P9-CQG-083

INFORMATION TRANSMISSION,
MODULATION, AND NOISE

McGRAW-HILL ELECTRICAL AND ELECTRONIC ENGINEERING SERIES

Frederick Emmons Terman, *Consulting Editor*
W. W. Harman and J. G. Truxal, *Associate Consulting Editors*

Angelakos and Everhart · Microwave Communications
Angelo · Electronic Circuits
Angelo · Electronics: BJTs, FETs, and Microcircuits
Aseltine · Transform Method in Linear System Analysis
Atwater · Introduction to Microwave Theory
Bennett · Introduction to Signal Transmission
Beranek · Acoustics
Bracewell · The Fourier Transform and Its Application
Brenner and Javid · Analysis of Electric Circuits
Brown · Analysis of Linear Time-invariant Systems
Bruns and Saunders · Analysis of Feedback Control Systems
Carlson · Communication Systems: An Introduction to Signals and Noise in Electrical Communication
Chen · The Analysis of Linear Systems
Chen · Linear Network Design and Synthesis
Chirlian · Analysis and Design of Electronic Circuits
Chirlian · Basic Network Theory
Chirlian and Zemanian · Electronics
Clement and Johnson · Electrical Engineering Science
Cunningham · Introduction to Nonlinear Analysis
D'Azzo and Houpis · Feedback Control System Analysis and Synthesis
Elgerd · Control Systems Theory
Eveleigh · Adaptive Control and Optimization Techniques
Feinstein · Foundations of Information Theory
Fitzgerald, Higginbotham, and Grabel · Basic Electrical Engineering
Fitzgerald and Kingsley · Electric Machinery
Frank · Electrical Measurement Analysis
Friedland, Wing, and Ash · Principles of Linear Networks
Gehmlich and Hammond · Electromechanical Systems
Ghausi · Principles and Design of Linear Active Circuits
Ghose · Microwave Circuit Theory and Analysis
Glasford · Fundamentals of Television Engineering
Greiner · Semiconductor Devices and Applications
Hammond · Electrical Engineering
Hancock · An Introduction to the Principles of Communication Theory
Harman · Fundamentals of Electronic Motion
Harman · Principles of the Statistical Theory of Communication
Harman and Lytle · Electrical and Mechanical Networks
Hayashi · Nonlinear Oscillations in Physical Systems
Hayt · Engineering Electromagnetics
Hayt and Kemmerly · Engineering Circuit Analysis
Hill · Electronics in Engineering
Javid and Brown · Field Analysis and Electromagnetics
Johnson · Transmission Lines and Networks
Koenig, Tokad, and Kesavan · Analysis of Discrete Physical Systems
Kraus · Antennas
Kraus · Electromagnetics

Kuh and Pederson · Principles of Circuit Synthesis
Kuo · Linear Networks and Systems
Ledley · Digital Computer and Control Engineering
LePage · Complex Variables and the Laplace Transform for Engineering
LePage and Seely · General Network Analysis
Levi and Panzer · Electromechanical Power Conversion
Ley, Lutz, and Rehberg · Linear Circuit Analysis
Linvill and Gibbons · Transistors and Active Circuits
Littauer · Pulse Electronics
Lynch and Truxal · Introductory System Analysis
Lynch and Truxal · Principles of Electronic Instrumentation
Lynch and Truxal · Signals and Systems in Electrical Engineering
McCluskey · Introduction to the Theory of Switching Circuits
Manning · Electrical Circuits
Meisel · Principles of Electromechanical-energy Conversion
Millman and Halkias · Electronic Devices and Circuits
Millman and Taub · Pulse, Digital, and Switching Waveforms
Minorsky · Theory of Nonlinear Control Systems
Mishkin and Braun · Adaptive Control Systems
Moore · Traveling-wave Engineering
Murdoch · Network Theory
Nanavati · An Introduction to Semiconductor Electronics
Oberman · Disciplines in Combinational and Sequential Circuit Design
Pettit · Electronic Switching, Timing, and Pulse Circuits
Pettit and McWhorter · Electronic Amplifier Circuits
Pfeiffer · Concepts of Probability Theory
Reza · An Introduction to Information Theory
Reza and Seely · Modern Network Analysis
Ruston and Bordogna · Electric Networks: Functions, Filters, Analysis
Ryder · Engineering Electronics
Schilling and Belove · Electronic Circuits: Discrete and Integrated
Schwartz · Information Transmission, Modulation, and Noise
Schwarz and Friedland · Linear Systems
Seely · Electromechanical Energy Conversion
Seely · Electron-tube Circuits
Seifert and Steeg · Control Systems Engineering
Shooman · Probabilistic Reliability: An Engineering Approach
Siskind · Direct-current Machinery
Skilling · Electric Transmission Lines
Stevenson · Elements of Power System Analysis
Stewart · Fundamentals of Signal Theory
Strauss · Wave Generation and Shaping
Su · Active Network Synthesis
Terman · Electronic and Radio Engineering
Terman and Pettit · Electronic Measurements
Thaler · Analysis and Design of Nonlinear Feedback Control Systems
Thaler · Elements of Servomechanism Theory
Thaler and Brown · Analysis and Design of Feedback Control Systems
Thaler and Pastel · Analysis and Design of Nonlinear Feedback Control Systems
Tou · Digital and Sampled-data Control Systems
Tou · Modern Control Theory
Truxal · Automatic Feedback Control System Synthesis

Tuttle · Electric Networks: Analysis and Synthesis
Valdes · The Physical Theory of Transistors
Van Bladel · Electromagnetic Fields
Weeks · Antenna Engineering
Weinberg · Network Analysis and Synthesis

BROOKLYN POLYTECHNIC INSTITUTE SERIES

Angelo · Electronic Circuits
Angelo · Electronics: BJTs, FETs, and Microcircuits
Levi and Panzer · Electromechanical Power Conversion
Lynch and Truxal · Signals and Systems in Electrical Engineering
Combining Introductory System Analysis *and* Principles of Electronic Instrumentation
Mishkin and Braun · Adaptive Control Systems
Schilling and Belove · Electronic Circuits: Discrete and Integrated
Schwartz · Information Transmission, Modulation, and Noise
Shooman · Probabilistic Reliability: An Engineering Approach
Strauss · Wave Generation and Shaping

INFORMATION TRANSMISSION, MODULATION, AND NOISE

A Unified Approach to Communication Systems

Second Edition

MISCHA SCHWARTZ
Professor of Electrical Engineering
Polytechnic Institute of Brooklyn

McGRAW-HILL BOOK COMPANY
New York / St. Louis / San Francisco
London / Sydney / Toronto
Mexico / Panama

Information Transmission,
Modulation, and Noise

Library of Congress Catalog Card Nunber 70-98490

ISBN 07-055761-6

89 KPKP 79876

To My Son David

PREFACE

A comparison with the first edition of this book will indicate that, except for Chapters 1 and 2, the book has undergone substantial revision for this second edition. Two trends in the past decade account for this: (1) The emphasis in communications has shifted dramatically to digital systems, and this trend can be expected to continue at a rapid rate in the near future. This is due, among other factors, to the increased use of integrated circuits with an associated stress on digital processing and the widespread use of computer systems. (2) The desire to constantly improve the performance of communication systems has led to an increased emphasis on optimization techniques, using modern statistical procedures.

Although the first edition contained much material on digital systems and the statistical analysis of communication systems, there is much more emphasis placed on both these modern trends in this edition. The book now breaks down naturally into two parts: Chapters 1 to 4 constitute a comprehensive introduction to modern communication systems and modulation theory; Chapters 5 to 8 constitute a self-contained introduction to modern statistical communication theory.

The enthusiastic reception accorded the first edition indicates that its special features of providing physical motivation for the mathematical material, of providing a blend of theory and practice, and of introducing examples of real systems where appropriate, were well received by the readers. An attempt has been made to retain all of these features in this edition.

The author was very pleased to find the first edition widely used, not only in regular electrical engineering courses throughout the world, but in various programs for continuing professional development offered both by industry and universities. It has also been used extensively for self-study by engineers working in the radar, sonar, and communications fields, as well as in other areas involving aspects of signal and data processing. He can only hope that this second edition is as well received.

As in the case of the first edition, the book has been written at the junior-senior level. It bridges the gap between the linear systems courses at the sophomore- and junior levels offered in most electrical engineering programs and the graduate-level courses in random processes and statistical communication theory given in graduate electrical engineering departments.

The book begins in Chapter 1 with an overall look at a typical digital communications system, showing the kinds of problems that may develop as a signal progresses through the system, and indicating the topics to be developed later. It continues by discussing in a qualitative way the information capacity of a system. Limitations in information capacity are shown to be due to the presence of energy-storage devices in the system and to unavoidable noise fluctuations. An exploration of the significance of these limitations is one of the major themes of the book.

The energy-storage or time-response limitation is shown in Chapter 2 to be expressible as a bandwidth limitation. For this purpose the Fourier series, Fourier integral, and concept of the frequency spectrum are developed and related to time response. The transmission through linear systems of elementary pulses, representative of simple binary signals, is then studied in detail.

Chapter 3 has been almost completely rewritten. It provides a comprehensive introduction to baseband digital communications, drawing on and applying the material of Chapter 2 to a study of analog-to-digital conversion, time-multiplexed signals, PAM systems, and pulse-code modulation. Waveshaping to avoid intersymbol interference is discussed in this chapter. New examples of digital systems, drawn from radio telemetry, space communication, and telephony, are included in this edition.

Chapter 4 extends the baseband considerations of Chapter 3 to high-frequency sine-wave transmission. Various types of high-frequency binary systems, including on-off-keyed systems and PSK and FSK systems, are introduced and compared on the basis of bandwidth. (Their comparative performance in noise is developed in Chapters 6 and 8.) Both synchronous, and envelope detection are considered. Analog modulation systems, including AM, single-sideband AM, and FM, are then studied in detail. The earlier discussion of binary FSK is found particularly useful in explaining analog FM systems.

Any further discussion of modern communication systems must of necessity include statistical concepts. To make the book as self-contained as possible, Chapter 5 introduces the necessary elements of probability theory. Many applications drawn from communication engineering are included to clarify the material and make it interesting. These include the calculation of error in PCM (done in more detail in Chapter 8), PCM repeater analysis, and a discussion of the statistics of fading media. Much of the material is new to this edition.

Chapter 6 on random signals and noise is essentially all new. As in Chapter 5, topics introduced are physically motivated where possible, providing a natural introduction to the more advanced treatments now commonly offered in first-year graduate courses. A discussion of the narrowband representation of noise enables the quantitative comparisons of high-frequency binary systems, noted above, to be carried out. It also leads directly into AM and FM receiver analysis, in a much more comprehensive fashion than was possible in the first edition.

Chapter 7 on Physical Sources of Noise has been completely rewritten, emphasizing noise sources appropriate to both semiconductors and lasers.

Thermal noise and shot noise are each treated in a twofold fashion; first, in a qualitative, physically motivated way; second, in the use of thermal noise, using a unique, modern-physics-oriented approach that brings out the quantum effects appropriate at light frequencies. In the case of shot noise, the derivation proceeds via the Poisson process, providing further application of the random processes of Chapter 6.

Chapter 8 on Statistical Communication Theory develops the basic aspects of optimum binary signalling in noise. With the help of some typical calculations drawn from space communications examples, it then points out how the rate of binary transmission, with error probability specified, is limited for a specified constraint on signal power and noise. Shannon's theorem is invoked to indicate that it may be possible to decrease the error probability even further by appropriate coding techniques. An information-feedback system is then introduced that actually performs, at least conceptually, at Shannon capacity. This provides the student with further insight into the significance of the famous Shannon expression, as well as enabling him to see how one may actually reduce error rates with specific systems. Error detection and coding are introduced at the end of this chapter.

With appropriate selection of material the book should lend itself to various types of courses. Thus, in those schools allowing a full year for the study of modern information-transmission systems, essentially the entire book can be studied. The first semester would then consist of a basic introduction to communication systems, using Chapters 1 to 4, while more quantitative analyses made possible by the introduction of statistical concepts in Chapters 5 to 8 would be treated in the second semester. For those schools in which Fourier analysis and/or probability theory are taught in other courses (either as prerequisites or concurrently), Chapters 2 and 5 can either be skipped or assigned for review.

In our own program at the Polytechnic we devote one semester to the course. The students have had a prior course in probability but very little material on Fourier analysis. We therefore devote substantial time to the material in Chapter 2 (following a short introduction to the course with the first few sections of Chapter 1), cover essentially all of Chapter 3, and then do selected portions of Chapter 4 (this includes Section 4-2 on binary communications, plus some additional material on analog AM and FM systems). The students are asked to quickly review the probability theory in Chapter 5, and we cover in class only the material in Section 5-6 on the detection of binary signals in noise. This provides the bridge between the digital systems of Chapter 3 and the study of random signals and noise in Chapter 6. We cover the first four sections of Chapter 6, then go on to Chapter 8, covering essentially all of that chapter. Time permitting, we may take up some of the material skipped over earlier in Chapters 4 and 6.

The author has profited immensely in the writing of this book from comments, criticisms, and suggestions provided by many readers, colleagues at other universities, and, most particularly, from colleagues at the Polytechnic. This second edition reflects in great measure the fine cooperative spirit of the members of the Communications Group at the Polytechnic who have been jointly responsible for the various courses and research effort in communications developed at the Polytechnic over the past decade. These colleagues include Professors R. R. Boorstyn, K. K. Clarke, D. T. Hess, R. L. Pickholtz, D. L. Schilling (presently at the City University of New York), and Jack K. Wolf. The author is grateful for their help.

MISCHA SCHWARTZ

CONTENTS

INFORMATION TRANSMISSION,
MODULATION, AND NOISE

CHAPTER ONE

INTRODUCTION TO INFORMATION TRANSMISSION

This book is devoted to a study of communication systems, or systems used to transmit information. In the course of the development we shall emphasize the limitations imposed on the information transmitted by the system through which it was passed and shall attempt some comparison of different systems on the basis of information-handling capabilities.

A complete system will generally include a transmitter, a transmission medium over which the information is transmitted, and a receiver which produces at its output a recognizable replica of the input information. In most communication work information transmission is closely related to the modulation or time variation of a particular sinusoidal signal called the *carrier*. A typical system diagram would thus appear as in Fig. 1-1.

The transmitter generally includes the source of the information to be transmitted—voice signals, TV signals, computer printouts, telemetry data in the case of space probes, or perhaps telemetry data transmitted from a remote automatically operated plant to a central control station.

As the signals traverse the transmission medium (or *channel* as it is most often referred to), they are distorted; noise and interfering signals are added, and it becomes a major task to correctly interpret the signals as finally received at the desired destination.

It is the purpose of this book to discuss in detail the effect of transmitting signals of various types over distorting channels—whether they be cables or wires for telephonic communication, space for radio transmission, or more exotic channels such as water (underwater communications) or the earth (seismic communications). We shall consider the limitations on information transmission introduced by transmission distortion and additive noise, using simple models for these effects, where possible, in this first treatment. We shall compare various ways of transmitting the desired information, as well as various techniques used at the receiver, to see how one goes about improving signal transmission in the presence of perturbing influences.

Although the stress in this book is placed on the type of communication systems shown in Fig. 1-1, much of the material is also particularly

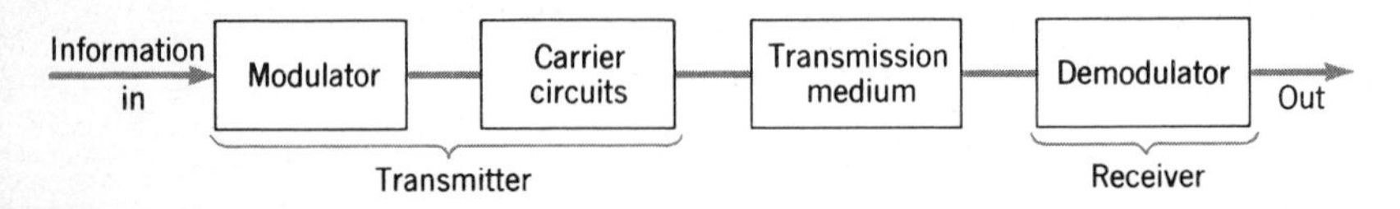

FIG. 1-1 Communication system.

appropriate to the broad area of *signal processing.* Here one is concerned less with the aspects of transmitting signals and more with the problem of *interpreting signals* once received. These signals could very well be the types of signals mentioned earlier, but might also include such types of data as biological signals (EEG and EKG data, for example), computer printouts of the results of scientific experiments, stock-market quotations, weather data, etc., etc. The basic problem here would be the analysis of the received data, that is, extracting the desired and pertinent information from the obscuring factors (or "noise") usually present. The techniques developed for the processing of these types of *random signals,* most commonly digital in form, draw heavily on principles developed in studies of information transmission.

1-1 TYPICAL DIGITAL COMMUNICATION SYSTEM DESIGN

To develop an overview to the general problem of information transmission and to pinpoint more specifically some of the problems encountered in communications, we shall detail in this section the problem of transmitting a typical digital data message from one point to another. Many of the problems raised will in turn be developed and studied in much more detail throughout this book. We consider a digital (or *binary*) data message since this is rapidly becoming the most common form of signal transmission, whether the message is directly digital in form, as in the case of computer printouts, or whether converted to a digital format, as is true with telemetered data and much of telephonic voice communication. The binary data message we introduce in this section will then be utilized, as a particularly simple form of signal, throughout this book.

By a binary message or signal we mean a sequence of two types of pulses of known shape, occurring at regularly spaced intervals, as shown in Fig. 1-2. Although the shape of the pulses is presumed known in advance, the occurrence of one or the other, say the 1 or 0 in Fig. 1-2, is not known beforehand, and the information carried is actually given by the particular sequence of binary 1's and 0's coming in. We shall most commonly use the rectangular pulse shapes of Fig. 1-2*a* and *b* in this book for simplicity's sake, but other types of signal shape could and are being used as well.

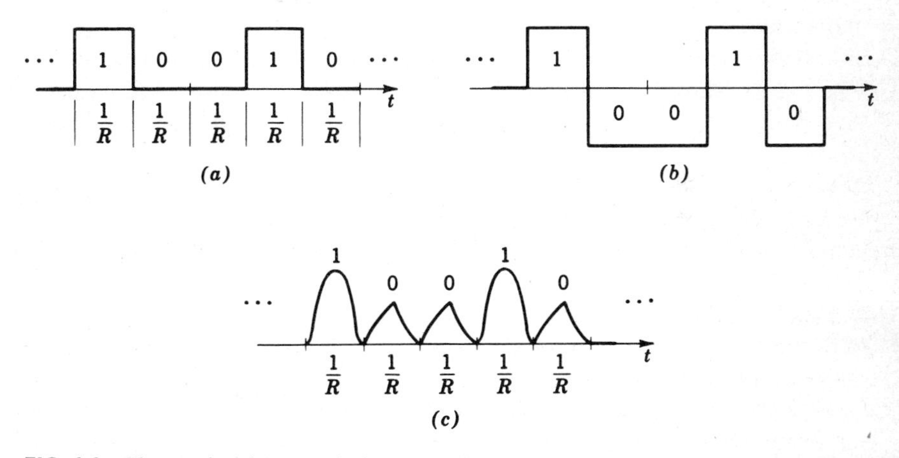

FIG. 1-2 Binary signal transmission. (a) On-off sequence; (b) bipolar sequence; (c) arbitrary waveshapes.

These pulses are shown occurring regularly every $1/R$ seconds, or at a rate of R/sec. $1/R$ is commonly referred to as the binary interval, and the signal source is said to be putting out R binary digits or *bits* per second (bits from *binary digits*). As an example, if $1/R = 10^{-3}$ sec, R is 1,000 bits/sec. If $1/R = 1$ μsec, $R = 10^6$ bits/sec.

We now assume that this sequence of binary symbols is to be transmitted to a distant destination. A typical system block diagram, a more detailed version of Fig. 1-1, is shown in Fig. 1-3. The two filters shown,

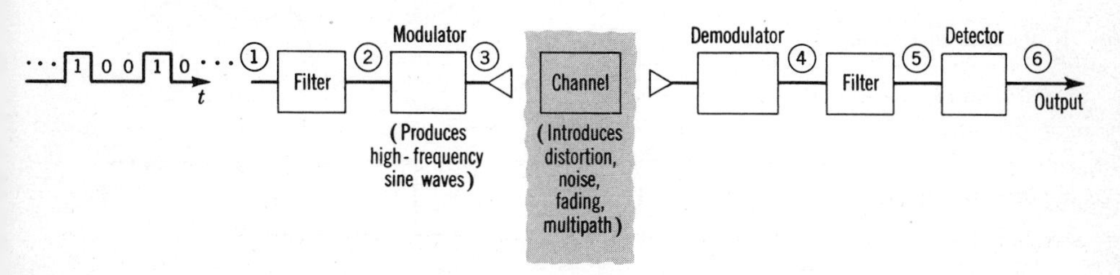

FIG. 1-3 Transmission of digital message.

one at the transmitter and the other at the receiver, represent the filtering of the signals either innately present in the system circuitry or purposely introduced as part of the design. The demodulator at the receiver serves to strip away the high-frequency sine-wave modulation introduced at the transmitter modulator. The modulation process is necessary to enable the signals to be effectively radiated into space (or whatever other medium is represented by the channel shown). The purpose of the detector at the receiver is to reproduce, "as best as possible," the original signal sequence representing the digital data to be transmitted.

Some typical waveshapes, corresponding to the numbered points in Fig. 1-3, are shown sketched in Fig. 1-4. Note that the filters cause excessive symbols to overlap into adjacent time slots. If carried too far, these

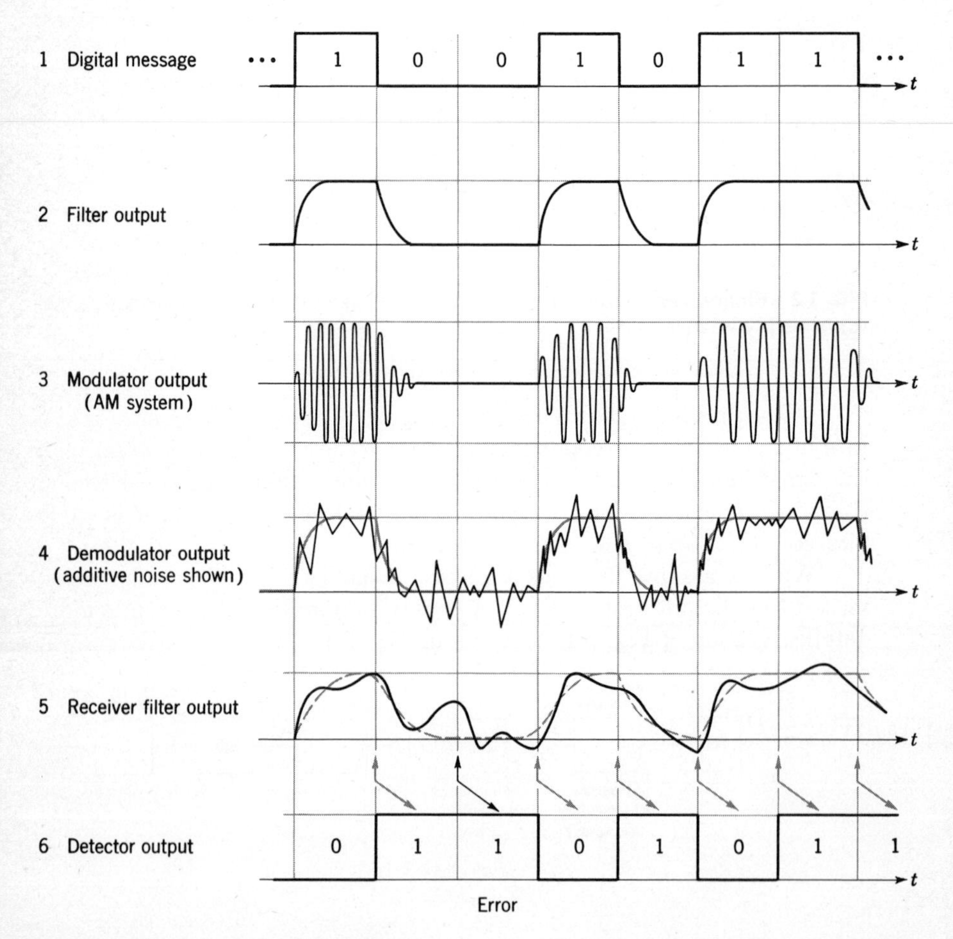

FIG. 1-4 Waveshapes in digital system.

interfere with symbols actually transmitted in the adjacent time slots, leading to confusion in symbol interpretation and possible errors at the system output. *Intersymbol interference* is a significant problem in many data communication systems, being particularly troublesome in the transmission of data via telephone lines. We shall have occasion to discuss this problem further in later chapters, and will indicate some design techniques used to overcome it.

The modulator chosen in this example happens to be of the amplitude-modulation (AM) type, in which a sine-wave oscillator adjusts its amplitude to the incoming signals. We could equally well have depicted the output of a frequency-shift-keyed (FSK) modulator, in which the carrier frequency alternates between two frequencies, depending on the symbol coming in (this is the digital version of an FM signal), or a phase-shift-keyed output (PSK), in which the polarity ($\pm$) of the sine wave depends on the incoming signal.

In this example the channel is shown as having introduced noise during transmission, so that the demodulator output represents the sum of signal plus noise waveshapes. Note that the noise introduced tends to obscure the signal if the two are comparable in magnitude. Some channels (the telephone line for example) introduce signal distortion as well (although this effect could be modeled by incorporating it in the filter following the demodulator). Others introduce signal fading, in which the received signal amplitude is found to fluctuate randomly, or so-called "multipath" effects, in which the radiated transmitter energy for one signal symbol, following several alternate paths to the receiver, appears at the receiver as a sequence of received symbols. Examples of this latter type of randomly fading channel include short-wave radio transmission via the ionosphere, as well as underwater and seismic communications, among others.

For simplicity's sake we shall stress additive noise effects in this book, although some reference will be made to fading channels.[1]

The receiver filter serves to eliminate some of the noise, at the expense, however, of further distorting the signal transmitted. This is shown pictorially in Fig. 1-4, in which the dashed lines of parts (4) and (5) represent the signal term, while the solid lines represent the composite sum of signal and noise.

To finally reproduce the original signal message, the detector must sample the receiver filter output once every bit interval and decide "as

[1] See M. Schwartz, W. R. Bennett, and S. Stein, "Communication Systems and Techniques," McGraw-Hill, New York, 1966, for a comprehensive treatment of such channels. An introductory treatment appears in Chap. 5 of our text.

best as possible" whether a 1 or 0 was transmitted. It is apparent from part (5) of Fig. 1-4 that the appropriate sampling times, in this example, occur at the end of each bit interval, since it is there that the filtered signals reach their maximum amplitudes. Mistakes can be made, however, with noise obscuring the correct symbol. An example of an error occurring is shown in part (6) of Fig. 1-4.

We have purposely used the words "as best as possible" twice in the discussion above because they play a key role in the overall design of any communication system. In the context of the simple digital communication system we have been describing, these words have a rather precise meaning. It is apparent that we would like to transmit any arbitrary sequence of data symbols with as small a number of errors occurring as is possible, costwise. The fact that errors will occur is apparent. Noise is always introduced in a system (we shall have a great deal to say on this subject later in the book), physical transmitters have power limitations so that one cannot generally increase the signal strength as much as necessary in order to overcome the noise, intersymbol interference is generally a problem with which to reckon, signal fluctuations during transmission give rise to possible errors, etc.

A comprehensive engineering design of a data system such as this one would have to take into account all possible sources of error and try to minimize their effect. This includes appropriate design of the signals at the transmitter. That is: How does one shape the signals or design the transmitter filter? Given various means of transmitting the binary symbols at the required high frequency—the AM technique shown in Fig. 1-4, FSK, PSK, etc., which is most appropriate for the problem at hand, including the particular channel over which transmission will take place? We shall find in discussing this particular question later in the book that there is the usual engineering tradeoff between the various modulation techniques. Thus PSK, although generally most effective in terms of conserving power or minimizing errors, is difficult to use over fading channels, and its use introduces severe phase control problems. FSK is generally less effective in the presence of noise and requires wider system bandwidths (a term to be defined in the next chapter), but is more effective over fading channels.

With a specific modulation scheme chosen how does one design the receiver? How does one minimize intersymbol interference and noise? How does one design the detector and the decision-making circuitry associated with it?

These are just a few of the design questions that arise in the actual development of a digital system of the type we have been considering. (We have, for example, ignored timing problems associated with the sequential transmission of 1's and 0's every $1/R$ sec. This is particularly important in the design of systems with the transmitter and receiver located thousands and even millions of miles apart, where the receiver must nonetheless always maintain the same timing as the transmitter.)

It is one of the purposes of this book to approach problems such as these systematically, indicating how one does go about designing a particular system. Although we shall stress for simplicity's sake the effects of noise and filtering, since their effect is important in every communication system, some of the other problems noted will be treated as well.

Now assume the system has been "optimumly" designed. We have minimized the number of errors that will occur, on the average, by appropriate design at both transmitter and receiver. How good is it? Is the resultant design appropriate for our needs (i.e., for the particular application for which intended)? This requires a quantitative system evaluation, modeling the presumed effect of the channel, noise introduced, etc., and such an evaluation will be one of the subjects treated in this book. For this purpose we shall find it necessary to introduce statistical concepts and approach the combined problems of system design and evaluation statistically.

If the error rate of the system, the average number of errors occurring per unit time, is too high, more complex system configurations, with their attendant increased cost, may be called for. These include signal coding techniques, sophisticated error detection and correction procedures, etc.

This raises another extremely important question: For a given channel over which we desire to communicate, is it possible to keep improving the system performance, i.e., reduce its error rate, as much as we like, with appropriate increased complexity of the system design? This is obviously a basic question in all communications design, for if the answer is *no*, there is no sense in even trying to design more complex systems.

To answer this question we choose to phrase it in somewhat more precise fashion. The question is now: With the rate of transmission of binary symbols, R bits/sec, fixed, as is the power available at the transmitter, is it possible, for a given channel, to reduce the error rate as much as desired (with appropriate system design and complexity)? The answer, as first established by Claude Shannon in 1948 to 1949 in a monumental piece of work,[1] is "yes," with one qualification. That is, his work was restricted to the study of a channel introducing noise only (intersymbol interference and fading effects were not included). This has since been extended to a few other channels by other investigators.

[1] C. E. Shannon, A Mathematical Theory of Communication, *Bell System Tech. J.*, vol. 27, pp. 379–423, July, 1948; pp. 623–656, October, 1948. C. E. Shannon, Communication in the Presence of Noise, *Proc. IRE*, vol. 37, pp. 10–21, January, 1949.

Shannon found that the chance of an error occurring may ideally be reduced as low as one likes by appropriate coding of the incoming signals, providing the binary signaling rate R, in bits per second, is less than a specified number determined by the transmitter power, channel noise, and channel response time or bandwidth (this latter concept will be discussed in detail in the next chapter). If one tries to push too many bits per second over the channel, the errors begin to mount up rapidly. The maximum rate of transmission of signals over the channel is referred to as the channel capacity.

Since the channel capacity is obviously an important concept in systems design (one can determine from this whether it pays to develop more complex systems), we shall devote some time to exploring its significance. The remainder of this chapter is devoted to a qualitative discussion of this concept, indicating why one physically expects a channel capacity to exist. There is a semantic difficulty that must be mentioned, however. We choose to use the word *channel* here most often to denote the physical medium over which transmission takes place. Many authors include in the channel various portions of the transmitter and receiver as well. Generally the meaning is clear from the context of the discussion. Shannon's channel capacity actually refers to this more general class of channels. To avoid confusion, we shall use the words *system capacity*, and, as we explain in the next section, all portions of the transmitter and receiver, as well as the physical channel, contribute to determining the capacity.

Since the maximum rate of transmission of binary symbols over a given channel is fixed, one would like to know what the binary rate R of a given signal source is. This is particularly true in the case of sources that are initially nonbinary in nature and that must be converted to binary symbols. These include speech, TV or facsimile, telemetry signals, etc. The concept of information content of a given message, in terms of the bits needed to represent it, is thus also explored in this chapter.

One last word before we close this section. We have emphasized digital communications here because of its relative simplicity as well as its technological importance. The questions raised here hold as well for other types of communication systems. Thus one is always interested in "optimizing" system performance. The major difficulty arising in much of communication system design and evaluation, however, is that no simple criterion exists as to "optimum performance." How does one determine whether a particular speech signal is reproduced as effectively as possible? When does a TV picture have an "optimum" appearance? It is apparent that for continuously varying (analog) signals such as these, simple performance measures may be hard to justify, since much of the essential system evaluation can only be done subjectively. Yet the techniques of minimizing noise or of appropriate signal filtering that we shall discuss are significant in these types of systems as well as in strictly digital ones, and we will be able to establish some measure of their performance.

1-2 INFORMATION AND SYSTEM CAPACITY

As we mentioned in the previous section, the information content of a message to be transmitted must be established in order to determine whether or not the message may be transmitted over a given channel. By the information content we mean the number of binary symbols that will ultimately be necessary, on the average, to transmit it. Although we may in actuality not transmit binary symbols, but choose to transmit a more complex signal pattern, we prefer to normalize all signal messages to their binary equivalents for simplicity's sake. (This conversion to a binary equivalent will be discussed further both in this chapter and in chapters that follow.)

Since all communication systems transmit *information* in one form or another and since we desire some measure of the information content of messages to be transmitted, it is important first to establish some measure of what is actually meant by the concept *information.* Although a precise mathematical definition can be set up for this concept, we shall rely on our intuitive sense in this introductory text.

Consider a student attending a class in which the teacher spends the entire time whistling one continuous note. Obviously, attending such a class would be a waste of time. What could one possibly learn from the one note? (Even if it were important for the student to repeat the sound exactly, he would be better off staying at home listening to a recording.) Perhaps in the next class the teacher chooses to devote the entire hour to a reading, word for word, from the text: no time for questions, no pauses, no original thoughts. Again why come to class (aside from the irrevelant fact that many persons might not then take the opportunity to read the book for themselves)?

What is the point to these hypothetical and obviously made-up stories? A student comes to class to receive *information.* That is, the teacher and student are in class to discuss new material or, at least, review old material in a new way.

The words and phrases used should thus be *changing* continuously; they should, in most cases, be *unpredictable* (otherwise—why come to class?).

The key phrase here is *unpredictable* change. If information is to be conveyed, we must presumably have sounds or, more generally, signals changing unpredictably with time. A continuous trilling of one note conveys no information to you. If the note is varied in a manner that you can interpret, however, the "signal" begins to convey meaning and information.

The binary sequence of the previous section was one simple example of a signal to be transmitted that is changing unpredictably in time. There the specific sequences of 1's and 0's were unknown beforehand and corresponded to the message to be transmitted.

So the transmission of information is related to signals changing with time, and changing in an unpredictable way. (For a well-known melody or old story conveys no new information although made up of changing notes or words.)

Why is it so important to stress these points? As noted previously, if we, as engineers, are to design systems to transmit information and are interested in the best possible type of system given practical equipment and a limited budget, we must know (at least intuitively) what it is we are transmitting and the effect of the system on this quantity.

To see how these concepts fit into our work in communication, consider the voltage-time diagram of Fig. 1-5. Assume that we have an

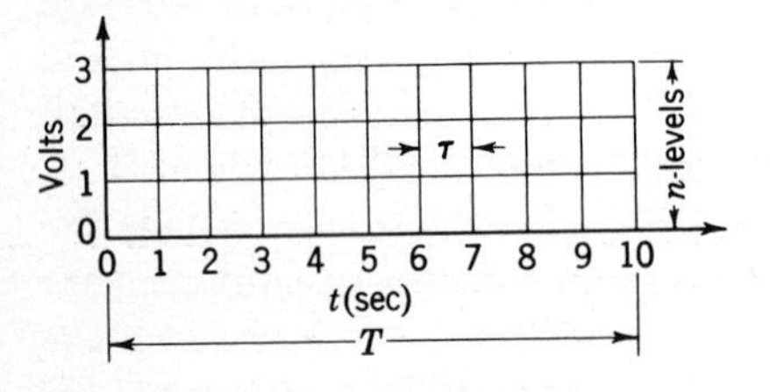

FIG. 1-5 Voltage-time diagram.

interval of time T sec long in which to transmit information and a maximum voltage amplitude (because of power limitations) that we can use. (In Fig. 1-5, T is 10 sec, and the maximum voltage is 3 volts.) A natural question to ask is: How much information can we transmit in this interval? Can we put a tag on the amount, and how does it depend on our system? (Note that the system has already introduced one limitation—that of power.)

The next question is: Why a limit to the amount of information? If information transmission is related to signals changing unpredictably with time, why not just change the signal as rapidly as we like and over as many subdivisions of the maximum amplitude as we like? This would imply increasing the information content indefinitely.

We deal with physical systems, however, and these systems do not allow us to increase indefinitely the rate of signal change and to distinguish indefinitely many voltage amplitudes, or levels:

1. All our systems have energy-storage devices present, and changing the signal implies changing the energy content. There is a limit on the rate of doing this determined by the particular system.
2. Every system provides inherent (even if small) variations or fluctuations in voltage, or whatever parameter is used to measure the signal amplitude, and we cannot subdivide amplitudes indefinitely. These unwanted fluctuations of a parameter to be varied are called noise. This noise is exactly the noise we noted as being introduced during transmission over a channel in the previous section. The channels discussed there are all examples of physical systems of the type with which we deal.

There is thus a *minimum* time τ required for energy change and a minimum detectable signal-amplitude change. As an example, τ is given as 1 sec in Fig. 1-5. If the inherent voltage fluctuations of the system may be assumed as an example to vary within ± 1 volt most of the time, the minimum detectable voltage change due to the signal is 1 volt. With a maximum voltage amplitude of 3 volts, there are thus four detectable levels of signal (0 voltage being assumed to be a possible signal value). For if the signal were to change by less than 1 volt, it could not be distinguished from the undesired noise fluctuations introduced by the system.

If the "amount of information" transmitted in T sec is related to the number of different and distinguishable signal-amplitude combinations we can transmit in that time—as we might intuitively feel to be the case—it is apparent that the information capacity of the system is limited. It is exactly these arguments, phrased in a much more quantitative manner, that were used by Shannon in developing his capacity expression referred to in the previous section. (Recall that the word *channel* was used in that section to refer to the physical medium over which signals are transmitted. Limitations on signal-time response may be produced anywhere in the systems of Figs. 1-1 or 1-3. The filters shown in Fig. 1-3, for example, definitely introduce specific time responses. Also, noise is often introduced at the receiver as well as during transmission. We shall thus refer here to *system capacity,* considering the effect of the overall system, rather than *channel capacity,* as commonly done in the literature.)

The system capacity, or maximum rate at which it can transmit information, should be measurable in terms of τ and n, the number of distinguishable amplitude levels. (Both limitations may be produced again anywhere in the system of Fig. 1-1, but we are speaking of the effect of the overall system.)

We can derive a more quantitative measure of system capacity in the following manner. We assume that the information transmitted in the 10-sec interval of Fig. 1-5 is directly related to the number of different signal-amplitude combinations in that time. For example, two different signals that might be transmitted are shown in Fig. 1-6. They differ over

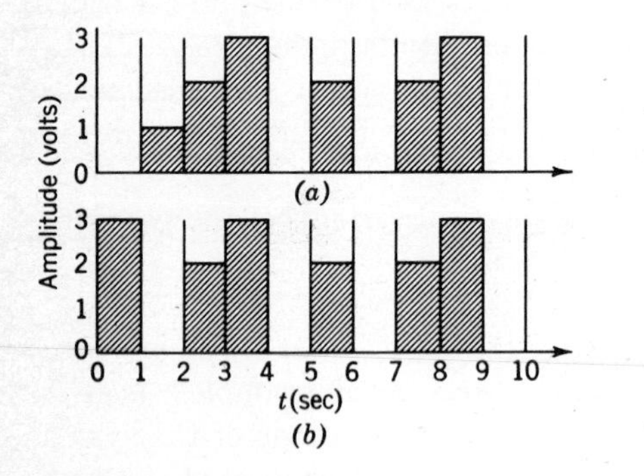

FIG. 1-6 Two different signals.

the first two intervals and have the same amplitudes over the remaining 8 sec. How many such combinations can we specifiy? There are four different possibilities in the first interval, and corresponding to each such possibility there are four more in the second interval, or a total of $4^2 = 16$ possibilities in two intervals. (The reader should tabulate the different combinations to check this result.) Repeating this procedure, we find that there are 4^{10} combinations of different signal amplitudes in 10 sec.

If, instead of 4, we had had n levels and, instead of 1-sec, τ-sec intervals, the number of combinations in T sec would have been

$$n^{T/\tau}$$

Under our basic assumption the information transmitted in T sec is related to this number of signal combinations. We might feel intuitively, however, that information should be proportional to the length of time of transmission. Doubling T (10 sec here) should double the information content of a message. The information content can be made proportional to T by taking the logarithm of $n^{T/\tau}$, giving us

$$\text{Information transmitted in } T \text{ sec} \propto \frac{T}{\tau} \log n \qquad [1\text{-}1]$$

The proportionality factor will depend on the base of logarithm used. The most common choice is the base 2, or

$$\text{Information} = \frac{T}{\tau} \log_2 n \qquad [1\text{-}2]$$

The unit of information defined in this manner is the *bit* (mentioned earlier and to be explained in the next section). The information content of the 10-sec strip of Fig. 1-5 is, for example,

$$10 \log_2 4 = 20 \text{ bits}$$

A 5-sec strip would have 10 bits of information. If there had been only two possible voltage levels (say 0 and 1), the information conveyed in 10 sec would have been 10 bits.

The system capacity can be defined as the maximum *rate* of transmitting information. From Eq. [1-2] this is simply

$$C = \frac{\text{information}}{T} = \frac{1}{\tau} \log_2 n \qquad [1\text{-}3]$$

and the units are given in bits per second.

System capacity is thus inversely proportional to the minimum interval τ over which signals can change and proportional to the logarithm of n.

We shall show in the next chapter, reviewing some simple concepts of networks, that there is an intimate inverse relationship between time response and frequency response. This will enable us to relate information transmitted and system capacity to the system "bandwidth."

These two parameters of system behavior, τ (or its inverse, bandwidth) and n (or, as we shall see, the signal-to-noise ratio in a system), are basic in any study of communication systems. Much of the material of this book will thus be devoted to a study of the time (or frequency) and noise characteristics of different networks and the frequency-noise characteristics of various practical communication systems.

1-3 BINARY DIGITS IN INFORMATION TRANSMISSION

The information content of a signal was defined in the previous section by Eq. [1-2],

$$\text{Information} = \frac{T}{\tau} \log_2 n \qquad \text{bits}$$

The use of the logarithm to the base 2 in defining the unit of information can be justified in an alternative and instructive way. Assume that a signal to be transmitted will vary anywhere from 0 to 7 volts with any one voltage range as likely as the next. Because of the system limitations described in Sec. 1-2, the signal can be uniquely defined only at the integral voltage values and will not change appreciably over an interval τ sec long. (The noise fluctuations are assumed to have the same magnitudes on the average as in the example of Sec. 1-2.)

The signal can thus be replaced by a signal of the type shown in Fig. 1-6; during any interval τ sec long it will occupy one of eight voltage levels (0 to 7 volts), each one equally likely to be occupied.

The process of replacing a continuous signal by such a discrete signal is called the *quantizing process*. A typical signal and its quantized equivalent are shown in Fig. 1-7.

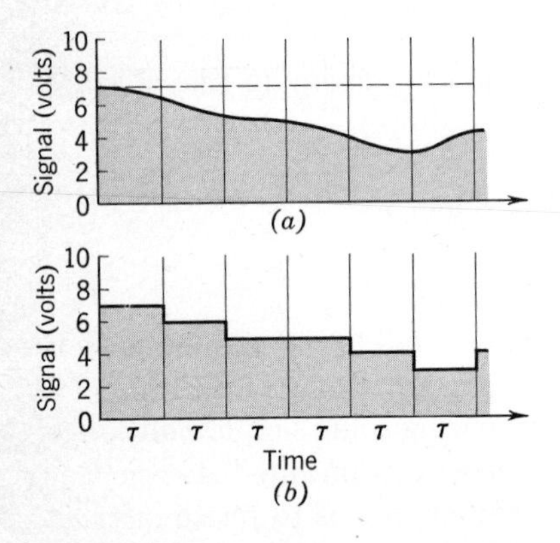

FIG. 1-7 Quantization of a signal. (a) Original signal; (b) quantized signal.

The signal can of course be transmitted by simply sending the successive integral voltage values as they appear. In any one interval any one of eight different voltages must be sent. The informational content of the signal is thus related to these eight different voltage levels (i.e., one of eight choices).

We ask ourselves, however: Is there another way of sending this information so that fewer than eight numbers are needed completely to specify the signal in any one interval? The informational content will then be assumed equal to the smallest number needed. The answer is yes: the simplest way uniquely to label a particular level and to indicate its selection is by means of a series of yes-no instructions. For this particular type of signal with eight levels three such yes-no instructions are needed.

To indicate the procedure followed, assume that the signal is at 7 volts during a particular instant. We first decide whether the proper level lies among the first four or the last four levels. If "yes," we use a designating symbol 1 for each level in the group of four; if "no," a designating symbol 0. In this case, then, levels 0 to 3 are labeled 0, and levels 4 to 7 are labeled 1 (see Fig. 1-8*a*). We can thus immediately reject the 0 levels and concentrate on choosing one of the remaining four. Our area of choice has been reduced considerably.

Again we separate the remaining levels into two parts. That half which does not contain the desired level (7 in this case) is labeled no, or 0; the other yes, or 1. This is shown in Fig. 1-8*b*. Again half the levels are eliminated, leaving only 6 and 7. This time we find level 7 uniquely singled out.

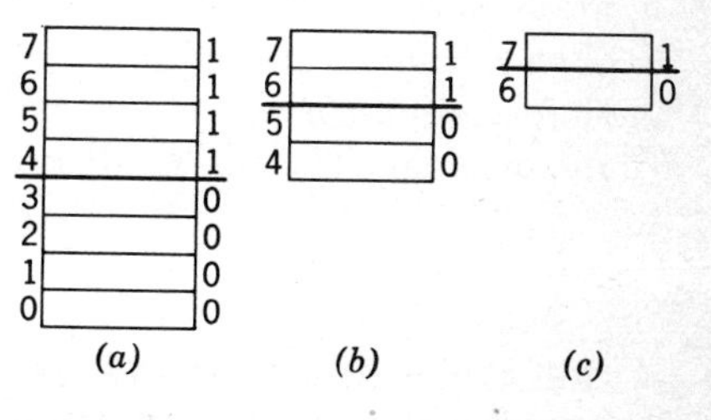

FIG.1-8 Binary selection of a signal. (a) First choice; (b) second choice; (c) third choice.

Note that this method required three consecutive yes-no responses.

Proceeding in a similar way, we could single out any one of the eight possible voltages. They can thus be uniquely identified by means of three 0 or 1 labels. This method of identification is called *binary coding*. A typical identification table would appear as follows:

	Binary coding
7	111
6	110
5	101
4	100
3	011
2	010
1	001
0	000

Instead of transmitting this signal as one of eight different voltage levels, we need transmit only three successive yes-no (voltage–no voltage) voltages during a particular interval. Any one yes-no label is a *bit*. Three bits are thus required to transmit the desired information as to a particular voltage level occupied for the eight-level signal under discussion.

This process of binary coding is the simplest one that can be devised for uniquely tagging a signal. With binary selection only three consecutive numbers, or 3 bits, are needed to transmit the informational content of this particular signal in any one time interval. For 16 levels 4 bits are required, for 32 levels 5 bits, and so on. For n levels $\log_2 n$ bits are required.

If the information in three successive intervals, each containing eight possible levels, is to be transmitted, then 3 bits for each interval, or 9 in all, are required for the signal transmission.

For T/τ intervals and n levels, $T/\tau \log_2 n$ bits must be transmitted. The information content of a signal is defined to be equal to the number of bits, or binary choices, needed for transmission.

1-4 RELATION BETWEEN SYSTEM CAPACITY AND INFORMATION CONTENT OF MESSAGES

In Sec. 1-2 we pointed out that the capacity (or ability) of a system to transmit information depended on the system time response and its ability to distinguish among different levels of a signal.

The capacity of a given system is defined as the maximum amount of information per second (in bits per second) that the system can transmit. The following chapters of this book will be devoted to relating the system limitations on information transmission to bandwidth and system noise properties. Here we take time to consider the relation between system capacity and the information content of signals to be transmitted.

Being able to measure the ability of a particular system to transmit information is not enough. The more basic question in information transmission should perhaps be phrased: Which system or group of systems will have sufficient capacity to transmit a specified class of signals or information-bearing messages? In order to answer this question, we must be able to measure the information content of a signal.

For example, assume that we are interested in transmitting a speech delivered in the English language. In order to choose a system for transmission, we must be able to determine the information content of the speech and the rate at which it is to be transmitted. The system chosen will then obviously have to accommodate the rate of information to be transmitted, or will have to possess a capacity greater than the rate of information transmission desired. The system must thus be "matched" to the class of signals being transmitted.

In order to choose a system or group of systems with the proper information capacity, we first determine or measure the information content of the signals that are to be transmitted. The detailed consideration of this point is in the realm of information theory, and the reader is referred to the ever-growing literature on this subject for thorough discussions.[1] We shall, however, indicate the approach used, and we shall find some of the material of value in our later work.

We have already shown, in Secs. 1-2 and 1-3 that, where different values of a signal are *equally likely* and where the signal appears at discrete voltage levels, the information content of a signal is readily evaluated. It is simply the logarithm of the number of equally likely signal combinations possible in a given interval. (In the example we considered, $n^{T/\tau}$ was the number of equally likely signal combinations in T sec, since each signal level was as likely to be occupied as any other. The information content of the strip T sec long was thus $\log_2 n^{T/\tau} = T/\tau \log_2 n$.)

The equally likely case is a very restricted one, however. For example, if we were transmitting a speech in the English language by transmitting the different letters in the successive words uttered, this would correspond to assuming that each letter was equally likely to occur. This assumption is obviously not true, since we know, for example, that the letter e occurs much more frequently than the letter z, or any other letter for that matter. We could guess that a particular letter to come would be an e and be much more sure that we were right than if we had guessed q, z, or u.

But this prior knowledge of the greater chance of e rather than z occurring reduces the information content of the speech being transmitted. For, as we noted in Sec. 1-2, the amount of information transmitted depends on the *uncertainty* of the message. In particular, if e were known to be the only letter occurring in this speech, no information at all would be transmitted, since all uncertainty as to the message would vanish.

Thus, although the different number of signals in any one interval is still the gamut of all letters from a to z, the fact that some occur more frequently than others reduces the information content of the message. In T sec the different signal combinations possible occur with differing relative frequencies, and the information content of the T-sec message is reduced as compared with the equally likely case.

The same considerations obviously hold true in our representation of different signals by different voltage levels in Sec. 1-2. If a 3-volt signal were to be the only signal transmitted, it would carry no information and transmission might just as well cease. If the 3-volt signal were to be

[1] Shannon, A Mathematical Theory of Communication, *op. cit.*

expected more often than any other, a message T sec long would carry information but the *information content* of the message would be less than if all voltage levels were equally likely. (Those signal combinations containing the 3-volt level would be expected to occur more frequently than any of the others.)

The information content of a message thus relates not only to the number of possible signal combinations in the message but also to their *relative frequency of occurrence.* This in turn depends on the source of the message. The information content of a message in the English language depends on a knowledge of the structure of the language and its alphabet: the relative frequency of occurrence of each letter, of different combinations of letters, of word combinations, of sentence combinations, etc. All these structural properties of English affect the different possible signal combinations and their relative frequency of occurrence, and hence the information content of a particular signal.

The decrease in the information content of a message due to unequally likely signals results in the requirement of a correspondingly reduced system capacity for information transmission. Telegraph transmission of messages in the English language has long taken this into account by coding the letter *e* with the shortest telegraph symbol. This thus reduces the average time required for transmission of a message.

But how do we *quantitatively* measure the information content of a message in this more general case of signals with differing frequencies of occurrence? To study this case, we shall first assume that successive signals (the individual letters in the case of English) are independent of one another and then attempt some further generalization. The assumption of *independence* implies that the occurrence of any one signal does not affect in any way the occurrence of any other signal. In the case of a message in the English language this assumption implies that the occurrence of one letter does not affect the occurrence of any other. (A q coming up could then be followed by an x or a z, as well as by a u. This assumption of independence is thus an oversimplification of a much more complicated situation in the case of English, but it does serve to simplify the analysis.)

To develop a quantitative measure of the information content of a message, we shall first rewrite our result for the equally likely case in a different form. We recall that we showed that $n^{T/\tau}$ was the total possible number of signal combinations in T sec if each signal lasted τ sec and there were n possible levels in each interval.

If we were to look at many messages, each T sec long, we would find that on the average each possible signal combination would occur with a relative frequency of $1/n^{T/\tau}$. For example, with $\tau = 1$ sec and $n = 4$, 64 different combinations are possible in an interval 3 sec long. The relative frequency of occurrence of each 3-sec message would be $\frac{1}{64}$. In 10,000 such 3-sec messages there would be approximately 10,000/64 messages of each of the 64 possibilities. The greater the number of 3-sec messages we were to look at, the more closely any one signal combination would approach a relative frequency of $\frac{1}{64}$.

The relative frequency of occurrence of any one combination, or *event*, we define to be its probability of occurrence, or, symbolically, P. Thus

$$P \equiv \frac{\text{number of times event occurs}}{\text{total number of possibilities}} \tag{1-4}$$

where the total number of possibilities must be very large compared with the number of possible events (10,000 as compared with 64 in the example just cited) if we are to have an accurate measure of the relative frequency.

For example, if we were interested in the probability of occurrence of a letter in the English alphabet, we would pick letters at random (say, from words on successive pages of a book to ensure independence of choice) and determine the number of times a given letter showed. We would have to pick many more than 26 total letters for our results to be valid, however.

If n possible events are specified to be the n possible signal levels, at any instant, of Sec. 1-2, then $P = 1/n$ for equally likely events. The information carried by the appearance of any one event in one interval is then

$$H_1 = \log_2 n = -\log_2 P \qquad \text{bits/interval} \tag{1-5}$$

Over m intervals of time (an interval is τ sec long) we should have m times as much information, assuming that each signal or event in time is independent. Therefore

$$H = mH_1 = -m \log_2 P \qquad \text{bits in } m \text{ intervals} \tag{1-6}$$

The information available in T sec is thus ($m = T/\tau$)

$$H = -\frac{T}{\tau} \log_2 P = \frac{T}{\tau} \log_2 n \qquad \text{bits in } T \text{ sec} \tag{1-7}$$

as in Sec. 1-2.

Now consider the case where the different signal levels (or events) are not equally likely. For the sake of simplicity we first assume just two levels to be transmitted, 0 or 1, the first with probability p, the second with probability q. Then

$$p \equiv \frac{\text{number of times 0 occurs}}{\text{total number of possibilities}} \tag{1-8}$$

and

$$q \equiv \frac{\text{number of times 1 occurs}}{\text{total number of possibilities}} \tag{1-9}$$

Since either a 0 or a 1 must always come up, $p + q = 1$. (The number of times 0 comes up plus the number of times 1 comes up equals the total number of possibilities.)

For example, say that the message to be transmitted by this two-level signal device represents the birth of either a boy or a girl in the United States; 1 corresponds to boy, 0 to girl. After counting 1,000,000 births, we find that 480,000 boys and 520,000 girls were born. Then we estimate $p = 0.52$, $q = 0.48$, and $p + q = 1$.

What is now the information content of a particular message consisting of a group of 0's and 1's? Each time a 0 appears, we should gain $-\log_2 p$ bits of information, and each time 1 appears we gain $-\log_2 q$ bits. If p and q are approximately each 0.5, either event occurring (0 or 1) carries almost the same amount of information. The two events are nearly equally likely. This is of course the case in the births of boys and girls in the United States to which we referred. But now assume $p \gg q$ (0 occurs more frequently, on the average). Since $-\log_2 q \gg -\log_2 p$, the occurrence of a 1, the more *rarely occurring* event, carries *more* information.

This seems to agree with our previous discussion, where we point out that, the greater the uncertainty of an event occurring, the more the information carried. Does this again agree with intuition? We use births as an example once more, but this time consider the case of a family with five sons and no daughters. The father has given up all hope of a daughter, especially since both his family and that of his wife have a long history of a preponderance of male children. The father, waiting expectantly for his wife to give birth again, receives word that his wife has given birth to—a *boy*. So? That is nothing new. A boy was expected. But had his wife given birth to—a *girl!* This news would be something tremendously different, it would carry much more information, it would be the completely unexpected!

More rarely occurring events thus carry more information than frequently occurring events in an intuitive sense, and our use of the $-\log_2 p$, $-\log_2 q$ formulation for information agrees with the intuitive concept.

The information carried by a group of 0 or 1 symbols should now be the sum of the bits of information carried by each appearance of 0 or of 1. If $p = 0.8$ and $q = 0.2$ and if p occurs 802 times in 1,000 possibilities, q occurring 198 times, the information content of the 1,000 appearances of a 0 or 1 is

$$\begin{aligned} H &= -(802 \log_2 0.8 + 198 \log_2 0.2) \\ &\doteq -1{,}000(0.8 \log_2 0.8 + 0.2 \log_2 0.2) \\ &= -1{,}000(p \log_2 p + q \log_2 q) \end{aligned}$$

(The dot over the equal sign indicates "approximately equal to.")

The information content of a longer message made up of many 0's and 1's thus depends on $p \log_2 p + q \log_2 q$, the information in bits per occurrence of a 0 or 1, times the relative frequency of occurrence of 0 or 1.

Generalizing for this case of two possible signals, we again consider a time interval T sec long, subdivided into intervals τ sec long. There are then $m = T/\tau$ possibilities for a 0 or 1 to occur. On the average ($m \gg 1/p$ and $1/q$) the 0 will appear $mp = (T/\tau)p$ times, the 1, $mq = (T/\tau)q$ times in the T-sec interval. (Remember again that q and p represent, respectively, the probability or relative frequency of occurrence of a 1 and a 0.)

The information content of a message T sec long is thus, on the average,

$$H = m(-p \log_2 p - q \log_2 q) \qquad \text{bits in } T \text{ sec} \tag{1-10}$$

The average information per interval τ sec long is

$$H_{\text{av}} = \frac{H}{m} = -p \log_2 p - q \log_2 q \qquad \text{bits/interval} \tag{1-11}$$

A communication system capable of transmitting this information should thus have an average capacity

$$C_{\text{av}} \geq \frac{H_{\text{av}}}{\tau} = \frac{1}{\tau}(-p \log_2 p - q \log_2 q) \qquad \text{bits/sec} \tag{1-12}$$

As a check consider the two possibilities, 0 or 1, equally likely. Then $p = q = 0.5$, and

$$H_{\text{av}} = \log_2 2 = 1 \text{ bit/interval}$$

or

$$\frac{T}{\tau} \log_2 2 = \frac{T}{\tau} \qquad \text{bits in } T \text{ sec}$$

(Note that $n = 2$ here.)

As a further check $p = 1, q = 0$ or $q = 1, p = 0$ gives $H_{\text{av}} = 0$. This corresponds to the case of a completely determined message (all 0's or 1's), which should carry no information according to our previous intuitive ideas.

Since $q = 1 - p$ in this simple case, H_{av} may be plotted as a function of p to give the curve of Fig. 1-9.[1] H_{av} reaches its maximum value of 1 bit per interval when $p = q = \frac{1}{2}$, or the two possibilities are equally likely. This is of course also in accord with our intuitive ideas that the

[1] Shannon, *op. cit.*

maximum information should be transmitted when events are completely random, or equally likely to occur.

We can now generalize to the case of n possible signals or signal levels in any one signal interval τ sec long. This could be the n levels of Sec. 1-2, the 26 possible letters in the English alphabet, any group of symbols or numbers of which one appears at a time, etc. We again seek an expression for the information content of a message T sec long ($T > \tau$) and the average information in bits for the interval τ sec long.

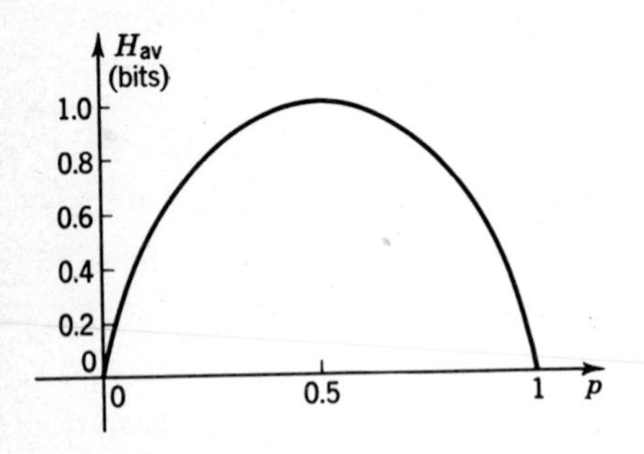

FIG. 1-9 Average information per interval with two possible signals. (From C. E. Shannon, A Mathematical Theory of Communication, *Bell System Tech. J.*, vol. 27, p. 394, fig. 7, July, 1948. Copyright, 1948, The American Telephone and Telegraph Co., reprinted by permission.)

Let the relative frequency of occurrence, or probability, of each possible signal level or symbol be $P_1, P_2, \ldots, P_n$, respectively,

$$P_1 + P_2 + \cdots + P_n = 1$$

and, for the jth symbol, $0 \leq P_j \leq 1$. If we again assume that occurrences in adjacent intervals are independent (i.e., that any particular symbol occurring in any one interval does not affect the relative rate of occurrence of any of the symbols in any other interval), we can find the information carried by a particular selection in any one interval, the total information (on the average) in T sec or $m = T/\tau$ intervals, and the average per τ-sec interval, as before.

As previously, if level (or symbol) j appears in any interval, it carries $-\log_2 P_j$ bits of information. In m intervals j will appear, on the average, mP_j times. The information in bits contributed, on the average, by each symbol appearing mP_j times in m intervals is then summed to give

$$H = -m \sum_{j=1}^{n} P_j \log P_j \qquad \text{bits in } m \text{ intervals}$$

$$H = -\frac{T}{\tau} \sum_{j=1}^{n} P_j \log P_j \qquad \text{bits in } T \text{ sec} \qquad [1\text{-}13]$$

The *average* information per single symbol interval (τ sec long) of a message with n possible symbols or levels, of probability P_1 to P_n, respectively, is then

$$H_{av} = -\sum_{j=1}^{n} P_j \log P_j \qquad \text{bits/interval} \qquad [1\text{-}14]$$

With τ-sec intervals the rate of transmission ~~of information~~ is $1/\tau$ symbols per second. The capacity required of a system to transmit this information would thus be

$$C_{av} \geq -\frac{1}{\tau} \sum_{j=1}^{n} P_j \log P_j \qquad \text{bits/sec} \qquad [1\text{-}15]$$

As a check let $P_1 = P_2 = \cdots = P_n = 1/n$ (equally likely events). Then

$$H = m \log_2 n \qquad \text{bits in } m \text{ intervals} \qquad [1\text{-}16]$$

as before.

As an example of the application of these results, we again consider the problem of determining the information content of a typical message of English speech. This is of course important knowledge required in determining the capacity of a communication system to be used for transmission of messages in English.

If we assume that the occurrence of any letter in the English alphabet is independent of preceding letters or words (a gross approximation), we may use a table of the relative frequency of occurrence of letters in the English alphabet to get an approximate idea of the information content of English speech or writing. This in turn can give us some estimate of the system capacity needed to transmit English at any specified rate (letters or words per second). Such a table appears in a book by Fletcher Pratt[1] and has been reprinted by S. Goldman[2] in his book "Information Theory." C. E. Shannon[3] in the article already referred to reproduces some of the information available.

[1] Fletcher Pratt, "Secret and Urgent," Garden City, New York, 1942.

[2] S. Goldman, "Information Theory," Prentice-Hall, Englewood Cliffs, N.J., 1953.

[3] Shannon, A Mathematical Theory of Communication, *op. cit.*

The relative frequency of occurrence of the letter *e* is found to be 0.131 (131 times in 1,000 letters), *t* occurs 0.105 of the time, *a* 0.086 of the time, etc., all the way down to *z* with a probability of 0.00077 (0.77 times in 1,000 letters). Equation [1-14] then gives, as a first approximation to the information content of English, 4.15 bits per letter. Had the letters been equally likely to occur ($P_j = 1/26$ for all 26 letters), we would have had $H_{av} = \log_2 26 = 4.7$ bits per letter. The fact that some letters are more likely to occur than others has thus *reduced* the information content of English from 4.7 bits to 4.15 bits per letter. If six letters are to be transmitted every second, we require a system with a capacity of at least $6 \times 4.15 = 25$ bits/sec. Doubling the number of letters to be transmitted doubles the required capacity as well.

The information content of English is actually much less than the 4.15 bits per letter figure because there is of course some dependence between successive letters, words, and even groups of words. Thus, if the letter *q* occurs, it is almost certain to be followed by a *u*. These two occurrences (*q* and then *u*) are thus not independent, and the uncertainty of a message with a *q* occurring is reduced, and with it the message information content. Similarly, *t* is frequently followed by an *h*, *r*, or *e* and almost never by a *q* or an *x*. Certain patterns of letters in groups of two thus occur much more frequently than others. The same holds true for groups of three letters. (The group *ter* frequently occurs in that order, while *rtn*, as an example, rarely appears.) Patterns also exist for four- and even five-letter combinations (*mani-*, *semi-*, etc.). In addition, there is dependence between successive words. The word *the* is almost always followed by a noun or adjective, a noun is frequently followed by a verb, etc. Various other words commonly occur together.

All these constraints on different letter and word combinations in English tend to reduce its information content. This reduction in information content of a message from the maximum possible (equally likely and independent symbols) is called the *redundancy* of the message. The redundancy of English, as an example, has been estimated to be considerably more than 50 percent.[1]

To calculate the information content of English more accurately, one would have to consider the influence of the different letters and words on one another. This requires additional knowledge of the statistics of the language, for example, the probability that *e* occurring would be followed by *a* or *b* or any of the other letters, the probability that a *th* would be followed by *a* or *b*, etc. These probabilities can also be calculated from the relative frequencies of occurrence of the different combinations.

[1] Goldman, *op. cit.*, p. 45.

They are called *conditional probabilities* because they relate the occurrence of an event to the previous occurrence of another event. We shall have occasion to discuss these probabilities in a later chapter in the book.

The existence of redundancy in message transmission can also be demonstrated very simply in the case of TV pictures. Here one signal element or interval corresponds to one spot on the screen, and the time taken for the electron beam to move one element corresponds to the time interval τ in our previous example. n, the number of levels, then corresponds to the number of intensity levels from white to gray to black that can be distinguished. In the case of TV one does not expect adjacent elements to change drastically very often from black to white as the beam sweeps across the screen (usually there is a gradual change from black through gray to white). In addition, a particular element will not be expected to change very much from one sweep interval to the next ($\frac{1}{30}$ sec). More than likely a given picture will persist for a while, backgrounds may remain the same for long intervals, small areas of black will remain black for a while, etc. The signal message considered as a time sequence, with various voltage levels corresponding to the different brightness levels and τ to the time to move across one element, will thus not consist of equally likely voltage levels, with the possibility of completely independent changes from one τ interval to the next. These constraints reduce the different number of signal combinations possible and thus quite markedly the average information content of a TV picture. (The limiting case again corresponding to the one in which the picture remains unchanged indefinitely. This implies no information content: you might as well turn off your set and go to bed.) Television scenes have a high percentage of redundancy and for this reason require (at least theoretically) much less system capacity than under the assumption of equally likely and independently varying signal levels. (Various estimates have indicated the redundancy of a typical TV pattern to be as high as 99.9 percent!)

PROBLEMS

1-1 In facsimile transmission, 2.25×10^6 square picture elements are needed to provide proper picture resolution. (This corresponds to 1,500 lines in each dimension.) Find the maximum information content if 12 brightness levels are required for good reproduction.

1-2 An automatic translator, with a capacity of 15×10^3 bits/sec, converts information from one coding system to another. Its input is a train of uniformly spaced variable-amplitude pulses, 2.71×10^5 pulses occurring each minute. Its output is another uniformly spaced variable-amplitude pulse train, with one-fifth the number of possible amplitude levels as in the input. Find the repetition rate of the output pulses.

1-3 (*a*) Find the capacity in bits per second that would be required to transmit TV picture signals if 500,000 picture elements were required for good resolution and 10 different brightness levels were specified for proper contrast. Thirty pictures per second are to be transmitted. All picture elements are assumed to vary independently, with equal likelihood of occurrence.

(*b*) In addition to the above requirements for a monochrome system a particular color TV system must provide 30 different shades of color. Show that transmission in this color system requires almost $2\frac{1}{2}$ times as much capacity as the monochrome system.

1-4 Refer to Prob. 1-3*b*. If 10 of the 30 color shades require only 7 brightness levels instead of 10, what is the capacity of the system? How many times greater is this capacity than that required for the monochrome system described in Prob. 1-3*a*?

1-5 Express the following decimal numbers in the binary system of notation: 6, 16, 0, 33, 1, 63, 127, 255, 117.

1-6 A system can send out a group of four pulses, each of 1 msec width, and each equally likely to have a height of 0, 1, 2, or 3 volts. The four pulses are always followed by a pulse of height -1, to separate the groups. A typical sequence of groups is shown in Fig. P 1-6. What is the average rate of information in bits per second that is transmitted with this system?

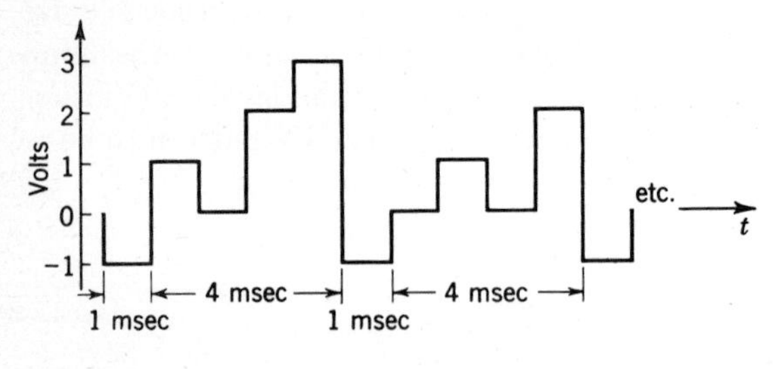

FIG. P 1-6

1-7 In the system of Prob. 1-6 the zero voltage level occurs one-half the time on the average, the 1-volt level occurs one-fourth the time on the average, and the remaining two levels occur one-eighth the time each on the average. Find the average rate of transmission of information.

1-8 An alphabet consists of the letters *A*, *B*, *C*, *D*. For transmission each letter is coded into a sequence of two binary (on-off) pulses. The *A* is represented by 00, the *B* by 01, the *C* by 10, the *D* by 11. Each individual pulse interval is 5 msec.

(*a*) Calculate the average rate of transmission of information if the different letters are equally likely to occur.
(*b*) The probability of occurrence of each letter is, respectively, $P_A = 1/5$, $P_B = 1/4$, $P_C = 1/4$, $P_D = 3/10$. Find the average rate of transmission of information in bits per second.

1-9 Repeat Prob. 1-8 with the letters coded into single pulses of 0, 1, 2, or 3 volts amplitude and of 10 msec duration.

1-10 A communication system is used to transmit one of 16 possible signals. Suppose that the transmission is accomplished by encoding the signals into binary digits.

(*a*) What will be the pulse sequence for the thirteenth symbol; for the seventh symbol?
(*b*) If each binary digit requires 1 μsec for transmission, how much information in bits does the system transmit in 8 μsec? Assume that the signals are equally likely to occur.
(*c*) If the symbols are sent directly without encoding, it is found that each symbol requires 3 μsec for transmission. What is the information rate in bits per second in this case?

CHAPTER TWO

TRANSMISSION THROUGH ELECTRIC NETWORKS

We indicated in the previous chapter that the system capacity, or rate of information transmission through a communication system, is related to the rapidity with which signals may change with time. From studies of the transient behavior of networks we know that in all networks with energy-storage elements (L and C) currents, or voltages as the case may be, cannot change instantaneously with time. A specified length of time is required (depending on the network) to reach a desired amplitude level.

In all networks inherent capacitance and inductance limit the time response. In many networks additional limitations are purposely imposed by adding filtering circuits which include inductance and capacitance. From our previous studies we also know that the time, or transient, response is inherently related to the familiar frequency, or steady-state sine-wave, response.

Since frequency concepts are widely used in radio and communication practice, and since frequency analysis of networks frequently simplifies the study of a system, we shall review and extend, in detail, the relation between frequency and time response. The importance of these concepts can be seen from some typical examples:

1. Radio-broadcasting stations are required to operate at their assigned frequency with very tight tolerances. Channels are spaced every 10 kHz in the amplitude-modulation (AM) broadcast band. This 10-kHz spacing is specified in order to prevent overlapping of stations and to allow as many stations as possible to be "squeezed into" the available frequency spectrum. As we shall see later on, these severe restrictions limit the maximum rate of information transmission.
2. Telephone cables are limited in their transient response (or, alternatively, in their frequency response). For a given cable to accommodate as many signal channels as possible, a limitation must be placed upon the frequency extent, or bandwidth, of each channel.

This again limits the rate of information transmission in a specified channel.

3. Television stations are limited to 6-MHz bandwidth, again to conserve available frequency space. This in turn imposes limitations on information-transmission capabilities.

Note that in all these simple examples it is the *frequency response* of the network that is specified. This has become common practice, and so it is important to study in detail the relation between frequency and time response, relating both in turn to the particular networks involved.

Consider the simplest set of examples of familiar time functions,

$$f(t) = a_1 \sin \omega_0 t,\ a_2 \sin 2\omega_0 t,\ a_3 \sin 3\omega_0 t,\ \ldots,\ a_n \sin n\omega_0 t \qquad [2\text{-}1]$$

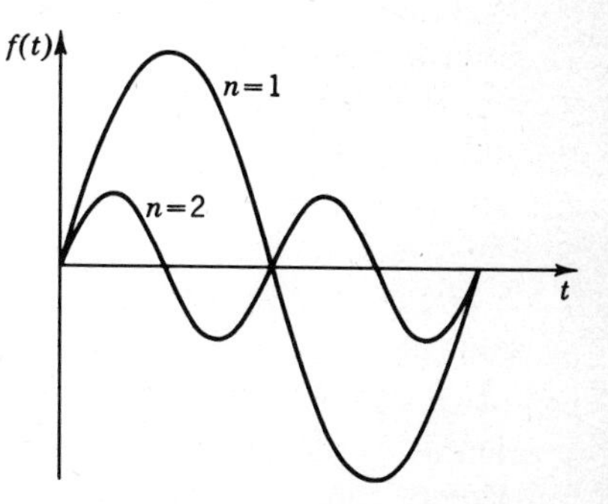

FIG. 2-1 Sinusoidal time functions.

The first two functions are shown in Fig. 2-1. As n increases, the rates of variation with time become more rapid. This may also be seen easily by comparing the different derivatives,

$$\frac{df(t)}{dt} = a_n n\omega_0 \cos n\omega_0 t \qquad [2\text{-}2]$$

As n increases, the maximum rate of change of $f(t)$ increases.

We can represent these functions in a different way by plotting the amplitude a_n versus angular frequency, as in Fig. 2-2. As the frequency increases, the time function varies more rapidly.

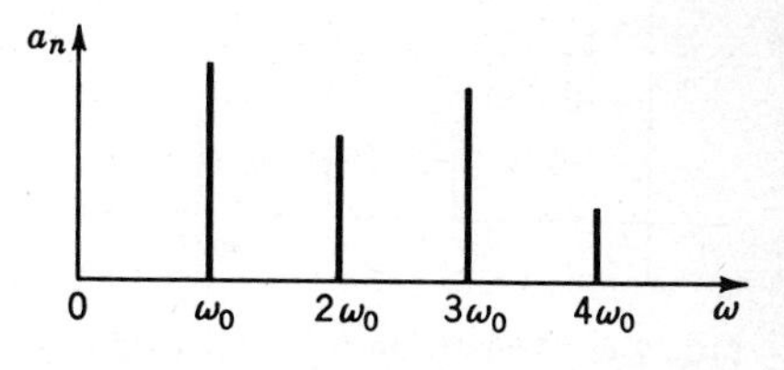

FIG. 2-2 Amplitude-frequency plot.

Now consider a simple example of an amplitude-modulated signal,

$$f(t) = A(1 + \cos \omega_m t) \cos \omega_0 t \qquad \omega_m \ll \omega_0 \tag{2-3}$$

The amplitude varies slowly (as compared with $\cos \omega_0 t$ time variations) between 0 and $2A$. Its rate of variation is given by ω_m, the modulating frequency. ω_0 is the carrier frequency. Figure 2-3 is a plot of one

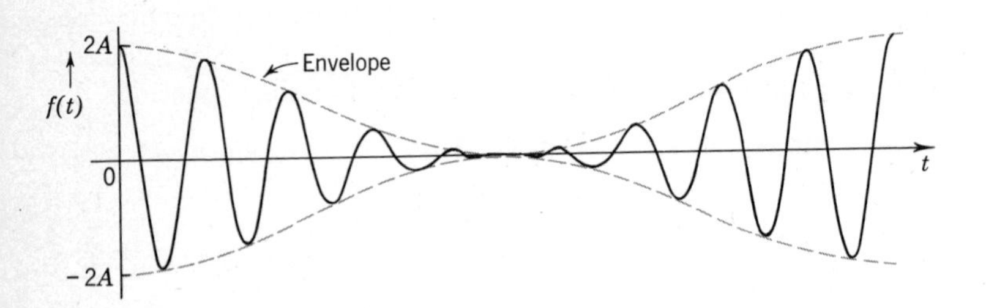

FIG. 2-3 Amplitude-modulated wave.

cycle of this function. The amplitude variations form the envelope of the complete signal and represent any information being transmitted. (In a practical situation the envelope would actually be a more complex function of time.)

A simple trigonometric manipulation of Eq. [2-3] gives

$$A(1 + \cos \omega_m t) \cos \omega_0 t = A \cos \omega_0 t + \frac{A}{2} [\cos (\omega_0 - \omega_m)t + \cos (\omega_0 + \omega_m)t] \tag{2-4}$$

The complete function can thus be represented as the sum of three sinusoidal functions and be plotted on an amplitude-frequency graph (Fig. 2-4). The two smaller lines are called the sideband frequencies;

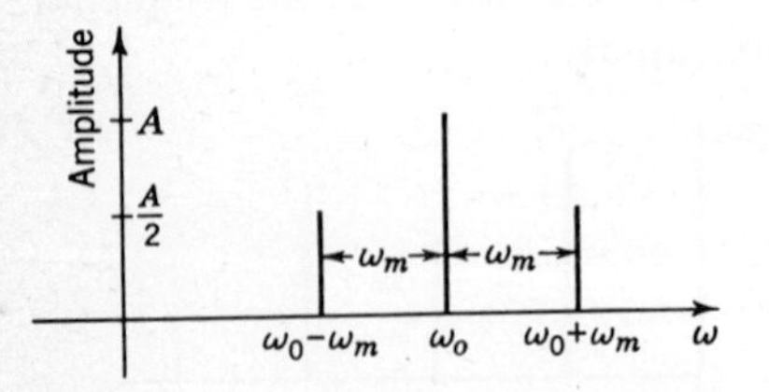

FIG. 2-4 Amplitude-frequency plot; modulated carrier.

the larger, central line, the carrier. As the amplitude-modulating signal varies more rapidly, ω_m increases and the sideband frequencies move farther away from the carrier. We therefore again have the notion that more rapid variations correspond to wider frequency swings.

These simple examples have all been cases of sinusoidal variations. As such, they are presumed to exist for all time and so carry no information. (Information-carrying signals are continuously varying in an unpredictable manner.) They are valuable in studying networks and systems, however, since any physical time function existing over a finite time interval can be expanded into a Fourier series of sinusoidal functions. Such a series will then represent that function over the interval desired. (Because of the periodic nature of sinusoidal functions the Fourier series representation will then repeat itself regularly outside the specified time interval.)

It is this concept that makes the frequency approach so useful in all electrical work. We are quite familiar with the steady-state sinusoidal response of networks. It is also a fact that sinusoidal analysis is frequently much simpler than a time, or transient, analysis. The ability to relate time to frequency response will thus simplify the solution of many problems.

The two examples previously discussed indicated that as the rate of time variation increased the frequency increased also. To generalize this concept somewhat, consider the series of periodic rectangular pulses shown in Fig. 2-5. Note that these pulses are related to the binary pulse

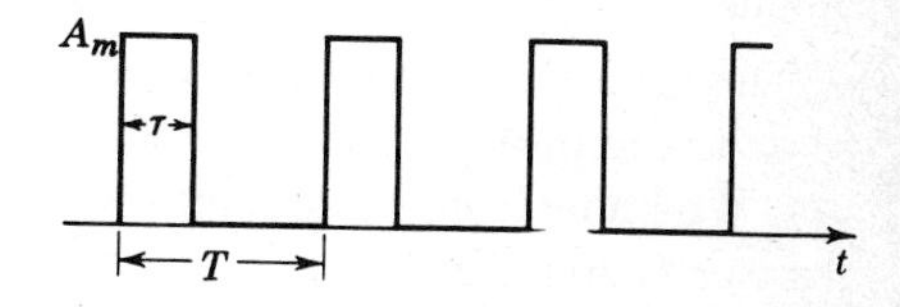

FIG. 2-5 Periodic pulse sequence.

sequence introduced in Sec. 1-1. We shall be dealing with such rectangular pulse sequences throughout the entire book because of their widespread occurrence in digital communications. Such a sequence of pulses can of course be expanded into a Fourier series. We ask the question: What is the amplitude-frequency plot of this sequence? This will further fix the time-frequency connection and will then enable us to go further into an examination of the effect of networks on time functions. Before treating this particular problem we review the Fourier-series concept.

2-1 REVIEW OF FOURIER SERIES

Let $f(t)$ be a periodic function of time with period T. Then $f(t)$ may be expanded into the following Fourier series[1] (we shall not deal here with the mathematical conditions necessary):

$$f(t) = \frac{a_0}{T} + \frac{2}{T} \sum_{n=1}^{\infty} (a_n \cos \omega_n t + b_n \sin \omega_n t) \qquad \omega_n = \frac{2\pi n}{T} \tag{2-5}$$

To find the constants a_n, we multiply through by $\cos \omega_n t$ and integrate over the period. All terms on the right-hand side vanish except the a_n term, since

$$\int_{-T/2}^{T/2} \cos \omega_j t \cos \omega_n t \, dt = 0 \qquad j \neq n$$
$$\int_{-T/2}^{T/2} \sin \omega_j t \cos \omega_n t \, dt = 0 \qquad \text{all } j$$

This gives us

$$\int_{-T/2}^{T/2} f(t) \cos \omega_n t \, dt = \frac{2a_n}{T} \int_{-T/2}^{T/2} \cos^2 \omega_n t \, dt = \frac{2a_n}{T} \frac{T}{2} = a_n \tag{2-6}$$

[Remember that $\cos^2 \omega_n t = (1 + \cos 2\omega_n t)/2$.] Thus

$$a_n = \int_{-T/2}^{T/2} f(t) \cos \omega_n t \, dt \qquad n = 1, 0, 2, 3, \ldots \tag{2-7}$$

Similarly, $$b_n = \int_{-T/2}^{T/2} f(t) \sin \omega_n t \, dt \qquad n = 1, 2, \ldots \tag{2-8}$$

The amplitude-frequency plot is proportional to a plot of $\sqrt{a_n^2 + b_n^2}$ versus ω_n. [The phase-frequency characteristic is a plot of $\tan^{-1} (-b_n/a_n)$ versus ω_n.] This plot of $\sqrt{a_n^2 + b_n^2}$ versus frequency will be referred to as the amplitude *spectrum* of the function.[2]

We know from our circuit analysis that, if the voltage across a 1-ohm resistor is given by

$$v_n(t) = A_n \cos \omega_n t + B_n \sin \omega_n t$$

[1] Many forms of the series may be written. For example,

$$f(t) = A_0 + \sum_{n=1}^{\infty} (A_n \cos \omega_n t + B_n \sin \omega_n t)$$

is commonly used. This is the same as Eq. [2-5] with

$$A_0 = \frac{a_0}{T} \qquad A_n = \frac{2a_n}{T} \qquad B_n = \frac{2b_n}{T}$$

[2] Note that this differs from the amplitude-frequency plot by the constant T.

the average power dissipated in the resistor is $(A_n{}^2 + B_n{}^2)/2$ watts. Alternatively, if $A_n = 2a_n/T$ and $B_n = 2b_n/T$, the average power dissipated can be written as

$$\left(\frac{2}{T}\right)^2 \frac{a_n{}^2 + b_n{}^2}{2}$$

The square of the amplitude spectrum is thus a measure of the power dissipated in a 1-ohm resistor at the different frequencies ($n = 0, 1, 2, \ldots$). By adding the power dissipated at each frequency we get the total average power dissipated when a periodic voltage is impressed across a resistor.

In our work to follow we shall be more interested in the amplitude spectrum $\sqrt{a_n{}^2 + b_n{}^2}$ and the phase angle $\tan^{-1}(-b_n/a_n)$ than in the individual Fourier coefficients a_n and b_n. (Note that in general two quantities must be specified at each frequency in order completely to specify the Fourier series.) This implies writing our Fourier series in the form

$$f(t) = \frac{a_0}{T} + \frac{2}{T} \sum_{n=1}^{\infty} \sqrt{a_b{}^2 + b_n{}^2} \cos(\omega_n t + \theta_n) \qquad \theta_n = \tan^{-1} \frac{-b_n}{a_n} \tag{2-9}$$

Equation [2-9] can of course be obtained from Eq. [2-5] by a simple trigonometric manipulation.

Since we shall frequently be interested in the amplitude and phase characteristics $\sqrt{a_n{}^2 + b_n{}^2}$ and θ_n of a periodic function, it would be much simpler to obtain these directly from $f(t)$, rather than by first finding a_n and b_n. We shall show that this can be done very simply by using yet another form of the Fourier series, the complex exponential form. This alternative form of the series may be written as

$$f(t) = \frac{1}{T} \sum_{n=-\infty}^{\infty} c_n e^{j\omega_n t} \tag{2-10}$$

The Fourier coefficient c_n is then a complex number defined as

$$c_n \equiv a_n - jb_n = \sqrt{a_n{}^2 + b_n{}^2}\, e^{j\theta_n} = \int_{-T/2}^{T/2} f(t) e^{-j\omega_n t}\, dt \tag{2-11}$$

$|c_n| = \sqrt{a_n{}^2 + b_n{}^2}$ is thus the desired amplitude spectrum and

$$\theta_n = \tan^{-1} \frac{-b_n}{a_n}$$

represents the phase characteristic. The coefficient c_n gives the complete frequency spectrum.

Equations [2-10] and [2-11] are completely equivalent to Eqs. [2-5], [2-7], and [2-8]. Not only does Eq. [2-11] give the amplitude and phase characteristics directly, but Eqs. [2-10] and [2-11] represent a much more compact form of the Fourier series,

$$f(t) = \frac{1}{T} \sum_{n=-\infty}^{\infty} c_n e^{j\omega_n t} \qquad [2\text{-}10a]$$

$$c_n = \int_{-T/2}^{T/2} f(t) e^{-j\omega_n t}\, dt \qquad [2\text{-}11a]$$

In rewriting Eqs. [2-5], [2-7], and [2-8] in this new form use has been made of the exponential form for the sine and cosine. Thus, in Eq. [2-5] let

$$\cos \omega_n t = \frac{e^{j\omega_n t} + e^{-j\omega_n t}}{2}$$

$$\sin \omega_n t = \frac{e^{j\omega_n t} - e^{-j\omega_n t}}{2j}$$

Regrouping terms in Eq. [2-5],

$$f(t) = \frac{a_0}{T} + \frac{1}{T} \sum_{n=1}^{\infty} [e^{j\omega_n t}(a_n - jb_n) + e^{-j\omega_n t}(a_n + jb_n)] \qquad [2\text{-}12]$$

If $c_n \equiv a_n - jb_n$, $c_n^* \equiv a_n + jb_n$, where c_n^* is the complex conjugate of c_n. But

$$a_n - jb_n = \int_{-T/2}^{T/2} f(t)(\cos \omega_n t - j \sin \omega_n t)\, dt = \int_{-T/2}^{T/2} f(t) e^{-j\omega nt}\, dt \qquad [2\text{-}13]$$

from Eqs. [2-7] and [2-8]. Since $\omega_n = 2\pi n/T$, $e^{-j\omega_n t} = e^{-j(2\pi nt/T)}$ and

$$e^{+j\omega_n t} = e^{-j(2\pi/T)(-n)t} = e^{-j\omega_{-n} t}$$

Therefore

$$c_n^* = a_n + jb_n = \int_{-T/2}^{T/2} f(t) e^{j\omega_n t}\, dt = \int_{-T/2}^{T/2} f(t) e^{-j\omega_{-n} t}\, dt \qquad [2\text{-}14]$$

[Remember again that $\omega_n = 2\pi n/T$, $\omega_{-n} \equiv 2\pi(-n)/T$.] From Eq. [2-14], then, assuming $f(t)$ real, $c_n^* = c_{-n}$. (That is, replacing n by $-n$ in c_n gives c_n^*.) Equation [2-12] can now be rewritten as

$$f(t) = \frac{a_0}{T} + \frac{1}{T} \sum_{n=1}^{\infty} (e^{j\omega_n t} c_n + e^{j\omega_n t} c_{-n}) \qquad [2\text{-}15]$$

But summing over $-n$ from 1 to ∞ is the same as summing over $+n$ from -1 to $-\infty$ ($c_{-3} \equiv c_n$, $n = -3$). Also

$$c_0 = \int_{-T/2}^{T/2} f(t) e^{j0}\, dt = a_0$$

Equation [2-15] can thus be further simplified to

$$f(t) = \frac{1}{T} \sum_{n=-\infty}^{\infty} c_n e^{j\omega_n t} \tag{2-10}$$

$$c_n = \int_{-T/2}^{T/2} f(t) e^{-j\omega_n t}\, dt \tag{2-11}$$

Although "negative" frequencies seem to appear in Eq. [2-10], they are actually fictitious. For if Eq. [2-10] is rewritten in real form, Eq. [2-9] is obtained, in which the only frequencies appearing ($\omega_n = 2\pi n/T$) are positive. This is very simply shown by writing $c_n = |c_n|e^{j\theta_n}$. Then Eq. [2-10] becomes

$$\begin{aligned} f(t) &= \frac{1}{T} \sum_{n=-\infty}^{\infty} |c_n| e^{j(\omega_n t + \theta_n)} \\ &= \frac{1}{T} \sum_{n=1}^{\infty} [|c_n|(e^{j(\omega_n t + \theta_n)} + e^{-j(\omega_n t + \theta_n)})] + \frac{c_0}{T} \\ &= \frac{a_0}{T} + \frac{2}{T} \sum_{n=1}^{\infty} |c_n| \cos(\omega_n t + \theta_n) \end{aligned}$$

(Remember again that $-\omega_n = \omega_{-n}$, $-\theta_n = \theta_{-n}$.)

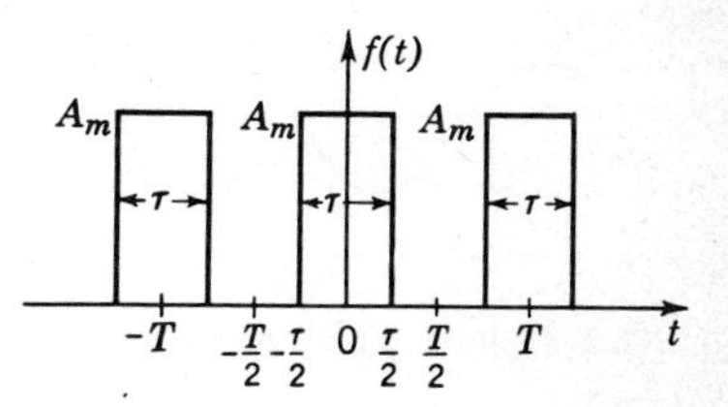

FIG. 2-6 Fourier analysis of periodic pulses.

As an example of the utility of this complex form of the Fourier series, consider the series of pulses of Fig. 2-6. (The origin has been chosen to coincide with the center of one pulse.) Then

$$\begin{aligned} c_n &= \int_{-\tau/2}^{\tau/2} A_m e^{-j\omega_n t}\, dt = -\frac{A_m}{j\omega_n} e^{-j\omega_n t} \Big]_{-\tau/2}^{\tau/2} \\ &= A_m \frac{e^{j\omega_n \tau/2} - e^{-j\omega_n \tau/2}}{j\omega_n} = \frac{2A_m}{\omega_n} \sin\frac{\omega_n \tau}{2} \end{aligned}$$

This may be written in the form

$$c_n = \tau A_m \frac{\sin (\omega_n \tau/2)}{\omega_n \tau/2} \tag{2-16}$$

If we define a normalized and dimensionless variable, $x = \omega_n \tau/2$,

$$c_n = \tau A_m \frac{\sin x}{x}$$

The $(\sin x)/x$ function will be occurring in many problems in the future and should be carefully studied. Note that it has its maximum value at $x = 0$, where $\sin x \to x$, $(\sin x)/x \to 1$. It approaches zero as $x \to \infty$, oscillating through positive and negative values. If x is a continuous variable, $(\sin x)/x$ has the form of Fig. 2-7.

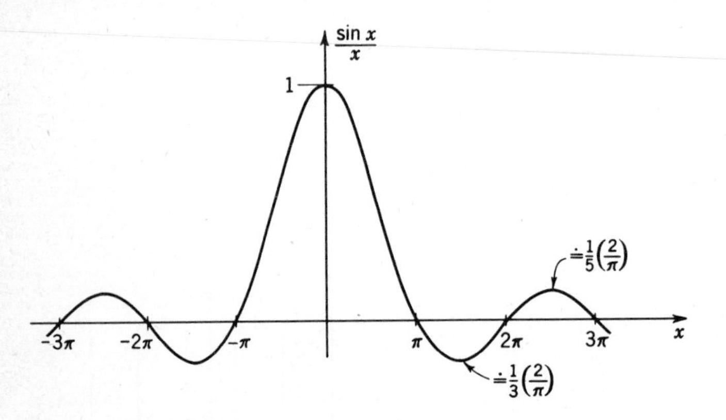

FIG. 2-7 $[(\sin x)/x]$ versus x.

In our particular problem n has discrete values only, ω_n takes on discrete values (harmonics of $\omega_1 = 2\pi/T$), and the normalized parameter x is thus also defined only at discrete points. The *envelope* of the plot of c_n will be exactly the curve of Fig. 2-7. The plot of c_n itself is shown in Fig. 2-8 ($\tau \ll T$). Since c_n is for this example a real number (alternately positive and negative), there is no need to find $|c_n|$ and θ_n and plot each separately. (Note again that the "negative" frequencies shown are just a mathematical artifice, since $\omega_n = 2\pi n/T$ and $\omega_{-n} = -2\pi n/T$.) The spacing between the successive lines is

$$\Delta\omega_n = \frac{2\pi}{T}(n + 1) - \frac{2\pi n}{T} = \frac{2\pi}{T}$$

or just the fundamental angular frequency. All lines of the frequency spectrum shown thus occur at multiples of this fundamental frequency. The "dc component" ($\omega_n = 0$) is of course just T times the average value.

TIME–FREQUENCY CORRESPONDENCE

Although the periodic function of Fig. 2-6 contains frequency components at all integral multiples of the fundamental frequency, the envelope of the amplitude decreases at higher frequencies. Note, however, that as the fundamental period T *decreases* (more pulses per second), the frequency lines move out farther. Again a *more rapid variation* in the time function corresponds to *higher-frequency* components. Alternatively, as T increases, the lines crowd in and ultimately approach an almost smooth frequency spectrum. Since the lines concentrated in the lower-frequency range are of higher amplitude, we note that most of the energy associated with this periodic wave is confined to the lower frequencies. As the function varies more rapidly (T decreases), the relative amount of the energy contained in the higher-frequency range increases.

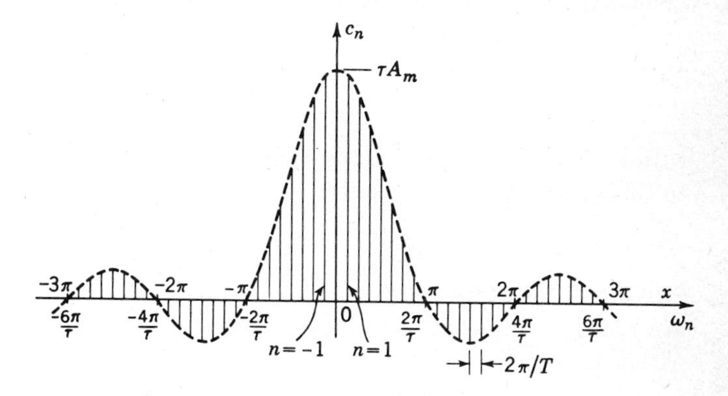

FIG. 2-8 Frequency spectrum, rectangular pulses ($\tau \ll T$).

Figure 2-8 and Eq. [2-16] emphasize another interesting phenomenon that will be very useful to us in later work. As the pulse width τ *decreases,* the frequency content of the signal extends out over a larger frequency range. The first zero crossing, at $\omega_n = 2\pi/\tau$, moves out in frequency. There is thus an *inverse relationship between pulse width, or duration, and the frequency spread of the pulses.*

If $\tau \ll T$ (that is, very narrow pulses), most of the signal energy will lie in the range

$$0 < \omega_n < \frac{2\pi}{\tau}$$

The first zero crossing is frequently a measure of the frequency spread of a signal (assuming, of course, that the envelope of the amplitude spectrum decreases with frequency). In keeping with the notation for networks we can talk of the bandwidth of the signal as being a measure of its frequency spread.

As in the case of the frequency response of networks the bandwidth occupied by the signal cannot be uniquely specified unless the signal is "band-limited" (i.e., occupies a finite range of frequencies with no frequency components beyond the range specified). However, some arbitrary (and frequently useful) criterion for bandwidth may be chosen to specify the range of frequencies in which most of the signal energy is concentrated. As an example, if the bandwidth B is specified as the frequency extent of the signal from zero frequency to the first zero crossing,

$$B = \frac{1}{\tau} \qquad [2\text{-}17]$$

where $\tau \ll T$. Any other criterion for bandwidth would still retain the inverse time-bandwidth relation, and, in general,

$$B = \frac{k}{\tau} \qquad [2\text{-}18]$$

with k a constant depending on the choice of criterion. We shall return to this important concept later.

POWER CONSIDERATIONS

We noted earlier that the average power dissipated in a 1-ohm resistor with a voltage

$$v_n(t) = A_n \cos \omega_n t + B_n \sin \omega_n t$$

impressed across it is just given by $(A_n^2 + B_n^2)/2$. If one deals with the complex exponential

$$v_n(t) = c_n e^{\omega_n t}$$

instead, it is apparent that the average power is given by

$$\bar{P} = \frac{1}{T}\int_0^T |v_n(t)|^2\, dt = |c_n|^2$$

This may be simply generalized to the case of an arbitrary periodic function $f(t)$ by using the Fourier-series expansion. Thus we recall that with an arbitrary (and possibly complex) voltage $f(t)$ impressed across a

1-ohm resistor the instantaneous power is just $|f(t)|^2$ and the average power is given by

$$\bar{P} = \frac{1}{T}\int_0^T |f(t)|^2\, dt \tag{2-19}$$

[If $f(t)$ is real, as in most of the examples of this book, $|f(t)|$ is simply replaced by $f(t)$.] Substituting for $f(t)$ its Fourier-series expansion of Eq. [2-10], we get

$$\bar{P} = \frac{1}{T^3}\int_0^T \left[\sum_m \sum_n c_m^* c_n e^{j(\omega_n - \omega_m)t}\right] dt \tag{2-19a}$$

Interchanging the order of summation and integration and noting that

$$\int_0^T e^{j(\omega_n-\omega_m)t}\, dt = T \qquad \omega_n = \omega_m$$
$$= 0 \qquad \text{elsewhere}$$

(this is simply checked by expanding in sines or cosines or by noting that $e^{j\omega t}$, although complex, is periodic, and averages to zero over a period), we get

$$\bar{P} = \frac{1}{T}\int_0^T |f(t)|^2\, dt = \sum_{n=-\infty}^{\infty} \left|\frac{c_n}{T}\right|^2 \tag{2-19b}$$

The average power is thus found by summing the power contribution at all frequencies in the Fourier-series expansion.

As an example, if we again consider the rectangular test pulses of Fig. 2-6, we have

$$\bar{P} = \frac{A_m{}^2\tau}{T} = \sum_{n=-\infty}^{\infty} \left(\frac{\tau A_m}{T}\right)^2 \frac{\sin^2(\omega_n\tau/2)}{(\omega_n\tau/2)^2} \tag{2-20}$$

This expression enables us to determine the relative power contributions at the various frequencies. For example the dc power is just $(c_o/T)^2$ or $(\tau A_m/T)^2$ in the rectangular pulse case. (This is of course easily checked by noting that the dc power is $\left[1/T \int_0^T f(t)\, dt\right]^2$.) The power in the first harmonic, that at $\omega_1 = 2\pi/T$, is just

$$2\left(\frac{\tau A_m}{T}\right)^2 \frac{\sin^2(\omega_1\tau/2)}{(\omega_1\tau/2)^2} \qquad \text{etc.}$$

(The factor of 2 is due to the equal power contributions at ω_1 and $\omega_{-1} = -\omega_1$.)

PERIODIC IMPULSES

A special case of the rectangular pulse train, of great utility in both digital-communication analysis and modern digital computation, is obtained by letting the pulse width τ in Fig. 2-6 go to zero and the amplitude $A_m \to \infty$, with $A_m\tau = 1$. This results in the extremely useful set of periodic impulses of infinite height, zero width, and unit area. These unit impulses are shown sketched in Fig. 2-9. The symbol usually adopted for an impulse of unit area and centered at $t = \tau$ is $\delta(t - \tau)$. It is represented by an arrow, as indicated in Fig. 2-9. If the area $(A_m\tau)$ is some number k, one writes $k\delta(t - \tau)$.

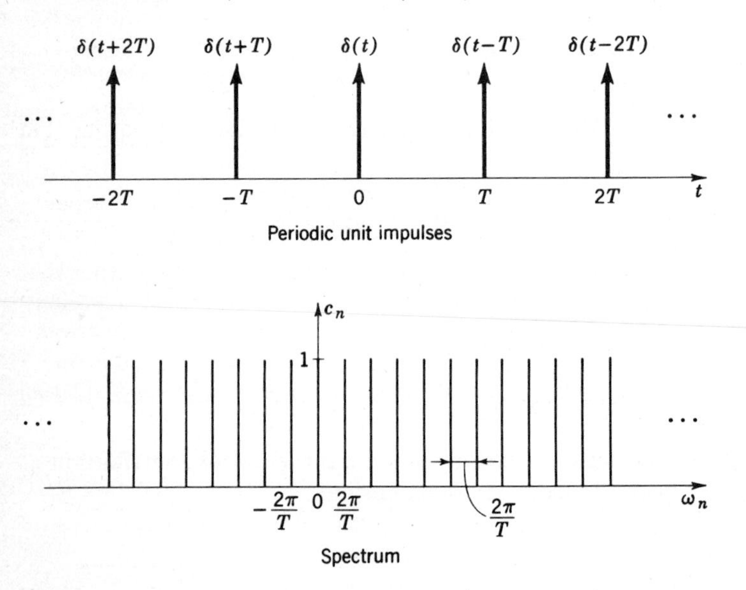

FIG. 2-9 Spectrum of periodic impulses.

We shall have more to say about the impulse function later, but note now that all spectral lines have the same height $A_m\tau$. The "bandwidth" thus approaches infinity. This agrees of course with the inverse time-bandwidth relation. As the width τ of the pulse $\to 0$, its bandwidth $1/\tau \to \infty$.

2-2 SIGNAL TRANSMISSION THROUGH LINEAR NETWORKS[1]

We have shown in the previous section that a series of rectangular pulses has an amplitude-frequency spectrum whose bandwidth is inversely pro-

[1] Although the emphasis throughout will be on electric networks, the conclusions drawn and analyses made will hold for all types of linear physical systems, assuming that the proper electrical analogs are found.

portional to the pulse width. Let us assume that these pulses will ultimately carry the information to be transmitted. In Chap. 1 it was pointed out that increased information transmission would correspond to narrowing these pulses down as much as possible. But this implies increasing the bandwidth—an intolerable situation in many communication systems. There is thus a minimum pulse width allowable in any particular case, depending on the system requirements. For example, if these pulses were to be used for TV transmission with a 6 MHz bandwidth specified, the minimum allowable pulse width would be $\frac{1}{6}$ μsec. As noted previously, these pulses contain considerable energy in the frequency range outside the bandwidth defined. This would likewise be intolerable in those systems in which different signals would be required to occupy different frequency channels, with no overlapping between channels. Frequency-selective filters must thus be utilized to sharpen the frequency characteristics of the signal (the series of pulses in this case). This will change the signal time response and, of course, the allowable information content of the signal. In addition, parasitic capacitance and inductance in any electrical system (the analogous quantities in nonelectrical systems) serve to alter the frequency characteristics of the signal and, correspondingly, the time response and information-handling capacity.

As a simple example of these interrelated concepts, consider a periodic series of pulses passed through the simple RC filter of Fig. 2-10 ($\tau \ll T$).

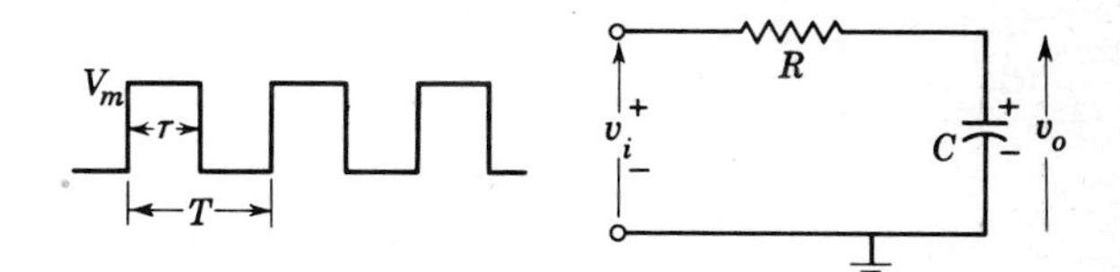

FIG. 2-10 Simple RC filter.

An analysis of the output voltage of this filter will enable us to draw simple conclusions as to the effect of the circuit on the time and frequency properties of the input pulses.

Assume that $RC \ll T$. The output voltage v_0 will then also be a series of nonoverlapping pulses. The spacing between successive pulses is so great that only the output due to one pulse at a time need be considered. (This is the practical situation, since large time constants RC lead to an overlapping of the output pulses and consequent jumbling of any information carried by the signal.) We know from transient analysis that the output and input waveshapes have the forms shown in Fig. 2-11. In

Fig. 2-11a, with $RC \geq \tau$

$$
\begin{aligned}
v_o(t) &= V_m(1 - e^{-t/RC}) && 0 < t < \tau \\
v_o(t) &= V'_m(e^{-(t-\tau)/RC}) && \tau < t,\ V'_m = V_m(1 - e^{-\tau/RC})
\end{aligned}
\qquad [2\text{-}21]
$$

In this case the RC network has considerably broadened the pulse, decreased its amplitude, and distorted its shape. But, provided that $RC \ll T$, we still obtain a unique output for each pulse input, with no mixing of the input pulses. (With $RC \ll T$ the output is negligibly small T sec later when the next pulse appears.)

The output in the case of Fig. 2-11b, with $RC \ll \tau$, differs considerably from that of Fig. 2-11a. Note that the output pulse is very much like the input, appearing almost rectangular in shape. The one notable difference between the input and output pulses, however, is the nonzero time required to reach the maximum output V_m, as well as the nonzero decay time as compared with the input.

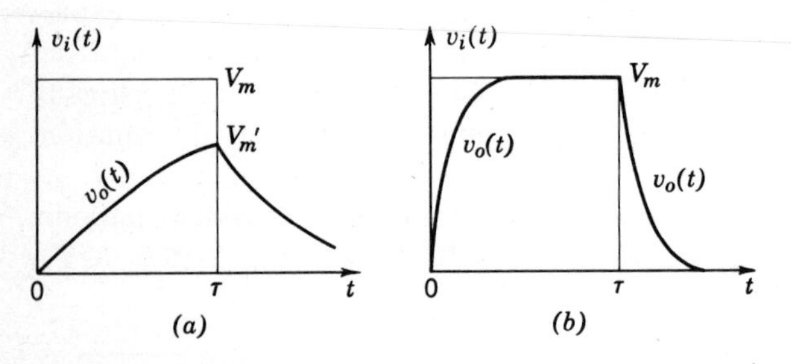

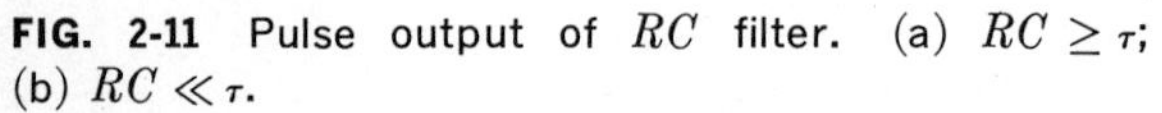

FIG. 2-11 Pulse output of RC filter. (a) $RC \geq \tau$; (b) $RC \ll \tau$.

The nonzero time required to reach a specified fraction of the peak voltage value V_m is commonly called the *rise time*. It is due, of course, to the presence of capacitance (the energy-storage element) in the circuit. The given input pulses were assumed to increase instantaneously from zero to maximum value—a highly idealized situation. The output pulses require a definite time to rise to any specified value. The rise time, as in the case of the bandwidth concept, is not uniquely defined (the rise here is exponential) but is frequently defined to be the time required for the pulse to rise from 10 to 90 percent of the final value.

As can be seen from Fig. 2-11b, the rise time is related to the time constant of the RC network. Very short rise times require small time constants. Figure 2-11a and b emphasizes two concepts that are extremely useful in all communication work. These concepts involve the relation between the input-pulse width and the time constant of the network through which the pulse is transmitted. (Later the time constant will be shown to be inversely proportional to the network bandwidth.) The two concepts may be phrased as follows:

1. If the pulse shape itself carries no information so that a somewhat "smeared-out" version of the pulse is allowable at the filter output (Fig. 2-11*a*), the filter time constant need be only approximately equal to the input-pulse width. Thus

$$RC \geq \tau \qquad [2\text{-}22]$$

in order to produce a pulse at filter output. But we still have $RC \ll T$ (interval between pulses) so that there will be no overlap between successive pulses.
2. If fidelity is required so that the output closely resembles the input (Fig. 2-11*b*), the filter time constant must be much less than the pulse width,

$$RC \ll \tau \qquad [2\text{-}23]$$

This implies rise times which are small compared with pulse width.

In more complicated filter networks it becomes difficult to talk of the network *time constant.* In these cases the filter-bandwidth concept becomes much more useful. We shall see very shortly that Eqs. [2-22] and [2-23] may be generalized to

$$B_f \leq \frac{1}{\tau}$$

for simple pulse reproduction with no special requirements for fidelity, and

$$B_f \gg \frac{1}{\tau}$$

where high fidelity (small rise time) is required. B_f is the normally defined 3-db filter bandwidth in hertz (i.e., the frequency at which the voltage amplitude is 0.707, or 3 db below the peak value in the case of a low-pass filter; for a bandpass filter the bandwidth would be the frequency difference between two 3-db points).

What effect does passing a series of pulses through a filter network have on the possible information content of the pulses, or, alternatively, how is the system information capacity (bits per second) limited by the network?

At the input to the filter the pulses were assumed rectangular, of duration τ sec, and appearing at intervals T sec apart. Additional pulses could thus have been inserted between the existing pulses. In particular, if a given pulse begins at $t = 0$ sec, a second pulse may begin and be inserted at $t = \tau$ sec. At the output of the filter this is no longer the case. The time insertion of the second pulse must be delayed somewhat in order to prevent overlap between two successive pulses. This is shown pictorially in Fig. 2-12. This undesired overlap of pulses is just the inter-

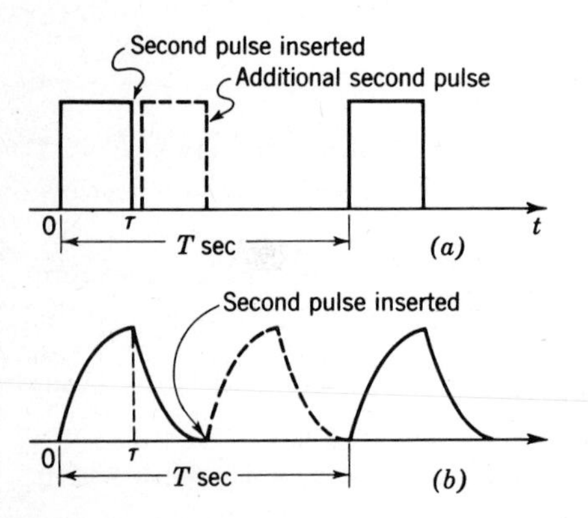

FIG. 2-12 Pulse lengthening due to RC filter. (a) Rectangular pulse train; (b) at filter output.

symbol interference mentioned briefly in Chap. 1. We shall discuss this concept in more detail in Chap. 3.

The number of pulses that could have been included, in a given time interval, has thus been limited by the network (containing an energy-storage device), limiting the corresponding amount of information that could have been transmitted in that interval. (Compare with the discussion in Sec. 1-2.)

Alternatively we can say that the bandwidth of the output set of pulses has been decreased by action of the RC filter. The RC network may be looked upon as either a band-limiting device or as an energy-storage device slowing down possible time changes.

FILTER-BANDWIDTH REQUIREMENTS

To demonstrate the band-limiting effect of the RC network, let us determine the frequency spectrum of the resultant output pulses. We shall do this formally by determining the complex Fourier coefficient c_n for the periodic series of output pulses.

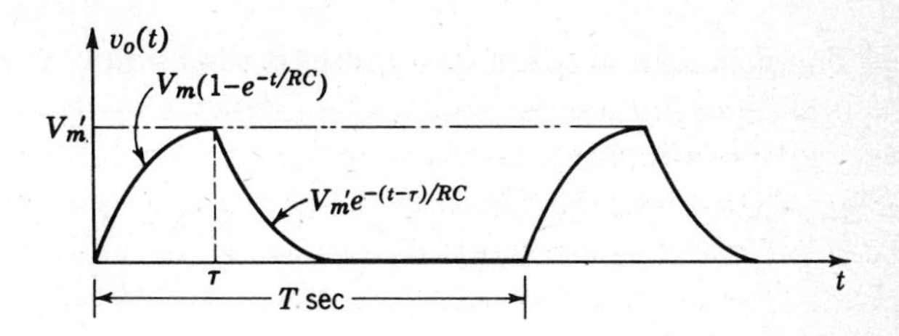

FIG. 2-13 Periodic pulsed output of RC network.

As the time origin is taken at $t = 0$, as shown in Fig. 2-13, the Fourier series and complex coefficients are then defined as follows:

$$v_o(t) = \frac{1}{T} \sum_{n=-\infty}^{\infty} c_n e^{j\omega_n t} \qquad \omega_n = \frac{2\pi n}{T} \tag{2-24}$$

$$c_n = \int_0^T v_o(t) e^{-j\omega_n t}\, dt \tag{2-25}$$

The evaluation of c_n is somewhat "messy" algebraically but quite straightforward.

$$c_n = \int_0^\tau V_m(1 - e^{-t/RC}) e^{-j\omega_n t}\, dt + \int_\tau^T V'_m e^{-(t-\tau)/RC} e^{-j\omega_n t}\, dt \tag{2-26}$$

with $V'_m = V_m(1 - e^{-\tau/RC})$. Since $RC \ll T$ (to prevent pulse overlapping), the second integral may be extended to an infinite limit without changing its value appreciably [$v_o(t)$ is assumed very nearly zero at $t = T$]. Then

$$c_n \doteq \int_0^\tau V_m(1 - e^{-t/RC}) e^{-j\omega_n t}\, dt + \int_\tau^\infty V''_m e^{-t/RC} e^{-j\omega_n t}\, dt \tag{2-27}$$

with $V''_m \equiv V'_m e^{\tau/RC} = V_m(e^{\tau/RC} - 1)$. Combining exponents in t, integrating the resultant exponentials, and collecting terms,

$$c_n = jV_m\tau e^{-j\omega_n\tau/2} \frac{e^{-j\omega_n\tau/2} - e^{j\omega_n\tau/2}}{\omega_n\tau} + \frac{(V_m + V''_m)e^{-\tau(1/RC + j\omega_n)}}{j\omega_n + 1/RC} - \frac{V_m}{j\omega_n + 1/RC}$$

But $\qquad V''_m = V_m(e^{\tau/RC} - 1) \qquad V_m + V''_m = V_m e^{\tau/RC}$

Therefore
$$\frac{c_n}{V_m\tau} = e^{-jx}\frac{\sin x}{x} + \frac{e^{-j2x} - 1}{\tau/RC + j2x} \qquad x \equiv \frac{\omega_n\tau}{2} \tag{2-28}$$

After some manipulation we finally obtain the desired result,

$$\frac{c_n}{V_m\tau} = \underbrace{e^{-jx}\frac{\sin x}{x}}_{\text{Spectrum of input pulses (including delay factor)}} \underbrace{\frac{1/RC}{1/RC + j\omega_n}}_{\text{Effect of filter}} \qquad x \equiv \frac{\omega_n\tau}{2} \tag{2-29}$$

The spectrum of the output pulses is thus simply the input-pulse spectrum multiplied by a complex function dependent upon the circuit constants.

The amplitude spectrum is

$$|c_n| = V_m\tau \left|\frac{\sin x}{x}\right| \frac{1/RC}{\sqrt{\omega_n^2 + (1/RC)^2}} \qquad x = \frac{\omega_n\tau}{2} \tag{2-30}$$

$|c_n|$ drops off more rapidly at higher frequencies than the corresponding coefficient for the input pulses, because of the filter term. The bandwidth of the pulse spectrum has thus been reduced by passing the pulses through the RC filter.

We normally define the bandwidth of the RC filter itself as being the frequency interval between zero frequency and the amplitude 3 db, or half-power point. For the RC network the 3-db bandwidth is simply

$$B_f = \frac{1}{2\pi RC} \qquad \text{Hz}$$

or

$$2\pi B_f = \frac{1}{RC} \qquad \text{radians/sec} \tag{2-31}$$

If we define the input-pulse bandwidth as the frequency interval to the first zero crossing of the amplitude spectrum,

$$B_p = \frac{1}{\tau} \tag{2-32}$$

Referring to Eq. [2-30], two special cases may be considered:

1. Assume $RC \gg \tau/\pi$. As ω_n increases from zero, the filter frequency term in Eq. [2-30] begins to decrease while the $(\sin x)/x$ term is still approximately equal to 1. When $\omega_n = 1/RC$, $\omega_n\tau/2 \ll \tau/2$, $(\sin x)/x \doteq 1$, and the overall 3-db bandwidth of the output-pulse spectrum is approximately that of the filter itself. From Eqs. [2-31] and [2-32] this corresponds to

$$B_f \ll B_p$$

or

$$B_f \ll \frac{1}{\tau} \tag{2-33}$$

If the filter bandwidth is less than the reciprocal of the input-pulse width, the filter limits the frequency spectrum of the pulses. Only those input-frequency components (Fourier-series components) within the passband of the filter are selected, and the output signal emerges a distorted replica of the input. This, of course, corresponds to the conclusions drawn from Fig. 2-11*a* and agrees with Eq. [2-22], derived from the transient analysis of the same problem.

2. $RC \ll \tau/2\pi$. In this case the overall bandwidth is dependent upon the original spectrum of the input pulses. For when $\omega_n = 2\pi/\tau$, $(\sin x)/x = 0$, but $(1/RC)/\sqrt{\omega_n^2 + (1/RC)^2} \doteq 1$. The output amplitude spectrum is changed only slightly by the filter from that of the input. The output pulses will thus look very much like the input pulses. This corresponds to the case

$$B_f \gg B_p$$

or

$$B_f \gg \frac{1}{\tau} \qquad [2\text{-}34]$$

The input pulses then appear as in Fig. 2-11*b*. This conclusion also agrees with that of Eq. [2-23], derived by the time analysis.

Conclusions as to system- (filter-) bandwidth requirements thus agree with the transient, or time, analysis. If $B_f \leq 1/\tau$, the output will be a distorted replica of the input, although the pulse character of the input will still be recognizable if B_f is not too small. Similarly, if $B_f \gg 1/\tau$, the output will be a reproduction of the input with high fidelity and the output rise time will be very much less than τ. These results are summarized pictorially in Fig. 2-14.

Summing up, the system bandwidth should be approximately the reciprocal of the pulse width to produce a recognizable pulse at the output. For small rise time and fidelity of reproduction, however, the bandwidth of the filter should be much greater than the reciprocal of the pulse width.

2-3 LINEAR NETWORKS, FILTER TRANSFER FUNCTIONS, AND COMPLEX FREQUENCY SPECTRA

Equation [2-29] was derived by expanding the pulse train at the output of the RC network in its complex Fourier series. In particular, the effect of the filter is given by the expression

$$\frac{1/RC}{1/RC + j\omega_n}$$

But this is just the ac complex transfer function (output voltage/input voltage) of the RC network at any sinusoidal frequency ω_n.

This suggests that Eq. [2-29] might have been written down directly, without the need actually to evaluate the Fourier coefficients by integration (a difficult job in this case). For consider one of the frequency components of the input-pulse spectrum. This is just a sine wave of fre-

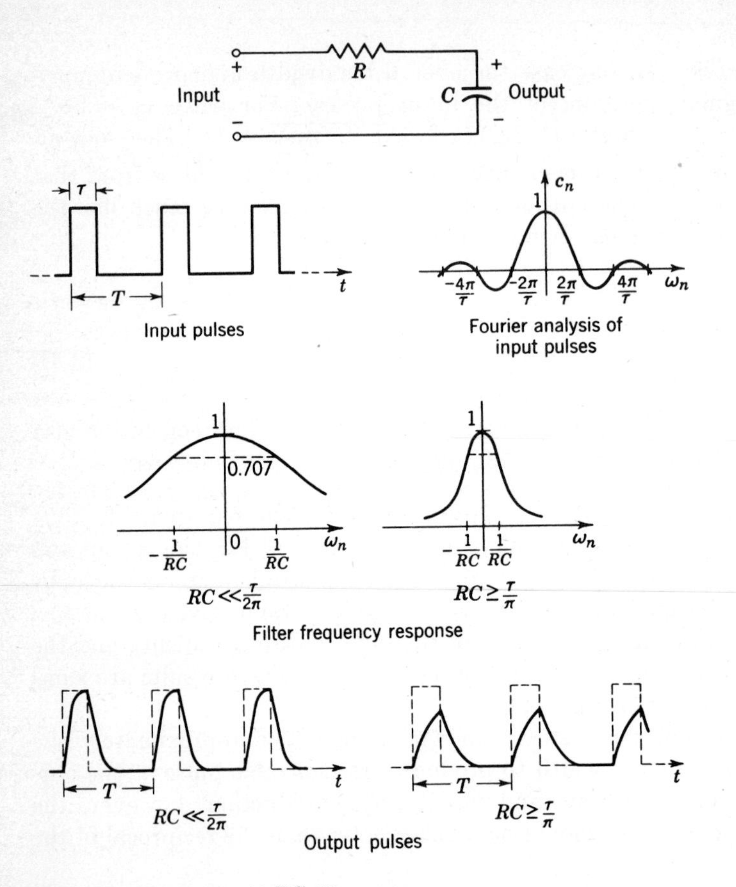

FIG. 2-14 Summary of RC filter response.

quency ω_n and amplitude $V_m\tau\{[\sin(\omega_n\tau/2)]/(\omega_n\tau/2)\}$. The complex filter output due to this one sine wave would simply be

$$V_m\tau \frac{\sin \omega_n\tau/2}{\omega_n\tau/2} \frac{1/j\omega_n C}{R + 1/j\omega_n C} = V_m\tau \frac{\sin x}{x} \frac{1/RC}{j\omega_n + 1/RC}$$

as in Eq. [2-29]. The total output due to all the input sine waves would be a superposition of an infinite number of terms of this type, one for each input frequency, or exactly the Fourier series of Eq. [2-24].

This result for the RC network is a basic one in our work and is a direct consequence of the assumed linearity, as well as time invariance (stationarity), of the network. For the concept of superposition, necessary in summing the response to all input sine waves, is in fact the defining relation of a linear network.

Specifically the linear stationary systems of which the RC network is just one simple example are defined in terms of two basic properties:

1. The response to a sum of excitations is equal to the sum of the responses to the excitations acting separately.
2. The relations between input and output are time-invariant or stationary.

The first condition is of course the statement of superposition noted above. The second condition implies that the system elements do not change with time.

Our familiar linear operations are those of multiplication by a constant, addition, subtraction, differentiation, and integration. The combination of these operations gives rise to a time-invariant linear system governed by constant-coefficient linear differential equations. The RC network of Fig. 2-14 is a particularly simple example of such a system.

The first condition cited above as defining a linear system may be summarized by saying that if an input $f_1(t)$ gives rise to an output $g_1(t)$, while an input $f_2(t)$ produces $g_2(t)$ at the output, the input $af_1(t) + bf_2(t)$ results in the output $ag_1(t) + bg_2(t)$. This is shown diagrammatically in Fig. 2-15. The time-invariant condition says that an input $f_1(t - \tau)$

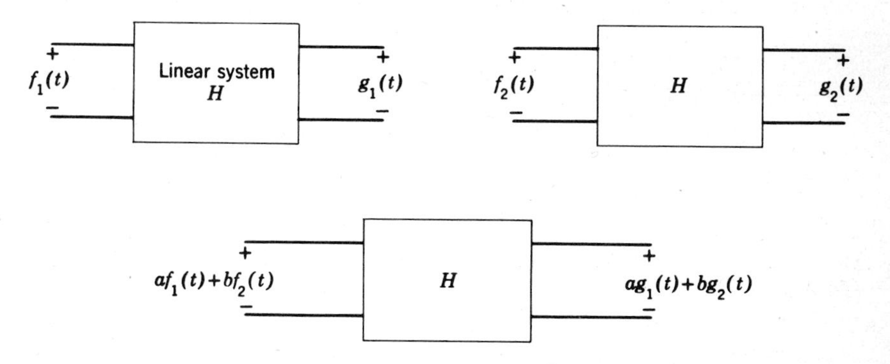

FIG. 2-15 Linear system, defining relations.

gives rise to an output $g_1(t - \tau)$. It is these simple conditions that enable Fourier analysis to play so significant a role in linear system analysis. For the system response to a number of sine waves is simply the sum of the responses to each individual sine wave. But any periodic function $f(t)$ may be decomposed into its Fourier series of sine waves (or complex exponentials). The sum of the responses to each sine wave is then just the Fourier-series representation of the output periodic function $g(t)$, providing a powerful tool in linear system analysis.

Specifically, let $H(\omega_n)e^{j\omega_n t}$ be the response of a linear network to the complex exponential $e^{j\omega_n t}$ applied at the input. $H(\omega_n)$ is termed the (complex) *frequency transfer response* (or *function*). With $c_n e^{j\omega_n t}$ applied at the input, $c_n H(\omega_n)e^{j\omega_n t}$ is the resultant output. If we now apply at the input

$$f(t) = \frac{1}{T} \sum_{n=-\infty}^{\infty} c_n e^{j\omega_n t} \qquad \omega_n = \frac{2\pi n}{T} \tag{2-35}$$

with

$$c_n = \int_{-T/2}^{T/2} f(t)e^{-j\omega_n t}\,dt \tag{2-36}$$

the output response $g(t)$ is, by the superposition condition,

$$\begin{aligned} g(t) &= \frac{1}{T} \sum_{n=-\infty}^{\infty} c_n H(\omega_n)e^{j\omega_n t} \\ &= \frac{1}{T} \sum_{n=-\infty}^{\infty} G(\omega_n)e^{j\omega_n t} \end{aligned} \tag{2-37}$$

with

$$G(\omega_n) = c_n H(\omega_n) \tag{2-38}$$

the complex frequency spectrum of the periodic output $g(t)$. Knowing the input Fourier coefficients c_n and the system transfer function $H(\omega_n)$ one may readily find the output response $g(t)$. This Fourier-series

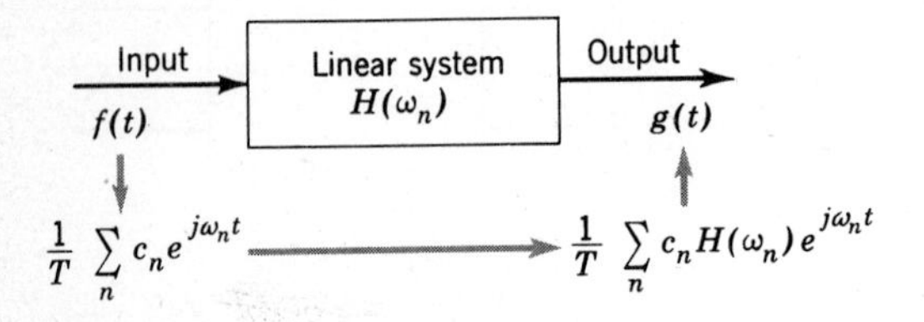

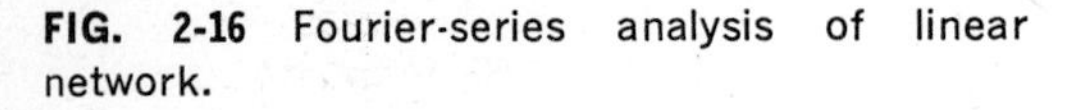

FIG. 2-16 Fourier-series analysis of linear network.

decomposition and superposition is diagrammed in Fig. 2-16. In the example of the RC filter

$$c_n = V_m \tau e^{-j\omega_n \tau/2} \frac{\sin(\omega_n \tau/2)}{\omega_n \tau/2}$$

$$H(\omega_n) = \frac{1/j\omega_n C}{R + 1/j\omega_n C}$$

$$G(\omega_n) = c_n H(\omega_n)$$

Thus

$$g(t) = \frac{1}{T} \sum_{n=-\infty}^{\infty} G(\omega_n)e^{j\omega_n t} = \begin{cases} V_m(1 - e^{-t/RC}) & 0 \le t \le \tau \\ V'_m e^{-(t-\tau)/RC} & \tau \le t \le T \end{cases} \tag{2-39}$$

and at all integral multiples of T sec. (Implicit of course is the assumption that $RC \ll T$. The more exact expression for the output voltage contains an exponential term involving T/RC. This is left as an exercise for the reader.)

It is really quite remarkable that the complex Fourier series on the left-hand side of Eq. [2-39] should reduce to the real periodic time function on the right-hand side.[1]

The frequency analysis of linear networks, using the Fourier-series expansion, is useful not only because it enables us, at least conceptually, to find the response to specific inputs, but also because it can be utilized to define the *bandwidth* of the linear system under study and then to study the band-limiting effects of the network on various signal inputs. This is just a generalization of the discussion carried out with the RC network.

Thus, using Eq. [2-38] as the defining relation of the effect of linear networks on input time functions, it is apparent that if $H(\omega_n)$ is nearly constant over a significant portion of the frequency range encompassed by c_n, the output signal will look very nearly like the input signal. Defining B_f as the filter bandwidth in Hz, we must thus have

$$B_f > B_p$$

if the output pulse train is to look very nearly like the input pulse train. Here B_p is, as before, some measure of the input signal bandwidth.

This concept will be reviewed and generalized later.

2-4 FOURIER INTEGRAL

The discussion thus far has been restricted to the case of periodic time functions passing through linear networks. Although such functions are commonly used for test purposes in many system studies, in practice they do not really represent the time functions occurring in communications practice. For as noted in Chap. 1, periodic functions carry no information. A closer approximation to the actual signals used in practice is provided by nonperiodic time functions. (A still better model will be the random signals discussed in later chapters.)

Here too we shall find the frequency response of signals and linear networks playing a key role in the discussion. In particular the inverse time-frequency relationship developed in the preceding section for periodic functions will still be found valid for nonperiodic functions.

[1] Those readers who disbelieve have only to sum the indicated Fourier series and show that it does indeed equal the indicated time function!

An extension of the time-frequency correspondence to nonperiodic time functions requires the introduction of the Fourier integral. This is simply done by recognizing that any time function (subject of course to certain broad mathematical definitions and restrictions) defined only over a specified time interval T sec long may be expanded in a Fourier series of base period T. The time function is then artificially made to repeat itself outside the specified time interval. As the time interval of interest becomes greater, the Fourier period is correspondingly increased. Ultimately, as the region of interest is made to increase beyond bound, the resultant Fourier series becomes, in the limit, the Fourier integral.

Consider a periodic function $f(t)$,

$$f(t) = \frac{1}{T} \sum_{n=-\infty}^{\infty} c_n e^{j\omega_n t} \qquad \omega_n = \frac{2\pi n}{T} \tag{2-40}$$

$$c_n = \int_{-T/2}^{T/2} f(t) e^{-j\omega_n t}\, dt \tag{2-41}$$

A typical amplitude-spectrum plot would appear as in Fig. 2-17. The spacing between successive harmonics is just

$$\Delta\omega = \omega_{n+1} - \omega_n = \frac{2\pi}{T} \tag{2-42}$$

Equation [2-40] may be written as

$$f(t) = \frac{1}{2\pi} \sum_{n=-\infty}^{\infty} c_n e^{j\omega_n t}\, \Delta\omega \tag{2-43}$$

Now consider the limiting case as $T \to \infty$. Then $\Delta\omega \to 0$, the discrete lines in the spectrum of Fig. 2-17 merge, and we obtain a continuous-fre-

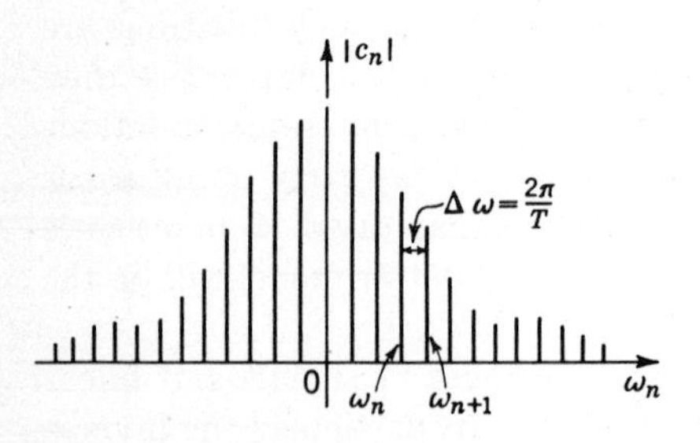

FIG. 2-17 Amplitude spectrum, periodic function.

quency spectrum. Mathematically, the infinite sum in Eq. [2-43] becomes the ordinary Riemann integral. c_n is now defined for *all* frequencies, not merely integral multiples of $2\pi/T$. In the limit, as $T \to \infty$, $\omega_n \to \omega$ and c_n becomes a continuous function $F(\omega)$.

$$F(\omega) = \lim_{T\to\infty} c_n \tag{2-44}$$

In the place of the Fourier series of Eq. [2-40] we now obtain as the Fourier-integral representation of a nonperiodic function $f(t)$

$$f(t) = \frac{1}{2\pi}\int_{-\infty}^{\infty} F(\omega)e^{j\omega t}\,d\omega \tag{2-45}$$

with

$$F(\omega) = \int_{-\infty}^{\infty} f(t)e^{-j\omega t}\,dt \tag{2-46}$$

from Eq. [2-41]. $F(\omega)$ is, in general, a complex function of ω and may be written

$$F(\omega) = |F(\omega)|e^{j\theta(\omega)} \tag{2-47}$$

A typical time function and its amplitude spectrum $|F(\omega)|$ are shown in Fig. 2-18. The periodic pulses previously considered serve as a good

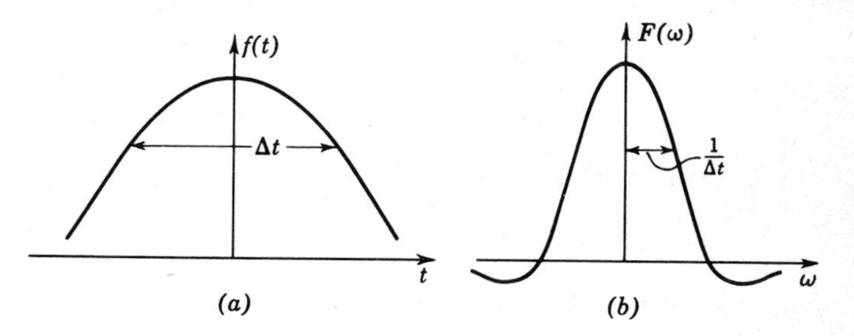

FIG. 2-18 A typical time function and its amplitude spectrum. (a) $f(t)$; (b) $F(\omega)$.

example of the transition from the Fourier series to the Fourier integral. The pulses are shown in Fig. 2-19*a* and the corresponding spectrum in Fig. 2-19*b*.

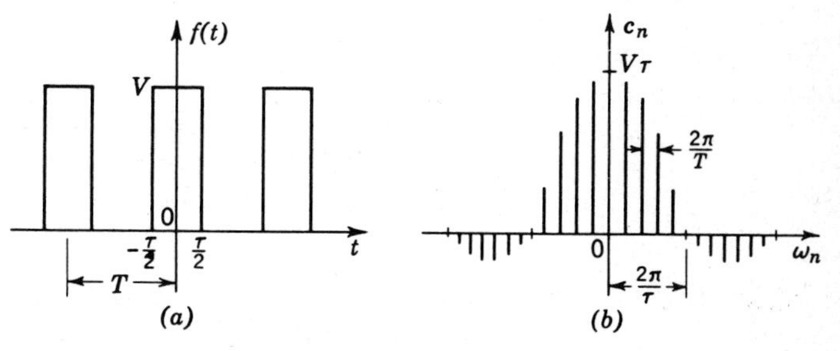

FIG. 2-19 Periodic pulses and their spectrum.

The frequency spectrum of the periodic pulses is of course a plot of the Fourier coefficient c_n (normally amplitude and phase plots).

$$c_n = V\tau \frac{\sin (\omega_n \tau/2)}{\omega_n \tau/2} \qquad \omega_n = \frac{2\pi n}{T} \tag{2-48}$$

As $T \to \infty$, all the pulses except for the one centered at $t = 0$ move out beyond bound and we are left, in the time plot, with a single pulse of amplitude V and width τ sec.

In the frequency plot $\omega_n \to \omega$ as $T \to \infty$, the lines move together and merge, and the spectrum becomes a continuous one (Fig. 2-20). Thus

$$F(\omega) = \lim_{T \to \infty} c_n = V\tau \frac{\sin (\omega\tau/2)}{\omega\tau/2} \tag{2-49}$$

The single pulse of Fig. 2-20*a* has the continuous-frequency spectrum of Fig. 2-20*b*, defined for all frequencies.

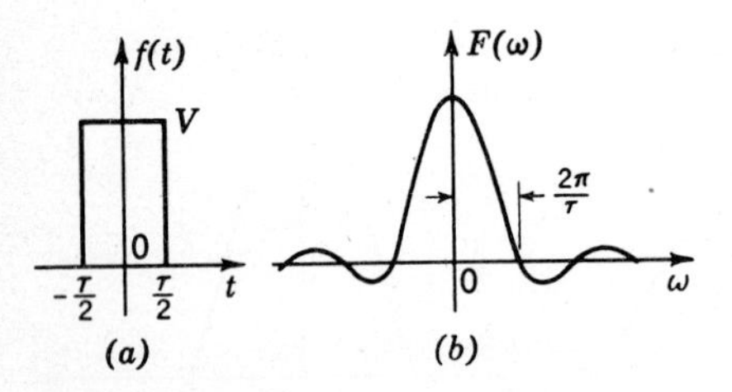

FIG. 2-20 Rectangular pulse and its spectrum. (a) Time plane; (b) frequency plane.

Equation [2-49] can of course be obtained directly from the defining relation for $F(\omega)$ (Eq. [2-46]).

$$F(\omega) = \int_{-\infty}^{\infty} f(t) e^{-j\omega t}\, dt$$

For a single pulse

$$f(t) = V \qquad |t| < \frac{\tau}{2}$$
$$f(t) = 0 \qquad |t| > \frac{\tau}{2} \tag{2-50}$$

Then $$F(\omega) = V \int_{-\tau/2}^{\tau/2} e^{-j\omega t}\, dt = \frac{V}{-j\omega}(e^{-j\omega\tau/2} - e^{j\omega\tau/2}) = V\tau \frac{\sin (\omega\tau/2)}{\omega\tau/2} \tag{2-51}$$

The reader may have noticed already that Eq. [2-46] is almost identical with the Laplace transform of $f(t)$,

$$F(p) = \int_0^\infty f(t)e^{-pt}\,dt \qquad [2\text{-}52]$$

Except for the lower limits of the integrals the two equations are the same, with p replacing $j\omega$, and vice versa. If $f(t)$ is zero for $t < 0$ (as is normally assumed to be the case in working most transient problems), the two integrals do become identical. $F(\omega)$ is frequently called the Fourier transform of $f(t)$.

The close connection between frequency and time response noted at the beginning of this section must exist because of the very definition of frequency response—given by the Fourier coefficients c_n in the case of periodic time functions and the Fourier transform in the case of nonperiodic time functions.

Although we talk of the "frequency properties" of networks and time functions as if they were physical entities in themselves, they are in reality mathematical abstractions that are utilized in determining the time response of linear networks. The concept, frequency response of a network, frequently serves to simplify many difficult problems, particularly since we have developed great facility in thinking in terms of amplitude and phase response of networks. It also gives us directly the response to an applied sine wave, thus serving to connect steady-state sine-wave and transient response.

The apparent slight difference between the Laplace transform and the Fourier transform due to the different lower limits in the integral may be eliminated by redefining either one of the two integrals. Thus, had we defined our original periodic function from 0 to T rather than from $-T/2$ to $T/2$, we would have obtained the so-called "one-sided" Fourier transform identical in form with the Laplace transform. This approach is useful in some cases and is discussed by S. Goldman.[1] Other authors have chosen to define the double-sided Laplace transform,[2] which has as its limits $-\infty$ and ∞.

Basically, however, the two transforms are the same, and both may be utilized in the solution of transient problems. Why, then, discuss the Fourier transform at all? Why not just use the Laplace transform?

In order to apply the Laplace transform to the solution of problems involving finite lumped-constant networks, we must be able to write the

[1] S. Goldman, "Frequency Analysis, Modulation, and Noise," McGraw-Hill, New York, 1948.

[2] B. Van der Pol and H. Bremmer, "Operational Calculus Based on the Two-sided Laplace Integral," 2d ed., Cambridge, New York, 1955.

network transfer functions as ratios of polynomials in p. Otherwise there is no way of determining the zeros and poles. The Fourier transform enables us to talk of the amplitude and phase characteristics of networks rather than the poles and zeros (the two concepts are of course directly related). Using the Fourier transform, we shall be able to relate the time response and amplitude-phase characteristics of networks directly. This is of tremendous value in those systems where the amplitude and phase characteristics are readily known or approximated.

One word of caution, however. Not all time functions may possess a Fourier transform. The condition for the existence of the Fourier transform is that the Fourier integrals exist in the limit as $T \to \infty$. For absolute convergence of the integral of Eq. [2-46]

$$\int_{-\infty}^{\infty} |f(t)|\, dt < \infty \qquad [2\text{-}53]$$

This is obviously not the case for such a simple time function as

$$\begin{aligned} f(t) &= t \qquad 0 < t \\ f(t) &= 0 \qquad t < 0 \end{aligned}$$

This integral does not even converge absolutely when $f(t)$ is a unit-step function, although the Fourier transform of the step may, under certain conditions, be defined and useful results obtained by its use. (Examples are given in the problems for this chapter.)

The Laplace transforms of many of these functions (i.e., those not possessing a Fourier transform) do exist, however, for the function e^{-pt} by which $f(t)$ is multiplied to obtain the Laplace transform of $f(t)$ can be made to provide suitable damping for the integral to converge. (The real part of p is chosen positive and large enough. In the case of the Fourier transform $p = j\omega$, and there is no damping.)

2-5 RESPONSE OF IDEALIZED NETWORKS

Judicious use of both Fourier series and the Fourier-integral representation of time functions will enable us to solve many problems relating to the transmission of signals through linear networks. In particular, assume a system frequency transfer function of the form

$$H(\omega) = A(\omega)e^{j\theta(\omega)} \qquad [2\text{-}54]$$

$A(\omega)$ thus represents the amplitude characteristic and $\theta(\omega)$ the phase characteristic of the system. Some important questions we shall consider include: What relations must exist between $A(\omega)$, $\theta(\omega)$, and the frequency spectrum of the input signal in order to preserve the signal fidelity at the output of the system? What effect do variations of $A(\omega)$ (network amplitude characteristic) have on the shape of the output signal as compared with the input? How do variations in $\theta(\omega)$ manifest themselves so far as the output signal is concerned? How is signal distortion related to system bandwidth?

We have already answered some of these questions for the special case of the transmission of a series of narrow rectangular pulses through an *RC* network. We shall now attempt to extend these results and others to more complicated systems.

Assume an input time function $f(t)$ with its associated Fourier transform $F(\omega)$. Then the Fourier transform of the output signal $g(t)$ is

$$G(\omega) = H(\omega)F(\omega) \tag{2-55}$$

{Equation [2-55] may be considered an extension of Eq. [2-38] to include continuous frequencies. $F(\omega)$ represents the amplitude and phase of a sinusoidal input of frequency ω. The network then introduces amplitude and phase changes given by $H(\omega)$.}

The output signal $g(t)$ is then given by

$$g(t) = \frac{1}{2\pi} \int_{-\infty}^{\infty} \underbrace{H(\omega)F(\omega)}_{G(\omega)} e^{j\omega t}\, d\omega \tag{2-56}$$

using Eq. [2-45].

Equation [2-56] will be the equation used to study the dependence of the output signal on network amplitude and phase characteristics. In using this equation many different kinds of practical circuits might be investigated. The mathematics would in most cases be quite cumbersome, however, and we would find it difficult to draw some general conclusions as to network amplitude and phase characteristics and their effect on signal transmission. We therefore resort to mathematical idealizations of practical circuits. Although these idealizations are physically unrealizable and therefore lead invariably to physically impossible results, they will serve to simplify the mathematics in many cases, will keep us from getting lost in a welter of algebraic quantities, and will produce results that can be interpreted quite usefully.[1]

[1] For many more examples than can be treated here see Goldman, *op. cit.;* T. Murakami and M. S. Corrington, Applications of the Fourier Integral in the Analysis of Color TV Systems, *IRE Trans. Circuit Theory,* vol. CT-2, no. 3, September, 1955; A. Papoulis, "The Fourier Integral and Its Applications," McGraw-Hill, New York, 1962.

As an example of such an idealization, consider the low-pass-filter characteristics of Fig. 2-21. Here the amplitude-frequency spectrum is a

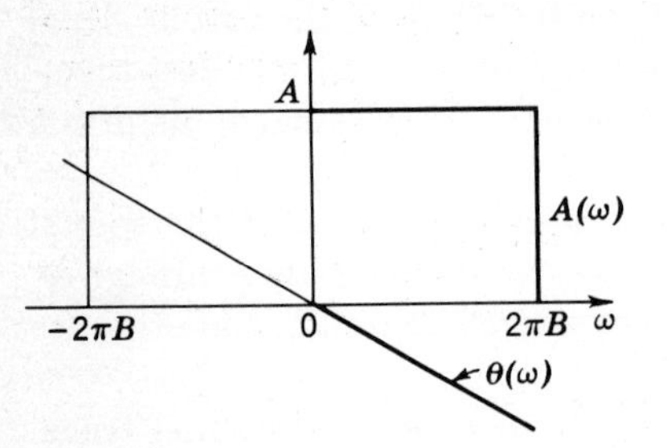

FIG. 2-21 Ideal low-pass filter.

constant for all frequencies below the cutoff frequency $2\pi B$. Thus, with $H(\omega) = A(\omega)e^{j\theta(\omega)}$,

$$\begin{aligned} A(\omega) &= A \qquad |\omega| \leq 2\pi B \\ A(\omega) &= 0 \qquad |\omega| > 2\pi B \end{aligned} \tag{2-57}$$

The phase shift $\theta(\omega)$ is assumed linearly proportional to frequency.

$$\theta(\omega) = -t_0\omega \tag{2-58}$$

(t_0 is a constant of this ideal network. Negative frequencies must be introduced in order to use the Fourier integrals, defined also for negative ω). This amplitude response is physically unattainable. The linear-phase-shift characteristic assumed is also physically impossible for finite lumped-constant networks. (Smooth transmission lines can have linear phase shift). In addition, the amplitude and phase characteristics of a given network are connected together by the pole-zero plot of the network. They are thus normally not chosen independently, as was done here. (There do exist so-called "all-pass" networks, synthesized from lattice structures, which provide phase variation with constant-amplitude response. These networks have their poles and zeros symmetrically arranged on either side of the $j\omega$ axis. Such networks can be included in an overall network to provide independent choice of amplitude and phase.)

The use of these idealizations to investigate the response of physical networks could thus lead to absurdities unless we are careful in interpreting our results.

If the input signal is a single rectangular pulse,

$$F(\omega) = V\tau \frac{\sin(\omega\tau/2)}{\omega\tau/2} \tag{2-59}$$

(The time origin is chosen at the center of the pulse.) The transform of the output signal is

$$G(\omega) = V\tau \frac{\sin(\omega\tau/2)}{\omega\tau/2} Ae^{-jt_0\omega} \qquad -2\pi B < \omega < 2\pi B \qquad [2\text{-}60]$$

$$G(\omega) = 0 \qquad \text{elsewhere}$$

Then $g(t)$ will be

$$g(t) = \frac{AV\tau}{2\pi} \int_{-2\pi B}^{2\pi B} \frac{\sin(\omega\tau/2)}{\omega\tau/2} e^{j\omega(t-t_0)}\, d\omega \qquad [2\text{-}61]$$

To evaluate this integral, we recall that

$$e^{j\theta} = \cos\theta + j\sin\theta$$

Then $\int_{-2\pi B}^{2\pi B} \frac{\sin(\omega\tau/2)}{\omega\tau/2} e^{j\omega(t-t_0)}\, d\omega$

$$= \int_{-2\pi B}^{2\pi B} \frac{\sin(\omega\tau/2)}{\omega\tau/2} [\cos\omega(t-t_0) + j\sin\omega(t-t_0)]\, d\omega$$

$$= \int_{-2\pi B}^{2\pi B} \frac{\sin(\omega\tau/2)}{\omega\tau/2} \cos\omega(t-t_0)\, d\omega + j\int_{-2\pi B}^{2\pi B} \frac{\sin(\omega\tau/2)}{\omega\tau/2} \sin\omega(t-t_0)\, d\omega \qquad [2\text{-}62]$$

The integrand of the first integral is an even function of ω. The integral is then just twice the integral from 0 to $2\pi B$. The integrand of the second integral is an odd function, and the integral, between equal negative and positive limits, vanishes. Equation [2-62] can thus be written

$$2\int_0^{2\pi B} \frac{\sin(\omega\tau/2)}{\omega\tau/2} \cos\omega(t-t_0)\, d\omega = \int_0^{2\pi B} \left[\frac{\sin\omega(t-t_0+\tau/2)}{\omega\tau/2} - \frac{\sin\omega(t-t_0-\tau/2)}{\omega\tau/2}\right] d\omega \qquad [2\text{-}63]$$

using the trigonometric relation for sum and difference angles.

Breaking the integral up into two integrals and changing variables [$x = \omega(t - t_0 + \tau/2)$ in the first integral, $x = \omega(t - t_0 - \tau/2)$ in the second], we get finally for $g(t)$

$$g(t) = \frac{AV}{\pi} \int_0^{2\pi B(t-t_0+\tau/2)} \frac{\sin x}{x}\, dx - \frac{AV}{\pi} \int_0^{2\pi B(t-t_0-\tau/2)} \frac{\sin x}{x}\, dx \qquad [2\text{-}64]$$

Unfortunately $\int_0^a [(\sin x)/x]\, dx$ cannot be evaluated in closed form but must be evaluated by expanding $(\sin x)/x$ in a power series in x and integrating term by term. Tables are available for the integral,[1] however, and it is called the sine integral of x,

$$\text{Si}\, x \equiv \int_0^x \frac{\sin x}{x}\, dx \qquad [2\text{-}65]$$

Equation [2-64] can thus be written

$$g(t) = \frac{AV}{\pi}\left\{\text{Si}\left[2\pi B\left(t - t_0 + \frac{\tau}{2}\right)\right] - \text{Si}\left[2\pi B\left(t - t_0 - \frac{\tau}{2}\right)\right]\right\} \qquad [2\text{-}66]$$

The sine integral appears very frequently in the literature pertaining to pulse transmission through idealized networks. It represents the area under the $(\sin x)/x$ curve plotted previously and reproduced in Fig. 2-22a.

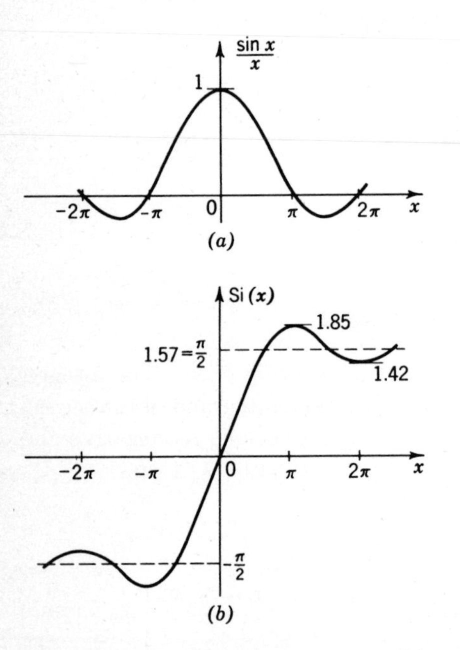

FIG. 2-22 (a) $(\sin x)/x$ versus x; (b) Si x versus x.

It thus has its maxima and minima at multiples of π (the points at which $\sin x$ changes sign). The first maximum is 1.85 at $x = \pi$. The curve is odd-symmetrical about $x = 0$ and approaches $\pi/2 = 1.57$ for large values of x.

[1] Goldman, *op. cit.;* E. Jahnke and F. Emde, "Tables of Functions," Dover, New York, 1945.

Since $(\sin x)/x$ is an even function and has zero slope at $x = 0$, the initial slope of $\text{Si}\, x$ is linear and $\text{Si}\, x \doteq x$, $x \ll 1$. The sine integral is plotted in Fig. 2-22*b*.

The response of an idealized low-pass filter to a rectangular pulse of width τ sec is given by Eq. [2-66] in terms of the sine integral. Figure 2-23 shows Eq. [2-66] plotted for different filter bandwidths B.

1. $$B = \frac{1}{5\tau} \qquad \left(B \ll \frac{1}{\tau}\right)$$
2. $$B = \frac{1}{\tau}$$
3. $$B = \frac{5}{\tau} \qquad \left(B \gg \frac{1}{\tau}\right)$$

The three cases are shown superimposed and compared with the rectangular-pulse input.

What conclusions can we draw from the curves of Fig. 2-23?

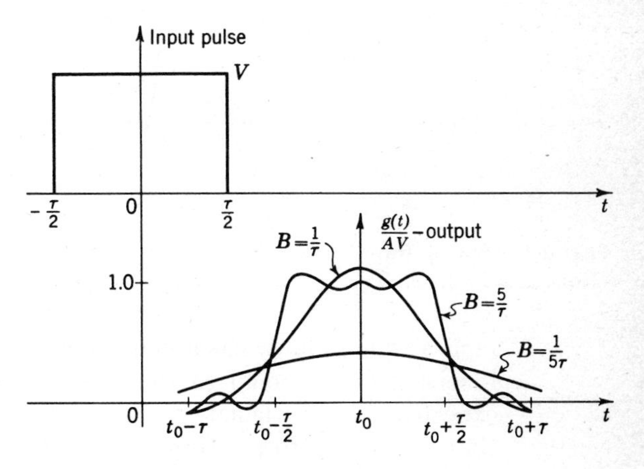

FIG. 2-23 Response of low-pass filter.

1. All three output curves are displaced t_0 sec from the input pulse and are symmetrical about $t = t_0$. The negative-linear-phase characteristic assumed for the filter has thus resulted in a *time delay* equal to the slope of the filter phase characteristic.
2. The curves bear out the filter-bandwidth–pulse-width relations we developed in the simple case of the *RC* filter:
   (*a*) $B \ll 1/\tau$. With the filter bandwidth much less than the reciprocal of the pulse width the output is much broader than the

input and peaks only slightly, i.e., is a grossly distorted version of the input.

(b) $B = 1/\tau$. Here the output is a recognizable pulse, roughly τ sec in width, but far from rectangular. The rise time is approximately half the pulse width.

(c) $B \gg 1/\tau$. The output resembles the input closely and has approximately the same pulse width. There are several marked differences between output and input, however. To point these out, the curve for this case is replotted on an expanded time scale in Fig. 2-24. Note that the output pulse, although a delayed

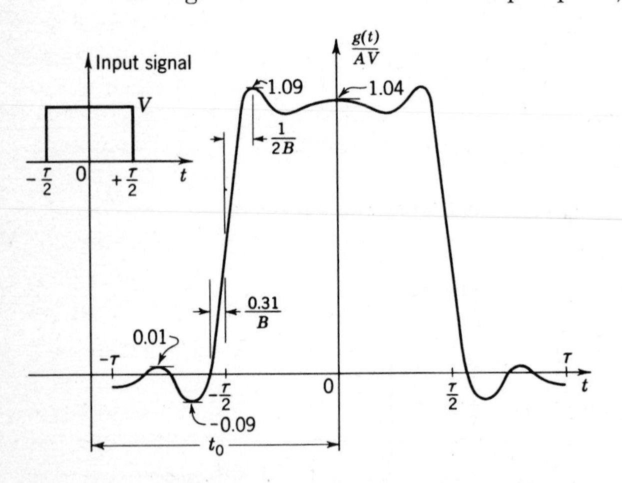

FIG. 2-24 Signal transmission through ideal low-pass filter ($B \gg 1/\tau$).

replica of the input, has a nonzero rise time. This rise time is inversely proportional to the filter bandwidth, as was to be expected. In particular, if the rise time here is defined as the time for a pulse to rise from zero to its maximum value of 1.09 AV,

$$\text{Rise time} = 0.8\,\frac{1}{B} \qquad [2\text{-}67]$$

Alternatively, if the rising curve is approximated by its tangent at the point $g(t) = 0.5AV$, the rise time of the resultant straight line (0 to AV) is

$$\text{Rise time} = 0.5\,\frac{1}{B} \qquad [2\text{-}68]$$

(Note that these results are valid only if $B \gg 1/\tau$, as is the case here.)

These results are in agreement with those obtained previously for the RC filter. If the object is merely to produce an output pulse which has about the same width as the input pulse, with fidelity unimportant, then the filter bandwidth required is approximately the inverse of the pulse width. This is the situation, for example, in a search radar system, where a recognizable signal pulse is required but its shape is of secondary interest.[1] If fidelity is required, then the bandwidth specified must be at least several times the reciprocal of the pulse width and is actually determined by rise-time considerations. For example, in tracking radars the time of arrival of individual pulses must be accurately known. The output pulse of the radar receiver must rise quite sharply so that the leading edge of the pulse may be accurately determined. The same considerations hold for loran navigational systems, where the time of arrival of each pulse, or its leading edge, must be accurately known. In pulse-modulation systems (to be considered in Chap. 3) the bandwidth is also determined by the specified pulse rise time. (Note that ordinarily i-f bandwidths, rather than a low-pass equivalent, are involved. As will be shown later, the i-f bandwidth is twice the low-pass equivalent, or $2B$. From Eqs. [2-67] and [2-68], then, rise time is $1/B$ or $1.6/B$, depending on the rise-time definition in this case. Defining rise time as the time required for a pulse to rise from 10 to 90 percent of its peak value leads essentially to the same result.)

These results, of course, account for the rule of thumb used in practice in most pulse system work that the system bandwidth should be the reciprocal of the pulse rise time. (Identical results are obtained, as is to be expected, from a transient analysis of typical wideband amplifier circuits.)

Some other interesting conclusions may be drawn from a study of Fig. 2-24. Note that the output pulse *overshoots*, or oscillates with damped oscillations, about the flat-top section of the pulse. This phenomenon is characteristic of filters with sharply cut-off amplitude response. (It is of course encountered in simple double-energy circuits which are underdamped.)

Figure 2-24 indicates that the output pulse actually has nonzero value for $t < -\tau/2$, *before* the input pulse has appeared. In fact Eq. [2-66] gives nonzero values for negative time as well as positive time (although centered about $t = t_0$). This appearance of an output before the input producing it has appeared is obviously physically impossible and is due

[1] Noise considerations lead to the same bandwidth requirement, as will be seen later on. Search radar specifications thus indicate that the overall receiver bandwidth should be at least the reciprocal of the pulse width. (Actually the i-f bandwidth is normally specified and is $2B$.)

specifically to the nonphysically realizable filter characteristics assumed. Thus the rectangular amplitude characteristic of the idealized low-pass filter can never be realized with physical circuits. (It can be approached closely, but the number of elements required increases as the approximation becomes better. An exact fit requires an infinite number of elements theoretically. The filter phase-shift constant t_0 then becomes infinite, and the filter produces the amplitude of the pulse after infinite delay.) As pointed out previously, however, network idealizations are valuable since they frequently provide insight into system performance and enable general conclusions as to network response to be drawn.

2-6 DISTORTIONLESS TRANSMISSION

The question may well be asked: Why were the "ideal" filter characteristics picked as constant amplitude and linear phase shift? These characteristics are of course rather simple to define and easy to talk about, but they are also important in a discussion of distortionless transmission.

Suppose that a given time function $f(t)$ is to be transmitted through a linear system and the resultant output is to "look" just like the input. Mathematically,

$$g(t) = Kf(t - t_0) \qquad K = \text{a constant} \tag{2-69}$$

The output $g(t)$ is a delayed replica of the input (delayed by t_0 sec) although of different magnitude (see Fig. 2-25).

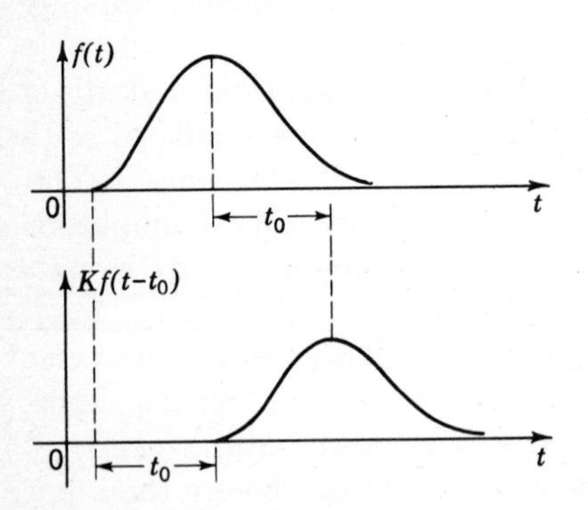

FIG. 2-25 Delayed replica of a signal.

The problem of distortionless transmission is a basic one in many fields of communication—telephony, radio, TV, etc.—in which the shape and appearance of the input must be retained after passage through electrical and electromechanical circuits, through wire or cable, through air, etc. A system producing such an output is said to be a "distortionless" system. This is indicated schematically in Fig. 2-26. What are the requirements for such a system in terms of frequency response?

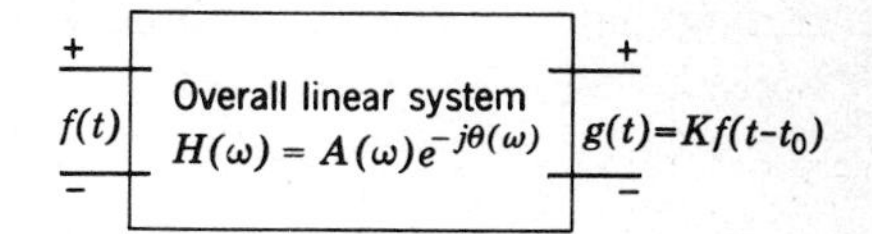

FIG. 2-26 Distortionless transmission.

Let $F(\omega)$ be the Fourier transform of $f(t)$; then

$$F(\omega) = \int_{-\infty}^{\infty} f(t)e^{-j\omega t}\,dt \qquad [2\text{-}70]$$

$$f(t) = \frac{1}{2\pi}\int_{-\infty}^{\infty} F(\omega)e^{j\omega t}\,d\omega \qquad [2\text{-}71]$$

$G(\omega)$ is the Fourier transform of the output function $g(t)$,

$$G(\omega) = \int_{-\infty}^{\infty} g(t)e^{-j\omega t}\,dt \qquad [2\text{-}72]$$

But $g(t) = Kf(t - t_0)$ by hypothesis. Therefore

$$\begin{aligned} G(\omega) &= K\int_{-\infty}^{\infty} f(t - t_0)e^{-j\omega t}\,dt \\ &= K\int_{-\infty}^{\infty} f(x)e^{-j\omega(x+t_0)}\,dx \qquad x = t - t_0 \qquad [2\text{-}73] \\ G(\omega) &= Ke^{-j\omega t_0}\int_{-\infty}^{\infty} f(x)e^{-j\omega x}\,dx \end{aligned}$$

But the integral in x is $F(\omega)$ (see Eq. [2-70]), or

$$G(\omega) = Ke^{-j\omega t_0}F(\omega) \qquad [2\text{-}74]$$

We know, however, that $G(\omega)$ and $F(\omega)$ are related by the network frequency transfer function $H(\omega)$,

$$G(\omega) = H(\omega)F(\omega) \qquad [2\text{-}75]$$

Comparing Eqs. [2-74] and [2-75] gives us the simple result for distortionless transmission,

$$H(\omega) = Ke^{-j\omega t_0} \qquad [2\text{-}76]$$

In words, if the signal is to be passed through a linear system without any resultant distortion, the overall system response must have a constant-amplitude characteristic over the frequency spectrum of the input and its phase shift must be linear over the same range of frequencies.

The words *overall* linear system are important here. A signal may well be distorted in passing through the different parts of the system, but phase- or amplitude-correction (equalization) networks may be introduced elsewhere to correct for distortion. It is the *overall* characteristic that determines the ultimate output. The system is assumed linear throughout, however, for otherwise the Fourier relations used do not apply.

2-7 EFFECT OF PHASE VARIATION ON SIGNAL TRANSMISSION

We found in the previous section that a distortionless-transmission system would require flat (uniform- or constant-) amplitude and linear-phase-shift characteristics over the range of frequencies covered by the signal.

In practice all linear systems introduce a certain amount of signal distortion because of bandwidth limitations and nonlinear phase characteristics. As pointed out previously, the amplitude and phase response of a given network are interrelated, and to treat the two characteristics separately in studying signal transmission could become an academic approach. We can gain some further insight into the effect of network characteristics on signal response, however, by treating amplitude and phase separately and attempting to superimpose the results.

We considered previously the effect of amplitude limiting on pulse transmission and arrived at the basic relation between system bandwidth and pulse rise time. The phase characteristic assumed was a linear one, and it is of interest to explore the question of network phase response further.

Note first that, although the effect of bandwidth limiting on an input pulse was to distort the pulse, the *distortion* was found to be *symmetrical:* the output waves were symmetrical about a delayed time $t = t_0$ (Fig. 2-23). This symmetry in the output pulses is a direct consequence of the linear-phase-shift characteristic assumed for the network. It is a general property of networks that where amplitude distortion occurs (as in band limiting) the output transient is symmetrically related to the input if the network phase shift is linear. Phase linearity in networks thus gives rise to the symmetrical pulse response and symmetrical step response and is particularly desirable in TV and radar systems.[1]

[1] T. Murakami and M. S. Corrington, Applications of the Fourier Integral in the Analysis of Color TV Systems, *IRE Trans. Circuit Theory*, vol. CT-2, no. 3, September, 1955.

Amplitude and phase distortion of a pulse input are illustrated in Fig. 2-27.[1]

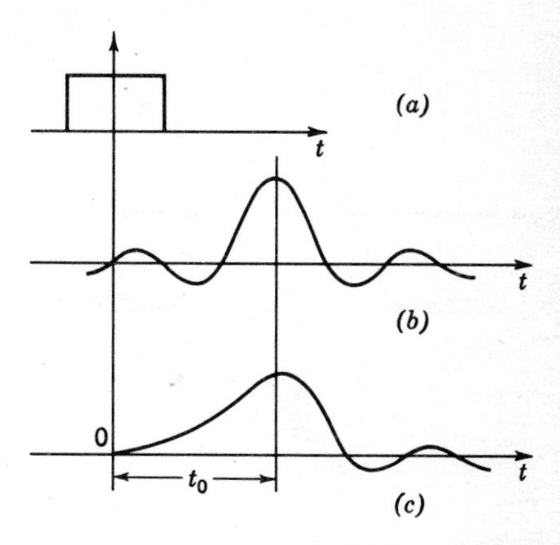

FIG. 2-27 Network amplitude and phase distortion. (a) Input pulse; (b) output response; amplitude distortion, linear phase shift (symmetrical); (c) phase distortion (nonsymmetrical). (From C. Cherry, "Pulses and Transients in Communication Circuits," fig. 60, p. 147, Chapman & Hall, London, 1949; Dover Publications, New York, 1950.)

The symmetrical-output property due to linear phase shift can be easily demonstrated. Assume an idealized linear network with frequency transfer function

$$H(\omega) = A(\omega)e^{-jt_0\omega} \tag{2-77}$$

The phase-shift characteristic is thus linear, with the amplitude characteristic $A(\omega)$ arbitrary. The input signal is $f(t)$, and the output response is $g(t)$. Then

$$G(\omega) = F(\omega)H(\omega) \tag{2-78}$$

with $F(\omega)$ the Fourier integral of $f(t)$ and $G(\omega)$ the Fourier integral of $g(t)$. We assume that $f(t)$ is a symmetrical function in time, either odd or even.

[1] C. Cherry, "Pulses and Transients in Communication Circuits," Chapman & Hall, London, 1949; Dover, New York, 1950.

Thus,

$$\text{If } f(t) = f(-t) \qquad f(t) \text{ is even}$$
$$\text{If } f(t) = -f(-t) \qquad f(t) \text{ is odd}$$

Examples of even and odd functions are shown in Fig. 2-28.

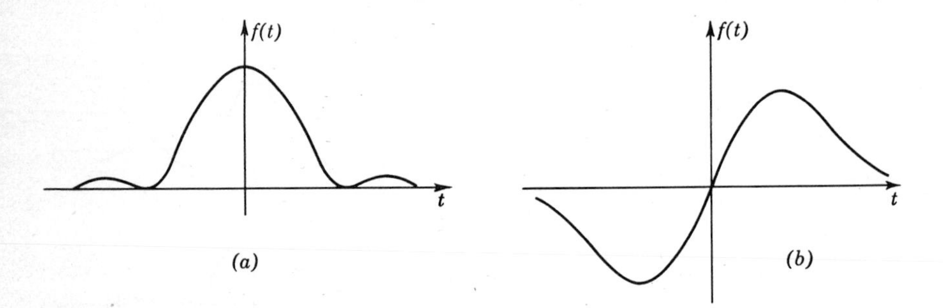

FIG. 2-28 Symmetrical time functions. (a) Even symmetry; (b) odd symmetry.

We should now like to prove that either type of symmetrical function produces a symmetrical function of the same type when passed through the idealized network with linear phase shift.

Case 1. $f(t)$ even. Then

$$\begin{aligned} F(\omega) &= \int_{-\infty}^{\infty} f(t)e^{-j\omega t}\,dt \\ &= \int_{-\infty}^{\infty} f(t)(\cos \omega t - j \sin \omega t)\,dt \\ &= 2\int_{0}^{\infty} f(t) \cos \omega t\,dt \end{aligned} \qquad [2\text{-}79]$$

since

$$\int_{-\infty}^{\infty} f(t) \sin \omega t\,dt = 0$$
$$\int_{-\infty}^{\infty} f(t) \cos \omega t\,dt = 2\int_{0}^{\infty} f(t) \cos \omega t\,dt$$

Equation [2-79] shows that $F(\omega)$ must also be an even function in ω since $\cos \omega t = \cos (-\omega t)$. The equation also indicates that $F(\omega)$ must be a *real* function of ω since the $f(t)$ is real. A simple example is of course the rectangular pulse of width τ, amplitude V, and centered about $t = 0$. For this pulse we have $F(\omega) = V\tau[\sin (\omega\tau/2)/(\omega\tau/2)]$, a real and even function of ω. Passing $f(t)$ through the idealized network, we obtain

$$\begin{aligned} g(t) &= \frac{1}{2\pi}\int_{-\infty}^{\infty} A(\omega)e^{-jt_0\omega}F(\omega)e^{jt\omega}\,d\omega \\ &= \frac{1}{2\pi}\int_{-\infty}^{\infty} A(\omega)F(\omega)[\cos \omega(t - t_0) + j \sin \omega(t - t_0)]\,d\omega \end{aligned} \qquad [2\text{-}80]$$

But $A(\omega)$ must be an even function of ω since it represents the magnitude of $H(\omega)$. [If $H(\omega) = (1/RC)/(1/RC + j\omega)$, $A(\omega) = (1/RC)/\sqrt{(1/RC)^2 + \omega^2}$.] $F(\omega)$ is even for $f(t)$ even. Therefore

$$A(\omega)F(\omega) \cos \omega(t - t_0)$$

is an even function, and

$$A(\omega)F(\omega) \sin \omega(t - t_0)$$

is an odd function. Or, from Eq. [2-80],

$$g(t) = \frac{1}{\pi} \int_0^\infty A(\omega)F(\omega) \cos \omega(t - t_0)\, d\omega \qquad [2\text{-}81]$$

an even function in *time* and symmetrical about $t = t_0$.

Case 2. $f(t)$ odd. Then

$$\begin{aligned} F(\omega) &= \int_{-\infty}^{\infty} f(t)e^{-j\omega t}\, dt \\ &= \int_{-\infty}^{\infty} f(t)(\cos \omega t - j \sin \omega t)\, dt \\ &= -2j \int_0^\infty f(t) \sin \omega t\, dt \end{aligned} \qquad [2\text{-}82]$$

$F(\omega)$ is then also odd in ω, since $\sin \omega t$ is odd. [$\sin(-\omega t) = -\sin \omega t$; so $F(-\omega) = -F(\omega)$.] As an example, let

$$\begin{aligned} f(t) &= e^{-at} & t > 0 \\ f(t) &= -e^{-a|t|} & t < 0 \end{aligned}$$

Then

$$F(\omega) = \frac{-2j\omega}{\omega^2 + a^2}$$

From Eq. [2-80]

$$\begin{aligned} g(t) &= \frac{1}{2\pi} \int_{-\infty}^{\infty} A(\omega)F(\omega)[\cos \omega(t - t_0) + j \sin \omega(t - t_0)]\, d\omega \\ &= \frac{j}{\pi} \int_0^\infty A(\omega)F(\omega) \sin \omega(t - t_0)\, d\omega \end{aligned} \qquad [2\text{-}83]$$

$g(t)$ is thus symmetrical about $t = t_0$. [$\sin \omega(t - t_0) = -\sin \omega(t_0 - t)$.]

Any signal input $f(t)$ beginning at time $t = 0$ can be shown to be decomposable into a sum of an even and an odd function. The above results may then be superimposed. The odd component produces an odd component of the output, the even component an even component of the output. The overall output function $g(t)$ is thus also symmetrical although symmetrical about its average value. For example, if $f(t)$ is a step, it may be decomposed into the odd and even functions shown in Fig. 2-29*b*. The response of the filter with linear phase shift to the unit step is symmetrical about $g(t) = 0.5$ as shown in Fig. 2-29*c*.

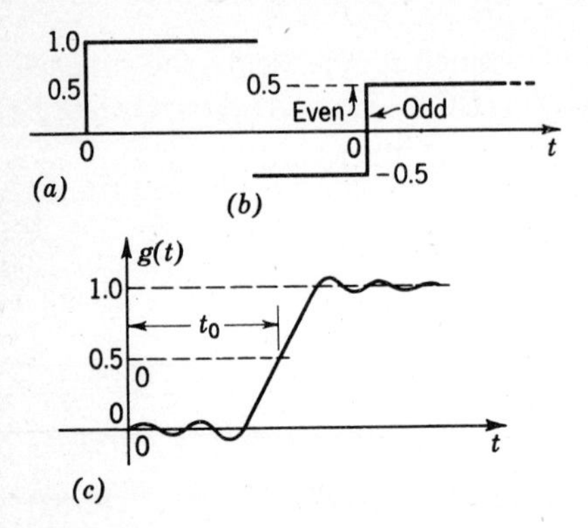

FIG. 2-29 Response of linear-phase network to a unit step. (a) Unit step; (b) even and odd components; (c) output response [symmetrical about $g(t) = 0.5$].

Although the linear-phase network is an idealization, the phase shift of a given network may be made very nearly linear by the addition of phase-correction networks. (The all-pass network mentioned previously may be used for this purpose.)

As an example of the effect of nonlinear phase characteristic on signals consider the step response of the simple RC low-pass filter of Fig. 2-30.

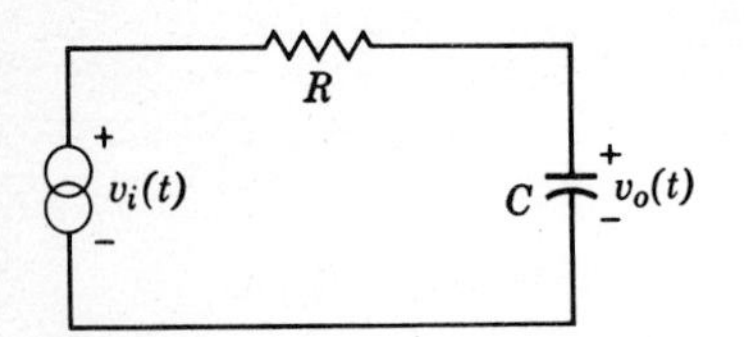

FIG. 2-30 RC network.

The step response is of course

$$v_o(t) = 1 - e^{-t/RC} \qquad v_i(t) = u(t) \tag{2-84}$$

This is obviously *not* symmetrical about the $v_o(t) = 0.5$ point. The nonlinearity of the filter phase characteristic is simply shown by writing the frequency transfer function:

$$H(\omega) = \frac{1}{1 + jRC\omega} = \frac{1}{\sqrt{1 + (\omega RC)^2}} e^{-j \tan^{-1} \omega RC} \tag{2-85}$$

The phase characteristic is thus

$$\theta(\omega) = \tan^{-1} \omega RC \qquad [2\text{-}85a]$$

A linear-phase network with the same amplitude characteristics would have as its transfer function

$$H(\omega)_{\text{linear phase}} = \frac{1}{\sqrt{1 + (\omega RC)^2}} e^{-j\omega t_0} \qquad [2\text{-}86]$$

The step response of a network with this phase characteristic has been calculated by Murakami and Corrington, in the paper previously referred to, and is shown plotted in Fig. 2-31, together with the step response of

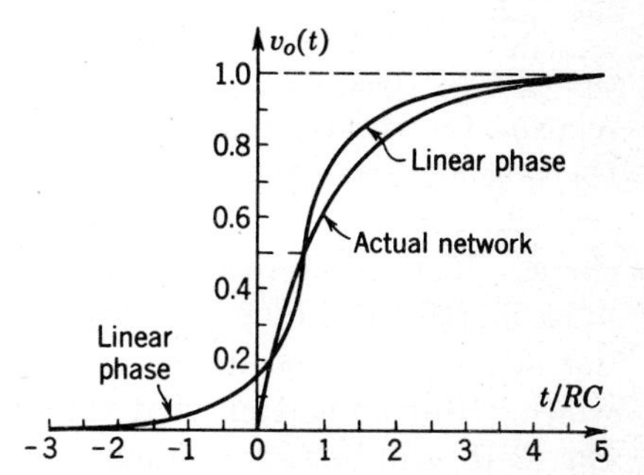

FIG. 2-31 Step response of RC network: actual and linear phase. (From T. Murakami and M. S. Corrington, Applications of the Fourier Integral in the Analysis of Color TV Systems, *IRE Trans. Circuit Theory*, vol. CT-2, no. 3, September, 1955, by permission.)

the ordinary RC network. The slope of the phase-shift curve, t_0, is chosen so that both curves coincide at the 50 percent point. Note the symmetrical response about the 50 percent point of the linear-phase network and the precursor or anticipatory transient ($t < 0$), the result of "noncompatible" amplitude and phase characteristics. The 10 to 90 percent rise time is 5 percent better for the linear-phase case.

2-8 BANDWIDTH-TIME INVERSE RELATIONSHIP FOR PULSES OF ARBITRARY SHAPE

We have shown, by example, in the last few sections that there exists an inverse relationship between pulse width and bandwidth. Thus the frequency spectrum of a rectangular pulse has a "bandwidth" (as arbitrarily defined) inversely proportional to the pulse width (Fig. 2-20). In order to pass this pulse through a given network with a minimum of distortion, then, the network bandwidth must be greater than the pulse "bandwidth." Alternatively, if a pulse is passed through a network with a bandwidth smaller than the pulse "bandwidth," a distorted pulse will emerge. This pulse will itself have a "bandwidth" (as found from its frequency spectrum or Fourier integral) very nearly that of the network.

Any pulse of specified width in time can be viewed as having come from a network of bandwidth inversely proportional to the pulse width. The smallest bandwidth anywhere in a linear system will determine the width (in time) of pulses emerging from that system.

The inverse frequency-time relationship was discussed in Sec. 2-1 for the specific case of a periodic set of rectangular pulses. The inverse relationship was expressed there by Eq. [2-18]. This inverse relationship is a specific property of the Fourier-series or Fourier-integral expansion of time functions and may be generalized to the case of a pulse of arbitrary shape. The exact relationship will again depend on the definition of pulse width (a unique quantity in the case of a rectangular pulse, but not so easily defined in the case of pulses with trailing edges) or bandwidth. But no matter what the definition, a relationship of the form of Eq. [2-18] will be found to hold.

A simple demonstration of the inverse pulse-width–bandwidth relationship may be carried out as follows. Assume that we have a *symmetrical positive pulse* $f(t)$ of arbitrary shape. (The only restriction on the shape is that the pulse have a definable width, i.e., that most of its energy be concentrated over a specified time interval. Any oscillations that are present are assumed to occur in the tail of the pulse.) We shall choose as our definition of pulse width (as the simplest choice and the one relating to our previous work) the width τ of an equivalent rectangular pulse of the same area and same amplitude, $f(0)$, at the point of symmetry, chosen to be $t = 0$. Figure 2-32 emphasizes this definition. Mathematically, we

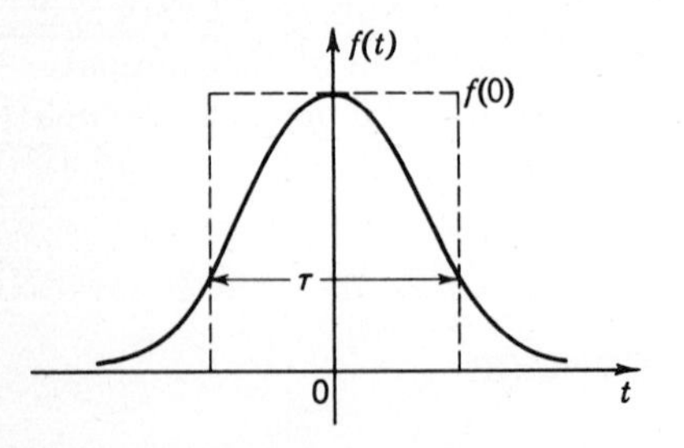

FIG. 2-32 Symmetrical pulse and its equivalent rectangular pulse.

have

$$\tau f(0) = \int_{-\infty}^{\infty} f(t)\, dt \qquad [2\text{-}87]$$

The frequency spectrum of $f(t)$ is found from the Fourier integral,

$$F(\omega) = \int_{-\infty}^{\infty} f(t) e^{-j\omega t}\, dt \qquad [2\text{-}88]$$

$F(\omega)$ will be real and symmetrical since $f(t)$ is assumed symmetrical (Eq. [2-79]).

We now define the pulse bandwidth $2\pi B$ as the width of the equivalent rectangular spectrum having the same area and same value at $\omega = 0$. (Note that this differs from our previous definitions.) This is shown in Fig. 2-33. Mathematically

$$2(2\pi B)F(0) = \int_{-\infty}^{\infty} F(\omega)\, d\omega \qquad [2\text{-}89]$$

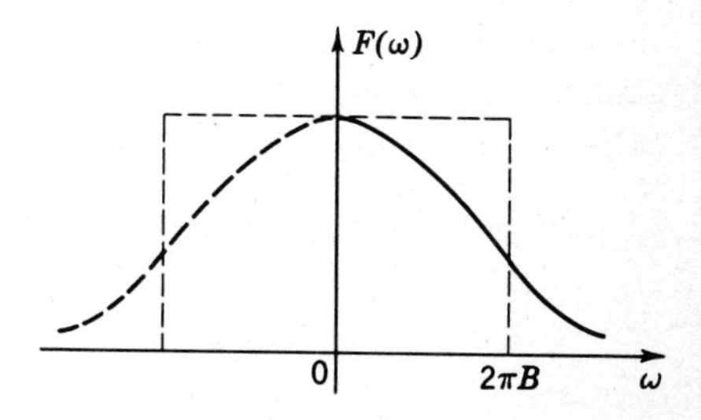

FIG. 2-33 Pulse spectrum and equivalent rectangular spectrum: $B = 1/2\tau$.

We must now relate B and τ. From Eq. [2-88]

$$F(0) = \int_{-\infty}^{\infty} f(t)\, dt \qquad [2\text{-}90]$$

[That is, the dc component represents the area under the pulse, and is thus proportional to the average value of $f(t)$, as expected.] Then

$$\tau f(0) = \int_{-\infty}^{\infty} f(t)\, dt = F(0) = \frac{1}{2B} \int_{-\infty}^{\infty} F(\omega) \frac{d\omega}{2\pi} \qquad [2\text{-}91]$$

from Eqs. [2-87], [2-89], and [2-90]. But

$$f(t) = \frac{1}{2\pi} \int_{-\infty}^{\infty} F(\omega) e^{j\omega t}\, d\omega \qquad [2\text{-}92]$$

or

$$f(0) = \frac{1}{2\pi} \int_{-\infty}^{\infty} F(\omega)\, d\omega \qquad [2\text{-}93]$$

Combining Eqs. [2-91] and [2-93],

$$B = \frac{1}{2\tau} \tag{2-94}$$

An inverse relationship between pulse rise time and bandwidth can also be demonstrated in a similar manner.

As has been noted several times previously the exact form of the inverse time-bandwidth relationship is not unique. Depending on the definition of pulse width (or rise time) and bandwidth one obtains somewhat different specific numbers, but the results are always of the form

$$B = \frac{K}{\tau} \tag{2-95}$$

with K a constant the order of unity.

Some examples of pulses, other than the rectangular pulse, encountered in practice are tabulated below together with their Fourier transforms. The inverse time-bandwidth relationship is apparent in all cases.

1. *Gaussian pulse* (Fig. 2-34)

$$f(t) = Ve^{-t^2/2\tau^2} \tag{2-96}$$

$$F(\omega) = \sqrt{2\pi}\,\tau V e^{-\tau^2\omega^2/2} \tag{2-96a}$$

Note that here τ is one possible measure of the width of the pulse. The width of $F(\omega)$, defined in an identical manner, is then $1/\tau$. The bandwidth of the signal thus increases inversely with its width, as expected.

2. *Cosine pulse* (Fig. 2-35)

$$f(t) = V\cos\frac{\pi t}{\tau} \qquad |t| \leq \frac{\tau}{2} \tag{2-97}$$

$$= 0 \qquad \text{elsewhere}$$

$$F(\omega) = \frac{2\tau V}{\pi}\,\frac{\cos \omega\tau/2}{1 - (\omega\tau/\pi)^2} \tag{2-98}$$

The pulse width here is exactly τ. The first zero crossing of $F(\omega)$ occurs at $\omega\tau = 3\pi$. If one accepts this as the pulse "bandwidth," one has

$$B = \frac{3}{2\tau}$$

in hertz.

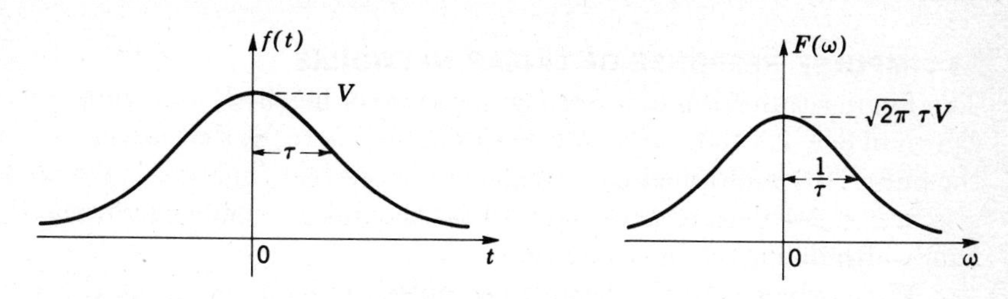

FIG. 2-34 Gaussian pulse and its Fourier transform.

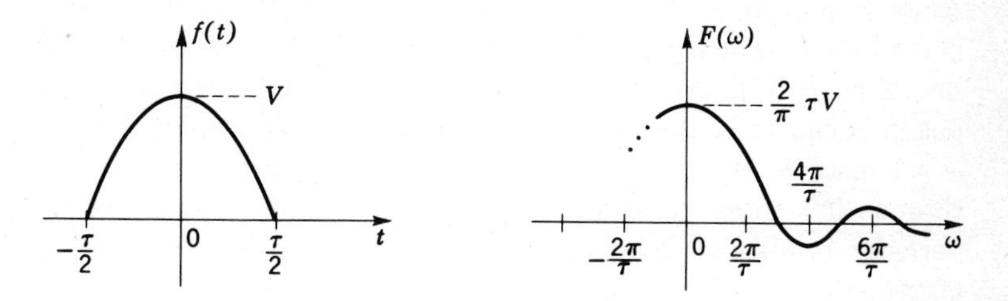

FIG. 2-35 Cosine pulse and its Fourier transform.

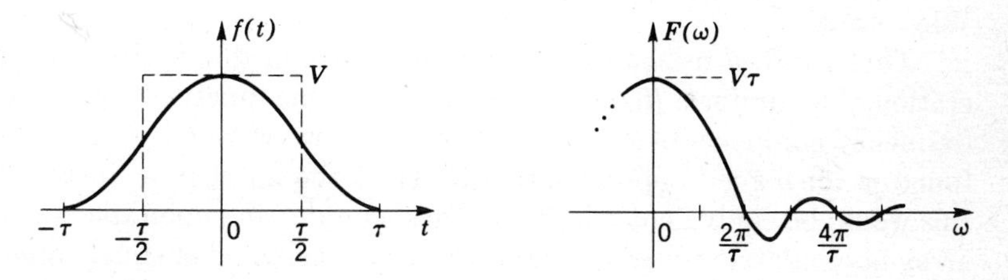

FIG. 2-36 Raised cosine pulse and its Fourier transform.

3. *Raised cosine pulse* (Fig. 2-36)

$$f(t) = \frac{V}{2}\left(1 + \cos\frac{\pi t}{\tau}\right) \qquad |t| \leq \tau \tag{2-99}$$

$$= 0 \qquad \text{elsewhere}$$

$$F(\omega) = V\tau\,\frac{\sin\omega\tau}{\omega\tau[1 - (\omega\tau/\pi)^2]} \tag{2-100}$$

Note here that although the full pulse width is 2τ, its width to the half-amplitude points (indicated in Fig. 2-36*a* by the dashed lines) is τ. The bandwidth to the first zero crossing is then $B = 1/\tau$.

2-9 IMPULSE RESPONSE OF LINEAR NETWORKS

The Fourier transform of a rectangular pulse of height V and width τ was shown in Sec. 2-4 to be given by $V\tau[\sin(\omega\tau/2)/(\omega\tau/2)]$; that is, the area of the pulse ($V\tau$) multiplied by a frequency-dependent function. We noted also in Sec. 2-4 that, as τ is decreased (as the pulse is made narrower), the bandwidth defined as K/τ increases.

We now consider the rectangular pulse to be one of extremely short duration and very large amplitude. In particular, we let $\tau \to 0$ and $V \to \infty$ such that $V\tau$, the area under the pulse, remains constant. If $V\tau = 1$ (unit area), we get, by definition, the unit impulse in time. This is the nonperiodic equivalent of the periodic pulses discussed at the end of Sec. 2-1 (see Fig. 2-9). The amplitude-frequency spectrum of this pulse becomes constant, or "flat," and equal to 1, at all frequencies. (Since $B\tau$ is a constant, B must become very large as τ goes to zero.) This continuous flat spectrum contrasts with the flat discrete line spectrum of the periodic impulses. The spectrum of three successively narrower and larger pulses is shown in Fig. 2-37. If the area under the pulse is a constant equal to A, the limiting function is an impulse of weight A. The amplitude spectrum equals A at all frequencies.

What is the significance of this impulse function? Why define it? Why use it?

One *practical* reason for using it appears immediately from a consideration of its uniform (flat) spectrum. This time function contains equal frequency components at all frequencies. If we were to use such a time function (or a good approximation to it) as the input to a linear system this would be equivalent to *simultaneously* impressing upon the system an array of oscillators covering all possible frequencies, all of equal amplitude and phase. We could determine with this one input the frequency response of the system at *all* frequencies. Any variations in amplitude and phase at the output of the system would be due to the system itself.

We can thus determine the frequency spectrum or frequency transfer function of a *linear* system by applying an impulse at the input and using, for example, a spectrum analyzer at the output. (Note the emphasis on the word *linear*. The superposition of many simultaneous outputs is the crucial concept here and is possible only with linear systems.)

In practice a periodic series of very narrow pulses, approaching the periodic impulses of Sec. 2-1, might be employed so that a repetitive pattern of the frequency spectrum would be obtained. Although the spectrum of the periodic impulses is discrete, with lines spaced at multiples of the repetition frequency (Fig. 2-9), a slow enough repetition rate will ensure that the spacing of the lines is close enough to approximate the continuous spectrum of a single narrow pulse or impulse. (This implies then that the repetition rate $1/T$ is very low compared with frequencies of interest in the spectrum under study.)

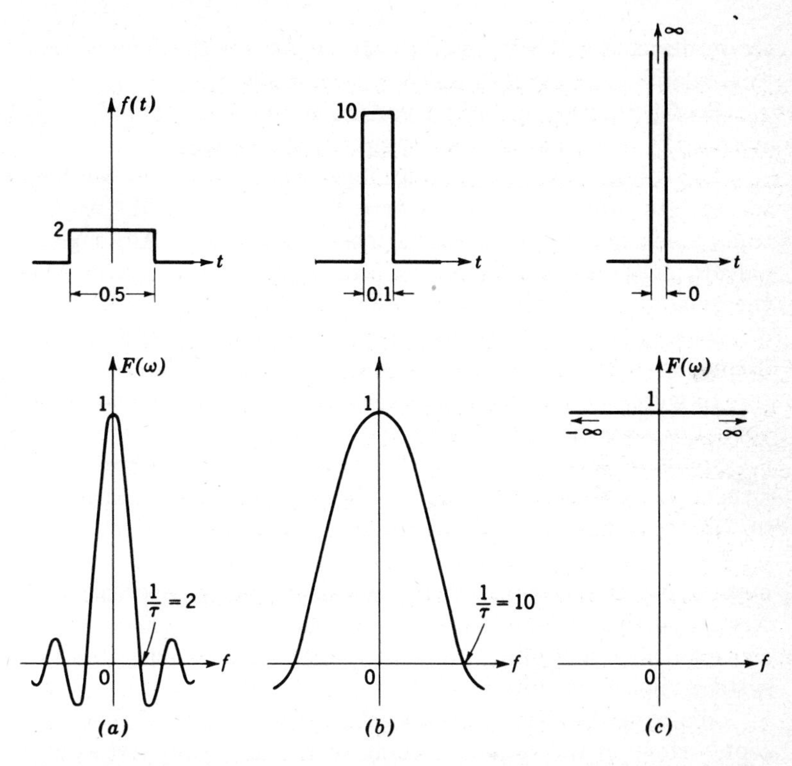

FIG. 2-37 Evolution of a unit impulse and its frequency spectrum. (a) $\tau = 0.5$; (b) $\tau = 0.1$; (c) $\tau = 0$.

The impulse response of a network represents one possible method of determining the network frequency or transfer function (sweep-generator and variable-frequency oscillators are of course also commonly used for this purpose), and this procedure has come into widespread use. There are some drawbacks, however. We can obviously only approximate an impulse function in practice; the width of an actual pulse used must be much less than the response time (reciprocal of the bandwidth) of the filter to be tested, for the impulse approximation to be valid. This requires some prior knowledge of the frequency properties of the network under test.

For example, if the network bandwidth is of the order of 100 kHz, a 1-μsec pulse applied at the input will look very nearly like an ideal impulse so far as the network is concerned. But if the network bandwidth is approximately 1 MHz, a 0.1-μsec pulse would have to be used. In mechanical systems an impulse can be approximated by an impact lasting a short interval of time. If the system response time is of the order of 0.1 sec (bandwidth of 10 Hz), the impact would have to be less than 0.01

sec in duration. These numbers are based on the inverse bandwidth-time relationship discussed in the previous section.

So the choice of the right approximation to an impulse depends on some prior knowledge of the response of the system.

In addition, the large amplitude pulses required may not be practical to use. In amplifier circuits, for example, they could easily drive the transistors used into nonlinear regions of operation. Although small but narrow pulses could be used as approximations, they contain very little energy and might not give the desired output amplitude.

The impulse response of networks is valuable in determining network frequency response, although with certain practical limitations.

Impulses have many other useful applications, however. They obviously provide us with information as to network time behavior (as is apparent from the close relationship between frequency and time response). We shall see shortly that the response of known networks to various other input functions can be readily determined from the impulse response. The use of the impulse also simplifies much of the mathematics of linear-system analysis, as we shall have occasion to demonstrate shortly. (Parenthetically, we shall see later on that impulse concepts play a leading role in the discussion of power and energy in systems, noise and interference, etc.)

We have discussed qualitatively the usefulness of the impulse concept. How do we formulate many of the useful properties of impulses quantitatively?

Our example of a unit impulse as the limiting case of a narrow rectangular pulse ($V \to \infty, \tau \to 0, V\tau = 1$) was only one of many that we might have chosen. Any other pulse shape would have done just as well. For example, a triangle of height V and base width 2τ has as its spectrum

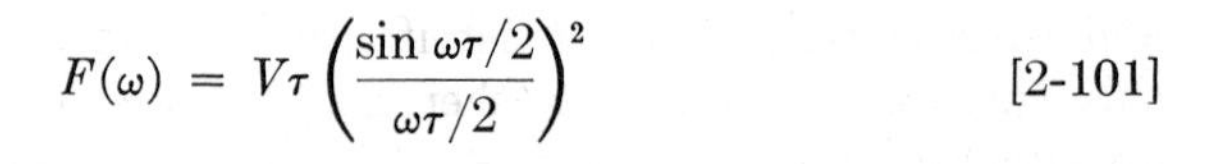

$$F(\omega) = V\tau \left(\frac{\sin \omega\tau/2}{\omega\tau/2} \right)^2 \qquad [2\text{-}101]$$

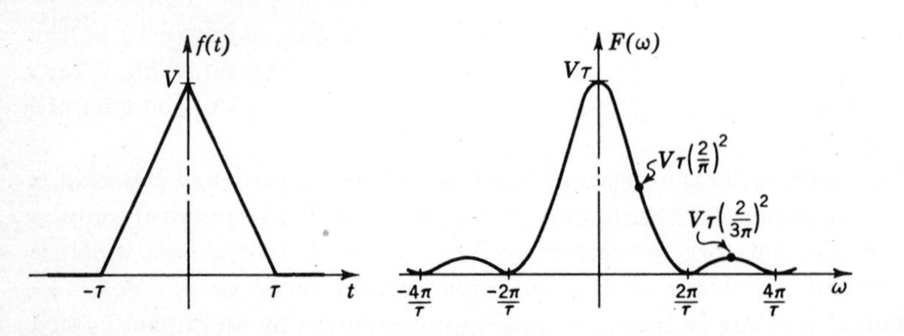

FIG. 2-38 Triangular pulse and its spectrum.

This is sketched in Fig. 2-38. Again the area of the pulse $V\tau$ appears multiplied by a frequency function. As $\tau \to 0$ and $V \to \infty$ such that $V\tau = 1$, we again get an infinitesimally narrow pulse of infinite height, while the frequency spectrum approaches a constant value of 1 for all frequencies. The pulses of Figs. 2-34 to 2-36 have the same property.

Other examples of unit impulses (i.e., unit area) can be generated from the following pulse-type time functions by considering the limit as $\tau \to 0$, $V\tau = 1$:

$f(t)$	$F(\omega)$
$\dfrac{Ve^{-\lvert t\rvert/\tau}}{2}$	$\dfrac{V\tau}{1 + (\tau\omega)^2}$
$\dfrac{V \sin(\pi t/\tau)}{\pi t/\tau}$	$V\tau,\ \lvert\omega\rvert < \pi/\tau$ $0,\ \lvert\omega\rvert > \pi/\tau$

Note that in the limit all these pulses tend to cluster about the point $t = 0$. Their amplitude V increases indefinitely, but their area $V\tau$ always remains 1. In the limit they all have a constant, or flat, frequency spectrum $[F(\omega) = 1]$.

To discuss these impulses quantitatively, we need a mathematical definition of an impulse that will include all the limiting cases mentioned, as well as many other possible ones. There are various acceptable definitions for the unit impulse, or *delta function*, as it is frequently called. One simple definition which satisfies the intuitive approach used above $(V \to \infty, \tau \to 0, V\tau = 1)$ is to define the unit-impulse function $\delta(t)$ by the following integral:

$$f(t) = \int_{-\infty}^{\infty} f(\tau)\, \delta(t - \tau)\, d\tau \qquad [2\text{-}102]$$

The notation $\delta(t - \tau)$ signifies that the impulse is peaked about the point $t = \tau$. The integral then implies that we take any function $f(\tau)$, multiply by a *weighting function* $\delta(t - \tau)$ which peaks strongly about the point $t = \tau$, and integrate the resulting function. We are thus weighting $f(\tau)$ strongly at $t = \tau$. If in particular $\delta(t - \tau)$ is infinite at $t = \tau$ and zero elsewhere, such that its area is normalized to 1, the integral of Eq. [2-102] will reproduce $f(t)$. $A\delta(t)$, with A a constant, represents an impulse function whose area is equal to A.

The integral and the weighting-function concept are demonstrated pictorially in Fig. 2-39. We assume there that $\delta(t - \tau)$ peaks sharply at $t = \tau$ and is zero outside an interval $t - \epsilon < \tau < t + \epsilon$. If ϵ is made small enough, $f(\tau)$ is very nearly constant and equal to $f(t)$ over this interval. As ϵ approaches zero, $f(\tau)$ approaches $f(t)$ exactly. Outside the interval the product $f(t)\, \delta(t - \tau)$ is zero. The integral can thus be

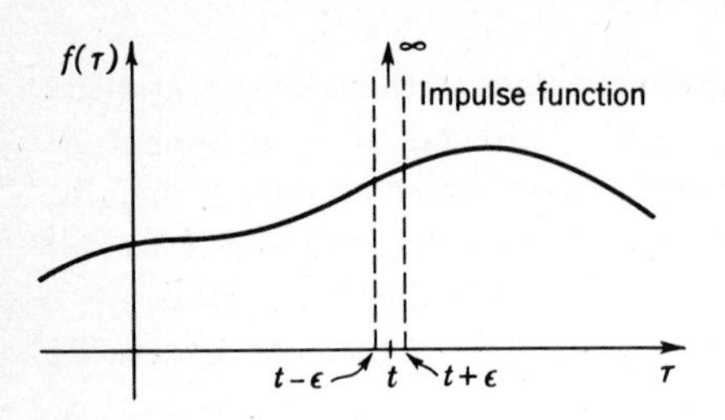

FIG. 2-39 Weighting-function property of impulse functions.

approximated by

$$f(t) \doteq f(t) \int_{t-\epsilon}^{t+\epsilon} \delta(t - \tau)\, d\tau = f(t) \qquad [2\text{-}103]$$

with $\delta(t)$ normalized so that the area under the curve is equal to 1 (*unit impulse*).

The integral thus satisfies our previous intuitive concept, namely, that $\delta(t - \tau) = 0$, $t \neq \tau$; $\delta(t - \tau) = \infty$, $t = \tau$; $\int_{-\infty}^{\infty} \delta(t - \tau)\, d\tau = 1$. It serves to generalize the concept of the delta, or impulse, function and enables us to treat it mathematically.

As an example of the use of the integral, consider in particular that the function $f(\tau)$ is the unit-step function $u(\tau)$. We shall show very simply that the integral does reproduce $u(t)$. Thus

$$\begin{aligned} u(t) &= \int_{-\infty}^{\infty} u(\tau)\delta(t - \tau)\, d\tau \\ &= \int_{0}^{\infty} \delta(t - \tau)\, d\tau \qquad [2\text{-}104] \end{aligned}$$

since

$$\begin{aligned} u(\tau) &= 0 \qquad \tau < 0 \\ u(\tau) &= 1 \qquad \tau > 0 \end{aligned}$$

For $0 < t < \infty$, $\int_{0}^{\infty} \delta(t - \tau)\, d\tau = 1$ and $u(t) = 1$, as expected. For $t < 0$, the integral is zero. [The point $\tau = t$ at which $\delta(t - \tau)$ peaks occurs outside the range of integration.] So we get of course $u(t) = 1$, $t > 0$; $u(t) = 0$, $t < 0$. This is nothing new but again demonstrates the agreement of the defining integral for $\delta(t)$ with our intuitive notions of $\delta(t)$.

In Eq. [2-104] let us now shift our origin of coordinates by defining $x = t - \tau$. The integral then becomes by substitution of variables in both the integrand and the limits of integration

$$u(t) = \int_{-\infty}^{t} \delta(x)\, dx \qquad [2\text{-}105]$$

The unit-step function is thus the integral of the unit impulse.

This is a fundamental relation between the impulse and step functions and is frequently also used as a definition of the impulse function.

To see that this agrees with the previous discussion of the impulse function, note that, if $\delta(x) = 0$ for $x \neq 0$ and $\delta(x) = \infty$ for $x = 0$, $u(t)$ will be zero for $t < 0$ and equals $\int_{-\infty}^{t} \delta(x)\, dx = 1$ if $t > 0$.

Alternatively, if $u(t) = 0$, $t < 0$; $u(t) = 1$, $t > 0$, $\delta(x)$ must be a function which is zero for $t < 0$. It must be infinite at $t = 0$ to make $u(t)$ jump by one unit in zero time at $t = 0$, and it must again be zero for $t > 0$. (Otherwise the integral could not be constant and equal to 1 for $t > 0$.) *The unit-step function is simply the integral of the δ function.* This property is a very useful one in studying linear systems. In a heuristic sense we might also consider $\delta(t)$ to be $du(t)/dt$. This is a troublesome concept mathematically, since the usual definition of a derivative implies continuity, while $u(t)$ is obviously a discontinuous function. For this reason the definition of $\delta(t)$ in terms of its integral form is to be preferred. We shall, however, have occasion to refer to $\delta(t)$ as du/dt and shall find this relation useful later in this section.

Now how do we utilize the impulse function in linear-system analysis?

We have shown, by example, that the unit impulse has a flat frequency spectrum of unit amplitude for all frequencies. That this is a necessary condition for *any* impulse function can be easily demonstrated by means of the integral definition for $\delta(t)$. Thus the Fourier transform of the impulse function is

$$F(\omega) = \int_{-\infty}^{\infty} \delta(t) e^{-j\omega t}\, dt \qquad [2\text{-}106]$$

But, from the integral definition of $\delta(t)$,

$$\int_{-\infty}^{\infty} \delta(t) e^{-j\omega t}\, dt = e^{-j\omega t}\Big|_{t=0} = 1$$

or

$$F(\omega) = 1 \qquad [2\text{-}107]$$

as expected.

For an impulse displaced t_0 sec in time, the Fourier transform is

$$F(\omega) = \int_{-\infty}^{\infty} \delta(t - t_0) e^{-j\omega t}\, dt = e^{-j\omega t_0} \qquad [2\text{-}108]$$

This again gives a flat-amplitude frequency spectrum but introduces the familiar phase factor to account for the time delay assumed. An impulse of weight A, or one whose area equals A, written as $A\delta(t)$, would have as its Fourier transform just A. If the function is $A\delta(t - t_0)$, the Fourier transform becomes $Ae^{-j\omega t_0}$.

We could of course have found the Laplace transform of $\delta(t)$ first and then, with $p = j\omega$, have obtained the Fourier transform. This would have given us

$$F(p) = e^{-pt_0} \qquad [2\text{-}109]$$

as the Laplace transform of $\delta(t - t_0)$ and

$$F(p) = 1 \qquad [2\text{-}110]$$

as the Laplace transform of $\delta(t)$. The corresponding Laplace transforms of $A\delta(t - t_0)$ and $A\delta(t)$ are, respectively, Ae^{-pt_0} and A.

The fact that the unit step is the integral of the impulse function now follows quite naturally. For we know from Laplace-transform theory that $(1/p)F(p)$ is the Laplace transform of $\int_0^t f(t)\,dt$, with $F(p)$ the Laplace transform of $f(t)$. In particular, since $(1/p)e^{-pt_0}$ is the Laplace transform of $u(t - t_0)$,

$$u(t - t_0) = \int_0^t \delta(t - t_0)\,dt \qquad [2\text{-}111]$$

using Eq. [2-109].

Now consider a linear network with transfer function $H(p)$. We know that the output $G(p)$ is given by

$$G(p) = H(p)F(p)$$

with $F(p)$ the Laplace transform of the input $f(t)$.

If $f(t) = \delta(t)$, $F(p) = 1$, and

$$G(p) = H(p) \qquad [2\text{-}112]$$

The unit-impulse response of the system is thus the inverse transform of the network transfer function.

Alternatively, if we determine the impulse response of a linear network, *the Laplace transform of the impulse response gives us the network transfer function.* Thus, with $f(t) = \delta(t)$, the resulting network response $g(t)$ is by definition the unit-impulse response $h(t)$, with $H(p)$ the Laplace transform of $h(t)$.

This agrees with our initial discussion of the impulse function from the frequency point of view. For if $f(t) = \delta(t)$, its Fourier transform is just 1 and the output frequency spectrum must be just the frequency spectrum of the network itself,

$$G(\omega) = H(\omega) \qquad [2\text{-}113]$$

If we measure the output frequency spectrum of the network (more generally, of any linear system) with an impulse applied at the input, we can determine the network transfer function. If we determine the impulse response in time and take its Laplace transform, we can also find the system transfer function. The frequency and time approaches are thus again shown to be related, tied together by the medium of the Fourier and Laplace transforms.

The impulse response of the RC filter of Sec. 2-2 is easily determined as an example. The transfer function of this filter is

$$H(p) = \frac{1/RC}{p + 1/RC}$$

This corresponds to a time function $h(t) = (1/RC)e^{-t/RC}$. $h(t)$ represents the unit-impulse response, or output voltage with a unit-impulse voltage applied to the input of the deenergized circuit. Note that this seems to contradict the usual assertion, based on the conservation of charge, that the voltage across the capacitor cannot change instantaneously. The voltage apparently jumps instantaneously from 0 to $1/RC$ volts. The apparent difficulty of course arises from the use of a nonphysical voltage pulse of infinite amplitude and zero duration. If we assume a voltage pulse of amplitude V volts ($V \to \infty$), a current of magnitude V/R amp begins to flow. In τ sec ($\tau \to 0$), $V\tau/R$ coulombs is deposited on the capacitor plates and the capacitor voltage charges up to $V\tau/RC$ volts. In the limit, as $\tau \to 0$ with $V\tau = 1$, the capacitor must charge from 0 to $1/RC$ volts in 0 time, checking the result obtained from the impulse response.

As a second example, we can calculate the impulse response of the ideal filter of Sec. 2-5. The frequency transfer function of this filter was

$$H(\omega) = Ae^{-j\omega t_0} \qquad |\omega| \leq 2\pi B$$

and $H(\omega) = 0$ elsewhere. With an impulse input voltage $f(t) = \delta(t)$, $F(\omega) = 1$ and $G(\omega) = H(\omega)$. The impulse response is then found from the inverse Fourier transform,

$$\begin{aligned} h(t) &= \frac{1}{2\pi} \int_{-\infty}^{\infty} H(\omega)e^{j\omega t}\, d\omega \\ &= 2BA \frac{\sin 2\pi B(t - t_0)}{2\pi B(t - t_0)} \end{aligned} \qquad [2\text{-}114]$$

The impulse response is simply the familiar $(\sin x)/x$ function in time. This is as might be expected, for the Fourier transform and its inverse are symmetric, respectively, in ω and t, as can be seen by a comparison of Eqs. [2-45] and [2-46]. A rectangular pulse in time has for its Fourier transform a $(\sin x)/x$ function in ω, while a rectangular pulse in ω has for its inverse transform a $(\sin x)/x$ function in time. This symmetry of course is the reason for the inverse frequency-time relationship demonstrated throughout this chapter.

CONVOLUTION INTEGRAL

Since the impulse response provides us with the network transfer function or frequency spectrum, it gives us a means of determining the response to any other input time function. This is immediately apparent from either of the two equivalent relations

$$G(p) = H(p)F(p)$$

or

$$G(\omega) = H(\omega)F(\omega)$$

Thus we can experimentally find $h(t)$ or $H(\omega)$ by applying an impulse to a given system and measuring the resultant time response $h(t)$ or the frequency response $H(\omega)$. For any other input $f(t)$ we can then predict the response $g(t)$ by finding the time function corresponding to $G(p)$. Since $g(t)$ is obviously related to $h(t)$ and $f(t)$ by the Laplace- or Fourier-transform relations, there should be a way of relating the three time functions directly.

We shall show that this relationship appears in the form of the so-called *convolution integral* given by

$$g(t) = \int_{-\infty}^{\infty} f(\tau)h(t - \tau)\, d\tau \qquad [2\text{-}115]$$

Note how similar this integral is to the integral definition for the delta function introduced previously as Eq. [2-102]. If the network impulse response happens to be an impulse function itself, or $h(t) = \delta(t - t_0)$ (a distortionless network is an example of a network with this property), Eq. [2-115] gives as the output a delayed version of the input:

$$g(t) = f(t - t_0)$$

Just as in the case of the defining integral for the delta function, Eq. [2-115] may be interpreted as a weighting-function integral. The network response to an arbitrary input $f(t)$ may be considered a weighted sum of different values of $f(t)$. The impulse response $h(t)$ acts as the weighting function in this case.

We shall first derive the convolution integral and then apply it to the solution of some simple problems. The significance of the integral will then be discussed further.

The derivation is demonstrated by first representing an arbitrary time function $f(t)$ by a series of impulses of varying weight. This we do by approximating the integral definition of the impulse function by a sum. Thus

$$f(t) = \int_{-\infty}^{\infty} f(\tau)\delta(t - \tau)\, d\tau$$

$$\doteq \sum_{n=-\infty}^{\infty} f(n\,\Delta\tau)\delta(t - n\,\Delta\tau)\,\Delta\tau \qquad [2\text{-}116]$$

This corresponds to breaking the time scale up into small intervals $\Delta\tau$ apart, letting the variable τ become $n\,\Delta\tau$, and summing over n instead of τ. This is shown in Fig. 2-40. For $\Delta\tau$ very small, $\delta(t - n\,\Delta\tau)\,\Delta\tau \doteq 1$ with a peak in the vicinity of $n\,\Delta\tau = t$, and the sum reproduces $f(t)$.

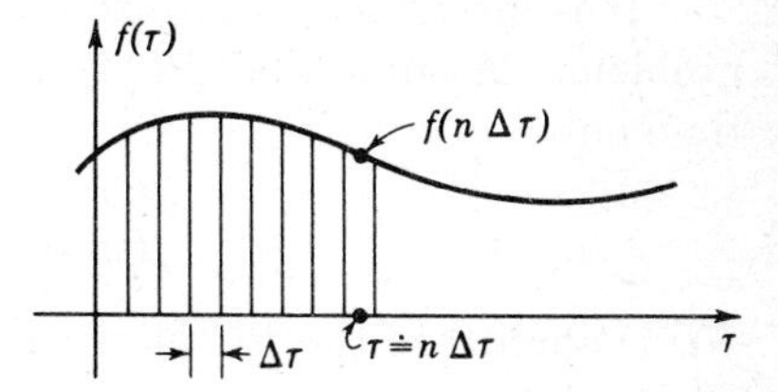

FIG. 2-40 Impulse representation of a function.

Now assume that $f(t)$ is applied to a *linear* system. The response to $f(t)$ may be found by superposing the response to each of the impulses of Eq. [2-116]. The response to an impulse $A\,\delta(t - n\,\Delta\tau)$ of weight A is by definition $Ah(t - n\,\Delta\tau)$, so that the output is given by

$$g(t) \doteq \sum_{n=-\infty}^{\infty} f(n\,\Delta\tau)h(t - n\,\Delta\tau)\,\Delta\tau \qquad [2\text{-}117]$$

In the limit, as $\Delta\tau \to 0$, $n\,\Delta\tau \to \tau$ and the sum becomes the integral

$$g(t) = \int_{-\infty}^{\infty} f(\tau)h(t - \tau)\,d\tau \qquad [2\text{-}115]$$

To show that this integral satisfies the transform relations

$$G(p) = F(p)H(p)$$

or $G(\omega) = F(\omega)H(\omega)$, we may take the Fourier (or Laplace) transform of each side of Eq. [2-115]. Thus

$$G(\omega) = \int_{-\infty}^{\infty} e^{-j\omega t}\left[\int_{-\infty}^{\infty} f(\tau)h(t - \tau)\,d\tau\right] dt \qquad [2\text{-}118]$$

If we now interchange the order of integration (the integrals must be assumed to converge absolutely),

$$G(\omega) = \int_{-\infty}^{\infty} f(\tau)\left[\int_{-\infty}^{\infty} e^{-j\omega t}h(t - \tau)\,dt\right] d\tau \qquad [2\text{-}119]$$

But

$$\int_{-\infty}^{\infty} e^{-j\omega t}h(t - \tau)\,dt = e^{-j\omega\tau}H(\omega) \qquad [2\text{-}120]$$

since a time delay of τ sec corresponds to a phase factor $e^{-j\omega t}$ in the Fourier transform. Using Eq. [2-120],

$$G(\omega) = H(\omega) \int_{-\infty}^{\infty} f(\tau) e^{-j\omega\tau} \, d\tau$$
$$= H(\omega) F(\omega) \qquad [2\text{-}121]$$

The transfer-function relations are thus seen to be satisfied by the convolution integral.

It is instructive to utilize the convolution integral to solve a simple problem. Assume that $f(t)$ is a simple exponential defined for positive time only.

$$\begin{aligned} f(t) &= e^{-at} & t &> 0 \\ f(t) &= 0 & t &< 0 \end{aligned} \qquad [2\text{-}122]$$

$f(t)$ is applied to the RC filter with transfer function

$$H(p) = \frac{1/RC}{p + 1/RC}$$

(assumed initially deenergized). What is the resulting output voltage $g(t)$?

We could use simple differential equations or the Laplace-transform method to find $g(t)$. Either way we find $g(t)$ to be given by

$$g(t) = \frac{e^{-at} - e^{-t/RC}}{1 - aRC} \qquad a \neq \frac{1}{RC} \qquad [2\text{-}123]$$

(This is the familiar response of an RC network to an exponential input and should be checked by the reader.)

Applying the convolution integral (Eq. [2-115]) to the solution of the same problem, we have

$$f(\tau) = e^{-a\tau} \qquad \tau > 0 \qquad [2\text{-}124]$$

and

$$h(t - \tau) = \frac{1}{RC} e^{-(t-\tau)/RC} \qquad t > \tau \qquad [2\text{-}125]$$

[Recall that $h(t) = (1/RC)e^{-t/RC}$ for the RC network.] Then

$$g(t) = \int_{-\infty}^{\infty} f(\tau) h(t - \tau) \, d\tau$$
$$= \int_0^t \frac{e^{-a\tau}}{RC} e^{-(1/RC)(t-\tau)} \, d\tau \qquad [2\text{-}126]$$

since $f(\tau) = 0$, $\tau < 0$; $h(t - \tau) = 0$, $\tau > t$. Performing the indicated integration, we get

$$g(t) = \frac{e^{-t/RC}}{RC} \left. \frac{e^{\tau(1/RC - a)}}{1/RC - a} \right|_0^t$$
$$= \frac{e^{-at} - e^{-t/RC}}{1 - aRC} \qquad [2\text{-}127]$$

The general form of the convolution integral (Eq. [2-115]) may be rewritten in somewhat different form for certain special cases. First, we note by a simple change of variables that $g(t)$ is symmetrical in $f(t)$ and $h(t)$. Thus

$$g(t) = \int_{-\infty}^{\infty} f(\tau)h(t - \tau)\, d\tau = \int_{-\infty}^{\infty} h(\tau)f(t - \tau)\, d\tau \qquad [2\text{-}128]$$

The reader should prove Eq. [2-128] for himself. Note that it is consistent with the transform relation since

$$G(p) = H(p)F(p) = F(p)H(p)$$

Interchanging H and F has no effect on G, and therefore $g(t)$ remains unchanged.

The convolution integral applies to both physically realizable and nonrealizable linear filters. (The ideal low-pass filter is an example of the latter.)

For physically realizable filters, i.e., those which can be synthesized in terms of R, L, C elements and active elements, we can develop an alternate form of the convolution integral by noting that $h(t) = 0$, $t < 0$. This is just a mathematical statement of the fact that no output can appear before the input is applied. [$\delta(t)$, the input, is applied at $t = 0$.] Then $h(t - \tau) = 0$, $t < \tau$, and

$$g(t) = \int_{-\infty}^{t} f(\tau)h(t - \tau)\, d\tau \qquad [2\text{-}129]$$

Equation [2-129] represents the convolution integral specialized to the case of realizable filters. If now $f(t) = 0$, $t < 0$, or the input is assumed applied at $t = 0$, we get

$$g(t) = \int_{0}^{t} f(\tau)h(t - \tau)\, d\tau \qquad [2\text{-}130]$$

This is actually the integral we used in finding the response of the simple RC filter to an exponential input. There we stated specifically that $h(t) = e^{-t/RC}/RC$, $t > 0$, and $h(t) = 0$, $t < 0$, and so rewrote the integral directly.

The ideal low-pass filter, on the other hand, was specifically noted to be nonrealizable, and we found in fact the existence of precursor, or anticipatory, transients. Thus, for the filter characteristic given by

$$H(\omega) = Ae^{-jt_0\omega}$$

where $|\omega| \leq 2\pi B$, and $H = 0$, $|\omega| > 2\pi B$,

$$h(t) = 2BA\, \frac{\sin 2\pi B(t - t_0)}{2\pi B(t - t_0)}$$

This impulse response is symmetrical about $t = t_0$, as expected, and has precursor transients. $h(t)$ is in fact defined for all t, $-\infty < t < \infty$. So for the general filter, including idealized networks, the general form of the convolution integral, as given by Eq. [2-115], must be used. If $f(t) = 0$, $t < 0$, however, Eq. [2-115] may be specialized to

$$g(t) = \int_0^\infty f(\tau)h(t - \tau)\,d\tau \qquad [2\text{-}131]$$

We have already indicated that the convolution integral contains as a special case the integral definition of the impulse, or delta, function. We can consider as another special case the step response of a linear network. Thus let $f(t) = u(t)$, a unit-step input. Then for the general linear filter (realizable and nonrealizable cases included)

$$g(t) = \int_0^\infty h(t - \tau)\,d\tau \equiv A(t) \qquad [2\text{-}132]$$

$A(t)$, the step response, is thus just the integral of the impulse response. A simple change of variables, with $x = t - \tau$, gives us also

$$A(t) = \int_{-\infty}^{t} h(x)\,dx \qquad [2\text{-}133]$$

In particular, if $h(t) = 0$, $t < 0$ (physically realizable filters),

$$A(t) = \int_0^t h(x)\,dx \qquad [2\text{-}134]$$

As an example, consider the RC filter again. The impulse response was found to be $h(t) = e^{-t/RC}/RC$, $t \geq 0$. The step response is then

$$\begin{aligned} A(t) &= \int_0^t h(x)\,dx = \frac{1}{RC}\int_0^t e^{-x/RC}\,dx \\ &= 1 - e^{-t/RC} \qquad [2\text{-}135] \end{aligned}$$

as expected.

The step response of the ideal filter is obtained using Eq. [2-133], since $h(t)$ is not zero for $t < 0$. Thus, with

$$h(t) = 2BA\,\frac{\sin 2\pi B(t - t_0)}{2\pi B(t - t_0)}$$

the step response of the ideal filter becomes

$$\begin{aligned} A(t) &= 2BA\int_{-\infty}^{t} \frac{\sin 2\pi B(x - t_0)}{2\pi B(x - t_0)}\,dx \\ &= \frac{A}{\pi}\int_{-\infty}^{2\pi B(t-t_0)} \frac{\sin y}{y}\,dy \qquad [2\text{-}136] \end{aligned}$$

with $y = 2\pi B(x - t_0)$. We again have the sine integral as a solution. From our previous discussion of the sine integral (Fig. 2-22) we recall that

$(1/\pi) \int_{-\infty}^{0} [(\sin x)/x]\, dx = \frac{1}{2}$. The step response becomes

$$A(t) = A\left[\frac{1}{2} + \frac{1}{\pi}\,\mathrm{Si}\; 2\pi B(x - t_0)\right] \qquad [2\text{-}137]$$

The impulse and step responses for both the RC filter and the ideal filter are sketched in Fig. 2-41.

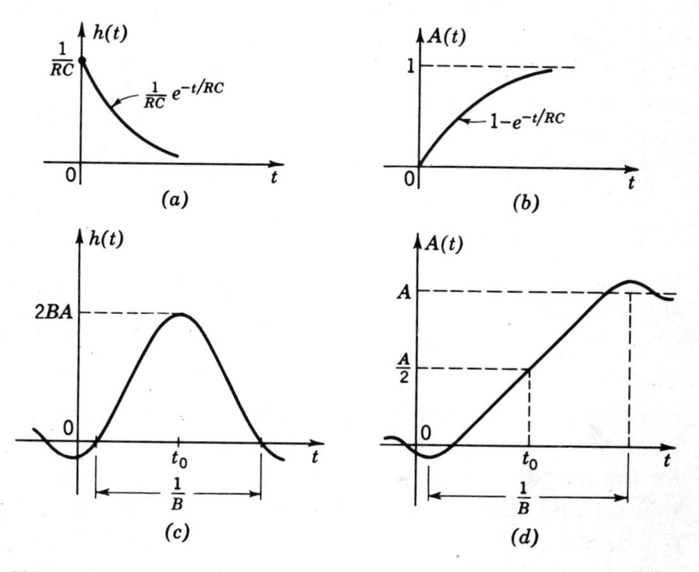

FIG. 2-41 Impulse and step responses. (a) Impulse response, RC filter; (b) step response, RC filter; (c) impulse response, ideal filter; (d) step response, ideal filter.

The step response of the ideal filter (Eq. [2-137]) may also be obtained from the previous result for the rectangular-pulse response of the same filter (Eq. [2-66] and Fig. 2-24) by letting the pulse width τ become very large. The input pulse then approaches a step, and the response approaches the step response. An alternative method of obtaining Eq. [2-137] is indicated in one of the problems for this chapter.

The convolution integral, in addition to providing us with another method of determining the response of a linear system to a specified input, enables us to draw a visual picture of the system behavior. This we can do by a qualitative discussion of the convolution integral for physically realizable filters,

$$g(t) = \int_0^t f(\tau)h(t - \tau)\, d\tau \qquad f(t) = 0,\ t < 0 \qquad [2\text{-}138]$$

Here t represents the time at which the filter response $g(t)$ is to be determined and might thus be called the *present*. The variable τ ranges over all values less than t, or values of time in the *past*. The variable $t - \tau$ which appears associated with the impulse response, thus measures time back into the past (see Fig. 2-42).

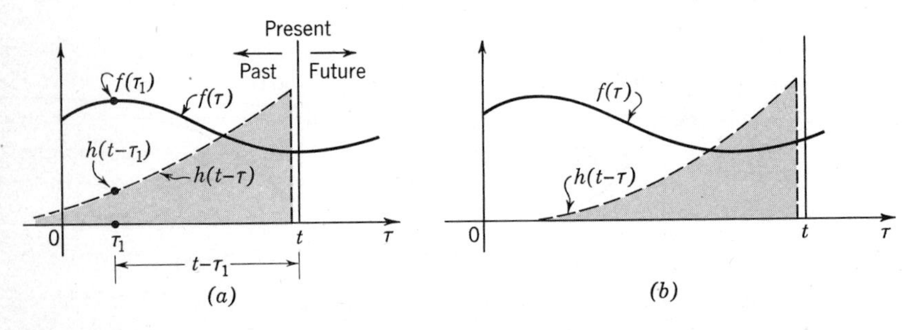

FIG. 2-42 Graphical interpretation of convolution integral. (a) A particular value of t; (b) t some time later.

For all realizable filters the impulse response $h(t)$ is zero for $t < 0$ and again goes to zero as $t \to \infty$ [for example, $h(t) = (1/RC)e^{-t/RC}$, $t > 0$, for the RC filter]. The function $h(t - \tau)$ which appears in the convolution integral then must be zero for $\tau > t$ and again approaches zero as $t - \tau \to \infty$, or $\tau \to -\infty$. $h(t - \tau)$ thus represents the impulse response turned round, or *folded over*, and extending back into the past. This is shown in Fig. 2-42.

We may thus interpret the filter response $g(t)$ as being the weighted superposition of past values of the input $f(\tau)$. $f(\tau)$, in the past, appears multiplied or weighted by $h(t - \tau)$.

Since $h(t - \tau)$ eventually goes to zero for $t - \tau$ large enough, or τ sufficiently "far back in the past," values of $f(\tau)$ that appeared "far back in the past" have a negligibly small influence on the present output of the network. The filter thus can be looked on as having a "memory" lasting only over the time interval $t - \tau$ for which $h(t - \tau)$ has significant values. This is just the response time of the filter, or the reciprocal of its bandwidth.

As an example, consider the RC filter again. The voltage across the capacitor is assumed to be the desired output. The charge on the capacitor at any given time ($\tau = t$, the *present*) is obviously due to all charge applied in the past. However, because of the resistor in series with the capacitor most of the charge applied in the past has been discharged and dissipated in the form of heat. The significant contribution to existing charge may be assumed to have been applied less than $2\pi RC$ sec or about 6 time constants before. ($B = 1/2\pi RC$.) Mathematically this is accounted for by the exponentially decreasing form of the impulse response.

As the present time t varies, the impulse response $h(t - \tau)$ scans the function $f(\tau)$, always producing a weighted sum of past inputs, always weighting most heavily values of $f(\tau)$ closest to the present ($t - \tau \doteq 0$). If the impulse response is quite narrow in width, the output comes close to being an undistorted replica of the input.

The filter weighting effect is analogous to the aperture effect of optical filters. The input $f(\tau)$ is "seen" through a "window," or aperture, whose width is that of the impulse-response function.

This analogy with optical filters is particularly illuminating if $h(t)$ happens to have the form of a rectangular pulse of width Δt. The filter output is then simply the input $f(\tau)$ averaged over the interval Δt. If Δt is small compared with time changes in the input function [if the filter time response is short compared with the rate of change of $f(\tau)$, or if the filter frequency response is wide compared with the input frequency spectrum], there is no aperture distortion and $f(t)$ is reproduced undistorted. This interpretation of course agrees with the previous bandwidth-time discussions of this chapter. Alternatively we can say that, as $\Delta t \to 0$, $h(t - \tau)$ approaches an impulse centered at $t = \tau$ and $f(t)$ is reproduced undistorted.

A simple example will demonstrate the usefulness of this graphical approach to the convolution integral. Assume that both $f(t)$ and $h(t)$ are unit-amplitude rectangular pulses of width Δt sec beginning at $t = 0$ (see Fig. 2-43*a* and *b*). The filter frequency response, or the Fourier transform of $h(t)$, is

$$H(\omega) = \Delta t\, e^{-j(\omega \Delta t/2)} \frac{\sin (\omega \Delta t/2)}{\omega \Delta t/2}$$

(The impulse response is symmetrical about the point $t = \Delta t/2$, and this gives rise to the phase term.) $F(\omega)$ is of course identical with $H(\omega)$.

We then find the output response $g(t)$ by convolving $f(\tau)$ with $h(t - \tau)$. The τ plane is depicted in Fig. 2-43*c* and *d*. Figure 2-43*c* shows the appropriate functions for $t < \Delta t$, Fig. 2-43*d* for $t > \Delta t$. In either case we are to multiply $f(\tau)$ by $h(t - \tau)$ and integrate over all values of τ up to t. Since $f(\tau) = 0$, $\tau < 0$, we need integrate only from $\tau = 0$. For $t < \Delta t$ the integral is simply the area under $f(\tau)$ and $h(t - \tau)$ included between 0 and t (see Fig. 2-43*c*). So

$$g(t) = t \qquad t < \Delta t$$

(Remember that both time functions have unit amplitude.)

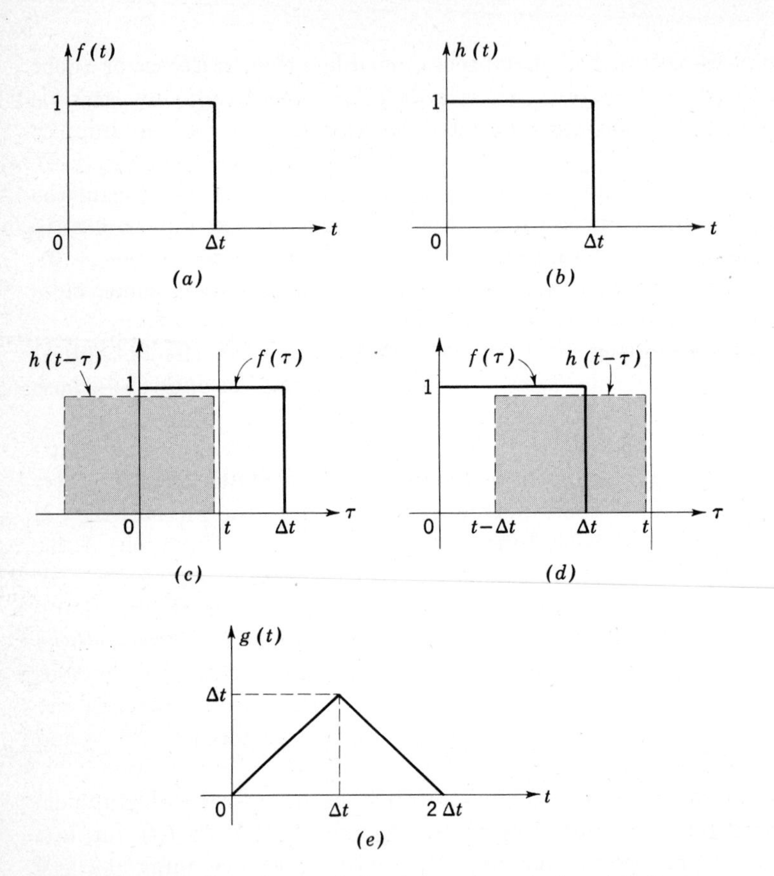

FIG. 2-43 Example of convolution-integral analysis. (a) System input; (b) system impulse response; (c) convolution integral, $t < \Delta t$; (d) convolution integral, $t > \Delta t$; (e) $g(t)$, output response.

For $t > \Delta t$ we can integrate only from $\tau = t - \Delta t$ to $\tau = \Delta t$, since $h(t - \Delta t)$ is zero for $\tau < t - \Delta t$. Again the integral is the area included between the two limits of τ, and

$$\begin{aligned} g(t) &= 2(\Delta t) - t \qquad & t > \Delta t \\ g(t) &= 0 & t > 2\,\Delta t \end{aligned}$$

The complete output function is thus a triangular pulse of base width $2\,\Delta t$ and height Δt. $g(t)$ is shown in Fig. 2-43*e*.

We can check this result quite easily from the frequency spectra. Since $F(\omega)$ and $H(\omega)$ are identical *in this case*,

$$G(\omega) = H^2(\omega) = (\Delta t)^2 e^{-j\omega\,\Delta t}\left[\frac{\sin\,(\omega\,\Delta t/2)}{\omega\,\Delta t/2}\right]^2$$

As indicated by Eq. [2-101], and as will be shown shortly by Eq. [2-142], this transform represents a triangular pulse in time, of width $2\,\Delta t$ sec, and whose amplitude is $(\Delta t)^2$. The pulse is symmetrical about the point $t = \Delta t$ from the phase factor. This is of course exactly the pulse of Fig. 2-43*e*.

Although we have discussed the convolution integral in terms of the network impulse response, it actually applies more generally to *any* three functions $a(t)$, $b(t)$, $c(t)$ whose transforms, respectively, are related by

$$A(p) = B(p)C(p) \tag{2-139}$$

Then

$$a(t) = \int_{-\infty}^{\infty} b(\tau)c(t-\tau)\,d\tau = \int_{-\infty}^{\infty} c(\tau)b(t-\tau)\,d\tau \tag{2-140}$$

For example, if we know the response to *any* input (a sine wave, for example), we can find the response to any other input in terms of the convolution integral.

We shall apply the convolution integral to the solutions of some problems in probability later in the book.

The convolution integral provides us with an additional tool in solving linear-system problems, and it enables us to describe the system behavior pictorially. Its application to particular problems is, however, sometimes quite complicated because of the integrals involved. In such cases a direct application of differential equations or transform theorem theory might provide the answer more readily. In the solution of a particular problem *all* available tools should be considered and the simplest and most direct eventually chosen.

2-10 USE OF IMPULSES IN FREQUENCY ANALYSIS

We have shown that the impulse response enables us to determine the response of linear networks to any other type of input. This gives us both an analytical tool (the convolution integral) and an experimental method for determining system response.

The impulse function is also particularly helpful in determining the Fourier transform or frequency spectrum of complicated waveshapes.

In particular, if a specified time function can be represented or approximated by a series of straight-line sections, or parabolic arcs, or in general by polynomials in t, its frequency spectrum can be found readily by using impulse functions. For we can successively differentiate the function over any section representable by a polynomial until an impulse or set of impulses remains. The Fourier transform of an impulse at $t = t_0$ is $e^{-j\omega t_0}$. Integrating back again to obtain the original function, we divide $e^{-j\omega t_0}$ by $1/j\omega$ for each integration. Adding up the resultant terms in $j\omega$ by superposition, we obtain the desired spectrum for the original function.

Note that this technique obviates the need for tedious and time-consuming integrations. It implies, however, that the time function can be accurately represented either by a polynomial in t or sectionally by polynomials in t.

Some examples will demonstrate the method:

1. $f(t)$ a rectangular pulse of width τ and height V (Fig. 2-44a). Differentiating once gives the two impulses of Fig. 2-44b. Each impulse

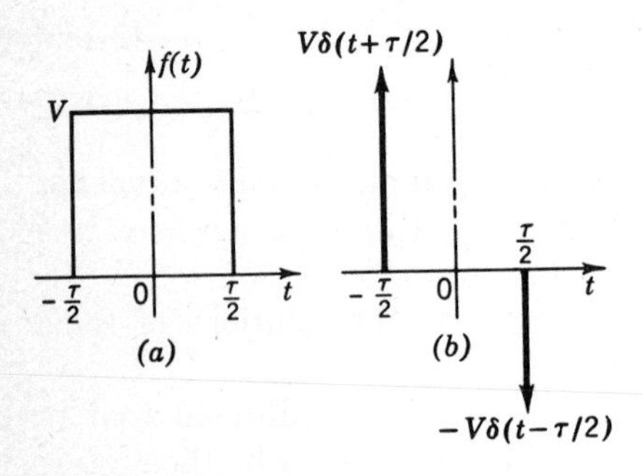

FIG. 2-44 Rectangular pulse and its derivative. (a) $f(t)$; (b) df/dt.

has a weight of V. Then

$$F(\omega) = \frac{V}{j\omega}(e^{j\omega\tau/2} - e^{-j\omega\tau/2}) = V\tau\,\frac{\sin(\omega\tau/2)}{\omega\tau/2} \qquad [2\text{-}141]$$

as previously.

2. $f(t)$ a triangle of height V and width 2τ (Fig. 2-45a). Here two successive differentiations are needed, and three impulses result. The

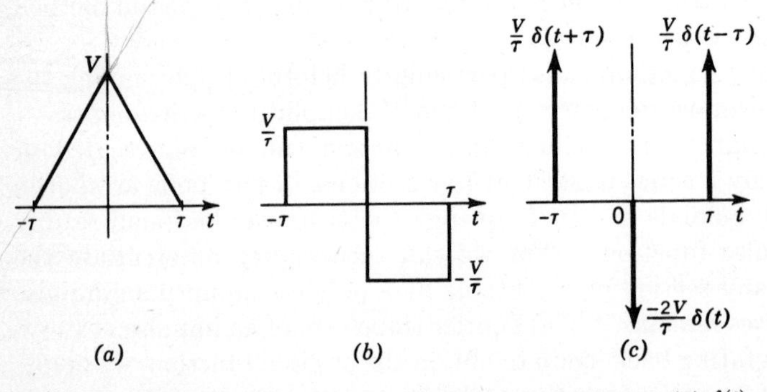

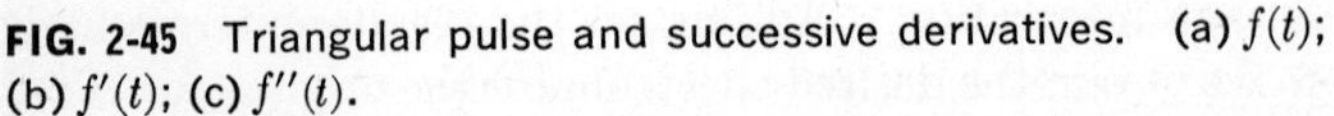
FIG. 2-45 Triangular pulse and successive derivatives. (a) $f(t)$; (b) $f'(t)$; (c) $f''(t)$.

middle impulse is twice the weight of the other two. Integrating twice to get $f(t)$ back again, with the appropriate phase factors for the impulses,

$$\begin{aligned} F(\omega) &= \frac{V}{\tau(j\omega)^2}(e^{j\omega\tau} - 2 + e^{-j\omega\tau}) \\ &= \frac{2V}{\tau(j\omega)^2}(\cos\omega\tau - 1) = \frac{4V}{\tau\omega^2}\left(\sin^2\omega\frac{\tau}{2}\right) \\ &= V\tau\left[\frac{\sin(\omega\tau/2)}{\omega\tau/2}\right]^2 \end{aligned} \qquad [2\text{-}142]$$

again, as obtained previously.

Note that we have to be careful to weight each impulse properly. Here the first derivatives had the magnitudes V/τ and $-V/\tau$, and these were used as the weight of the impulses [that is, $(V/\tau)\,\delta(t-\tau)$, $(-2V/\tau)\,\delta(t)$, $(V/\tau)\,\delta(t+\tau)$].

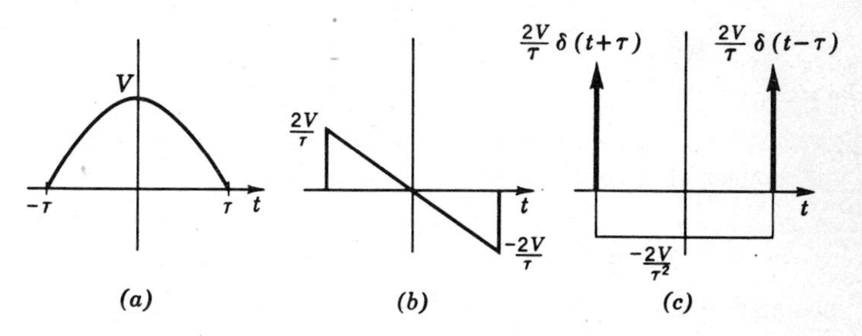

FIG. 2-46 Parabolic pulse and its derivatives. (a) $f(t)$; (b) $f'(t)$; (c) $f''(t)$.

3. $f(t)$ a parabolic pulse of height V and width 2τ (Fig. 2-46*a*).

$$\begin{aligned} f(t) &= V\left[1 - \left(\frac{t}{\tau}\right)^2\right] \qquad |t| \le \tau \\ f(t) &= 0 \qquad |t| > 0 \end{aligned} \qquad [2\text{-}143]$$

Differentiating once, we get the triangle of Fig. 2-46*b*. Differentiating a second time, we get two impulses, $(2V/\tau)\,\delta(t+\tau)$, $(2V/\tau)\,\delta(t-\tau)$, and a pulse of amplitude $-2V/\tau^2$, width 2τ. We could differentiate the pulse once more to reduce it to two impulses. Instead we apply superposition at this point, combining the known spectra for the impulses and pulse, and then dividing by $(j\omega)^2$ to obtain $F(\omega)$. This gives

$$F(\omega) = \frac{2V}{\tau(j\omega)^2}(e^{j\omega\tau} + e^{-j\omega\tau}) - \frac{2V}{\tau^2(j\omega)^2} 2\tau \frac{\sin \omega\tau}{\omega\tau}$$
$$= \frac{4V\tau}{(\omega\tau)^2}\left(\frac{\sin \omega\tau}{\omega\tau} - \cos \omega\tau\right) \quad [2\text{-}144]$$

The method of superposition is applicable in all our problems and can be used very effectively to find the transforms of various simple shapes made up of impulses, pulses, and triangles. For example, the Fourier transform of the raised triangle of Fig. 2-47*a* can be obtained by super-

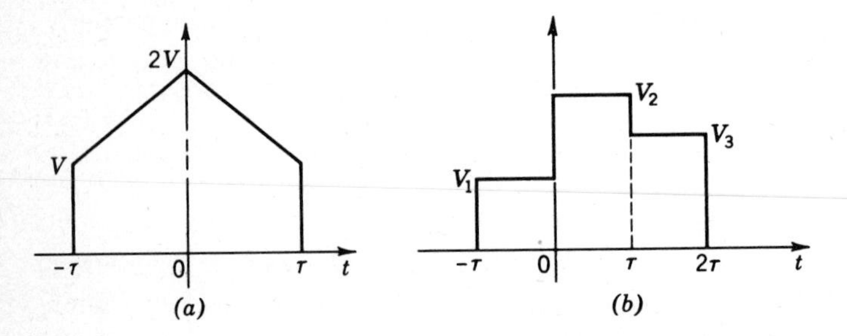

FIG. 2-47 Complex pulse shapes. (a) Another triangular pulse; (b) complex of rectangular pulses.

posing the transforms of a pulse and a triangle. Thus

$$F(\omega) = V\tau\left[\frac{\sin(\omega\tau/2)}{\omega\tau/2}\right]^2 + 2V\tau\frac{\sin \omega\tau}{\omega\tau} \quad [2\text{-}145]$$

The pulse of Fig. 2-47*b* is obviously the superposition of three pulses, each of width τ, with appropriate delays. The Fourier transform of the group is thus

$$F(\omega) = \frac{\sin(\omega\tau/2)}{\omega\tau/2}(V_1\tau e^{j\omega\tau/2} + V_2\tau e^{-j\omega\tau/2} + V_3\tau e^{-j(3\omega\tau/2)}) \quad [2\text{-}146]$$

Other simple examples are included among the problems for this chapter.

Note that throughout this section, as in the previous sections of this chapter, runs the theme of superposition. This is a basic concept of linear systems and often serves to simplify complex-looking problems. Superposition techniques break down in nonlinear systems, however, as will be noted in the sections on frequency modulation in Chap. 4.

The use of impulse, or delta, functions also serves to simplify the mathematics of linear systems, as has been demonstrated in the last two sections. We shall utilize the impulse function in the discussion of various topics throughout this book.

2-11 TELEVISION BANDWIDTH REQUIREMENTS

Before concluding this chapter on signal transmission through linear networks, we give one further example—the calculation of the bandwidth needed to transmit a typical TV test pattern.[1]

The pattern we choose consists of a series of alternating black and white spots. The pattern is 6 in. high by 8 in. wide, and the spots are 0.0121 in. high and 0.0188 in. wide. (These dimensions have been chosen as a compromise between the ability of the eye to resolve detail and the frequency bandwidth required.) There are then $6/0.0121 = 495$ horizontal lines that must be covered by a scanning beam of electrons. The equivalent of 30 additional lines is allowed to enable the beam to retrace from the final line back to the first again. (Actually interlaced scanning is used in which even-numbered lines are first covered, then odd-numbered lines. This increases the time allotted for scanning the picture. The persistence of the eye is then relied on to give the effect of a more rapid scanning rate.) The total number of lines scanned is thus 525 (the time for the 30 additional lines is blanked out). The standard scanning rate is 30 frames per second, so that each line is scanned in

$$\frac{1}{30 \times 525} = 63.5\ \mu\text{sec}$$

This represents the time in which the beam must sweep from left to right through the 8 in. of the pattern and return to the left side again. Allowing 10 μsec for the return, or horizontal, trace interval leaves 53.5 μsec for the actual sweep. In this time $8/0.0188 = 425$ alternating black and white spots will be swept through. The electrical output will thus consist of a series of square waves of $53.5/425 = 0.125\ \mu$sec width.

Our problem is to determine how much bandwidth is needed to resolve these square-wave pulses of 0.125 μsec duration, separated by the same time interval. We can relate this to the problem of resolving two pulses of width τ, separated by τ sec (τ is 0.125 μsec). These pulses are shown in Fig. 2-48.

The response of the idealized low-pass filter to a single pulse was given by Eq. [2-66]. Since the system is assumed linear, superposition

[1] S. Deutsch, "Theory and Design of Television Receivers," McGraw-Hill, New York, 1951.

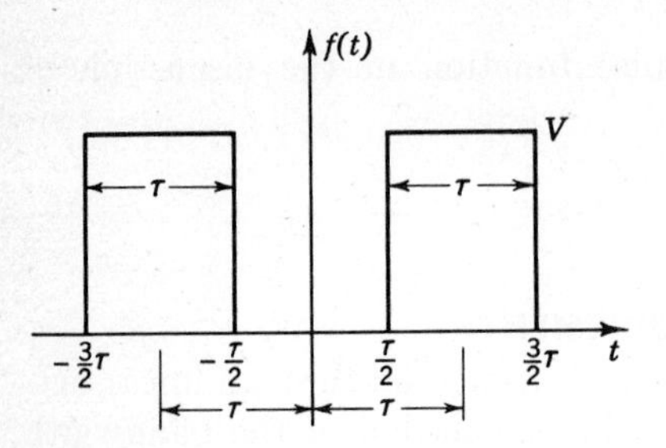

FIG. 2-48 Two test pulses.

may be applied. The two pulses applied to the low-pass filter with bandwidth B Hz then produce a response at the output given by

$$g(t) = \frac{AV}{\pi}\left\{\text{Si}\left[2\pi B(t - t_0) + \frac{3\tau}{2}\right] - \text{Si}\left[2\pi B(t - t_0) + \frac{\tau}{2}\right] + \text{Si}\left[2\pi B(t - t_0) - \frac{\tau}{2}\right] - \text{Si}\left[2\pi B(t - t_0) - \frac{3\tau}{2}\right]\right\} \quad [2\text{-}147]$$

This equation has been plotted by S. Goldman in his book,[1] and some of these curves are reproduced in Fig. 2-49. (t_0 is assumed zero so that the

[1] Goldman, *op. cit.*

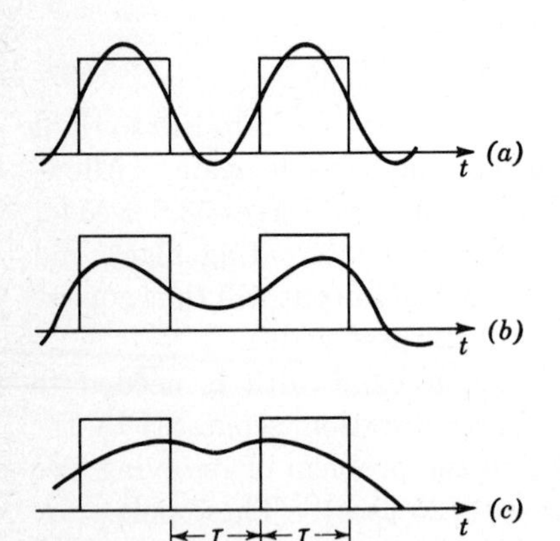

FIG. 2-49 Low-pass filter response to TV test pattern (two pulses). (a) $B = 1/\tau$; (b) $B = 1/2\tau$; (c) $B = 1/3\tau$. (From S. Goldman, "Frequency Analysis, Modulation, and Noise," McGraw-Hill, New York, 1948.)

input and output curves are superimposed for comparison.) These results indicate that for resolution of (the ability to separate) the two pulses

$$B \geq \frac{1}{2\tau} \tag{2-148}$$

Note that this is a different situation from the one previously stressed in this chapter, in which it was desired to reproduce the pulse faithfully ($B \gg 1/\tau$). Here the problem is to distinguish between black and white spots or between an on and off signal. The signal detail as such is not important in this application.

If $B < 1/2\tau$, the two pulses merge into one and the resolution disappears. If $B > 1/2\tau$, the sides of the pulses are sharpened, the rise times decreasing.

For the TV case outlined $\tau = 0.125$ μsec, and

$$B \geq 4 \text{ MHz}$$

It will be shown later that in order to transmit these signals at high frequencies bandwidths of $2B = 8$ MHz will be required. In practice some bandwidth compression is utilized (vestigial sideband transmission), and 6-MHz bandwidths are prescribed for home TV receivers.

2-12 SUMMARY

Major emphasis has been placed in this chapter on developing, through different examples, the inverse frequency-time relationship of signal transmission.

We demonstrated, first, through the use of the Fourier-series representation for periodic functions and then through the Fourier-integral representation of nonperiodic functions that more rapid time variations in a signal give rise to the higher-frequency components in the signal spectrum.

A general conclusion drawn from an analysis of signal transmission through linear networks was that the system bandwidth had to be approximately the reciprocal of the signal duration in order to produce at the system output a signal of the same general form as the input. For a symmetrical output signal the system phase characteristics had to be linear.

High-fidelity signal reproduction, or reproduction of detail, required bandwidths in excess of the reciprocal of the signal duration. The bandwidths needed were found to be about the reciprocal of the rise time, or the time taken for the signal to change from one level to another.

These conclusions are of importance in the design of practical system circuitry and are also of considerable theoretical importance in the study of information transmission through communication systems.

We recall that in the introductory chapter we discussed in a qualitative way the two system limitations on the amount of information per unit time (system capacity) a system could transmit:

1. Inability of the system to respond instantaneously to signal changes (due to the presence of energy-storage devices)
2. Inability of the system to distinguish infinitesimally small changes in signal level (due to inherent voltage fluctuations or noise)

These two limitations were tied together in a simple expression developed for system capacity,

$$C = \frac{1}{\tau} \log_2 n \qquad \text{bits/sec} \tag{2-149}$$

where τ was the minimum time required for the system to respond to signal changes and n the number of distinguishable signal levels.

In this chapter we have found that this minimum response time is proportional to the reciprocal of the system bandwidth (two alternative ways of referring to the same phenomenon). System capacity thus could be written

$$C = B \log_2 n \qquad \text{bits/sec} \tag{2-150}$$

where B is the system bandwidth in hertz.

In the next chapters we shall discuss some common communication systems and their bandwidth requirements. We shall then return to a discussion of the second limitation on the system capacity—inherent noise.

PROBLEMS

2-1 (*a*) Find the Fourier-series representations of each of the pulse trains in Fig. P 2-1. Choose the time origin so that a cosine series is obtained in each case.

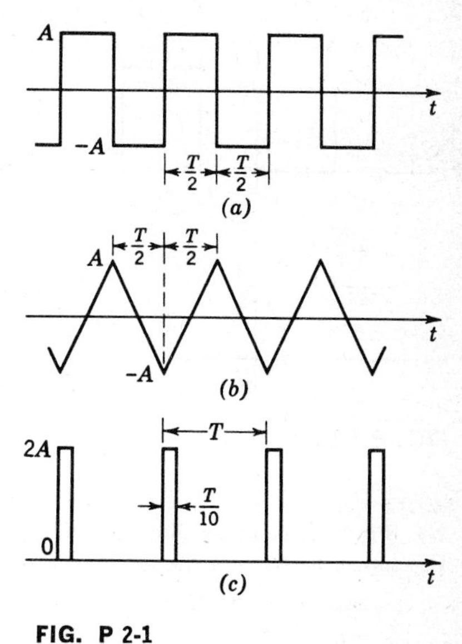

FIG. P 2-1

(*b*) Plot the first 10 Fourier coefficients vs. frequency for each pulse train. Compare the plot of Fig. P 2-1*a* with each of the other two, paying particular attention to the rate of decrease of the higher-frequency components. Note that Fig. P 2-1*b* represents in form the integral of Fig. P 2-1*a* and has no discontinuities in the function. The pulses of Fig. P 2-1*c* are much narrower than those of Fig. P 2-1*a*.

2-2 (*a*) Find the cosine Fourier-series representation of the half-wave rectified sine wave of Fig. P 2-2.

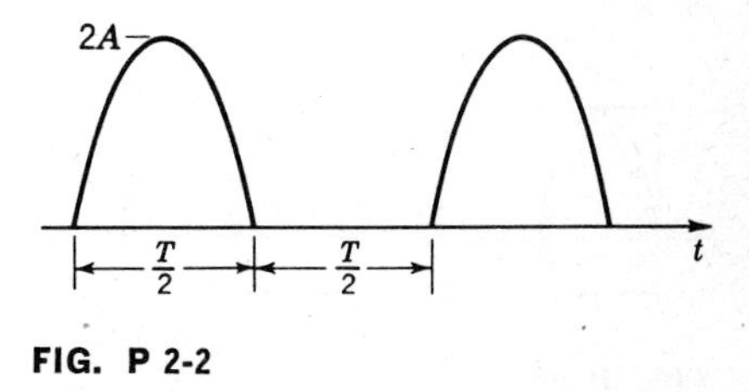

FIG. P 2-2

(*b*) Compare the successive Fourier coefficients and their rate of decrease with those of Fig. P 2-1*a*.

2-3 Find the complex Fourier series for the two pulse trains of Fig. P 2-3. Plot and compare the two amplitude spectra. What is the significance of the term $e^{-i\omega_n t_0}(\omega_n = 2\pi n/T)$ in the expression for the complex Fourier coefficient for the rectangular pulses?

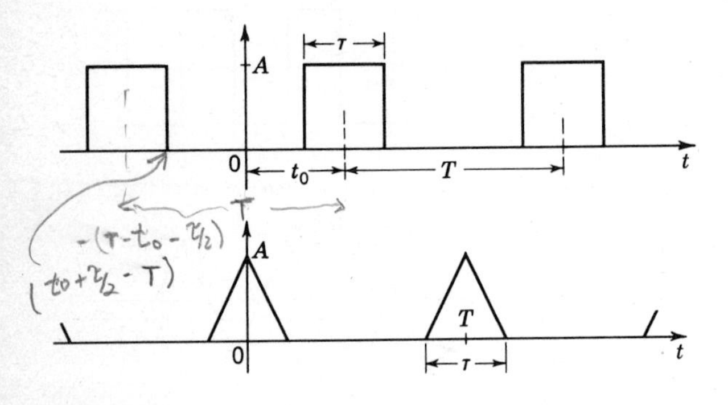

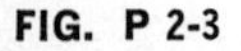
FIG. P 2-3

2-4 Find the complex Fourier series for the periodic function of Fig. P 2-4. *Hint:* Use superposition and the result of Prob. 2-3.

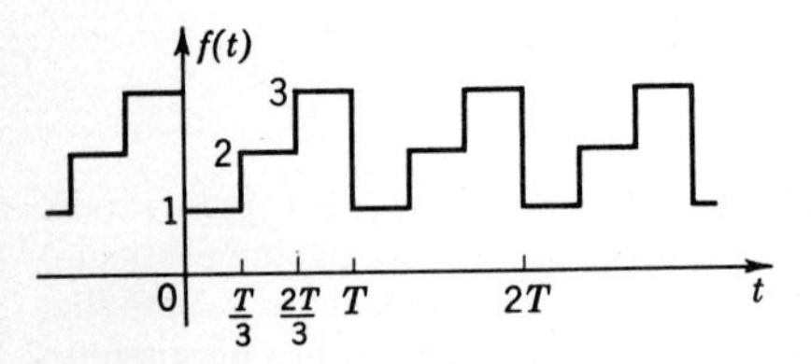

FIG. P 2-4

2-5 Consider the three cases of a rectangular, a triangular, and a half cosine pulse, all with $\tau/T = 1/10$ (Fig. P 2-5). Plot the amplitude spectra to the first zero crossing in each. What is the percent of the total power in the first 10 frequency components?

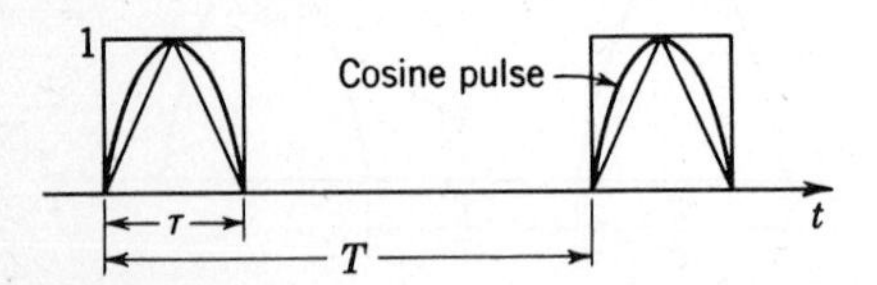

FIG. P 2-5

2-6 Find the complex Fourier series for the periodic function of Fig. P 2-6. Find the percent power in the first six components.

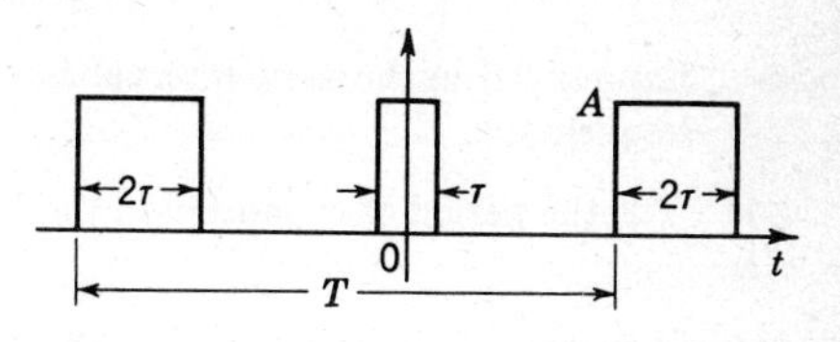

FIG. P 2-6

2-7 Plot the frequency spectra for the pulses of Fig. P 2-7.

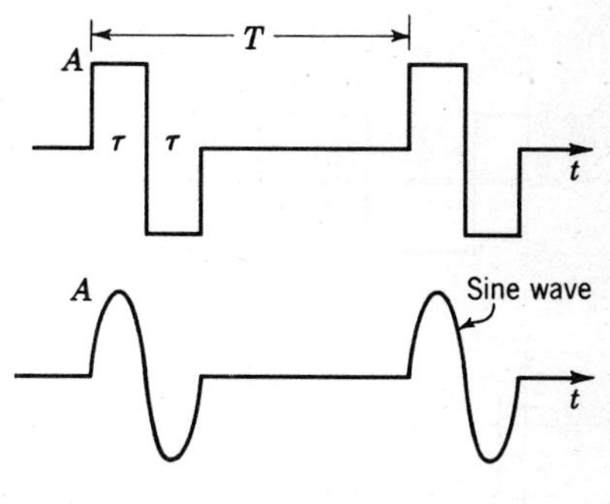

FIG. P 2-7

2-8 Find the complex Fourier series for the two periodic functions of Fig. P 2-8.

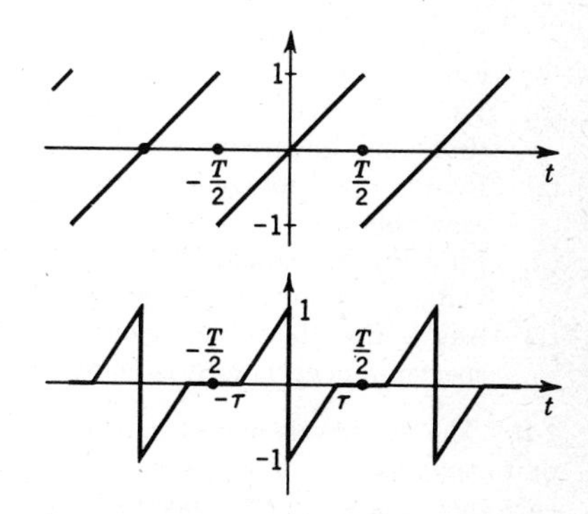

FIG. P 2-8

2-9 $f(t) = \sin \omega t$ for $0 \leq t \leq \pi/\omega$. It is undefined for values of t outside this interval.

(*a*) Express $f(t)$ inside the given interval as a sum of cosine terms with fundamental radian frequency ω.

(*b*) Express $f(t)$ in the same interval as a sum of sine terms with fundamental radian frequency ω.

2-10 T is the period of expansion in the Fourier-series representation for the function $f(t)$.

(*a*) Show that if $f(t) = f(t + T/2)$ the Fourier series will contain no odd harmonics.
(*b*) Show that if $f(t) = -f(t + T/2)$ the Fourier series will contain no even harmonics.

2-11 Each of the pulse trains shown in Fig. P 2-11 represents a voltage $v(t)$ appearing across a load resistor of 1 ohm.

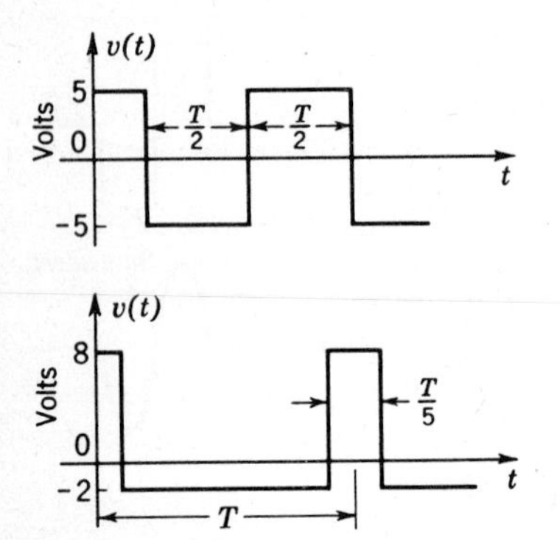

FIG. P 2-11

(*a*) Find the total average power dissipated in the resistor by each voltage.
(*b*) Find the percentage of the total average power contributed by the first-harmonic (fundamental) frequency of each voltage.
(*c*) Find the percentage of the total average power due to the first 10 harmonics of each voltage.
(*d*) Find the percentage of the total average power due to those harmonics of the voltage within the first zero-crossing interval of the amplitude spectrum.
(*e*) Add a fixed-bias (dc) level E volts to each of the pulse trains. Sketch the spectrum of each as E is increased from zero to 10 volts.

2-12 The duty cycle of a train of rectangular pulses is defined to be the ratio of time on to time off, or τ/T in the notation of this book. Sketch the pulse trains and their corresponding frequency spectra for the following cases:

(*a*) 0.1 duty cycle; (1) pulse width τ of 1 μsec; (2) pulse width of 10 μsec. Compare the frequency spectra.
(*b*) Repetition period T is 1 msec; (1) τ is 10 μsec; (2) τ is 1 μsec. What is the effect of varying τ, with T fixed, on the frequency scale? Compare the time and frequency plots.
(*c*) 10-μsec pulses; (1) 0.1 duty cycle; (2) 0.001 duty cycle. What is the effect of varying T, with τ fixed, on the frequency spectrum?

2-13 Find the complex Fourier coefficient c_n, and write the Fourier series for the function shown in Fig. P 2-13. Leave the coefficients in complex form.

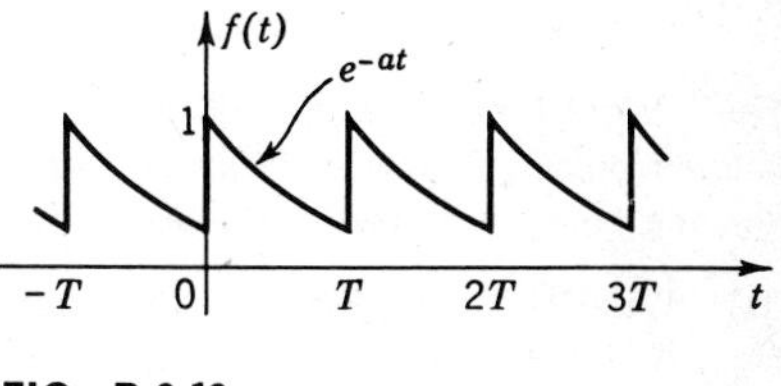

FIG. P 2-13

2-14 (*a*) Show that the complex Fourier coefficient c_n of the pulse train of Fig. P 2-14 is given by $c_n = 2A\tau[\sin(\omega_n\tau/2)/(\omega_n\tau/2)] \cos \omega_n\tau$. *Hint:* Use the result of Prob. 2-3 and the principle of superposition.

(*b*) Sketch the spectrum envelope and indicate the location of the spectral lines for $\tau/T = 0.1$.

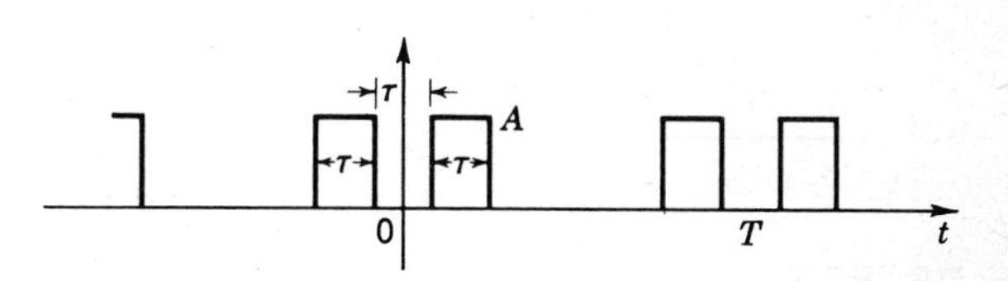

FIG. P 2-14

2-15 The teletype printer uses a code consisting of seven units or time intervals per character (letter). This includes start and stop pulses during the first and last time interval corresponding to each letter. During the remaining five intervals the signal may be either on (a 1) or off (a 0). The average character has the form shown in Fig. P 2-15.

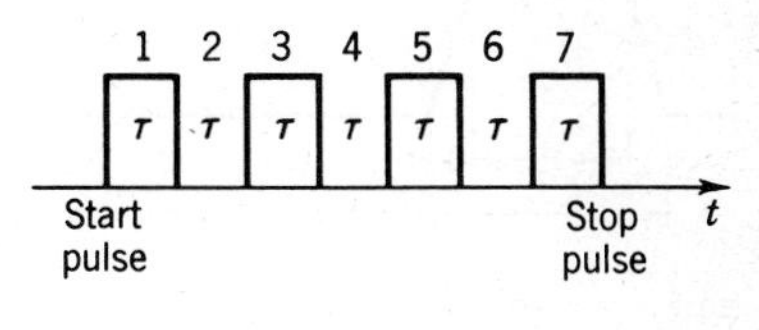

FIG. P 2-15

Assuming that the average word in the English language contains five letters and that one character is needed to transmit a space between words, there are on the average six characters per word. The rate of transmission is 25 words per minute.

(*a*) Find the length of one time interval.
(*b*) Find the bandwidth needed to transmit the first five harmonics of the pulse train of Fig. P 2-15. What bandwidth would be needed if the rate of transmission were increased to 100 words per minute? Why are these bandwidths the maximum required to transmit a telegraph signal?

2-16 A pulse-position-modulation (PPM) system produces symmetrical trapezoidal-shaped pulses, 5 μsec wide at the top and 5.5 μsec wide at the base. The average spacing between pulses is 125 μsec.

(*a*) What is the approximate system bandwidth required?
(*b*) If the average pulse spacing were halved, what bandwidth would be required?
(*c*) The pulses are passed through a low-pass filter cutting off at 200 kHz. Sketch the waveshape at the output of the filter.

2-17 (*a*) At what frequency does the amplitude spectrum of the four-pulse group of Fig. P 2-17 first go to zero?
(*b*) The pulse group in (*a*) is put through a single low-pass *RC* section having as its half-power frequency the frequency found in (*a*). Sketch the output waveshape.

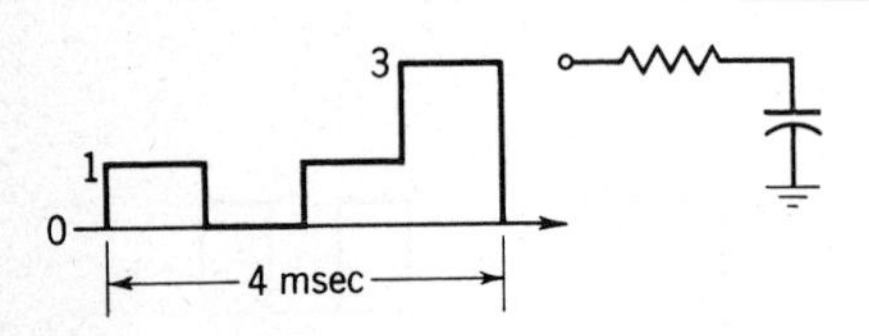

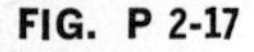
FIG. P 2-17

2-18 (*a*) What is the bandwidth in which 90 percent of the power in the trapezoidal pulses of Fig. P 2-18 is contained?
(*b*) What is the effect of passing these pulses through a single *RC* filter with a 3-db (half-power) frequency of 318 kHz?

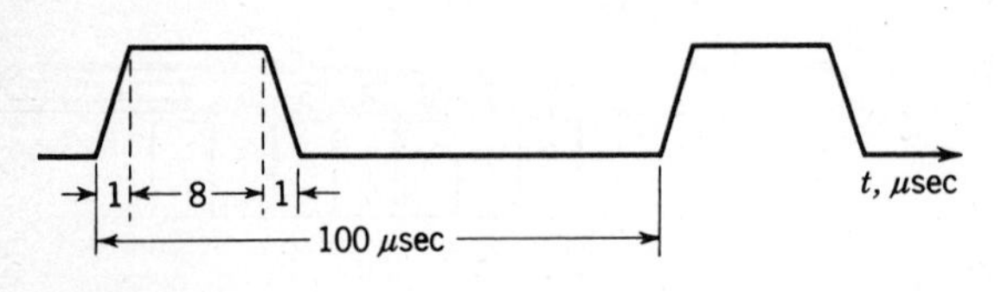

FIG. P 2-18

2-19 Find the exact expression for the repetitive pulse response of the *RC* filter. Show that this contains a term identical with Eqs. [2-24] and [2-29], plus an exponential term involving T/RC.

2-20 Find the Fourier transform of each of the pulses shown in Fig. P 2-20. Compare with the complex Fourier coefficients of the corresponding pulse trains obtained in Prob. 2-5.

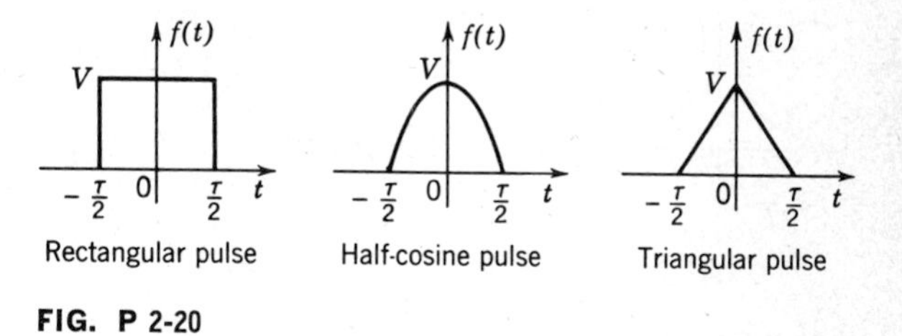

FIG. P 2-20

2-21 The Fourier transform of $f(t)$ is $F(\omega)$, and the Fourier transform of $g(t)$ is $G(\omega)$. For each of the following cases, find $G(\omega)$ in terms of $F(\omega)$:

(*a*) $g(t) = f(3t)$.
(*b*) $g(t) = f(t + a)$.

2-22 (*a*) Show that for each of the pulses of Prob. 2-20 any "bandwidth" definition (e.g., the first zero crossing of the spectrum) would give $B = K/\tau$, K a constant.

(*b*) Superimpose sketches of $|F(\omega)|$ for each of these pulses, and compare the spectrum curves. Focus attention particularly on the l-f and h-f ends of the spectra.

2-23 Find the Fourier transform of each of the pulses of Fig. P 2-23 in two different ways:

(*a*) From the Fourier integral directly.
(*b*) From the Laplace transform of the appropriate time function.

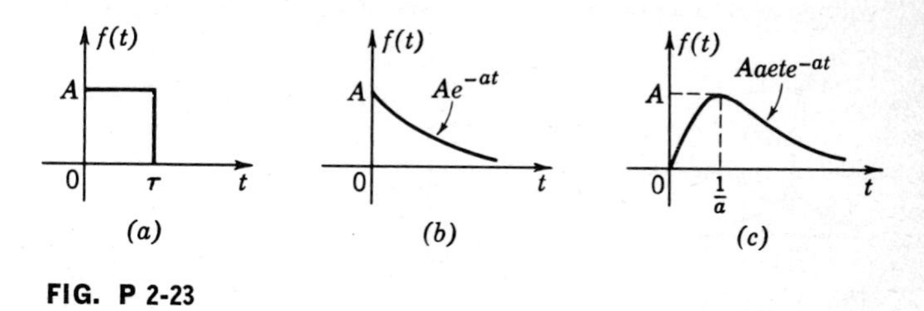

FIG. P 2-23

2-24 Find the Fourier transform of each of the functions of Fig. P 2-24. The last two time functions are even and odd, respectively. What, therefore, is to be expected of their Fourier transforms?

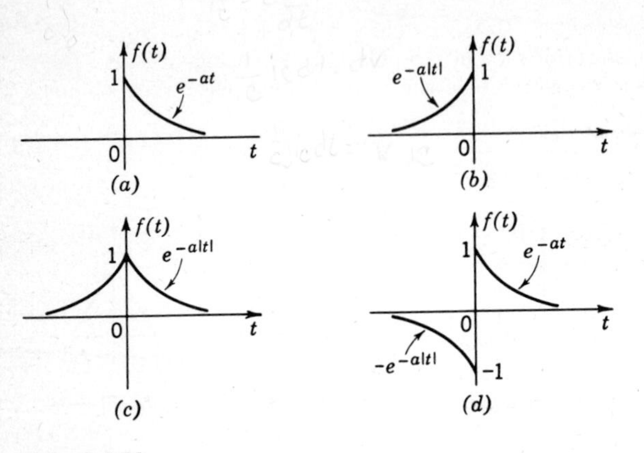

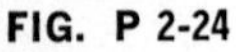
FIG. P 2-24

2-25 Find the Fourier transform of $f(t) = e^{-a|t|} \cos \omega_0 t$.

2-26 Find the frequency spectrum of the output voltage of Fig. P 2-26. Leave the answer in complex form.

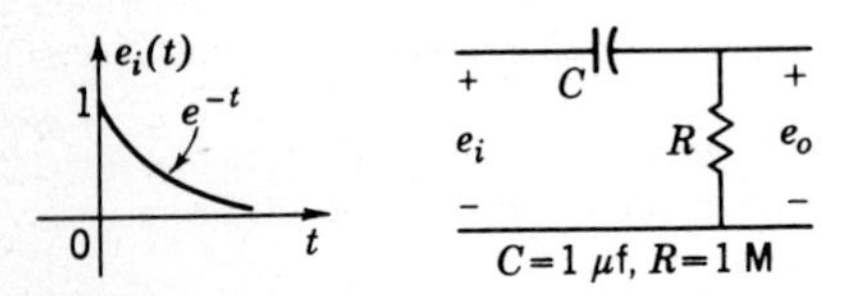

FIG. P 2-26

2-27 For the circuit and the input current shown in Fig. P 2-27 find the capacitor voltage as a function of time. Leave the result in integral form.

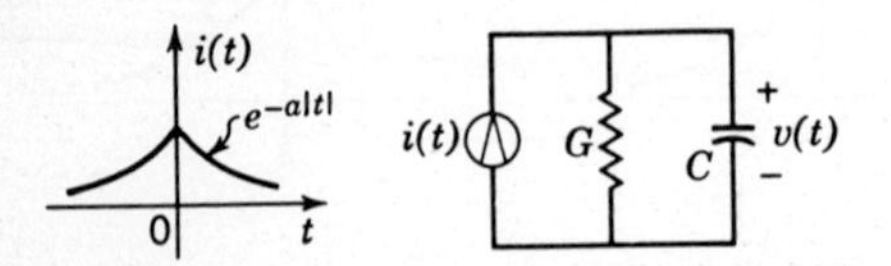

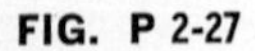
FIG. P 2-27

2-28 "Integration smoothes out time variations; differentiation accentuates time variations." Verify this statement for the two circuits shown in Fig. P 2-28, by comparing the input and output spectra with an arbitrary input $v_i(t)$.

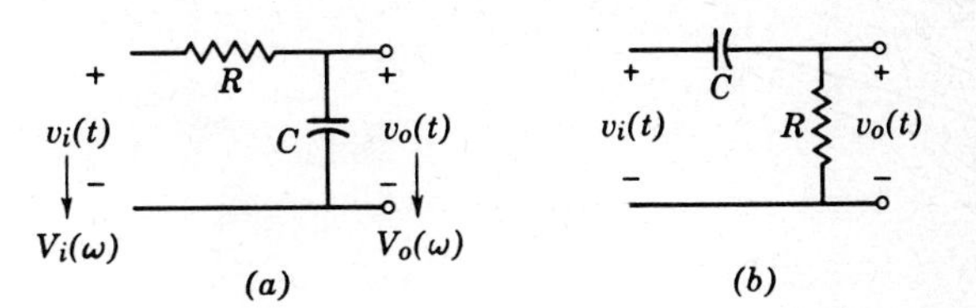

FIG. P 2-28 (a) Integrator ($R \gg 1/j\omega C$); (b) differentiator ($R \ll j\omega C$).

2-29 A voltage of the form $v(t) = Ae^{-a|t|}$ is applied to a series RLC circuit. Find an expression in integral form for the voltage across the capacitor.

2-30 A delay line and integrating circuit, combined as shown in Fig. P 2-30, are one example of a "holding circuit" commonly used in radar work, sampled-data servo systems, and pulse-modulation systems (see Chap. 3).

(*a*) Show that the Laplace-transform transfer function is $H(p) = V_o(p)/V_i(p) = (1/\tau p)(1 - e^{-\tau p})$.

(*b*) Using the results of (*a*), show that the frequency transfer function is $H(\omega) = e^{-j\omega\tau/2} \sin (\omega\tau/2)/(\omega\tau/2)$.

(*c*) If $v_i(t)$ is a rectangular pulse of width τ sec, show that the output is a triangular pulse of width 2τ sec, by (1) actually performing the successive operations indicated in the figure; (2) using the spectrum approach, the results of (*b*), and the results of Prob. 2-20.

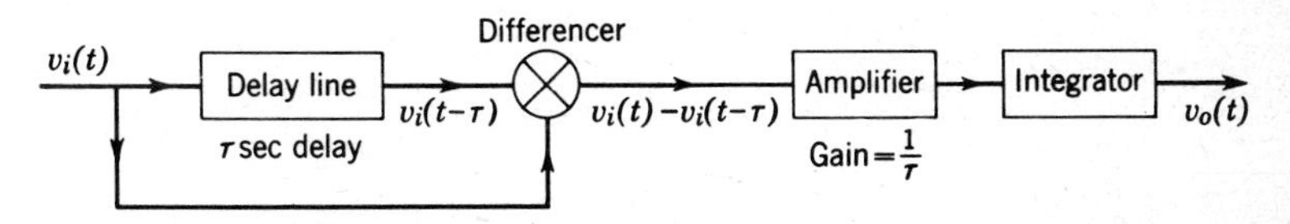

FIG. P 2-30 Holding circuit, (sin x)/x filter.

2-31 Two identical low-pass RC sections are connected in tandem. Find the overall frequency transfer function. Why can this not be obtained by multiplying together the transfer functions of the individual sections?

2-32 A pulse of the form $e^{-at}u(t)$ with $u(t)$ the unit-step function is applied to an idealized network with transfer characteristic $H(\omega) = Ae^{-jt_0\omega}, |\omega| \leq \omega_c$; $H(\omega) = 0$, $|\omega| > \omega_c$.

(*a*) Show that the output time response is given by

$$g(t) = \frac{A}{\pi} \int_0^{\omega_c} \left[\frac{a \cos \omega(t - t_0)}{a^2 + \omega^2} + \frac{\omega \sin \omega(t - t_0)}{a^2 + \omega^2} \right] d\omega$$

(*b*) Show that, for $\omega_c \gg a$,

$$g(t) \doteq \frac{A}{\pi} \text{Si}\, \omega_c(t - t_0) + \frac{A}{2} e^{-a|t-t_0|}$$

$$\text{Si}\, x \equiv \int_0^x \frac{\sin x}{x}\, dx$$

Note that $\int_0^\infty [(\cos mx)/(1 + x^2)]\, dx = (\pi/2)e^{-a|m|}$.

2-33 Using the results of Prob. 2-32, show that the unit-*step response* of an idealized network with amplitude A, cutoff frequency ω_c, and phase constant t_0 is given by

$$A(t) = A\left[\frac{1}{2} + \frac{1}{\pi} \text{Si}\, \omega_c(t - t_0)\right]$$

Sketch $A(t)/A$.

2-34 The unit-step response of the idealized network of Prob. 2-33 can also be obtained by first finding the response to a pulse of unit amplitude and width τ. As τ is allowed to get very large, the pulse approaches a unit step. Show that the network response approaches the unit-step response of Prob. 2-33.

2-35 A filter has the following amplitude and phase characteristics:

$$A(\omega) = 1 - \alpha + \alpha \cos 2\pi n \frac{\omega}{\omega_c} \qquad |\omega| \le \omega_c$$

$$A(\omega) = 0 \qquad |\omega| > \omega_c$$

$$\theta(\omega) = -t_0\omega$$

(This represents a filter with ripples in the passband.)

(*a*) Show that the step response has the form

$$A(t) = \frac{1}{2} + \frac{1}{\pi}\left\{(1 - \alpha)\, \text{Si}\, \omega_c(t - t_0) + \frac{\alpha}{2} \text{Si}\, [\omega_c(t - t_0) + 2\pi n] + \frac{\alpha}{2} \text{Si}\, [\omega_c(t - t_0) - 2\pi n]\right\}$$

(Use the approach of either Prob. 2-33 or Prob. 2-34.)

(*b*) Sketch $A(t)$ if $\alpha = \frac{1}{4}$, $n = 2$. Use the straight-line approximation for Si x indicated in Fig. P 2-35.

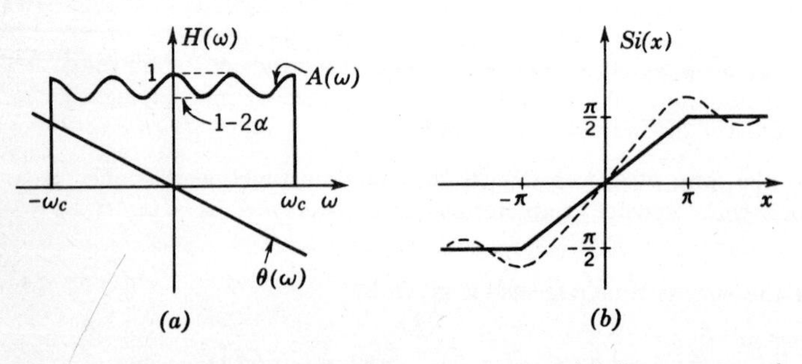

FIG. P 2-35 (a) Low-pass filter with ripples; (b) linear approximation to Si x.

2-36 In the filter characteristic of Prob. 2-35 let $2n = 1$ and $\alpha = \frac{1}{2}$. Sketch the amplitude characteristic for this case. Find the step response of this filter, and compare with the step response of the idealized filter of Prob. 2-33. This shows the effect of rounding off the sharp corners of the ideal low-pass filter.

2-37 The Butterworth filter is one example of a physically realizable low-pass filter. It has an amplitude transfer characteristic given by

$$A(\omega) = |H(\omega)| = \frac{1}{\sqrt{1 + (\omega/\omega_c)^{2n}}}$$

The constant n is a positive integer whose value depends on the complexity of the filter. Two examples of such filters, for $n = 3$ and $n = 4$, respectively, are shown in Fig. P 2-37. They are each normalized to a half-power frequency ω_c of 1 radian/sec

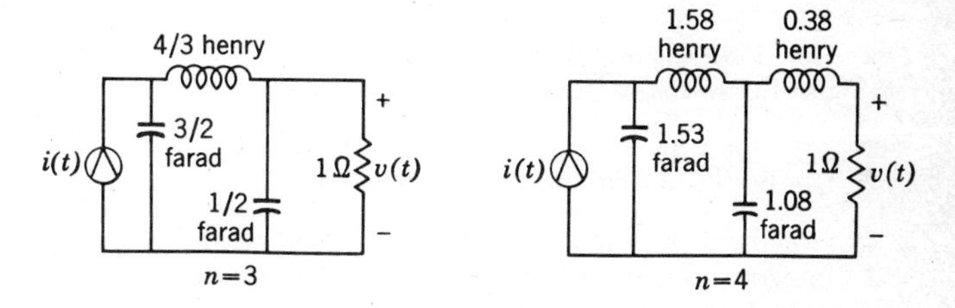

FIG. P 2-37 Butterworth filters.

and with load resistor of 1 ohm. (Note that n represents the number of reactive elements in each filter.)

(*a*) Find the complex frequency transfer function $H(\omega) = V(\omega)/I(\omega)$ for each filter, and show that $|H(\omega)|$ is given by the Butterworth expression noted above. Sketch this for both cases.

(*b*) Show that the Laplace-transform transfer function $H(p)$ for each of the two filters has poles lying on a unit circle centered about the origin in the p plane and spaced π/n radians apart. The two (complex conjugate) poles closest to the imaginary axis are located $\pi/2n$ radians away from this axis.

(*c*) Find the step response for each of the two filters, and compare with the step response of the ideal low-pass filter (Prob. 2-33).

2-38 Two students were asked the effect of passing the square wave of Fig. P 2-38 through the high-pass RC network shown. John reasoned that since the fundamental frequency of the square wave was four octaves above the 3-db fall-off frequency of the filter its attenuation and that of all the harmonics should be negligible. The output should thus look like the input. Conrad disagreed and to prove his point set up an experiment and obtained the oscilloscope trace for the output shown in the lower part of Fig. P 2-38.

Explain the fallacy (or fallacies) in John's argument. Describe the characteristics of a network to be connected in tandem with the RC network so that no overall distortion will result.

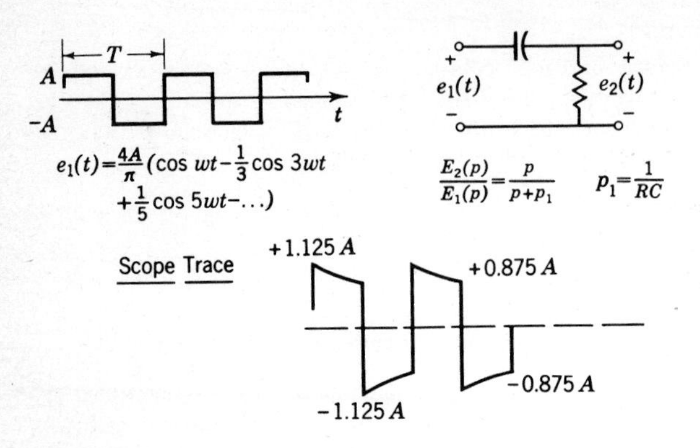

FIG. P 2-38

2-39 (*a*) Find the impulse response of the Butterworth filters of Prob. 2-37.
(*b*) Use Eq. [2-134] to find the step response.

2-40 (*a*) Find the impulse response of the idealized filter of Prob. 2-35.
(*b*) Use Eq. [2-133] to find the step response.

2-41 (*a*) Find the impulse response of the high-pass RC filter of Fig. P 2-38.
(*b*) Integrate the impulse response to find the step response. (As a check, determine the step response by inspection.)
(*c*) Use the convolution integral to find the response of the same filter to an input exponential $e_1(t) = e^{-at}u(t)$.

2-42 Check Eq. [2-123], the response of a low-pass RC filter to an exponential input by:

(*a*) Solving the differential equation for the circuit.
(*b*) Using Laplace transforms.

2-43 Prove Eq. [2-128].

2-44 A ramp voltage beginning at time $t = 0$, $tu(t)$, is applied to a low-pass RC section. Find the output voltage, using the convolution integral.

2-45 (*a*) Repeat Prob. 2-44 if the ramp voltage is suddenly reduced to zero T sec after being applied.
(*b*) Check the result of (*a*) by superposing the responses to a ramp voltage $tu(t)$ applied at time $t = 0$ and a negative ramp $-tu(t - T)$ applied T sec later.

2-46 (*a*) Use Laplace transforms to prove the relation

$$h(t) = A(0) + \frac{dA}{dt}$$

$h(t)$ and $A(t)$ represent the unit-impulse and unit-step responses, respectively, of a physically realizable linear system.
(*b*) As an example, find the impulse response of the high-pass RC filter of Prob. 2-41 from the given step response.
(*c*) Show that the relation stated in (*a*) agrees with Eq. [2-134].

2-47 Show that a train of unit impulses, periodically spaced T sec apart, may be represented by the Fourier series

$$\frac{1}{T} + \frac{2}{T} \sum_{n=1}^{\infty} \cos \omega_n t \qquad \omega_n = \frac{2\pi n}{T}$$

Sketch the frequency spectrum.

2-48 Find the Fourier transform of a function $f(t) = \delta(t) - \delta(t - \tau) + \delta(t - 2\tau) - \delta(t - 3\tau)$. Compare $|F(\omega)|$ with the Fourier coefficient obtained in Prob. 2-14.

2-49 Obtain the Fourier transform of the functions shown in Fig. P 2-49 by successive differentiation to obtain impulses or by breaking the function up into simpler forms for which the Fourier transform is readily available.

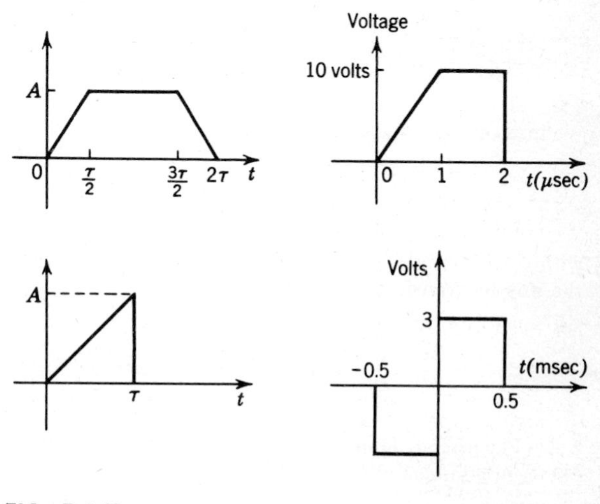

FIG. P 2-49

2-50 A linear system has a transfer function

$$H(\omega) = \tau \left[\frac{\sin (\omega\tau/2)}{\omega\tau/2} \right] e^{-j5\omega\tau}$$

(a) The input is $f(t) = \delta(t)$. Find the output $g(t)$ and sketch.

(b) The input is a rectangular pulse of unit amplitude and width τ. Find $g(t)$ and sketch.

(c) The input is a rectangular pulse of unit amplitude and width 10τ. Provide an approximate sketch of the output. (No calculations are needed here.)

(d) Compare the results of (a), (b), and (c) by relating relative signals and system bandwidths.

2-51 (a) Show the Fourier transform of $\delta(t - kt_0)$ is periodic in ω with period $2\pi/t_0$.
(b) Show the Fourier transform of

$$\sum_{k=-N}^{N} \delta(t - kT)$$

is given by

$$\frac{\sin \omega (N + \frac{1}{2})T}{\sin (\omega T/2)}$$

Sketch the spectrum and show it is periodic with period $2\pi/T$. *Hint:* This is a geometric series of $2N + 1$ terms, with the common ratio $e^{-j\omega T}$.
(c) Let N in part (b) $\rightarrow \infty$. Show the Fourier transform of

$$\sum_{k=-\infty}^{\infty} \delta(t - kT)$$

is given by

$$\frac{2\pi}{T} \sum_{l=-\infty}^{\infty} \delta\left(\omega - \frac{2\pi l}{T}\right)$$

2-52 Show that a *periodic* function $f(t)$ can be written as the convolution of $f(t)$ (defined over one period only) with

$$\sum_{k=-\infty}^{\infty} \delta(t - kT)$$

with T the period. Use this and the result of Prob. 2-51c to obtain an expression for the Fourier transform of a periodic signal.
Ans.:

$$F(\omega) = \frac{2\pi}{T} \sum_{n=-\infty}^{\infty} c_n \, \delta(\omega - \omega_n) \qquad \omega_n = \frac{2\pi n}{T} \qquad c_n = \int_{-T/2}^{T/2} f(t) \, e^{-j\omega_n t} \, dt$$

2-53 The *discrete* Fourier transform is defined as follows:
Given N numbers, $x(l)$, $l = 0, 1, 2, \ldots, N - 1$, then

$$X(n) \equiv \frac{1}{N} \sum_{l=0}^{N-1} x(l) e^{-2j\pi ln/N}$$

is the discrete Fourier transform of $x(l)$. This is used in digital-computer calculations of Fourier transforms.[1]

(a) Show that with $X(n)$ given $x(l)$ may be recovered by using the inverse discrete transform

$$x(l) = \sum_{n=0}^{N-1} X(n) e^{2\pi jln/N}$$

[1] J. W. Cooley, P. A. W. Lewis, and P. D. Welch, Application of the Fast Fourier Transform to Computation of Fourier Integrals, Fourier Series, and Convolution Integrals, *IEEE Trans. Audio and Electroacoustics*, vol. AU-15, no. 2, pp. 79–84, June, 1967.

Hint: substitute for $X(n)$ its defining series in terms of $x(l)$, interchange the order of summation, and sum the resultant geometric series. See Prob. 2-51*b*.

(*b*) Show that $X(n)$, and $x(l)$ obtained from $X(n)$ in (*a*) above, are periodic with period N: $X(x + kN) = X(n)$, k an integer; $x(l + kN) = x(l)$, k an integer.

(*c*) $x(l) = 1$, $l = 0, 1, \ldots, N - 1$. Show $X(n) = 1$, $n = 0, \pm N, \pm 2N, \ldots$; $X(n) = 0$, all other values of n. Sketch $X(n)$, and compare with the continuous Fourier transform $F(\omega)$ of a constant.

(*d*) $x(l) = 1$, $l = 0, 1, 2, \ldots, m - 1$; $x(l) = 0$, $m < l < N - 1$. This is then the discrete equivalent of the rectangular pulse used throughout this chapter

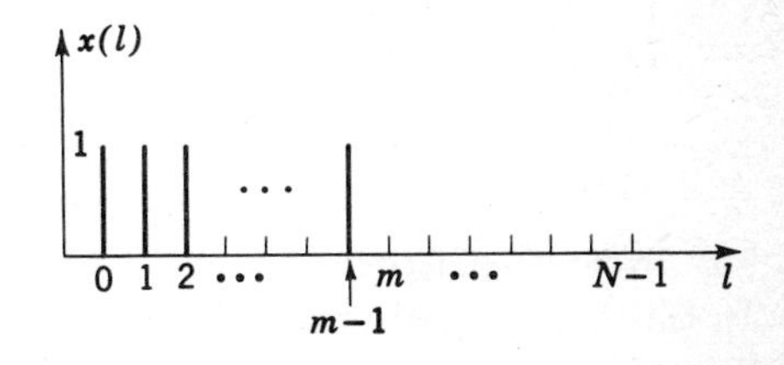

FIG. P 2-53

(see Fig. P 2-53). Show

$$X(n) = \frac{m}{N} \exp\left(-2\pi j \frac{n}{N} \frac{m-1}{2}\right) \frac{\sin(\pi nm/N)}{m \sin(\pi n/N)}$$

Interpret each of the terms in $X(n)$ and compare with the transform $F(\omega)$ of a rectangular pulse, $f(t) = 1$, $0 \le t \le \tau$; $f(t) = 0$, elsewhere.

(*e*) In (*d*) above, let $N = 100$, $m = 10$. Sketch $|X(n)|$ for $-2N \le n \le 2N$, and compare with $|(\sin x)/x|$, $x = \pi n/10$. Repeat for $N = 1{,}000$. Again compare with $(\sin x)/x$, $x = \pi n/100$.

CHAPTER THREE

DIGITAL COMMUNICATION SYSTEMS

3-1 INTRODUCTION

In this book, we are basically interested in investigating the information-handling capabilities of different communication systems. We showed in the previous chapter that one important parameter determining the system information capability is the system bandwidth. In Chaps. 5 and 6 we shall discuss the limitations on system performance due to noise.

We pause at this point in our discussion to describe some typical digital communication systems, to compare them on the basis of bandwidth requirements, and to develop further the inverse frequency-time relationship of the previous chapter.

The stress here will be on digital systems because of their paramount importance in modern-day technology and because the concepts important to information transmission are so easily developed through the study of digital systems. This was first pointed out in Chap. 1 where the transmission of digital signals step-by-step through a typical system was discussed.[1]

The current widespread use of digital signaling is the result of many factors.

1. The relative simplicity of digital circuit design and the ease with which one can apply integrated circuit techniques to digital circuitry

[1] Good overall references on digital systems, covering much of the material of this chapter in more detail, are the books "Data Transmission," W. R. Bennett and J. R. Davey, McGraw-Hill, New York, 1965 and "Principles of Data Communication," R. W. Lucky, J. Salz, and E. J. Weldon, McGraw-Hill, New York, 1968. The emphasis in these books is on digital systems developed primarily for telephone usage. Various aspects of digital space systems are considered in S. W. Golomb (ed.), "Digital Communications with Space Applications," Prentice-Hall, New Jersey, 1964; A. V. Balakrishnan (ed.), "Space Communications," McGraw-Hill, New York, 1965; and E. L. Gruenberg (ed.), "Handbook of Telemetry and Remote Control," McGraw-Hill, New York, 1967.

2. The ever-increasing use and availability of digital processing techniques
3. The widespread use of computers in handling all kinds of data
4. The ability of digital signals to be coded to minimize the effects of noise and interference

Although some communications signals are inherently digital in nature—e.g., teletype data, computer outputs, pulsed radar and sonar signals, etc., many signals are analog, or smooth, functions of time. If these signals are to be transmitted digitally, they first have to be sampled at a periodic rate and then further converted to discrete amplitude samples by quantization. The sampling procedure in this analog-to-digital (A/D) conversion process will be the first topic treated in this chapter. Speech, TV, facsimile, and telemetered data signals are all examples of analog signals that are often transmitted digitally.

We shall discuss time multiplexing of digital signals, in which signals from different sources are sequentially transmitted through the system. (Sine-wave carrier modulation by these signals for remote transmission will be discussed in Chap. 4.) Some of the problems arising in the use of digital systems will also be discussed. These include time synchronization, intersymbol interference, quantization noise, transmission bandwidth requirements, etc. Examples will be given of various types of digital systems in use.

3-2 NYQUIST SAMPLING

Consider a continuous varying signal $f(t)$ that is to be converted to digital form. We do this simply by first sampling $f(t)$ periodically at a rate of f_c samples per second. Although in practice this sampling process would presumably be carried out electronically by gating the signal on and off at the desired rate,[1] we show the sampling process conceptually in Fig. 3-1 using a rotating mechanical switch.

[1] See Sec. 3-10 for some specific techniques.

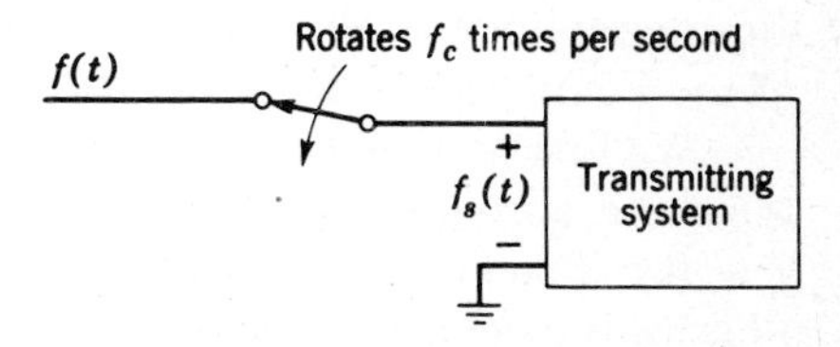

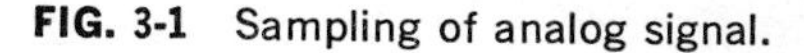
FIG. 3-1 Sampling of analog signal.

Assume that the switch remains on the $f(t)$ line τ sec, while rotating at the desired rate of $f_c = 1/T$ times per second ($\tau \ll T$). The switch output $f_s(t)$ is then a sampled version of $f(t)$. A typical input function $f(t)$, the sampling times, and the sampled output $f_s(t)$, are shown in Fig. 3-2. f_c is called the sampling rate, and T is the sampling interval.

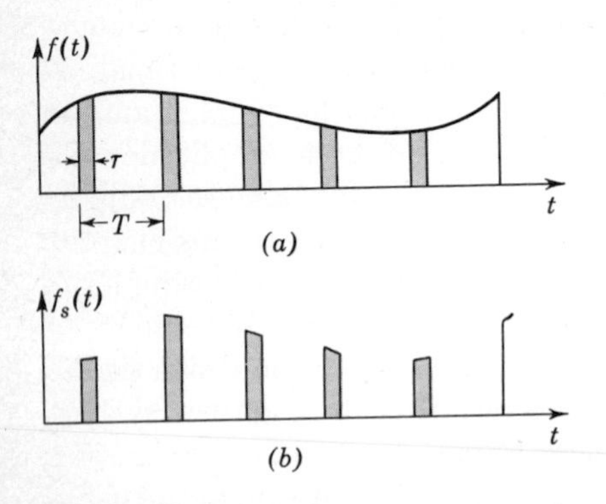

FIG. 3-2 The sampling process. (τ = sampling time; $T = 1/f_c$: sampling interval.) (a) Input $f(t)$; (b) sampled output $f_s(t)$.

The question that immediately arises is: What should the sampling rate be? Are there any limits to the rate at which we sample? One might intuitively feel that the process of sampling has irretrievably distorted the original signal $f(t)$. The process of sampling has been introduced to convert the signal $f(t)$ to a digital form for further processing and transmission. Yet eventually, and at a remote location, in most cases, one would like to retrieve $f(t)$ again. Have we lost valuable information in the sampling process?

The answer to this last question, and from this, to the other questions asked, is that under a rather simple assumption (closely approximated in practice), the sampled signal $f_s(t)$ contains within it *all* information about $f(t)$! Further, $f(t)$ can be uniquely extracted from $f_s(t)$! This rather amazing and not at all obvious result can be demonstrated through the use of the Fourier analysis of Chap. 2.

We first assume the signal $f(t)$ is *band-limited* to B Hz. This means there are absolutely no frequency components in its spectrum beyond $f = B$. The Fourier transform $F(\omega)$ of such a signal is shown in Fig. 3-3. Physically occurring signals generally do not have the sharp frequency cutoff implied by the concept of bandlimitedness. Except for a few

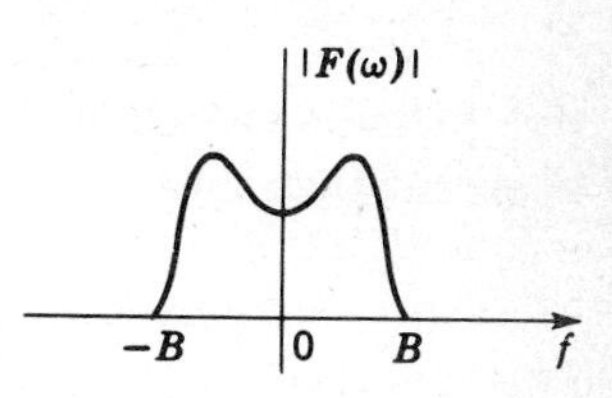

FIG. 3-3 Band-limited signal.

singular cases, examples of which were given in the previous chapter, real signals contain frequency components out to all frequencies. Yet we know from our discussion in the previous chapter that the frequency content drops rapidly beyond some defined bandwidth. This approximation of real signals by band-limited ones introduces no significant error in the analysis and will therefore henceforth be assumed. (In practice, sharp-cutoff low-pass filters are frequently introduced before the sampling process to ensure that the band-limited condition is obeyed to the approximation desired.)

With the signal $f(t)$ band-limited to B Hz, it is then readily shown that sampling the signal does *not* destroy any information content, provided that the sampling rate $f_c \geq 2B$. The minimum sampling rate of $2B$ times per second is called the *Nyquist*[1] sampling rate and $1/2B$ the Nyquist sampling interval.

To demonstrate this result through Fourier analysis we use a simple strategem. It is apparent that the sampled signal $f_s(t)$ may be represented in terms of $f(t)$ by the simple relation

$$f_s(t) = f(t)S(t) \tag{3-1}$$

with $S(t)$ a periodic series of pulses of unit amplitude, width τ, and period $T = 1/f_c$. This so-called switching or gating function is sketched in Fig. 3-4. (Note the recurrence of our familiar periodic pulses of Chap. 2.) If

[1] After H. Nyquist of the Bell Telephone Laboratories.

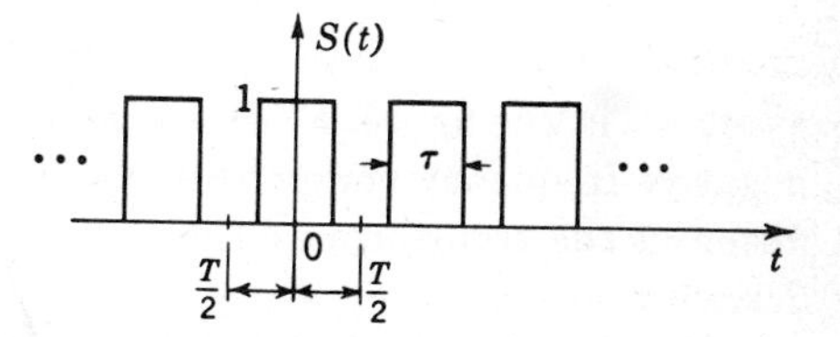

FIG. 3-4 Periodic switching function.

this periodic switching function is expanded in its Fourier series, it is apparent that the sampled signal $f_s(t)$ can be written in the form

$$f_s(t) = df(t)\left(1 + 2\sum_{n=1}^{\infty} \frac{\sin n\pi d}{n\pi d} \cos 2\pi n f_c t\right) \qquad [3\text{-}2]$$

Here $d = \tau/T$, the so-called duty cycle.

We now determine the frequency spectrum of $f_s(t)$ by taking its Fourier transform $F_s(\omega)$. But by the linearity of the Fourier-transform operation this is the same as taking the Fourier transform of each term in the series of Eq. [3-2] and then superposing the results. Consider a typical time function $f(t) \cos n\omega_c t$ in the series. Call its Fourier transform $F_c(\omega)$. Then

$$F_c(\omega) = \int_{-\infty}^{\infty} f(t) \cos n\omega_c t e^{-j\omega t}\, dt \qquad [3\text{-}3]$$

Writing the cosine term as the sum of two exponentials, we then get

$$\begin{aligned} F_c(\omega) &= \frac{1}{2}\int_{-\infty}^{\infty} f(t)e^{-j(\omega - n\omega_c)t}\, dt + \frac{1}{2}\int_{-\infty}^{\infty} f(t)e^{-j(\omega + n\omega_c)t}\, dt \\ &= \frac{1}{2}F(\omega - n\omega_c) + \frac{1}{2}F(\omega + n\omega_c) \end{aligned} \qquad [3\text{-}4]$$

since each integral in Eq. [3-4] is identically the Fourier transform of $f(t)$, although referenced or shifted to a different frequency.

The process of multiplication of $f(t)$ by $\cos n\omega_c t$ has resulted in a shift of its frequency spectrum, positively and negatively, by the frequency $n\omega_c$. This is shown in Fig. 3-5. If $F(\omega)$ was originally centered about 0,

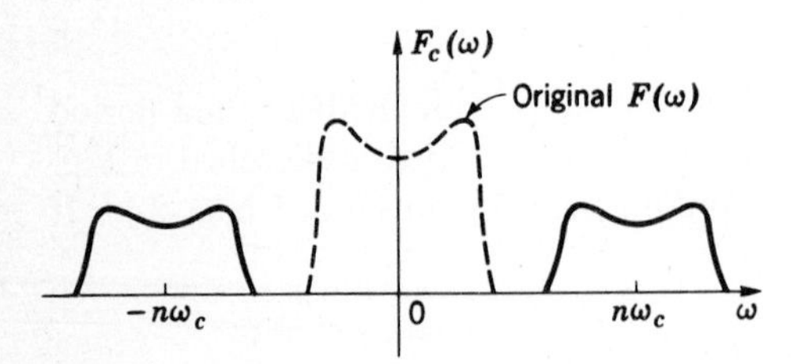

FIG. 3-5 Spectrum of $f(t) \cos n\omega_c t$.

as shown in the figure, $F_c(\omega)$ is centered at $\pm n\omega_c$. (Recall again that the negative-frequency components are due to the complex exponential definition of the transform and may be combined with the positive terms if desired.)

We shall see shortly that $f(t) \cos n\omega_c t$ is an example of an amplitude-modulated (AM) signal, with $f(t)$ the modulating signal and $n\omega_c$ the carrier frequency. This technique of switching or gating $f(t)$ on and off at the carrier rate is in fact one common method of generating AM signals.

Superposing all the Fourier transforms in Eq. [3-2], we get for the Fourier transform of $f_s(t)$,

$$F_s(\omega) = dF(\omega) + d \sum_{\substack{n=-\infty \\ n \neq 0}}^{\infty} \frac{\sin n\pi d}{n\pi d} F(\omega - n\omega_c) \qquad [3\text{-}5]$$

This sum of individual Fourier transforms, each centered at a multiple of the sampling frequency, is shown sketched in Fig. 3-6. Note that

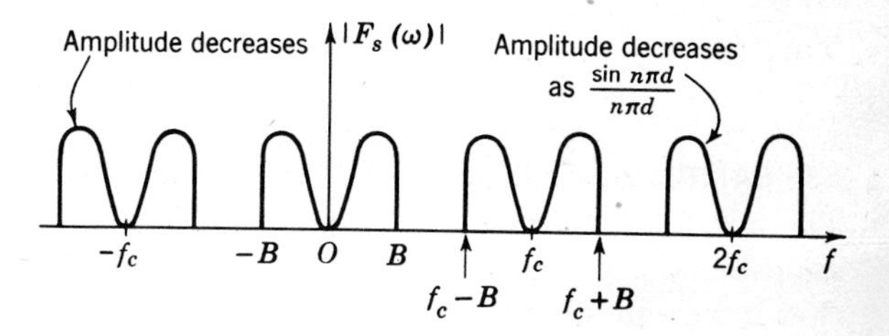

FIG. 3-6 Amplitude spectrum, sampled input.

the amplitude of each successive component decreases as $(\sin n\pi d)/n\pi d$. The effect of sampling $f(t)$ has thus been to shift its spectrum up to all harmonics of the sampling frequency. Alternately, we may say that the effect of multiplying the periodic sampling function $S(t)$ by the nonperiodic $f(t)$ has been to broaden its discrete-line spectrum into a continuous spectrum symmetrically situated about the original frequency lines.

So long as the different spectra in Fig. 3-6 are separated, it is apparent that $f(t)$ can be filtered out from $f_s(t)$. This requires a low-pass filter, passing $F(\omega)$ but cutting off sharply before reaching the frequency spectrum component centered at $\pm f_c$. This answers the question previously raised about the possibility of retrieving $f(t)$, *undistorted*, from its sampled version $f_s(t)$. Systems transmitting these sampled values of the signal $f(t)$ are commonly called *sampled-data* or *pulse-modulation* systems.

We now visualize the switch rotation rate slowing down. The frequency f_c and all its harmonics start closing in on one another. It is apparent that eventually the spectral components in Fig. 3-6 will overlap and merge. The component $F(\omega - \omega_c)$ centered about $\pm f_c$ will in particular merge with the unshifted $F(\omega)$ term, centered about the origin. It is then impossible to separate out $F(\omega)$, and hence $f(t)$, from $f_s(t)$. The limiting frequency at which $F(\omega)$ and $F(\omega - \omega_c)$ merge is, from Fig. 3-6, given by $f_c - B = B$, or

$$f_c = 2B \tag{3-6}$$

which is just the Nyquist sampling rate introduced earlier.

Generally one samples at a somewhat higher rate to ensure separation of the frequency spectra and to simplify the problem of low-pass filtering to retrieve $f(t)$. As an example, speech transmitted via telephone is generally filtered to $B = 3.3$ kHz. The Nyquist rate is thus 6.6 kHz. For digital transmission the speech is normally sampled at an 8-kHz rate, however. As another example, if the signal $f(t)$ to be sampled contains frequency components as high as 1 MHz, at least 2 million samples per second are needed.

3-3 SAMPLING THEOREM—FURTHER DISCUSSION

The minimum Nyquist sampling rate, $2B$ samples per second, for signals band-limited to B Hz, is so important a concept in digital communications that it warrants further discussion and extension. In particular we shall demonstrate rather simply that any $2B$ independent samples per second (not necessarily periodically obtained) suffice to represent the signal uniquely. One may thus sample aperiodically, if so desired; one may also sample a signal $f(t)$ and its successive derivatives, etc.

The lower, Nyquist, limit on the sampling rate in the case of periodic sampling is highly significant. Once we have satisfied ourselves that the sampled values of a signal $f(t)$ do in fact contain complete information about $f(t)$ we might logically feel that there must be a *minimum* value of f_c in order not to lose information or to be able eventually to reconstruct the continuous input signal. For we note that if we sample at too low a rate, the signal may change radically between sampling times. We thus lose information and eventually produce a distorted output. A sampling rate too low for the signal involved is shown in Fig. 3-7. In order not to lose the signal dips and rises, additional sampling pulses must be added as shown.

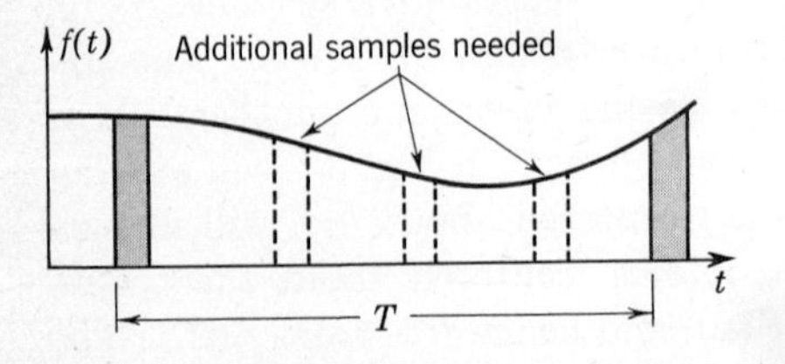

FIG. 3-7 Sampling frequency too low.

There is obviously a relation between the rate at which a signal varies and the number of pulses needed to reproduce it exactly. The rate at which a signal varies is of course related to its maximum frequency component, or bandwidth, B. Equation [3-6] tells us that at least $2B$ uniformly spaced samples are needed every second in order eventually to reproduce the signal without distortion.

This statement, arising quite naturally out of a consideration of the frequency spectrum of a periodically sampled signal, is the famous *sampling theorem*[1] of modern communication theory.

The theorem is of course particularly important in sampled-data and pulse-modulation systems, where sampling is inherent in the operation of the system. But it has deep significance in the modern concepts of information theory. For any measure of the information content of a specified signal must be related to the number of *independent* quantities needed to describe that signal completely (see Chap. 1). If the number is written in the form of binary units, the information content is measured in bits.

Although our statement of the sampling theorem, arising from Eq. [3-6], relates to the periodic sampling of a band-limited signal, the theorem can be generalized to *any* group of independent samples. Thus the more general theorem states that *any $2B$ independent samples per second will completely characterize a band-limited signal. Alternatively any $2BT'$ unique (independent) pieces of information are needed to completely specify a signal over an interval T' sec long.*

These statements should come as no surprise to us, for we use similar results constantly in applying the Taylor's-series expansion of calculus: a function, obeying certain conditions, may be expressed and completely specified at any point in terms of the value of the function and its successive derivatives at another point. All polynomials, for example $y = t, t^2, t^3$, etc., possess a finite number of nonzero derivatives, and so only a finite number of pieces of information are needed to describe these functions. Here the number of independent pieces of information are measured in the vicinity of one point and are not periodically spaced samples.

The proof of the more general form of sampling theorem, that *any* $2BT'$ pieces of information are needed to characterize a signal over a T'-sec interval, follows readily from an application of the Fourier series. Thus assume that we are interested in a band-limited function $f(t)$ over an

[1] H. S. Black, "Modulation Theory," chap. 4, Van Nostrand, Princeton, N.J., 1953; A. Papoulis, "The Fourier Integral and Its Applications," McGraw-Hill, New York, 1962, and "Systems and Transforms with Applications in Optics," McGraw-Hill, New York, 1968.

interval T' sec long. We may then expand $f(t)$ in a Fourier series with T' as the base period. But with $f(t)$ band-limited to B Hz we get only a finite number of terms in the Fourier series,

$$f(t) = \frac{c_0}{T'} + \frac{2}{T'} \sum_{n=1}^{BT'} |c_n| \cos (\omega_n t + \theta_n) \qquad \omega_n = \frac{2\pi n}{T'} \tag{3-7}$$

or

$$f(t) = \frac{c_0}{T'} + \frac{2}{T'} \sum_{n=1}^{BT'} (a_n \cos \omega_n t + b_n \sin \omega_n t) \tag{3-8}$$

with $|c_n| = \sqrt{a_n^2 + b_n^2}$, $\theta_n = -\tan^{-1}(b_n/a_n)$. [Since B is the maximum-frequency component of $f(t)$, ω_n has a maximum value

$$2\pi B = \frac{2\pi n}{T'}$$

The maximum value of n is thus BT'.]

The c_0 term is the dc term. It merely serves to shift the level of $f(t)$ and does not provide any new information. (Information implies signals *changing* with time, as pointed out in Chap. 1.) There are thus $2BT'$ independent Fourier coefficients (the sum of the c_n's and θ_n's or a_n's and b_n's), and any $2BT'$ independent samples of $f(t)$ are needed to specify $f(t)$ over the T'-sec interval.

It is of interest to note at this point the connection between the sampling theorem just discussed and the equation for channel capacity developed in Chap. 1 and rephrased at the end of Chap. 2,

$$C = B \log_2 n \qquad \text{bits/sec} \tag{3-9}$$

where B is the channel bandwidth in hertz. In T' sec the channel will allow the transmission of

$$BT' \log_2 n \qquad \text{bits}$$

This expression for system capacity includes possible effects of noise ($\log_2 n$), but, aside from noise considerations, it says simply that the information that can be transmitted over a band-limited system is proportional to the product of bandwidth times the time for transmission. The concept embodied in this last sentence was first developed by R. V. L. Hartley of Bell Laboratories in 1928 and is called *Hartley's law*.[1] Modern communication theory has extended Hartley's law to include the effects of noise, giving rise to Eq. [3-9] the expression for system capacity.

Hartley's law and the sampling theorem (as developed by Nyquist in 1928) are in essence the same. For a band-limited signal may be viewed as having been "processed," or emitted, by a band-limited system. The

[1] H. S. Black, *op. cit.*

information carried by this signal must thus be proportional to BT' by Hartley's law. This of course is the same result as that obtained from the sampling theorem. (The factor of 2 that appears to distinguish the expression for channel capacity and the results of the sampling theorem can be accounted for by using $\tau = 1/2B$ instead of $1/B$ in the development of the channel-capacity expression of Chap. 2.)

3-4 DEMODULATION OF SAMPLED SIGNALS

If $2BT'$ samples completely specify a signal, it should be possible to recover the signal from the samples. This is the demodulation process required for sampled-data or pulse-modulation systems. How do we accomplish it?

Note that the sampled output was expressed by Eq. [3-2] in the form

$$f_s(t) = df(t)\left(1 + 2\sum_{n=1}^{\infty} \frac{\sin n\pi d}{n\pi d} \cos \frac{2\pi nt}{T}\right) \qquad [3\text{-}10]$$

As noted earlier, the simplest way to demodulate this output signal would be to pass the sampled signal through a low-pass filter of bandwidth B Hz. This is shown in Fig. 3-8.

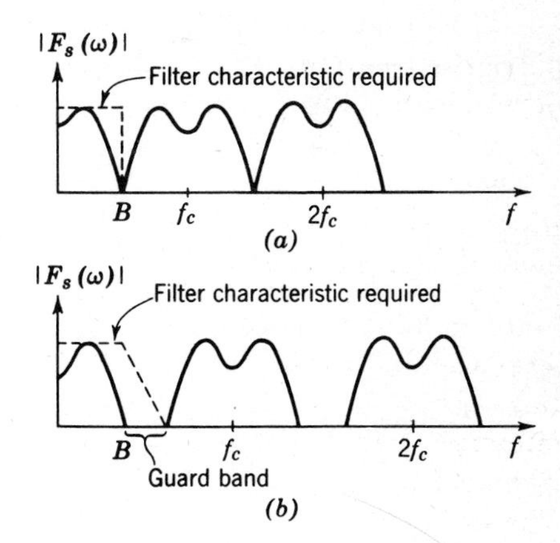

FIG. 3-8 Sampled-data demodulation using low-pass filter. (a) $f_c = 2B$; (b) $f_c > 2B$.

If we sample at exactly the Nyquist rate ($f_c = 2B$), the filter required must have infinite cutoff characteristics, as shown in Fig. 3-8*a*. This requires an ideal filter, an impossibility in practice. A practical low-pass filter with sharp cutoff characteristics could of course be used, with resulting complexity in filtering and some residual distortion (part of the lower sideband about f_c would be transmitted). This situation can of course be relieved somewhat by sampling at a higher rate, as shown in Fig. 3-8*b*. A guard band is thus made available, and filter requirements are less severe. The filter must cut off between B and $f_c - B$, its attenuation at $f_c - B$ being some prescribed quantity measured with respect to the passband.

In the example given previously, that of transmitting speech signals digitally, it was noted that the voice transmission is commonly band-limited to 3.3 kHz. The Nyquist sampling rate is then 6.6 kHz. A sampling rate of 8 kHz is most frequently used, however, so that the filter guard band is 1.4 kHz (from 3.3 to 8.0 − 3.3 = 4.7 kHz).

It is instructive to consider a simple proof for the low-pass filter demodulation of periodic samples.[1] This will fill out in a more quantitative way our rather qualitative discussion based on Eq. [3-10] and Fig. 3-8. It also helps to clarify further the actual mechanism of filtering.

Thus assume that a signal $f(t)$ band-limited to B Hz has been sampled at intervals of $1/2B$ sec. We shall first show that $f(t)$ may be reconstructed from these samples (a necessary result from the sampling theorem) and shall then demonstrate that an ideal low-pass filter is called for in the reconstruction or demodulation process.

To show that the given samples suffice to reproduce $f(t)$, take the Fourier transform $F(\omega)$ of $f(t)$,

$$F(\omega) = \int_{-\infty}^{\infty} f(t)e^{-j\omega t}\,dt \qquad [3\text{-}11]$$

Then

$$F(\omega) = 0 \qquad |\omega| > 2\pi B \qquad [3\text{-}12]$$

by virtue of the band-limited assumption on $f(t)$.

$F(\omega)$ can be arbitrarily made periodic with a period of $4\pi B$, as shown in Fig. 3-9. It can then be expanded in a Fourier series of period $4\pi B$ to

[1] B. M. Oliver, J. R. Pierce, and C. E. Shannon, Philosophy of PCM, *Proc. IRE*, vol. 36, no. 11, p. 1324, November, 1948.

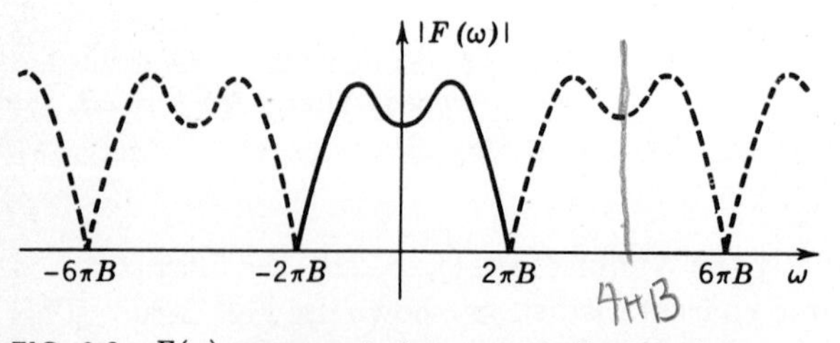

FIG. 3-9 $F(\omega)$ represented as a periodic function.

be used within the interval $|\omega| \leq 2\pi B$. Thus

$$\begin{aligned} F(\omega) &= \frac{1}{4\pi B} \sum_{-\infty}^{\infty} c_n e^{j(2\pi n/4\pi B)\omega} \\ &= \frac{1}{4\pi B} \sum_{n=-\infty}^{\infty} c_n e^{jn\omega/2B} \qquad |\omega| < 2\pi B \qquad [3\text{-}13] \\ F(\omega) &= 0 \qquad |\omega| > 2\pi B \end{aligned}$$

with c_n defined by

$$c_n = \int_{-2\pi B}^{2\pi B} F(\omega) e^{-jn\omega/2B} \, d\omega \qquad [3\text{-}14]$$

But, since $F(\omega)$ is the Fourier transform of $f(t)$, $f(t)$ can be written

$$f(t) = \frac{1}{2\pi} \int_{-\infty}^{\infty} F(\omega) e^{j\omega t} \, d\omega = \frac{1}{2\pi} \int_{-2\pi B}^{2\pi B} F(\omega) e^{j\omega t} \, d\omega \qquad [3\text{-}15]$$

In particular, at time $t = -n/2B$

$$f\left(-\frac{n}{2B}\right) = \frac{1}{2\pi} \int_{-2\pi B}^{2\pi B} F(\omega) e^{-jn\omega/2B} \, d\omega = \frac{c_n}{2\pi} \qquad [3\text{-}16]$$

from Eq. [3-14].

This means that if we are given $f(t)$ at the various sampling intervals (for example, $t = -3/2B$, $-2/2B$, $-1/2B$, 0, $1/2B$, etc.), we can find the corresponding Fourier coefficient c_n. But, knowing c_n, we can in turn find $F(\omega)$ from the Fourier series of Eq. [3-13]. Knowing $F(\omega)$, we find $f(t)$ for *all* possible times by Eq. [3-15]. A knowledge of $f(t)$ at sampling intervals $1/2B$ sec apart thus suffices to determine $f(t)$ at all times. This completes the first part of our proof. We have demonstrated that $f(t)$ may be reproduced completely, solely from a knowledge of $f(t)$ at the periodic sampling intervals. How do we now actually reconstruct $f(t)$ from these samples?

If we substitute the Fourier-series expansion for $F(\omega)$ (Eq. [3-13]) into Eq. [3-15], the Fourier-integral representation of $f(t)$, we get

$$\begin{aligned} f(t) &= \frac{1}{2\pi} \int_{-2\pi B}^{2\pi B} F(\omega) e^{j\omega t} \, d\omega \\ &= \frac{1}{2\pi} \int_{-2\pi B}^{2\pi B} \frac{1}{4\pi B} \left(\sum_n c_n e^{jn\omega/2B} \right) e^{j\omega t} \, d\omega \qquad [3\text{-}17] \end{aligned}$$

If the order of integration and summation are now interchanged (the integral is finite and no difficulties can arise because of improper integrals; the c_n coefficients also approach zero for n large enough), the resulting integral may be readily evaluated. We get

$$
\begin{aligned}
f(t) &= \sum_n \frac{c_n}{2\pi} \frac{1}{4\pi B} \int_{-2\pi B}^{2\pi B} e^{j\omega(t+n/2B)}\, d\omega \\
&= \sum_n \frac{c_n}{2\pi} \frac{\sin 2\pi B(t + n/2B)}{2\pi B(t + n/2B)} \qquad [3\text{-}18]
\end{aligned}
$$

But $c_n/2\pi = f(-n/2B)$, from Eq. [3-16]. Therefore

$$
\begin{aligned}
f(t) &= \sum_{n=-\infty}^{\infty} f\left(-\frac{n}{2B}\right) \frac{\sin 2\pi B(t + n/2B)}{2\pi B(t + n/2B)} \\
&= \sum_{n=-\infty}^{\infty} f\left(\frac{n}{2B}\right) \frac{\sin 2\pi B(t - n/2B)}{2\pi B(t - n/2B)} \qquad [3\text{-}19]
\end{aligned}
$$

since all positive and negative values of n are included in the summation. Mathematically, then, Eq. [3-19] indicates that we are to take each sample, multiply it by a $(\sin x)/x$ weighting factor centered at the sample's time of occurrence, and sum the resultant terms. This is exactly what is done, however, when we pass the samples through an ideal low-pass filter cutting off at B Hz.

We can demonstrate this very simply in the following manner: Assume $f(t)$ again limited to B Hz. We sample $f(t)$ for a length of time τ (the sampling time) periodically at intervals of $1/2B$ sec. If $\tau \ll 1/2B$, $f(t)$ may be assumed very nearly constant over the sampling time (Fig. 3-10).

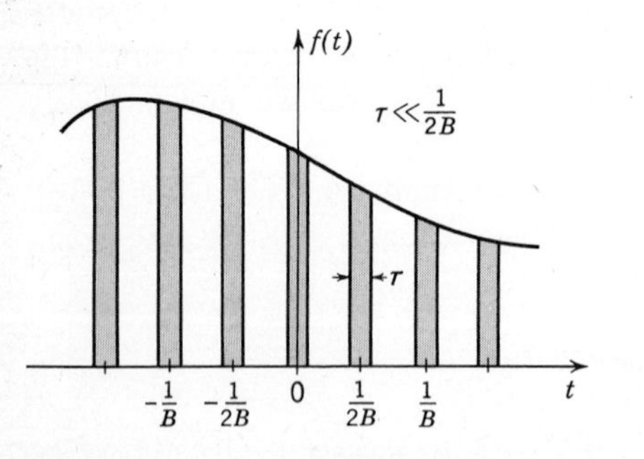

FIG. 3-10 The sampling process.

The individual sample $f(n/2B)$ has a Fourier transform (or frequency spectrum) given by

$$F_n(\omega) = \int_{-\infty}^{\infty} f(t)e^{-j\omega t}\,dt \doteq \tau f\left(\frac{n}{2B}\right) e^{-jn\omega/2B} \qquad [3\text{-}20]$$

since $f(t)$ is assumed constant over the τ-sec interval and zero elsewhere.

But note that this is just the Fourier transform of an impulse (delta function) of amplitude $\tau f(n/2B)$ and located at $t = n/2B$ in time. [It should be recalled from Eq. [2-108] that the Fourier transform of $K\delta(t - n/2B) = Ke^{-jn\omega/2B}$.]

The amplitude factor $\tau f(n/2B)$ represents the area under the curve of the individual sample. By assuming the sample duration τ to be very small, we have effectively approximated the sample by an impulse of the same area.

Assume that this impulse is now passed through an ideal low-pass filter of bandwidth B Hz with zero phase shift and unity amplitude assumed for simplicity. (The idealized linear phase shift just serves to delay the occurrence of the output signal, as pointed out in Chap. 2.) The response of this idealized filter to an impulse $K\delta(t - t_0)$ was shown in Eq. [2-114] to be given by

$$2KB\,\frac{\sin 2\pi B(t - t_0)}{2\pi B(t - t_0)}$$

The output response $g_n(t)$ of the ideal low-pass filter to the individual sample $f(n/2B)$ must thus be given by

$$g_n(t) = 2\tau f\left(\frac{n}{2B}\right) B\,\frac{\sin 2\pi B(t - n/2B)}{2\pi B(t - n/2B)} \qquad [3\text{-}21]$$

This result can of course also be obtained directly from Eq. [3-20] and the assumed filter characteristic. Thus, for an ideal low-pass filter with zero phase shift and unit amplitude, the output response to the sample $f(n/2B)$ applied at the input is

$$\begin{aligned} g_n(t) &= \frac{1}{2\pi}\int_{-2\pi B}^{2\pi B} F_n(\omega)e^{j\omega t}\,d\omega \\ &= f\left(\frac{n}{2B}\right)\frac{\tau}{2\pi}\int_{-2\pi B}^{2\pi B} e^{j\omega(t-n/2B)}\,d\omega \qquad [3\text{-}22] \end{aligned}$$

Integrating, as indicated, we get for $g_n(t)$

$$g_n(t) = 2B\tau f\left(\frac{n}{2B}\right)\frac{\sin 2\pi B(t - n/2B)}{2\pi B(t - n/2B)}$$

as in Eq. [3-21]. $g_n(t)$ is plotted in Fig. 3-11. Note that $g_n(t)$ has its maximum value at $t = n/2B$ (the filter was assumed to have zero phase shift), and precursors again appear because of the idealized filter characteristics assumed. The output is a maximum at the given sampling point, $t = n/2B$, and zero at all the other sampling points. At the sampling point $g_n(n/2B) = 2B\tau f(n/2B)$, or g_n is just the input $f(n/2B)$ to within a constant. The next sample, occurring $1/2B$ sec later, or at $t = (n + 1)/2B$, likewise produces an output given by Eq. [3-21], but delayed $1/2B$ sec, and proportional to $f[(n + 1)/2B]$.

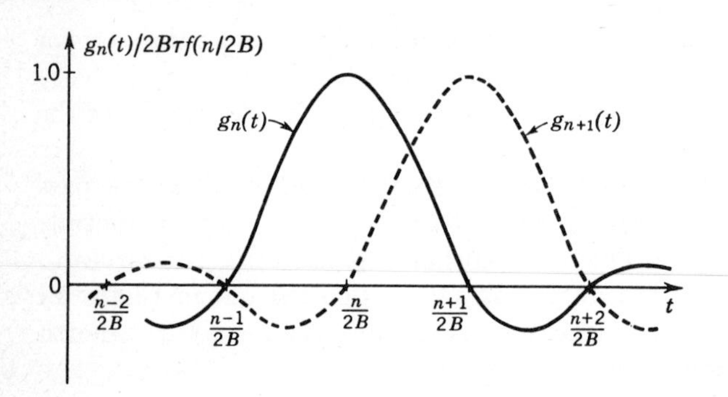

FIG. 3-11 Filter response to sampled inputs.

The peak of each $(\sin x)/x$ term occurs at a sampling point where all other outputs are zero (Fig. 3-11), and the output at each of the sampling points is exactly proportional to the magnitude of the input sample at that point.

Since we have been assuming a linear ideal filter, the complete output is just the superposition of the individual sample outputs, or

$$\begin{aligned} g(t) &= \sum_{n=-\infty}^{\infty} g_n(t) \\ &= 2B\tau \sum_{n=-\infty}^{\infty} f\left(\frac{n}{2B}\right) \frac{\sin 2\pi B(t - n/2B)}{2\pi B(t - n/2B)} \\ &= 2B\tau f(t) \end{aligned} \qquad [3\text{-}23]$$

by comparison with Eq. [3-19]. The output of the low-pass filter, $g(t)$, is thus identically proportional to the original signal $f(t)$ at *all* instants of time, not only at the sampling points. [Had we included a linear-phase-shift characteristic, we would of course have obtained a constant time delay for the output function. The output $g(t)$ would thus be a delayed replica of $f(t)$.]

The original input $f(t)$ may thus be reproduced from the samples by passing them through an ideal low-pass filter of bandwidth B Hz. The $2B\tau$ proportionality factor represents the ratio of the filter and sampling pulse bandwidths. Since we assumed $\tau \ll 1/2B$, and since the effective pulse "bandwidth" is proportional to $1/\tau$ (Chap. 2), the pulse spectral width is much greater than that of the filter. (We in fact assumed the pulse spectrum to be constant over the filter bandwidth, as shown by Eq. [3-20].) Most of the energy of the input sample thus lies outside the filter spectrum, as shown in Fig. 3-8. If the pulse width is so narrow that its spectrum is almost constant (again the justification for representing it by an impulse of the same area), the output amplitude is reduced by the ratio of bandwidths, or just $2B\tau$.

These considerations of filter operation from a frequency point of view of course agree with the time approach. For, as shown by Eq. [3-10], an ideal low-pass filter of unit amplitude and bandwidth B would produce $g(t) = df(t)$ at the output. But the pulse duty cycle d is simply τ/T, or $2B\tau$, with $T = 1/2B$, agreeing with the result of Eq. [3-23].

The ideal filter is of course a mathematical artifice and can only be approximated in practice. Actual filter outputs will thus only approximate the actual input $f(t)$. An example of a low-pass filter frequently used in practice will be described in the next section.

3-5 TIME-DIVISION MULTIPLEXING AND PAM SYSTEMS

We have indicated in the previous sections that one may convert band-limited analog data to a sampled form by sampling at least at the Nyquist rate. These sampled values of the signal carry the original intelligence, and demodulation may be carried out by low-pass filtering. A system transmitting these sampled values of the signal is commonly called a *pulse-amplitude-modulation* (PAM) system. For the sequence of samples may alternatively be visualized as a periodic sequence of pulses (the *carrier*) whose amplitude is modulated (or varied) in accordance with the intelligence to be transmitted. This is in fact apparent from the form of the sampled data expression $f_s(t) = f(t)S(t)$. The switching function $S(t)$ represents the unmodulated pulse carrier and $f(t)$ the intelligence modulating the carrier. This concept of amplitude modulation will be pursued further in the next chapter.

Most PAM systems in use transmit many signals simultaneously, rather than just one. It is apparent from the sampling process, with a very narrow sample τ sec wide taken every T sec, that much of the time no information is being transmitted through the system (see Fig. 3-2). It is thus possible to transmit other information from other sources in the vacant intervals. The transmission of samples of information from several signal channels simultaneously through one communication system with different channel samples staggered in time is called *time-division multiplexing* or *time multiplexing* for short.

In a typical time-multiplexing scheme, the various signals to be transmitted are sequentially sampled and combined for transmission over the one channel. Figure 3-12 shows a time diagram of a time-multiplexed

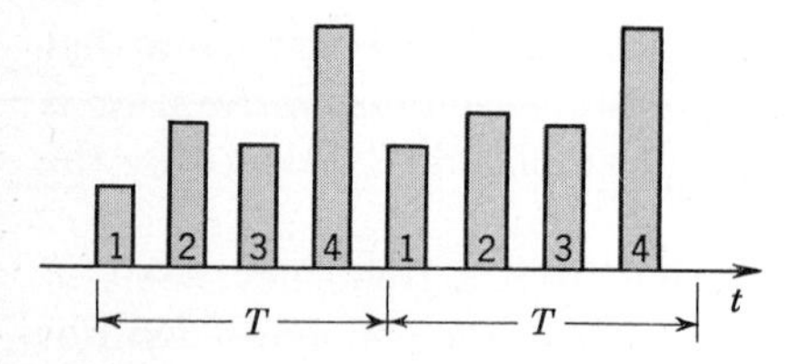

FIG. 3-12 Time multiplexing.

signal system having four information carriers. It is apparent that all signals to be multiplexed must either be of the same bandwidth, or, if such a scheme is to be used, sampling must take place at a rate determined by the maximum bandwidth signal. (Alternatively, relatively low-bandwidth signals may be first combined before sampling, as will be noted later.) A sampler time multiplexing a multichannel input sequentially into the transmission channel is shown in Fig. 3-13. A mechanical switch is again

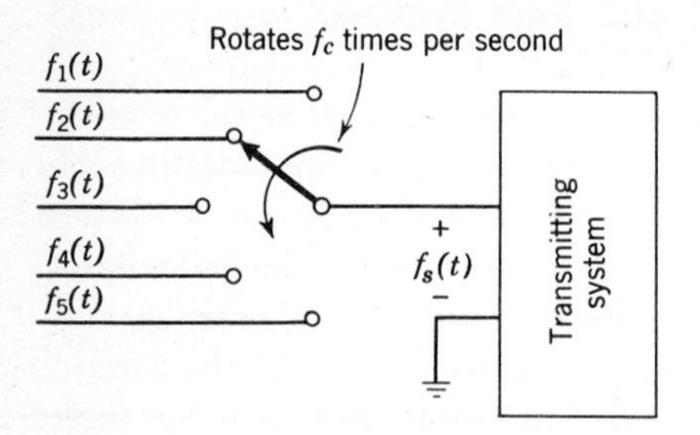

FIG. 3-13 Sampler for time multiplexing.

shown for simplicity, but in practice electronic switching would normally be used.

Time multiplexing has been widely used in the radio, teletype, and telephone fields, and for telemetering purposes. Experimental data from space probes are commonly sampled and transmitted sequentially, for example. A space vehicle transmitting back vital measured information such as temperature, electron density, magnetic intensity, and a multitude of other sensor outputs need only utilize one transmission channel, providing the sampling is carried out rapidly enough to accommodate the most rapidly varying signal to be transmitted.

If the various signals to be multiplexed have widely differing bandwidths, it is possible to first combine the low-bandwidth signals into a single wider-bandwidth analog signal by *frequency-multiplexing* techniques. (This is described in more detail in the next chapter. Here the different signals are mixed into successively higher and contiguous frequency bands, and then combined.) Teletype signals with 100-Hz bandwidth are commonly combined in this way for final time multiplexing with 3.2-kHz voice channels.

As is true with all engineering developments, one pays a price for introducing time multiplexing into a system. First it is apparent that the necessary transmission bandwidth increases with the number of signals multiplexed, for the transmission bandwidth is proportional to the reciprocal of the width of the pulses transmitted (Chap. 2). Thus, in the case of the single-signal pulses of Fig. 3-2, the bandwidth required to transmit the pulses shown would be approximately $1/\tau$ Hz. However, it is possible to widen the pulses substantially, just to the point where they begin to overlap say, and so require approximately $1/T$-Hz bandwidth. The pulses of Fig. 3-12 can only be widened to the point where they begin to interfere with the next adjacent pulses, however. In this case the minimum bandwidth is four times that of the single-signal case of Fig. 3-2. In general, then, if N signals are time multiplexed, the required transmission bandwidth is N times that for single-signal samples.

As an example, assume that 10 voice channels, each band-limited to 3.2 kHz are sequentially sampled at an 8-kHz rate and time multiplexed on one channel. The successive pulses are then spaced 12.5 μsec apart, as shown in Fig. 3-14. The bandwidth required to transmit these pulses is

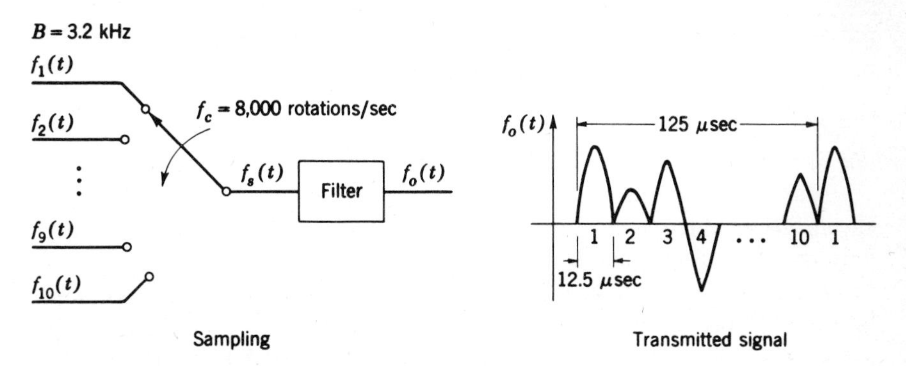

FIG. 3-14 Time-multiplexed voice channels.

roughly 80 kHz. (A more accurate determination of the filtering and bandwidth necessary to prevent pulse overlap will be considered later in Sec. 3-9 in the discussion of intersymbol interference.) The filter shown in Fig. 3-14 is used to widen the pulses as required. Filtering could be incorporated in the sampling operation as well, or it could be carried out further along the transmission path.

The other problem introduced by time multiplexing involves proper synchronization and registration of the successive pulses at the receiver. For it is apparent that the successive pulses must on reception be delivered to the appropriate destination. This implies that a switch is available at the receiver, is synchronized to the original transmitter switch, and deposits each sample in its appropriate signal channel. This is no mean task with high-speed data systems, and with receiver and transmitter located thousands or millions of miles apart. Figure 3-15 shows trans-

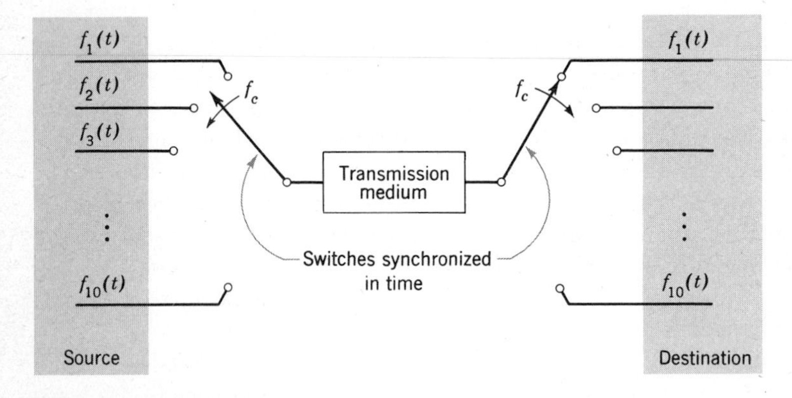

FIG. 3-15 Switching in a PAM system.

mitter and receiver switching in a time-multiplexed PAM system. The registration problem, with sample 1 placed in the 1 line at the destination and not in line 2 or 3, is particularly acute. (Although you might not mind eavesdropping on someone else's conversation, you in turn would not like to have your conversation picked up by a stranger!)

Various techniques have been utilized in practice to perform synchronization and registration of signals. This is a major problem not only with PAM systems, but with all other digital systems as well. The techniques have included: (1) the use of special marker pulses, tagged to be easily differentiable from regular signal pulses, and sent periodically at prescribed intervals; (2) continuous sine waves of known phase and frequency which can be filtered out at the receiver to provide the necessary timing information; and (3) schemes which derive timing information from the transmitted signal pulses themselves by averaging over long periods of time, etc. Reference is made to the literature for details of various synchronizing techniques.[1] Some specific methods adopted for particular systems are included in Sec. 3-10 in the description of some actual operating systems.

When a large number of signals are time multiplexed together in a PAM system, it is apparent that the individual signal pulses as transmitted are of necessity narrow compared with the sampling interval T. These narrow samples do not provide very much signal power when ultimately received at the appropriate receiver signal channel for demodulation. This problem is commonly handled by following the receiver channel-selecting switch by a capacitor fed from a low-impedance source. The switch-capacitor combination is called a *sample-and-hold* or *holding circuit*, and it also provides the low-pass filtering necessary to finally demodulate the signal samples.

A simple diagram of a PAM system using such a holding circuit for demodulation is shown in Fig. 3-16. As the demodulator switch reaches a particular channel position, the capacitor rapidly charges up (at the rate determined by the capacitance and the internal impedance of the switch circuit) to the voltage at the switch. If the switch is properly synchro-

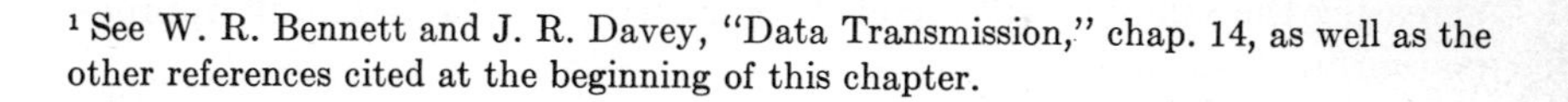
[1] See W. R. Bennett and J. R. Davey, "Data Transmission," chap. 14, as well as the other references cited at the beginning of this chapter.

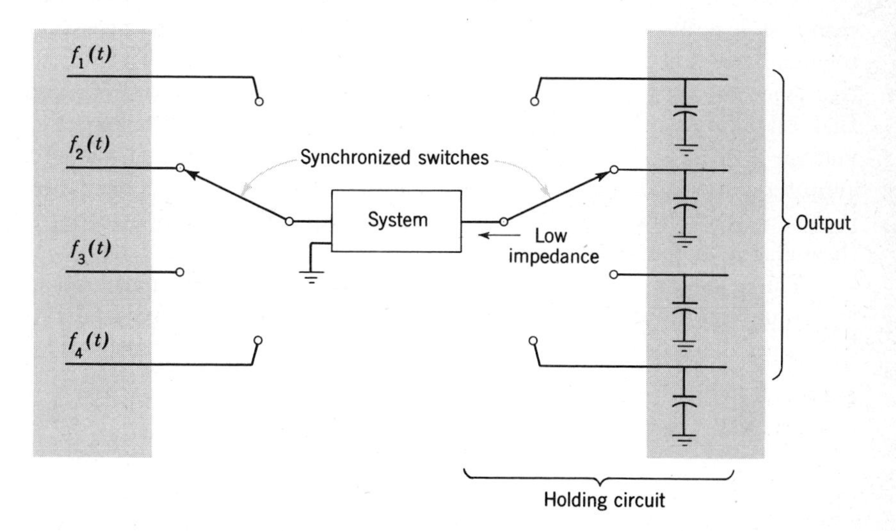

FIG. 3-16 PAM system using holding circuit.

nized, this voltage is just the amplitude of the pulse corresponding to that channel. As the switch leaves the particular position, the capacitor "holds" the charge until the next sampling interval. The holding circuit thus serves to "stretch" the narrow input impulses into rectangular waves of width T—the Nyquist interval.

The holding-circuit operation is shown graphically in Fig. 3-17. It is apparent that the holding-circuit output is a step approximation to the

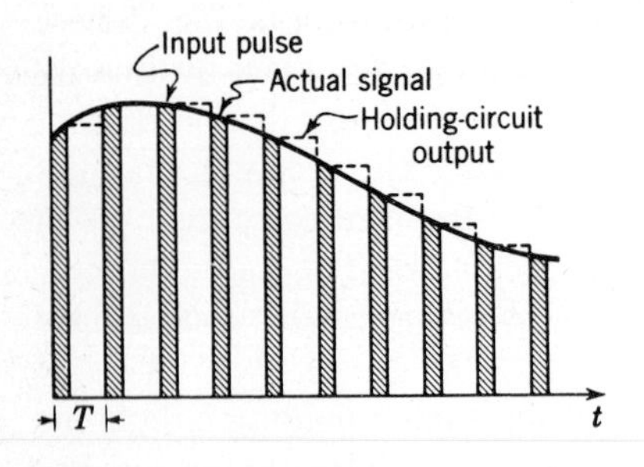

FIG. 3-17 Holding-circuit output.

actual signal. Further filtering is frequently used to produce a smoothed and improved version of the desired signal. It is also apparent that increasing the sampling rate tends to improve the holding-circuit approximation to the signal.

That this circuit, described above from the point of view of time response, is actually a crude low-pass filter, of the type deemed necessary for demodulation in Fig. 3-8, may easily be demonstrated. To find the frequency response of this circuit, we first find the Laplace transform and then set $p = j\omega$. This is done very simply by noting that if the applied voltage is a pulse of unit amplitude the capacitor charges up to unit amplitude and holds the charge until the switch again closes, T sec later. If the applied voltage is zero at that time, the capacitor quickly discharges through the low-impedance source.

The capacitor voltage may thus be represented as the superposition of two unit steps, one positive step and one negative step delayed in time by T sec. This provides a delay factor e^{-pT} in the Laplace transform, or

$$V_o(p) = \frac{1 - e^{-pT}}{p} \qquad [3\text{-}24][1]$$

[1] If the input voltage is a unit pulse of width τ, $\tau \ll T$, its Laplace transform is approximately τ. The transfer function of the filter is then $V_o(p)/\tau$, with $V_o(p)$ given by Eq. [3-24]. The transfer function or frequency response is thus proportional to $V_o(p)$.

With $p = j\omega$,

$$V_o(\omega) = Te^{-\frac{1}{2}j\omega T} \frac{\sin(\omega T/2)}{\omega T/2}$$

$$= Te^{-j\pi f/f_c} \frac{\sin(\pi f/f_c)}{\pi f/f_c} \qquad [3\text{-}25]$$

with $f_c = 1/T$ the sampling frequency. The holding circuit thus behaves as a $(\sin x)/x$ filter. The frequency response (amplitude only) of such a filter is shown in Fig. 3-18*a*. (Compare with the filtering in Fig. 3-8.)

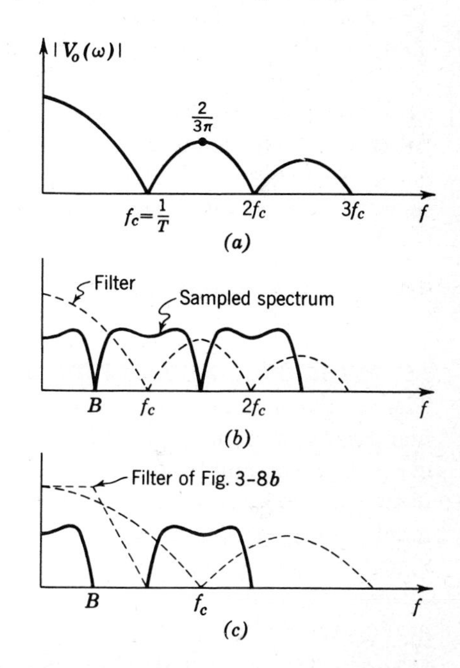

FIG. 3-18 Holding circuit as low-pass filter. (a) $(\sin x)/x$ filter; (b) demodulation with $(\sin x)/x$ filter; $f_c = 2B$; (c) $f_c > 2B$.

It is apparent from Fig. 3-18 that the holding circuit, by itself, serves as a rather poor low-pass filter if the minimum Nyquist sampling rate is used. It gives zero output at f_c, instead of cutting off sharply at $f_c/2$. In fact it is down to $2/\pi$, or 0.636 of the peak value at $f = f_c/2$. The filtering properties may of course be improved by increasing the sampling rate, as shown in Fig. 3-18*c*.

As an example of the improvement possible by increasing the sampling rate, assume $B = 100$ Hz for the particular signal and $f_c = 400$ Hz (the minimum Nyquist rate is 200 Hz). Ideally we would want a filter having unity response up to and including B and tapering off to zero at $f = f_c - B$ (the guard band). With these numbers the filter is down to

$$\frac{\sin (\pi 100/400)}{\pi 100/400} = 0.9 \ (-1 \text{ db}) \qquad \text{at } f = B = 100 \text{ Hz}$$

and
$$\frac{\sin (\pi 300/400)}{\pi 300/400} = 0.3 \ (-10 \text{ db}) \qquad \text{at } f = f_c - B = 300 \text{ Hz}$$

This may be satisfactory in some applications, but ordinarily additional filtering is inserted to improve the holding-circuit response.

The switch is necessary in understanding the operation of the holding circuit, so that the circuit is essentially a linear time-varying one. Linear circuits with the same transfer function may also be developed. These do not require the use of switches for their operation. One example of such a linear circuit is shown in Fig. P 2-30 in the problem section for Chap. 2.

3-6 PULSE CODE MODULATION: A/D CONVERSION COMPLETED

The relative simplicity of PAM systems makes them quite attractive for many data-handling applications. They are not completely digital, however, since the amplitudes of the pulses transmitted may vary continuously in accordance with the original analog signal variations. Very commonly in modern communication technology the PAM signals are further digitized before transmission. The digital signals may then be encoded into any equivalent form desired. Systems embodying the transmission of digitized and coded signals are commonly called *pulse-code-modulation* (PCM) systems. Binary digital systems constitute the most frequently encountered form of PCM systems.

There are many advantages to using PCM systems.

1. The signals may be regularly reshaped or regenerated during transmission since information is no longer carried by continuously varying pulse amplitudes but by discrete symbols.
2. All-digital circuitry may be used throughout the system.
3. Signals may be digitally processed as desired.
4. Noise and interference may be minimized by appropriate coding of the signals, etc.

Some of these factors will be described in detail later in the book.

The process of digitizing the original PAM signals is called the *quantization* process. It consists of breaking the amplitudes of the PAM signals up into a prescribed number of discrete amplitude levels. The resultant signals are said to be *quantized*. Unlike the sampling process this results in an irretrievable loss of information since it is impossible to reconstitute the original analog signal from its *quantized* version. However, as first noted in the discussion of the concept of information in Chap. 1, there is actually no need to transmit all possible signal amplitudes as done in PAM systems. Because of noise introduced during transmission and at the receiver, the demodulator or detector circuit will not be able to distinguish fine variations in signal amplitude. In addition the ultimate recipient of the information—our ears in the case of sound or music, our eyes in the case of a picture—is limited with regard to the fine gradation of signal it can distinguish.

This ultimate limitation in distinguishing among all possible amplitudes thus makes quantization possible. In a specific system the sampled pulses may be quantized, or both quantization and sampling may be performed simultaneously. This latter process is portrayed in Fig. 3-19. The total amplitude swing of $A_0 = 7$ volts is divided into equally spaced amplitude levels $a = 1$ volt apart. There are thus $M = A_0/a + 1$ possible amplitude levels, including zero. In Fig. 3-19 samples are shown taken every second and the nearest discrete amplitude level is selected as the one to be transmitted. The resultant quantized and sampled version of the signal of the smooth signal of Fig. 3-19*a* thus appears in Fig. 3-19*b*. (The signal of 0.3 volts at 0 sec is transmitted as 0 volts, etc.)

Although the level separation is shown here as uniform, the separation is often tapered in practice to improve the system noise performance.[1] In particular the spacing of levels is decreased at low amplitude levels. We shall assume equal spacing for simplicity, however.

Obviously the quantizing process introduces some error in the eventual reproduction of the signal. The demodulated signal will differ somewhat from the derived signal, as noted earlier. The overall effect is as if additional noise had been introduced into the system. (In the case of sound transmission this manifests itself as a background cackle. In the case of picture transmission the continuous gradation of grays from black to white is replaced by a discrete number of grays, and the picture also looks somewhat noisy.) This *quantization noise* may of course be reduced by decreasing the level separation a or by increasing the number of levels M used. Experiment has shown[2] that 8 to 16 levels are sufficient for good

[1] See M. Schwartz, W. R. Bennett, and S. Stein, "Communication Systems and Techniques," p. 264, McGraw-Hill, New York, 1966. See also Sec. 3-10, in which compandors, providing the same tapering, are described.

[2] H. F. Mayer, Pulse Code Modulation, summary chapter in L. Martin (ed.), "Advances in Electronics," vol. III, pp. 221–260, Academic, New York, 1951.

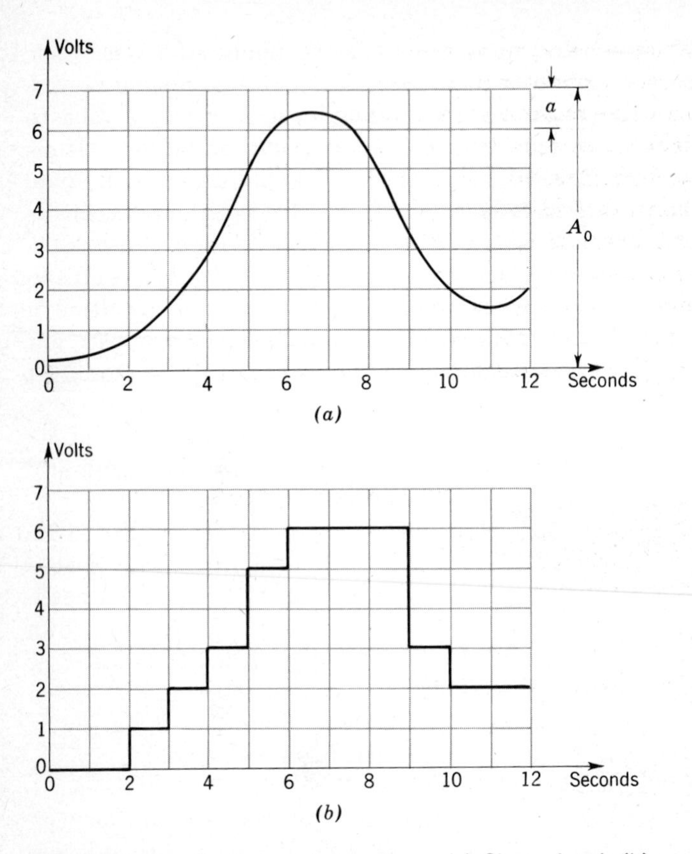

FIG. 3-19 Quantization and sampling. (a) Given signal; (b) quantized and sampled version.

intelligibility of speech. (Two-level PCM is even understandable, although quite noisy![1])

If the quantized signal samples were transmitted directly as pulses of varying (although quantized) heights, the resultant system would simply be quantized PAM. (This is also often referred to as *M*-ary PAM.) But with discrete or numbered voltage levels each level can be coded in some arbitrary form before transmission. It is this possibility of coding discrete sample levels that gives quantized signals much greater flexibility in transmission than the continuous varying pulses of PAM.

Most commonly the quantized and sampled signal pulse is encoded into an equivalent group or packet of binary pulses of fixed amplitude.

[1] J. S. Mayo, Pulse Code Modulation, *Electro-Technol.*, pp. 87–98, November, 1962.

This finally provides the binary signals first encountered in Sec. 1-1, typical examples of which are shown in Fig. 1-2. The encoding of amplitude levels into binary form follows the usual decimal-binary conversion, tabulated for the numbers 0 to 7 in Sec. 1-3, and repeated in the table on page 142. Thus 8-level PCM requires three binary digits for transmission, 16 levels require four binary digits, etc. Since these binary digits must be transmitted in the sampling interval originally allotted to one quantized sample, the binary pulse widths are correspondingly narrower and the transmission bandwidth goes up proportionately to the number of binary pulses needed. An example of the binary encoding process is shown in Fig. 3-20. Here three binary pulses are transmitted in the original sam-

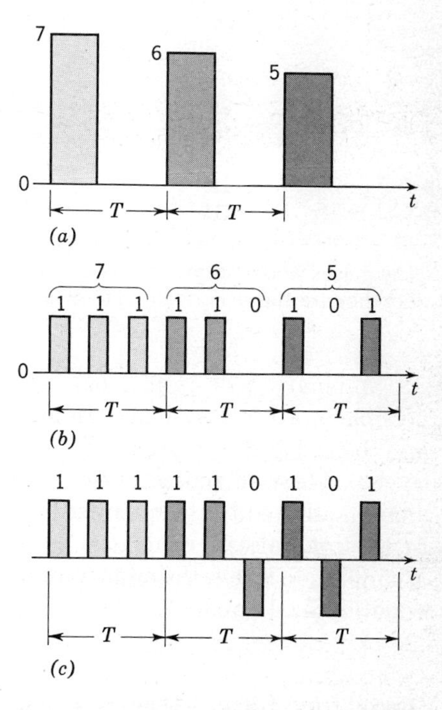

FIG. 3-20 Binary coding of samples. (a) Given signal (already sampled and quantized); (b) coded samples; (c) another form of binary code.

pling interval, so that the bandwidth is increased by a factor of 3. (Note that if this represented a time-multiplexed system, the three samples of amplitudes 7, 6, and 5, respectively, would represent three separate signals.)

Two examples of binary signals are shown in Fig. 3-20: (1) an on-off signal, in which the 1 bit is represented by the presence of a pulse, the 0 bit by its absence; and (2) a so-called bipolar signal, with the 1 represented by a positive pulse, the 0 by a negative pulse. The two sets of binary signals are tabulated in the following table, again for the case of $M = 8$ possible signal levels.

BINARY SIGNALING (AMPLITUDE NORMALIZED TO 1)†

Amplitude level	On-off pulses	Bipolar pulses
0	000	−1 −1 −1
1	001	−1 −1 1
2	010	−1 1 −1
3	011	−1 1 1
4	100	1 −1 −1
5	101	1 −1 1
6	110	1 1 −1
7	111	1 1 1

†The pulses would normally have any desired amplitude. Here they are shown normalized to an amplitude level of 1.

A binary code is just one special case of the coding theoretically possible in a PCM system. In general any one quantized signal sample may be coded into a group of m pulses, each with n possible amplitude levels. These m pulses must be transmitted in the original sampling interval allotted to the quantized sample. Since the information carried by these m pulses is equivalent to the information carried by the original M amplitude levels, the number of possible amplitude combinations of the m pulses must equal M. Thus

$$M = n^m \qquad [3\text{-}26]$$

(Recall from Chap. 1 that m pulses, each with n possible heights, may be combined in n^m different ways. Each combination must correspond to one of the original M levels.) If there are two possible levels, $n = 2$, and we have the binary code just mentioned. With $M = 8$, three binary pulses are necessary. If $n = 3$, a ternary code results. Obviously, if $m = 1$, $n = M$ and we are back to our original uncoded but quantized samples. As the number n of levels chosen for the coded pulses increases, m decreases, as does the bandwidth required for transmission.

This ability to code back and forth is one of the reasons for the increased usage of PCM systems. Although most PCM systems in use are binary in form, so that the bandwidth required for transmission is wider than that required before binary encoding, bandwidth *reduction* schemes have been suggested in which successive M-level samples would be further collapsed into much higher-level samples. This is a reversal of the binary encoding process. For example, two successive binary pulses provide a total of four possible combinations, and so need four numbers to represent them. Similarly, 2 eight-level samples require 64 numbers to represent them. If combined into one *wider* pulse, covering the two adjacent time slots, the required bandwidth could be reduced by a factor of 2, but the one pulse used for transmission would take on 64 possible levels. This process of combining or collapsing successive pulses into one much wider pulse with many more amplitude levels can of course be continued indefinitely. (The m in Eq. [3-26] then represents the number of adjacent pulses, each with n amplitude levels, that are collapsed into one pulse with M levels. The bandwidth is then reduced by a factor of m.) There is one major difficulty, however; if the spacing between levels remains fixed, the required peak power goes up *exponentially* with the number of pulses combined. Alternately, if the peak power or amplitude swing is to remain fixed, the levels must be spaced closer and closer together. This then makes it easier for noise in the system to obscure adjacent levels. Such bandwidth reduction techniques are thus only possible in a relatively low-noise environment.

The binary form of transmission provides, on the other hand, the most noise immunity, and this is therefore the digital communication scheme with which we shall concern ourselves most in the chapters to follow. For, as indicated by the examples of Fig. 3-20, the information carried by the binary signals is represented either by the absence or presence of a pulse, or by its polarity. All the receiver has to do is to then correspondingly recognize the absence or presence of a pulse, or the polarity (plus or minus) of a pulse, and then decode into the original quantized form to reconstruct the signal. The pulse shape or exact amplitude is not significant as in the case of the original analog signal, or a PAM signal. By transmitting binary pulses of high enough amplitude, we can ensure correct detection of the pulse in the presence of noise with as low an error rate (or possibility of mistakes) as required. We shall have more to say about this noise-improvement capability of PCM systems in the next section, as well as in later chapters of the book.[1]

[1] In Chap. 8 we consider encoding blocks of m successive binary digits into $M = 2^m$ possible signal symbols or code words for the purpose of reducing the probability of error. The resultant bandwidths then generally *increase* with m, or at best remain fixed. These encoding schemes for error reduction should not be confused with the binary to M-level amplitude transmission discussed here for bandwidth reduction.

As noted earlier binary PCM lends itself readily as well to signal reshaping at periodic intervals during transmissions. This is a common practice in digital communication over telephone circuits. The reshaping of signals at these intermediate *repeaters* enhances the signal decisions when finally received at the receivers.[1]

It is now of interest to tie together the processes of sampling, quantization, and binary encoding, as well as their reverse processes at the receiver. Figure 3-21 shows these various operations in block diagram

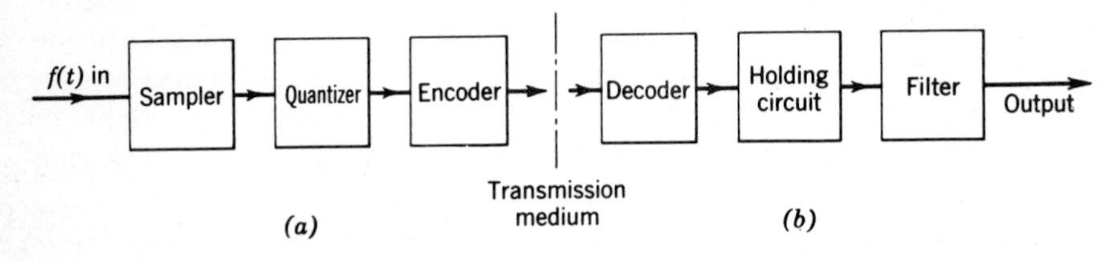

FIG. 3-21 One-channel PCM system. (a) Transmitter; (b) receiver.

form for a complete one-channel PCM system. (Synchronizing circuitry, as well as modulation and demodulation circuits to be discussed later, are not shown.) Note that this is the same form as a PAM system, with the addition of a quantizer and encoder at the transmitter and a decoder at the receiver. In Fig. 3-22 we take a 10-channel PCM system and show the timing and pulse widths required as we progress through the system. Again, for simplicity's sake, necessary synchronizing pulses are not shown. (The insertion of additional synchronizing bits would of course require narrower pulses and hence wider bandwidths.) Thus if the original analog channels each have 3.2-kHz bandwidth, as shown, and a sampling rate of 8,000 samples per second is used, the time-multiplexed pulses at the sampler output [point (1), part (*a*)] appear at 12.5-μsec intervals. Assuming an eight-level quantizer, with integer steps for simplicity, the quantized output pulses also appear at 12.5-μsec intervals, but with integer amplitudes only as shown in part (*c*), point (2), of Fig. 3-22. Each quantized pulse is then encoded into three binary pulses, each occupying a 4.2-μsec time slot, as shown.

[1] W. R. Bennett and J. R. Davey, "Data Transmission," pp. 128–131, McGraw-Hill, New York, 1965. Since most telephone systems provide relatively low-noise transmission, eight-level PCM rather than binary is often used to provide a higher rate of information transmission over the channel. Section 5-9 gives a brief analysis of repeaters.

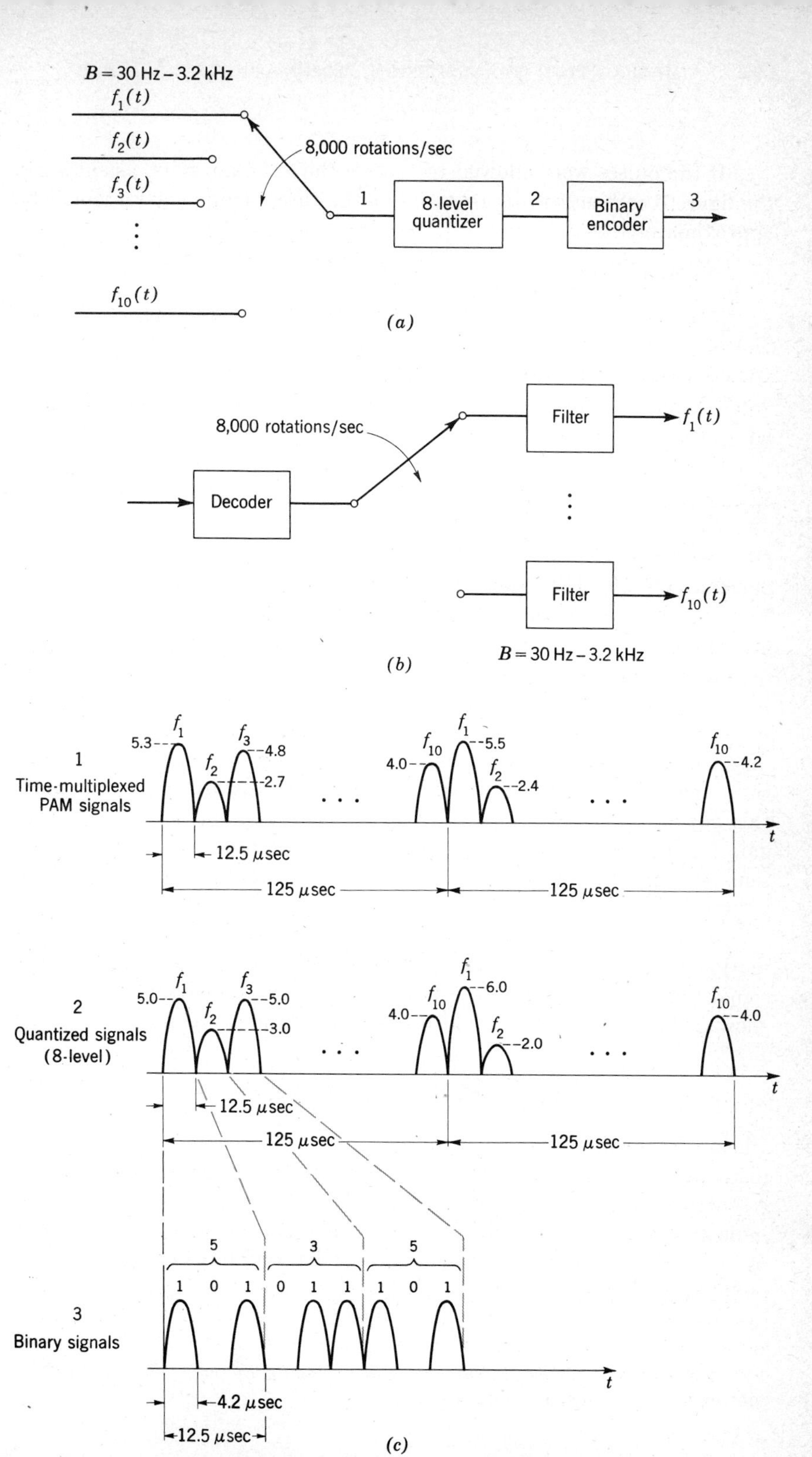

FIG. 3-22 Ten-channel PCM system. (a) Transmitter; (b) receiver; (c) signal shapes.

If the pulses were allowed to occupy the full time slots, as shown in the figure, the bandwidths required at the three points shown would be approximately

1. 80 kHz (1/12.5 μsec)
2. 80 kHz
3. 240 kHz

(1) is the PAM bandwidth; (3) is the binary PCM bandwidth. (As noted earlier we shall discuss requisite bandwidths more quantitatively later after a discussion of intersymbol interference.)

3-7 INFORMATION CAPACITY OF PCM SYSTEMS:[1] SIGNAL POWER AND BANDWIDTH EXCHANGE

We noted in the previous section that by encoding M amplitude levels into their binary equivalent, we improve the detectability of the signal, but at the expense of increased bandwidth. Alternately, by collapsing several successive binary signal (or bit) intervals into one interval with increased amplitude levels, we reduce the bandwidth required for transmission, but at the expense of increased signal power required. There is thus an exchange possible between power and bandwidth. We can make this power-bandwidth exchange more quantitative and at the same time gain further insight into the noise-improvement properties of PCM systems by discussing it in the context of the information capacity of PCM systems.

Recall that in Chap. 1 we developed a relation for the rate of information transmission, or the information capacity, of a system. For signals given in the form of equally likely voltage levels (Fig. 1-5), we showed that the capacity could be defined by the expression

$$C = \frac{1}{\tau} \log_2 n \qquad \text{bits/sec} \tag{3-27}$$

where τ is the minimum response time of the system and n is the number of distinguishable voltage levels. In Chap. 2 we further showed that τ is essentially the inverse of the system bandwidth. Some thought will indicate that the quantized PAM or PCM signals are already in the form of Eq. [3-27] so that we can quantitatively determine the information rate of PCM systems.

[1] B. M. Oliver, J. R. Pierce, and C. E. Shannon, The Philosophy of PCM, *Proc. IRE*, vol. 36, p. 1324, November, 1948.

Thus assume that we have quantized our signal to M possible amplitude levels. Assuming that all levels are equally likely (see Sec. 1-4 for a discussion of unequally likely signals), each sample carries $\log_2 M$ bits of information (Chap. 1).

Assume further that the original signal was sampled at the minimum Nyquist rate. For a band-limited signal of B-Hz bandwidth, this corresponds to $2B$ samples per second, or one sample every $T = 1/2B$ sec. The rate of information transmission, or system capacity, must thus be

$$C = \frac{\log_2 M}{T} = 2B \log_2 M \qquad \text{bits/sec} \qquad [3\text{-}28]^1$$

This is of course just in the form of Eq. [3-27], although rewritten with different symbols.

The encoding process does not change the rate of transmission of information but only converts the M quantized levels to a code group of m pulses of n levels each. The capacity can therefore be written, with $M = n^m$,

$$C = 2mB \log_2 n \qquad \text{bits/sec} \qquad [3\text{-}29]$$

This is of course in agreement with the comments of the previous section stating that by sending m pulses in the time previously allotted to one, the transmission bandwidth is in turn increased by a factor of m. Here mB represents the effective system bandwidth.

Alternatively we can say that we now send m pulses for each previous pulse sample. The pulse rate at the output of the encoder is thus $2mB$ pulses per second. But with n equally likely levels in each pulse the average information content of each pulse is $\log_2 n$ bits. The capacity is the number of pulses per second times the average number of bits per pulse, or $C = 2mB \log_2 n$ bits/sec, as before.

We can now relate the number of levels n to the average power required for transmission. The capacity will thus be shown to depend on both bandwidth and signal power, and the exchange possible between the two will be made quantitative. Thus, assume as in the previous section that the n amplitude levels are equally spaced. If, as shown in Fig. 3-23, the spacing is a, while the total amplitude swing is A_0, we must have $A_0 = (n - 1)a$, with the zero level included. From this we can calculate the average signal power.

[1] This is the capacity per signal channel transmitted. In the case of a time-multiplexed system, we would of course sum the individual capacities, or multiply by the number of signals time multiplexed, to determine the total system capacity.

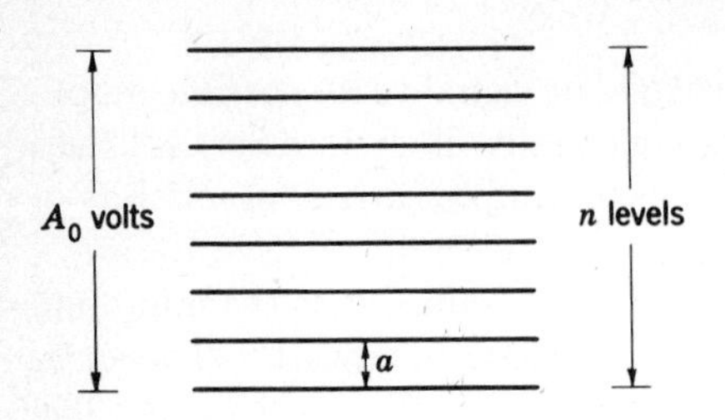

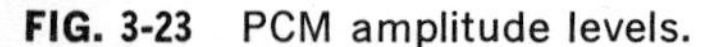

FIG. 3-23 PCM amplitude levels.

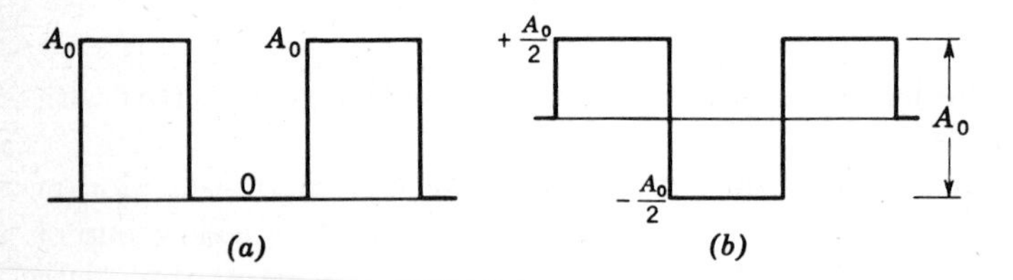

FIG. 3-24 Binary-code pulses at the decoder. (a) On-off pulses; (b) bipolar pulses.

Consider first the binary pulse group $n = 2$, with the two possible signal types noted in the previous section, and shown again in Fig. 3-24.

1. With on-off signals used, the on signal is a pulse of A_0 volts amplitude, and the off signal is zero. If the signal is assumed equally likely to be on or off, the peak power is A_0^2 and the average power is

$$S = \frac{A_0^2}{2} \qquad [3\text{-}30]$$

(Signal is on half the time and off half the time, on the average; a 1-ohm resistor is assumed.)

2. If bipolar pulses are sent instead, the pulses will be of amplitude $+A_0/2$ and $-A_0/2$, respectively. The total voltage swing is still A_0 volts. Again assuming either polarity of pulse equally likely, the peak power is $A_0^2/4$ and is the same as the average power

$$S = \frac{A_0^2}{4} \qquad [3\text{-}31]$$

Less power, on the average, is thus required to send bipolar pulses.

For the code with n possible levels, each spaced a volts apart, $A_0 = (n - 1)a$. For double-polarity pulses the various pulse heights possible will be $\pm a/2$, $\pm 3a/2$, . . . , $\pm(n - 1)a/2$. Again assuming the different levels equally likely, the average power over a large time interval will be

$$S = \frac{2}{n}\left\{\left(\frac{a}{2}\right)^2 + \left(\frac{3a}{2}\right)^2 + \cdots + \left[\frac{(n-1)a}{2}\right]^2\right\}$$
$$= (a)^2 \frac{n^2 - 1}{12} \qquad [3\text{-}32]$$

(The reader should check this relation for himself.)

Solving for n^2 in terms of the average power S and substituting into Eq. [3-29] for the capacity, we have finally

$$C = W \log_2 \left(1 + \frac{12S}{a^2}\right) \qquad [3\text{-}33]$$

with $W = mB$ the transmission bandwidth.

For a given capacity, then, one may reduce the transmission bandwidth mB by increasing the average signal power S. This is just the procedure noted in the previous section where successive pulses are combined into one wider pulse with increased amplitude levels. But note how inefficient this bandwidth-power exchange is. One must increase the power *exponentially* to determine a corresponding linear decrease in bandwidth. As an example, assume $12S/a^2 \gg 1$. Then if the power is increased eightfold, the bandwidth may be reduced by a factor of 3. Similarly, by a linear *increase* in bandwidth, the required signal power may be *reduced exponentially*.

How does one now determine the level spacing a? This ultimately depends on the noise encountered while attempting to decode the received signals at the receiver. The presence of noise at the receiver makes it difficult to distinguish adjacent amplitude levels until they are spaced "far enough apart." We shall see in later chapters, after discussing the statistical properties of noise, that the chance of an error at the decoder (i.e., mistaking one level for another) is related directly to the rms noise level at that point. Calling this rms noise voltage σ, the spacing a between levels must then be some constant K times σ to keep the average number of errors per unit time within desired limits. (The actual quantitative relationship between $K\sigma$ and the error rate will be developed in a later chapter.) As the spacing increases, the chance of an error obviously decreases.

The capacity expression of Eq. [3-33] may be rewritten to emphasize the role noise plays in the system by letting $a = K\sigma$. Then with $N = \sigma^2$, the mean noise power at the decoder, we have

$$C = W \log_2 \left(1 + \frac{12}{K^2} \frac{S}{N}\right) \qquad [3\text{-}34]^1$$

The quantity S/N, average signal power to average noise power ratio, or SNR as it is commonly abbreviated, will appear quite often in later discussions of the effect of noise on systems. At this point we simply note, as first pointed out in Chap. 1, that noise plays a role in determining system capacity through its effect on limiting the number of amplitude levels that may be used. This is exactly the reason it appears in the capacity expression of Eq. [3-34]. Equation [3-34] thus emphasizes the point made in Chap. 1 that the capacity is limited by bandwidth and noise.

The capacity expression of Eq. [3-34] is interesting for another reason as well. Shannon[2] has shown that there is a maximum possible rate of transmission of binary digits over a channel limited in bandwidth to W Hz, with mean noise power N and mean signal power S. This rate of transmission, or capacity C, is maximum in the sense that if one tries to transmit information at a higher rate, the number of errors made in decoding the signals at the receiver begins to mount rapidly. In fact the chance of an error in a code word of block length m can be shown to approach certainty as $m \to \infty$. On the other hand, if the information rate in bits per second is *less* than C, the chance of an error goes rapidly to zero. This maximum rate of transmission was found by Shannon to be given by

$$C = W \log_2 \left(1 + \frac{S}{N}\right) \qquad [3\text{-}35]$$

Note that this is in precisely the same form as the PCM capacity expression of Eq. [3-34]! Shannon's maximum-capacity expression provides an upper bound on the rate at which one can communicate over a channel of

[1] We are assuming throughout this discussion that the power level remains fixed at all points from transmitter to receiver. In actual practice, of course, it will vary from point to point. For example, it will normally be much less at the receiver antenna, or receiver input, than at the transmitter output. But amplifiers may of course be used and normally *are* used to bring the signal up to the desired power level. The *relative* signal amplitude compared to the level spacing a remains unchanged throughout transmission and reception, however, so that the ratio S/a^2 of Eq. [3-33] remains fixed. One may, therefore, just as well consider *relative* signal power, and assume it unchanging throughout the transmission path. Although S and N are, strictly speaking, mean-squared voltages, we use the common term power to represent both. They would represent the power dissipated in a 1-ohm resistor.

[2] C. E. Shannon, Communication in the Presence of Noise, *Proc. IRE*, vol. 37, pp. 10–21, January, 1949.

bandwidth W, and signal-to-noise ratio SNR.[1] His original derivation indicated that it was theoretically possible to transmit at almost this rate with the error rate made to approach zero as closely as desired, but he did not specify any particular system for so communicating. He showed only that complex encoding and indefinitely large time delays in transmission would be needed. Comparing Eqs. [3-34] and [3-35] we are now in a position to compare the PCM system with this hypothetical one transmitting at a maximum possible rate over a given channel.[2]

In particular note that the PCM system requires $K^2/12$ times the signal power of the optimum Shannon system for the same capacity, bandwidth, and noise power. We shall show in a later chapter that the transmission of binary digits, with an average error rate of 1 digit in 10^5 digits, requires a peak signal-to-noise ratio, A_0/σ, of 9.2. The bipolar pulses of Fig. 3-24 require a peak signal-to-rms-noise ratio $(A_0/2\sigma)$ of 4.6. Increasing the signal above this value reduces the error rate quite rapidly.

Since $A_0 = K\sigma$ for the binary pulses (there are $n = 2$ levels, so that the spacing between the two levels is just $a = K\sigma = A_0$), K must have the value 9.2 for an average error rate of 10^{-5}. For this value of K, $K^2/12 = 7$, and the PCM system requires seven times as much power (8.5 db) as the theoretically optimum one for the same channel capacity. (Note however, that the optimum system can theoretically be made to transmit error free at the cost of large time delays in transmission, while the PCM system, in this example, has the nonzero error rate of 10^{-5}.)

3-8 QUANTIZATION NOISE IN PCM[3]

As has been previously pointed out, noise added during transmission and at the receiver may cause errors in recognition of the code symbols being transmitted. If the signal pulses transmitted are above a specified threshold level, however (about which we shall have more to say in later chapters), the average error rate can be kept quite low. Aside from causing errors occasionally, additive noise will have no other effect on the

[1] A detailed discussion of this equation appears in Chap. 8.

[2] A particular digital system that does have the properties of Shannon's optimum system has recently been described in the literature. It uses M-ary PAM transmission and relies on the availability of a noise-free feedback channel between receiver and transmitter. The system is described in detail in Chap. 8. For the original description see J. P. M. Schalkwijk, A Coding Scheme for Additive Noise Channels with Feedback, Part II, Bandlimited Signals, *IEEE Trans. Information Theory,* vol. IT-12, no. 2, pp. 183–189, April, 1966.

[3] Oliver, Pierce, and Shannon, *op. cit.*; H. F. Mayer, Pulse Code Modulation, in L. Martin (ed.), "Advances in Electronics," vol. III, pp. 227–260, Academic, New York, 1951.

output signal. This is the basic reason for transmitting coded signals. The problem becomes one of recognizing the presence or absence of a pulse (binary code) or of the amplitude level of a pulse ($n > 2$). So long as the noise does not cause a 0 symbol to be mistaken for a 1 symbol, for example, the effect of additive noise can be completely removed. With the signal power great enough, mistakes can be made to occur rather infrequently. So, aside from these occasional errors, additive noise does not appear at the output.

But, in order to code a continuous signal, we must first quantize it into discrete steps of amplitude. Once quantized, the instantaneous values of the continuous signal can never be restored exactly. This, as we have pointed out previously, gives rise to random error variations which are called quantization noise. Instead of additive noise, we are now faced with the problem of artificially introduced quantization noise. This noise can be reduced to any desired degree, however, by choosing the quantum steps or level separations fine enough.

To calculate the rms quantization noise, assume equal spacing between levels. Let the signal at the transmitter be initially quantized into M levels, with a the spacing in volts between adjacent levels. With a maximum plus-minus signal excursion of P volts

$$a = \frac{P}{M} \tag{3-36}$$

(The continuous signal is assumed to have 0 average value, or no dc component.) The quantized amplitudes will be at $\pm a/2$, $\pm 3a/2$, . . . , $\pm(M - 1)(a/2)$, and the quantized samples will cover a range

$$A = (M - 1)a \qquad \text{volts}$$

(See Fig. 3-25. Do not confuse this with the previous discussion of the *coded* samples, leading to Eq. [3-32], and illustrated by Fig. 3-23.)

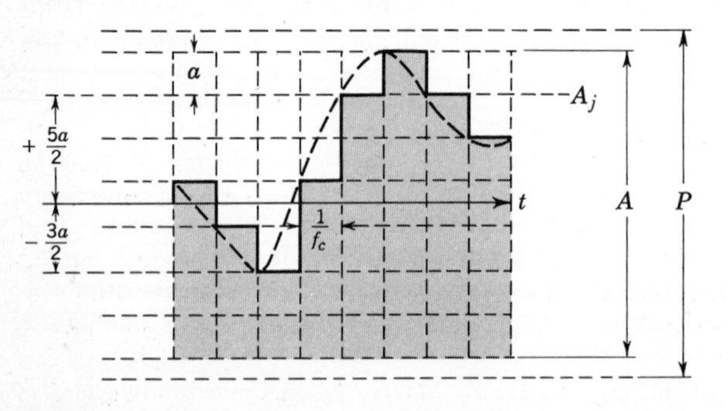

FIG. 3-25 Quantized approximation to a signal: eight levels.

The quantization process introduces an irreducible error, since a sample appearing at the receiver output at quantized voltage A_j volts could have been due to any signal voltage in the range $A_j - a/2$ to $A_j + a/2$ volts. This region of uncertainty is shown in Fig. 3-26. So far as the

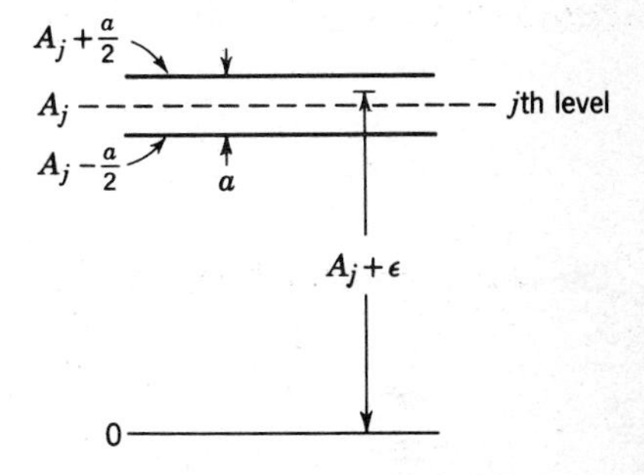

FIG. 3-26 Region of uncertainty at system output.

ultimate recipient of the message is concerned, this region of uncertainty could just as well have been due to additive noise masking the actual signal level. The one difference is that additive noise, as we shall see in Chap. 5, can theoretically take on all possible voltage values. The quantization noise, on the other hand, is limited to $\pm a/2$ volts. The distinction between these two will become clearer in Chap. 5.

We can calculate a mean-squared-error voltage due to quantization and treat this just like the mean additive noise power of the last section. To do this, assume that over a long period of time all voltage values in the region of uncertainty eventually appear the same number of times. The instantaneous voltage of the signal will be $A_j + \epsilon$, with $-a/2 \leq \epsilon \leq a/2$. ϵ represents the error voltage between the instantaneous (actual) signal and its quantized equivalent. Under our assumption all values of ϵ are equally likely. The mean-squared value of ϵ will then be

$$\overline{\epsilon^2} = \frac{1}{a}\int_{-a/2}^{a/2} \epsilon^2 \, d\epsilon = \frac{a^2}{12} \qquad [3\text{-}37]$$

The average value of the error is zero with the assumption made. The rms error is then $a/\sqrt{12} = a/(2\sqrt{3})$ volts, and this represents the rms "noise" at the system output.

Just as a signal-to-noise ratio appeared useful in considering the effect of additive noise in the last section, it is desirable to introduce a signal-to-noise (SNR) expression for the quantization noise. We shall define two different SNR ratios, one in terms of the peak signal excursion, P volts, the other in terms of the mean signal power, $S_o = (M^2 - 1)a^2/12$ (by comparison with Eq. [3-32]).

Peak signal Since $P = aM$ is the peak-to-peak signal excursion, the ratio of peak signal voltage to rms noise will be

$$\frac{S_{ov}}{N_{ov}} = \frac{P}{a/(2\sqrt{3})} = 2\sqrt{3}\,M \qquad [3\text{-}38]$$

The corresponding power ratio is

$$\frac{S_o}{N_o} = 12M^2 \qquad [3\text{-}39]$$

or, in decibels,

$$\left(\frac{S_o}{N_o}\right)_{db} = 10.8 + 20 \log_{10} M \qquad [3\text{-}40]$$

The power ratio thus goes up as the square of the number of levels. The SNR decibel improvement with M is given in the accompanying table. Also indicated is the relative bandwidth as obtained from the discussion following.

QUANTIZATION SNR IMPROVEMENT
WITH NUMBER OF LEVELS

S_o/N_o, db	M	Relative bandwidth
17	2	1
23	4	2
29	8	3
35	16	4
41	32	5
47	64	6
53	128	7

Since M, the number of levels used, determines the number of pulses into which the quantized signal is encoded before transmission, increasing M increases the number of code pulses and hence the bandwidth. We can thus relate SNR to bandwidth. This is easily done by noting that $M = n^m$, with m the number of pulses in the code group and n the number of code levels. With this relation, Eqs. [3-39] and [3-40] become, respectively,

$$\frac{S_o}{N_o} = 12n^{2m} \qquad [3\text{-}41]$$

and

$$\left(\frac{S_o}{N_o}\right)_{db} = 10.8 + 20m \log_{10} n \qquad [3\text{-}42]$$

In particular, for a binary code ($n = 2$),

$$\left(\frac{S_o}{N_o}\right)_{db} = 10.8 + 6m \qquad [3\text{-}43]$$

Since the bandwidth is proportional to m, the number of pulses in the code group, the output SNR increases exponentially with bandwidth. The decibel SNR increases linearly with bandwidth (Eq. [3-43]). These results are of course similar to those noted in the previous section with additive noise.

For a 128-level system $S_o/N_o = 53$ db, and seven-pulse binary-code groups are transmitted, requiring a sevenfold bandwidth increase.

Mean signal power Essentially similar results are obtained upon defining a mean power SNR. With a quantized level spacing of a volts and signal swings of $\pm P/2$ volts the mean signal power is

$$S_o = \tfrac{1}{12}(M^2 - 1)a^2$$

assuming all signal levels equally likely. (This result is obtained in the same way as Eq. [3-32]. Recall that Eq. [3-32] gave the mean power in the transmitted code group. The present calculation refers to the original quantized signal both at the transmitter, before being encoded, and at the receiver, after decoding.)

Since $N_o = a^2/12$, the mean power output SNR is

$$\frac{S_o}{N_o} = M^2 - 1 \qquad [3\text{-}44]$$

For $M \gg 1$ this differs only by a constant from the peak S_o/N_o relation given by Eq. [3-39]. For a system with 128 levels the quantization SNR is 42 db. A binary-code group requires seven pulses ($2^7 = 128$), or seven times the bandwidth of the original quantized signal.

3-9 SIGNAL SHAPING, INTERSYMBOL INTERFERENCE, AND BANDWIDTH[1]

In discussing the transmission bandwidth of digital systems thus far, we have indicated that the bandwidth is roughly given by the reciprocal of the time slot or interval in which the pulse is constrained to lie. Thus if 10 signals are sampled and time multiplexed every 125 μsec, each signal sample is constrained to lie within its 12.5 μsec time slot (Fig. 3-22*c*). The minimum bandwidth required to transmit the time-multiplexed signal train is then roughly 1/12.5 sec or 80 kHz. (Wider bandwidths may of course be used if it is desired to send pulses narrower than 12.5 μsec.) If eight-level quantization is now used, and binary signals sent, the bandwidth increases to 240 kHz.

The transmission bandwidths necessary may be made more precise by introducing the concept of *intersymbol interference*. By choosing signal waveshapes to minimize or eliminate this phenomenon, we shall find the bandwidth necessary for their transmission. For practical waveshapes the bandwidth necessary will be precisely that given by our usual rule of thumb; that is, it is *the reciprocal of the time interval.*

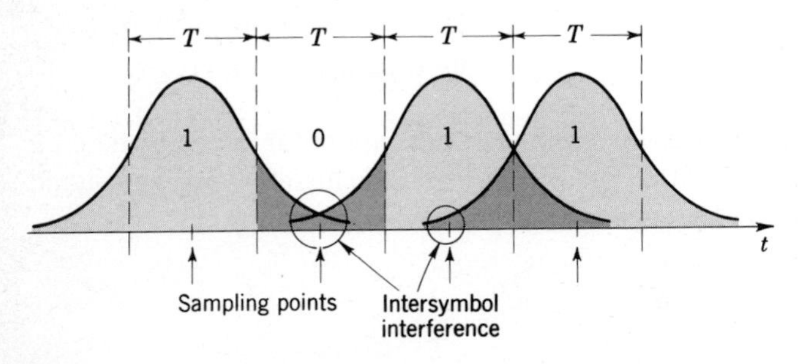

FIG. 3-27 Intersymbol interference in digital transmission.

Consider the sequence of pulses shown in Fig. 3-27. Although these are shown as binary pulses, they could equally well be pulses of identical shape, but of arbitrary height (PAM or quantized PAM). They are shown recurring at T-sec intervals. T is the sampling interval in the PAM or quantized PAM case, the binary interval in the case of binary encoded symbols. System filtering causes these pulses to spread out as

[1] W. R. Bennett and J. R. Davey, *op. cit.*, chap. 5; R. W. Lucky, J. Salz, and E. J. Weldon, "Principles of Data Communication," McGraw-Hill, New York, 1968. This latter book contains a thorough discussion of adaptive equalizers that automatically shape signals to minimize intersymbol interference.

they traverse the system, and they overlap into adjacent time slots as shown. At the receiver the original pulse message may be derived by sampling at the center of each time slot as shown, and then basing a decision on the amplitude of the signal measured at that point.

The signal overlap into adjacent time slots may, if too strong, result in an erroneous decision. Thus, as an example, in the case of Fig. 3-27 the 0 transmitted may appear as a 1 if the tails of the adjacent pulses add up to too high a value. (In practice there may be contributions due to the tails of several adjacent pulses rather than the one pair shown in Fig. 3-27.) This phenomenon of pulse overlap and the resultant difficulty with receiver decisions is termed *intersymbol interference.*

Note that this interference may be minimized by purposely widening the transmission bandwidth as much as desired. This is unnecessarily wasteful of bandwidth, however, and if carried too far may introduce too much noise into the system (see Chap. 1). Instead we seek a way of *purposely* designing the signal waveshapes and hence transmission filters used to minimize or eliminate this interference with as small a transmission bandwidth as possible. One obvious signal waveshape to use is one that is maximum at the desired sampling point, yet goes through zero at all adjacent sampling points, multiples of T sec away. This ideally provides zero intersymbol interference. With such a waveshape, chosen at the receiver, it should then be possible to design the overall system, back to the original sampling point at the transmitter, to provide this desired waveshape. (Recall that the original sampled pulses are essentially impulses if the time τ, during which the analog signal is sampled, is small compared to the sampling interval. The pulse broadening and shaping is then due to innate system filtering as well as filters purposely put in to achieve the final desired shape.)

One signal waveshape producing zero intersymbol interference is just the $(\sin x)/x$ pulse introduced in Chap. 2 as the impulse response of an ideal low-pass filter. Specifically, if the filter has a flat amplitude spectrum to B Hz, and is zero elsewhere, the impulse response is just $(\sin 2\pi Bt)/2\pi Bt$. This pulse is shown sketched in Fig. 3-28. Note that the pulse goes through zero at equally spaced intervals, multiples of $1/2B$ sec away from the peak at the origin. If $1/2B$ is chosen as the sample interval T, it is apparent that pulses of the same shape and *arbitrary amplitude* that are spaced $T = 1/2B$ sec apart will not interfere. This is shown in Fig. 3-29.

There are practical difficulties with this particular waveshape, however:

1. It implies that the overall characteristic between transmitter sampling and receiver decision point is that of an ideal low-pass filter. As noted

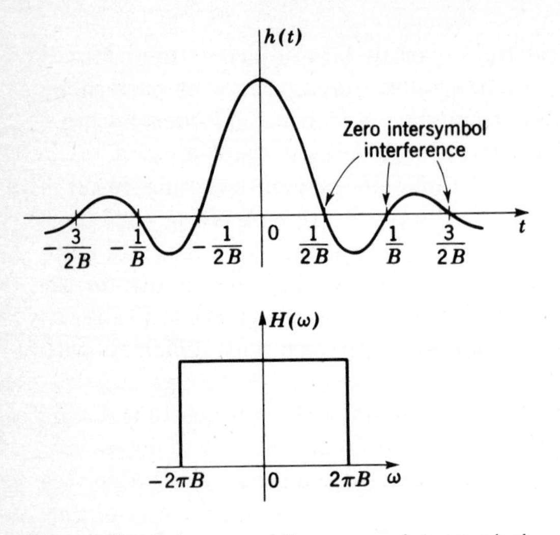

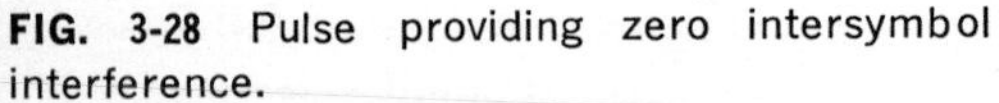
FIG. 3-28 Pulse providing zero intersymbol interference.

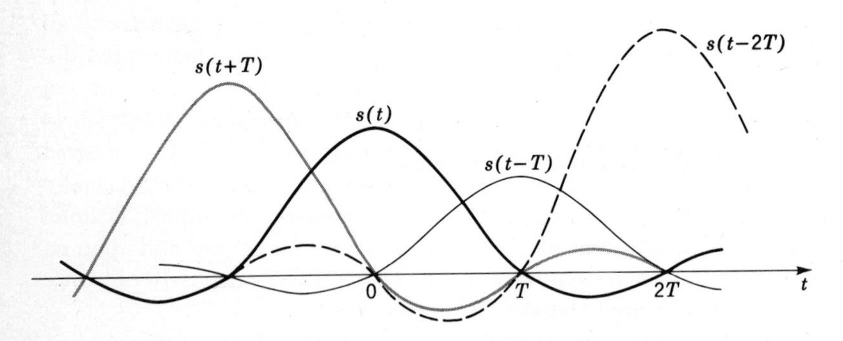

FIG. 3-29 Sequence of pulses: zero intersymbol interference.

in the previous chapter this is physically unrealizable, and very difficult to approximate in practice because of the sharp cutoff in its amplitude spectrum at B Hz.

2. This particular pulse, if attainable, would require extremely precise synchronization. If the timing at the receiver varies somewhat from exact synchronization, the zero intersymbol interference condition disappears. In fact, under certain signal sequences, the tails of all adjacent pulses may add up as a divergent series, causing possible errors! Since some timing jitter will inevitably be present even with the most sophisticated synchronization systems, this pulse shape is obviously not the one to use.

It is, however, possible to derive from this waveshape related waveshapes with zero intersymbol interference that do overcome the two difficulties mentioned. They are much simpler to attain in practice, and the effects of timing jitter may be minimized. The particular class of such waveshapes we shall discuss is one of several first described by Nyquist.[1]

To attain this particular class we start with the ideal low-pass filter of Fig. 3-28, but modify its characteristics at the cutoff frequency to attain a more gradual frequency cutoff, which is hence more readily realized. In particular, if the new frequency characteristic is designed to have odd symmetry about the low-pass cutoff point, it is readily shown, following Nyquist, that the resultant impulse response retains the derived property of having zeroes at uniformly spaced time intervals. An example of such a spectrum often approximated in practice is the so-called raised cosine amplitude spectrum. This particular spectrum and the original ideal low-pass spectrum are shown in Fig. 3-30. To avoid con-

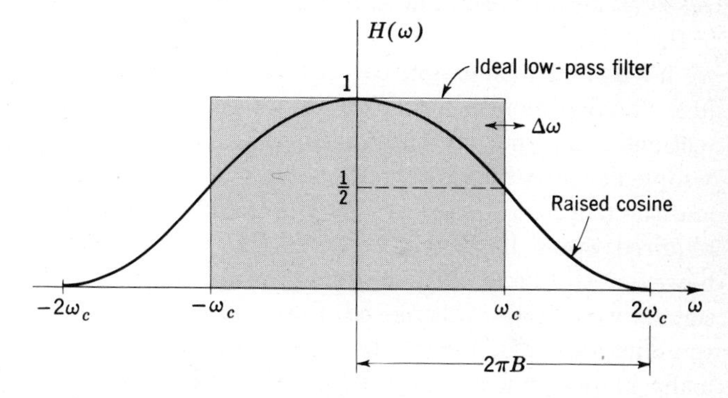

FIG. 3-30 Raised cosine spectrum.

fusion with the transmission bandwidth B, we henceforth label the ideal low-pass cutoff frequency ω_c. The raised cosine spectrum is then given by

$$\begin{aligned} H(\omega) &= \frac{1}{2}\left(1 + \cos\frac{\pi\omega}{2\omega_c}\right) \qquad |\omega| \leq 2\omega_c \\ &= 0 \qquad \text{elsewhere} \end{aligned} \tag{3-45}$$

The bandwidth in this case is just $2\omega_c = 2\pi B$. If we now measure frequency with respect to ω_c by letting $\omega = \omega_c + \Delta\omega$ (see Fig. 3-30), we have

$$H(\omega) = \frac{1}{2}\left[1 + \cos\frac{\pi}{2}\left(1 + \frac{\Delta\omega}{\omega_c}\right)\right] = \frac{1}{2}\left(1 - \sin\frac{\pi}{2}\frac{\Delta\omega}{\omega_c}\right) \tag{3-46}$$

[1] H. Nyquist, Certain Topics in Telegraph Transmission Theory, *Trans. AIEE*, vol. 47, pp. 617–644, April, 1928. See Bennett and Davey, *op. cit.*, for further discussion.

Since the sine term has odd symmetry [sin $(-x) = -\sin x$], it is apparent that the raised cosine spectrum displays the odd symmetry about ω_c noted above. This is also apparent from Fig. 3-30.

The impulse response of a filter with this frequency characteristic is readily shown to be given by

$$h(t) = \frac{\omega_c}{\pi} \frac{\sin \omega_c t}{\omega_c t} \frac{\cos \omega_c t}{1 - (2\omega_c t/\pi)^2} \tag{3-47}$$

It has the $(\sin x)/x$ term of the ideal filter multiplied by an additional factor that decreases with increasing time. The $(\sin x)/x$ term ensures zero crossings of $h(t)$ at precisely the same equally spaced time intervals as the ideal low-pass filter. The additional factor multiplying the $(\sin x)/x$ term reduces the tails of the pulse considerably below that of the $(\sin x)/x$ term, however, so that such pulses when used in digital transmission are relatively insensitive to timing jitter.

Letting the sampling interval $T = 1/2f_c = \pi/\omega_c$, so that the zeroes of the pulse occur at T-sec intervals, as in the previous ideal low-pass case, we have the transmission bandwidth given by $B = 2f_c = 1/T$. This is just the bandwidth criterion adopted rather arbitrarily in previous sections. As an example, consider again an analog signal of 3.2 kHz sampled at an 8-kHz rate. If this signal only were transmitted by PAM, and the $(\sin x)/x$ ideal waveshape were used, the transmission bandwidth required would be $B = 1/2T = 4$ kHz. Using a raised cosine spectrum, however, the bandwidth required would be $B = 1/T = 8$ kHz. If 10 signals were time multiplexed, the bandwidth would increase by a corresponding factor of 10. Thus although the $(\sin x)/x$ signal shape theoretically allows transmission at very nearly the original analog signal bandwidth (it would be the original bandwidth if the Nyquist sampling rate were used), the use of more realistic waveshapes results essentially in a doubling of the required bandwidth.

The fact that the raised cosine spectrum is just one example of a class of spectra with odd symmetry about ω_c providing zero crossings at equally spaced sampling intervals is demonstrated as follows:

Assume a low-pass filter to have the characteristic

$$\begin{aligned} H(\omega) &= 1 + H_1(\omega) && |\omega| < \omega_c \\ &= H_1(\omega) && \omega_c < |\omega| < 2\omega_c \\ &= 0 && \text{elsewhere} \end{aligned} \tag{3-48}$$

(If H_1 is zero, we have just the ideal low-pass filter.) As an example, let

$$\begin{aligned} H_1(\omega) &= \frac{1}{2}\left(\cos \frac{\pi}{2}\frac{\omega}{\omega_c} - 1\right) && |\omega| < \omega_c \\ &= \frac{1}{2}\left(\cos \frac{\pi}{2}\frac{\omega}{\omega_c} + 1\right) && \omega_c < |\omega| < 2\omega_c \end{aligned} \tag{3-49}$$

This is of course just the raised cosine case.

Assume now for simplicity's sake that the overall filter has zero phase shift. (A linear phase term of course results in a corresponding pulse time delay.) Assume further that $H_1(\omega)$ has *odd symmetry* about ω_c. Then

$$H_1(\omega_c + \Delta\omega) = -H_1(\omega_c - \Delta\omega) \tag{3-50}$$

The raised cosine has this property. Another arbitrarily chosen example is shown in Fig. 3-31*a*, with the overall characteristic sketched in part (*b*) of the figure.

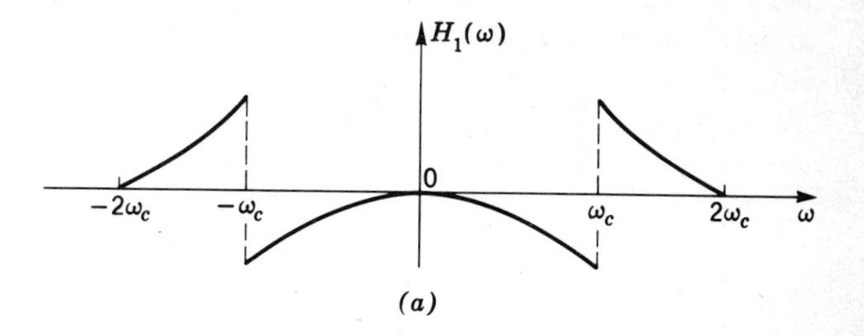

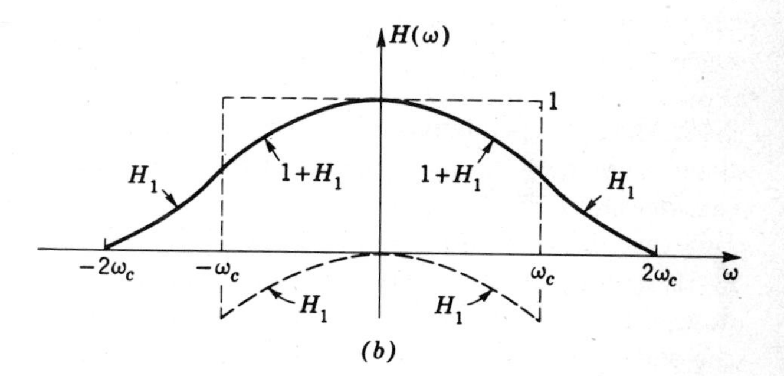

FIG. 3-31 Nyquist filter. (a) Odd symmetry about ω_c; (b) overall filter.

Taking the Fourier transform of Eq. [3-48] to find the impulse response $h(t)$, we have, by superposition,

$$h(t) = \frac{\omega_c}{\pi}\frac{\sin \omega_c t}{\omega_c t} + h_1(t) \tag{3-51}$$

where
$$h_1(t) = \frac{1}{2\pi}\int_{-\infty}^{\infty} H_1(\omega)e^{j\omega t}\,d\omega \tag{3-52}$$

But $H_1(\omega)$, having zero (or linear) phase shift, must be even in ω (see Fig. 3-31*a*). (Recall that all *amplitude* spectra are symmetrical about $\omega = 0$.)

Then

$$h_1(t) = \frac{1}{\pi}\int_0^\infty H_1(\omega)\cos \omega t\, d\omega$$

$$= \frac{1}{\pi}\int_0^{\omega_c} H_1(\omega)\cos \omega t\, d\omega + \frac{1}{\pi}\int_{\omega_c}^{2\omega_c} H_1(\omega)\cos \omega t\, d\omega \qquad [3\text{-}53]$$

using the fact that $H_1 = 0$, $\omega > 2\omega_c$.

We now make use of the property of odd symmetry about ω_c by letting $\omega = \omega_c - x$ in the first integral of Eq. [3-53], and $\omega = \omega_c + x$ in the second integral. The new dummy variable x ranges between 0 and ω_c in both integrals, and the two may be combined into the following one integral, after using the odd symmetry property:

$$h_1(t) = \frac{1}{\pi}\int_0^{\omega_c} H_1(\omega_c - x)\,[\cos(\omega_c - x)t - \cos(\omega_c + x)t]\,dx \qquad [3\text{-}53a]$$

Using now the trigonometric identity

$$\cos(a - b) - \cos(a + b) = 2\sin a \sin b$$

we finally obtain the following interesting result

$$h_1(t) = \frac{2}{\pi}\sin\omega_c t\int_0^{\omega_c} H_1(\omega_c - x)\sin xt\,dx \qquad [3\text{-}53b]$$

Note that independently of the precise value of the integral (this will depend on the particular characteristic chosen for H_1), the $\sin \omega_c t$ preceding *guarantees* that $h_1(t)$ will be *zero* at intervals spaced $T = \pi/\omega_c$ sec apart. But this is just the original sampling interval T, so that $h(t)$ from Eq. [3-51] does go through zero at all intervals multiples of T away from the desired sampling point $t = 0$. This property of $h(t)$ is of course due to the odd symmetric choice $H_1(\omega)$.

We have thus shown that one may use shaped pulses to attain zero intersymbol interference. In practice, of course, one never quite attains these desired shapes, and timing jitter introduces some intersymbol interference. But an appropriate choice of spectrum (with the raised cosine as one prominent example) may be used to minimize this effect. The transmission bandwidth required is then just $B = 1/T$, as noted earlier.

The problem of minimizing intersymbol interference with digital transmission is a major one in the telephone plant where the effects of additive noise are generally minimal. In digital systems designed for space communications, however, noise considerations play a highly significant role. Receiver design and signal shaping must take the effects of noise into account. We shall see, in considering this problem later, that the minimization of errors due to additive noise leads to so-called *matched-filter* receivers. In an environment involving the minimization of both intersymbol interference and errors due to noise, some compromise between the two effects is generally necessary. Yet our discussion later will show that the receiver and signal shaping problem is not too critical; in the matched-filter case, for example, we shall again find that transmission bandwidths equal to the reciprocal of the interval T are close to optimum, so that one may simultaneously combat both intersymbol interference and noise without too much difficulty.[1]

3-10 EXAMPLES OF PAM AND PCM SYSTEMS

In this section we present some examples of specific PAM and PCM systems. The examples have been chosen to further focus attention on the elements of digital communications discussed earlier, as well as to present some typical engineering solutions to problems raised earlier. Thus, in the context of summarizing the characteristics of these systems we will indicate some of the methods actually used to perform the sampling and time-multiplexing operations discussed earlier. Some methods of providing time synchronization, as well as the A/D conversion necessary in PCM systems, will be summarized. The purpose here is to enhance the understanding of these various system operations and functions by presenting them in the context of typical systems.

As noted earlier digital signaling finds applications in a variety of fields: teletype and telephonic communications, industrial telemetry, space communications and telemetry, biomedical telemetry, etc. We shall confine ourselves here to discussing one PAM system typical of those used in radio telemetry, one PCM system developed for space telemetry, and one PCM system in use for short-haul telephone communication. Many other examples may be found in the references cited, as well as in the current literature.[2]

[1] The minimization of errors due to both intersymbol interference and noise is considered in Lucky, Salz, and Weldon, *op. cit.*, chap. 5. They show that if the channel may be assumed to introduce no distortion other than noise, the effect of both intersymbol interference and noise may be simultaneously minimized by splitting the Nyquist filter $H(\omega)$ equally into two parts, $H^{1/2}(\omega)$ at the transmitter, $H^{1/2}(\omega)$ at the receiver. As will be seen in Chap. 6, the receiver filter is then exactly the *matched filter* that minimizes errors due to noise.

[2] See for example, E. L. Gruenberg (ed.), "Handbook of Telemetry and Remote Control," McGraw-Hill, New York, 1967, and A. V. Balakrishnan (ed.), "Space Communications," McGraw-Hill, New York, 1963. The *IEEE Transactions on Communication Technology*, the *IEEE Transactions on Aerospace and Electronic Systems*, the annual *Proceedings of the IEEE National Telemetering Conference*, and the *Proceedings of the IEEE International Communications Conference* generally contain examples of current system designs and applications.

A RADIO TELEMETRY PAM SYSTEM[1]

The PAM system to be discussed is typical of those used for radio telemetry purposes. It is of interest because it provides for the multiplexing of many data channels of greatly varying data rates or bandwidths. This is done by performing the multiplexing operations in two steps: low data rate channels are first time multiplexed to form composite data channels of much wider bandwidths; the composite channels are then in turn time multiplexed with wider bandwidth channels to form the main multiplexed signal.

Specifically, in a typical application 318 different data channels are to be sampled, time multiplexed, and transmitted via radio. The data rates involved range from a bandwidth of 1 Hz to one of 2 kHz. The main multiplexer is designed to handle 16 channels, with 2,500 samples per second for each. The low data rate signals must thus be combined in groups to provide the 2,500 samples per second required for any one channel. The various data channels, and their bandwidths, sampling rates, and group assignments, where necessary, are indicated in the following table:[2]

Group	No. of data channels	Bandwidth per channel	Sampling rate (samples/sec)	Required accuracy, %	Main multiplexer position
1	3	2 kHz	5,000	10	2 and 10, 4 and 12, 5 and 13
2	2	1 kHz	2,500	5	3, 6
3	5	100 Hz	312.5	2	7
4	28	25 Hz	78	2	8
5	55	5 Hz	39	1	9
6	115	5 Hz	19.5	2	14
7	110	1 Hz	19.5	2	15

The two 1-kHz channels, sampled at a 2,500 samples per second rate, obviously can be tied directly to the main multiplexer. The channels in groups 3 to 7 must be combined, however. For ease in timing this combining is done by sampling at binary submultiples of the final 2,500 samples per second rate. The required sampling rates can then be obtained by successively dividing down by factors of two from a master clock. As

[1] "Handbook of Telemetry and Remote Control," *op. cit.*, pp. 9-11 to 9-18.

[2] *Ibid.*, p. 9-11.

an example, the 312.5 samples per second rate is obtained by dividing down by 8 from 2,500, and the 19.5 samples per second rate by dividing down by 128. Figure 3-32 shows the preliminary multiplexing at the

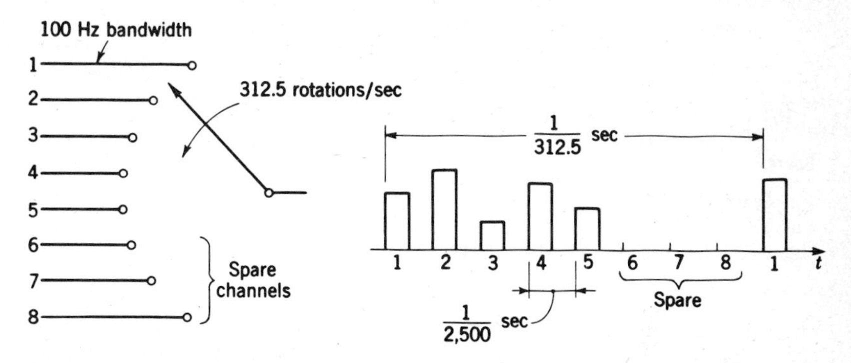

FIG. 3-32 Preliminary multiplexing, group 3, PAM system

312.5 sample rate. Note that five data channels only are combined, leaving three time slots available for synchronization, calibration, or inclusion of additional data channels later, if so desired.

It is apparent that such spare capacity exists in each of the low data rate groupings. Thus, a sampling rate of 78 samples per second makes 32 time slots available. In this application there are only 28 channels (group 4) requiring this sampling rate. The 55 data channels of 5-Hz bandwidth (group 5) are grouped together and shown sampled at almost four times the Nyquist rate to meet the higher accuracy requirements in that case. Here nine spare time slots are available. The 110 channels of 1-Hz bandwidth are obviously highly oversampled, but system simplicity results.

The three high-bandwidth channels indicated as group 1 require 5,000 samples per second each. Two time slots, spaced $1/5{,}000$ sec apart, must be thus set aside, of the 16 available in the main multiplexer, for each of the three channels. These then use six time slots. Adding the two used by the two 1-kHz channels, and the five time slots required by the premultiplexed groups 3 to 7, we have thirteen used in all. Two additional time slots in the final multiplexed signal are used for synchronization purposes, and one is left available as a spare. The main multiplexer assignments are indicated in the table.

The final multiplexed signal, following these assigned time slots, is shown in Fig. 3-33.[1] The slots occupied by the three wide-bandwidth

[1] *Ibid.*, p. 9-13, fig. 12*a*.

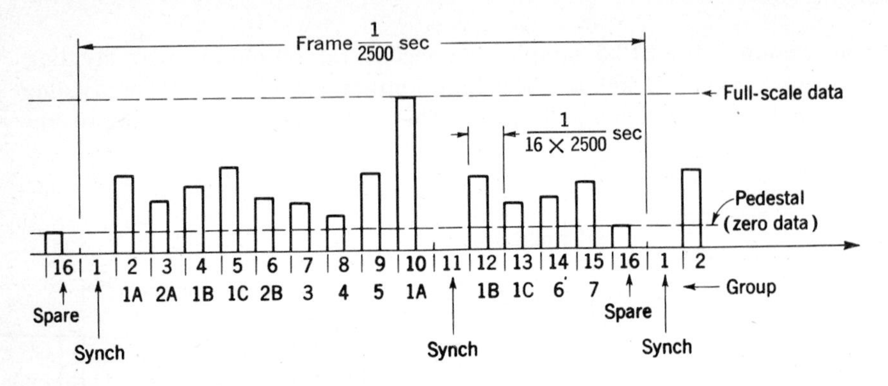

FIG. 3-33 Composite signal format.

channels in group 1 are indicated by the labels 1A, 1B, 1C, respectively, while the two 1-kHz channels are indicated by 2A and 2B. The entire sequence of 16 time slots shown is called a *frame*. In the format of Fig. 3-33 the composite data samples occupy 50 percent of each time slot. To provide synchronization all data samples are placed on a pedestal of about 20 percent of the full scale data level. The absence of pulses in slots 1 and 11 then provides the necessary synchronization. Zero data corresponds to the pedestal height, as shown by the spare data slot 16. (In other systems synchronization may be provided by a specified digital code word inserted in the appropriate time slot.[1])

The block diagram of the entire PAM system (transmitter only) is shown in Fig. 3-34.[2] The programmer shown provides the clock pulses for all multiplexers, starting with a clock rate of 40 kHz, and counting down by 2's, as indicated earlier.

The receiver for this PAM radio telemetry system performs the necessary functions of demultiplexing the data samples, low-pass filtering them, and sending them to the appropriate receiving channel. We shall not discuss the demultiplexer here, but refer the reader to the reference for a detailed discussion.[3]

COMMUTATION

The sampling and time-multiplexing operations represented conceptually here and in previous sections by a rotating mechanical switch represent the heart of the PAM system. Many electronic schemes have been devised to perform these operations. The process of sequentially sam-

[1] *Ibid.*, p. 9-13, fig. 12; pp. 5-34 and 5-35, fig. 5*a*, *b*, *c*, *d*.

[2] *Ibid.*, p. 9-12, fig. 11

[3] *Ibid.*, pp. 9-13 to 9-18.

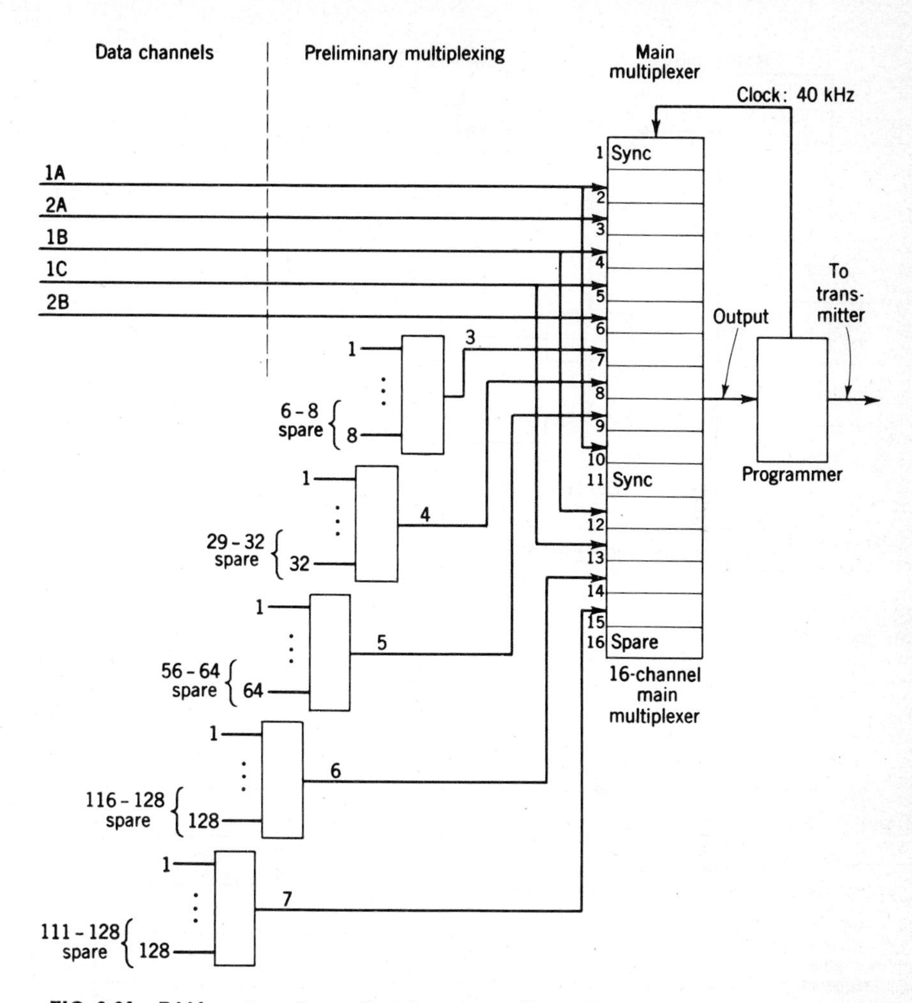

FIG. 3-34 PAM system for radio telemetry. [From E. L. Gruenberg (ed.), "Handbook of Telemetry and Remote Control," chap. 9, fig. 11, simplified, McGraw-Hill, New York, 1967.]

pling a group of input data channels is frequently referred to as a *commutation* process, and both mechanical and electronic commutators have been devised to perform this function.[1] The commutator is commonly divided into two parts: channel gates that sequentially open to allow one channel at a time through, performing the sequential sampling operation, and a sequential gate control generator, operated by a clock, that actually operates the gates. One common example is shown in Fig. 3-35*a*.[2] The

[1] *Ibid.*, pp. 4–53 to 4–92.

[2] *Ibid.*, fig. 78*a*, p. 4–79.

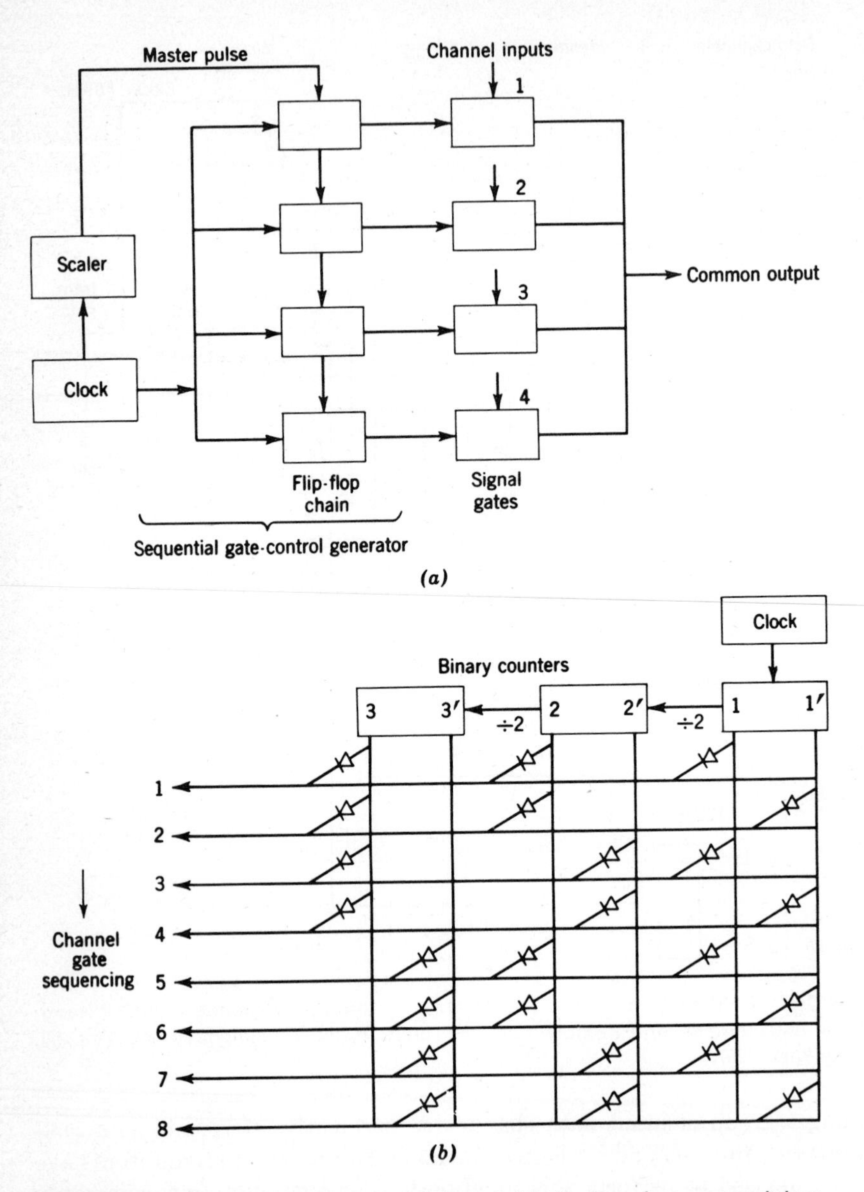

FIG. 3-35 Typical electronic commutators. (a) Broken-ring commutator: four data channels; (b) counter-matrix gate-control generator: eight channels. [From E. L. Gruenberg, *op. cit.*, chap. 4, figs. 78(a) and 79, simplified.]

sequential gate control generator here consists of a chain of flip-flops. The master pulse turns on the first element of the chain. The next channel pulse turns off the first flip-flop, and this in turn turns on the second element. This stepping continues to the end of the chain, when the master pulse again turns on the first element.

The counter-matrix arrangement of Fig. 3-35*b* provides another common form of commutation. The outputs shown open the corresponding gates sequentially. To perform the necessary sequencing, counter waveforms are combined appropriately. As an example consider the binary waveforms of Fig. 3-36. These correspond respectively to the

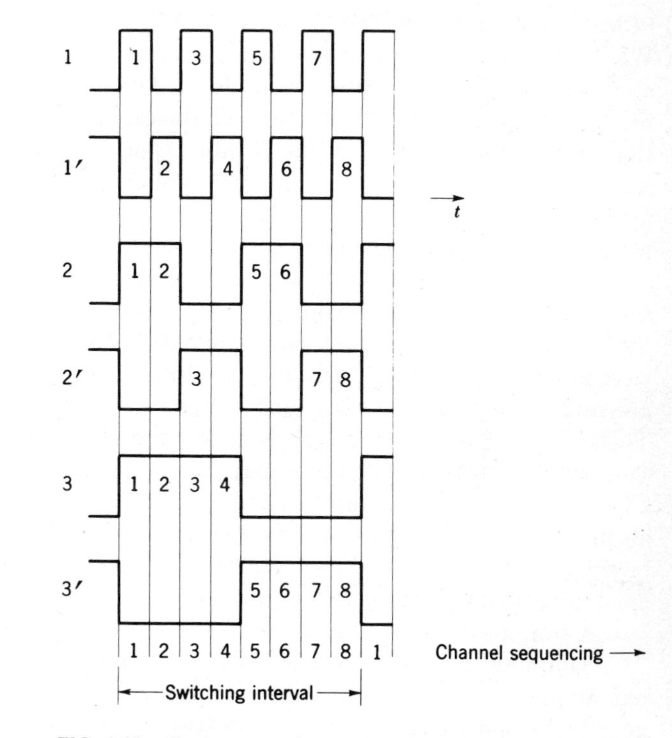

FIG. 3-36 Timing waveforms for channel sequencing.

appropriate binary counters of Fig. 3-35*b*. Thus the basic binary sequence 1 and its phase-shifted version 1′ correspond to 1 and 1′, respectively, of Fig. 3-35*b*. Dividing down successively by 2, as indicated in Fig. 3-35*b*, we get the waveforms shown in Fig. 3-36. The three pairs of waveforms shown provide the $2^3 = 8$ possible connections needed for the eight channels indicated. Specifically, note from Fig. 3-36 that outputs

1, 2, and 3 are all simultaneously positive during the first ⅛ of the switching cycle of Fig. 3-36. Connecting these outputs to channel 1 as shown in Fig. 3-35*b*, channel 1 is gated on during the first ⅛ of the cycle only, being turned off during the remainder of the cycle. It is apparent from Fig. 3-36 that connecting the other seven channels appropriately, as shown in Fig. 3-35*b*, three simultaneously positive pulses are obtained in time sequence, providing the desired sequential gating action.

Many other types of electronic commutators are available and have been used in practice.[1]

PCM TELEMETRY FOR THE NIMBUS SATELLITE

We now turn to an example of a PCM system developed for space applications. The particular example we have chosen to describe very briefly is the telemetering system used aboard the NASA second-generation meteorological satellite, the Nimbus spacecraft. This spacecraft is designed to provide appropriate meteorological data (cloud measurements, distribution of rainfall, the earth's outgoing thermal radiation, heat balance, etc.) of the entire earth at least once a day.[2]

Two independent PCM telemetering systems are used. One, the so-called A system, records continuously on an endless loop tape recorder; the other, the B system, may be commanded at any time. The systems were produced for NASA by Radiation, Inc. In the A system 544 inputs are multiplexed for recording and ultimate transmission on command. Thirty-two of these input channels, in two groups of 16 channels each, are sampled once per second; the remaining 512 channels, in two groups of 256 channels each, are sampled once every 16 sec. Premultiplexing, just as in the PAM systems described in the previous section, must thus be carried out on the two 256 channel groups, combining them first before combining with the 16 channel groups.

A simplified block diagram of the system is shown in Fig. 3-37.[3]

Sampling and multiplexing are first carried out by means of the timer and 16-position shift registers shown. A pulse of the 64 pulse per second generator shown triggers the first position of the first (uppermost) shift register shown, gating on the first of the 16 upper channels. The second pulse, $\frac{1}{64}$ sec later, triggers the first position of the second shift register, gating on the first of the 16 channels in the next group. One of the 256

[1] *Ibid.*, pp. 4-53 to 4-92. Included is an extensive bibliography.

[2] R. Stampfl, in A. V. Balakrishnan (ed.), "Space Communications," chap. 18, McGraw-Hill, New York, 1963. The telemetry systems are described on pp. 370–376, and pp. 404–406.

[3] *Ibid.*, fig. 18-4, pp. 372, 373.

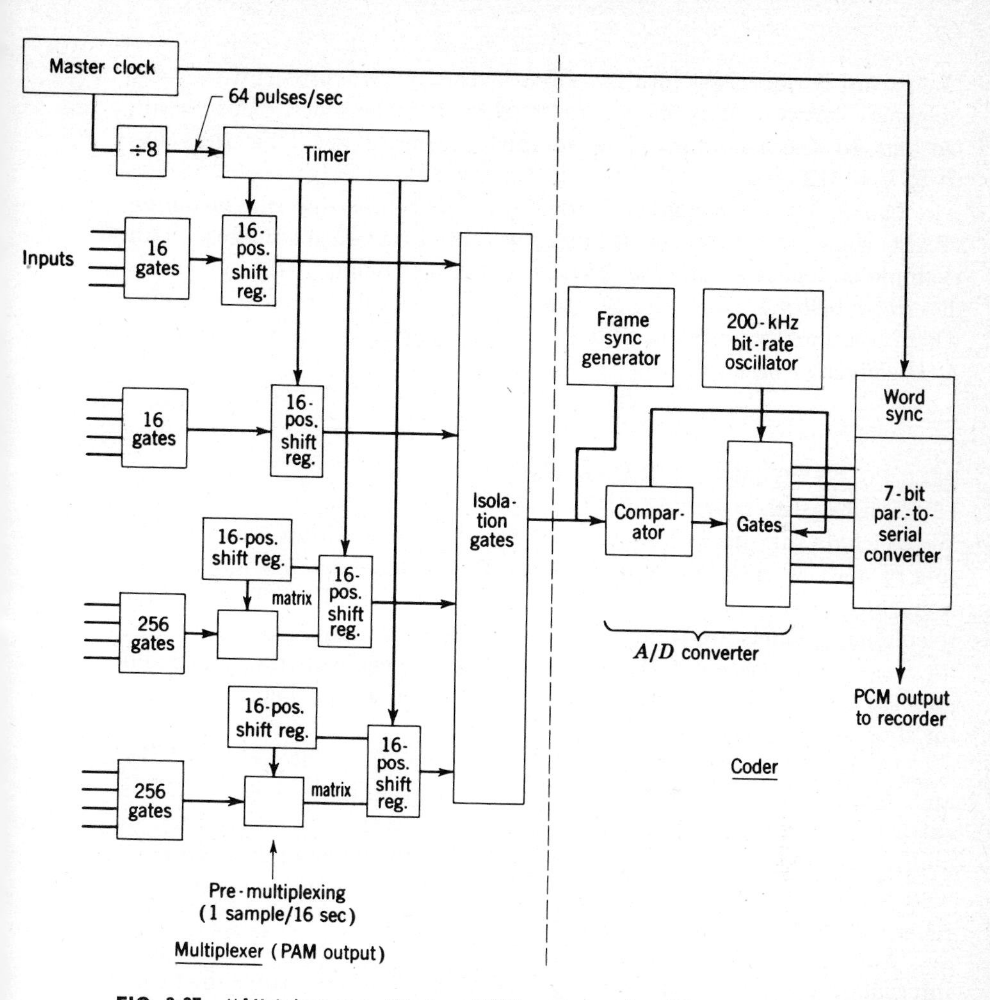

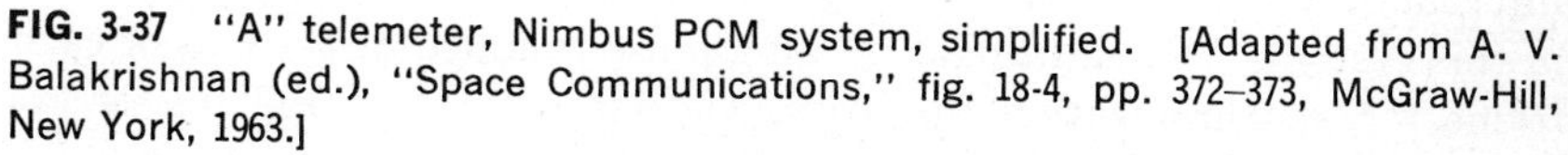

FIG. 3-37 "A" telemeter, Nimbus PCM system, simplified. [Adapted from A. V. Balakrishnan (ed.), "Space Communications," fig. 18-4, pp. 372–373, McGraw-Hill, New York, 1963.]

multiplexed group is then gated on, etc. The fifth pulse returns to the first shift register, opening its second position. The cycle thus continues for the full 64 pulses, returning to the first of the 16 upper channels 1 sec later, as desired.

Premultiplexing of the 256 channel groups is done by using an additional 16-position shift register for each group to form a matrix with the shift registers mentioned above.

Each of the 64 outputs of the multiplexer constitutes a word $\frac{1}{64}$ sec long. The 64 consecutive words in turn make up a frame 1 sec long.

Each word is quantized into $2^7 = 128$ levels and converted into a 7-bit sequence. A word sync bit (a zero) is added and the coded PCM stream fed out to the recorder. The complete frame 1 sec long thus has $64 \times 8 = 512$ bits.

The A/D conversion is carried out by comparing the incoming (PAM) signal to a sequence of binary weighted reference voltages. (An example of such a scheme, as used in a Bell Telephone PCM system, is described below.) The binary coded signal is stored in a core buffer where the word sync bit is added. Frame sync is also provided at the A/D converter input.

BELL PCM SYSTEM FOR SHORT-RANGE TELEPHONE COMMUNICATION[1]

We conclude this discussion of some typical systems with a brief description of a PCM system adopted by the Bell System for telephone communication over short-haul distances of 10 to 50 miles. Twenty-four telephone channels are time multiplexed, sampled, and coded into PCM for carrier transmission in this system. It was specifically developed and adopted by the telephone system to provide an economical carrier system for short distance, exchange area, telephony.

In this system the 24 channels are each low-pass filtered to 4 kHz, and then sampled at an 8-kHz rate. Quantization to $2^7 = 128$ levels[2] is carried out, and, as in the Nimbus telemetry system just described, an additional eighth time slot is inserted following the seven binary symbols. This additional time slot in this case is used to carry the exchange area signaling information necessary to connect subscribers.

There are thus $24 \times 8 = 192$ bits transmitted in the 125-μsec interval between individual samples. This interval constitutes the frame in this system. An additional framing pulse is added at the end of each frame for synchronization purposes, so that a total of 193 pulses are actually transmitted in the 125-μsec frame interval. A typical frame is shown in Fig. 3-38[3]. The transmitted binary pulses occupy 50 percent of their assigned 0.65-μsec time slots. Pulses are thus 0.325-μsec wide.

In the actual system *compressors* are used following the sampling and multiplexing operations to improve the signal-to-quantization noise ratio.

[1] C. G. Davis, An Experimental Pulse Code Modulation System for Short-haul Trunks, *Bell System Tech. J.*, vol. 41, pp. 1–24, January, 1962.

[2] Actually 127 levels only are used. The all-zero code sequence is not utilized because of timing difficulties it introduces at the repeaters.

[3] C. G. Davis, *op. cit.*, fig. 2, p. 4.

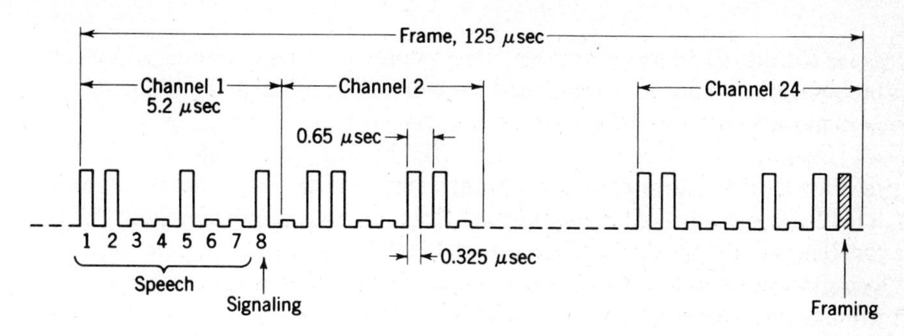

FIG. 3-38 System time assignment. Bell PCM system. (From C. G. Davis, An Experimental Pulse Code Modulation System for Short-haul Trunks, *Bell System Tech. J.*, vol. 41, fig. 2, p. 4, January, 1962. Copyright, 1962, The American Telephone and Telegraph Co., reprinted by permission.)

The compressor is essentially a nonlinear device that restricts or compresses high-amplitude excursions. The logarithmic output-input characteristic of Fig. 3-39 serves as a particularly simple (and practical)

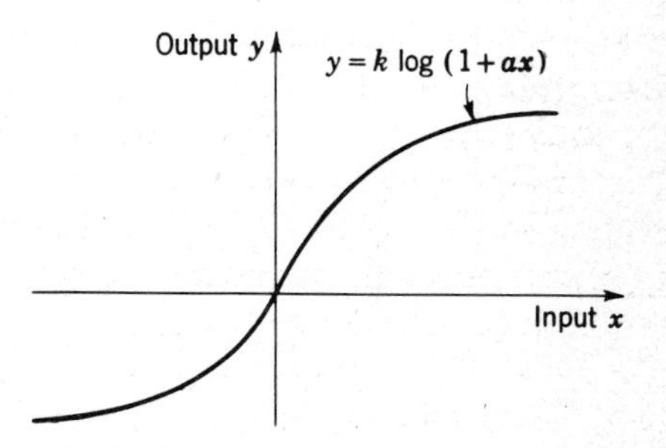

FIG. 3-39 Typical compressor characteristics.

example. (Parallel combinations of diode-resistor networks are used to provide a piecewise approximation to this characteristic.) An *expandor* at the receiving terminal provides the inverse characteristic necessary to undo the nonlinear signal distortion. The combination of compressor and expandor is often referred to as a *compandor*. The effect of the signal compression is to provide a large number of the 128 quantizing levels at the lower signal levels, reducing the number correspondingly at the higher amplitude levels. This is equivalent to introducing a nonlinear or tapered quantization characteristic, with smaller step sizes used at the lower signal levels and larger steps used at the higher signal levels. The quantization noise is thus effectively smaller at the low signal levels, obviously a desirable property, while the corresponding increase in noise at the higher levels is relatively unnoticed.[1,2]

A simplified block diagram of the Bell PCM system is shown sketched in Fig. 3-40.[3] Note that *two* compressors rather than one are shown. This lessens the speed of response requirements on the compressor. The 24 channels are divided into groups of 12 each, successive channel samples being fed alternately to the two compressors. The compressor outputs are then combined sequentially at the encoder.

[1] B. Smith, Instantaneous Companding of Quantized Signals, *Bell System Tech. J.*, vol. 36, pp. 653–709, May, 1957.

[2] H. Mann, H. M. Strube, and C. P. Villars, A Companded Coder for an Experimental PCM Terminal, *Bell System Tech. J.*, vol. 41, pp. 173–226, January, 1962.

[3] C. G. Davis, *op. cit.*, fig. 1, p. 3.

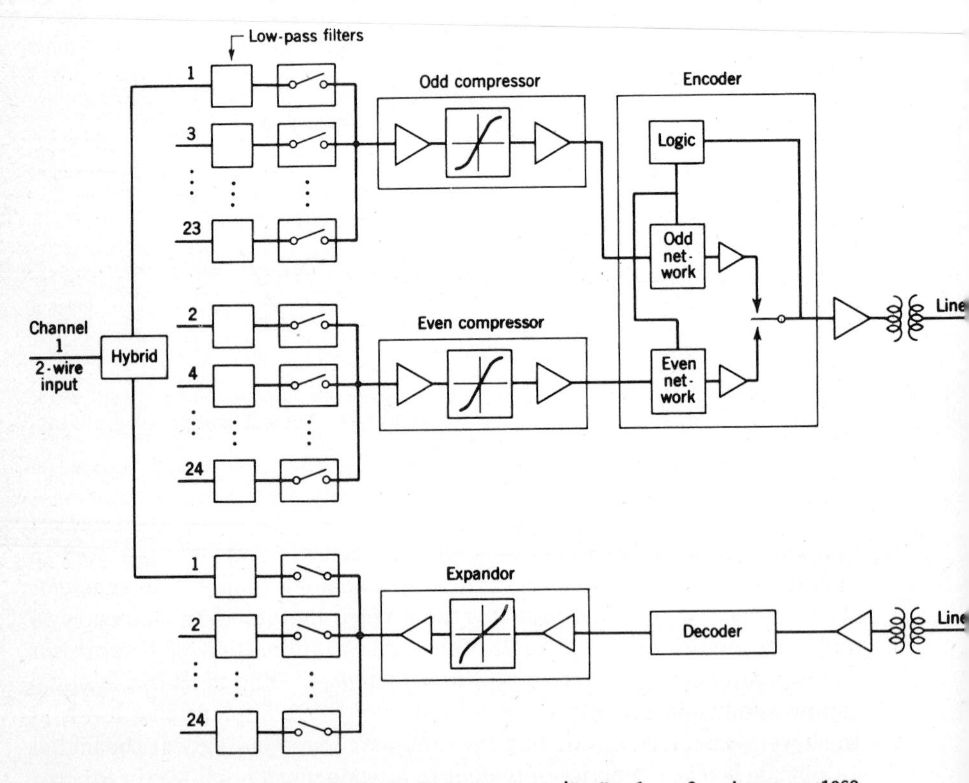

FIG. 3-40 Bell PCM system (From C. G. Davis, *op. cit.*, fig. 1, p. 3. January, 1962. Copyright, 1962, The American Telephone and Telegraph Co., reprinted by permission.)

The encoder used is similar to the one mentioned earlier for the Nimbus PCM telemetry system. The incoming PAM current samples, 5.2-μsec wide, are compared with a sequence of binary weighted currents. A simplified block diagram is shown in Fig. 3-41.[1] The particular binary

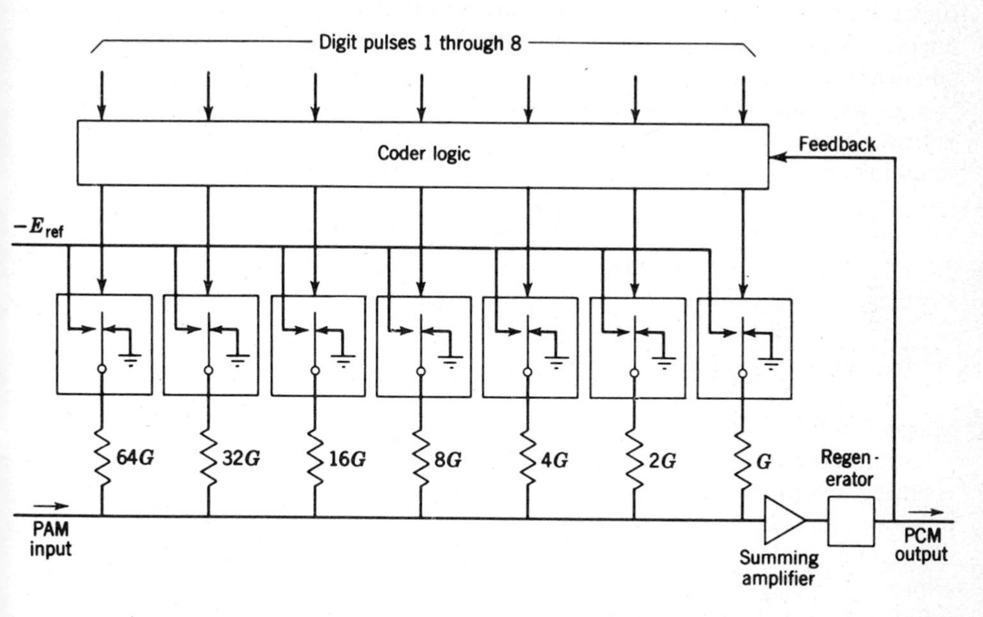

FIG. 3-41 Simplified block diagram of network coder, Bell PCM system. (From C. G. Davis, *op. cit.*, fig. 4, p. 7. January, 1962. Copyright, 1962, The American Telephone and Telegraph Co., reprinted by permission.)

code corresponding to a given analog signal sample at the input is generated by comparing the incoming signal current with a locally generated fixed reference current that is sequentially stepped through the required binary range. Thus, with conductance $64G$ connected to $-E_{ref}$, $2^6 = 64$ units of current are subtracted from the incoming signal current. If $32G$ is switched in, $2^5 = 32$ units are subtracted. If both $64G$ and $32G$ are present, $64 + 32 = 96$ units are subtracted. If $16G$ alone is switched in, 16 units of current are subtracted, etc. The net current drives the summing amplifier.

The specific coding logic used in the encoder of Fig. 3-41 first switches $64G$. If the resultant input to the summing amplifier is negative, this indicates that the incoming (PAM) sample is less than 64 units in ampli-

[1] *Ibid.*, fig. 4, p. 7.

tude. The decision circuit then transmits a pulse, switching $64G$ back to ground at the same time. If the summing amplifier input is positive, the incoming sample is greater than 64 units in amplitude. The decision unit then transmits no pulse, and $64G$ is left connected. Conductance $32G$ is then connected in and the same procedure repeated, continuing until conductance G is switched in. The circuit thus sequences through the seven possible binary choices.

As an example say the incoming signal has a 61.2 units amplitude. Subtracting 64 units by switching in $64G$ results in a negative input to the summing amplifier and a pulse output of the decision circuit. With $32G$ now switched in and $64G$ switched out, 32 units are subtracted. This obviously results in a positive input to the summing amplifier and no pulse output of the decision circuit. $32G$ is left switched in, and $16G$ now switched in. An additional 16 units for a total of 48 are now subtracted. Since this is still less than the input 61.2, again the decision circuit does not transmit a pulse. Both $32G$ and $16G$ are left switched in, and $8G$ now switched in. Continuing in this manner it is apparent that not until $2G$ is switched in, at which time a total of $32 + 16 + 8 + 4 + 2 = 62$ units are subtracted, does the summing amplifier input become negative again, resulting in a pulse output. The final conductance G again provides a no pulse output. If a pulse out of the decision circuit is used to denote a 0, no pulse a 1 (this is easily done by adding the output to a continuous sequence of pulses in mod 2 addition), the sequence of outputs in this case is 0111101, just the binary form for 61. The reader can easily check that all values of input ≥ 61 and < 62 are transmitted as a binary 61. The circuit thus performs both quantization (quantizing to the nearest integer below an incoming sample value), and binary encoding.

The decoding operation at the receiver is essentially the inverse of the encoding scheme just described. Incoming pulses are stored until all seven are in, and then summed through appropriate binary weighting resistors to provide the desired decimal number. (As an example, the first binary digit in, representing 64 units, is weighted by 64, the second by 32, etc.) The decoder is shown in Fig. 3-42.[1]

3-11 PULSE-POSITION MODULATION (PPM)

We have discussed in this chapter thus far PAM and PCM systems. Pulse-amplitude modulation is but one example of several types of

[1] *Ibid*, fig. 5, p. 8. This paper describes the system in detail.

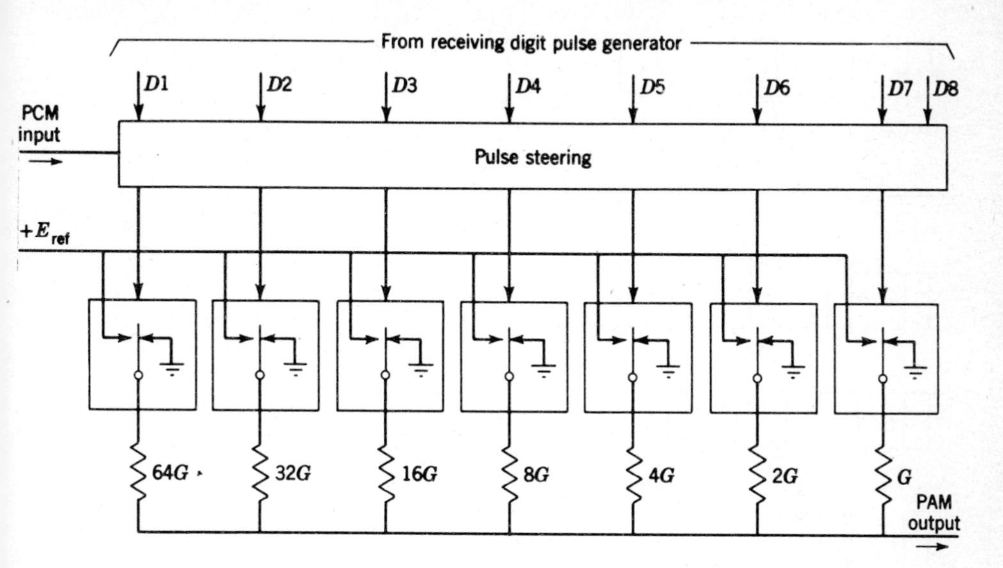

FIG. 3-42 Simplified block diagram of decoder. (From C. G. Davis, *op. cit.*, fig. 5, p. 8. Copyright, 1962, The American Telephone and Telegraph Co., reprinted by permission.)

uncoded pulse-modulation systems used in practice. In these systems a periodic sequence of pulses constitutes the carrier. Depending upon the type of system, the input signal may modulate or vary the pulse amplitude (producing PAM), the pulse width, or duration [producing pulse-width modulation (PWM) or pulse-duration modulation (PDM)—both terms have been used], and the pulse position [producing pulse-position modulation (PPM)]. The carrier, a typical input signal, and the three types of modulation output are shown in Fig. 3-43. In the PDM output the normal pulse width increases or decreases depending on the signal amplitude (positive or negative). In the PPM case the width and the amplitude of the pulses are unaltered, but the pulse occurrence is delayed or advanced in accordance with the input signal.

These are the basic types of pulse-modulation systems. Variations of these schemes are of course possible and have been used, for example, double-polarity PAM, PDM with the position of the pulse leading edge fixed and the trailing edge varied, pulse modulation with the period T modulated in accordance with the signal, etc.[1]

[1] See Schwartz, Bennett, and Stein, *op. cit.*, chap. 6, for a discussion of some of these systems.

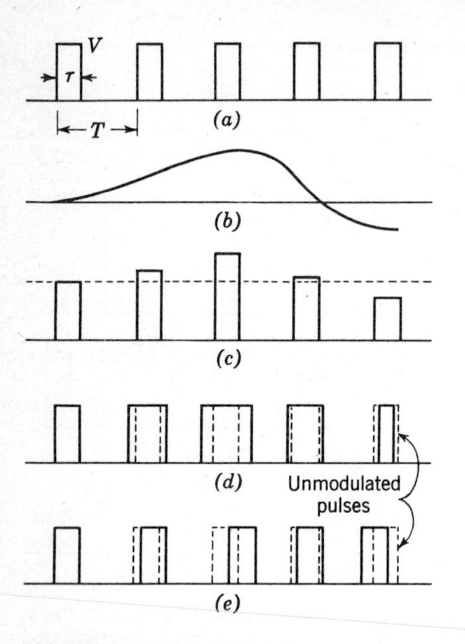

FIG. 3-43 Pulse-modulation systems. (a) Carrier; (b) signal input; (c) PAM output; (d) PDM output; (e) PPM output.

Pulse-position-modulation and pulse-duration-modulation systems require substantially more bandwidth than equivalent PAM systems, for they depend on the accurate location of the pulse edges. This increased bandwidth in the case of PPM will be shown in Chap. 6 to lead to signal-to-noise improvement, just as wideband FM provides signal-to-noise improvement. (There is no signal-to-noise improvement in PAM, just as AM provides no improvement.) This improvement is not as efficient as that of PCM, however, which is discussed earlier in this chapter.

The simplicity of PPM, PDM, and other types of pulse-modulation systems, as contrasted to the coding techniques required to generate binary PCM signals, suggests their use in many applications. In those cases of analog signal transmission particularly, where quantization and binary encoding are not desirable or are unwarranted (the accuracy requirements may require an inordinate number of quantization levels), one may prefer to use PPM or PDM rather than PCM transmission. Since the modulated pulses of PDM and PPM are of uniform amplitude, they may be reshaped periodically during transmission, just as in the case of PCM pulses, to avoid excessive distortion.

Rather than discuss the elements of each of these various systems in detail we focus attention on PPM only, describing its basic elements by considering a specific commercial system developed many years ago by the Federal Telephone Laboratories.[1]

This system accommodates 23 voice channels, each of frequency range 100 to 3,400 Hz. The Nyquist sampling rate used is again 8 kHz (greater than $2B = 6.8$ kHz), and the Nyquist sampling interval is thus again 1/(8 kHz), or 125 μsec. This leaves about 5 μsec per channel to be utilized for varying pulse position (plus and minus about the mean position) in accordance with the voice input. The timing diagram for several such channels is shown in Fig. 3-44. The marker pulses (M) in this system are two identical pulses, spaced by the pulse width. (In other systems single marker pulses of larger amplitude or greater width than the channel pulses have been used.) The pulses actually used are trapezoidal in shape, with a base width of 0.8 μsec and rise and decay time of 0.15 μsec. The maximum pulse time shift (corresponding to maximum signal input) is ± 1 μsec away from normal (zero input) signal. The total space required by each channel is 2.3 μsec (2 μsec plus an allowance for the pulse rise and decay time at either end of the maximum displacement points). A guard time of 2.7 μsec is thus made available between the channels. These numbers are all indicated in Fig. 3-44 and repeated in Fig. 3-45, which shows the multiplexed pulse train and the details of multiplexing. The trapezoidal pulse used is shown in Fig. 3-46.

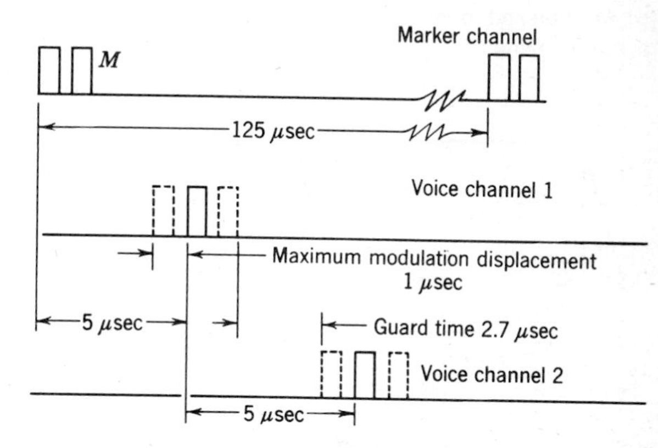

FIG. 3-44 Federal Telephone Laboratories PPM system (individual channels).

[1] D. D. Greig and A. M. Levine, Pulse Time Modulated Multiplex Radio Relay System Terminal Equipment, *Elec. Commun.*, vol. 23, p. 159, 1946.

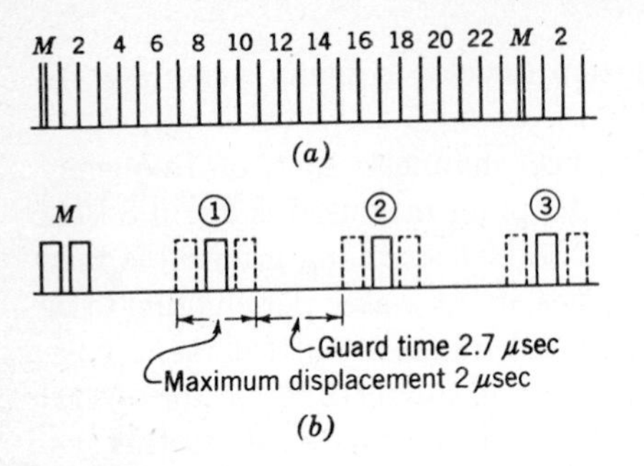

FIG. 3-45 The 23 channels multiplexed. (a) Multiplexed pulse train; (b) details of multiplexing.

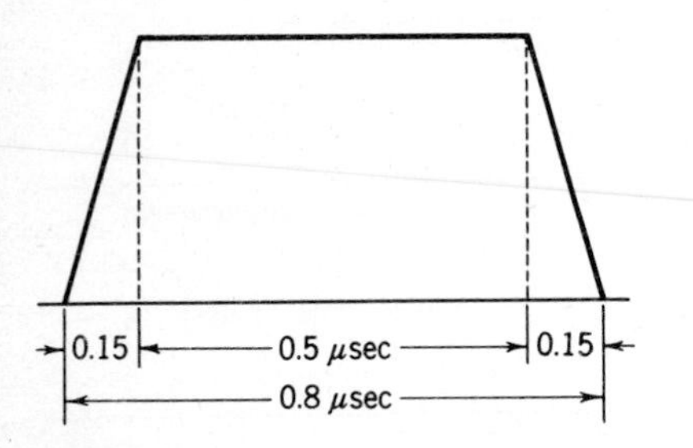

FIG. 3-46 System pulse shape.

From our time-frequency discussions of Chap. 2 the system bandwidth is inversely proportional to the rise time. Using $B = 1/(2 \text{ rise time})$, $B = 3.3$ MHz. The bandwidth actually used in this system is 2.8 MHz. Note then that the requirement of 23 channels to be multiplexed in 125 μsec, with appropriate provision allowed for maximum pulse displacement and guard time, has led to the use of rather narrow pulses, with very small rise times and correspondingly wide bandwidth. This is of course different from the situation in PAM, where the pulse shape is not of extreme importance and lower bandwidths can be used. The sharp pulses obtained with this bandwidth, however, and the uniform amplitudes and shapes used serve to minimize possible difficulties with noise. The noise-improvement possibilities of wideband PPM will be discussed specifically in Chap. 6.

Just as in the case of PAM, there are many possible ways of generating and detecting time-multiplexed PPM signals. One particularly simple method of generating PPM is outlined in the article by Grieg and Levine already referred to. The method is indicated in block-diagram

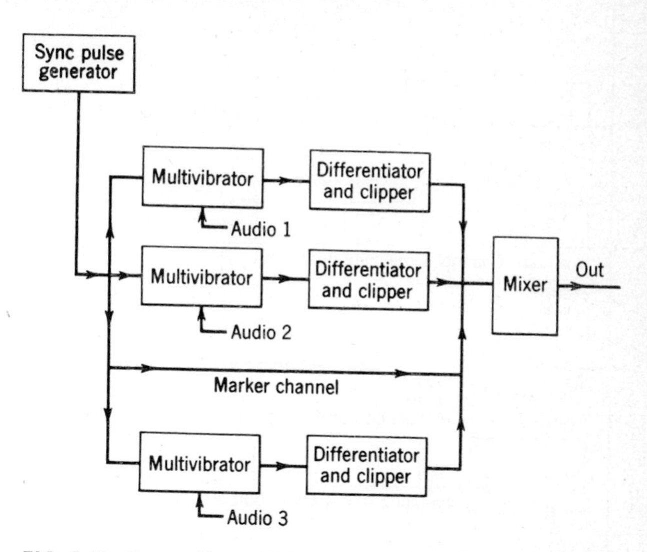

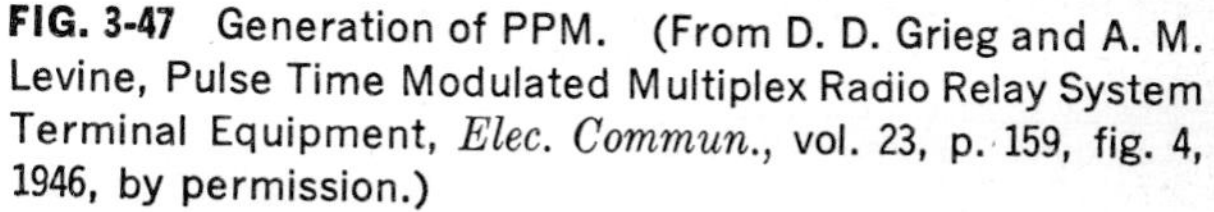

FIG. 3-47 Generation of PPM. (From D. D. Grieg and A. M. Levine, Pulse Time Modulated Multiplex Radio Relay System Terminal Equipment, *Elec. Commun.*, vol. 23, p. 159, fig. 4, 1946, by permission.)

form in Fig. 3-47. The multivibrators shown are all of different time constant and provide successively increasing output pulse widths. They are all turned on periodically at the same time, however, by the synchronizing pulse generator. The multivibrator pulses, of varying widths, are then differentiated and clipped and combined in the mixer. The synchronizer output is fed directly to the mixer and serves as a marker at the beginning of each cycle. (These marker pulses can be made of greater amplitude or width to distinguish them from the channel pulses.) The successive pulse outputs and the final multiplexed output are shown in Fig. 3-48.

A method of demodulating multiplexed PPM pulses using a tapped delay line is shown in Fig. 3-49. The taps on the delay line are connected 5 μsec apart. The two taps corresponding to the two marker pulses (a and b) are a pulse width apart. The gate generator can be turned on only by marker pulses a and b in coincidence. This prevents other pulses moving down the delay line from turning on the gate generator. (If single marker pulses were used, the gate generator could be made to respond to pulses of the marker height or duration only.) The gate-generator pulses last the duration of the maximum plus or minus displacement of the PPM pulses in any one channel (2.3 μsec for the Federal Telephone Laboratories system) and serve to open all channel outputs connected to the respective taps.

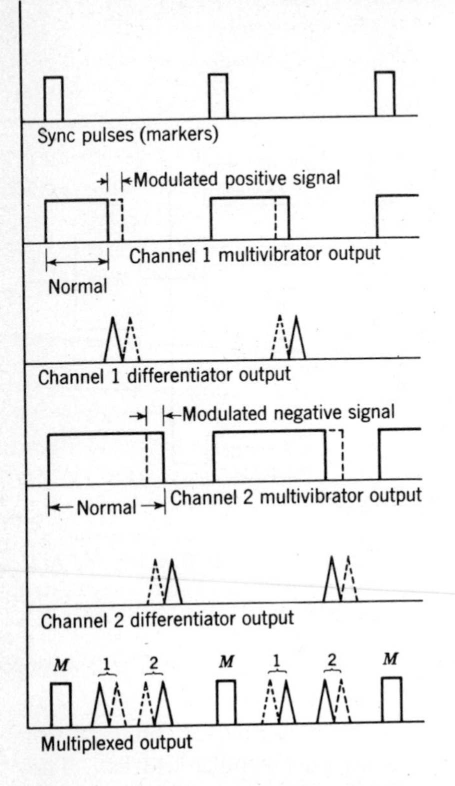

FIG. 3-48 Pulse-position-modulation generator pulse diagrams. (From D. D. Grieg and A. M. Levine, *op. cit.*, by permission.)

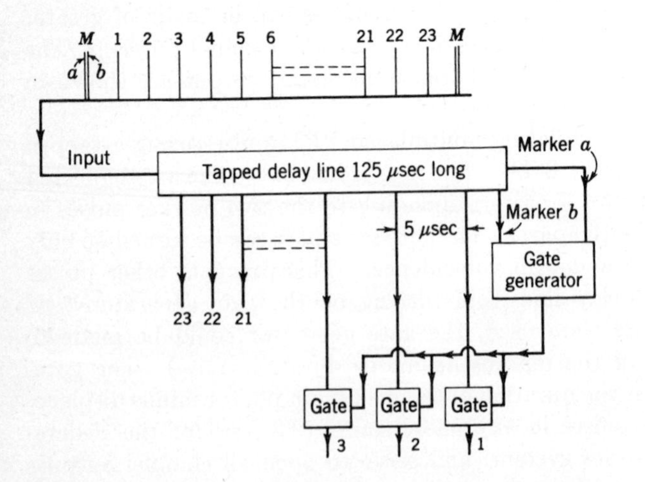

FIG. 3-49 Pulse-position-modulation system demodulator.

During the gate time a pulse arriving at any particular channel turns on a linear sweep in that channel. The sweep is turned off at the end of the gate interval. At the time the sweep is turned off its voltage level is sampled and stored in a capacitor. Pulses advanced in time thus produce greater voltages; those delayed in time provide smaller voltages. Because of the linearity of the sweep the voltage sampled is proportional to the pulse position and, after possible inversion and further filtering, reproduces the original audio signal.

3-12 SUMMARY

The digital systems studied in this chapter are of ever-increasing technological importance, and, because of their relative conceptual simplicity, serve as a good introduction to the whole field of information transmission and signal processing. As digital processing assumes an even greater role in modern technology, the digital techniques such as sampling, A/D conversion, etc., discussed in this chapter will become even more fundamental in systems design and development.

We began the chapter by pointing out that fundamental to digital transmission is the necessity of sampling analog waveshapes. We then showed, using the Fourier analysis of the previous chapter, that for signals band-limited to B Hz, at least $2B$ samples per second were necessary to retain all the information in the original (analog) signal. Sampling plays a key role in data analysis and processing by computer, as well as in preparing signals for digital transmission.

By leaving sufficient space between samples additional signals can be *time multiplexed* for transmission on one channel. This is of course at a price—the resultant increased bandwidth requirements. For the transmission of *digital data*, that is, discrete numbers rather than continuous (analog) waveshapes, the signal samples must further be quantized into a specified number of amplitude levels. This results in so-called quantization noise—the difference between the actual sample amplitude and its quantized approximation. The use of an appropriate number of levels makes this effect tolerable. The resultant stream of signal digits may then be coded into any number system desired. The binary system is the one most commonly used, since the resultant stream of binary pulses is least susceptible to noise and interference during transmission. There is a tradeoff between signal power and bandwidth that occurs with the coding. The power requirements are reduced while the bandwidth is increased with the use of binary digits. Conversely, successive signal samples may be combined into increasingly more amplitude levels to decrease bandwidth, but at the cost of exponentially increasing signal power.

The question of minimum bandwidth needed to transmit signal samples was investigated in the context of intersymbol interference. We found that pulse overlap may be eliminated (at least ideally) by appropriately shaping the pulses transmitted. The bandwidth required to transmit these pulses was found to be the reciprocal of the pulse width, in agreement with the qualitative discussions of Chap. 2. This question of filtering will be pursued further in Chap. 6, in studying the detection of signals in noise.

The treatment in this chapter was devoted entirely to so-called *baseband* systems, those that transmit the basic signals with no frequency translation carried out. In Chap. 4, we point out the common need for transmission of these digital signals at high frequencies, and discuss various ways of performing the requisite frequency translation up at the transmitter (*modulation*), and down at the receiver (*demodulation*). Again Fourier analysis and bandwidth concepts will be found to play a key role.

In Chap. 6 we return to these systems, comparing them on their relative performance in the presence of noise.

PROBLEMS

3-1 A sinusoidal input at 1 Hz, sin $2\pi t$, is to be sampled periodically.

(*a*) Find the maximum allowable interval between samples.
(*b*) Samples are taken at intervals of ⅓ sec. Perform the sampling operation graphically, and show to your satisfaction that no other sine wave (or any other time function) of bandwidth less than 1.5 Hz can be represented by these samples.
(*c*) The samples are spaced ⅔ sec apart. Show graphically that these may represent another sine wave of frequency less than 1.5 Hz.

3-2 A function $f_1(t)$ is band-limited to 2,000 Hz, another function $f_2(t)$ to 4,000 Hz. Determine the maximum sampling interval if these two signals are to be time multiplexed.

3-3 Two signals, $f_1(t) = 10 \sin 2\pi \times 1{,}000t$ and $f_2(t) = 5 \sin 2\pi \times 2{,}000t$, are sampled at a rate of 10,000 samples per second and are then time multiplexed.

(*a*) Sketch the resulting multiplexed function.
(*b*) After transmission the two sampled signals are separated out and fed to holding circuits. Sketch the holding-circuit output in the $f_1(t)$ channel.

3-4 Explain the $e^{-j\omega T/2}$ factor appearing in Eq. [3-25] for the output of the holding circuit. Does this agree with the sketch of a typical output shown in Fig. 3-17?

3-5 Twenty-four signal channels plus one synchronization (marker) channel, each band-limited to 3,300 Hz, are sampled and time multiplexed at an 8-kHz rate. Calculate the minimum bandwidth needed to transmit this multiplexed signal in a PAM system.

3-6 A four-channel time-multiplexed PAM system is to be designed for voice signal transmission. The system is to accept voice signals in the range 300 to 3,300 Hz.

Outline a possible block diagram for such a system. Indicate, in the block diagram, a possible method of sampling and then multiplexing the four channels. Show a possible method of separating the different signals back into their respective channels. (The methods used may incorporate multivibrators, switching circuits, delay lines, counters, gate circuits, etc.)

Include any filters that may be needed in the system. Indicate the bandwidths required for transmission at various points in the system, and show sketches of pulse shapes at different points.

3-7 Five signal channels are sampled and time multiplexed. The multiplexed signal is then passed through a low-pass filter. Three of the channels handle signals covering the frequency range 300 to 3,300 Hz. The other two carry signals covering the range 50 Hz to 10 kHz. Included is a synchronization signal.

(*a*) What is the minimum sampling rate?
(*b*) For this sampling rate what is the minimum bandwidth of the low-pass filter?

3-8 Channel 1 of a two-channel PAM system handles 0- to 5-kHz signals; the second channel handles 0- to 10-kHz signals. The two channels are sampled by a pulse generator that puts out square waves at the lowest frequency that is theoretically adequate. The sampled signals are then added and passed through a low-pass filter before transmission. After demodulation at the receiver the pulses in each of the two channels are passed through appropriate holding circuits.

(*a*) What is the frequency of the square-wave generator? What is the minimum cutoff frequency of the low-pass filter that will preserve the amplitude of the output pulses?

(*b*) The signal in channel 1 is $2 \sin 2\pi \times 2{,}500t$, and that in channel 2 is $\sin 2\pi \times 5{,}000t$. Sketch these signals, and below them draw the waveshapes at the input to the low-pass filter, at the filter output, and at the output of the holding circuit in channel 2.

3-9 A PAM system has been designed to accommodate 10 voice channels with frequency range 300 to 3,300 Hz. The channels are sampled at an 8-kHz rate.

(*a*) Calculate the frequency response (in decibels) of a holding circuit in one of the channels at $f = 0, B, f_c - B$ Hz.

(*b*) Design a constant-K and m-derived matching section filter to follow the holding circuit so that the combination of filters, including the holding circuit, meets the following specifications:

f	Attenuation, db
0	0
B	≤ 3
$\geq f_c - B$	≥ 25

Assume that the relative frequency response of the holding circuit in decibels and the attenuation due to each filter section are additive.

3-10 A single information channel carries voice frequencies in the range 50 to 3,300 Hz. The channel is sampled at an 8-kHz rate, and the resulting pulses are transmitted over either a PAM system or a PCM system.

(*a*) Calculate the minimum bandwidth of the PAM system.

(*b*) In the PCM system the sampled pulses are quantized into eight levels and transmitted as binary digits. Find the transmission bandwidth of the PCM system, and compare with that of (*a*).

(*c*) Repeat (*b*) if 128 quantizing levels are used. Compare the rms quantization noise in the two cases if the peak-to-peak voltage swing at the quantizer is 2 volts.

3-11 Repeat Prob. 3-10 for the case of 12 voice channels, each carrying frequencies of 50 to 3,300 Hz, which are sampled and time multiplexed. Include block diagrams of both the PAM and PCM systems.

3-12 The sinusoidal voltage $10 \sin 6{,}280t$ is sampled at $t = 0.33$ msec and thereafter periodically at a 3-kHz rate. The samples are then quantized into eight voltage levels and coded into binary digits. Draw the original voltage and below it to scale the outputs of the sampler, the quantizer, and the encoder. Calculate the rms quantization noise.

3-13 A signal voltage in the frequency range 100 to 4,000 Hz is limited to a peak-to-peak swing of 3 volts. It is sampled at a uniform rate of 8 kHz, and the samples are quantized to 64 evenly spaced levels. Calculate and compare the bandwidths and ratios of peak signal to rms quantization noise if the quantized samples are transmitted either as binary digits or as four-level pulses.

3-14 Calculate the capacity in bits per second of the PCM systems of Probs. 3-10*b* and *c*, and Probs. 3-11 and 3-13.

3-15 Consider the so-called *sinusoidal roll-off* spectrum of Fig. P 3-15[1]: measuring from ω_c as shown,

$$|H(\Delta\omega)| = \frac{1}{2}\left(1 - \sin\frac{\pi}{2}\frac{\Delta\omega}{\omega_x}\right) \qquad |\Delta\omega| < \omega_x$$

$$|H(\Delta\omega)| = 0 \qquad |\Delta\omega| > \omega_x$$

$$|H(\Delta\omega)| = 1 \qquad -\omega_c < \Delta\omega < -\omega_x$$

Assume linear phase shift, $\theta(\omega) = -j\omega t_0$.

(*a*) Show the impulse response is given by

$$h(t) = \frac{\omega_c}{\pi}\frac{\sin\omega_c(t - t_0)}{\omega_c(t - t_0)}\frac{\cos\omega_x(t - t_0)}{1 - [2\omega_x(t - t_0)/\pi]^2}$$

(*b*) Sketch $|H(\omega)|$ and $h(t)$ for $\omega_x/\omega_c = 0.5, 0.75, 1.0$ (this is the so-called *roll-off factor*) and compare. Pay specific attention to the bandwidth and pulse tails in each case.

(*c*) The raised cosine spectrum is given by $\omega_x/\omega_c = 1.0$. Show that the impulse response in this case has an additional zero half-way between the usual zero crossings. Discuss the possibility in this case of detecting successive pulses by "slicing" the received signal at the half-amplitude levels.

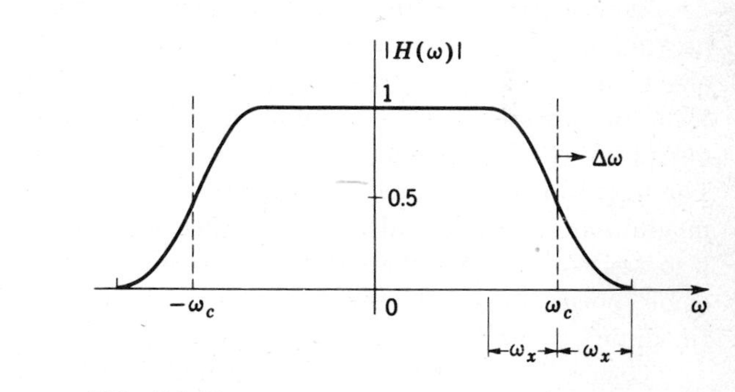

FIG. P 3-15

3-16 Indicate, in block-diagram form, a method of generating and multiplexing PPM pulses other than the one shown in the text. Show a possible method of demodulating the multiplexed PPM signals.

3-17 Repeat Prob. 3-6 for a PPM system. Compare the bandwidths required in the two cases.

[1] W. R. Bennett and J. R. Davey, *op. cit.*, pp. 55, 56.

CHAPTER FOUR

MODULATION TECHNIQUES

4-1 INTRODUCTION

All information-bearing signals must ultimately be transmitted over some medium (channel) separating the transmitter and the receiver. The medium may in some cases be the air, in others, a set of wires or a hollow conducting tube (a waveguide). Efficiency of transmission—whether measured in terms of bandwidth, power required, complexity of the circuitry, etc.—requires that this information be processed in some manner before being transmitted over an intervening medium.

The step of processing the signal for more efficient transmission is called the *modulation process.*

We have already discussed one particular form of modulation in the previous chapter: *pulse modulation,* in which the height, width, or position of a set of pulses (the carrier) is altered in a definite pattern corresponding to the information to be transmitted. This form of modulation enables one to time multiplex many information channels for sequential transmission over a single channel. If A/D conversion (quantization) is adopted in addition, the signal pulses may further be processed by binary encoding if so desired. This serves the purpose of coding the signals to better overcome noise and distortion ultimately encountered on the channel (medium), as well as to enable digital circuitry to be utilized.

Generally, however, the pulse signals must themselves modulate in turn a high-frequency sine-wave carrier in order to be transmitted over the medium. In this case of *continuous-wave* (*c-w*) modulation, the amplitude, phase, or frequency of a specified sine wave (the carrier) is altered in accordance with the information to be transmitted.

Sine-wave modulation enables signals to be transmitted at frequencies much higher than the signal-frequency components. It is a well-known phenomenon of electromagnetic theory that an efficient radiator of electric energy (the antenna) must be at least the order of magnitude of a wavelength in size. Since the wavelength of a 1-kHz tone is 300 km, radio transmission is hardly practicable at audio frequencies. But a 1-MHz carrier wave could be (and is) transmitted efficiently with an antenna 300 ft high.

In this chapter we shall study various types of c-w modulation, with emphasis primarily on amplitude modulation (including both normal and single-sideband modulation) and frequency modulation. We have indicated above that pulse-modulation systems generally require some form of c-w modulation to finally transmit the signals over the desired channel; this is also true of analog signals that are to be transmitted, without conversion to pulse or digital forms, to remote points. Examples, of course, include the usual radio and TV broadcasts, as well as many telemetered signals where it is more expedient to transmit the data directly, without digitizing them.

Not only is c-w modulation necessary to launch signals effectively over a desired medium, but it also leads to the possibility of *frequency multiplexing*, or staggering of frequencies over the specified band. This is directly analogous to the *time-multiplexing* technique introduced in the previous chapter, in which many signal channels are sequentially sampled and transmitted serially in time. In the frequency-multiplexed case signal channels are transmitted over adjacent, nonoverlapping frequency bands. They are thus transmitted in parallel, simultaneously in time. This means that many telephone conversations can be transmitted over a single pair of wires (depending of course on the bandwidth allowed for each conversation and the total bandwidth of the system including the wires). Very often a group of digital channels, each incorporating many time-multiplexed signals, may further be combined by frequency-multiplexing techniques.

Operation at higher carrier frequencies makes more bandwidth available and hence leads to the possibility of frequency multiplexing more signals or transmitting wider-band signals than is possible at lower frequencies. As an example, the amplitude-modulation (AM) broadcast band in this country is fixed at 550 to 1,600 kHz. This provides for 100 channels, spaced 10 kHz apart. This entire band, however, is just a fraction of the band, 6 MHz, needed for one TV channel. Operating TV broadcasts at the 60-MHz range and up (the VHF band) makes several channels available. Increasing the carrier frequency to 470 MHz and up (the UHF band) makes many more TV channels available. This is one reason for the current interest in millimeter and optical frequencies for communication purposes.

In this chapter we shall concentrate, as in the previous chapter, on the spectral (bandwidth) characteristics of the various types of c-w systems, as well as on general systems aspects of their generation and reception. This will further solidify the Fourier analysis of Chap. 2. It parallels the approach used in the previous chapter in discussing pulse-modulation systems. After introducing statistical techniques and noise in the next chapter, we shall be in a position to further compare the various c-w systems, as well as the pulse systems already studied, on the basis of their performance in noise.

Although the techniques of c-w modulation are essentially common to all types of signals transmitted—digital as well as analog—we shall begin the discussion in this chapter by considering digital (binary) transmission via high-frequency carrier. We do this for several reasons:

1. Because of the increasing importance of the transmission of digital data
2. Because the resultant carrier systems are sufficiently distinct to warrant separate categorization in the literature
3. Because the spectral properties of such systems, which can be obtained very simply for these types of signals, may be readily extended to more complex types of signals

In essence, then, the use again of the binary pulse as a typical signal enables us to more readily understand the performance of c-w systems with more complex test signals, while being of sufficient technological importance to warrant discussion in its own right.

4-2 BINARY COMMUNICATIONS[1]

As noted earlier there are essentially three ways of modulating a sine-wave carrier: variation of its amplitude, phase, and frequency in accordance with the information being transmitted. In the binary case this corresponds to switching the three parameters between either of two possible values. Most commonly the amplitude switches between zero (*off* state) and some predetermined amplitude level (the *on* state). Such systems are then called *on-off-keyed* (OOK) systems. Similarly, in *phase-shift keying* (PSK), the phase of a carrier switches by π radians or 180°. Alternately, one may think of switching the polarity of the carrier in

[1] W. R. Bennett and J. R. Davey, "Data Transmission," McGraw-Hill, New York, 1968, discuss these systems in much more detail than is possible here. They include a thorough treatment of the spectral characteristics of these systems and summarize the characteristics of many of the data sets in actual use. M. Schwartz, W. R. Bennett, and S. Stein, "Communications Systems and Techniques," McGraw-Hill, New York, 1966, also discuss these systems in more detail than is possible here, although their emphasis is primarily on the relative noise immunity of the systems. Some of this material will be discussed in Chap. 6 of this text after the basic aspects of noise phenomena are introduced.

accordance with the binary information stream. In the *frequency-shift-keyed* (FSK) case, the carrier switches between two predetermined frequencies, either by modulating one sine-wave oscillator or by switching between two oscillators locked in phase. Although other types of binary sine-wave signaling schemes are in use as well,[1] we shall concentrate here on the basic schemes.

ON–OFF KEYING

Assume a sequence of binary pulses, as shown in Fig. 4-1*a*. The 1's turn on the carrier of amplitude A, the 0's turn it off (Fig. 4-1*b*). It is

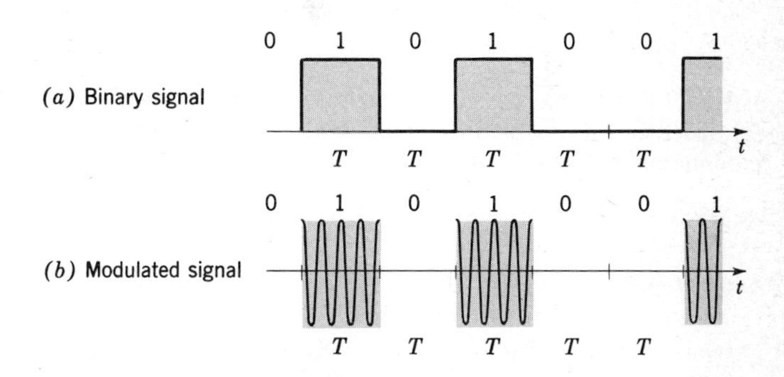

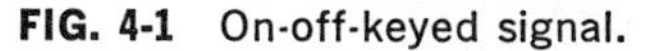

FIG. 4-1 On-off-keyed signal.

apparent that the spectrum of the OOK signal will depend on the particular binary sequence to be transmitted. Call a particular sequence of 1's and 0's $f(t)$. Then the amplitude-modulated or OOK signal is simply

$$f_c(t) = Af(t) \cos \omega_c t \qquad [4\text{-}1]$$

where $f(t) = 1$ or 0, over intervals T sec long. But note that this is in exactly the form of a typical component of the sampled signals of Chap. 3 (see Eq. [3-3]). As shown there, upon taking the Fourier transform of the amplitude-modulated (OOK) signal $f_c(t)$, we have

$$F_c(\omega) = \frac{A}{2} [F(\omega - \omega_c) + F(\omega + \omega_c)] \qquad [4\text{-}2]$$

(See Eq. [3-4].)

The effect of multiplication by $\cos \omega_c t$ is simply to shift the spectrum of the original binary signal (the so-called baseband signal) up to fre-

[1] One common example is *differential phase-shift keying* (DPSK). See Bennett and Davey, *op. cit.;* Schwartz, Bennett, and Stein, *op. cit.*

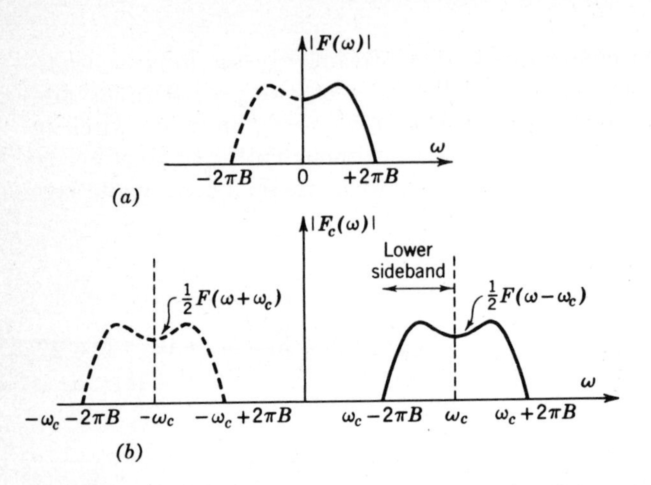

FIG. 4-2 Amplitude spectrum—amplitude-modulated wave. (a) Spectrum of modulating signal; (b) spectrum of amplitude-modulated wave.

quency ω_c (Fig. 4-2). This is the general form of an AM signal. It contains upper and lower sidebands symmetrically distributed about the carrier or center frequency ω_c. Note the important fact that with an initial baseband bandwidth of $2\pi B$ rps (B Hz), the AM or transmission bandwidth is twice that, that is, $\pm 2\pi B$ rps or $\pm B$ Hz about the carrier, for a total bandwidth of $2B$ Hz.

Note again that this is a general result for AM waves that is true for all modulating signals $f(t)$ and not just for the binary case we are in the process of considering. As an example, let $f(t) = \cos \omega_m t$, a single sine wave of frequency ω_m. Then, by trigonometry,

$$\cos \omega_m t \cos \omega_c t = \tfrac{1}{2} \cos (\omega_m + \omega_c)t + \tfrac{1}{2} \cos (\omega_m - \omega_c)t$$

The single-line spectral plot representing $\cos \omega_m t$ is thus replaced by *two* lines, symmetrically arrayed about ω_c (Fig. 4-3). Similarly, if $f(t)$ is a finite sum of sine waves, each sine wave is translated up in frequency by ω_c, resulting in the AM spectrum of Fig. 4-4.

As another special case assume the signal $f(t)$ to be the single rectangular pulse of Chap. 2. (This is then the special case of a binary train in which all symbols are 0, except for one 1.) For a pulse of amplitude A and width T (the binary interval), the spectrum of the AM signal becomes simply

$$\frac{AT}{2}\left[\frac{\sin (\omega - \omega_c)T/2}{(\omega - \omega_c)T/2} + \frac{\sin (\omega + \omega_c)T/2}{(\omega + \omega_c)T/2}\right]$$

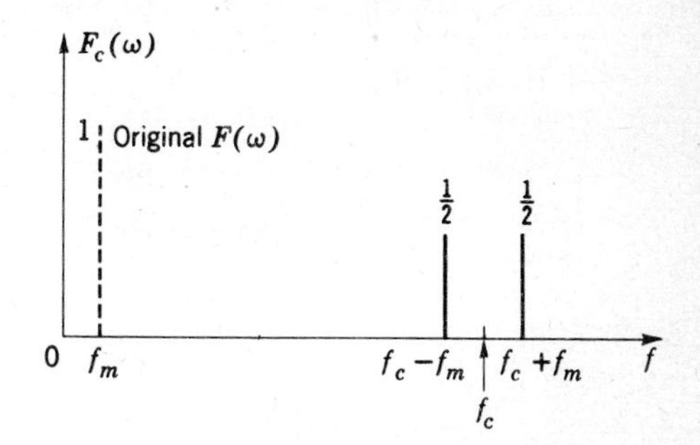

FIG. 4-3 Frequency-shift effect of AM (positive frequencies only are shown here).

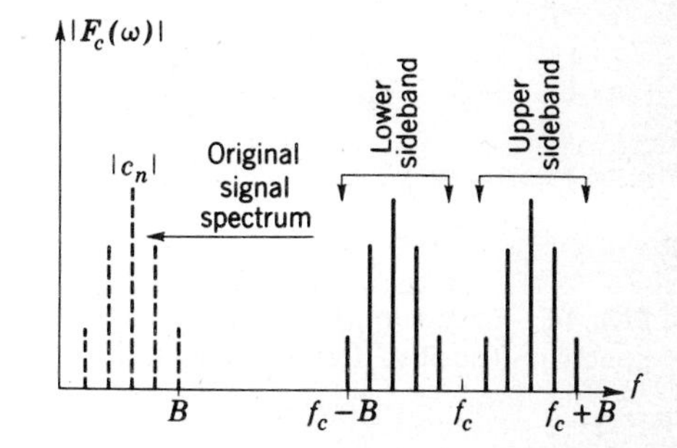

FIG. 4-4 Amplitude-modulation spectrum—periodic modulating signal.

With an initial bandwidth of approximately $1/T$ Hz (from 0 frequency to the first zero crossing), we now have a transmission bandwidth of $2/T$ ($\pm 1/T$ about the carrier). Another special case is a binary train of alternating 1's and 0's, resulting in a periodically alternating OOK signal. The spectrum of this signal is just the $(\sin x)/x$ line spectrum of a pulse of width T, periodic with period $2T$, translated up to frequency f_c. This is shown in Fig. 4-5.

FREQUENCY-SHIFT KEYING

Here

$$\left.\begin{aligned} f_c(t) &= A \cos \omega_1 t \\ \text{or} \qquad f_c(t) &= A \cos \omega_2 t \end{aligned}\right\} \qquad -\frac{T}{2} \leq t \leq \frac{T}{2} \tag{4-3}$$

A 1 corresponds to frequency f_1, a zero to frequency f_2 (Fig. 4-6). (*Note:* Generally, f_1 and $f_2 \gg 1/T$. In some systems, particularly over

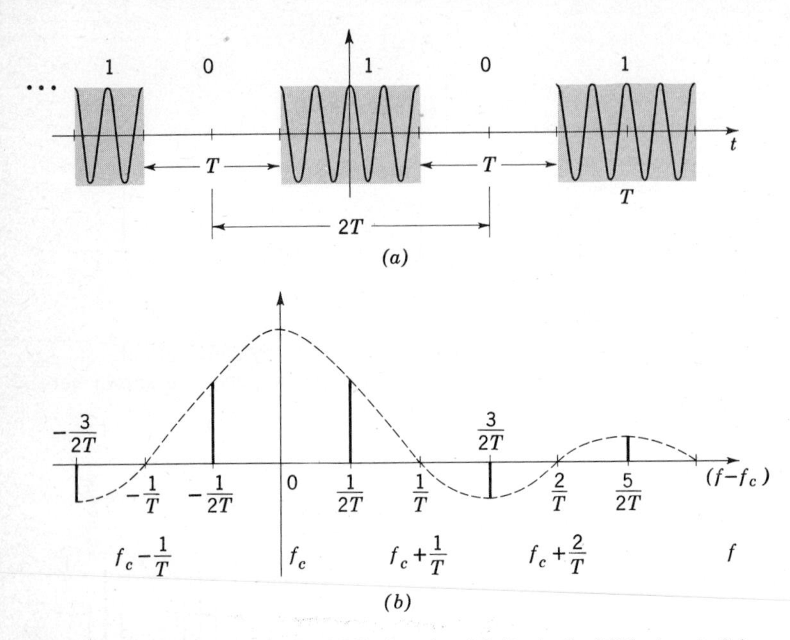

FIG. 4-5 Spectrum, periodic OOK signal. (a) Periodic OOK signal; (b) spectrum (positive frequencies only).

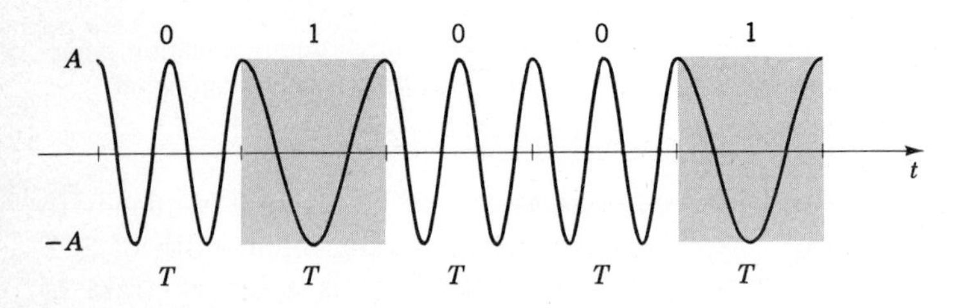

FIG. 4-6 FSK wave.

telephone lines, f_1 and $f_2 \sim 1/T$, as shown here.) An alternate representation of the FSK wave consists of letting $f_1 = f_c - \Delta f$, $f_2 = f_c + \Delta f$. Tho two frequencies then differ by $2\,\Delta f$ Hz. Then

$$f_c(t) = A \cos (\omega_c \pm \Delta\omega)t \qquad -\frac{T}{2} \leq t \leq \frac{T}{2} \tag{4-3a}$$

The frequency then deviates $\pm\Delta f$ about f_c. Δf is commonly called the *frequency deviation*. The frequency spectrum of the FSK wave $f_c(t)$ is in general difficult to obtain. We shall see this is a general characteristic of FM signals. However, one special case which provides insight into the spectral characteristics of more complex FM signals, and leads to a good rule of thumb regarding FM bandwidths, may be readily evaluated. Assume the binary message consists of an alternating sequence of 1's and 0's. If the two frequencies are each multiples of the reciprocal of the binary period T (that is, $f_1 = m/T$, $f_2 = n/T$, m and n integers), and are synchronized in phase, as assumed in Eq. [4-3], the FSK wave is the periodic function of Fig. 4-7. Note, however, that this may also be visualized as

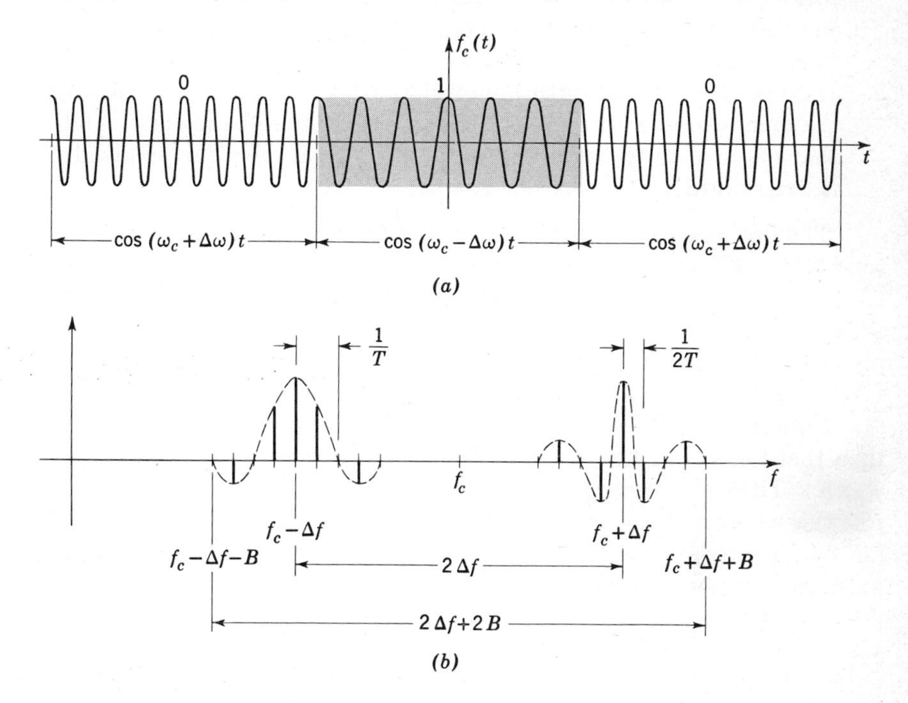

FIG. 4-7 Spectrum, periodic FSK wave. (a) Periodic FSK signal; (b) spectrum (positive frequencies only).

the linear superposition of two periodic OOK signals such as the one of Fig. 4-5, one delayed T sec with respect to the other. The spectrum is then the linear superposition of two spectra such as the one in Fig. 4-5. Specifically, it is left for the reader to show that the positive frequency spectrum is of the form

$$\frac{\sin [(\omega_1 - \omega_n)T/2]}{(\omega_1 - \omega_n)T/2} + (-1)^n \frac{\sin [(\omega_2 - \omega_n)T/2]}{(\omega_2 - \omega_n)T/2}$$

with $\omega_n = \pi n/T$, $\omega_1 = \omega_c - \Delta\omega$, $\omega_2 = \omega_c + \Delta\omega$. This spectrum is shown sketched in Fig. 4-7 for the special case $\Delta f \gg 1/T$. Note that if the original baseband binary signal had bandwidth $B = K/T$ ($K = 2$, or the second zero crossing, would appear to be a good choice here), the bandwidth of the composite signal would appear to be $2\,\Delta f + 2B$, as indicated in Fig. 4-7.

Two extreme cases are of interest:

1. If $\Delta f \gg B$, the bandwidth approaches $2\,\Delta f$. Thus, if one uses a wide separation of tones in an FSK system, the bandwidth is essentially just that separation. It is virtually independent of the bandwidth of the baseband binary signal. *This is distinctly different from the AM case.*
2. If $\Delta f \ll B$, the bandwidth approaches $2B$. In this case, even with the tones chosen very close together, the minimum bandwidth is still that required to transmit an OOK (AM) signal; here the bandwidth *is* determined by the baseband signal.

The first case is commonly called *wideband FM*, the second *narrowband FM*. We shall see shortly that the bandwidth $2\,\Delta f + 2B$ and its two extreme values are quite good approximations to FM bandwidths with complex modulating signals. This analysis through the use of simple binary signals provides, in addition, much more physical insight into FM bandwidth determination than is possible with complex signals.

Note that the FM transmission bandwidth is generally much greater than that for AM, which is always $2B$, that is, twice the baseband bandwidth. Then why use FM? We shall show in a later chapter that it is just this wideband property of FM that makes its performance generally far superior to AM in a noisy environment. This is analogous to the pulse modulation results noted in the previous chapter. Encoding of pulse-amplitude-modulation (PAM) signals into binary pulse-code modulation (PCM) results in an expansion of the system bandwidth but the noise immunity increases considerably. A general characteristic of communication systems to which we shall refer after discussing noise in systems is that one can generally improve system performance in the presence of noise by encoding or modulating signals into equivalent wideband forms. Binary PCM and FM are examples of such wideband signals.

It is common in FM analysis to denote the dependence of transmission bandwidth on the relative magnitudes of the frequency deviation Δf baseband bandwidth B by defining a parameter β, the *modulation index*, as the ratio of the two. Thus,

$$\beta \equiv \frac{\Delta f}{B} \tag{4-4}$$

In terms of β the FM transmission bandwidth is

$$\begin{aligned} \text{FM bandwidth} &= 2\,\Delta f + 2B \\ &= 2B(1 + \beta) \end{aligned} \tag{4-5}$$

Narrowband FM systems correspond to $\beta \ll 1$, wideband systems to $\beta \gg 1$. We shall find the modulation index β playing a significant role throughout our discussion of FM.

PHASE-SHIFT KEYING

In this case we have the phase-shift-keyed signal given as

$$f_c(t) = \pm \cos \omega_c t \qquad -\frac{T}{2} \leq t \leq \frac{T}{2} \tag{4-6}$$

Here a 1 in the baseband binary stream corresponds to positive polarity, and a 0 to negative polarity. The PSK signal thus corresponds essentially to a bipolar binary stream (Fig. 3-24), translated up in frequency. An example is shown in Fig. 4-8. The discontinuous-phase transitions shown

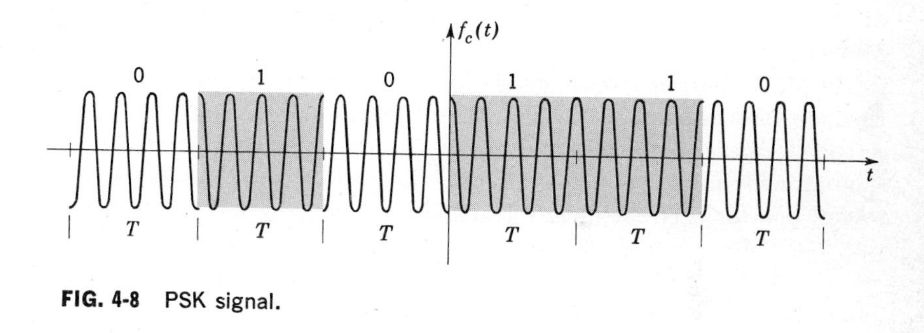

FIG. 4-8 PSK signal.

at the beginning and end of each bit interval, whenever a transition from 1 to 0 or 0 to 1 takes place, are actually smoothed out during transmission (recall that abrupt signal changes require infinite bandwidths to be transmitted). The information regarding polarity is, however, retained in the center of each interval, so that decoding at the receiver is normally timed to be carried out in the vicinity of the center. This is also true for OOK and FSK signals, since the abrupt switches at the beginning and end of the pulses in those cases will also not be transmitted. The spectrum of the PSK signal is essentially that of the bipolar stream, translated up to the carrier frequency f_c.

DETECTION OF BINARY SIGNALS

One may very well ask at this point which of the three binary signaling techniques discussed (and other possible ones as well) one would prefer using in practice. What are the relative advantages and disadvantages of the different techniques? We have already noted that FSK systems perform better in a noisy environment than do OOK systems. We shall show later in this book that PSK systems perform still better and, in fact, may be shown to be the optimum possible for binary signaling in the presence of additive noise. (This ignores intersymbol interference and other types of distortion; it assumes additive noise is the sole form of disturbance during transmission.)

Then why not always use PSK techniques? The answer lies in the detection process at the receiver. Recall that we modulate a sine-wave carrier with the baseband binary stream of necessity to shift the resultant modulated signal to an appropriate frequency for transmission. At the receiver we must undo this process or *demodulate* the signal to recover the original binary stream. This process of demodulation is often also called *detection*. (The binary stream must then be further processed as in Chap. 3 to eventually retrieve the individual signals contained within it.) There are essentially two common methods of demodulation. One, called *synchronous* or *coherent detection*, simply consists of multiplying the incoming signal by the carrier frequency, as locally generated at the receiver, and then low-pass filtering the resultant multiplied signal. The other method is called *envelope detection*. The synchronous detection procedure is diagrammed in Fig. 4-9. Note that the FSK signals require *two* sine waves, one for each frequency transmitted. This is just the reverse of

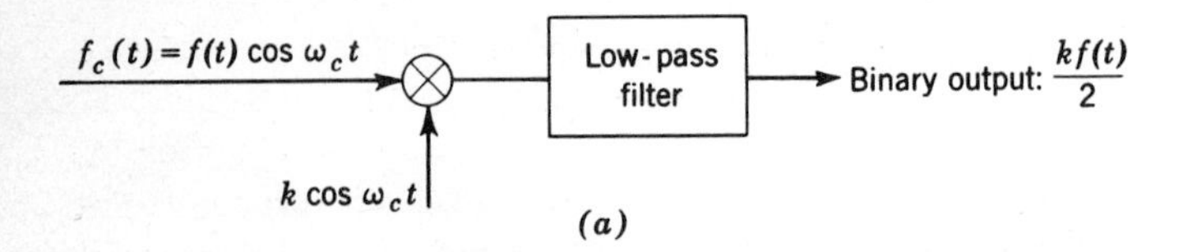

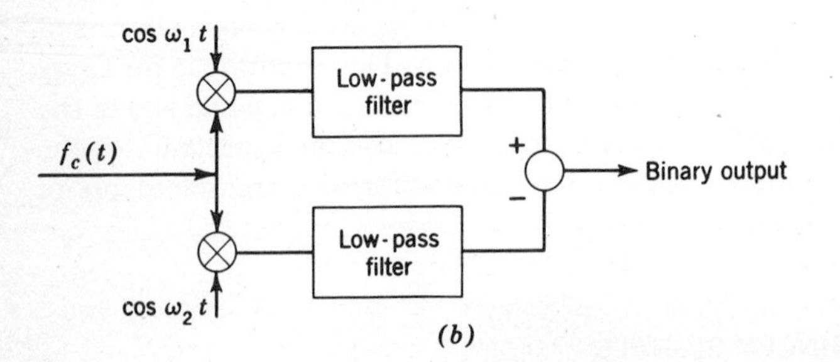

FIG. 4-9 Synchronous detection. (a) OOK or PSK signals; (b) FSK signals.

the original modulation process at the transmitter and serves to translate the binary signals back down to baseband.

To demonstrate the synchronous method, assume the high-frequency binary signal to have the AM form $f_c(t) = f(t) \cos \omega_c t$. [If $f(t) = \pm 1$, we have the PSK signal; if it is 1 or 0, we have the OOK case.] If we multiply this by $k \cos \omega_c t$ as indicated (k is an arbitrary constant of the multiplier), we get $kf(t) \cos^2 \omega_c t = (k/2)(1 + \cos 2\omega_c t)f(t)$. But the term $f(t) \cos 2\omega_c t$ represents $f(t)$ translated up to frequency $2f_c$, the second harmonic of the carrier frequency f_c. This is rejected by the low-pass filter and the output is $(k/2)f(t)$, just the desired baseband binary sequence. (The constant factor has no significance since the output signal can always be amplified or attenuated by any derived amount.) So the synchronous detector does the desired job of reproducing the signal $f(t)$.

But we have assumed the local carrier $\cos \omega_c t$ is exactly at the same frequency as the incoming carrier term $\cos \omega_c t$, and in phase with, or synchronized to, it as well. If the locally generated sine wave were at a frequency $\cos (\omega_c + \Delta\omega)t$, the multiplication would generate $kf(t) \cos (\omega_c + \Delta\omega)t \cos \omega_c t = (k/2)[\cos (2\omega_c + \Delta\omega)t + \cos \Delta\omega t]f(t)$. The output of the low-pass filter would then be $[kf(t)/2] \cos \Delta\omega t$ if $\Delta\omega$ were within the filter passband, which is not at all the desired signal! (This technique for shifting carrier frequencies or superheterodyning AM signals will be discussed later.) Alternately, if the local signal were at the right frequency ω_c, but θ radians out of phase with the incoming carrier, that is, $\cos (\omega_c t + \theta)$, it is apparent that the low-pass filter output would be $[kf(t)/2] \cos \theta$. This is the desired baseband output but is attenuated in amplitude. In particular, as θ increases, $\cos \theta$ decreases. For θ close to $\pi/2$ the output is very nearly zero. As θ increases beyond $\pi/2$, the output signal reverses sign. If the baseband signal is a bipolar sequence, the entire signal reverses polarity and all 1's become 0's, all 0's become 1's! So not only must the locally generated carrier be at the same frequency, but also synchronized in phase as well. This is the reason for the term *synchronous* detection.

Phase synchronism is quite difficult to attain, particularly if transmission takes place over long distances. This means that a receiver clock which provides the synchronism must be synchronized or slaved to the transmitter clock to within a fraction of a carrier cycle, no mean task. As an example, if transmission is at $f_c = 3$ MHz with a period $1/f_c = 0.3$ μsec (this is in the so-called h-f band, used for short-wave broadcasting), a phase difference $\theta \ll \pi/2$ radians implies clock synchronization to within much less than a quarter of a period, or much less than 0.07 μsec! At a carrier frequency of 100 MHz, $\pi/2$ radians corresponds to 2.5 nanosec while at a carrier frequency of 1,000 MHz, this is 0.25 nanosec. Phase synchronization is thus quite a difficult task. It is particularly difficult if either the transmitter or receiver is mounted on a rapidly moving vehicle introducing doppler phase shifts proportional to the relative velocity between transmitter and receiver. If the relative velocities change rapidly enough, the phase shifts in turn change rapidly and synchronism cannot be maintained. The same problem arises if the signal transmission takes place through a fading medium in which randomly moving scatterers also introduce random doppler phase shifts.

The reader may have recognized already that this problem of maintaining synchronism is similar to that encountered in baseband binary transmission in the previous chapter. The problem here, however, is a much more difficult one; the synchronism previously considered was, for example, from bit interval to bit interval. With high carrier frequencies, $1/f_c \ll T$, the bit interval, so that the problem is compounded. As with the baseband digital synchronization case, various methods are available for obtaining the required phasing information.

1. A pilot carrier may be transmitted, and superimposed on the high-frequency binary signal stream, which may be extracted at the receiver and used to synchronize the receiver local oscillator (this is similar to the marker or timing pulses transmitted with the baseband data stream to maintain bit interval, word, and frame synchronism).
2. A so-called phase-locked loop, locking on either the data stream or a pilot tone, may be used at the receiver to drive the phase difference to zero,[1] etc.

We shall have more to say about synchronous detection later in this chapter, as well as in the discussions on binary transmission in the presence of noise. It is interesting to note, however, that it is not really necessary to physically multiply by a pure sine wave to obtain the desired demodulated form $f(t)\cos^2 \omega_c t$. Switching or gating $f_c(t)\cos \omega_c t$ on and off at a rate of f_c times per second and then using low-pass filtering will accomplish the same job (see Fig. 4-10). This is apparent from a consideration of the switching function of Chap. 3. All-digital gating circuits are in fact often simpler to design and use for this purpose than is multiplication by a pure

[1] A. J. Viterbi, "Principles of Coherent Communication," chap. 2, McGraw-Hill, New York, 1966.

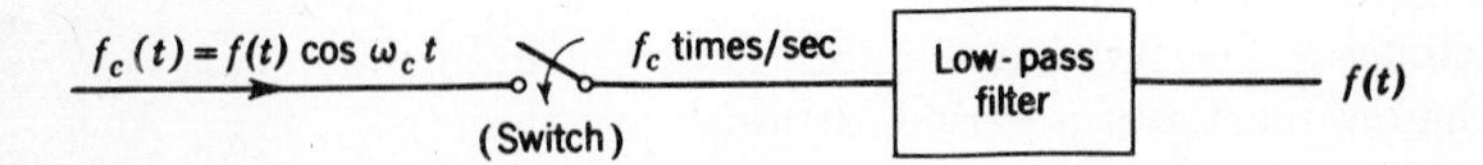

FIG. 4-10 Another synchronous detector.

sine wave. [If the baseband binary stream contains time-multiplexed information, further switching (see Chap. 3) is then necessary to separate out the individual signals. But this switching, gating, or commutating function is of course carried out at the much slower, baseband, rate.]

The difficulty of maintaining phase synchronism notwithstanding, PSK transmission and phase-locked loop synchronous detection has been successfully used in space communications. The Pioneer V deep-space probe, as an example, used biphase modulation in its telemetry system. The Pioneer IV system also used phase-coherent techniques.[1]

The other common form of detection, *envelope detection*, avoids the timing and phasing problems of synchronous detection. Here the incoming high-frequency signal is passed through a nonlinear device and low-pass filter. Envelope detection will be considered later in this chapter in discussing more general AM signals and receivers. Suffice it to say at this point that one common form of envelope detector is a diode half-wave rectifier (the nonlinear device) followed by a RC low-pass filter (Fig. 4-11).

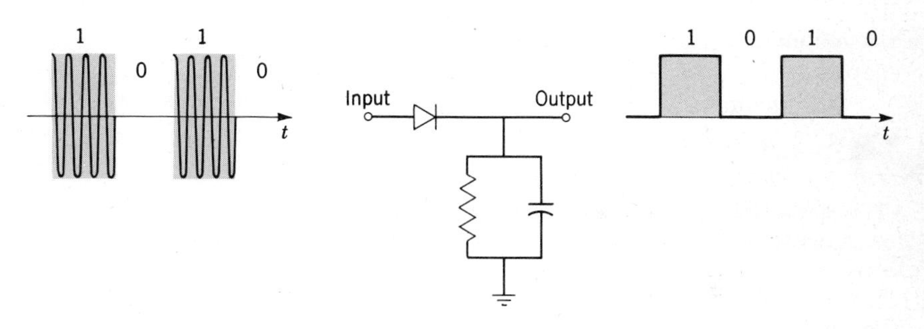

FIG. 4-11 Envelope detector.

As the name indicates the output of the detector represents the envelope of the incoming high-frequency wave. The RC time constant is long enough to hold the incoming amplitude over many carrier cycles, yet short enough compared to a binary period to discharge once the binary signal changes.

Notice one difficulty, however. The PSK signal has a *constant* envelope (Fig. 4-8) so that it cannot be used with an envelope detector. Thus the PSK system *requires* synchronous detection. As usual there is a tradeoff in economics. One may avoid using synchronous detection by adopting envelope detection, but at the price of requiring

[1] See, for example, A. V. Balakrishnan (ed.), "Space Communications," chaps. 14 and 15, McGraw-Hill, New York, 1963.

FSK or OOK transmission, rather than the more optimum (in the presence of noise) PSK system.[1] We shall also show in discussing the effects of noise, in later chapters, that envelope detection of OOK or FSK signals is somewhat inferior to synchronous detection of these signals. An additional price must thus be paid for this much simpler circuit, although in this case it can be demonstrated that the cost will not be too great.

The envelope-detection process requires the presence of an unchanging carrier term in addition to the varying high-frequency binary signal. This is the reason envelope detection cannot be used with PSK signals. To demonstrate this point consider an OOK signal $f_c(t) = Af(t) \cos \omega_c t$. Recall that $f(t)$ is a random sequence of 1's and 0's. Squaring the expression (this is the simplest nonlinear operation that demonstrates the envelope-detection process), we get $A^2 f^2(t) \cos^2 \omega_c t$. But $f(t) = 1$ or 0, so $f^2(t) = 1$ or 0 as well. The output of a low-pass filter is then $A^2/2$ or 0, reproducing the derived 1,0 sequence. In the case of the PSK signal, however, $f(t) = \pm 1$, $f^2(t) = 1$, and the output is *always* $A^2/2$. There are other ways of analyzing the envelope detector, and these are considered in discussions of AM systems, as well as in the problems at the end of this chapter.

To conclude the discussion of binary signaling at this point, we show in Fig. 4-12 a diagram of a complete time-multiplexed PCM system.

[1] This is why the space communication systems have used synchronous detection, the cost notwithstanding. The reduction possible in transmitter power is worth the price there.

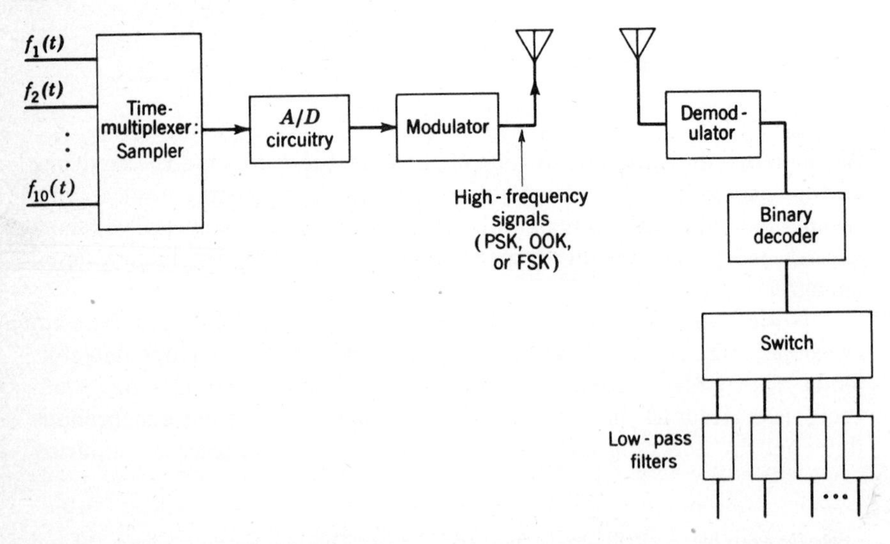

FIG. 4-12 Complete PCM system.

This includes the initial time-multiplexing and A/D circuitry (the latter includes the quantization and binary encoding operations); the modulator which produces the high-frequency binary signals; then, at the receiver, the demodulator, which includes a synchronous or envelope detector, binary decoder, a switch, or commutator circuitry for unscrambling the time-multiplexed signals; and finally low-pass filters in each output channel for providing the final output signals. Note that this is essentially the same set of blocks discussed in Chap. 3, but with the addition of the high-frequency modulator and demodulator.

4-3 AMPLITUDE MODULATION

In the previous section we discussed sine-wave or c-w modulation systems in which the modulating signal consisted of a digital pulse train. Such systems are important in their own right because of the continuously increasing use of digital systems. The discussion was also important because it enabled us to introduce the basic concept of sine-wave modulation and demodulation (or detection), as well as to introduce spectrum considerations for AM and FM systems in which the modulating signal takes a particularly simple form. We now generalize our approach somewhat, and consider AM and FM systems designed to handle more complex signals than a sequence of binary digits. The signals could be voice, TV, as encountered in standard broadcasting, or complex combinations of both digital and analog waveshapes, obtained by time multiplexing and frequency multiplexing many individual signal channels into one master signal. This composite signal would in turn amplitude modulate or frequency modulate a sine-wave carrier.

In this section, we concentrate on AM systems. Much of the material is identical to that already discussed previously, so that our treatment will be rather brief. We shall then go on to single-sideband (SSB) systems, concluding with a discussion of FM systems. Here too the material, particularly that relating to spectral considerations, will be similar to that presented previously.

Recall that when we talk of sine-wave modulation we imply that we have available a source of sinusoidal energy with an output voltage or current of the form

$$v(t) = A \cos (\omega_c t + \theta) \tag{4-7}$$

This sinusoidal time function is called a carrier. Any one of the three quantities A, ω_c, or θ may be varied in accordance with the information-carrying, or modulating signal. We restrict ourselves in this section to AM systems in which A only is assumed to be varied. A basic assumption to be adhered to in all the work to follow will be that the modulating signal varies slowly compared with the carrier. This then means that we can talk of an envelope variation, or variation of the locus of the carrier peaks. Figure 4-13 shows a typical modulating signal $f(t)$ and the carrier

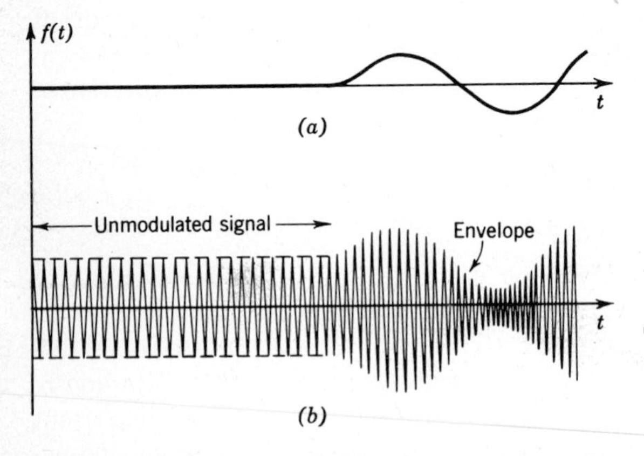

FIG. 4-13 Amplitude modulation of a carrier. (a) Modulating signal; (b) amplitude-modulated carrier.

envelope variation corresponding to it. The amplitude-modulated carrier of Fig. 4-13 can be described in the form

$$f_c(t) = K[1 + mf(t)] \cos \omega_c t \qquad [4\text{-}8]$$

(We arbitrarily choose our time reference so that θ, the carrier phase angle, is zero.) Obviously, $|mf(t)| < 1$ in order to retain an undistorted envelope, in which case the envelope is a replica of the modulating signal. Note that in both Chap. 3 and the previous section, we discussed modulated carrier terms of the form

$$f_d(t) = f(t) \cos \omega_c t \qquad [4\text{-}9]$$

The present expression, Eq. [4-8], differs by the addition of the carrier term $K \cos \omega_c t$. This term is necessary to ensure the existence of an envelope, as shown in Fig. 4-13. This was no problem in the case of the OOK signals of the previous section, since the on-off sequence of pulses ensured the presence of an envelope. With more complex modulating signals $f(t)$, one must ensure this by adding the appropriate carrier term.

Some systems actually transmit signals of the form of Eq. [4-9]. These are called *double-sideband* (DSB) or *suppressed-carrier* systems. As indicated earlier (Figs. 4-2 to 4-4), the effect of multiplying an arbitrary signal $f(t)$ by $\cos \omega_c t$ is to translate or shift the spectrum up to the range of frequencies surrounding the carrier frequency f_c. The further addition of the carrier term, as in Eq. [4-8], provides a discrete spectral line at frequency f_c as well. Figures 4-3 and 4-4, for the case of a single-sine-wave modulating signal $f(t) = \cos \omega_m t$ and a group of sine waves, respectively, then become Figs. 4-14 and 4-15, with the additional carrier line shown.

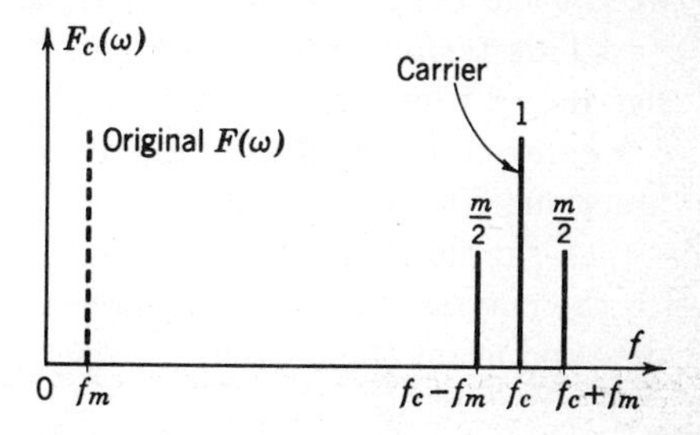

FIG. 4-14 AM spectrum, including carrier. (Positive frequencies only are shown here.)

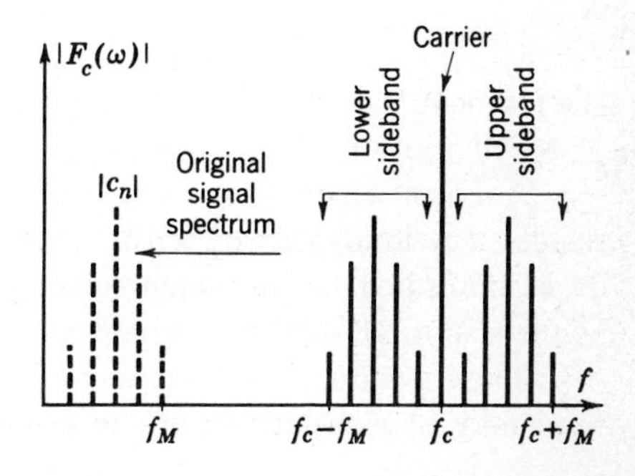

FIG. 4-15 Amplitude-modulation spectrum-periodic modulating signal.

Similar results of course hold for an aperiodic $f(t)$. The only stipulation here, as is apparent from Fig. 4-2, is that the bandwidth B of the modulating signal be less than the carrier frequency f_c. Two sets of side frequencies appear, as noted earlier: an upper and a lower set. Each set contains a band of frequencies corresponding to the band of frequencies covered by the original signal and is called a sideband. Each sideband contains all the spectral lines—both amplitude and phase—of the original signal and so presumably contains all the information carried by $f(t)$. It is left for the reader to show that the phase angles of the upper and lower sidebands differ in sign. [As a hint, recall from Chap. 2 that for real $f(t)$, the phase of the Fourier transform $F(\omega)$ is *odd* symmetrical about the origin.]

We now ask ourselves the very pertinent question: How do we physically produce an AM signal? One method indicated in Chap. 3 and referred to again in the previous section was that of gating or switching $f(t)$ on and off at the carrier rate. This translates $f(t)$ up to all harmonic multiples of frequency f_c. Bandpass filtering at one of these frequencies, we get the DSB form of the AM signal.

This technique is one of several that can be used to generate AM signals. From the form of Eqs. [4-8] and [4-9] it is apparent that we need some device that will provide at its output the product of the two input functions $f(t)$ and $\cos \omega_c t$.

A simple example might be a galvanometer with separate windings for the magnet and the coil. The force on the coil is then proportional to the product of the currents in each winding. If

$$F \propto i_1 i_2 \qquad [4\text{-}10]$$

$$i_1 = a[1 + mf(t)] \qquad [4\text{-}11]$$

(that is, a dc bias plus the modulating signal) and

$$i_2 = b \cos \omega_c t \qquad [4\text{-}12]$$

the force on the coil, as a function of time, is of the same form as Eq. [4-8]. The coil motion would then appear as an amplitude-modulated signal.

A motor with stator and rotor separately wound would give the same results. A loudspeaker, with the modulating signal and an appropriate dc bias applied to the voice coil and the carrier to the field winding, is again a similar device. These are all examples of *product* modulators.

These product modulators and other types of modulators commonly used may be classified in one of two categories:

1. Their terminal characteristics are nonlinear.
2. The modulation device contains a switch, as just described, which changes the system from one linear condition to another.

A *linear* time-invariant device by itself can never provide the product function needed. We may recall from our discussion of linear systems in Chap. 2 that *a non-time-varying linear system can generate no new frequencies.* The response to all sine-wave inputs may be superposed at the output. Such a system can therefore never be used as a modulator.

If we now add a switch to a linear system which switches the system in a specified manner from one linear condition to another (a simple example would be a switch turning the system on and off) and which does this independently of the signal, the system is still basically linear; the two basic characteristics of a linear system still apply. Such a system is still governed by linear differential equations, but the coefficients of the equations now vary with time according to the prescribed switching action.

The system is now a linear time-varying system. As we shall see shortly, such a system can be used for modulation and will generate new frequencies. This is then a generalization of our previous switching process.

If instead of a switch we now introduce a device whose static terminal characteristics are nonlinear, the differential equations governing the system become nonlinear differential equations. Again, as we shall see, modulation or generation of new frequencies becomes possible. Two typical kinds of nonlinear characteristics are shown in Fig. 4-16. Actually all physical devices have some curvature or nonlinearity in their static terminal characteristics, but we usually assume the region of operation to be small enough so that the characteristics are very nearly linear over this region.

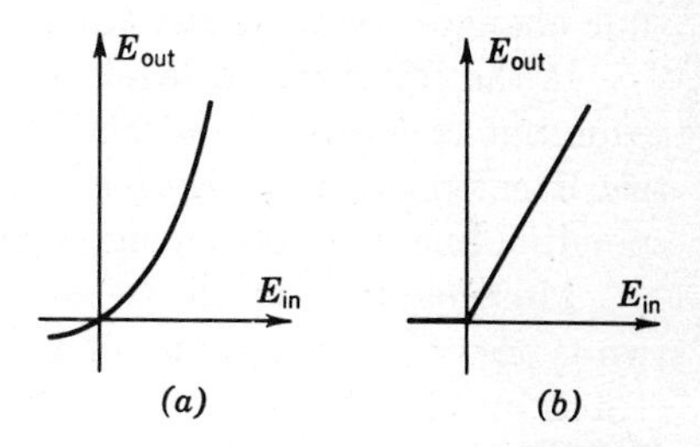

FIG. 4-16 Nonlinear characteristics. (a) Nonlinearity due to curvature of characteristics; (b) "strong" discontinuity; piecewise-linear characteristic.

Figure 4-16*a* might be the characteristic of a typical square-law diode or transistor. Figure 4-16*b* might represent an approximation to a linear rectifier. The two curves differ in the type of nonlinearity. The linear rectifier of Fig. 4-16*b* has a strong nonlinearity at one point and is approximated by linear characteristics elsewhere. (The use of the term linear rectifier and the piecewise-linear characteristics assumed have sometimes led to the mistaken conclusion that the device is linear.) The characteristics of Fig. 4-16*a* have no one point of strong nonlinearity but are everywhere nonlinear owing to the curvature of the characteristics.

Figure 4-16*b* can also be used as the representation of the characteristics of a switching device and, in fact, serves as a good example of the distinction between linear time-invariant, nonlinear, and switching characteristics. Thus, if the device is operated in one of the linear portions of the curve only, it behaves as a linear time-invariant device. Typically, the device would be biased to operate in a region far from the discontinuity, and the signals applied would be small enough to keep it in that region. The output is then a linear replica of the input, and if two signals are applied, the output is a superposition of the response to each. If the signal applied to the device is now increased to the point where the discontinuity is crossed and the two linear regions are involved, the output is no longer a simple linear function of the input. If one signal is applied, the output is a distorted version, and if two signals are applied, the output no longer can be found by superposing the two input responses separately. The output is then a nonlinear function of the input; as the input amplitude changes, the output does not change proportionally.

If the input-signal level is now dropped back to its original small value but is somehow switched in a predetermined fashion between the two linear regions of Fig. 4-16*b*, we have a linear switch with the corresponding linear time-varying equations.

Most modulators in use can thus be classified as having one of the two types of characteristics of Fig. 4-16 if switches are included as represented by Fig. 4-16*b*.

We shall first analyze simple modulators incorporating the nonlinear characteristics of Fig. 4-16*a*. The analysis of the piecewise-linear characteristic of Fig. 4-16*b* will be very similar to that describing the switching operation in Chap. 3.

SQUARE-LAW MODULATOR

The simplest form of this type of modulator appears in Fig. 4-17. e_o and e_i represent the incremental output and input variations, respectively, away from some fixed operating point (Fig. 4-17*b*). The nonlinear characteristics of this case are assumed continuous so that e_o may be expanded in a power series in e_i,

$$e_o = a_1 e_i + a_2 e_i^2 + a_3 e_i^3 + \text{(higher-order terms)} \qquad [4\text{-}13]$$

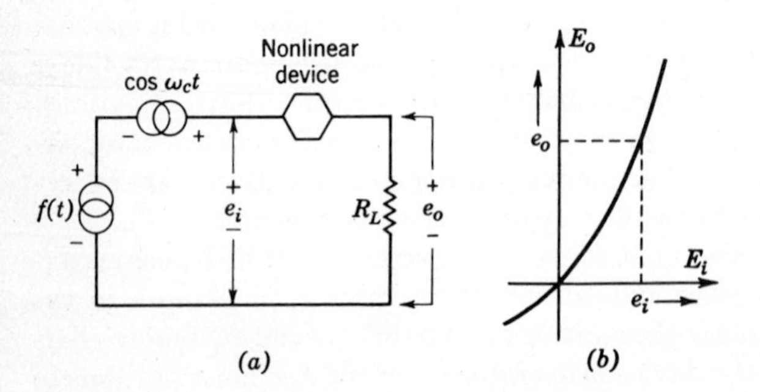

FIG. 4-17 Simple modulator. (a) Simple circuit; (b) terminal characteristics.

The quadratic term in the power series denotes the presence of curvature in the characteristics and is all-important in the modulation process.

For the circuit in Fig. 4-17a

$$e_i = \underbrace{\cos \omega_c t}_{\text{Carrier}} + \underbrace{f(t)}_{\text{Modulating signal}} \qquad [4\text{-}14]$$

Then, retaining just the first two terms of the power series of Eq. [4-13] we get

$$\begin{aligned} e_o &= a_1 \cos \omega_c t + a_1 f(t) + a_2[\cos^2 \omega_c t + 2f(t) \cos \omega_c t + f^2(t)] \\ &= \underbrace{a_1 f(t) + a_2 \cos^2 \omega_c t + a_2 f^2(t)}_{\text{Unwanted terms}} + \underbrace{a_1 \cos \omega_c t \left[1 + \frac{2a_2}{a_1}\right] f(t)}_{\text{Amplitude-modulated terms}} \end{aligned} \qquad [4\text{-}15]$$

The second term thus contains the desired AM signal. (Let $m \equiv 2a_2/a_1$, $K \equiv a_1$.) The first term contains unwanted terms that can be filtered out. $\mathrm{Cos}^2\, \omega_c t$ gives $\frac{1}{2}(1 + \cos 2\omega_c t) = dc$ and twice the carrier frequency. $f^2(t)$ contains frequency components from dc to twice the maximum frequency component of $f(t)$, that is, $2B$. This is left as an exercise for the reader. It may be noted in passing that this device could also be used as a second-harmonic generator if $\cos \omega_c t$ only were introduced and its second harmonic retained.

PIECEWISE-LINEAR MODULATOR (STRONG NONLINEARITY)

A typical circuit (identical with the circuit of Fig. 4-17a) and the piecewise-linear characteristics for such a device (here chosen as a "linear" rectifier) are shown in Fig. 4-18. R_L is the load resistance; R_d, the diode forward resistance.

We should like to demonstrate that the piecewise-linear characteristics can also be used for simple AM. The discontinuity assumed in the terminal characteristics prevents the use of the power-series expansion as was done in the previous modulator.

We could, of course, approximate the characteristics by a polynomial. This would convert them to the continuous form of Fig. 4-17 with the discontinuity more pronounced at the origin. Both types of nonlinearity could thus be treated simultaneously. We prefer, however, to use a different approach similar to that used in discussing the switching operations in Chap. 3.

We shall assume that the carrier amplitude is much greater than the maximum value of $f(t)$,

$$|f(t)| \ll a$$

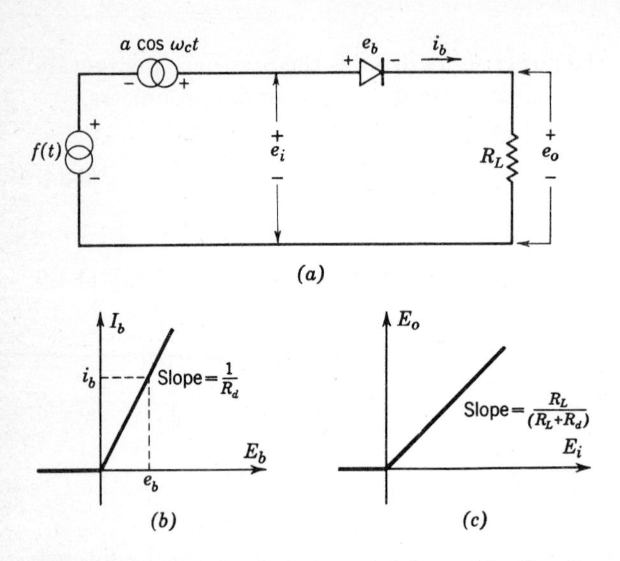

FIG. 4-18 Piecewise-linear modulator. (a) Simple modulator; (b) diode characteristic; (c) circuit characteristic.

If the carrier alone were present, the rectifier would clip the negative halves of the carrier. Under the assumption of a strong carrier the clipping of the carrier plus modulating signal will occur approximately at the same point in the cycle as the carrier alone. We thus have

$$
\begin{aligned}
e_i &= a \cos \omega_c t + f(t) & |f(t)| \ll a \\
e_o &\doteq \frac{R_L e_i}{R_L + R_d} = b e_i & a \cos \omega_c t > 0 \\
e_o &\doteq 0 & a \cos \omega_c t < 0
\end{aligned}
\qquad [4\text{-}16]
$$

Since the output now varies between two values (be_i and 0) periodically, at the frequency of the carrier, the input may be looked on as having been switched between two regions of the diode operation. By the simple expedient of assuming a weak signal compared with the carrier we have converted the nonlinear device into a linear switching device. Since the transitions from one region of operation to another are now variables of time independent of the signal $f(t)$, we have effectively replaced a nonlinear equation by a linear time-varying one. This approach is the one actually used in solving some types of nonlinear differential equations. It is to be emphasized, however, that this is an approximate technique, valid only for small signals. Equation [4-16] can be written mathematically as

$$e_o \doteq [a \cos \omega_c t + f(t)]S(t)$$

with

$$S(t) = b \qquad -\tfrac{1}{4}T < t < \tfrac{1}{4}T,\ T = \frac{1}{f_c} \qquad [4\text{-}17]$$

$$S(t) = 0 \qquad t \text{ elsewhere}$$

and repeating at multiples of $T = 1/f_c$ sec.

$S(t)$ is again our familiar periodic switching function as shown in Fig. 4-19. Again using the Fourier representation,

$$S(t) = b\left[\frac{1}{2} + \sum_{n=1}^{\infty} \frac{\sin (n\pi/2)}{n\pi/2} \cos n\omega_c t\right] \qquad [4\text{-}18]$$

But

$$e_o = [a \cos \omega_c t + f(t)]S(t)$$

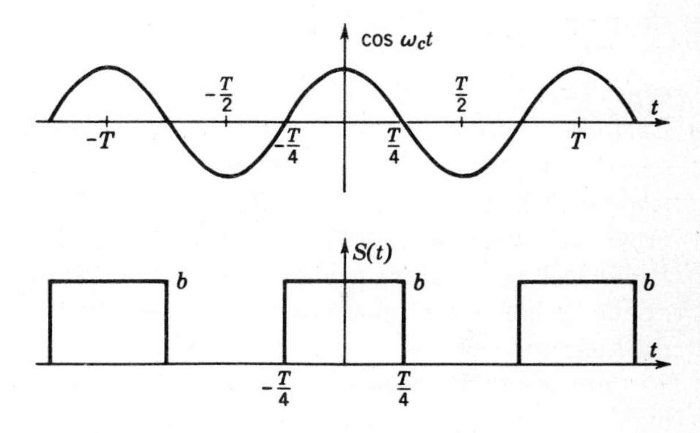

FIG. 4-19 Switching function.

By using the Fourier-series expansion for $S(t)$, performing the indicated multiplication, and collecting terms, e_o can be written in the form

$$e_o(t) = b\Bigg(\Bigg\{\tfrac{1}{2}f(t) + \frac{2}{\pi} a \cos^2 \omega_c t + \sum_{n=3}^{\infty} \frac{\sin (n\pi/2)}{n\pi/2} [f(t) + a \cos \omega_c t] \cos n\omega_c t\Bigg\} + \frac{a}{2} \cos \omega_c t \left[1 + \frac{4}{\pi a} f(t)\right]\Bigg) \qquad [4\text{-}19]$$

Again, we can filter this output. The terms in the braces give direct-current, low frequencies up to B [the maximum frequency component of $f(t)$], and then higher frequencies: $2f_c$, $3f_c - B$, $3f_c$, $3f_c + B$, etc. After filtering,

$$e_o(t) \doteq K[1 + mf(t)] \cos \omega_c t$$

where

$$K = \frac{ab}{2} \qquad [4\text{-}20]$$

$$m = \frac{4}{\pi a}$$

The piecewise-linear modulator of Fig. 4-18 thus produces an amplitude-modulated output upon proper filtering.

We have actually demonstrated this modulation property for the device considered as a switch rather than as a nonlinear device, under the assumption of a small signal (relative to the carrier). The truly nonlinear solution leads to harmonics of the signal $f(t)$ and higher-order modulation products of the carrier and its harmonics. These additional terms would be filtered out in an actual modulator, leaving Eq. [4-20] as the solution in the more exact analysis also.

The two nonlinear devices just discussed are representative of many types that are used for low-level (i.e., low-power) modulation purposes. Examples include nonlinear-resistance modulators, semiconductor diode modulators, mechanical-contact modulators (also called "choppers," or *vibrators*, which are basically switching devices of the piecewise-linear type previously discussed), semiconductor modulators such as photocells and Hall-effect modulators using germanium crystals, magnetic modulators, nonlinear-capacitor modulators, vacuum-tube modulators, transistor modulators, etc.

For standard radio-broadcast work these low-level modulators are highly inefficient. (Their use in balanced modulator circuits for SSB transmission or as parts of servo systems is discussed later in this chapter.) Class C amplifier operation provides higher efficiency of power generation and is commonly used for standard broadcast transmitters. The modulation process here involves the gross nonlinearity due to class C operation. A tuned circuit at the amplified output then provides the necessary filtering to obtain the desired amplitude-modulated wave: a carrier plus sidebands.

A simplified example of a tube-type plate modulated class C amplifier is shown in Fig. 4-20. With E_{cc} highly negative the tube is normally cut off. Conduction thus occurs only at the peak of the carrier input signal. The output is then a series of pulses at the carrier frequency. (The carrier "switches" the tube on, just as in the case of the piecewise-linear rectifier previously analyzed as one basic prototype for many nonlinear circuits.) The magnitude of the plate current varies very nearly linearly with the plate voltage. Varying E_{bb} by means of $f(t)$ therefore produces pulses of plate current whose amplitude is proportional to $f(t)$—the modulating

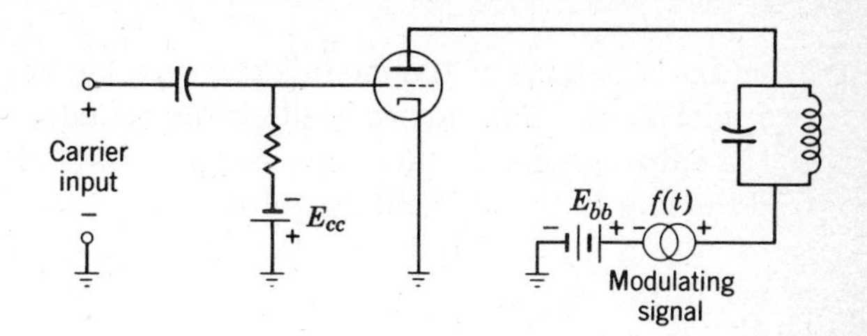

FIG. 4-20 Simplified version: plate-modulated class C amplifier.

signal. The presence of the desired carrier and sidebands, the amplitude-modulated signal, can, of course, be demonstrated just as in the case of the piecewise-linear rectifier by expanding the mathematical expression for the modulated pulses in a Fourier series. Transistor class C amplifiers, of course, function the same way.

A block diagram of a typical standard broadcast transmitter is shown in Fig. 4-21. This is a simplified version with only the essential elements shown.

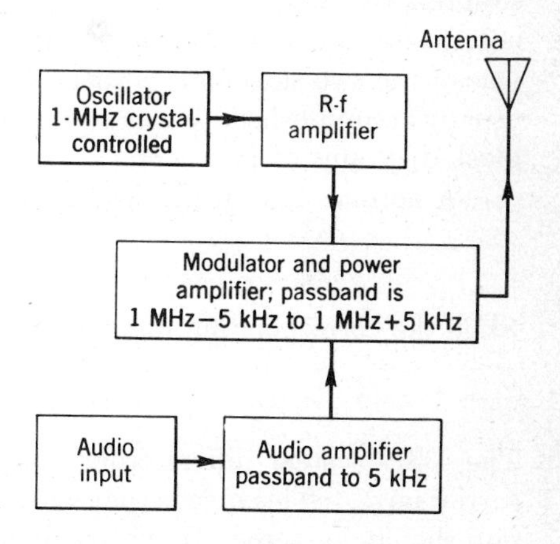

FIG. 4-21 Standard high-level broadcast transmitter.

4-4 SINGLE-SIDEBAND TRANSMISSION[1] AND BALANCED MODULATORS

We note from the frequency plot of the amplitude-modulated carrier (Fig. 4-15) that the desired information to be transmitted [as originally

[1] Single-sideband issue, *Proc. IRE*, vol. 44, no. 12, December, 1956.

given by variations of the modulating signal $f(t)$] is carried in one of the two sidebands. The carrier itself carries no information. Transmission of the entire spectrum thus represents a waste of frequency space since transmitting both sidebands requires double the bandwidth needed for transmitting one sideband. It also represents a waste of power (especially power in the carrier). This has resulted in the widespread use of SSB transmission for transoceanic radiotelephone circuits and wire communications. In this type of transmission the carrier and one sideband are suppressed and only the remaining sideband transmitted.

There is also experimental evidence to the effect that SSB transmission is less susceptible to fading than AM transmission.

With the apparent advantages of SSB over AM, why is it that SSB is not used for standard radio broadcasting? The answer is that the circuitry *required at the receiver* is too complex *at present.* Compatible SSB techniques have been proposed to alleviate this problem.[1] To demonstrate these statements more quantitatively, we must discuss methods of producing and then detecting SSB signals.

A method commonly used to generate an SSB signal is first to suppress the carrier of the AM signal. The resulting transmission is, as noted earlier, *suppressed-carrier,* or *double-sideband* (DSB), transmission. One of the two sidebands is then filtered out. The carrier suppression is usually accomplished by means of a *balanced modulator,* as indicated in the block diagrams of Fig. 4-22.

A normal AM output has the form

$$f_c(t) = K[1 + mf(t)] \cos \omega_c t \qquad [4\text{-}21]$$

while the corresponding mathematical expression for a DSB output is

$$f_d(t) = K'f(t) \cos \omega_c t \qquad [4\text{-}22]$$

The DSB system differs from normal AM simply by suppressing the carrier term. This means that with no modulating signal applied the output should be zero. If we recall the nonlinear devices discussed previously, the carrier term arose because of a quiescent condition or dc bias somewhere in the modulator. If we can balance out this quiescent condition, we have a simple product modulator with the output given by Eq. [4-22]. (In the case of the magnetic product modulators mentioned —the galvanometer, motor, loudspeaker—the assumption was that a dc

[1] L. R. Kahn, Compatible Single Sideband, *Proc. IRE,* vol. 49, pp. 1503–1527, October, 1961. The article describes a compatible SSB system developed for use with normal AM radio sets. See also Schwartz, Bennett, and Stein, *op. cit.,* pp. 193–198, for a discussion of this and other compatible SSB systems.

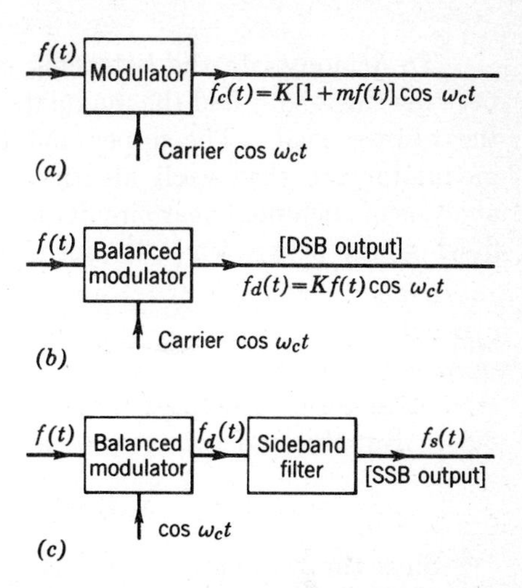

FIG. 4-22 Amplitude-modulation systems. (a) Normal AM; (b) suppressed carrier, or DSB; (c) SSB.

term appeared in one of the currents. If we balance out this dc current, we have the desired suppressed-carrier output.)

One possible scheme for suppressing or balancing out the carrier term is shown in Fig. 4-23. This method uses two nonlinear elements of the types previously considered in a balanced arrangement. The resulting device is then an example of a *balanced modulator*.

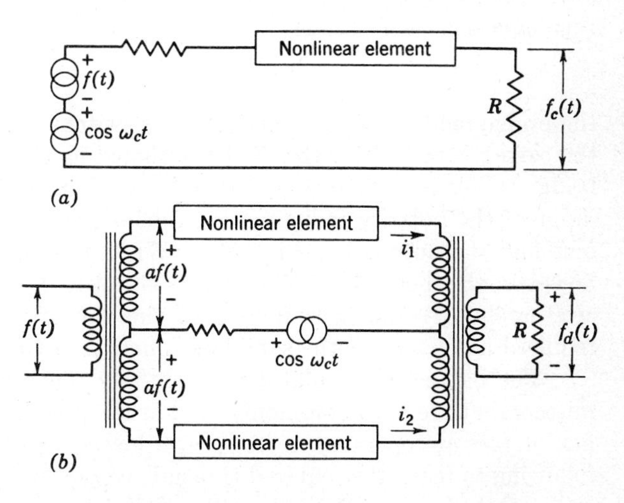

FIG. 4-23 Balanced modulator. (a) Modulator; (b) balanced modulator.

To demonstrate the balancing effect, assume that the transformer between the load R and the modulator is ideal. The modulator then sees the resistor load. The upper and the lower sections of the balanced modulator are then each identical with Fig. 4-23a, and the previous analysis of such nonlinear circuits holds. In particular, the two currents flowing may be written in the form

$$i_1 = K[1 + mf(t)] \cos \omega_c t \qquad [4\text{-}23]$$

and

$$i_2 = K[1 - mf(t)] \cos \omega_c t \qquad [4\text{-}24]$$

(the other terms generated in the nonlinear process are assumed filtered out). But the load voltage $f_d(t)$ is proportional to $i_1 - i_2$, or

$$f_d(t) = K'f(t) \cos \omega_c t \qquad [4\text{-}25]$$

Since the switching operation discussed in Chap. 3 in connection with sampling was shown to produce terms of exactly the DSB form, these switches may also be considered balanced modulators. One such switch, the shunt-bridge diode modulator, is shown in Fig. 4-24. The diodes in

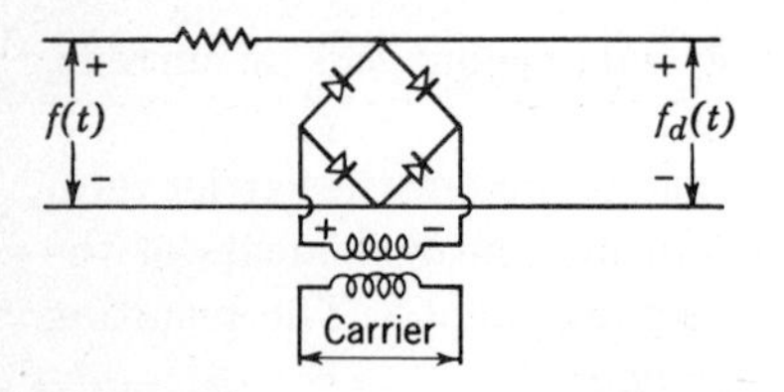

FIG. 4-24 Balanced modulator: shunt-bridge diode modulator.

the bridge modulator may be treated as switches, switching on and off at the carrier rate. With the diodes assumed identical (i.e., balanced) the bridge is balanced when $f(t) = 0$ and there is no output; with the carrier input of the polarity indicated the diodes are essentially short-circuited out, and the output is again zero. When the carrier reverses polarity, however, the diodes open and $f_d(t)$ is equal to $f(t)$. The output is thus alternately $f(t)$ and 0, switching at the carrier-frequency rate, just as required here, and for the sampling function in Chap. 3 as well.

The balanced modulator is essentially the heart of the SSB transmitters that are most commonly used at the present time. The balanced modulators are operated at low power levels, with linear amplifiers then following to reach the required transmitted power. This is in contrast to ordinary AM (Fig. 4-21), where the modulation is frequently carried out at high power levels, with class C amplifiers.

In addition to the balanced modulator, the SSB system requires a sideband filter with sharp cutoff characteristics at the edges of the passband. (Attenuation at the carrier frequency must be especially high.) Because of the stringent requirements on the filter, the modulation is commonly performed at a relatively low fixed frequency at which it is possible to design the required sideband filter.

A typical SSB transmitter incorporating the two features mentioned —low-level balanced modulator and low-frequency (l-f) filtering—is shown in Fig. 4-25.[1] The modulation process in this transmitter is carried

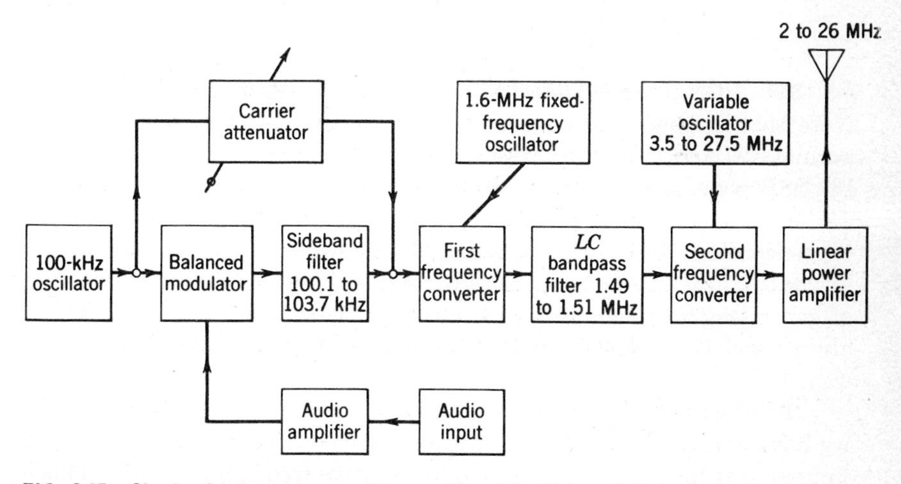

FIG. 4-25 Single-sideband transmitter. (From B. Fisk and C. L. Spencer, Synthesizer Stabilized Single-sideband Systems, *Proc. IRE*, vol. 44, no. 12, p. 1680, December, 1956, by permission.)

out at 100 kHz. Two levels of frequency conversion, or mixing, are employed, with the signal successively shifted up in frequency by the frequency of the two injected signals (1.6 MHz and the final variable frequency).

Frequency converters are basically similar to the modulators previously discussed, with a large injected signal of frequency close to that of the incoming signal superimposed on an incoming wave. The combined signal is then put through a nonlinear device and sum and difference frequencies again obtained. In the transmitter shown in Fig. 4-25 the difference frequencies of the second converter are rejected and sum fre-

[1] B. Fisk and C. L. Spencer, Synthesizer Stabilized Single-sideband Systems, *Proc. IRE*, vol. 44, no. 12, p. 1680, December, 1956.

quencies transmitted. This serves to raise the frequency of the incoming signal by the frequency of the injected signal. In a receiver the reverse process takes place: The difference frequencies are transmitted and the incoming signal thus lowered in frequency.

In the discussion of SSB detection to follow, the need for reinjecting the suppressed carrier will be discussed. To ensure that the reinjected carrier is of precisely the right frequency a pilot carrier of known amplitude is transmitted (usually 10 to 20 db below the carrier in normal AM). The carrier attenuator in Fig. 4-25 proves for the pilot carrier.

4-5 SSB SIGNAL REPRESENTATION: HILBERT TRANSFORMS[1]

Note that although we have written mathematical expressions for both normal AM and DSB (suppressed-carrier) signals (Eqs. [4-21] and [4-22]), the SSB generation has been described only in terms of filtering out one of the sidebands of the DSB signal. It is of interest to develop an explicit expression for a SSB signal not only for its own sake, but because it also serves two other purposes: (1) as a byproduct it indicates an alternate way of generating SSB signals (the so-called *phase-shift method* of SSB generation); and (2) it enables us to discuss SSB detection in a quantitative way.

To obtain the desired expression for a single-sideband signal assume we have available a baseband or low-frequency time function $z(t)$ whose Fourier transform $Z(\omega)$ is nonzero for positive frequencies only. This is admittedly not a real or physical possible time function, since we showed in Chap. 2 that all real time functions had Fourier transforms defined over both negative and positive frequency ranges. [In fact recall that $|F(\omega)| = |F(-\omega)|$ for real $f(t)$, while the phase angle of $F(\omega)$ is odd symmetric about zero frequency.] By definition, then,

$$Z(\omega) = 0 \qquad \omega < 0 \tag{4-26}$$

An example of such a Fourier transform is shown in Fig. 4-26*a*. Now assume $z(t)$ multiplied by $e^{j\omega_c t}$. A little thought will indicate that this corresponds to a translation up by frequency ω_c, so that the transform of $z(t)e^{j\omega_c t}$ is just $Z(\omega - \omega_c)$. This is shown in Fig. 4-26*b*. It is apparent from the figure that this is just a SSB signal with *upper* sideband only present. A function $z(t)$ with the peculiar one-sided spectral property is called an *analytic signal*.[2] Since $z(t)e^{j\omega_c t}$ is complex, we simply take its

[1] Schwartz, Bennett, and Stein, *op. cit.*, pp. 29–35.

[2] Schwartz, Bennett, and Stein, *ibid.*

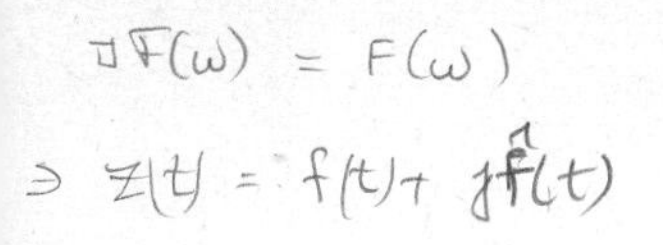

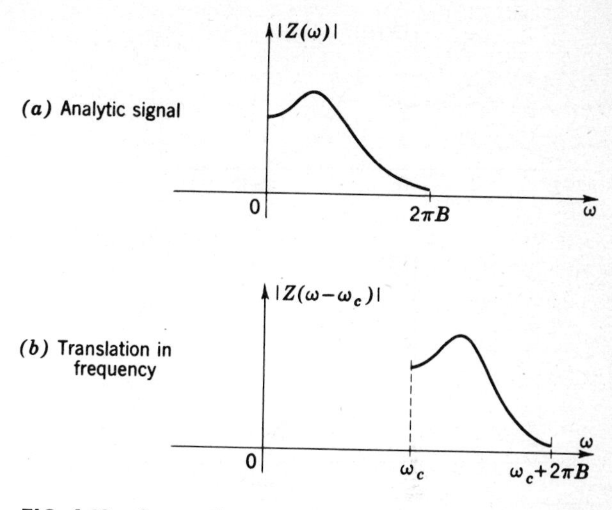

FIG. 4-26 Generation of a SSB signal.

real part to obtain the SSB signal $f_s(t)$:

$$f_s(t) = \text{Re}\,[z(t)e^{j\omega_c t}] \tag{4-27}$$

Now given a modulating or information-bearing signal $f(t)$, how does one generate the analytic signal $z(t)$ from it? Once we show this we have in essence closed the loop, indicating how one goes from $f(t)$ to its SSB version $f_s(t)$. Consider $f(t)$ passed through a $-90°$ phase shifter, a device which shifts the phase of all positive frequency components of $f(t)$ by $-90°$ and all negative components by 90°. (The phase characteristic is always an *odd* function of frequency.) Call the output of the phase shifter $\hat{f}(t)$. We then have

$$\begin{aligned} \hat{F}(\omega) &= -jF(\omega) \qquad \omega \geq 0 \\ &= +jF(\omega) \qquad \omega < 0 \end{aligned} \tag{4-28}$$

Note now that $j\hat{F}(\omega) = F(\omega)$, $\omega \geq 0$; $= -F(\omega)$, $\omega < 0$. This indicates that we can generate the analytic signal $z(t)$ by adding $j\hat{f}(t)$ to $f(t)$:

$$z(t) = f(t) + j\hat{f}(t) \tag{4-29}$$

For we then have

$$\begin{aligned} Z(\omega) &= 2F(\omega) \qquad \omega \geq 0 \\ &= 0 \qquad\qquad \omega < 0 \end{aligned} \tag{4-30}$$

just as desired (Fig. 4-27). By this expedient of defining an analytic signal $z(t)$ and then showing how it may be generated from $f(t)$ and its 90° phase-shifted version $\hat{f}(t)$, we have finally the desired SSB representation

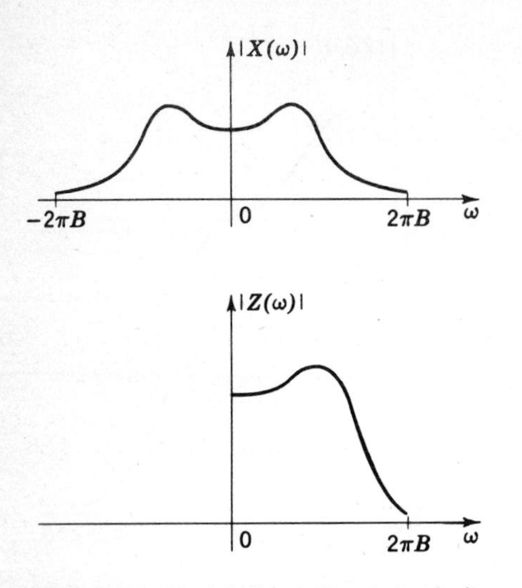

FIG. 4-27 Modulating signal and its analytic signal.

in terms of $f(t)$:

$$f_s(t) = \text{Re}\,[z(t)e^{j\omega_c t}] = \text{Re}\,\{[f(t) + j\hat{f}(t)]e^{j\omega_c t}\}$$
$$= f(t)\cos\omega_c t - \hat{f}(t)\sin\omega_c t \qquad [4\text{-}31]$$

Equation [4-31] is not only the explicit expression for a SSB signal in terms of the baseband signal $f(t)$; it tells us in addition how to generate $f_s(t)$; that is, we generate a DSB signal $f(t)\cos\omega_c t$ by using a product or balanced modulator, generate another DSB signal $\hat{f}(t)\sin\omega_c t$ by phase shifting $f(t)$ by 90° and then multiplying by $\sin\omega_c t$ and then subtracting the two. This so-called phase-shift method of generating SSB signals is diagrammed in Fig. 4-28.[1]

It is left to the reader to show that the *lower* sideband only may be similarly generated by writing

$$f_s(t) = f(t)\cos\omega_c t + \hat{f}(t)\sin\omega_c t \qquad [4\text{-}32]$$

As a simple example of these results, and to develop somewhat more insight into the significance of the operations indicated by Eqs. [4-31] and [4-32], consider $f(t)$ to be just the single-frequency term $\cos\omega_m t$. The DSB signal is then

$$f_{d1}(t) = \cos\omega_m t\cos\omega_c t = \tfrac{1}{2}[\cos(\omega_m + \omega_c)t + \cos(\omega_c - \omega_m)t] \qquad [4\text{-}33]$$

[1] D. E. Norgaard, The Phase-shift Method of Single-sideband Signal Generation, *Proc. IRE*, vol. 44, no. 12, p. 1718, December, 1956.

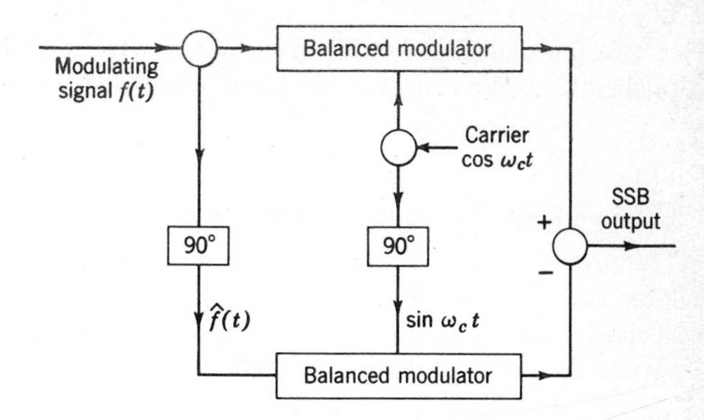

FIG. 4-28 Phase-shift method of generating SSB (upper-sideband case).

containing the expected two sideband frequencies. How do we now cancel one of these sideband terms? If it is desired to retain the upper sideband term only, we must subtract from Eq. [4-33] an expression of the form

$$f_{d2}(t) = \tfrac{1}{2}[\cos(\omega_c - \omega_m)t - \cos(\omega_m + \omega_c)t] = \sin \omega_m t \sin \omega_c t \tag{4-34}$$

Upon subtracting Eqs. [4-33] and [4-34], the lower sideband cancels, leaving the desired upper sideband

$$\cos(\omega_c + \omega_m)t$$

Similarly, upon *adding* Eq. [4-33] and [4-34], the upper sideband cancels, leaving the lower sideband

$$\cos(\omega_c - \omega_m)t$$

These are, of course, the operations indicated by Eqs. [4-31] and [4-32].

The 90° phase shift of all frequency components of $f(t)$ is commonly called a Hilbert transformation, and $\hat{f}(t)$ is called the *Hilbert transform* of $f(t)$. We shall use this transform in the next section in discussing the detection of SSB signals.

One major problem in the design of phase-shift SSB system is the practical realization of the wideband 90° phase-shift network, for *all* the frequency components of the modulating signal $f(t)$ must be shifted by 90°.

A modified version of the phase-shift method of generating SSB signals consists in inserting a phase shift $\beta(f)$ in the $f(t)$ branch of Fig. 4-28 and a phase shift $\beta(f) + 90°$ in the $\hat{f}(t)$ branch. $\beta(f)$ varies in an arbitrary manner with frequency. The details of the analysis are left to the reader as an exercise. The design requirements in this case are for a set of two parallel phase-shift curves 90° apart. This is more easily accomplished than the constant 90° curve required for the method of Fig. 4-28.

4-6 DEMODULATION, OR DETECTION

The process of separating a baseband or modulating signal from a modulated carrier is called demodulation, or detection. Detection is, of course, necessary in all radio receivers where the information to be received is carried by the modulating signal.

We have discussed the demodulation process previously in Sec. 4-2, in connection with high-frequency digital signal detection. We considered there two basic methods of demodulating high-frequency signals: envelope detection and synchronous (coherent) detection. Both methods again appear here in discussing AM and SSB demodulation. In particular, we shall find that synchronous detection is generally required for SSB (or DSB) systems, while normal AM may use either detection method.[1] Since, as noted earlier, envelope detection is much more simply instrumented, this is the preferred method of AM detection.

Demodulation is basically the inverse of modulation and requires nonlinear or linear time-varying (switching) devices also. Since the nonlinear circuits used are essentially the same, the details of detector operation are left as exercises for the reader. (Again, two types of nonlinear detector may be considered: the so-called "square-law" detector with the current-voltage characteristics represented by a power series, and the piecewise-linear detector with a nonlinearity concentrated at one point.)

Recall from Sec. 4-2 that envelope detection consisted of passing the amplitude-modulated carrier through a nonlinear device and then low-pass filtering the nonlinear output. One common configuration, already noted in Sec. 4-2, consists of a half-wave diode rectifier followed by a parallel *RC* circuit. The circuit is diagrammed in Fig. 4-29. The rectified output is shown in the two cases of filtering and no filtering.

The analysis of the detector of Fig. 4-29*a* is carried out very simply by either treating the diode as a square-law device (as in Sec. 4-2),[2] or as a piecewise device. In the former case, it is apparent that with the input given by

$$f_c(t) = K[1 + mf(t)] \cos \omega_c t \tag{4-35}$$

[1] Some thought will convince the reader that amplitude modulation is the generalized form of on-off keying, while double-sideband modulation corresponds to phase-shift keying in the binary case.

[2] Other nonlinear representations lead to essentially the same results, but with much more algebraic complexity and manipulation.

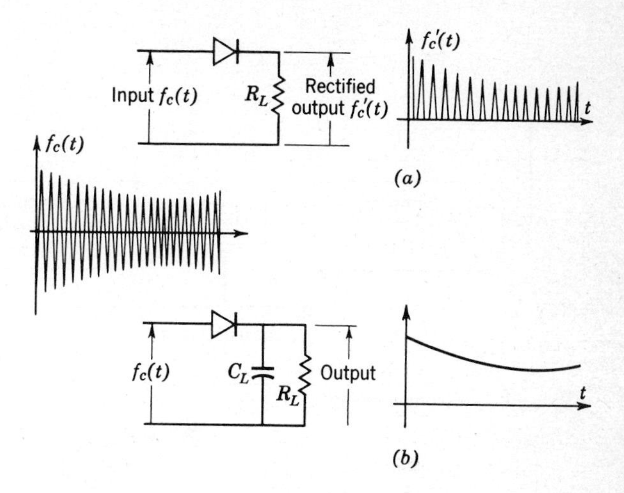

FIG. 4-29 Amplitude-modulation detector. (a) Detection, no filtering; (b) detection with filtering.

the output may be written

$$f_c'(t) = [f_c(t)]^2 = K^2[1 + mf(t)]^2 \cos^2 \omega_c t \qquad [4\text{-}36]$$

Expanding the expression and selecting the low-pass terms, it is easily shown that these latter contain the desired $f(t)$ plus distortion $[f^2(t)]$ components. Similarly, if the diode is approximated as a piecewise-linear device, switching at the carrier-frequency rate, the output may be written

$$f_c'(t) = K[1 + mf(t)] \cos \omega_c t \, S(t) \qquad [4\text{-}37]$$

with $S(t)$ the switching function previously defined. By again expanding $S(t)$ in its Fourier series, it can be shown very simply that the output contains a component proportional to $f(t)$ plus higher-frequency terms (sum and difference frequencies of carrier and modulating signal).

The capacitor of Fig. 4-29*b* serves to filter out these higher-frequency terms. Looked at another way, C_L is chosen so as to respond to envelope variations, but the circuit time constant (including the diode forward resistance) does not allow the circuit to follow the high-frequency (h-f) carrier variations. A block diagram of a typical superheterodyne radio receiver incorporating such a detector is shown in Fig. 4-30.

The incoming signal passes first through a tuned radio-frequency (r-f), amplifier which can be tuned variably over the radio band 550 to 1,600 kHz. This signal is then mixed with a locally generated signal. The sum

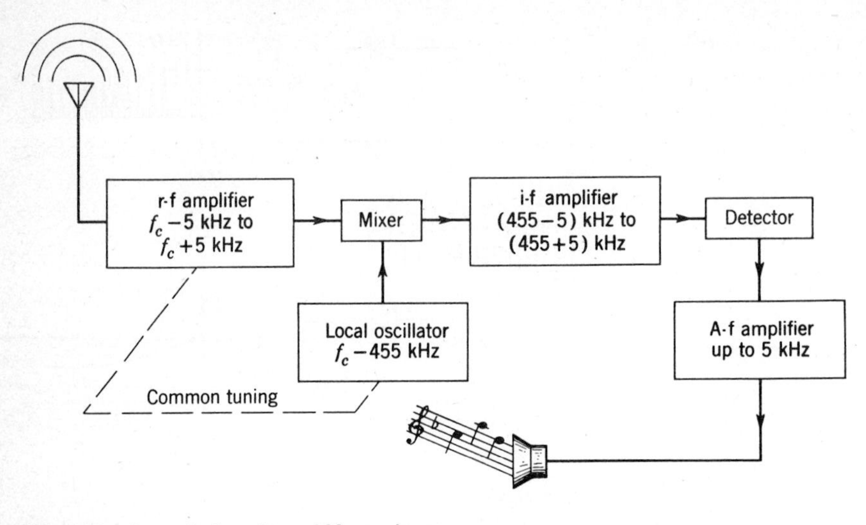

FIG. 4-30 Superheterodyne AM receiver.

and difference frequencies generated contain a term centered about 455 kHz. (The local oscillator and r-f amplifier are tuned together so that there is always a difference frequency of 455 kHz between them.) The mixer acts as a frequency converter, shifting the incoming signal down to the fixed intermediate frequency of 455 kHz. Several stages of amplification are ordinarily used, with double-tuned circuits providing the coupling between stages. The intermediate-frequency (i-f) signal is then detected as described above, amplified further in the audio-frequency (a-f) amplifiers and applied to the loudspeaker. The superheterodyning operation refers to the use of a frequency converter and fixed, tuned i-f amplifier before detection.

The synchronous detection operation as indicated in Sec. 4-2 consists simply of multiplying the incoming carrier signal by a locally generated carrier $\cos \omega_c t$ and low-pass filtering the resultant. This is shown again in Fig. 4-31. Based on our discussion in the previous sections, it is appar-

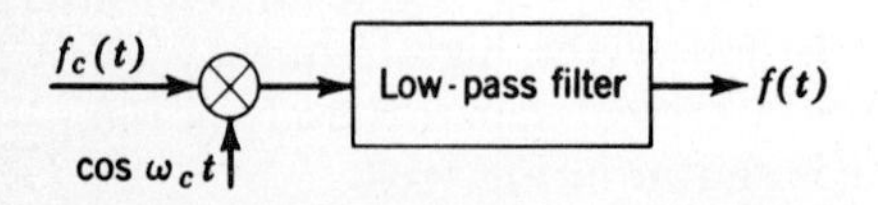

FIG. 4-31 Synchronous detection (frequency conversion).

ent that the multiplication by $\cos \omega_c t$ serves to translate the frequency components in the incoming-modulated carrier up and down by the carrier signal. The low-pass filter rejects the terms centered about $2f_c$, and passes those at baseband frequencies. The synchronous detection process can thus equally well be considered to consist of a frequency converter plus filter. This conversion process is diagrammed in Fig. 4-32. The

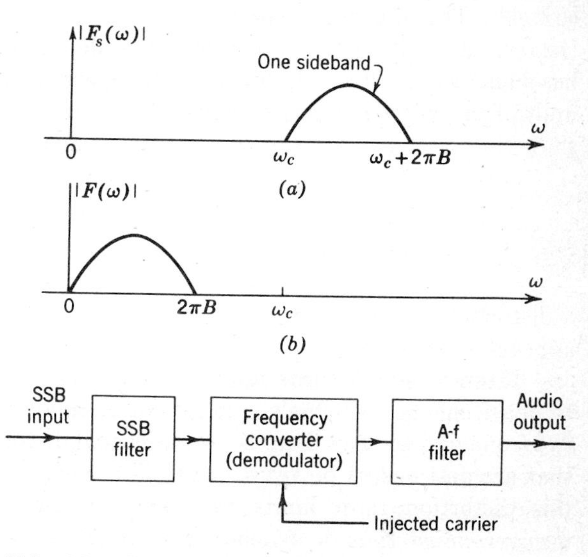

FIG. 4-32 Single-sideband demodulation. (a) Before frequency conversion; (b) after frequency conversion.

multiplication by $\cos \omega_c t$ may be accomplished in innumerable ways: by the use of nonlinear devices, switching, etc. The key factor, however, is the necessity of having available at the receiver the carrier frequency term $\cos \omega_c t$.

In the case of normal AM it is apparent that multiplication by $\cos \omega_c t$ results in the prefiltered expression

$$K[1 + mf(t)] \cos^2 \omega_c t$$

For a DSB signal the equivalent expression is

$$f(t) \cos^2 \omega_c t$$

while for the SSB signal the prefiltered term is, using Eq. [4-31],

$$f(t) \cos^2 \omega_c t - \hat{f}(t) \sin \omega_c t \cos \omega_c t$$

It is apparent that the desired term $f(t)$ appears at the filter output in all three cases.

For most types of baseband signals the local carrier must not only be of the right frequency but must be synchronized in phase with the carrier as well. This was emphasized in Sec. 4-2. If the carrier shifts in phase the resultant output signal may be a considerably distorted version of the baseband signal $f(t)$. To demonstrate this for the SSB case consider multiplication by a carrier term $\cos(\omega_c t + \theta)$. It is left for the reader to show that in this case the filtered output signal is given by

$$2f_o(t) = f(t) \cos \theta + \hat{f}(t) \sin \theta \qquad [4\text{-}38]$$

Note that not only is the desired output reduced as θ increases from 0, but a distortion term $\hat{f}(t) \sin \theta$ appears. If $\theta = \pi/2$, in particular, only $\hat{f}(t)$ appears at the output.

Interestingly, it turns out that the human ear is relatively insensitive to phase changes in signals. It cannot distinguish the Hilbert transform $\hat{f}(t)$ from $f(t)$, so that there is no noticeable distortion in received signals that are destined to be *heard*. For all other signals—digital, TV, etc.—this distortion term limits the operation of the SSB system. SSB receivers must thus be synchronized in phase, which again presents the problem noted in Sec. 4-2.

The kind of distortion to be expected with lack of phase synchronization may be determined by evaluating the Hilbert transform $\hat{f}(t)$. This is readily accomplished by using the 90° phase-shift definition. As an example, assume the signal $f(t)$ is an *even* function of time. (This simplifies the discussion.) Test pulses of the kind considered in Chap. 2 are examples. Then $F(\omega)$ is an even function of frequency. The Hilbert transform $\hat{f}(t)$ is then given by

$$\begin{aligned} \hat{f}(t) &= \frac{1}{2\pi} \int_{-\infty}^{\infty} \hat{F}(\omega) e^{j\omega t} \, d\omega \\ &= \frac{1}{\pi} \int_0^{\infty} F(\omega) \sin \omega t \, d\omega \end{aligned} \qquad [4\text{-}39]$$

Notice that $\hat{f}(t)$ is then an odd function of time. Using Eq. [4-39] we may calculate $\hat{f}(t)$ quite readily. Two simple examples are tabulated below, and sketched in Fig. 4-33.

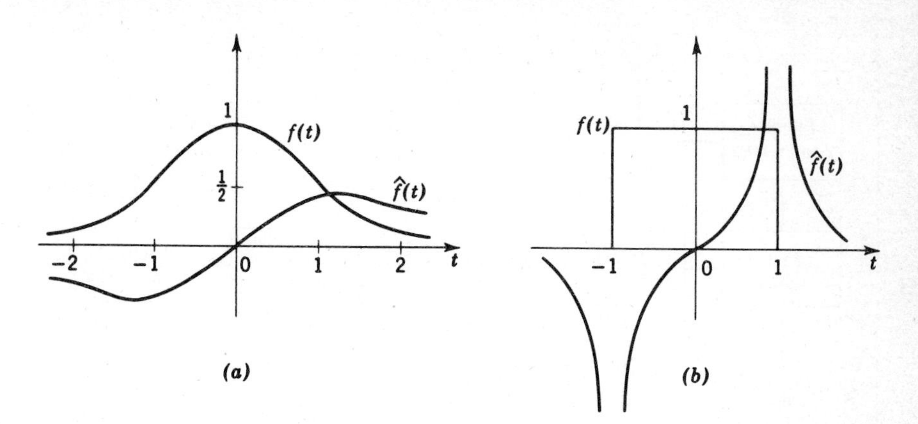

FIG. 4-33 Examples of Hilbert transforms. (a) $f(t) = 1/(1 + t^2)$; (b) $f(t)$, a unit pulse.

1. $f(t) = 1/(1 + t^2)$. Then

$$F(\omega) = \pi e^{-|\omega|} \qquad \hat{f}(t) = \frac{t}{1 + t^2}$$

2. $f(t)$ = the unit pulse of Fig. 4-33*b*. Then

$$F(\omega) = 2\frac{\sin \omega}{\omega} \qquad \hat{f}(t) = \frac{1}{\pi} \ln \frac{|t + 1|}{|t - 1|}$$

Note that in the case of the sharply varying unit pulse the Hilbert transform becomes infinite at the points of sharp variation. In the case of the more realistic pulse of Fig. 4-33*a*, these peaks in the Hilbert transform are less pronounced but still quite high. This distortion becomes quite noticeable in TV and other signal receivers when the phase shift θ approaches $\pi/2$.

Various methods of synchronizing the local carrier exist, as noted in Sec. 4-2. A small amount of unmodulated carrier energy is commonly sent as a pilot signal, and after extraction at the receiver, is used to provide the necessary synchronization.

A typical SSB-receiver block diagram is shown in Fig. 4-34.

The carrier filter and sideband filter are used to separate the sideband and the pilot carrier. Double-frequency conversion (the first and the second mixers) is used, just as in the case of the SSB transmitter (Fig. 4-25), to obtain a convenient range of frequency for the filters. The output of the carrier filter is amplified and limited and used for automatic

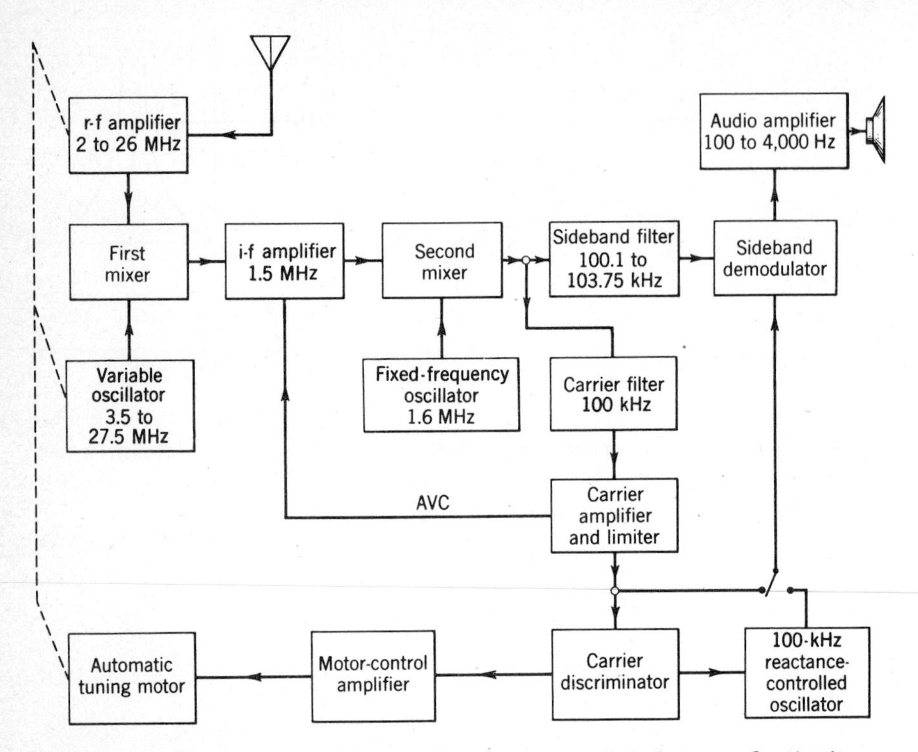

FIG. 4-34 Single-sideband receiver. (From B. Fisk and C. L. Spencer, Synthesizer Stabilized Single-sideband Systems, *Proc. IRE*, vol. 44, no. 12, p. 1680, December, 1956, by permission.)

tuning controls, either as a reinsertion carrier directly or to lock the frequency of a local oscillator that furnishes a suitable insertion carrier.

4-7 FREQUENCY MODULATION

We have investigated thus far in the past few sections the effect of slowly varying the amplitude of a sinusoidal carrier in accordance with some information to be transmitted. The desired information is then found to be concentrated in sidebands about the carrier frequency. By choosing the carrier frequency high enough, information transmission by means of radio (through the air) becomes practicable. Alternatively, with AM the signal-frequency spectrum may be shifted to a frequency range where circuit design becomes more feasible, where circuit components are more readily obtained and more economically built, or where equipment size and weight can be reduced.

In addition, many information channels may be transmitted simultaneously by means of frequency-multiplexing techniques.

For these reasons and others, AM systems of the standard, SSB, or DSB type are commonly used in both the communication and the control field.

Amplitude modulation is, however, not the only means of modulating a sine-wave carrier. We could just as well modulate the phase and frequency of a sine wave in accordance with some information-bearing signal. And such frequency-modulation (FM) systems are of course also utilized quite commonly.

We first discussed all three methods of modulating a sine-wave carrier in Sec. 4-2 in the special case of binary modulating signals. Just as we then extended the AM analysis to more general forms of information-bearing signals in the sections that followed, we propose to investigate FM in more generality in this section and in those that follow.

We noted previously, in discussing binary FM or FSK systems, that FM transmission requires wider transmission bandwidths than the corresponding AM systems. We shall find this to be generally true for all types of modulating signals. Why then use FM? Again, as noted earlier, FM provides better discrimination against noise and interfering signals. This fact will be demonstrated in Chap. 6 after we have studied noise in systems.

Interestingly, the expression found for FM transmission bandwidth in the binary case, $2\,\Delta f + 2B$, with Δf the frequency deviation away from the carrier and B the baseband bandwidth, will be found to serve as a quite useful rule of thumb for more complex signals as well.

Although we shall briefly consider phase modulation (PM), the stress will be placed on FM. Actually, for complex modulating signals, one form is easily derived from the other, and both are encompassed in the class of *angle-modulation* systems, as will be demonstrated below.

Before going further we need first to specify what we mean by *frequency modulation.* We might start intuitively by saying that we shall consider a frequency-modulation system one in which the frequency of the carrier is caused to vary in accordance with some specified information-carrying signal. Thus, we might write the frequency of the carrier as $\omega_c + Kf(t)$, where $f(t)$ represents the signal and K is a constant of the system. This is, of course, analogous to the AM case, and was possible in the binary FM case. We run into some difficulty, however, when we attempt to express the more general frequency-modulated carrier mathematically. For we can talk about the frequency of a sine wave only when the frequency is constant and the sine wave persists for all time. Yet here we are attempting to discuss a variable frequency!

The difficulty lies in the fact that, strictly speaking, we can talk only of the sine (or cosine) of an *angle.* If this angle varies linearly with time, we can specifically interpret the frequency as the derivative of the angle. Thus, if

$$f_c(t) = \cos \theta(t) = \cos (\omega_c t + \theta_0) \tag{4-40}$$

the usual expression for a sine wave of frequency ω_c, we are implicitly assuming $\theta(t)$ to be linear with time, with ω_c its derivative.

When $\theta(t)$ does not vary linearly with time, we can no longer write Eq. [4-40] in the standard form shown, with a specified frequency term. To obviate this difficulty, we shall define an *instantaneous radian frequency* ω_i to be the derivative of the angle as a function of time. Thus, with

$$f_c(t) = \cos \theta(t) \tag{4-41}$$

we have

$$\omega_i \equiv \frac{d\theta}{dt} \tag{4-42}$$

(This then agrees, of course, with the usual use of the word *frequency* if $\theta = \omega_c t + \theta_0$.)

If $\theta(t)$ in Eq. [4-41] is now made to vary in some manner with a modulating signal $f(t)$, we call the resulting form of modulation *angle modulation*. In particular, if

$$\theta(t) = \omega_c t + \theta_0 + K_1 f(t) \tag{4-43}$$

with K_1 a constant of the system, we say we are dealing with a *phase-modulation* system. Here the phase of the carrier wave varies linearly with the modulating signal. Binary signaling is a special case here.

Now let the *instantaneous frequency*, as defined by Eq. [4-42], vary linearly with the modulating signal,

$$\omega_i = \omega_c + K_2 f(t) \tag{4-42a}$$

Then

$$\theta(t) = \int \omega_i \, dt = \omega_c t + \theta_0 + K_2 \int f(t) \, dt \tag{4-44}$$

This, of course, gives rise to an FM system. As an example, if we again consider binary FM (FSK), we have the baseband signal $f(t)$ switching between either of two states. Then $\omega_i = \omega_c \pm \Delta\omega$, and the instantaneous frequency switches correspondingly between two frequencies. The phase angle $\theta(t)$ increases linearly with time in any one binary interval T, switching back to its initial value θ_0 as a new interval begins.

Both phase modulation and frequency modulation are seen to be special cases of angle modulation. In the phase-modulation case the phase of the carrier varies with the modulating signal, and in the frequency-modulation case the phase of the carrier varies with the integral of the modulating signal. If we first integrate our modulating signal $f(t)$ and then allow it to phase modulate a carrier, this gives rise to a frequency-modulated wave. This is exactly the method used for producing a frequency-modulated carrier in the Armstrong indirect FM system, as we shall see shortly.

A frequency-modulated carrier is shown sketched in Fig. 4-35c. The modulating signal is assumed to be a repetitive sawtooth of period T

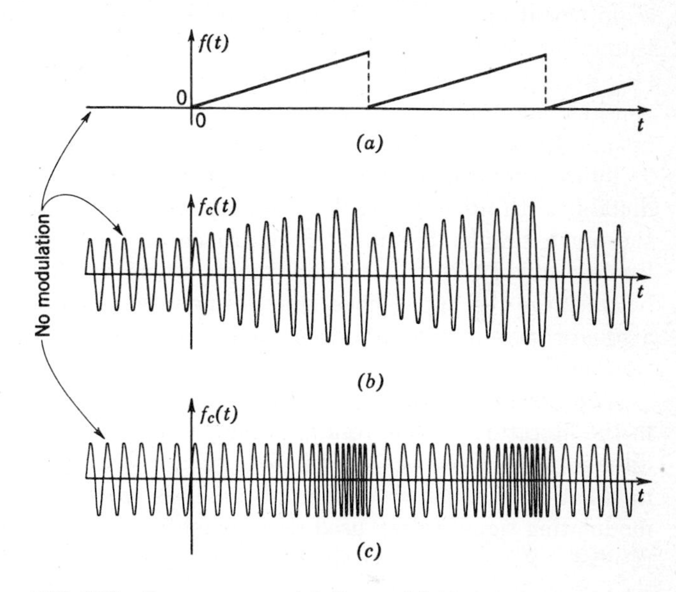

FIG. 4-35 Frequency modulation. (a) Modulating wave; (b) AM carrier; (c) FM carrier.

($2\pi/T \ll \omega_c$). Compare this with Fig. 4-7a, in which the modulating signal is a periodic square wave. As the sawtooth modulating signal increases in magnitude, the FM oscillates more rapidly. Its amplitude remains unchanged, however.

Frequency modulation is a nonlinear process, and so, as pointed out in previous sections, we would expect to see new frequencies generated by the modulation process. As indicated by Eqs. [4-42a] and [4-44] the FM signal oscillates more rapidly with increasing amplitude of the modulating signal. We would, therefore, expect the frequency spectrum of the FM wave, or its bandwidth, to widen correspondingly.

The analysis of the FM process is inherently much more complicated than that for AM, particularly in dealing with a general modulating signal. This is due to the nonlinearity of the FM process. Superposition cannot be used and so the analysis for one particular type of modulating signal cannot be readily applied to another. However, at the two extremes of small-amplitude modulating signal (narrowband FM) and large-amplitude modulating signal (wideband FM) it turns out that spectral analyses for different modulating signals produce the transmission bandwidth requirements: $2B$ Hz and $2\,\Delta f$ Hz, respectively, found in the binary FSK case, while at intermediate amplitudes of modulating signal the bandwidth requirement obtained for the binary FSK signal serves as a good rule of thumb for all types of signals.

Because of the difficulty of analyzing general FM signals, we shall consider here only one nonbinary modulating signal; namely, we shall assume $f(t)$ a single sine wave. Although this is a poor "approximation" to information-carrying signals, which may vary randomly and unpredictably with time, the results obtained are in substantial agreement with the much simpler binary FSK approach of Sec. 4-2, and will serve to further solidify the discussion there. (One would expect agreement between the two approaches, since a sine wave is similar in appearance to a square wave of the same frequency. The sine wave is in fact the first term in the square-wave Fourier series.)

Spectral analyses of other classes of modulating signals may be found in the literature.[1] The results obtained are, as noted, not substantially different than those obtained here for square-wave (FSK) and sine-wave modulating signals. FM spectral analysis with a random or noiselike modulating signal (often used as a model for real signals) leads to similar results.[2] Assume then a sinusoidal modulating signal at frequency f_m:

$$f(t) = a \cos \omega_m t \tag{4-45}$$

The instantaneous radian frequency ω_i is

$$\omega_i = \omega_c + \Delta\omega \cos \omega_m t \qquad \Delta\omega \ll \omega_c \tag{4-46}$$

where $\Delta\omega$ is a constant depending on the amplitude a of the modulating signal and on the circuitry converting variations in signal amplitude to corresponding variations in carrier frequency.

The instantaneous radian frequency thus varies about the unmodulated carrier frequency ω_c, at the rate ω_m of the modulating signal and with a maximum deviation of $\Delta\omega$ radians. Just as in the binary case $\Delta f = \Delta\omega/2\pi$ gives the maximum frequency deviation away from the carrier frequency and is called the *frequency deviation.*

[1] H. S. Black, "Modulation Theory," Van Nostrand, Princeton, N.J., 1955; S. Goldman, "Frequency Analysis, Modulation, and Noise," McGraw-Hill, New York, 1948; D. Middleton, "An Introduction to Statistical Communication Theory," chap. 14, McGraw-Hill, New York, 1960.

[2] Schwartz, Bennett, and Stein, *op. cit.*, sec. 3.10; D. Middleton, *op. cit.*

The phase variation $\theta(t)$ for this special case is given by

$$\theta(t) = \int \omega_i \, dt = \omega_c t + \frac{\Delta\omega}{\omega_m} \sin \omega_m t + \theta_0 \qquad [4\text{-}47]$$

θ_0 may be taken as zero by referring to an appropriate phase reference, so that the frequency-modulated carrier is given by

$$f_c(t) = \cos(\omega_c t + \beta \sin \omega_m t) \qquad [4\text{-}48]$$

with

$$\beta \equiv \frac{\Delta\omega}{\omega_m} = \frac{\Delta f}{f_m} \qquad [4\text{-}49]$$

Again, as in Sec. 4-2, for binary FSK, β is called the *modulation index* and is by definition the ratio of the frequency deviation to the baseband bandwidth. For a single sine wave of frequency f_m, the baseband bandwidth B is just f_m.

We noted previously that increasing the amplitude of the modulating signal should increase the bandwidth occupied by the FM signal. Increasing the modulating-signal amplitude corresponds to increasing the frequency deviation Δf or the modulation index β. We would thus expect the bandwidth of the FM wave to depend on β. This will be demonstrated in the sections to follow. The average power associated with the frequency-modulated carrier is independent of the modulating signal, however, and is in fact the same as the average power of the unmodulated carrier. This is again in contrast to the AM case, where the average power of the modulated carrier varies with the modulating-signal amplitude.

That this statement is true may be demonstrated by using Eq. [4-48] for a sinusoidal modulating signal. Assuming that $f_c(t)$ represents the instantaneous voltage impressed across a 1-ohm resistor, the average power over a cycle of the modulating frequency is given by

$$\frac{1}{T}\int_0^T f_c^2(t)\, dt = \frac{1}{T}\int_0^T \cos^2(\omega_c t + \beta \sin \omega_m t)\, dt$$

where $T = 1/f_m$. This expression may be rewritten as

$$\frac{1}{T}\int_0^T \frac{1 + \cos(2\omega_c t + 2\beta \sin \omega_m t)}{2}\, dt$$

The second term of the integral goes to zero since it is periodic in T. This assumes $\omega_m = 2\pi/T \ll \omega_c$. The first term gives ½ watt. If the amplitude of the carrier had been written as A_c volts, the average power would have been found to be $A_c^2/2$ watts for a 1-ohm resistor.

Although shown only for a sinusoidal modulating signal, the above result is true for any modulating signal whose highest frequency component B Hz is small compared with the carrier frequency f_c. This is left as an exercise for the reader.

NARROWBAND FM

To simplify the analysis of FM, we shall treat it in two parts. We shall first consider the sinusoidally modulated carrier with $\beta \ll \pi/2$, and then with $\beta > \pi/2$. Small β corresponds to narrow bandwidths, and FM systems with $\beta \ll \pi/2$ are thus called *narrowband FM systems.* The equations for *narrowband FM* appear in the form of the equations of the product modulator of the previous sections on AM and so give rise to sideband frequencies equally displaced about the carrier, just as in the case of AM.

To demonstrate this point, consider the sinusoidally modulated carrier of Eq. [4-48], and assume $\beta \ll \pi/2$. (This implies that the maximum phase shift of the carrier is much less than $\pi/2$ radians. This is ordinarily taken to mean $\beta < 0.2$ radian, although $\beta < 0.5$ radian is sometimes used as a criterion.) We have

$$\begin{aligned} f_c(t) &= \cos(\omega_c t + \beta \sin \omega_m t) \\ &= \cos \omega_c t \cos(\beta \sin \omega_m t) - \sin \omega_c t \sin(\beta \sin \omega_m t) \end{aligned} \qquad [4\text{-}50]$$

But, for $\beta \ll \pi/2$, $\cos(\beta \sin \omega_m t) \doteq 1$, and $\sin(\beta \sin \omega_m t) \doteq \beta \sin \omega_m t$.

The frequency-modulated wave for small modulation index thus appears in the form

$$f_c(t) \doteq \cos \omega_c t - \beta \sin \omega_m t \sin \omega_c t \qquad \beta \ll \frac{\pi}{2} \qquad [4\text{-}51]$$

Note that this expression for $f_c(t)$, the frequency-modulated carrier, has a form similar to that of the output of a product modulator; it contains the original unmodulated carrier term plus a term given by the product of the modulating signal and carrier. For the sinusoidal modulating signal, $\beta \sin \omega_m t$, this product term provides sideband frequencies displaced $\pm\omega_m$ radians from ω_c. The bandwidth of this narrowband FM signal is thus $2f_m = 2B$ Hz, agreeing with the FSK result.

If we had assumed a general modulating signal $f(t)$ instead of the sinusoidal modulating signal used here, we would have obtained similar results. For, as shown by Eqs. [4-42*a*] and [4-44], we then have

$$\begin{aligned} \omega_i &= \omega_c + K_2 f(t) \\ \theta(t) = \int \omega_i\, dt &= \omega_c t + \theta_0 + K_2 \int f(t)\, dt \end{aligned}$$

Suppressing the arbitrary phase angle θ_0, and defining a new time function $g(t)$ to be the integral of $f(t)$, $g(t) \equiv \int f(t)\, dt$, we have, as the frequency-modulated carrier,

$$f_c(t) = \cos [\omega_c t + K_2\, g(t)] \qquad [4\text{-}52]$$

If K_2 and the maximum amplitude of $g(t)$ are now chosen small enough so that $|K_2 g(t)| \ll \pi/2$,

$$f_c(t) = \cos \omega_c t - K_2\, g(t) \sin \omega_c t \qquad [4\text{-}53]$$

Our previous discussion of product modulators indicates that the spectrum of this general case of narrowband FM consists of the carrier plus two sidebands, one on each side of the carrier and each having the form of the spectrum of the function $g(t)$. Narrowband FM is thus equivalent, in this sense, to AM. The bandwidth of a narrowband FM signal, in general, is $2B$ Hz, where B is the highest-frequency component of either $g(t)$ or its derivative $f(t)$, the original modulating signal. (Remember that the linear process of integration adds no new frequency components. The lower-frequency components, however, are accentuated in comparison with the higher-frequency components.) The FSK result is, of course, a special case here.

Although AM and narrowband FM have similar frequency spectra and their mathematical representations both appear in the product-modulator form, they are distinctively different methods of modulation. In the AM case we were interested in variations of the carrier envelope, its frequency remaining unchanged; in the FM case, we assumed the carrier amplitude constant, its phase (and effectively the instantaneous frequency also) varying with signal. This distinction between the two types of modulation must be retained in the narrowband FM case also (here maximum phase shift of the carrier is assumed less than 0.2 radian).

To emphasize this distinction in the two types of modulation, note that the product modulator or sideband term in either Eq. [4-51] or Eq. [4-53] appears in phase quadrature with the carrier term ($\sin \omega_c t$ as compared with $\cos \omega_c t$). In the AM case, we had both carrier and sideband terms in phase,

$$f_c(t) = \cos \omega_c t + m f(t) \cos \omega_c t$$

For the narrowband FM case, we have

$$f_c(t) = \cos \omega_c t - K_2 g(t) \sin \omega_c t$$

That the inphase, or phase-quadrature, representation is fundamental in distinguishing between AM and narrowband FM (or, alternatively, small-angle phase modulation) is demonstrated very simply by the use of rotating vectors.[1]

[1] Black, *op. cit.*, pp. 186–187.

We rewrite Eq. [4-51] in its sideband frequency form,

$$\begin{aligned} f_c(t) &= \cos \omega_c t - \beta \sin \omega_m t \sin \omega_c t \\ &= \cos \omega_c t - \frac{\beta}{2} [\cos (\omega_c - \omega_m)t - \cos (\omega_c + \omega_m)t] \end{aligned} \qquad [4\text{-}54]$$

This can also be written in the form

$$f_c(t) = \text{Re}\left[e^{j\omega_c t} \left(1 - \frac{\beta}{2} e^{-j\omega_m t} + \frac{\beta}{2} e^{j\omega_m t}\right)\right] \qquad [4\text{-}55]$$

where Re[ ] represents the real part of the expression in the square brackets.

$e^{j\omega_c t}$ may be represented as a unit vector rotating counterclockwise at the rate of ω_c radians/sec. Superimposed on this rotation are changes in the vector due to the three terms in parentheses. We then take the real part of the resultant vector, or its projection on the real axis. If we suppress the continuous ω_c rotation and concentrate solely on the three terms in parentheses, we may plot these as the three vectors of Fig. 4-36*a*. Note that the resulting vector deviates in phase from the unmodulated vector, while its amplitude is very nearly unchanged. (This is for the case $\beta \ll \pi/2$. Here β is shown larger for the sake of clarity.)

We can use the same vector approach for the AM case and get

$$\begin{aligned} f_c(t) &= \cos \omega_c t + m \cos \omega_m t \sin \omega_c t \\ &= \cos \omega_c t + \frac{m}{2} [\cos (\omega_c + \omega_m)t + \cos (\omega_c - \omega_m)t] \\ &= \text{Re}\left[e^{j\omega_c t} \left(1 + \frac{m}{2} e^{j\omega_m t} + \frac{m}{2} e^{-j\omega_m t}\right)\right] \end{aligned} \qquad [4\text{-}56]$$

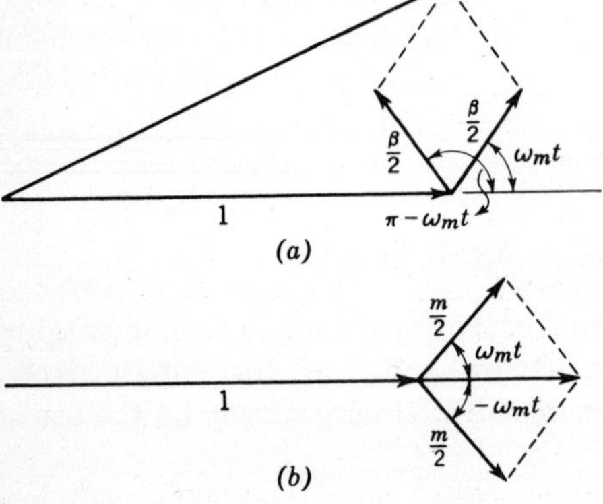

FIG. 4-36 Vector representation of (a) narrowband FM and (b) AM.

The three quantities in parentheses here are likewise shown plotted in Fig. 4-36*b*. Note that the two sideband vectors have been rotated by $\pi/2$ radians as compared with the FM case. The resultant vector here varies in amplitude but remains in phase with the unmodulated carrier term.

We can summarize by saying that the resultant of the two sideband vectors in the FM case will always be perpendicular to (or in phase quadrature with) the unmodulated carrier, while the same resultant in the AM case is colinear with the carrier term. The FM case thus gives rise to phase variations with very little amplitude change ($\beta \ll \pi/2$), while the AM case gives amplitude variations with no phase deviation.

The distinction and similarity between AM and narrowband FM (or phase modulation as well) leads us to a commonly used method of generating a frequency-modulated wave with small modulation index ($\beta < 0.2$ radian).

We demonstrated in our discussion of AM systems that the output of a balanced modulator provides just the product or sideband term required by Eq. [4-53]. For an *amplitude-modulated* output we would then add to this output the *inphase* carrier term. This is shown in block diagram form in Fig. 4-37*a*.

To obtain a *phase-modulated* output, according to our previous discussion, we must add a *phase-quadrature* carrier term to the balanced-modulator output. Such a system is shown in Fig. 4-37*b* in block-diagram form. (Remember that this is restricted only to small phase variations of 0.2 radian or less.)

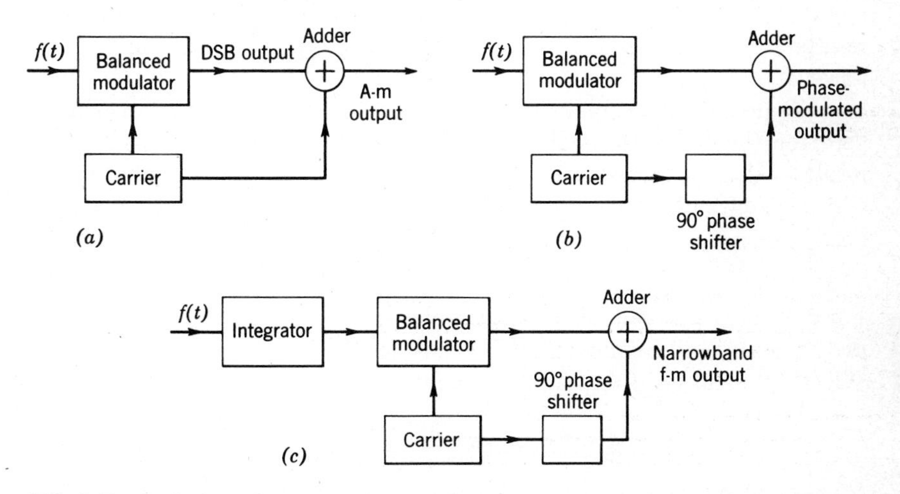

FIG. 4-37 Evolution of a narrowband FM system. (a) Possible AM system; (b) phase-modulation system (small phase deviations <0.2 radian); (c) narrowband FM system ($\beta < 0.2$ radian).

As we demonstrated previously, however, phase modulation and FM differ only by a possible integration of the input modulating signal (Eq. [4-44]). To obtain a narrowband frequency-modulated output, then, we need only integrate our input signal and then apply it to the phase-modulation system of Fig. 4-37*b*. The resultant narrowband FM system ($\beta < 0.2$ radian) is shown in Fig. 4-37*c*.

The narrowband FM system of Fig. 4-37*c* is used in the Armstrong indirect FM system, which is described in a later section.

WIDEBAND FM

We have just shown that a frequency-modulated signal with small modulation index ($\beta \ll \pi/2$) has a frequency spectrum similar to that of an amplitude-modulated signal. The significant distinction between the two cases arises from the fact that in the FM case the sidebands are in phase quadrature with the carrier, while in the AM case they are in phase with the carrier. The bandwidth of the narrowband FM signal, just like the AM signal, is thus $2B$, with B the maximum-frequency component of the modulating signal.

The noise and interference reduction advantages of FM over AM, mentioned previously, become significant, however, only for large modulation index ($\beta > \pi/2$). The bandwidths required to pass this signal become correspondingly large, as first noted in Sec. 4-2. Most FM systems in use are of this wideband type.

Here the comparison between FM and AM that was valid for small modulation index ($\beta \ll \pi/2$) ends. We shall show that the previous results for narrowband FM can be extended to the wideband case, however.

We demonstrate the increase in signal bandwidth with increasing β here for the idealized model of a sinusoidal modulating signal. The results, as noted earlier, are similar to those obtained in the binary FM case. The frequency-modulated carrier is again written in the expanded form

$$f_c(t) = \cos \omega_c t \cos (\beta \sin \omega_m t) - \sin \omega_c t \sin (\beta \sin \omega_m t) \qquad [4\text{-}57]$$

For $\beta \ll \pi/2$ we would, of course, get our previous result of a single carrier and two sideband frequencies. But now let β be somewhat larger at first. We can expand $\cos (\beta \sin \omega_m t)$ in a power series to give us

$$\cos (\beta \sin \omega_m t) \doteq 1 - \frac{\beta^2}{2} \sin^2 \omega_m t \qquad \beta^2 \ll 6 \qquad [4\text{-}58]$$

If we assume $\beta \ll \sqrt{6}$ and retain just the first two terms in the power series for the cosine, we get the additional term $\sin^2 \omega_m t \cos \omega_c t$ in the expression for $f_c(t)$. This term gives, upon trigonometric expansion, additional sideband frequencies spaced $\pm 2\omega_m$ radians from the carrier and also contributes a term $-\beta^2/4$ to the carrier. Sin ($\beta \sin \omega_m t$) can still be represented by $\beta \sin \omega_m t$, the first term in its power-series expansion, for $\beta^2 \ll 6$, so that $f_c(t)$ becomes

$$f_c(t) \doteq \left(1 - \frac{\beta^2}{4}\right) \cos \omega_c t - \frac{\beta}{2} [\cos (\omega_m - \omega_c)t - \cos (\omega_m + \omega_c)t] + \frac{\beta^2}{8} [\cos (\omega_c + 2\omega_m)t + \cos (\omega_c - 2\omega_m)t] \qquad \beta^2 \ll 6 \quad [4\text{-}59]$$

Note that the carrier term has now begun to decrease somewhat with increasing β, the first-order sidebands at $\omega_c \pm \omega_m$ increase with β, and a new set of sidebands (the second-order sidebands) appear at $\omega_c + 2\omega_m$. The spectrum of this signal is shown in Fig. 4-38*b*, while the narrowband case is plotted in Fig. 4-38*a*. This appearance of new sidebands with

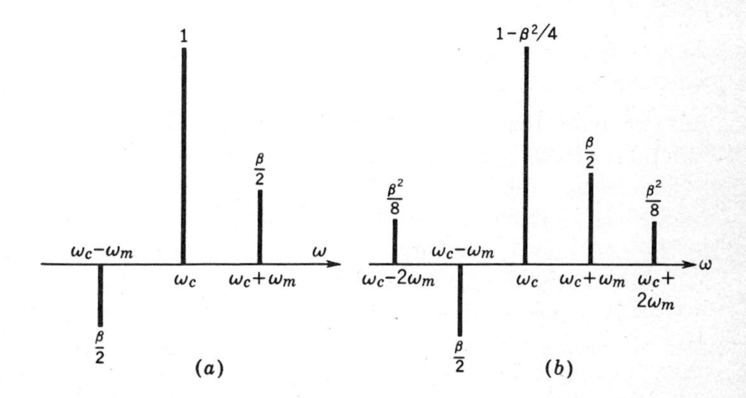

FIG. 4-38 Effect of increasing β on FM spectrum ($\beta^2 \ll 6$). Sine-wave modulation. (a) Narrowband case, $\beta \ll \pi/2$; (b) increasing β.

increasing modulation index corresponds to the widening of the spectrum in the binary FM case as the two frequencies there are deviated further away from the carrier. This is distinctly different from the AM case, where the number of sideband frequencies was dependent solely on the number of modulating frequencies, and not on the amplitude of the modulating signal. Here new sets of significant sidebands appear as the modulation index increases. For a fixed modulating frequency, β is proportional to the amplitude of the modulating signal, so that increases in signal amplitude generate new sidebands. This produces a corresponding increase in bandwidth. (The bandwidth has been doubled in this case, going from $2f_m$ to $4f_m$.)

Since the average power in the frequency-modulated wave is independent of the modulating signal, increasing power in the sideband frequencies must be accompanied by a corresponding decrease in the power associated with the carrier. This accounts for the decrease in carrier amplitude that we have already noted.

As β increases further, we require more terms in the power-series expansion for both $\cos(\beta \sin \omega_m t)$ and $\sin(\beta \sin \omega_m t)$. This gives rise to increasingly more significant sideband components, and the bandwidth begins to increase with β, or with increasing amplitude of the original modulating signal. This power-series approach can be used to explore the characteristics of wideband FM. It becomes a tedious job of trigonometric manipulation to determine the significant sideband frequencies and their associated amplitudes, however, so that we shall resort to a somewhat different approach.

We are basically interested in determining the frequency components of the frequency-modulated carrier given by

$$f_c(t) = \cos(\omega_c t + \beta \sin \omega_m t) = \cos \omega_c t \cos(\beta \sin \omega_m t) - \sin(\beta \sin \omega_m t) \sin \omega_c t$$

But we note that both $\cos(\beta \sin \omega_m t)$ and $\sin(\beta \sin \omega_m t)$ are periodic functions of ω_m. As such, each may be expanded in a Fourier series of period $2\pi/\omega_m$. Each series will contain terms in ω_m and all its harmonic frequencies. Each harmonic term multiplied by either $\cos \omega_c t$ or $\sin \omega_c t$, as the case may be, will give rise to two sideband frequencies symmetrically situated about ω_c. We thus get a picture of a large set of sideband frequencies in general, all displaced from the carrier ω_c by integral multiples of the modulating signal ω_m. The sidebands corresponding to the $\sin \omega_c t$ term will be quadrature sidebands, while those corresponding to the $\cos \omega_c t$ term will be inphase sidebands. For small values of β, $\cos(\beta \sin \omega_m t)$ and $\sin(\beta \sin \omega_m t)$ vary slowly, and so only a small number of the sidebands about ω_c will be significant in amplitude. As β increases, these two terms vary more rapidly, and the amplitudes of the higher-frequency terms become more significant. This picture of course agrees with our previous conclusion that increasing β produces a wider-bandwidth signal and agrees as well with the binary FM analysis. It is again in contrast to the AM case, where only the carrier and a single set of sidebands appear.

These remarks are given more significance by considering some plots of $\cos(\beta \sin \omega_m t)$. These are shown in Fig. 4-39 for various values of β. The four curves drawn demonstrate some interesting points:

1. For $\beta < 0.5$ the curve can be represented approximately by a dc component plus a small component at twice the fundamental frequency

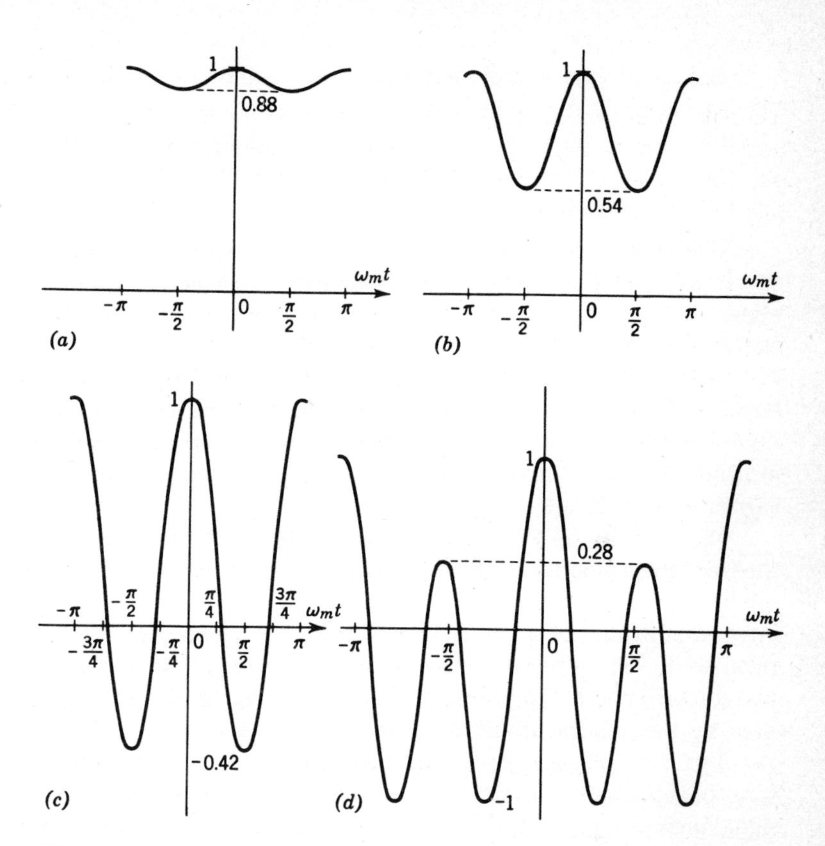

FIG. 4-39 Plots of cos (β sin $\omega_m t$) for various β. (a) $\beta = 0.5$, cos (0.5 sin $\omega_m t$); (b) $\beta = 1$, cos (sin $\omega_m t$); (c) $\beta = 2$, cos (2 sin $\omega_m t$); (d) $\beta = 5$, cos (5 sin $\omega_m t$).

ω_m. But these terms multiplied by $\cos \omega_c t$ give just the carrier and the second-order sideband terms that we obtained previously for $\beta \ll \sqrt{6}$.

2. For $\beta < \pi/2 = 1.57$ the function remains positive and appears as a dc component with some ripple superimposed. One would thus expect a Fourier analysis to give a large dc component plus successively decreasing harmonics. This dc component, or average value of the function, decreases with β, however. Again, if we multiply this by $\cos \omega_c t$, we obtain a picture of a carrier term decreasing with β, while the sidebands increase with β. These sidebands are all displaced from the carrier by even integral values of $\omega_m(\pm 2\omega_m, \pm 4\omega_m$, etc.).
3. For $\beta > \pi/2 = 1.57$ the function takes on negative values. As β increases, the positive and negative excursions become more rapid and

one would expect a greatly increased harmonic content, with much less energy at direct current. Again converting these results to variations about the carrier frequency, we would expect a rapid decrease in the energy content of the carrier for $\beta > \pi/2$ and increased energy in the sideband components.

The value $\beta = \pi/2$ thus represents the transition from a more or less slowly varying periodic time function with most of the spectral energy appearing in the carrier to a rapidly varying function with the spectral energy spread out over a wide range of frequencies.

We would have obtained similar results had we plotted the quadrature term $\sin (\beta \sin \omega_m t)$, with the one difference that the average value of this function is always zero. Its frequency components are all odd integral multiples of ω_m, so that they give rise to odd-order sidebands about the carrier when multiplied by $\sin \omega_c t$.

We can sum up these qualitative results by stating that for $\beta < \pi/2$ the frequency-modulated wave consists of a large carrier term plus smaller sideband terms. The carrier and even-order sidebands (those at even integral multiples of ω_m away from ω_c: $\pm 2\omega_m$, $\pm 4\omega_m$, $\pm 4\omega_m$, etc.) are contributed by the inphase term of the expression for the frequency-modulated carrier. The odd-order sidebands ($\pm \omega_m$, $\pm 3\omega_m$, etc.) are contributed by the quadrature-phase term.

For $\beta > \pi/2$, we get a picture of a wave with only a small carrier term plus increased energy in the sidebands. The bandwidth of the FM signal increases rapidly with $\beta > \pi/2$. This implies wide bandwidths and is, of course, as pointed out previously, the desirable situation in most FM systems used for their good noise- and interference-rejection properties.

By wideband FM, we shall thus mean a frequency-modulated signal with $\beta > \pi/2$.

To demonstrate these conclusions more quantitatively, we must actually determine the frequency components and their amplitudes for the frequency-modulated signal of arbitrary β and sinusoidal modulating signal. As was noted previously, this may be done by expanding both $\cos (\beta \sin \omega_m t)$ and $\sin (\beta \sin \omega_m t)$ in their respective Fourier series. Multiplying the cosine term by $\cos \omega_c t$ and the sine term by $\sin \omega_c t$ then gives us our FM signal.

Both series may be found simultaneously by considering the periodic complex exponential

$$v(t) = e^{j\beta \sin \omega_m t} \qquad -\frac{T}{2} < t < \frac{T}{2} \tag{4-60}$$

The real part of this function gives us our cosine function; the imaginary part, the sine function. If we expand this exponential in its Fourier series, we can expect to get a real part consisting of even harmonics of ω_m and an imaginary part consisting of the odd harmonics. [This is deduced from our previous discussion of cos $(\beta \sin \omega_m t)$ and the associated Fig. 4-39.] By equating reals and imaginaries, we shall then obtain the desired Fourier expansions for cos $(\beta \sin \omega_m t)$ and sin $(\beta \sin \omega_m t)$, respectively.

The Fourier coefficient of the complex Fourier series for the exponential of Eq. [4-60] is given by

$$c_n = \int_{-T/2}^{T/2} e^{j(\beta \sin \omega_m t - \omega_n t)}\, dt \qquad \omega_m = \frac{2\pi}{T},\ \omega_n = \frac{2\pi n}{T} = n\omega_m \tag{4-61}$$

Normalizing this integral by letting $x = \omega_m t$, we get

$$\frac{c_n}{T} = \frac{1}{2\pi} \int_{-\pi}^{\pi} e^{j(\beta \sin x - nx)}\, dx \tag{4-62}$$

This integral can be evaluated only as an infinite series (as was the case in Chap. 2 for the sine integral Si x). It occurs very commonly in many physical problems, however, and so has been tabulated in many books.[1] It is called the *Bessel function of the first kind* and is denoted by the symbol $J_n(\beta)$. (Note from Eq. [4-62] that c_n is a function of both β and n.)

In particular,

$$J_n(\beta) \equiv \frac{1}{2\pi} \int_{-\pi}^{\pi} e^{j(\beta \sin x - nx)}\, dx \tag{4-63}$$

so that

$$c_n = TJ_n(\beta) \tag{4-64}$$

For $n = 0$, we get $c_0 = TJ_0(\beta)$, the dc component of the Fourier-series representation of the periodic complex exponential of Eq. [4-60]. Increasing values of n give the corresponding Fourier coefficients for the higher-frequency terms of the Fourier series. The spectrum of the complex exponential of Eq. [4-60] (and ultimately that of the frequency-modulated signal) will thus be given by the value of the Bessel function and will depend on the parameter β. Thus

$$e^{j\beta \sin \omega_m t} = \frac{1}{T} \sum_{n=-\infty}^{\infty} c_n e^{j\omega_n t} = \sum_{n=-\infty}^{\infty} J_n(\beta) e^{j\omega_n t} \qquad \omega_n = n\omega_m \tag{4-65}$$

As an example, let $n = 0$. Then

$$J_0(\beta) \equiv \frac{1}{2\pi} \int_{-\pi}^{\pi} e^{j\beta \sin x}\, dx \tag{4-66}$$

[1] See, for example, E. Jahnke and F. Emde, "Tables of Functions," Dover, New York, 1945.

is the dc component of the Fourier series of Eq. [4-65]. But

$$e^{j\beta \sin x} = 1 + (j\beta \sin x) + \frac{(j\beta \sin x)^2}{2!} + \frac{(j\beta \sin x)^3}{3!} + \cdots \quad [4\text{-}67]$$

using the series expansion for the exponential.

The power terms in $\sin x$ may be rewritten as sines and cosines of integral multiples of x, so that Eq. [4-67] can also be written

$$e^{j\beta \sin x} = 1 + j\beta \sin x - \frac{\beta^2}{2} \frac{1 - \cos 2x}{2} + \cdots \quad [4\text{-}68]$$

If we now integrate over a complete period of 2π radians (as called for by Eq. [4-66]), the terms in $\sin x$, $\cos 2x$, etc., vanish, leaving us with

$$J_0(\beta) = 1 - \frac{\beta^2}{4} + \cdots \quad [4\text{-}69]$$

an infinite series in β. (This is, of course, one method of evaluating the integral of Eq. [4-63].) But a comparison with Eq. [4-59] shows that this is exactly the coefficient of the carrier term of our FM signal, obtained there by a power-series expansion of $\cos(\beta \sin \omega_m t)$. The advantage of the present approach, using a Fourier-series expansion, is that the Bessel function is already tabulated so that we do not have to repeat the evaluation of Eq. [4-63] for different values of n by means of an infinite series.

From Eq. [4-65], the Fourier-series expansion of the complex exponential, we can obtain our desired Fourier series for $\cos(\beta \sin \omega_m t)$ and $\sin(\beta \sin \omega_m t)$. It can be shown, either from the integral definition of $J_n(\beta)$ (Eq. [4-63]) or from the power-series expansion of $J_n(\beta)$, that

$$\begin{aligned} J_n(\beta) &= J_{-n}(\beta) \qquad n \text{ even} \\ \text{and} \qquad J_n(\beta) &= -J_{-n}(\beta) \qquad n \text{ odd} \end{aligned} \quad [4\text{-}70]$$

Writing out the Fourier series term by term, and using Eq. [4-70] to combine the positive and negative terms of equal magnitude of n, we get

$$\begin{aligned} e^{j\beta \sin \omega_m t} = J_0(\beta) &+ 2[J_2(\beta) \cos 2\omega_m t + J_4(\beta) \cos 4\omega_m t + \cdots] \\ &+ 2j[J_1(\beta) \sin \omega_m t + J_3(\beta) \sin 3\omega_m t + \cdots] \end{aligned} \quad [4\text{-}71]$$

But
$$e^{j\beta \sin \omega_m t} = \cos(\beta \sin \omega_m t) + j \sin(\beta \sin \omega_m t) \quad [4\text{-}72]$$

Equating real and imaginary terms, we get

$$\cos(\beta \sin \omega_m t) = J_0(\beta) + 2J_2(\beta) \cos 2\omega_m t + 2J_4(\beta) \cos 4\omega_m t + \cdots \quad [4\text{-}73]$$

and
$$\sin(\beta \sin \omega_m t) = 2J_1(\beta) \sin \omega_m t + 2J_3(\beta) \sin 3\omega_m t + \cdots \quad [4\text{-}74]$$

Equations [4-73] and [4-74] are the desired Fourier-series expansions for the cosine and sine terms. As noted previously, the cosine term has only the even harmonics of ω_m in its series, the sine term containing the odd harmonics.

The spectral distribution of the frequency-modulated carrier is now readily obtained. As previously written,

$$f_c(t) = \cos \omega_c t \cos (\beta \sin \omega_m t) - \sin \omega_c t \sin (\beta \sin \omega_m t) \qquad [4\text{-}75]$$

Using the Fourier-series expansions for the cosine and sine terms, and then utilizing the trigonometric sum and difference formulas (as in the AM analysis), we get

$$\begin{aligned} f_c(t) = J_0(\beta) \cos \omega_c t &- J_1(\beta)[\cos (\omega_c - \omega_m)t - \cos (\omega_c + \omega_m)t] \\ &+ J_2(\beta)[\cos (\omega_c - 2\omega_m)t + \cos (\omega_c + 2\omega_m)t] \\ &- J_3(\beta)[\cos (\omega_c - 3\omega_m)t - \cos (\omega_c + 3\omega_m)t] \\ &+ \cdots \end{aligned} \qquad [4\text{-}76]$$

We thus have a time function consisting of a carrier and an infinite number of sidebands, spaced at frequencies $\pm f_m$, $\pm 2f_m$, etc., away from the carrier. This is in contrast to the AM case, where the carrier and only a single set of sidebands existed. The odd sideband frequencies arise from the quadrature term of Eq. [4-75]; the even sideband frequencies arise from the inphase ($\cos \omega_c t$) term. This, of course, agrees with our previous qualitative discussion based on Fig. 4-39.

The magnitudes of the carrier and sideband terms depend on β, the modulation index, this dependence being expressed by the appropriate Bessel function. Again, this is at variance with the AM case, where the carrier magnitude was fixed and the two sidebands varied only with the modulation factor.

We showed previously, from qualitative considerations of the time variation of the function $f_c(t)$, that for $\beta \ll \pi/2$, we should have primarily a carrier and one or two sideband pairs. For $\beta > \pi/2$, we should have increasingly more significant sideband pairs as β increases. The magnitude of the carrier should also decrease rapidly.

We can now verify these qualitative conclusions by referring to plots or tabulations of the Bessel function.[1] As an example, a graph of the functions $J_0(\beta)$, $J_1(\beta)$, $J_2(\beta)$, $J_8(\beta)$, and $J_{16}(\beta)$ is shown in Fig. 4-40. Note that for $\beta > \pi/2$ the value of $J_0(\beta)$ decreases sharply. $J_0(\beta)$ represents the magnitude of the carrier term, so that this result agrees with that obtained from the curves of Fig. 4-39.

For β small ($\beta \ll \pi/2$), the only Bessel functions of significant magnitude are $J_0(\beta)$ and $J_1(\beta)$. The FM wave thus consists essentially

[1] See, for example, the various curves and tables in A. Hund, "Frequency Modulation," chap. 1, McGraw-Hill, New York, 1942.

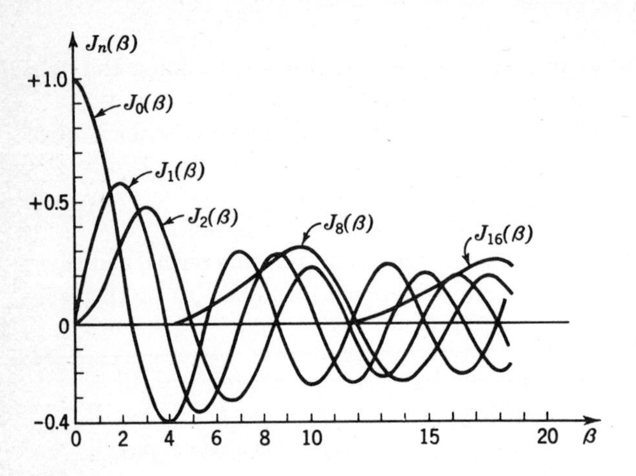

FIG. 4-40 Examples of Bessel functions of the first kind.

of the carrier and the two first-order sideband frequencies. This is, of course, the narrowband FM case considered previously. As β increases, however, the magnitudes of the higher-order sidebands begin to increase, the carrier magnitude begins to decrease, and the bandwidth required increases. The significant number of sideband terms depends on β, the modulation index. Note that $J_8(\beta)$ is essentially zero up to $\beta = 4$ and then begins to increase to a peak value at $\beta = 9.5$. $J_{16}(\beta)$ begins to increase significantly for $\beta > 12$ and peaks at about $\beta = 18$.

It is apparent from the curves of Fig. 4-40, and can be shown in general, that the higher-order Bessel functions, $J_n(\beta)$ with $n \gg 1$, are essentially zero to the point $\beta = n$. They then increase to a peak, decrease again, and eventually oscillate like damped sinusoids. This means that for $\beta \gg 1$ the number of significant sideband frequencies in the frequency-modulated wave is approximately equal to β. [$J_n(\beta) \doteq 0$, $n > \beta$, so that the corresponding sideband terms are negligible.] Since the sidebands are all f_m Hz apart, and since there are two sets of them, on either side of the carrier, the bandwidth B_T of the FM signal is approximately

$$B_T \doteq 2\beta f_m = 2\frac{\Delta f}{f_m} f_m = 2\,\Delta f \qquad \beta \gg 1 \tag{4-77}$$

This agrees, of course, with the results of the binary FM case. This assumes a sinusoidal modulating signal of frequency f_m Hz. Δf represents the maximum-frequency deviation away from f_c, the unmodulated carrier frequency, and depends on the amplitude of the modulating signal. So for large β the bandwidth is directly proportional to the amplitude of the modulating signal. This is again to be compared with the AM or narrowband FM case where the bandwidth is $2B$, B being the baseband bandwidth (f_m in the case of a single sine wave at that frequency).

The bandwidth is equal to $2\,\Delta f$ only for very large modulation index. For smaller values of β, we can determine the bandwidth by counting the significant number of sidebands. The word significant is usually taken to mean those sidebands which have a magnitude of at least 1 percent of the magnitude of the unmodulated carrier. We have been assuming an unmodulated carrier of unit amplitude ($\cos \omega_c t$), so that the significant sidebands will be those for which $J_n(\beta) > 0.01$. The number will vary with β and can be determined readily from tabulated values of the Bessel function.

An example of such a determination is shown by the accompanying table giving values of β up to 2. Using such a table, we can plot band-

TABLE OF SIGNIFICANT SIDEBANDS

β	$J_0(\beta)$	$J_1(\beta)$	$J_2(\beta)$	$J_3(\beta)$	$J_4(\beta)$	No. of side-bands	Band-width
0.01	1.00	0.005	....	....	....	1	$2f_m$
0.20	0.99	0.100	....	....	....	1	$2f_m$
0.50	0.94	0.24	0.03	....	....	2	$4f_m$
1.00	0.77	0.44	0.11	0.02	....	3	$6f_m$
2.00	0.22	0.58	0.35	0.13	0.03	4	$8f_m$

width versus the modulation index β. Such a curve has been drawn in Fig. 4-41. The transmission bandwidth B_T is shown normalized to both Δf, the frequency deviation, and to the baseband bandwidth B(f_m in the sine-wave case). Also shown for reference is the rule-of-thumb relation

$$B_T = 2\,\Delta f + 2B = 2B(1 + \beta) \qquad [4\text{-}77a]$$

as well as a sine-wave analysis curve with sideband frequencies having amplitudes 10 percent or greater of the carrier included. The bandwidth in this case is, of course, less than that for the 1 percent amplitude case, and agrees closely with the rule-of-thumb relation.

All curves show B_T equal to $2B$ for small β, and approaching $2\,\Delta f$ for large β.

We are now in a position to make some calculations of the required bandwidth for typical sinusoidal signals. Since the bandwidth ultimately varies with Δf, the frequency deviation, or the amplitude of the modulating signal, some limit must be put on this amplitude to avoid excessive

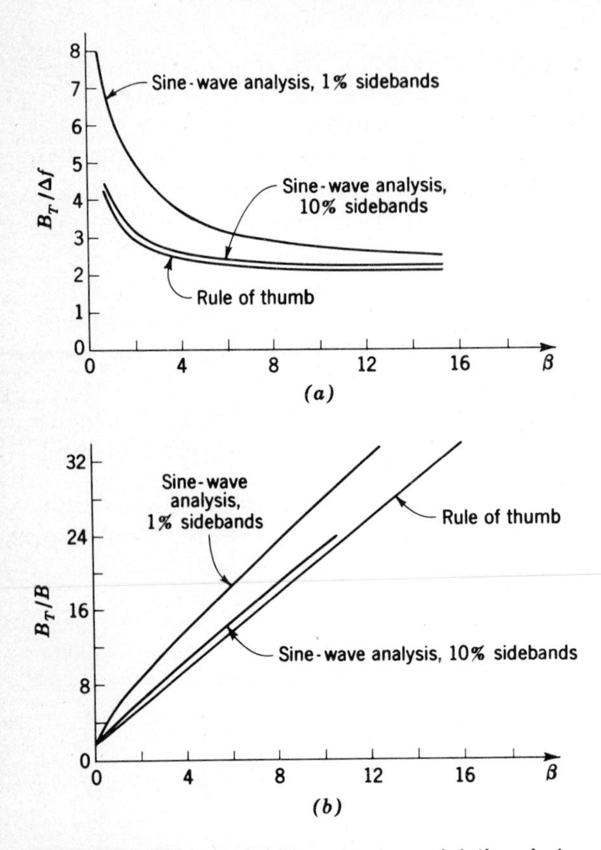

FIG. 4-41 FM bandwidth versus modulation index. (a) Normalized to frequency deviation; (b) normalized to baseband bandwidth.

bandwidth. The FCC has fixed the maximum value of Δf at 75 kHz for commerical FM broadcasting stations. What does this imply in the way of required bandwidth? If we take the modulating frequency f_m to be 15 kHz (typically the maximum audio frequency in FM transmission), $\beta = 5$, and the required bandwidth is 240 kHz, from Fig. 4-41. (Alternatively, for $\beta = 5$, there are eight significant sideband frequencies, or $2 \times 8 \times 15 = 240$ kHz is the bandwidth required.) For $f_m < 15$ kHz (the lower audio frequencies) β increases above 5, and the bandwidth eventually approaches $2\,\Delta f = 150$ kHz. So it is the *highest* modulating frequency that determines the required bandwidth. (These are the extreme cases, since $\Delta f = 75$ kHz corresponds to the maximum possible amplitude of the modulating signal.) The corresponding rule-of-thumb result is 180 kHz for the $\beta = 5$ case. The difference is, of course, due to the difference in definition of bandwidth. For the 10 percent sideband level case, the bandwidth is 190 kHz, which is in close agreement with the rule of thumb.

Frequency modulation is used for transmission of the sound channel in commercial TV, and the maximum-frequency deviation there (corresponding to the maximum amplitude of the modulating signal) has been fixed at 25 kHz by the FCC. For a 15-kHz audio signal this gives $\beta = 1.7$, or a bandwidth of 110 kHz. The lower audio frequencies require bandwidths well within this limit, for with f_m decreasing and $\beta > 15$, the bandwidth required approaches $2\,\Delta f = 50$ kHz.

The relatively large bandwidths required for commercial FM, as compared with AM sound transmission, are the price one has to pay to obtain significant improvement in noise and interference rejection. This question will be discussed in detail in Chap. 6.

The amplitude spectra of a frequency-modulated signal are shown plotted in Fig. 4-42 for $\beta = 0.2, 1, 5, 10$. The sinusoidal modulating

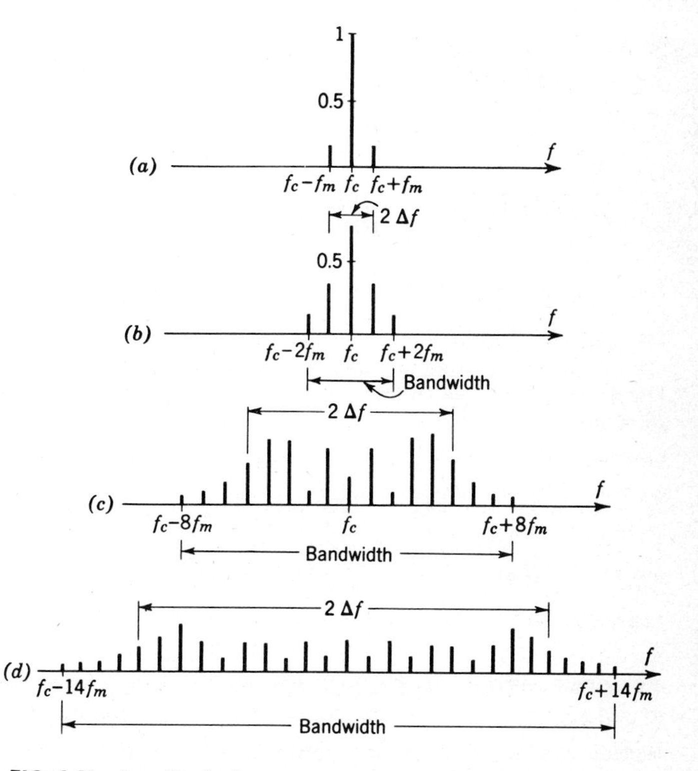

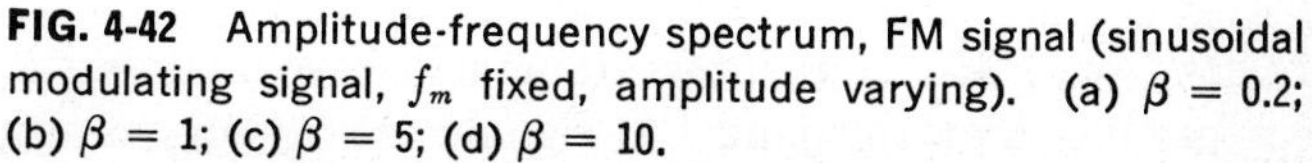

FIG. 4-42 Amplitude-frequency spectrum, FM signal (sinusoidal modulating signal, f_m fixed, amplitude varying). (a) $\beta = 0.2$; (b) $\beta = 1$; (c) $\beta = 5$; (d) $\beta = 10$.

signal is assumed to be of constant modulation frequency f_m, so that β is proportional to the signal amplitude. The amplitude of the spectral line at frequency $f_c \pm nf_m$ is given by $J_n(\beta)$. If the amplitude of the modulating signal is fixed (for example, $\Delta f = 75$ kHz for all signals) and different audio frequencies are considered, β increases as f_m decreases and we get spectrum plots similar to those of Fig. 4-42. For these plots Δf is fixed, however, so that we get a picture of more and more spectral lines crowding into a fixed-frequency interval. An example of such a plot is shown in Fig. 4-43. Δf has been chosen as 75 kHz, with f_m varying from

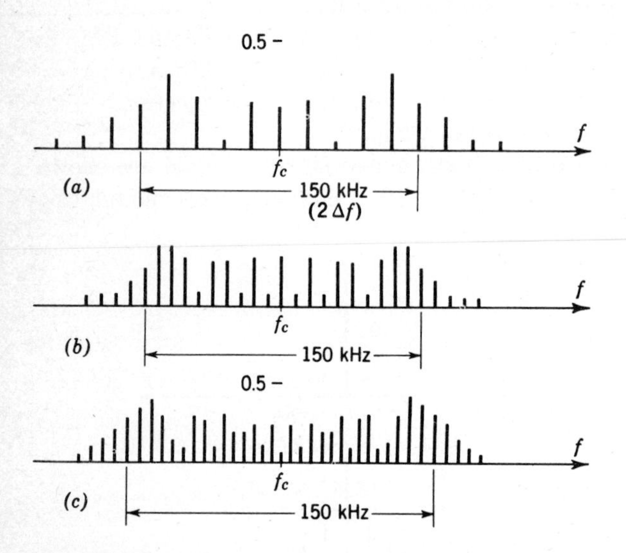

FIG. 4-43 Amplitude-frequency spectrum, FM signal (amplitude of Δf fixed, f_m decreasing). (a) $f_m = 15$ kHz ($\beta = 5$); (b) $f_m = 7.5$ kHz ($\beta = 10$); (c) $f_m = 5$ kHz ($\beta = 15$).

15 kHz down to 5 kHz. This plot shows clearly the lines concentrating within the 2 Δf or 150-kHz points as β increases. (Remember again that the bandwidth approaches 2 Δf for large β.)

4-8 GENERATION OF FREQUENCY-MODULATED SIGNALS

The methods of generating the wideband FM signals discussed in the last section can be grouped essentially into two types:

1. *Indirect FM* Here integration and phase modulation are first used to produce a narrowband FM signal. Frequency multiplication is then

utilized to increase the modulation index to the desired range of values.

2. *Direct FM* Here the carrier frequency is directly modulated or varied in accordance with the input modulating signal (hence, the classification *direct*).

INDIRECT FM

A simple block diagram of an indirect FM system is shown in Fig. 4-44. The modulating signal is first integrated and then used to phase modulate

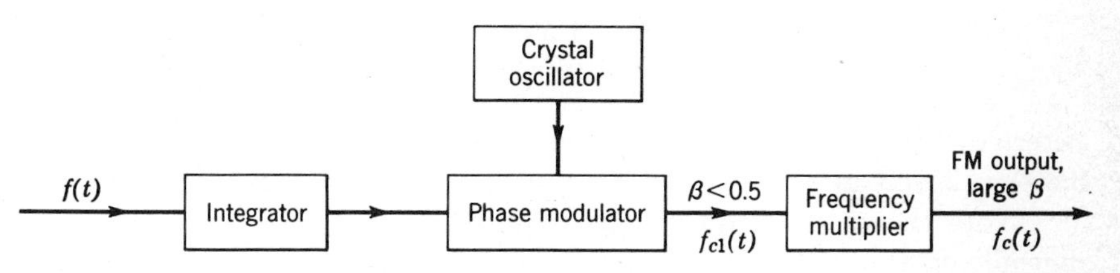

FIG. 4-44 Indirect FM system.

a crystal-controlled carrier frequency. The maximum phase variation is kept very small (< 0.5 radian in practice), so that the output of the phase modulator is a narrowband FM signal. This signal is then multiplied in frequency to produce the desired wideband FM signal.

At the output of the phase modulator the FM signal has the form

$$f_{c1}(t) = \cos\,[\omega_{c1}t + \theta_1(t)] \qquad \theta_1(t) = k\int f(t)\,dt \tag{4-78}$$

For a sine-wave modulating signal this would be

$$f_{c1}(t) = \cos\,(\omega_{c1}t + \beta_1 \sin \omega_m t) \qquad \beta_1 = \frac{\Delta f}{f_m} \tag{4-79}$$

The signal frequency is multiplied n times by the frequency multiplier so that at its output the signal appears in the form

$$f_c(t) = \cos n[\omega_{c1}t + \theta_1(t)] \tag{4-80}$$

For the sine-wave modulating signal, then,

$$\begin{aligned} f_c(t) &= \cos\,(n\omega_{c1}t + n\beta_1 \sin \omega_m t) \\ &= \cos\,(\omega_c t + \beta \sin \omega_m t) \qquad \beta = n\beta_1,\ \omega_c = n\omega_{c1} \end{aligned} \tag{4-81}$$

By picking n appropriately, β may be set at any desired value.

As an example, assume that the audio signal contains frequencies of 50 Hz to 15 kHz. The maximum-frequency deviation Δf is 75 kHz. β then ranges from 1,500 ($f_m = 50$ Hz) down to 5 ($f_m = 15$ kHz).

We indicated in Sec. 4-7 that $\beta_1 < 0.2$ radian corresponds to a narrowband FM signal. In practice $\beta_1 < 0.5$ radian is used. The resultant signal has some AM associated with it (5 percent residual AM distortion for $\beta = 0.5$), but amplitude limiting in the frequency multipliers is utilized to eliminate most of this undesired AM.

Since a *maximum* modulation index of 0.5 radian is called for at the phase-modulator output, this must be associated with the *lowest-frequency* modulating signal. For example, if $\beta_1 = 0.5$ radian for a 50-Hz signal, $\beta_1 = 0.5/300$ radian for a 15-kHz signal. ($\beta_1 = \Delta f/f_m$; $\Delta f = 75$ kHz.) The narrowband FM signal must then be multiplied in frequency 3,000 times to produce the desired modulation index at the output:

$$n = \frac{\beta}{\beta_1} = \frac{1{,}500}{0.5} = 3{,}000$$

Alternatively, the maximum-frequency deviation Δf_1 will be 75 kHz/3,000 = 25 Hz at the input to the multiplier and 75 kHz at the output.

Note then that although the required *bandwidth* of an FM signal depends on the *highest* modulating frequency (as shown in the previous section), the choice of *modulation index* and required frequency multiplication depend on the *lowest* modulating frequency. For this reason, 50 Hz is usually chosen as the lowest audio frequency to be passed in an indirect FM system. Still lower modulating frequencies would require excessively large frequency multiplication and attendant complications in the circuitry of the system.

A straight frequency multiplication of 3,000 would produce inordinately high carrier frequencies at the output of the multipliers (for example, if $f_{c1} = 200$ kHz, $f_c = 600$ MHz). In practice, then, a combination of both frequency multiplication and frequency conversion is used to produce a wideband FM signal at the desired carrier frequency. The frequency converters employed are essentially those discussed previously in connection with AM demodulators. They produce at their output the difference frequencies between the input signal and an injected sinusoidal signal. This serves to shift the FM spectrum downward to a specified carrier frequency but keeps the multiplied modulation index intact.

The points discussed above are all incorporated in the Armstrong indirect FM transmitter. E. H. Armstrong was one of the first engineers to recognize the possible merits of FM broadcasting, and he pioneered in many of the important developments in the field. As A. Hund[1] points out, it was Armstrong's demonstration of the merits of FM in a paper presented at an IRE meeting in 1935 that stirred up interest in the field and eventually resulted in the recognition of the capabilities of FM transmission.

[1] A. Hund, "Frequency Modulation," p. 231, McGraw-Hill, New York, 1942.

A block diagram of an Armstrong-type transmitter for standard FM broadcast use is shown in Fig. 4-45. (A detailed diagram of the original Armstrong transmitter and a thorough discussion of its component parts may be found in Hund's book and in the literature cited there.) The

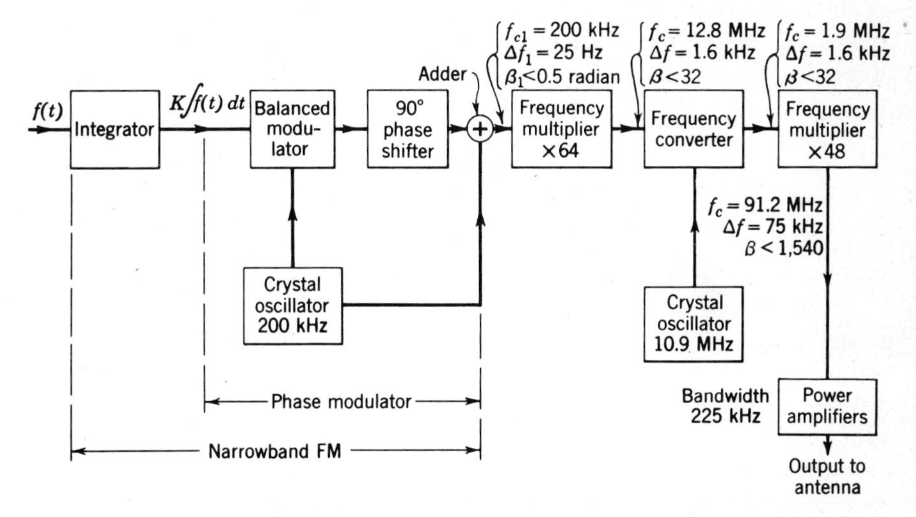

FIG. 4-45 An Armstrong-type indirect FM transmitter (simplified diagram).

transmitter shown operates at an unmodulated carrier frequency of 91.2 MHz with a maximum-frequency deviation of 75 kHz. The transmitter is crystal-controlled, starting at the controlled carrier frequency of 200 kHz. There is a total frequency multiplication of $64 \times 48 = 3{,}072$, so that the maximum modulation index at the multiplier input can be fixed at 0.5 radian. An injected frequency of 10.9 MHz is used to shift the carrier frequency down from 12.8 to 1.9 MHz between multiplier sections.

DIRECT FM

As apparent from the terminology used, direct FM systems are those in which the modulating signal directly controls the carrier frequency. These direct FM systems normally require a much smaller amount of frequency multiplication than the indirect system.

At the heart of the direct FM transmitter is the device employed to vary the carrier frequency. The carrier frequency is normally generated by an oscillator whose frequency-determining circuit is a high-Q resonant circuit or crystal. Variations in the inductance or capacitance of this resonant circuit will then change the resonant frequency or the oscillator frequency. Thus, assume the capacitance of the resonant (tank) circuit is made proportional to the baseband modulating signal $f(t)$. Then

$$C = C_0 + \Delta C = C_o + Kf(t) \qquad [4\text{-}82]$$

Here C_0 is the zero-signal capacitance. If the change ΔC in capacitance is small compared with C_0, it is simple to show that the instantaneous frequency ω_i of the tuned circuit becomes linearly proportional to $f(t)$, as desired. Specifically,

$$\begin{aligned}\omega_i &= \frac{1}{\sqrt{LC}} = \frac{1}{\sqrt{LC_0}}\frac{1}{\sqrt{1 + \Delta C/C_0}} \\ &\doteq \omega_c\left(1 - \frac{\Delta C}{2C_0}\right) \\ &= \omega_c\left[1 - \frac{Kf(t)}{2C_0}\right] \\ &= \omega_c - K_2 f(t) \qquad \omega_c^2 \equiv \frac{1}{LC_0}, \ \Delta C \ll C_0 \end{aligned} \qquad [4\text{-}83]$$

and

$$\Delta f = \frac{\Delta C}{2C_0} f_c = \frac{Kf_c}{2C_0} f(t) \qquad [4\text{-}84]$$

It is apparent that small variations in the inductance L produce the same result.

How small ΔC should be compared with C_0 depends on how accurate the linear approximation to the square-root term in Eq. [4-83] must be. With $\Delta C/C_0 < 0.013$, the distortion due to this approximation is less than 1 percent. (The reader can check this for himself by evaluating the next, nonlinear term, in the power-series expansion of the square root.) Although the change in capacitance is of necessity small, the frequency deviation Δf may be quite large if the zero-signal resonant frequency f_c is large enough. As an example, if $\Delta C/2C_0 = 0.005$, and $f_c = 15$ MHz, $\Delta f = 75$ kHz, which is the maximum deviation specified for standard radio broadcasting. Generally, then, this method of *directly* varying the instantaneous frequency requires substantially less additional frequency multiplication and conversion than does the indirect method.

There are various ways of obtaining a capacitance (or inductance) variation proportional to the signal intelligence $f(t)$. The most common involve the use of a capacitance plus controlled source from an active circuit element. A Miller-effect capacitance is one example, a reactance tube another.[1] One common method is to use as the capacitance in the

[1] M. Schwartz, "Information Transmission, Modulation, and Noise," 1st ed., McGraw-Hill, New York, 1959, pp. 136–141.

circuit a back-biased *varactor* diode. The modulating signal $f(t)$ controls the back-bias voltage, varying in turn the *varactor* capacitance.

FM transmitters have been used quite commonly in space communications and telemetry. A typical block diagram of a direct-type FM transmitter for space applications is shown in Fig. 4-46.[1] This all-solid

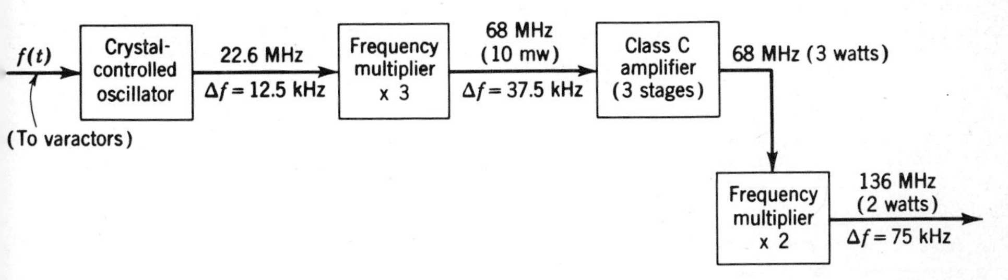

FIG. 4-46 Typical direct FM transmitter, space telemetry. [From A. V. Balakrishnan (ed.), "Space Communications," p. 190, McGraw-Hill, New York, 1963.]

state system provides 2 watts output power at 136 MHz. A series reactive network consisting of varactor diodes and an inductor provides the required variable capacitance in the 22.6 MHz crystal-controlled oscillator. The FM signal is then tripled in frequency and amplified to provide 3 watts output at 68 MHz. A frequency doubler then provides the desired output. The initial frequency deviation is 12.5 kHz. The final Δf, after frequency multiplication by 6, is 75 kHz.

Frequency multiplexing of several data channels is quite commonly used in space telemetry.[2] In this case the individual data channels each frequency modulate a carrier, the resultant FM signals being arranged to occupy adjacent frequency bands. These FM signals are then summed and the composite complex signal used to frequency modulate a very high sine-wave carrier for final transmission. An example of such an *FM-FM system* is shown in Fig. 4-47. (The notation FM-FM is commonly used to indicate two steps of FM. AM-FM similarly implies initial frequency multiplexing of multiple data channels using AM techniques, with the composite set of AM signals then used to frequency modulate a high-frequency carrier.) In this example, the ratio of frequency deviation to

[1] W. B. Allen, in A. V. Balakrishnan (ed.), "Space Communications," pp. 190–192, McGraw-Hill, New York, 1963.

[2] See FM-FM Telemetry Systems, in E. L. Gruenberg (ed.), "Handbook of Telemetry and Remote Control," chap. 6, McGraw-Hill, New York, 1967. Also see R. Stampfl, The Nimbus Satellite Communication System, in A. V. Balakrishnan, *op. cit.*, chap. 18.

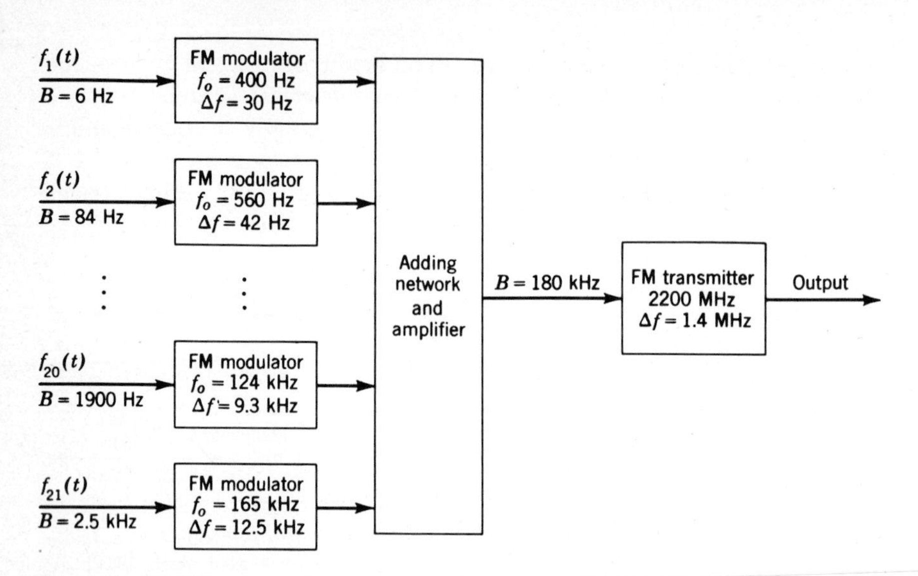

FIG. 4-47 FM-FM system for space telemetry (proportional-bandwidth subcarrier channels).

subcarrier frequency in each subchannel is held fixed at 7.5 percent. For a modulation index of 5 for each subchannel, the corresponding signal bandwidths increase with the channel number. Low-bandwidth signals would thus be used in the low-numbered channels, wider-bandwidth signals in the higher-number channels.

Very often one or more of the subchannels may be used for the transmission of time-multiplexed digital data, or several subchannels may be combined to accommodate wider-band data signals.

4-9 FREQUENCY DEMODULATION

Frequency demodulation, the process of converting a frequency-modulated signal back to the original modulating signal, can be carried out in a variety of ways. Ultimately, however, the process used must provide an output voltage (or current) whose *amplitude* is *linearly* proportional to the *frequency* of the input FM signal. The term *frequency discriminator* is commonly used to characterize a device providing this frequency-amplitude conversion.

Various schemes have been proposed to accomplish the task of frequency demodulation. We shall discuss two types of frequency discriminators in this section: a balanced discriminator involving the use of tuned circuits and so-called zero-crossing detectors.

To help us understand the functioning of these devices, consider again the FM signal given by

$$f_c(t) = A \cos \theta(t) = A \cos [\omega_c t + K\int f(t)\, dt] \qquad [4\text{-}85]$$

Here $f(t)$ is the baseband or modulating signal. It is apparent that to extract $f(t)$ from this expression, we must somehow generate the instantaneous frequency

$$\omega_i(t) = \frac{d\theta(t)}{dt} = \omega_c + Kf(t) \qquad [4\text{-}86]$$

If we then balance out the constant term ω_c, we have the desired output $f(t)$.

We shall show later that zero-crossing detectors do provide a measure of ω_i and, hence, $f(t)$. The approach we consider now is that of studying Eq. [4-85] for the FM signal. Some thought indicates that ω_i may be generated from that equation by differentiating $f_c(t)$ with respect to time. Specifically, if we assume the amplitude A constant (we shall have to include a limiter to ensure this later), the derivative of $f_c(t)$ is just

$$\frac{df_c(t)}{dt} = -A \sin \theta(t) \left[\frac{d\theta}{dt}\right]$$
$$= -A[\omega_c + Kf(t)] \sin \theta(t) \qquad [4\text{-}87]$$

Note that this expression is precisely in the form of an AM signal, whose *envelope* is given by

$$A[\omega_c + Kf(t)] = A\omega_c\left[1 + \frac{Kf(t)}{\omega_c}\right]$$

Since we have been assuming the frequency deviation $\Delta f = Kf(t) \ll \omega_c$, the envelope never goes to zero, and, in fact, varies only slightly about the average quantity $A\omega_c$. It is then apparent that we may now obtain $f(t)$ by envelope detecting $df_c(t)/dt$ (see Fig. 4-48).

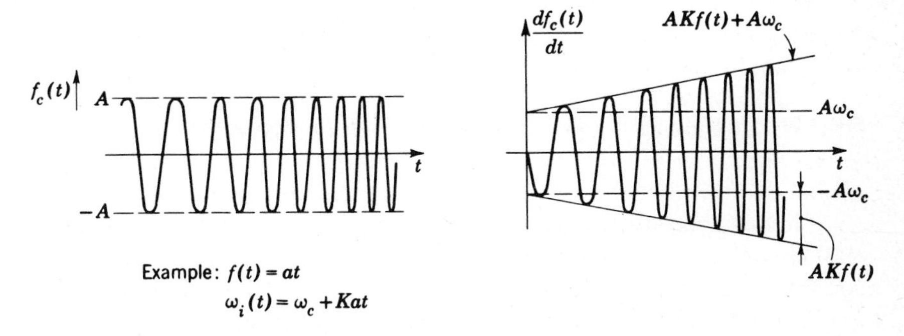

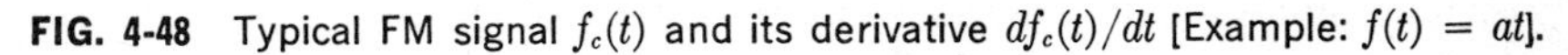
FIG. 4-48 Typical FM signal $f_c(t)$ and its derivative $df_c(t)/dt$ [Example: $f(t) = at$].

The FM detector thus consists of a differentiating circuit followed by an envelope detector (Fig. 4-49). To ensure that the amplitude A is truly constant (otherwise additional, distortion, terms involving dA/dt appear at the output), one normally inserts a *limiter* prior to differentia-

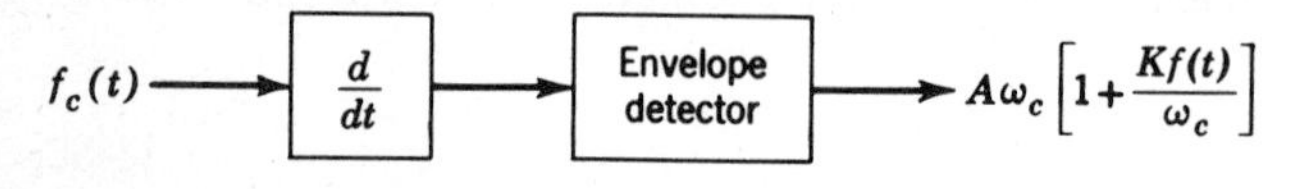

FIG. 4-49 FM detector.

tion. The limiter serves to keep amplitude variations from appearing in the output. If one uses hard clipping to keep the amplitude invariant, square waves at the varying instantaneous frequency $\omega_i(t)$ are produced. It is then necessary to follow the limiter with a bandpass filter centered about ω_c to convert the square waves back to the cosinusoidal form of Eq. [4-85]; i.e., terms centered about $2\omega_c$ and the other higher harmonics of ω_c are filtered out. This is commonly incorporated in the differentiator, as will be shown in the paragraphs that follow.

There are various ways of performing the necessary differentiation. We consider here only one, that involving the single-tuned circuit of Fig. 4-50.

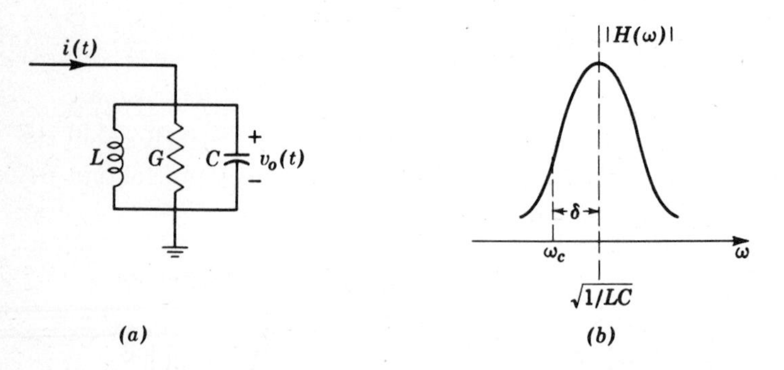

FIG. 4-50 Single-tuned circuit. (a) Circuit; (b) amplitude response.

To demonstrate that this circuit does provide the required differentiation and envelope detection, we first note that if the circuit is detuned so that the unmodulated carrier frequency ω_c lies on the sloping part of the amplitude-frequency characteristic and the frequency variations occur within a small region about the unmodulated carrier, the amplitude of the output wave will follow the instantaneous frequency of the input. This is shown in both Figs. 4-50 and 4-51. The region over which the characteristics are very nearly linear must be wide enough to cover the maximum-frequency deviation.

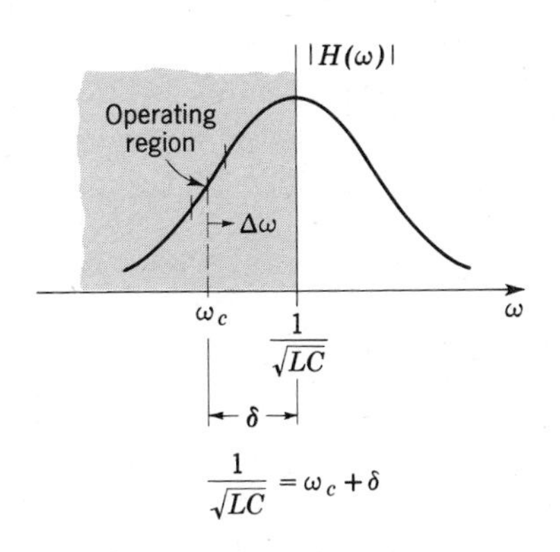

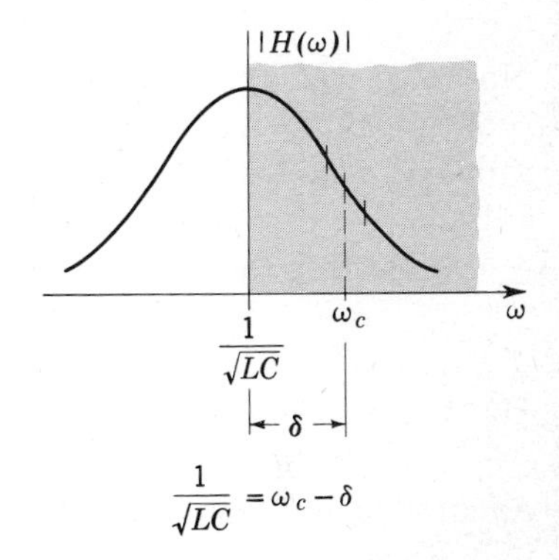

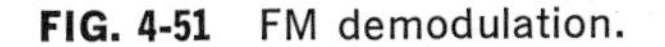

FIG. 4-51 FM demodulation.

We demonstrate this quantitatively by calculating the frequency response of the circuit. It is apparent that the magnitude of the transfer impedance is given by

$$|H(\omega)| = \left|\frac{V_o(\omega)}{I(\omega)}\right| = \frac{1/G}{\sqrt{1 + (\omega C - 1/\omega L)^2/G^2}} \qquad [4\text{-}88]$$

We now define $\Delta\omega = \omega - \omega_c$ as the frequency deviation away from the carrier frequency ω_c, and let $1/\sqrt{LC} = \omega_c \pm \delta$, as in Fig. 4-51. (The sign depends on the slope along which we operate.) To ensure linearity of the operating region, assume

$$\Delta\omega \ll \omega_c$$

(This is of course normally true anyway.) Also, let $\delta \ll \omega_c$. (The carrier frequency ω_c is then close to the center frequency $1/\sqrt{LC}$.) It is then easily shown that Eq. [4-88] reduces to

$$|H| \doteq \frac{1/G}{\sqrt{1 + 4(C/G)^2(\delta - \Delta\omega)^2}} \qquad [4\text{-}89]$$

in the case of operation along the left-hand slope.

Now assume further that $\delta \gg \Delta\omega$. This implies, of course, that although ω_c is close to $1/\sqrt{LC}$, we never deviate far enough away from ω_c

to approach $1/\sqrt{LC}$. This is necessary to ensure linearity of the operating region. Further, assume that $\delta \ll G/2C \equiv \alpha$. Equation [4-89] then reduces to

$$|H| \doteq \frac{1}{G}\left[\left(1 - \frac{\delta^2}{2\alpha^2}\right) + \frac{\delta}{\alpha^2}\Delta\omega\right] \tag{4-90}$$

Note that with the assumptions made $|H|$ does have a component linearly proportional to $\Delta\omega$, as desired. This shows the single-tuned circuit *has* provided the desired differentiation, while the envelope of the output voltage *does* provide the desired output $f(t)$.

Collecting the assumptions made, we have

$$\Delta\omega \ll \delta \ll 1/\sqrt{LC}$$
$$\Delta\omega \ll \omega_c$$
$$\delta \ll \alpha$$

As an example, assume we are to demodulate an FM signal with a carrier frequency $f_c = 10$ MHz. The maximum-frequency deviation $\Delta f = 75$ kHz. We then use as the differentiating circuit a single-tuned circuit centered at $1/2\pi\sqrt{LC} = 10$ MHz + 200 kHz, with $\alpha/2\pi = 10^6$/sec. (Then $\delta/2\pi = 200$ kHz.)

The requirements on the inequalities to obtain linearity may be relaxed somewhat and the $\delta^2/2\alpha^2$ term in Eq. [4-90] (which is large compared with the $\Delta\omega$ term) eliminated by using a *balanced demodulator*, that is, two single-tuned circuits, one tuned to $\omega_c + \delta$ and the other to $\omega_c - \delta$. If the individual outputs are envelope detected and subtracted, the final output is

$$|H|\ \omega_{c+\delta} - |H|\ \omega_{c-\delta} = \frac{2\delta}{G\alpha^2}\Delta\omega \tag{4-91}$$

as desired. An example of such a balanced demodulator is shown in Fig. 4-52.

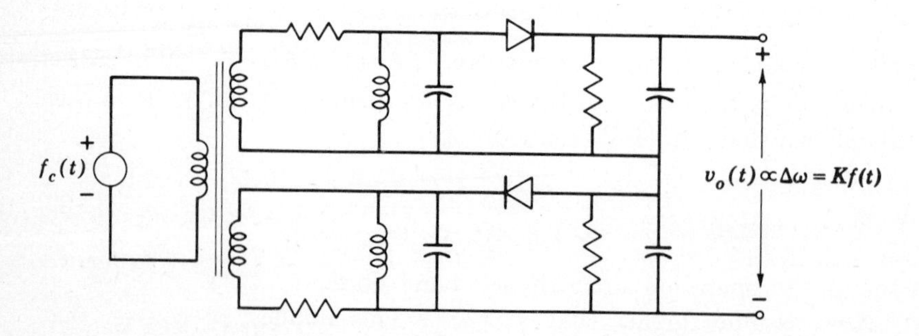

FIG. 4-52 Balanced FM demodulator.

A block diagram of a typical FM receiver, covering the broadcast range of 88 to 108 MHz, is shown in Fig. 4-53. Note that except for the limiter and discriminator circuits, the form of the receiver is similar to that of a conventional AM receiver. All h-f circuits prior to the dis-

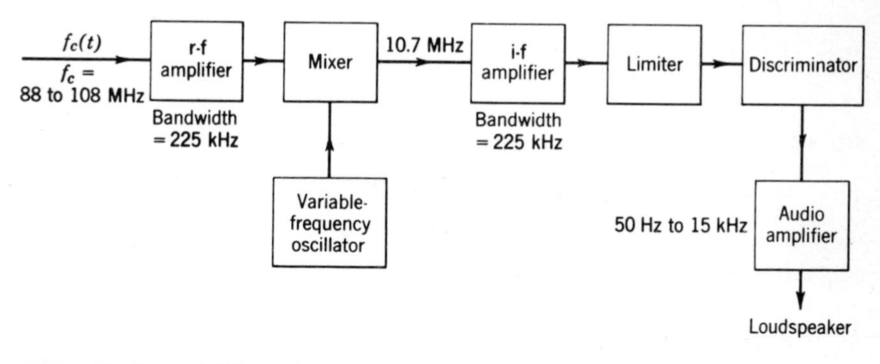

FIG. 4-53 Typical FM receiver.

criminator must be designed for the FM bandwidth of 225 kHz, however, while the audio amplifier which amplifies the recovered modulating signal need cover only the 50- to 15-kHz range. The i-f amplifiers are tuned to a center frequency of 10.7 MHz. The audio amplifier normally includes a deemphasis circuit. This circuit, in conjunction with a preemphasis circuit in the transmitter, provides additional discrimination against noise and interference. It is discussed in detail in Chap. 6.

ZERO-CROSSING DETECTORS

It was noted above that FM detectors normally include hard limiters to eliminate any amplitude fluctuations that could then be converted (erroneously) to a detected FM output. It is thus apparent that the FM information must be contained in the points at which the FM signal $f_c(t)$ crosses the origin, the so-called *zero crossings*. We shall demonstrate this statement below, providing another means of detecting FM signals.

Consider then again the FM signal given by

$$f_c(t) = A \cos \theta(t) = A \cos [\omega_c t + K \int f(t)\, dt] \qquad [4\text{-}85]$$

Again, as an example, let $f(t) = at$, $0 \leq t \leq T$, a repetitive ramp. Then $\omega_i = \omega_c + Kat$, $\theta(t) = \omega_c t + Kat^2/2$. Let t_1 be a zero crossing, as shown in Fig. 4-54, with $t_2 = t_1 + \Delta t$ the next zero crossing. Then $\theta(t_2) - \theta(t_1) = \pi$. Assume the bandwidth B of $f(t)$ is much less than f_c, the carrier frequency. The information-bearing signal $f(t)$ changes much

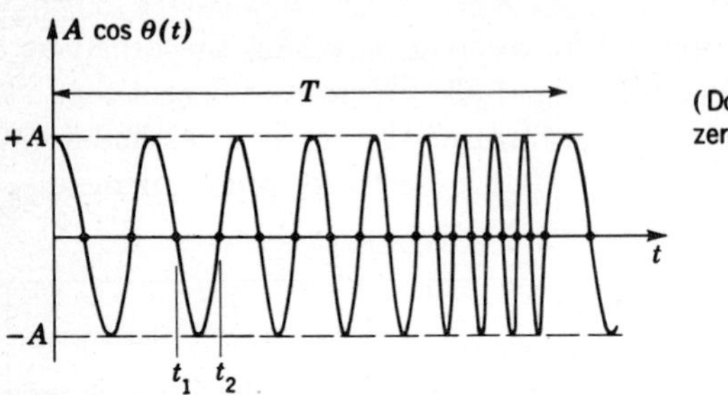

FIG. 4-54 Zero-crossing determination.

more slowly than f_c. In the interval $(t_2 - t_1)$, then, $f(t)$ may be assumed effectively constant, so that

$$\begin{aligned}\theta(t_2) - \theta(t_1) &= \pi = \omega_c(t_2 - t_1) + K\int_{t_1}^{t_2} f(t)\,dt \\ &\doteq \omega_c(t_2 - t_1) + Kf(t_1)(t_2 - t_1) \\ &= \underbrace{[\omega_c + Kf(t_1)]}_{\frac{d\theta}{dt} \equiv \omega_i}(t_2 - t_1)\end{aligned} \qquad [4\text{-}92]$$

From Eq. [4-92] we thus have

$$\omega_i = \omega_c + Kf(t) \doteq \frac{\pi}{t_2 - t_1}$$

and

$$f_i = f_c + \frac{K}{2\pi} f(t) \doteq \frac{1}{2(t_2 - t_1)} \qquad [4\text{-}93]$$

The desired output $f(t)$ may thus be found by measuring the spacing between zero crossings.

If positive-going zero crossings only are considered [those at which the slope of $f_c(t)$ is positive], we get

$$f_i \doteq \frac{1}{t_2 - t_1} \qquad [4\text{-}93a]$$

A simple way of measuring the spacing of zero crossings is to actually *count* the number of zero crossings in a given time interval.

Thus, consider a counting interval T_c long enough so that it counts a significant number of zero crossings, yet short enough compared with $1/B$ so that $f(t)$ still does not change too much in this interval (see Fig. 4-55). Then

$$\frac{1}{f_c} < T_c \ll \frac{1}{B} \qquad [4\text{-}94]$$

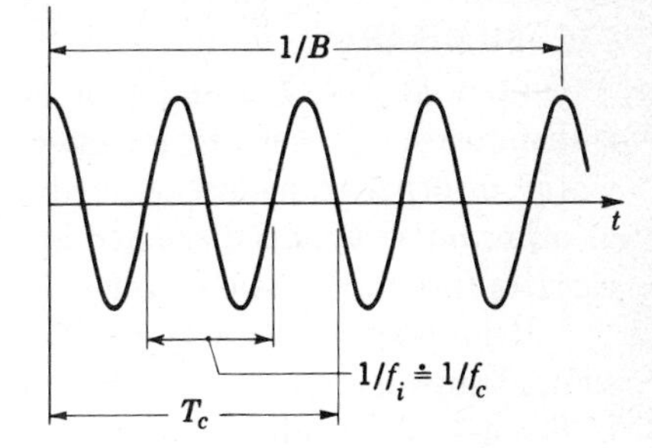

FIG. 4-55 Counting intervals.

(As an example, let $f_c = 10$ MHz, $T_c = 1$ μsec, $B = 20$ kHz.) Let n_c be the number of positive zero crossings in T_c sec. We then have

$$n_c \doteq \frac{T_c}{t_2 - t_1} \qquad [4\text{-}95]$$

or

$$f_i = \frac{n_c}{T_c} \qquad [4\text{-}96]$$

(n_c is approximately 10 in the example noted.) The instantaneous frequency f_c, and from this the derived $f(t)$, are thus found directly in terms of the measured count n_c. [In practice, one would half-wave rectify $f_c(t)$, differentiate to accentuate the zero-crossing points, again rectify to eliminate the negative pulses due to the negative-going zero crossings, and pass the resultant sequence of positive pulses into a low-pass device with time constant T_c. The output would then be a direct measure of $f_i = f_c + Kf(t)/2\pi$. By using a balanced device one may again obtain $f(t)$ directly.]

Other types of FM detectors in addition to those described above have been developed and used in practice. One particular class, which is of particular importance in deep-space communications where power is at a premium, uses feedback techniques to improve signal detectability in the presence of noise. Among the different types of detectors in this group one may consider the phase-locked loop, the FM demodulator with feedback (FMFB),[1] and the frequency-locked loop FM demodulator.[2]

[1] Schwartz, Bennett, and Stein, *op. cit.*, sec. 3-9, pp. 157–163. (The phase-locked loop is treated in detail, both as a tracking and acquisition device and as a demodulator, in A. J. Viterbi, "Principles of Coherent Communications," McGraw-Hill, New York, 1966.)

[2] K. K. Clarke and D. T. Hess, Frequency Locked Loop Demodulator, *IEEE Trans. Commun. Technol.* pp. 518–524, August, 1967.

4-10 SUMMARY

As noted at the beginning of this chapter, high-frequency sine-wave transmission is necessary for effective radiation of signals through space, water, or other transmission media. The PCM, PAM, and PPM, as well as other pulse signals discussed in Chap. 3, must thus modulate sine-wave carriers for transmission over a desired channel.

Beginning with the types of sine-wave modulation peculiar to binary signal transmission—PSK, FSK, and OOK, among others—we proceeded to discuss amplitude and angle modulation of sine-wave carriers. Stress was placed on methods of generating and detecting AM and FM signals, as well as on bandwidth considerations for various systems. This latter discussion served to further solidify the frequency analysis of Chap. 2.

We indicated that the bandwidth of normal AM signals was essentially twice that of the modulating (baseband) signals (single-sideband signals require only the baseband bandwidth, however), while the bandwidth of FM signals was approximately given by $2\,\Delta f + 2B = 2B(1 + \beta)$. Here Δf is the frequency deviation, proportional to the modulating signal amplitude, B is the modulating signal bandwidth, and $\beta \equiv \Delta f/B$, the modulation index. For wideband FM the transmission bandwidth is substantially larger than that required for AM.

Both AM and narrowband FM systems were found to require some type of product modulator for the modulation process. Both nonlinear and piecewise-linear devices were shown to provide the necessary product term. Switches similar to those introduced in Chap. 3 are equivalent to the piecewise-linear characteristic.

We shall return to these various modulation schemes in Chap. 6, comparing them there on the basis of both bandwidth and their performance in the presence of noise.

PROBLEMS

4-1 The output of a PCM system consists of a binary sequence of pulses, occurring at the rate of 2×10^6 bits/sec. Compare the approximate transmission bandwidths required in the two cases of:

(*a*) OOK transmission, amplitude modulation of a sine-wave carrier.
(*b*) FSK, switching between two sine waves of frequencies 100 and 104 MHz. Repeat the FSK calculation if the two frequencies are 100 and 120 MHz. Sketch the spectra in all cases and indicate assumptions made.

4-2 A binary message consists of an alternating sequence of 1's and 0's. FSK transmission is used, with the two frequencies each multiples of the *binary* period T and synchronized in phase (see Fig. 4-7*a*). Find the spectrum of the FSK wave and sketch, verifying the expression given in the text and sketched in Fig. 4-7*b*.

4-3 Twenty-four signal channels, plus one synchronization (marker) channel, each band-limited to 3,300 Hz, are sampled and time multiplexed at an 8-kHz rate.

(*a*) Calculate the minimum bandwidth needed if the multiplexed signal used in a PAM system amplitude modulates an h-f sine-wave carrier.
(*b*) Calculate the bandwidth needed to *frequency multiplex* the same 24 signal channels, using normal AM techniques. (The spacing between frequencies is the minimum possible.) What bandwidth would be needed if SSB AM were used in both the modulation and frequency-multiplexing processes?
(*c*) Repeat (*b*) if FM-FM multiplexing is used. Each of the 24 channels first frequency modulates a carrier ($\beta = 1$), with the carrier spacing chosen as small as possible. The composite group of 24 FM waves in turn frequency modulates a high-frequency carrier with $\beta = 5$.

4-4 A 1-Hz sinusoidal input and a 60-Hz sinusoidal carrier are combined in a product modulator to give $\sin 2\pi t \sin 2\pi 60t$. Sketch this function, and show why simple envelope detection will not reproduce $\sin 2\pi t$.

4-5 Consider the *bipolar chopper* of Fig. P 4-5.

(*a*) Sketch an arbitrary time function $f(t)$ and just below it the chopper output $f_d(t)$.
(*b*) Show $f_d(t) = f(t)\, S(t)$, with $S(t)$ a square-wave switching or sampling function of alternate values $+1$ and -1. Use this formulation, as was done in Chap. 3, to find the spectrum of $f_d(t)$. Assume $f(t)$ is band-limited to B Hz.

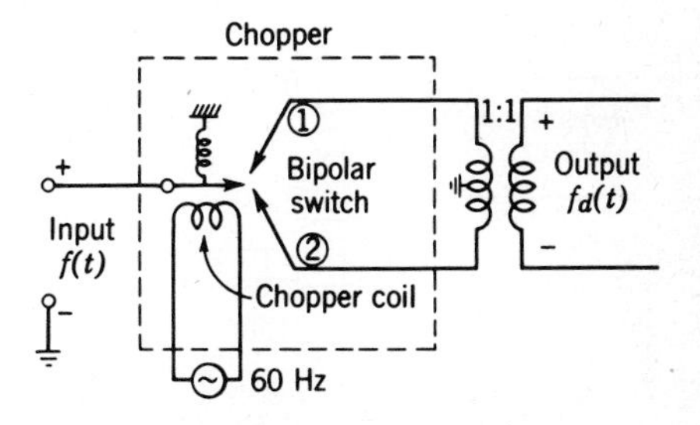

FIG. P 4-5 Mechanical balanced modulator, using bipolar chopper.

(*c*) The chopper output is fed into an ac amplifier centered at 60 Hz, with passband $\pm B$ about 60 Hz. What is the maximum bandwidth B if the amplifier output is to be $kf(t) \cos 2\pi 60t$, k a constant? (Note that the chopper-amplifier combination acts as a *balanced modulator*.)

4-6 The 1-Hz input of Prob. 4-4 is applied to a 60-Hz bipolar chopper.

(*a*) Sketch the output wave and compare with the sketch of Prob. 4-4.
(*b*) The chopped signal is now demodulated by another 60-Hz chopper. Sketch the resultant output signal for the three cases of: (1) the two choppers in synchronism; (2) the two choppers out of phase by ⅛ cycle; and (3) the two choppers out of phase by ¼ cycle. Which of these cases will reproduce $\sin 2\pi t$ if filters are available?

4-7 A signal $f(t)$ is first modulated by being passed through a bipolar chopper. After amplification by a constant A the chopped signal is demodulated by another chopper, which lags the first chopper by a fraction αT of the chopper period. The resultant demodulated signal may be written analytically in the form $Af(t)S(t)S(t - \alpha T)$, with $S(t)$ a switching function. This signal is filtered to reproduce $f(t)$. Compare the outputs of the filter for different values of α. Show that the results are in agreement with those of Prob. 4-6*b*.

4-8 A transformer is arranged to allow the secondary winding to be rotated mechanically with respect to the primary. Show that with the primary input a 60-Hz sine wave and the secondary rotated at a constant rate of 1 rps, the output voltage is of the DSB (suppressed-carrier) type.

4-9 Shown in Fig. P 4-9 are two examples of phase-sensitive diode demodulators used in servomechanism practice. The input in each case is either $f(t)\ S(t)$ or $f(t) \cos \omega_c t$.

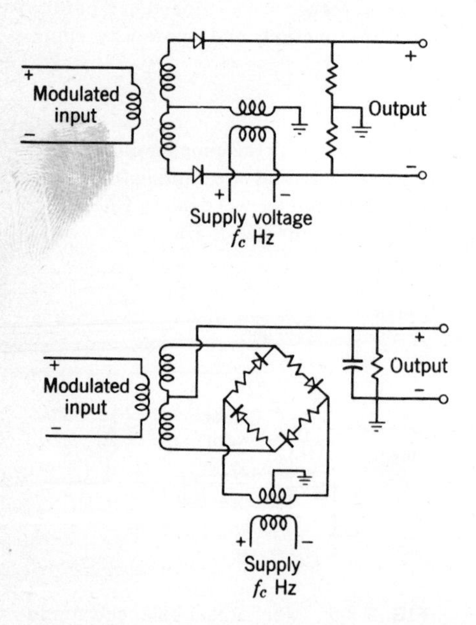

FIG. P 4-9 Diode demodulators.

The power-supply frequency f_c Hz is in each case the same as the frequency of the square wave $S(t)$. Explain the operation of these demodulators. Compare with the bipolar chopper of Prob. 4-5. Why are these devices phase-sensitive?

4-10 An amplifier has a slight nonlinearity, so that its output-input characteristic is given by

$$e_o = A_1 e_i + A_2 e_i^2$$

(*a*) $e_i = \cos \omega_m t$. Find all the frequencies present in the output. How could this device be used as a second-harmonic generator?
(*b*) $A_1 = 10$, $A_2/A_1 = 0.05$. Determine the ratio of second-harmonic amplitude to fundamental amplitude. (This is frequently called the *second-harmonic distortion*.)

4-11 (*a*) Let $e_i = 0.1 \cos 500t$ in the amplifier of Prob. 4-10. Determine the frequency terms present in the output and the amplitude of each. Repeat for $e_i = 0.2 \cos 3{,}000t$.
(*b*) $e_i = 0.1 \cos 500t + 0.2 \cos 3{,}000t$. Determine the different frequency terms present in the output and the amplitude of each. What new frequency terms appear that were not present in the two cases of (*a*)? These are known as the *intermodulation* frequencies and are due to the nonlinear mixing of the two frequency terms.

4-12 A voltage $e_i = 0.1 \cos 2{,}500t + 0.2 \cos 3{,}000t$ is applied at the input to the amplifier of Prob. 4-10. The output voltage e_o is in turn applied to an ideal rectangular filter passing all frequencies between 50 Hz and $4{,}000/2\pi$ Hz with unity gain, rejecting all others. Calculate the power at the output of the filter in the frequency terms generated by the amplifier nonlinearity. Determine the ratio of power in these terms to the power in the 2,500 and 3,000 radians/sec terms at the filter output.

4-13 A semiconductor diode has a current-voltage characteristic at room temperature given by

$$i = I_0 \left(e^{40v} - 1\right)$$

(*a*) What is the maximum variation in the voltage v to ensure i linearly proportional to v, with less than 1 percent quadratic distortion?
(*b*) Let $v = 0.01 \cos \omega_1 t + 0.01 \cos \omega_2 t$. Expand i in a power series in v, retaining terms in v, v^2, and v^3 only. Tabulate the different frequencies appearing in i, and the relative magnitudes of each.

4-14 A band-limited signal $f(t)$ may be expressed in terms of the finite Fourier series

$$f(t) = \frac{c_0}{T} + \frac{2}{T} \sum_{n=1}^{M} |c_n| \cos (\omega_n t + \theta_n) \qquad \omega_n = \frac{2\pi n}{T}$$

The maximum-frequency component of $f(t)$ is $f_M = M/T$.

(*a*) Sketch a typical amplitude spectrum for $f(t)$. Then indicate the frequency components present in $f^2(t)$.
(*b*) What is the highest-frequency component of $f^2(t)$?
(*c*) Find the spectrum of $f(t) \cos \omega_c t$ ($\omega_c \gg \omega_M$), and compare with that of $f(t)$.

4-15 For each network shown in Fig. P 4-15, calculate the maximum-frequency component in the output voltage.

(*a*) e_{in} has a single component of frequency f Hz.
(*b*) e_{in} has a single component of frequency $2f$ Hz.
(*c*) e_{in} has two frequency components, f and $2f$.

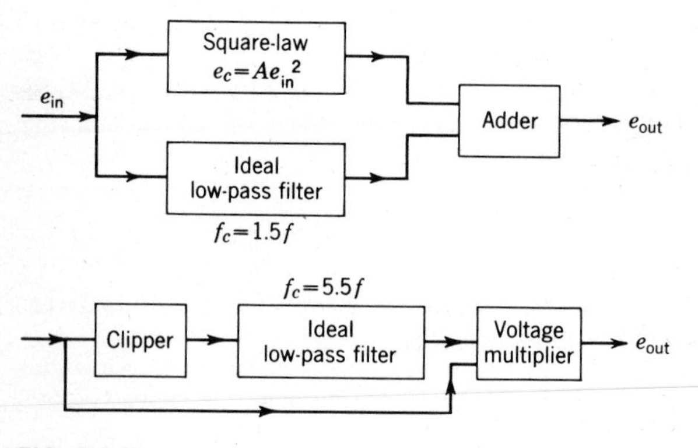

FIG. P 4-15

4-16 Spectrum of coherent radar pulses. These pulses may be generated by turning a sine-wave generator of frequency f_c on and off periodically at the same phase points. For the example shown in Fig. P 4-16,

$$f_c(t) = V \cos \omega_c t \qquad \frac{-\tau}{2} < t < \frac{\tau}{2}$$

$$f_c(t) = 0 \qquad \text{elsewhere in a period}$$

(*a*) Find the spectrum of $f_c(t)$ by expanding the pulsed carrier in a Fourier series. (Use the exponential form of $\cos \omega_c t$.)
(*b*) Write $f_c(t) = V \cos \omega_c t \cdot S(t)$, with $S(t)$ a rectangular pulse train. Expand $S(t)$ in a Fourier series. Find the spectrum of $f_c(t)$, and compare with the result of (*a*).

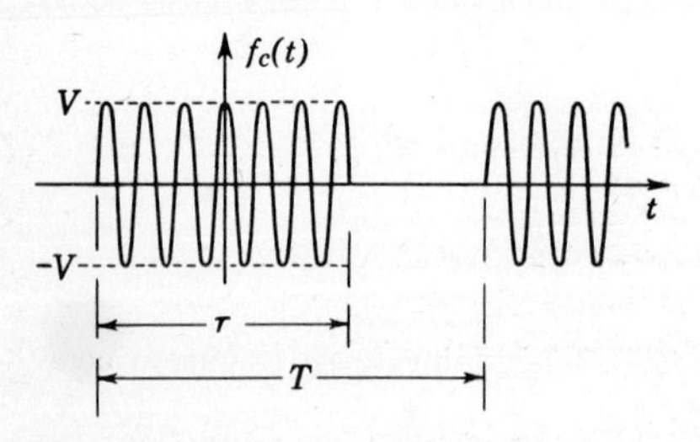

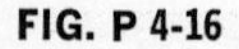
FIG. P 4-16

4-17 A periodic function $f(t)$ is band-limited to 10 kHz and has a uniform or flat amplitude spectrum from 0 to 10 kHz. It is given by

$$f(t) = \frac{1}{2} + \sum_{n=1}^{10} \cos n\omega t \qquad \omega = 2\pi \times 1{,}000$$

Show that the function $f^2(t)$ has an amplitude spectrum decreasing linearly from 0 to 20 kHz. The envelope of the spectrum of $f^2(t)$ is thus triangular in shape.

4-18 A function $f(t)$ has as its Fourier transform

$$F(\omega) = K \qquad |\omega| < \omega_M \qquad F(\omega) = 0 \qquad |\omega| > \omega_M$$

If $f(t)$ is the input voltage to a square-law device with the characteristics $e_o = Ae_i^2$, show that the output has the triangular spectrum of Fig. P 4-18 (compare with Prob. 4-17).

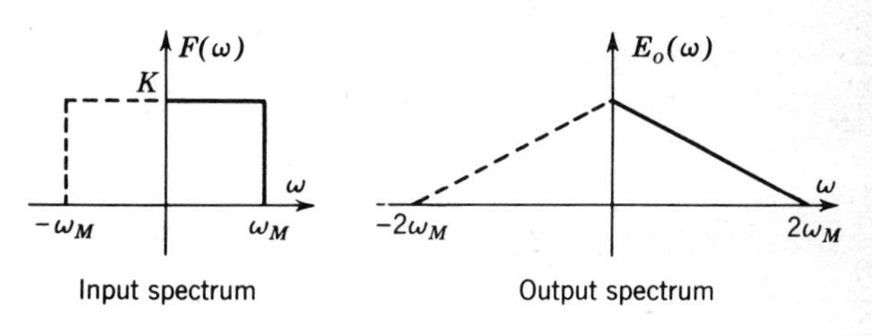

FIG. P 4-18

4-19 An AM signal consists of a carrier voltage 100 sin $(2\pi \times 10^6 t)$ plus the voltage $(20 \sin 6{,}280t + 50 \sin 12{,}560t) \sin (2\pi \times 10^6 t)$.

(*a*) Draw the amplitude-versus-frequency plot of the modulated signal.
(*b*) How much peak power will this signal deliver to a 100-ohm load?
(*c*) What is the average power (over a modulation cycle) delivered to a 100-ohm load?
(*d*) Sketch the envelope over a modulation cycle. (Indicate numerical values at key points.)

4-20 Assume that the carrier in Prob. 4-19 is suppressed after modulation. Repeat Prob. 4-19 under this condition.

4-21 *Frequency conversion.* An amplitude-modulated carrier

$$f_c(t) = A[1 + mf(t)] \cos \omega_c t$$

is to be raised or lowered in frequency to a new carrier frequency $\omega_c' = \omega_c \pm \Delta\omega$. The resultant time function will appear in the form $f_c'(t) = A'[1 + mf(t)] \cos \omega_c' t$.

(*a*) Show that a product modulator producing an output $f_c(t) \cos \Delta\omega t$ will accomplish this frequency conversion. Draw a simple block diagram for this frequency converter. Specify the filtering required in the two cases of frequency raised by $\Delta\omega$ and lowered by $\Delta\omega$ radians/sec.
(*b*) Show that a nonlinear device with terminal characteristics $e_o = \alpha_1 e_i + \alpha_2 e_i^2$ will perform this frequency conversion if $f_c(t) + \cos \Delta\omega t$ is applied at the input.

Sketch a typical amplitude spectrum of the output signal, and specify the filtering necessary to produce the frequency conversion.

4-22 A general nonlinear device has output-input characteristics given by $e_o = \alpha_1 e_i + \alpha_2 e_i^2 + \cdots$ (this could represent a diode, transistor, vacuum triode, etc.).

(*a*) Show that all combinations of sum and difference frequencies appear in the output if e_i is a sum of cosine waves.
(*b*) Show that this device can be used as a modulator if $e_i = f(t) + \cos \omega_c t$.
(*c*) The input is a normal AM wave given by $e_i = K[1 + mf(t)] \cos \omega_c t$. Show that this nonlinear device will demodulate the wave, i.e., reproduce $f(t)$.
(*d*) Show that this device can act as a frequency converter if

$$e_i = K[1 + mf(t)] \cos \omega_c t + \cos \Delta\omega t$$

In particular, show that with proper filtering the AM wave spectrum can be shifted up or down by $\Delta\omega$ radians/sec.

4-23 A function appears in the form

$$f_c(t) = \cos \omega_c t - K \sin \omega_m t \sin \omega_c t$$

Using block diagrams, show how to generate this, given $\cos \omega_c t$ and $\sin \omega_m t$ as inputs.

4-24 Amplitude-modulation detector. An amplitude-modulated voltage

$$f_c(t) = K[1 + mf(t)] \cos \omega_c t$$

is applied to the diode-resistor combination of Fig. P 4-24. The diode has the piecewise-linear characteristics shown. (It is then called a linear detector.) Show that for $|mf(t)| < 1$ the l-f components of the load voltage e_L provide a perfect replica of $f(t)$.

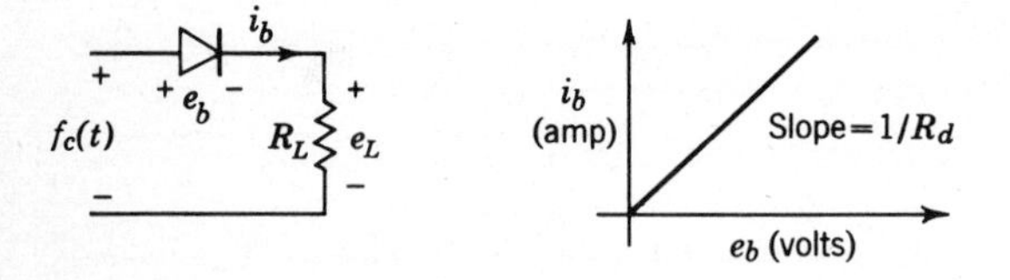

FIG. P 4-24 Linear detector.

4-25 Repeat Prob. 4-24 for the case where the diode-resistor combination of Fig. P 4-24 has the square-law characteristic $e_L = \alpha_1 e_i + \alpha_2 e_i^2$. Show that e_L contains a component proportional to $f(t)$.

4-26 (*a*) A DSB (suppressed-carrier) signal of the form $f(t) \cos \omega_c t$ is to be demodulated. Show that multiplying this signal by $\cos \omega_c t$ in a product modulator reproduces $f(t)$.
(*b*) "The DSB demodulator is phase-sensitive." Demonstrate the validity of this statement by multiplying $f(t) \cos \omega_c t$ by $\cos (\omega_c t + \theta)$ and letting θ vary from 0 to π radians.

4-27 Prove the following properties of Hilbert transforms:

(*a*) If $f(t) = f(-t)$, then $\hat{f}(t) = -\hat{f}(-t)$.
(*b*) If $f(t) = -f(-t)$, then $\hat{f}(t) = \hat{f}(-t)$.

(*c*) The Hilbert transform of $\hat{f}(t)$ is $-f(t)$.
Hint: these are easily proven using the 90° phase-shift description of the transform process.

4-28 Define a function $y(t) = f(t) - j\hat{f}(t)$, with $\hat{f}(t)$ the Hilbert transform of $f(t)$.

(*a*) Show

$$\begin{aligned} Y(\omega) &= 2F(\omega) \qquad \omega \leq 0 \\ &= 0 \qquad \omega > 0 \end{aligned}$$

(*b*) Show Re $[y(t)e^{j\omega_c t}]$ is a *lower-sideband* SSB signal and is given by Eq. [4-32] in the text.

4-29 Find the Hilbert transforms of the two pulses shown in Fig. 4-33 in the text, verifying the results shown there. *Hint:* Find the Fourier transforms of the two pulses, introduce the necessary 90° phase shift, and then take *inverse* transforms.

4-30 Show that detection of a SSB signal requires multiplication by a carrier term precisely in phase with the incoming signal. Show that a phase difference of θ radians between the SSB carrier and the carrier term at the receiver results in the output signal given by Eq. [4-38]. Sketch the output function of Eq. [4-38] for the two pulses of Fig. 4-33 for $\theta = 0, \pi/4, \pi/2$.

4-31 *Single-sideband detection.* An audio tone $\cos \omega_m t$ and a carrier $\cos \omega_c t$ are combined to produce an SSB signal given by $\cos (\omega_c - \omega_m)t$.

(*a*) Show that $\cos \omega_m t$ can be reproduced from the SSB signal by mixing with $\cos \omega_c t$ as a local carrier.
(*b*) The local carrier drifts in frequency by $\Delta\omega$ radians to $\omega_c + \Delta\omega$. What is the demodulated signal now?
(*c*) The local carrier shifts in phase to $\cos (\omega_c t + \theta)$. Find the audio output signal.

4-32 (*a*) Refer to Fig. 4-28. Find the output signal if a phase-shift network providing $\beta(f)$ deg is inserted in the $f(t)$ branch and a network giving $\beta(f) + 90°$ is inserted in the $f_{f2}(t)$ branch. The input signal $f(t)$ may be assumed a sine wave.
(*b*) How may $f_{f2}(t)$ be reproduced at the receiver?

4-33 The 90° phase shift in the $f_{f2}(t)$ branch of Fig. 4-28 is replaced by θ deg of phase shift, while the carrier shift of 90° is replaced by $180 - \theta$ deg. Find the output signal. Is it in the form of an SSB signal? How does it compare with the case of $\theta = 90°$?

4-34 (*a*) An amplitude-modulated voltage given by

$$v_c(t) = V(1 + m \cos \omega_m t) \cos \omega_c t \qquad \text{volts}$$

is applied across a resistor R. Calculate the percentage average power in the carrier and in each of the two sideband frequencies. Calculate the peak power.
(*b*) The voltage $v_c(t)$ is now a suppressed-carrier voltage $v_c(t) = V \cos \omega_m t \cos \omega_c t$ applied across R ohms. What is the fraction of the total average power in each of the sideband frequencies?

4-35 An SSB radio transmitter radiates 1 kw of average power summed over the entire sideband. What total average power would the transmitter have to radiate if it were operating as a DSB (suppressed-carrier) system and the same distance coverage were required?

4-36 An AM transmitter is tested by using the dummy load and linear narrowband receiver of Fig. P 4-36. The r-f amplifier portion of the receiver is swept succes-

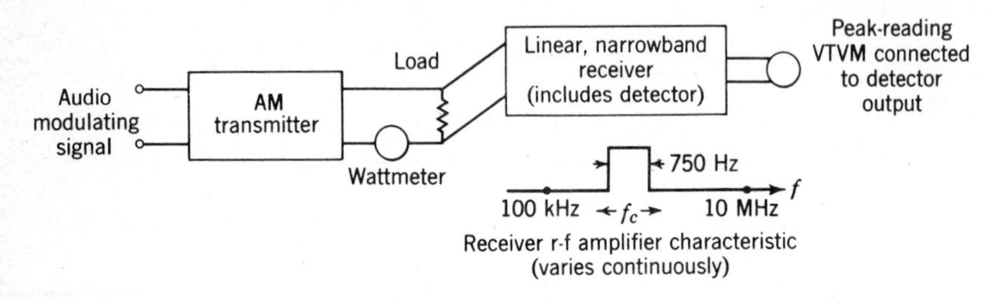

FIG. P 4-36

sively and continuously over the range 100 kHz to 10 MHz. Its frequency characteristic at any frequency f_c in this range is shown in the figure.

With no audio input the wattmeter reads 100 watts average power. The peak-reading vacuum-tube voltmeter (VTVM) and r-f amplifier tuning characteristic indicate an output of 10 volts peak at 1 MHz.

(*a*) With an audio input signal of 10 volts peak at 1 kHz the wattmeter reads 150 watts. At what frequencies are there receiver outputs? What are the various amplitudes, as read on the VTVM?

(*b*) The 1-kHz modulating signal is replaced by the composite signal $2 \cos 12{,}560t + 3 \cos 18{,}840t$. What is the new wattmeter reading? At what frequencies will there be receiver outputs? What are the various amplitudes?

(*c*) The transmitter modulator is modified so that the carrier is suppressed but the other characteristics are unchanged. An audio signal of 10 volts peak at 1 kHz is applied. What is the wattmeter reading? What are the receiver output frequencies and amplitudes? Sketch the pattern that would be seen on an oscilloscope connected across the load. (The left-to-right sweep period is 2,000 μsec; the return takes negligible time.)

4-37 The radio receiver of Fig. P 4-37 has the r-f amplifier and i-f amplifier characteristics shown. The mixer characteristic is given by $e_3 = e_2(\alpha_0 + \alpha_1 e_{01})$, with α_0 and α_1

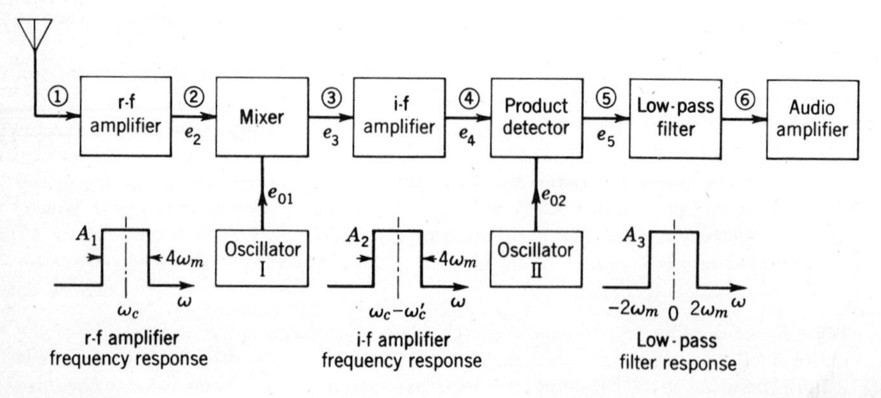

FIG. P 4-37

constants, and $e_{01} = A \sin \omega_c' t$. The product-detector output is $e_5 = e_4 e_{02}$, with $e_{02} = B \cos (\omega_c - \omega_c')t$.

Find the signals at points 4 and 6 for each of the following signals at point 1:

(a) Normal AM, $(1 + \sin \omega_m t) \sin \omega_c t$
(b) Double sideband, suppressed carrier, $\sin \omega_m t \sin \omega_c t$
(c) Single sideband, $\cos (\omega_c - \omega_m)t$

4-38 The product detector and oscillator II in Fig. P 4-37 are replaced by an envelope detector with the characteristic $e_5 = be_4^2$. The r-f amplifier can be continuously tuned over the range $f_c = \omega_c/2\pi = 535$ to 1,605 kHz. The tuning capacitor of oscillator I is ganged to the tuning capacitor of the r-f amplifier so that $f_c' = \omega_c'/2\pi = f_c + 455$ kHz at all times. The circuit is then that of a superheterodyne radio receiver. The passband of the amplifiers is ± 5 kHz about the center frequency.

(a) Determine all frequencies present at points 3 and 4 if the signal at 1 is a 1-MHz unmodulated carrier.
(b) A 1-kHz audio signal amplitude modulates the 1-MHz carrier. Determine the frequencies present at points 2 and 4.
(c) The r-f amplifier is tuned to 650 kHz. To your amazement, however, a radio station at 1,560 kHz (WQXR in the New York City area) is heard quite clearly. Can you explain this phenomenon?
(d) The input at point 1 is the DSB signal of Prob. 4-37, with $f_c = 1$ MHz and $f_m = \omega_m/2\pi = 1$ kHz. Find the frequencies present at points 5 and 6. Compare with the product-detector output of Prob. 4-37.

4-39 Consider the system shown in Fig. P 4-39. Show that the output $g(t)$ is a SSB signal. Do this by assuming $F(\omega)$ as shown ($B \ll f_c$) and carefully sketching the transforms at the output of each device. Preserve the distinction between the shaded and unshaded halves of $F(\omega)$. Is the lower or upper half of a conventional DSB signal retained?

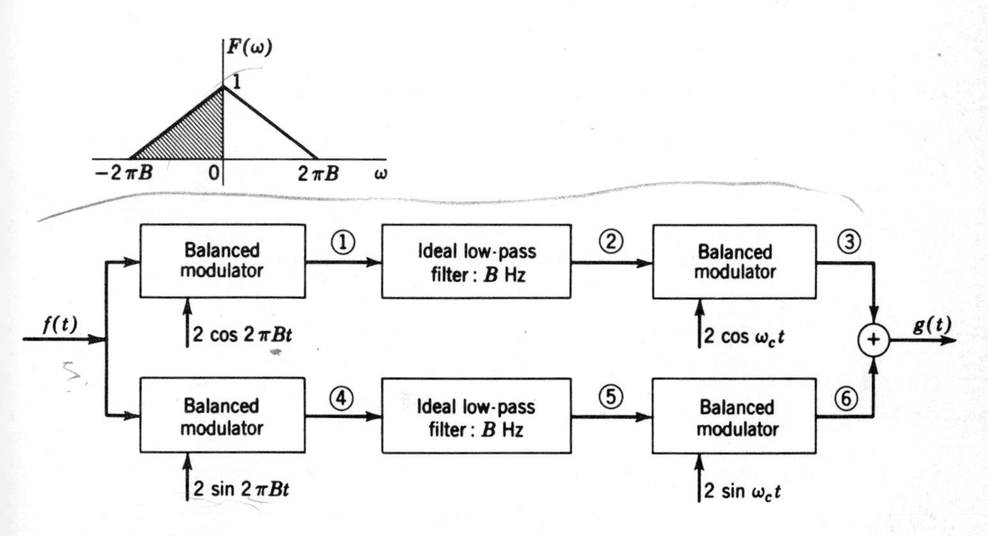

FIG. P 4-39

4-40 A DSB signal $(\cos \omega_m t + a \cos 2\,\omega_m t) \cos \omega_c t$ is applied to the receiver shown in Fig. P 4-40. Find the signals shown at points 1, 2, and 3 if

(a) $\omega_1 = \omega_c - 3\omega_m$
(b) $\omega_1 = \omega_c$

Discuss.

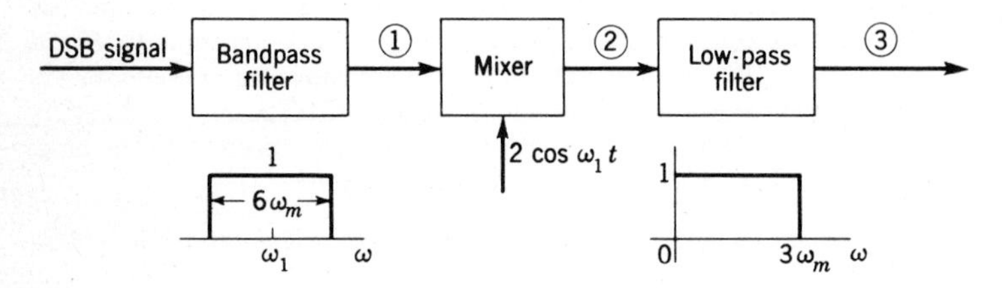

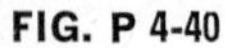
FIG. P 4-40

4-41 An AM signal $r(t) = a[1 + f(t)] \cos(\omega_0 t + \theta)$, where θ is a constant and $f(t)$ has a Fourier transform which is zero for $|\omega| > \Omega$. Assume that $\Omega \ll \omega_0$ and $|f(t)| \leq 1$. Consider the receiver shown in Fig. P 4-41. The low-pass filters have transfer functions

$$H(\omega) = \begin{cases} 1 & |\omega| < \Omega \\ 0 & |\omega| > \Omega \end{cases}$$

By calculating the signals at points 1 through 6, show that the receiver can be used to demodulate $f(t)$.

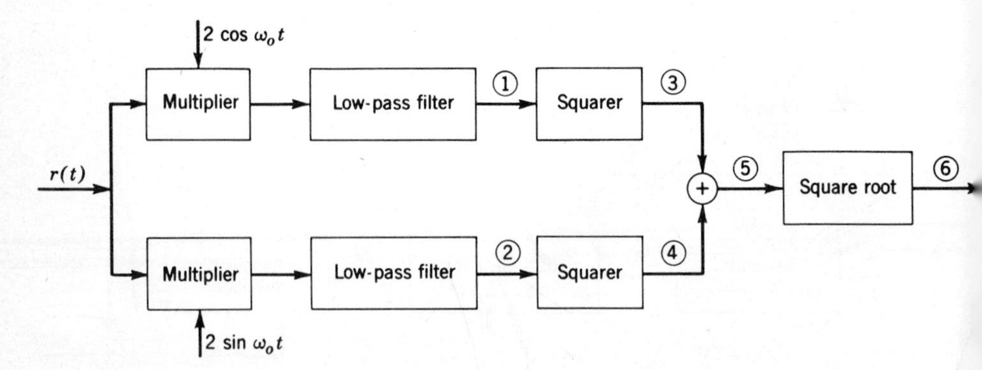

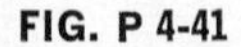
FIG. P 4-41

4-42 Consider a stereo broadcasting station that desires to transmit the two signals:

$$x(t) \rightarrow \text{left channel}$$
$$y(t) \rightarrow \text{right channel}$$

Both $x(t)$ and $y(t)$ have frequency components only in the range $(2\pi)\,30 \leq \omega \leq (2\pi)10{,}000$. The station actually broadcasts the signal $[A + x(t) + y(t)]\cos\omega_0 t + [A + x(t) - y(t)]\sin\omega_0 t$, where $\omega_0 = 2\pi(10^6)$ and $A \gg |x(t)| + |y(t)|$.

(*a*) Give and discuss a block diagram of a receiving system which provides as separate outputs $x(t)$ and $y(t)$. Assume the phase and frequency of the transmitted wave is known exactly at the receiver. You may use mixers, multipliers, adders, filters, etc.

(*b*) What would be the output of a receiver which merely consisted of an envelope detector and a low-pass filter? (Would it be intelligible?) *Hint:* Approximate for large A.

4-43 (*a*) A 1-kHz sine-wave frequency modulates a 10-MHz carrier. The amplitude of this audio signal is such as to produce a maximum-frequency deviation of 2 kHz. Find the bandwidth required to pass the frequency-modulated signal.

(*b*) Repeat (*a*), if the modulating signal is a 2-kHz sine wave.

(*c*) The 2-kHz sine wave of (*b*) is doubled in amplitude, so that $\Delta f = 4$ kHz. Find the required bandwidth.

4-44 A 100-MHz sine-wave carrier is to be frequency modulated by a 10-kHz sine wave. An engineer designing the system reasons that he can minimize bandwidth by decreasing the audio amplitude. He arranges for a maximum-frequency deviation Δf of only ± 10 Hz away from the 100-MHz carrier frequency and assumes that he requires only 20-Hz bandwidth.

(*a*) Find the fallacy in his reasoning, and specify the actual bandwidth required.

(*b*) Would he have been right to assume a bandwidth of 2 MHz if the audio amplitude had been chosen to produce a Δf of 1 MHz? Explain the difference between the results of (*a*) and (*b*).

4-45 An FM receiver similar to that of Fig. 4-53 is tuned to a carrier frequency of 100 MHz.

(*a*) A 10-kHz audio-signal frequency modulates a 100-MHz carrier, producing a β of 0.1. Find the bandwidths required of the r-f and i-f amplifiers and of the audio amplifier.

(*b*) Repeat (*a*) if $\beta = 5$.

(*c*) Two signals at 100 MHz are tuned in alternately. The carriers are of equal intensity. One is modulated with a 10-kHz signal and has $\beta = 5$; the other is modulated with a 2-kHz signal and has $\beta = 25$. Which requires the larger bandwidth? Explain. Compare the audio-amplifier outputs in the two cases.

(*d*) Two other signals are tuned in alternately. The carriers are again of equal intensity. One has a frequency deviation of 10 kHz with $\beta = 5$, the other a deviation of 2 kHz with $\beta = 25$. Which requires the larger bandwidth? Which gives the larger audio output?

4-46 A general frequency-modulated voltage has the form $f_c(t) = A_c \cos\,[\omega_c t + K \int f(t)\,dt]$. Show that the average power dissipated in a 1-ohm resistor is $A_c^2/2$ averaged over a modulating cycle.

4-47 (*a*) A modulating signal $f(t) = 0.1 \sin 2\pi \times 10^3 t$ is used to modulate a 1-MHz carrier in both an AM and an FM system. The 0.1-volt amplitude produces a 100-Hz frequency deviation in the FM case. Compare the receiver r-f amplifier and audio-amplifier bandwidths required in the two systems.

(*b*) Repeat (*a*) if $f(t) = 20 \sin 2\pi \times 10^3 t$.

4-48 A sine wave is switched periodically from 10 to 11 MHz at a 5-kHz rate. Sketch the resultant waveshape. What is the form of the modulating signal if this frequency-shifted wave is considered an FSK signal? What is the approximate transmission bandwidth? Compare with the bandwidth required if the modulating signal is approximated by a 5-kHz sine wave of the same amplitude.

4-49 The tuned circuit of an oscillator is shown in Fig. P 4-49. The current source $g_m e_g$ is controlled by the voltage $e_g(t)$ across R as shown.

(*a*) Show that with $1/\omega_c \gg RC$, the total effective capacitance of the tuned circuit is given by

$$C_T = C_0 + C + g_m RC$$

(*b*) What is the oscillator frequency if $g_m = 4{,}000$ μmhos?
(*c*) What is the modulation index β if g_m varies from 3,000 to 5,000 μmhos at a 1-kHz rate?

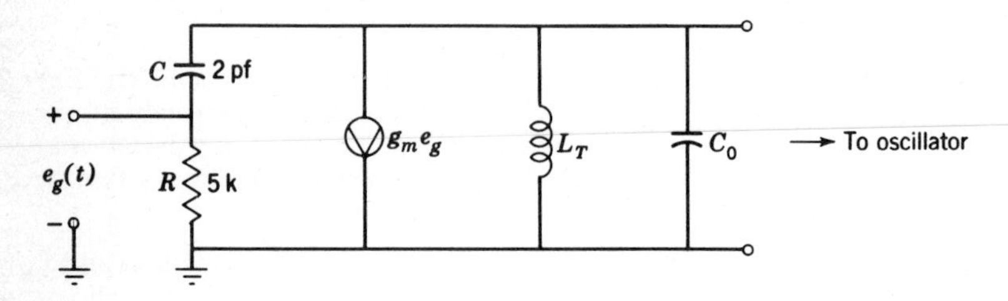

FIG. P 4-49

4-50 Consider a circuit such as the one in Fig. P 4-49, but with R and C interchanged. Calculate the effective inductance introduced by the controlled source. Repeat Prob. 4-49.

4-51 A controlled source circuit similar to that of Fig. P 4-49 provides an effective capacitance $C' = [-2e_g(t) + 6]10^{-12}$ farad appearing in parallel with C_0 and L_T. For this case $C_0 = 14$ $\mu\mu$f, $L_1 = 50$ μhenrys, and $e_g(t) = 0.5 \sin 2\pi \times 500t$.

(*a*) What is the carrier frequency?
(*b*) What is the frequency deviation Δf?
(*c*) Determine the modulation index β. Show a rough sketch of the frequency spectrum of the FM wave indicating the approximate bandwidth.
(*d*) Repeat (*c*) if the frequency deviation is reduced to 0.01 of its value in (*b*), with the modulating frequency unchanged.

4-52 The capacitance of a PN junction is given in terms of its reverse bias voltage V by

$$C = \frac{C_0}{(1 + V/\psi)^K}$$

with ψ the contact potential, and K and C_0 known constants. Such a reversed-bias diode is to be used as a variable capacitor.

(*a*) $K = 0.5$, $\psi = 0.5$, $C_0 = 300$ pf. Plot C versus the reversed-bias voltage V.
(*b*) The diode is to be used as the sole source of capacitance in a tuned circuit. Assume the diode is biased at -6 volts and the center frequency is to be adjusted

to 10 MHz. Plot frequency deviation away from 10 MHz versus voltage deviation away from -6 volts. Calculate the modulation sensitivity in hertz per volt in the vicinity of -6 volts (this is the initial slope of the curve plotted), and the maximum voltage excursion in either direction for a maximum of 1 percent deviation away from a linear frequency-voltage characteristic.

(*c*) A fixed capacitance of 100 pf is added in series with the diode. Repeat part (*b*) and compare.

4-53 *FM Stereo.* Call the left-speaker output L, and the right-speaker output R. For stereo broadcasting the two signals are first combined (*matrixed*) to form composite signals $L + R$, and $L - R$. (As in monaural FM the nominal bandwidth of each signal is 15 kHz.) The $L - R$ signal is multiplied by a 38-kHz carrier in a balanced modulator to form a DSB signal, as shown in Fig. P 4-53. The DSB signal,

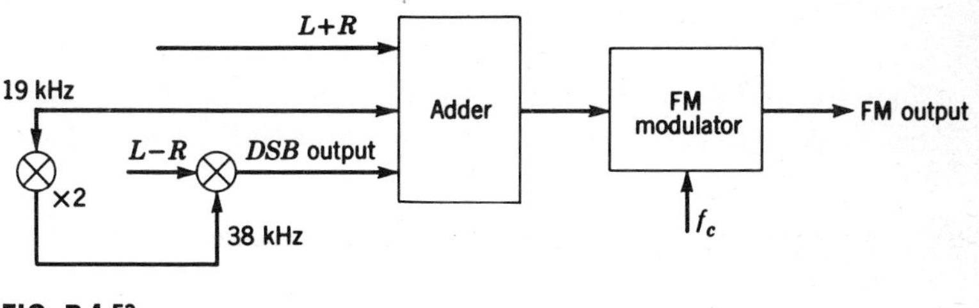

FIG. P 4-53

the baseband $L + R$ signal, and a 19-kHz tone from which the 38-kHz carrier is obtained are then summed to provide a frequency-multiplexed signal that in turn modulates the FM carrier. (The 19-kHz tone provides the pilot carrier for receiver synchronization.)

(*a*) Sketch a typical spectrum for the frequency-multiplexed signal.

(*b*) Estimate the frequency deviation, modulation index, and transmission bandwidth required of the FM signal. Compare with the monaural case ($L + R$ above).

CHAPTER FIVE

STATISTICAL METHODS IN ANALYSIS OF INFORMATION-TRANSMISSION SYSTEMS

Chapter 1 of this book stressed the fact, in a somewhat qualitative way, that there were two fundamental limitations to the transmission of information through a communication system:

1. Nonzero response time of the system (inversely related to the bandwidth)
2. Presence of noise in the system

Much of the material of the preceding chapters has been devoted to a discussion of the inverse time-frequency relationship of linear systems and to the time-frequency properties of some typical systems.

Although we have commented frequently on the fact that noise plays a basic role in communication systems studies, we have not specifically taken it into account up to this point. In discussing digital systems, we indicated that binary pulse-code modulation (PCM) is more noise immune than multilevel PCM. In comparing FM and AM systems we noted that "we would show later" that FM provides noise improvement over AM. In comparing various methods of transmitting digital information by high-frequency carrier, we indicated that "phase-shift keying (PSK) can be shown to be relatively more noise immune than on-off keying (OOK) and frequency-shift keying (FSK)."

We would now like to explore the effects of noise in systems in detail, actually verifying these statements and others. In order to do this we first have to introduce the basic tools of probability theory. For, as we shall see in studying noise present in systems, the noise we deal with is generally random and unpredictable in nature. Any thorough study of the effects of noise in systems must thus utilize aspects of probability theory, the mathematical discipline dealing with random phenomena.

The study of probability theory is also extremely important for an understanding of information-transmission systems because of the inherently statistical nature of signal messages to be transmitted. The statistical nature of signal messages was discussed in an elementary fashion in Chap. 1. We shall be referring to random signals over and over again in this chapter and those to follow.

The introduction to probability in the next few sections will of necessity be brief. We shall apply concepts as soon as they are developed to interesting problems in information transmission–binary transmission in noise, PCM repeater analysis, transmission over fading channels, radar, etc. Our approach will in general be nonrigorous. We shall often try to motivate some of the statistical properties of random signals and noise in an intuitive sense. Our goal is to develop a feeling for the terms used and the ability to utilize the theory of probability in studying various communication systems.

The reader is referred to some more advanced books for a deeper study of random signals and noise and applications to communications systems than will be possible in this book.[1]

5-1 INTRODUCTION TO PROBABILITY[2]

The need for statistical analysis arises in many branches of science. In many cases, measurements of various parameters may be deterministic in nature. They follow classical laws, and results may presumably be predicted exactly if all pertinent information is known. If a great many variables are involved, however, it frequently becomes extremely difficult to analyze the problem exactly. Instead, various average properties may be defined.

For example, if we are interested in investigating various properties of a gas trapped in a container, we could conceivably do this analytically by following the path of each molecule as it moves along, colliding with its neighbors, with the walls of the container, etc. Application of the simple

[1] M. Schwartz, W. R. Bennett, and S. Stein, "Communication Systems and Techniques," McGraw-Hill, New York, 1966; J. M. Wozencraft and I. M. Jacobs, "Principles of Communication Engineering," Wiley, New York, 1965; D. Middleton, "An Introduction to Statistical Communication Theory," McGraw-Hill, New York, 1960; W. B. Davenport, Jr. and W. L. Root, "Introduction to Random Signals and Noise," McGraw-Hill, New York, 1958.

[2] A. Papoulis, "Probability, Random Variables, and Stochastic Processes," McGraw-Hill, New York, 1965; A. M. Mood and F. A. Graybill, "Introduction to the Theory of Statistics," McGraw-Hill, New York, 1963; H. Cramer, "Mathematical Methods of Statistics," Princeton, Princeton, N.J., 1946; E. Parzen, "Modern Probability Theory and Its Applications," Wiley, New York, 1960; W. Feller, "An Introduction to Probability Theory and Its Applications," 2d ed., Wiley, New York, 1957.

laws of mechanics would presumably tell us all we want to know about each gas molecule, and therefore about the gas as a whole. But with millions of molecules to treat simultaneously, each moving with a possibly different velocity from the next, this theoretically possible calculation becomes practically impossible. Instead, because we deal with large numbers of molecules, we can determine average values for the velocity, force, momentum of the molecules, etc., and from these determine such average properties of the gas as temperature and pressure. (We shall define the concept of average more precisely later on, but we rely on our intuition now.)

The crucial point here is the phrase "large numbers." Whenever we deal with large numbers of variables (whether the millions of gas molecules in a small container, the millions of possible messages in a particular language to be transmitted, or the millions of different pictures that could possibly be seen on a small TV screen), we can talk about the average properties, obtained by applying statistical concepts to the variables in question.

These large numbers we deal with and the statistics based on them can be generated in different ways. For example, we could conceivably set up 1,000 blackboards and ask 1,000 persons to start simultaneously writing anything that came into their minds. If at a given instant we scanned the particular letter being written by each, we could determine the relative frequency of occurrence of letters in the language (say, English) being used. (This assumes that the persons were selected randomly, have no connection with one another, were not briefed beforehand, etc.) Alternatively we could watch one typical person writing and perhaps pick out the first letter in each new paragraph as he writes. This, too, could be used to determine the relative frequency of occurrence of letters in the language. But we would need many paragraphs of writing here (say, 1,000 or more) to get a good approximation to our desired frequencies. The same type of experiments could be used to determine the average properties of a gas; one could either measure simultaneously the velocity of each molecule in a container or follow one molecule about over a long period of time, and from this determine its average velocity and other statistical properties.

The distinction between these two methods of determining average or statistical properties of a particular quantity will be considered later on. Now we simply emphasize that they both involve dealing with large numbers of events.

How do we now make use of large numbers of measurements to determine the statistical properties of a particular occurrence?

Assume, as the simplest type of experiment, that we are engaged in repeatedly tossing a coin. Can we determine whether heads or tails will appear in any particular throw? Of course, given enough information about the way in which the coin is dropped, initial velocity, etc., we could predict heads or tails. But this is extremely difficult. Instead we say ordinarily that we *do not know exactly* which side will appear on any one throw but that the odds are 50–50 either way; i.e., in the *long run* (large number of tries), as many heads will appear as tails. Either event (heads or tails) is equally likely.

How do we arrive at this conclusion? We may perhaps say intuitively that by virtue of complete symmetry, assuming an unloaded coin, there can be no preference as to heads or tails. Or perhaps we have come to the same conclusion after long experience with coin tossing (a valuable and instructive use of time!).

Thus, although in only a few throws the frequency of occurrence of heads or tails may not be the same, we feel sure that over many repetitions (tosses) of the experiment, either event will occur very nearly the same number of times. If we plotted, for example, the ratio of the number of heads H to the total number of tosses N, we might obtain the curve of Fig. 5-1.

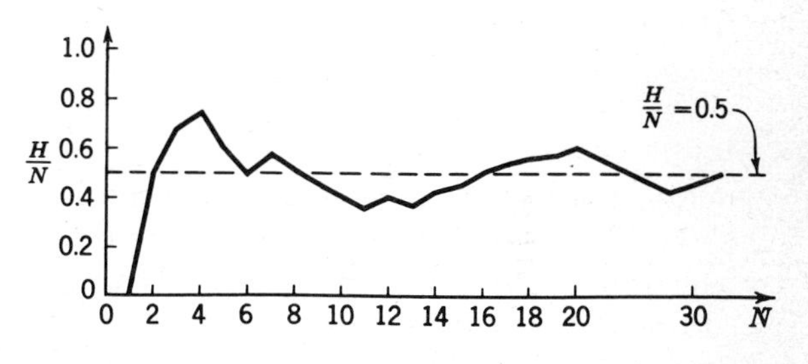

FIG. 5-1 Coin-tossing experiment. Fraction of heads thrown.

Initially (N small) H/N takes on greatly differing values. As N increases, however, H/N approaches 0.5 more and more closely. For large N, then, we are presumably justified in saying that heads will occur as often as tails. We then generalize by saying that in the next throw there is a 50 percent chance of heads occurring.

The same conclusions would of course be drawn from an experiment with signals consisting just of on-off pulses. If a long series of tests with all types of messages show that pulses appear as frequently as spaces (no pulses), we would presumably say there is a 50 percent chance of getting a pulse (or a space) in any one interval. The pulses and spaces are then equally likely.

We can consider similarly the tossing of a die: we feel intuitively, or might perhaps demonstrate by actual tosses of a die, that in many tosses each face should come up very nearly as often as any other. We say that, with six faces, each face has a $\frac{1}{6}$ chance of coming up on any one toss. Any two faces have a $\frac{1}{3}$ chance; any three faces, a $\frac{1}{2}$ chance; etc.

These intuitive ideas of chance as associated with many repetitions of an experiment may be formalized somewhat, as in Sec. 1-4, by defining the probability of one event out of several possible occurring as follows: Say there are n possible outcomes of an experiment given by $A_1, A_2, \ldots, A_n$. (In the case of the coin, A_1 is H, A_2 is T, n is 2. In the case of the die, n is 6, and the A's represent different faces of the die.) If the experiment is repeated N times and the relative frequency of occurrence of event A_K is found to approach a limiting value for N much larger than n, this limiting value is defined to be the probability of occurrence, $P(A_K)$, of event A_K. Thus

$$P(A_K) = \frac{\text{number of times } A_K \text{ occurs in } N \text{ tries}}{N} \qquad N \gg n \qquad [5\text{-}1]$$

If the occurrence of any one event excludes the occurrence of any others (a head excludes the occurrence of a tail), the events are said to be *mutually exclusive*. Then, if all possible events A_1 to A_n are included,

$$P(A_1) + P(A_2) + \cdots + P(A_n) = 1 \qquad [5\text{-}2]$$

Equation [5-2] is an obvious statement of the fact that some one of the n events must occur in the N tries, and the sum of all the events must be equal to N. In probability terms, then, the probability of *any* event occurring is 1. (The probability of a head or a tail is 1; the probability of any one of the six faces of a die coming up is 1; etc.)

We can now generalize our definition of probability a bit. If the n events are mutually exclusive, the probability of one of K events of the n occurring ($K \leq n$) must be the sum of the probabilities of each event occurring. Thus

$$P(A_{j+1} \text{ or } A_{j+2} \text{ or } \cdots \text{ or } A_{j+k}) = P(A_{j+1}) + P(A_{j+2}) + \cdots + P(A_{j+k}) \leq 1 \qquad [5\text{-}3]$$

For the relative frequency of the K events is simply

$$\frac{N_{j+1} + N_{j+2} + \cdots + N_{j+k}}{N} = \frac{N_{j+1}}{N} + \frac{N_{j+2}}{N} + \cdots + \frac{N_{j+k}}{N}$$

or the sum of the probabilities, as above.

The simplest examples of the calculation of probabilities relate to games of chance. We shall refer to some of these as we proceed, for the very same ideas involved appear in the calculation of noise and signal statistics.

As an example, assume that a box contains three white and seven black balls. What is the probability of drawing a white ball? If we were repeatedly to draw one ball at a time from the box, replacing each ball after it was drawn, we would again expect intuitively to find the white balls appearing 30 percent of the time, the black balls 70 percent of the time. There are thus two possible events here: A_1 is a white ball, A_2 is a black ball, and $P(A_1) = 0.3$, $P(A_2) = 0.7$.

Alternatively we could tag each white and black ball separately. Any one ball would thus appear very nearly $1/10$ of the time in many ($N \gg 10$) drawings of a ball. But we can lump together the chances of drawing any one of the three white balls and get a probability of 0.3 of drawing a white ball. (Here we could also say that each ball is equally likely to be drawn. Any one ball must thus have a $1/10$ chance of being drawn.) This latter approach is simply one example of adding the probabilities of each of a group of mutually exclusive events to determine the probability of occurrence of the overall group.

For another example, consider two dice to be thrown. What is the probability of getting a 6? Again two approaches are possible:

1. Since each die has six faces, a total of $6 \times 6 = 36$ possible outcomes exists. Each possibility is equally likely to occur (as determined by experiment or, more probably, by intuition) so that the probability of drawing any combination of two faces is $1/36$. But there are five different face combinations that give a 6 (5,1; 4,2; 3,3; 2,4; 1,5). The probability of the overall event is $1/36 + \cdots + 1/36 = 5/36$.
2. More directly, of 36 possible outcomes, 5 are the favorable ones corresponding to a 6 occurring. We can call the drawing of a 6 the desired event rather than the drawing of any one of the combinations giving 6 as above. Then 6 will occur $5/36$ of the time in many repeated tosses. [As a check, $P(2) = 1/36$, $P(3) = 2/36$, $P(4) = 3/36$, $P(5) = 4/36$, etc. Then

$$P(2) + P(3) + \cdots + P(11) + P(12) = 1$$

as expected.]

For the third example, say two coins are thrown. What is the probability of one head and one tail? Here there are four possible outcomes: H,H; H,T; T,H; T,T. Each outcome is equally likely and has a probability of $1/4$ of appearing. Head-tail can occur in two ways if no distinction is made as to which coin turns up head. So

$$P(H,T) = 1/4 + 1/4 = 1/2$$

Alternatively we can say that the head-tail combination is one of three possible: HT, TT, HH. But in many throws we would expect HT to come up 50 percent of the time.

Note that from our definition of probability as the relative frequency of occurrence of a specified event we have gradually and almost unconsciously moved to another interpretation of probability:

We enumerate the total number r of possible outcomes of an experiment that are *equally likely* and *mutually exclusive* (i.e., in a long series of repetitions of the experiment each outcome would occur on the average as often as any other). Then the probability of one of these events occurring is simply $1/r$. Our specified event could be just one of these equally likely events, in which case its probability of occurrence would be $1/r$. Or it might comprise a group, say, K ($K < r$), of the r events. Its probability would then be K/r by summing the probabilities of the mutually exclusive events.

For example, in the case of the two dice thrown, all face combinations are equally likely. Then $r = 36$, and the probability of any one face combination occurring is $1/36$. But of these 36 combinations $K = 5$ constitutes our desired event (a 6 appearing); so $5/36$ is the probability of throwing a 6.

By reducing the calculation of probability to the determination of the equally likely events first we can often avoid the necessity of actually performing an experiment many times to determine relative frequencies of occurrence. Thus, in calculating the probabilities of getting various head-tail combinations in the tossing of a coin, there is usually no need actually to carry out a coin-tossing experiment. We know intuitively that tossing a head or a tail on an unbiased coin is equally likely and begin our calculations from that point.

This is not possible in many cases, however. The calculation of the probability of occurrence of the different letters in the English alphabet cannot be based on any "equally likely" argument. We must actually calculate the relative frequency of occurrence of a particular letter in a long series of trials.

Where the "equally likely" approach is fruitful, however, use can sometimes be made of some of the fundamental notions of permutations and combinations. As an example, say a set of 2 cards is to be drawn from a deck of 52. What is the probability of drawing one spade and one heart?

The total number of ways in which two successive cards may be drawn is obviously 52×51. (For each card drawn, 51 remain. But 52 different cards may be drawn on the first try.) This number is just the permutations of 52 elements taken 2 at a time. Each one of these 52×51 possibilities or permutations is equally likely, so that the probability of drawing any one pair of cards is $1/(52 \times 51)$. But 13 different spades could have been drawn as the first card and for each such spade 13 different hearts as the second card. 13×13 is thus the number of spade-heart possibilities. But a heart followed by a spade would have done just as well; so $2(13 \times 13)$ is the number of favorable possibilities of the 52×51 possible outcomes. The probability of a spade and a heart is thus $2(13 \times 13)/(52 \times 51)$.

In general we may have n items (say, n books, n symbols of an alphabet, n voltage levels, etc.) which can be selected. m of these items are to be selected successively, with no duplication allowed. Thus, if a particular book or symbol is selected as the first of m selections, it can no longer be used in further selections. (If the 2-volt level is picked in the first interval, it can no longer be chosen the second or succeeding intervals.) What is the probability of selecting a particular symbol combination, book combination, or voltage-level combination? This probability will depend on the number of permutations or combinations of n items, m at a time.

If we pay attention to the *order* of the selection, the number of ways of picking m of n events is the number of *permutations* of n items taken m at a time. Thus, if two of the letters a, b, c, d, e are to be selected and *order* is important, ab differs from ba.

The number of ways of selecting m of n events is called the number of *combinations* if the order is unimportant. Thus, if an a and a b are all that are desired, ab is the same as ba.

If the m elements selected are *ordered*, the first may be selected in n ways, the second in $(n - 1)$ ways, etc. (Remember that once a particular item is chosen it cannot be selected again.) Then the number of possible selections is $P_{n,m}$ (number of permutations of n items taken m at a time).

$$P_{n,m} = n(n-1)(n-2) \cdots [n-(m-1)] = \frac{n!}{(n-m)!} \qquad [5\text{-}4]$$

As an example, if two of the five letters a, b, c, d, e are to be selected, there are $5 \times 4 = 20$ possibilities, with order important. (The reader may wish to tabulate these to convince himself of this result.) In particular, if n selections are made,

$$P_{n,n} = n! \qquad [5\text{-}5]$$

For example, there are $3! = 6$ permutations of the three letters a, b, c selected in three successive tries: abc, acb, bac, bca, cab, cba.

How do we now find the number of unordered selections or combinations? Call this number $C_{n,m}$. Then, for example, $abcd$ and $abce$ are different combinations, but $abcd$ and $abdc$ are not to be distinguished. There are $m!$ ways of obtaining each particular combination of m units, or, alternatively, $m!$ ways of rearranging a combination of m items, once obtained. (For example, *abc* can be rearranged in $3! = 6$ ways.) Then $C_{n,m}m!$ must equal the total number of possible selections of $P_{n,m}$.

$$C_{n,m} = \frac{P_{n,m}}{m!} = \frac{n!}{m!(n-m)!} \qquad [5\text{-}6]$$

Very commonly the symbol $\binom{n}{m}$ is used to denote the number of combinations of n objects taken m at a time. We shall follow this practice in writing equations where required. By definition, we then have

$$\binom{n}{m} \equiv \frac{n!}{m!(n-m)} \qquad [5\text{-}6a]$$

The card problem we discussed previously is a good example of using permutations and combinations to calculate probabilities: We draw 2 cards from a 52-card pack. The number of combinations possible is $C_{52,2} = (52 \times 51)/2$. Each of these is equally likely. But of these the possible number of ways for a spade-heart to appear is 13×13. Then $P(\text{heart,spade}) = [2(13 \times 13)]/(52 \times 51)$, as before.

Alternatively there are $P_{52,2} = 52 \times 51$ permutations of the cards taken two at a time. Of these, $2(13 \times 13)$ represent possible ways of getting a spade and a heart, and we get the same result for the probability of getting a spade and a heart.

5-2 CONDITIONAL PROBABILITY AND STATISTICAL INDEPENDENCE

Up to this point we have discussed the probability of one particular event occurring: a head or a tail in coin tossing, a particular number in the tossing of dice, a particular letter (say, e or g, for example) in the English alphabet, etc. We shall have occasion in future sections to discuss the probability of occurrence of two or more events, and it is appropriate at this point to extend our definition of probability as the relative frequency of occurrence to include the *joint probability* of two or more events occurring.

Random signals and noise offer many examples of the joint occurrence of two random events. Consider, for example, a binary pulse train (Fig. 5-2*a*). What is the probability that two pulses spaced T sec apart will both be 0's (or 1's)? Figure 5-2*b* represents the voltage at the output of an amplifier. What is the joint probability that the noise voltage at time t_1 exceeds 4 volts (or some other specified voltage), and the voltage at time t_2 exceeds 5 volts? (We shall relate the joint probability in this case to the system bandwidth later in the next chapter.)

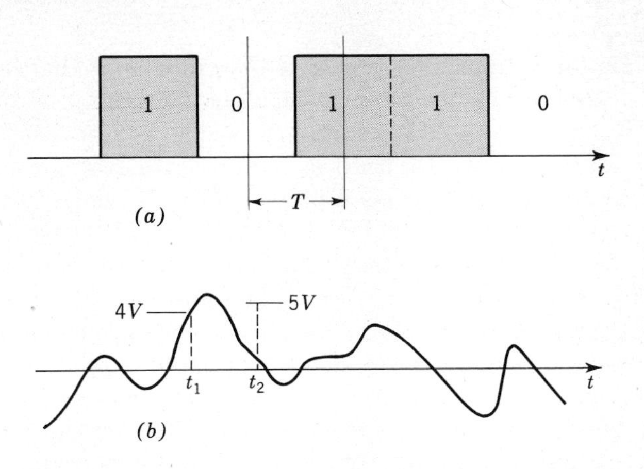

FIG. 5-2 Joint occurrence of two random events. (a) Binary signals; (b) noise voltage at output of an amplifier.

In Sec. 1-4 we indirectly hinted at the need for determining such probabilities. There we pointed out that in determining the statistical structure of a language such as English, a count of the relative frequency of occurrence of different letters of the alphabet does not provide full information about the language statistics. Letters commonly occur in pairs (for example, *qu*, *th*, *tr*, *st*) or even in groups of three or more. Some word combinations occur more frequently than others. All these joint occurrences affect the overall statistics and must be included in an accurate determination, for example, of the language information content.

The fact that some events occur *jointly* indicates that there may be some dependence of one event on another: the letter u obviously follows q much more frequently than does the letter e, for example. So we should be able to determine the independence of several events by measuring the frequency of joint occurrence. We shall restrict ourselves here to two events, A and B, only. The reader is referred to the literature on probability and noise statistics previously cited for an extension to more than two events of interest.

A simple example of the calculation of a joint probability will clarify these ideas and enable us to generalize quite readily. We determined previously, as an example in the calculation of probability, the chance of drawing a heart and a spade from a deck of cards. Using permutations and combinations, we were able to show that

$$P(\text{hearts,spades}) = \frac{2(13 \times 13)}{52 \times 51} = 2\frac{13}{4 \times 51}$$

Although we considered the heart-spade drawing at that time to be a single event of interest, we could just as well have treated it as a problem in joint probability: What is the probability that in the drawing of two cards one will be a heart, the other a spade?

To answer this question, consider first the case in which we are interested in drawing a heart as the first card, a spade as the second. Let A represent the drawing of a heart, B the drawing of a spade. Then $P(AB)$ represents the desired probability of drawing first a heart, then a spade.

The chance of drawing a heart as the first card is $P(A) = 1/4$, since 13 of the 52 cards are hearts. The probability of drawing a spade as the second card is not $1/4$, however, but $13/51$, since 51 cards are left. The second drawing is thus dependent upon, or *conditioned* by, the first drawing. We designate this possible dependence of the second of the two events on the first by the symbol $P(B|A)$: the probability of event B occurring, it being known that A has occurred. This is called a *conditional probability*. Here $P(B|A) = 13/51$. [Had the first card been replaced, we would have had $P(B) = 1/4$, independent of the first card.]

The chance of drawing the desired sequence heart-spade (AB) is then

$$P(AB) = 1/4 \times 13/51 = P(A)P(B|A)$$

since, for every heart we draw ($1/4$ of the time on the average), there is a $13/51$ chance of drawing a spade.

But we could just as well have asked for a spade first, then a heart in this particular problem. (The original problem specified, not a particular sequence of cards, but merely a heart and a spade.) This sequence obviously gives $P(BA) = 1/4 \times 13/51 = P(B)P(A|B)$ again. Since the probabilities of the two sequences are mutually exclusive (*either* a heart *or* a spade is drawn as the first card), the chance of a heart-spade combination being drawn is the sum of the two probabilities, or

$$1/4 \times 13/51 + 1/4 \times 13/51 = 2 \times 1/4 \times 13/51$$

This agrees of course with the result using permutations and combinations. To recapitulate, there we considered the heart-spade drawing as the one desired event and calculated the corresponding probability of this event occurring from a consideration of its relative frequency of occurrence. Here we have chosen to treat it as two events (one dependent on the other in this case).

We can now generalize the concepts of joint and conditional probability and their relation to the dependence of two events on one another. Assume that we perform an experiment and look for the occurrence of two events A and B in that order. We repeat this experiment many times and measure the relative frequency of occurrence of each event separately and as a pair (AB). (We shall actually give an example of such an experiment, and calculations based on it, shortly.)

Our problem is to determine the relative dependence (or independence) of B on A. The conditional probability $P(B|A)$, or the probability of B occurring, given A having occurred, will serve as a measure of this dependence. How do we then determine $P(B|A)$ from our measurements?

Let n_{AB} represent the number of times in n repetitions of the experiment that the combination AB appears. For n a "large number," the joint probability of first A and then B occurring is

$$P(AB) = \frac{n_{AB}}{n} \tag{5-7}$$

by using our previous definition of probability. In the same n trials the outcome A is found to appear n_A times, the outcome B n_B times. n_A must include n_{AB}, since some of the times that A appears it is followed by a B. The ratio $n_{AB}/n_A \leq 1$ represents the relative frequency of occurrence of B preceded by the event A and is just the desired conditional probability $P(B|A)$ (n_A "large" enough). Thus

$$P(B|A) = \frac{n_{AB}}{n_A} \leq 1 \tag{5-8}$$

As an example, n might represent a large number of drawings of two successive cards, n_A the number of times a heart appears, n_B the number of times a spade appears, n_{AB} the number of times a heart is followed by a spade. Dividing the numerator and denominator of Eq. [5-8] by n, we get

$$P(B|A) = \frac{n_{AB}/n}{n_A/n} = \frac{P(AB)}{P(A)} \tag{5-9}$$

from Eq. [5-7] and the definition of $P(A)$. Equation [5-9] is the defining equation relating conditional and joint probabilities. Multiplying through by $P(A)$ we get

$$P(AB) = P(A)P(B|A) \tag{5-10}$$

This is the relation used in the two-card problem given as an example.

Now assume that $P(B|A) = P(B)$. This implies that the probability of event B happening is independent of A. Such a situation would be true in the two-card problem if the first card were immediately replaced after having been drawn. In this case Eq. [5-10] gives

$$P(AB) = P(A)P(B) \tag{5-11}$$

We thus multiply the two separate probabilities together to find the probability of the event AB, if B is independent of A.

Equation [5-11] can also be written

$$P(AB) = P(B)P(A)$$

This implies that $P(AB)$ is independent of the order of occurrence of A and B in this case. It must thus be the same as $P(BA)$, the probability of A following B. But we also write

$$\begin{aligned} P(BA) &= P(B)P(A|B) \\ &= P(B)P(A) \end{aligned} \qquad [5\text{-}12]$$

in this case. Then $P(A|B) = P(A)$, and A is independent of the occurrence of B.

Two events A and B are said to be *statistically independent* if their probabilities satisfy the equations

$$P(AB) = P(BA) = P(A)P(B) \qquad [5\text{-}13]$$

and

$$P(B|A) = P(B) \qquad P(A|B) = P(A) \qquad [5\text{-}14]$$

Some examples of conditional probability are in order at this point.

1. *An urn containing two white balls and three black balls* Two balls are drawn in succession, the first one not being replaced. What is the chance of picking two white balls in succession? (Note that this is similar to the two-card problem.)

Letting event A represent a white ball on draw 1, event B a white ball on draw 2,

$$P(AB) = P(A)P(B|A) = \tfrac{2}{5} \times \tfrac{1}{4} = \tfrac{1}{10}$$

Here two of the five balls are white so that the chance of drawing a white is $\frac{2}{5}$. Once a white is drawn, however, only one white ball remains among the four balls left. The chance of drawing a white ball now, assuming a white drawn on the first try, is $\frac{1}{4}$.

Alternatively we can say that there are $C_{5,2} = 5!/3!2! = 10$ possible combinations of five balls arranged in groups of two. Of these only one combination, two white balls, is of interest. The chance of drawing this combination is thus $P(AB) = 1/C_{5,2} = \frac{1}{10}$ again. (Combinations are used here instead of permutations because the two white balls drawn cannot be distinguished from one another.)

If the first ball were replaced before drawing the second, the two events would be *independent:*

$$P(AB) = P(A)P(B) = (\tfrac{2}{5})^2 = \tfrac{4}{25}$$

2. *Two urns contain white and black balls* Urn A contains two black balls and one white ball; urn B contains three black balls and two white balls (Fig. 5-3). One of the urns is selected at random, and one of the balls in it chosen. What is the probability $P(W)$ of drawing a white ball?

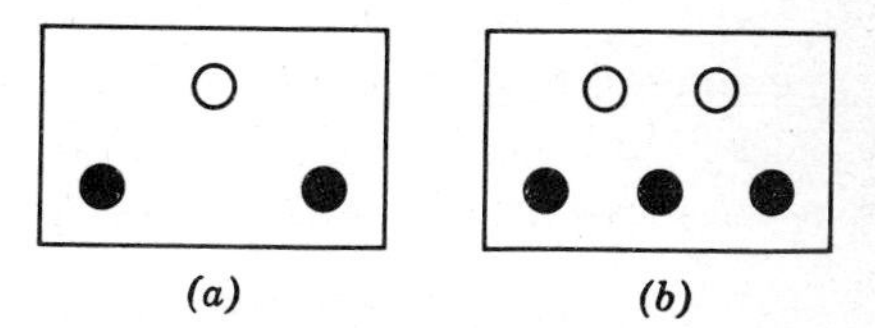

FIG. 5-3 Two-urn problem. (a) Two black, one white; (b) three black, two white.

There are two ways of satisfying the desired event W—the drawing of one white ball:

(*a*) Pick urn A; draw W.
(*b*) Pick urn B; draw W.

These two subevents are mutually exclusive, so that

$$P(W) = P(AW) + P(BW)$$

But
$$P(AW) = P(A)P(W|A) = \tfrac{1}{2} \times \tfrac{1}{3} = \tfrac{1}{6}$$

Thus the probability of drawing a white ball from urn A is the probability $P(A) = \frac{1}{2}$, of first picking A, times the probability $P(W|A) = \frac{1}{3}$, that a white ball will be selected once A is chosen. [$P(A) = P(B) = \frac{1}{2}$, since the urns are selected at random.] Similarly

$$P(BW) = P(B)P(W|B) = \tfrac{1}{2} \times \tfrac{2}{5} = \tfrac{1}{5}$$

Therefore
$$P(W) = \tfrac{1}{6} + \tfrac{1}{5} = \tfrac{11}{30}$$

[Note that if the balls were in one box $P(W)$ would be $\frac{3}{8}$. Because of the two urns the balls are not equally likely to be drawn.]

3. *Color blindness*[1] Assume that 5 men out of 100 and 25 women out of 10,000 are color-blind. A color-blind person is chosen at random from a representative sample of 10,000 men and 10,000 women. What is the probability that he will be male?

It may be assumed that 500 men and 25 women of this sample are color-blind. This gives a total of 525 color-blind persons. The probability that the person chosen will be male is thus $\frac{500}{525} = \frac{20}{21}$.

Alternatively let $N = 20{,}000$, $N_M = N_W = 10{,}000$ be the number of men or women in the sample, $N_C = 525$ be the total number of color-blind persons. $N_{MC} = 500$ represents the number of color-blind men. Then the conditional probability that a man is selected if a color-blind

[1] W. Feller, "An Introduction to Probability Theory and Its Applications," 2d ed., Wiley, New York, 1957.

person is chosen is

$$P(M|C) = \frac{N_{MC}}{N_C} = \frac{500}{525}$$

As a check, let $P(MC)$ be the probability of selecting a color-blind man, $P(C)$ the probability of selecting a color-blind person.

$$P(MC) = P(C)P(M|C) = {}^{525}\!/_{20,000} \times {}^{500}\!/_{525} = {}^{500}\!/_{20,000}$$

as expected. We could also determine the probability of selecting a color-blind man by first picking a man at random. The chance of doing this is $P(M) = 1/2$. Once a man is selected, the probability that he is color-blind is $P(C|M) = 5/100$. The probability of picking a color-blind man is again $5/200$.

4. *Statistics of three-letter alphabets* Suppose that we have an alphabet containing three letters A, B, C. We wish to determine the statistics of messages using this alphabet. In particular what is the relative frequency of occurrence, or probability, of each letter, and the probability of two-letter groups such as AA, AB, BC, CA, etc., occurring? From this we can determine the statistical dependence (or independence) of successive letters.

We take a typical example of a message using this three-letter alphabet and proceed to count the frequency of occurrence of the individual letters and groups of two successive letters. This is then an example of the repeated experiment mentioned previously as a method of determining the different probabilities of interest. A typical message of 50 letters appears as follows:

CBACABCABABCCBBABBCABAAAACACCBBCC
ACCBBBBACCAACBABC

Letting n = the total number of letters, n_A the number of A's, n_{AA} the number of pairs (diads) of AA, n_{AB} the number of pairs of AB, n_{BA} the pairs of BA, etc., we get $n = 50$: $n_A = 16$, $n_B = 17$, $n_C = 17$, and

$$\begin{array}{lll} n_{AA} = 4 & n_{AB} = 6 & n_{AC} = 6 \\ n_{BA} = 6 & n_{BB} = 6 & n_{BC} = 5 \\ n_{CA} = 6 & n_{CB} = 5 & n_{CC} = 5 \end{array}$$

From the almost equal number of times that the different letters appear we can conclude that the letter probabilities are equal. The different letters are thus equally likely in this particular three-letter language.

$$P(A) = \frac{n_A}{n} = 0.32 \qquad P(B) = 0.34 = P(C)$$

are the calculated probabilities, and 0.33 would be the actual probabilities if the letters were equally likely to occur.

The conditional probabilities can be calculated quite easily from the number of times the different pairs appear. For example,

$$P(B|A) = \frac{P(AB)}{P_A} = \frac{n_{AB}}{n_A} = \frac{6}{16}$$

Thus $P(B|A)$ represents the number of times that an A is followed by a B. Repeating this calculation for the different pair combinations—nine in number—we get the following conditional-probability table.[1]

CALCULATED PROBABILITIES

$P(j\|i)$		j		
		A	B	C
i	A	4/16	6/16	6/16
	B	6/17	6/17	5/17
	C	6/17	5/17	5/17

THEORETICAL PROBABILITIES, ASSUMING STATISTICAL INDEPENDENCE

$P(j\|i)$		j		
		A	B	C
i	A	1/3	1/3	1/3
	B	1/3	1/3	1/3
	C	1/3	1/3	1/3

Both the table and the list of the number of times each pair appears lead us to conclude that the letters in the pair combinations are *statistically independent.* For statistical independence $P(j|i) = P(j)$, or the probability of occurrence of any letter is independent of the letter preceding. Here, theoretically, we would expect all the $P(j|i)$'s to be 1/3. Because of the relatively short length of the message used (small sample, small number of experiment repetitions) the actual frequencies calculated differ somewhat.

C. E. Shannon has published an example of a three-letter alphabet in which any letter in a particular sequence is dependent on the letter immediately preceding.[2] (There is no dependence on letters before that one.) This alphabet consists of the letters A, B, C with the following

[1] C. E. Shannon, A Mathematical Theory of Communication, *Bell System Tech. J.*, vol. 27, pp. 379–423, July, 1948.

[2] Shannon, *ibid.*

probability tables:

i	$P(i)$
A	$9/27$
B	$16/27$
C	$2/27$

$P(j\|i)$		j: A	B	C
i	A	0	$4/5$	$1/5$
	B	$1/2$	$1/2$	0
	C	$1/2$	$2/5$	$1/10$

The letter B should thus occur most frequently, the letter C only occasionally. The letter A has 0 probability of being followed by another A, as is true also for C following a B. Each time A appears there is a $4/5$ probability that a B will follow, a $1/5$ probability that a C will follow. A C is followed by a B $2/5$ of the time, by an A $1/2$ of the time, by another C $1/10$ of the time.

A typical message given by Shannon for this three-letter language is

ABBABABABABABABBBABBBBBAB
ABABABABBBACACABBABBBBABB
ABACBBBABA

Calculating the number of times each letter and each pair of letters occurs, just as we did previously, we get $n = 60$:

$$n_A = 22 \qquad n_B = 35 \qquad n_C = 3$$
$$n_{AA} = 0 \qquad n_{AB} = 18 \qquad n_{AC} = 3$$
$$n_{BA} = 19 \qquad n_{BB} = 16 \qquad n_{BC} = 0$$
$$n_{CA} = 2 \qquad n_{CB} = 1 \qquad n_{CC} = 0$$

Again calculating a conditional probability table based on the actual relative frequencies, we get:

CALCULATED

$P(j\|i)$		j: A	B	C
i	A	0	$18/22$	$3/22$
	B	$19/35$	$16/35$	0
	C	$2/3$	$1/3$	0

THEORETICAL

$P(j\|i)$		j: A	B	C
i	A	0	$4/5$	$1/5$
	B	$1/2$	$1/2$	0
	C	$1/2$	$2/5$	$1/10$

Note again that the calculated probabilities, based on relative frequencies of occurrence, and the theoretical probabilities agree reasonably well. The reader should check these results for himself. He should also calculate the joint probabilities $P(ij)$ and compare for both the theoretical and actual cases.

5-3 AXIOMATIC APPROACH TO PROBABILITY[1]

The relative-frequency approach to probability discussed in the last two sections lends itself to experimental measurements of probability. It also helps one develop a "physical feel" for some of the concepts introduced. The theory of probability has achieved its greatest impetus, however, by being developed on an axiomatic basis. Although this approach using set theory is more abstract, it is at the same time much more general than the relative-frequency approach, and hence more readily extendable to more complex situations.

We shall outline this approach very briefly, contenting ourselves with paralleling the equations developed using the relative-frequency approach. The reader is referred to the references cited on probability at the beginning of Sec. 5-1 for much more comprehensive discussions.

We begin with some simple definitions from set theory, and then follow by setting up the axioms in terms of these. Assume that a typical experiment among those discussed earlier (coin and die tossing, cards, balls in urns, three-letter alphabets, etc) has S possible outcomes. We call S the set and each outcome an element in it. For example, assume that three balls a, b, c, are to be placed into two urns. There are then $2^3 = 8$ possible elements in the space S: a may be in either urn, b may be in either urn, and c may be in either urn. The eight elements are tabulated in the table below:

Element	Urn I	Urn II
1	abc	...
2	...	abc
3	ab	c
4	ac	b
5	bc	a
6	a	bc
7	b	ac
8	c	ab

[1] A Papoulis, "Probability, Random Variables, and Stochastic Processes," chap. 2, McGraw-Hill, New York, 1965.

This total set may in turn be split into possibly overlapping subsets. Call these A, B, C, ... As examples, four such subsets are defined below and the elements contained within them shown bracketed.

A: all the balls appear in 1 urn	{1,2}
B: ball a is in urn I	{1,3,4,6}
C: at least two balls in urn I	{1,3,4,5}
D: no balls in urn II	{1}

Many more subsets may obviously be defined.

Note that subset D is contained within A (all elements in D are in A), that A and B overlap by having element 1 in common, and that B and C overlap by having 1, 3, 4 in common.

We define the set $A + B$ (or *union* of A and B) to be another set containing all elements in either A or B or both. In the example above $A + B$ contains the elements {1,2,3,4,6}; $B + C$ contains {1,3,4,5,6}.

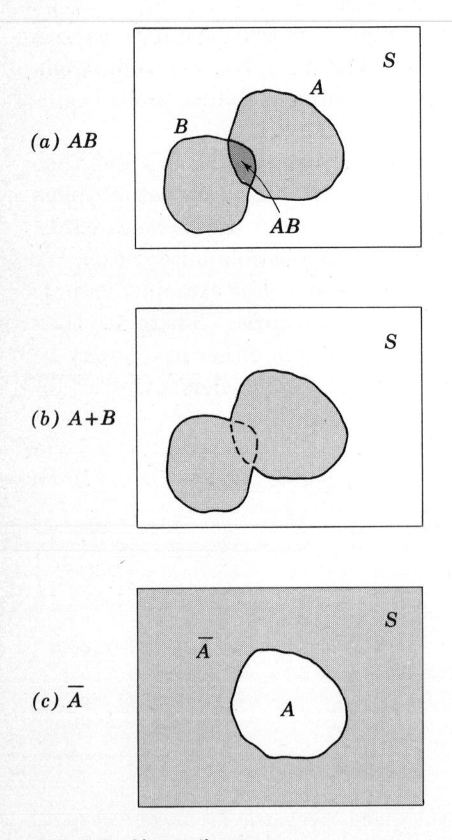

FIG. 5-4 Venn diagrams.

The set AB (or *intersection* of A and B) contains elements common to A and B. Again in this example AB contains the element $\{1\}$, while BC contains $\{1,3,4\}$. $\bar{A}$ is defined to be the *complement* of A: this is the set containing all elements *not* in A. Again in this example $\bar{A}$ has the elements 3, 4, 5, 6. It is apparent that $A + \bar{A} = S$. The zero or null set 0 contains no elements. These set-theory definitions and relations are easily visualized by using two-dimensional diagrams called *Venn diagrams*. S, A, B, $A + B$, $\bar{A}$, and AB are thus diagrammed in Fig. 5-4. The set S has been chosen rectangular here solely for simplicity's sake.

As a second example, indicating that the elements of a set do not necessarily have to be related to *discrete* outcomes, let x represent the rectified (positive) voltage at the output of an amplifier and y the rectified (positive) voltage at the output of another amplifier. The set S is given by all possible combinations of x and y, or by the elements $\{x,y > 0\}$. Let A be the subset $\{x > 10\}$, that is, all x voltages greater than 10 volts; B the subset $\{y > 10\}$ (Fig. 5-5). AB is then the set "both voltages greater than 10," as shown. It is apparent that $S = AB + \overline{AB}$. Define the set C as $\{x > y\}$. This is the set for which all x voltages are greater than y voltages. On the Venn diagram in Fig. 5-5 this is shown bounded

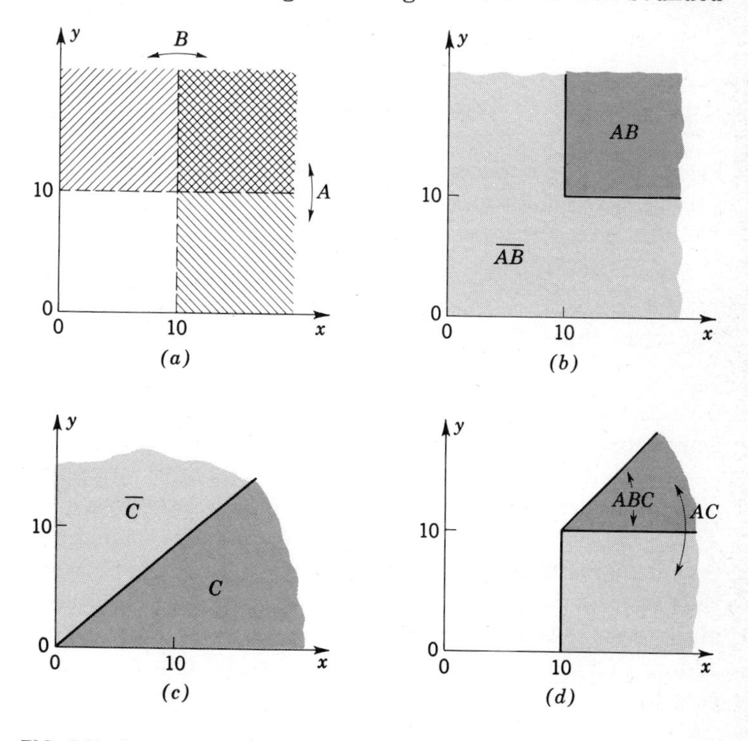

FIG. 5-5 Sets and various subsets.

by the 45° line (Fig. 5-5*c*). It is left for the reader to show that AC and ABC are the subsets indicated in Fig. 5-5*d*.

The Venn diagrams may be used to prove the following set identities:

$$\begin{aligned} AS &= A \\ A\bar{B} + AB &= A(\bar{B} + B) = A \\ A(B + C) &= AB + AC \\ \overline{A + B} &= \bar{A}\bar{B} \\ \overline{AB} &= \bar{A} + \bar{B} \end{aligned}$$

We now introduce probability by assigning to each subset or *event* a number we *define* to be the probability associated with event A. This number must have the property that it is positive. Since the set S contains all possible outcomes, $P(S) = 1$ is an obvious condition. This simply states the truism that some event must occur if an experiment is performed. To complete the basic definitions and make them consistent with the relative-frequency approach we have already discussed, we also require that if two sets A and B have no elements in common ($AB = 0$), their probabilities add. (Such events are said to be *mutually exclusive*.) We summarize by tabulating three basic axioms:

1: $$P(S) = 1 \qquad [5\text{-}15]$$

2: $$P(A) \geq 0 \qquad [5\text{-}16]$$

3: If $AB = 0$ $$P(A + B) = P(A) + P(B) \qquad [5\text{-}17]$$

From these three axioms other relations among the probabilities follow quite simply and often almost mechanically.

For example, since $A + \bar{A} = S$, and $A\bar{A} = 0$, by definition,

$$P(A + \bar{A}) = P(A) + P(\bar{A}) = P(S) = 1$$

But $P(A)$ and $P(\bar{A})$ are both positive by hypothesis. Then

$$P(A) \leq 1 \qquad [5\text{-}18]$$

Our probabilities as defined thus range between zero and one, as expected.

Similarly, we can calculate the probability of overlapping events by using the three axioms and referring to an appropriate Venn diagram. Thus, consider A and B overlapping as in Figs. 5-4 and 5-5. What is the probability of event $(A + B)$? It is apparent that we cannot simply add the two probabilities, since $AB \neq 0$. A and B are *not* mutually exclusive in this case. As a matter of fact, from the Venn diagrams we note that by adding $P(A)$ and $P(B)$ we have included AB twice. It is thus apparent that we must have

$$P(A + B) = P(A) + P(B) - P(AB) \qquad [5\text{-}19]$$

Here $P(AB)$ is exactly the kind of joint probability discussed earlier. As a check, note that $A + B = A + \bar{A}B$. But A and $\bar{A}B$ have no elements in common and are thus mutually exclusive events. $[A(\bar{A}B) = 0.]$ Then

$$P(A + B) = P(A) + P(\bar{A}B)$$

But $B = \bar{A}B + AB$, with $(\bar{A}B)(AB) = 0$. Hence

$$P(B) = P(\bar{A}B) + P(AB)$$

Finally, then, with $P(\bar{A}B) = P(B) - P(AB)$, we get Eq. [5-19].

Although the manipulations to determine probabilities of various events follow readily from the basic axioms, the actual *numbers* used must still be provided from other sources. Thus one either guesses at the probabilities of various outcomes of experiments, assumes them, or calculates them, as in the previous sections. The numerical part of probability does *not* come automatically.

As an example assume in Fig. 5-5 that $P(A) = 0.5$ and $P(B) = 0.3$. Thus we have good reason to believe that the rectified voltage of amplifier x has a 50 percent chance of exceeding 10 volts, while that of amplifier B has a 30 percent chance. If we *know* the joint probability $P(AB)$ is 0.4, (this is the probability that x and y will both exceed 10 volts when measured), the total probability $P(A + B) = 0.5 + 0.3 - 0.4 = 0.4$. In other words, the probability that either amplifier A output or amplifier B output exceeds 10 volts is 0.4.

Conditional probabilities are introduced by axiom as well. Thus we define the conditional probability of event A occurring, given B having occurred, as given by

$$P(A|B) = \frac{P(AB)}{P(B)} \tag{5-20}$$

This is, of course, in agreement with Eq. [5-9], obtained from relative-frequency considerations. Here we do not derive it, we simply take it as an axiom. The motivation is apparent from the Venn diagram of Fig. 5-4. With B having first occurred, it is apparent that the compound region of occurrence is immediately restricted to AB. Thus $P(A|B)$ must be proportional to $P(AB)$. Division by $P(B)$ serves to *normalize* $P(A|B)$ so that it is always ≤ 1. As a check let A be wholly in B. Then $AB = A$, and $P(A|B) = P(A)/P(B) \geq P(A)$. If we *know* we are in region B, our chance of further being in A (a subset of B) is greater than if we did not possess this additional information.

As an example of the use of these conditional probabilities, in addition to the examples discussed in Sec. 5-2, assume that one out of four symbols is transmitted in a communication system. The four symbols could be four levels in a four-level digital system, or simply four letters in a four-letter alphabet. Call these symbols A_1, A_2, A_3, and A_4. Call the four corresponding symbols received at the receiver B_1, B_2, B_3, B_4. Ideally A_1 is received as B_1, A_2 as B_2, etc. Because of noise or other distortion during transmission, however, A_1 as received may be mistaken for B_2, or B_3, or B_4. The same is true for the other transmitted symbols. Alternately, at the receiver, B_1 as received may actually have come from any one of the four transmitted symbols. We would like to determine the probability that B_1, as received, actually came from A_1. Some of the transitions that may occur are shown in Fig. 5-6.

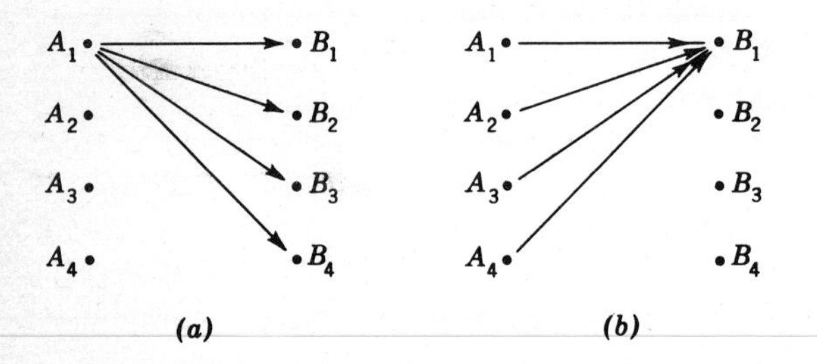

FIG. 5-6 Four-level transmission. (a) As referred to transmitter; (b) as referred to receiver.

Specifically, we would like to find the conditional probability $P(A_1|B_1)$, that is, the probability that A_1 was transmitted, given B_1 received. (This is often called the a posteriori probability, the probability *after the fact.* We shall have occasion to refer to such probabilities in much more detail later in the book.) To calculate this we first have to specify some probabilities. Let $P(A_1) = P(A_2) = \frac{1}{8}$, $P(A_3) = \frac{1}{4}$, $P(A_4) = \frac{1}{2}$. (These are called the *a priori* probabilities.) Then level A_4 is the most probable one to be transmitted, followed by A_3, A_1 and A_2 in that order. Note that A_1, A_2, A_3, and A_4 are all mutually exclusive events and incorporate the complete transmission set. Thus

$$\sum_{j=1}^{4} P(A_j) = P(S) = 1$$

We must also specify the various probabilities that transitions will take place. Specifically, let

$$P(B_1|A_1) = \tfrac{1}{2} \qquad P(B_1|A_2) = \tfrac{1}{4} \qquad P(B_1|A_3) = P(B_1|A_4) = \tfrac{1}{8}$$

From Eq. [5-20] the desired a posteriori probability $P(A_1|B_1)$ is given by

$$P(A_1|B_1) = \frac{P(A_1B_1)}{P(B_1)} \qquad [5\text{-}21]$$

This is easily written, but how do we find the two terms $P(A_1B_1)$ and $P(B_1)$? The first term, the joint probability of transmitting A_1 and receiving B_1, is easily calculated from the known conditional probabilities:

$$P(A_1B_1) = P(A_1)P(B_1|A_1) = (\tfrac{1}{8})(\tfrac{1}{2}) = \tfrac{1}{16} \qquad [5\text{-}22]$$

The second term, the probability $P(B_1)$ of receiving B_1, is obviously related to the transitions shown in Fig. 5-6. To calculate it consider the Venn diagram shown in Fig. 5-7. The set S contains the mutually exclu-

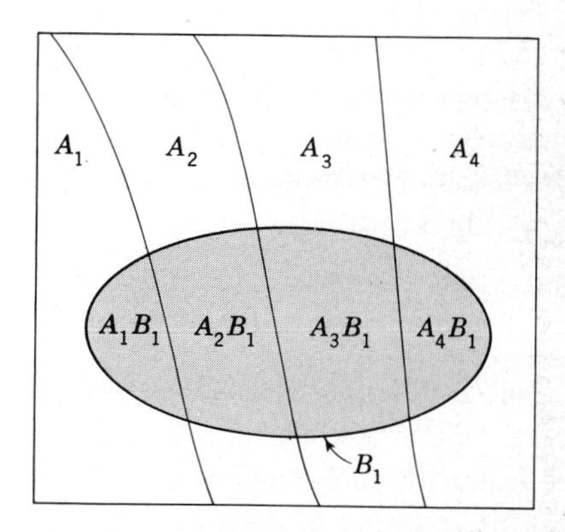

FIG. 5-7 Venn diagram for four-level transmission.

sive events A_1, A_2, A_3, A_4, as noted above. It must also contain the four mutually exclusive events B_1, B_2, B_3, B_4. In particular, event B_1 is shown overlapping with all four transmitted events. It is apparent from the figure that B_1 is the sum of four mutually exclusive events:

$$B_1 = A_1B_1 + A_2B_1 + A_3B_1 + A_3B_1 \qquad [5\text{-}23]$$

The probabilities then add, and we have

$$\begin{aligned} P(B_1) &= \sum_{j=1}^{4} P(A_jB_1) \\ &= \sum_{j=1}^{4} P(A_j)P(B_1|A_j) \qquad [5\text{-}24] \end{aligned}$$

using the conditional probability axiom to find each term $P(A_jB_1)$. In words: The probability of jointly transmitting A_j and receiving B_1 is the probability of first transmitting A_j, and then receiving B_1 with A_j transmitted. Summing over the mutually exclusive events A_j, $j = 1, 2, 3, 4$, we find the probability of receiving B_1. In particular, it is really found that $P(B_1) = 3/16$. The desired a posteriori probability $P(A_1|B_1)$ is then found from Eq. [5-21] to be $1/3$.

Note that nowhere was it necessary to discuss the relative frequency of occurrence of the various events. The form of the expression $P(A_1)$ was obtained solely from the use of axioms of probability. In actually evaluating this probability numerically, however, we had to assume a knowledge of the various a priori and transition probabilities. These would normally be estimated from repeated experiments, guessed, or calculated. (Generally, the a priori probabilities are estimated from many repeats of the experiment. One then falls back on the relative-frequency approach to actually evaluate probabilities numerically. The transition probabilities, however, are generally calculated from models of noise on the transmission channel. Examples of this type of calculation will be given later.)

5-4 DISCRETE AND CONTINUOUS PROBABILITY DISTRIBUTIONS

In this introductory discussion of probability we have thus far been primarily concerned with calculating the probability of occurrence of a finite number of discrete events. For instance, in the coin-tossing example there were two possible events: head or tail. In the case of a die we might be interested in the probability of one of the six possible faces coming up. In the case of the English alphabet there are 26 letters, and we may ask for the probability that one of them, say f, will occur. (Only in the two-amplifier example of Fig. 5-5 did we discuss continuous outcomes.) In the discrete examples we can represent the different possible outcomes by a discrete variable x. If x has n distinct values $x_1, x_2, \ldots, x_n$, each of which has probability $P_1, P_2, P_3, \ldots, P_n$, x is called a discrete chance variable, or discrete stochastic variable, or *discrete random variable.* As an example, say that we toss two dice. Let x represent the number coming up on any throw. Then x takes on the discrete values 2, 3, 4, . . . , 12, each one with its associated probability.

If the variable x is now allowed to take *any* value in a whole interval, however, and to each subinterval in the overall interval there is associated a probability of occurrence, x is called a *continuous random variable.* For example, if we start a pointer on a wheel spinning, the pointer will stop at any position on the wheel's circumference. It is not limited to discrete positions only.

We shall be primarily concerned with continuous chance variables because of our interest in the statistics of noise and signals in noise. If we were to plot the noise voltage at the output of an amplifier, for example, it might have the continuous appearance of the curve of Fig. 5-2*b*. All values of voltage are possible, not just discrete values.

How do we now compute probabilities in the case of continuous chance variables?

We shall start by first using the relative-frequency approach. In dealing with continuous variables, we shall find it convenient to introduce a *probability-density function*. We shall compare this probability-density function with the analogous mass density with which we are presumably quite familiar. Charge-density functions occur of course in electric-field problems also and are used to compute the field due to a continuous, or smoothed-out, array of charge, rather than discrete charges.

As an example of the determination of probability for continuous chance variables, say that we are interested in determining the distribution of height among American males. The height of a man may take on *any* value within a specified interval and thus represents a *continuous* random variable. Once we have such a height distribution, we may use this to calculate the probability that the height of a given male will lie between 5 ft 7 in. and 5 ft 9 in. We may use this to determine an average height, etc.

Assume that we select for this determination a representative sample of 1,000 men. We measure their heights (this is an example of the repeated experiment of the previous sections) and group them according to the nearest even inch (5 ft 0 in., 5 ft 2 in., 5 ft 4 in., etc.). All heights from 4 ft 11 in. to 5 ft 1 in. will, for example, be grouped in the 5-ft 0-in. category. The relative frequency of men found in each height interval is then the number grouped in that interval divided by the total number (1,000). A typical plot of such a height distribution is shown in Fig. 5-8. x represents the height, and n_x/n the relative number of men grouped in the interval Δx in. about x. n is the 1,000-man sample here. Since all heights in a 2-in. interval about the even heights are grouped together, they are shown as horizontal lines covering the 2-in. grouping. For a

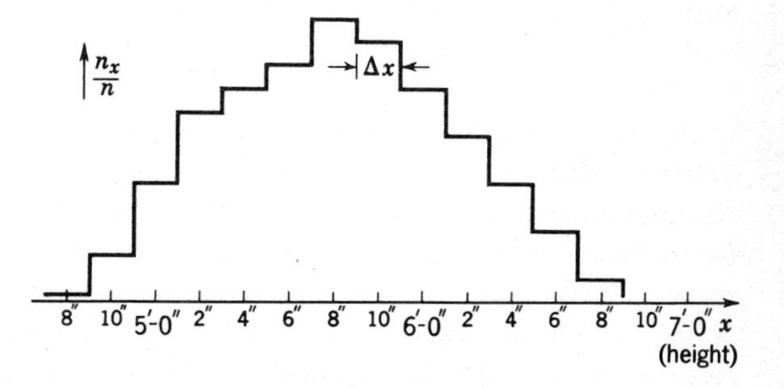

FIG. 5-8 Relative distribution of heights.

large enough sample we can say that n_x/n represents the probability $P(x_j)$ that the height of an American male will lie between $x_j - \Delta x/2$ and $x_j + \Delta x/2$.

This probability $P(x_j)$ obviously depends on the choice of interval size Δx (2 in. here). For if we decrease Δx, say, to 1 in., the number of men in this height interval decreases also. In particular, if we begin using smaller and smaller intervals, the ordinate will become smaller and smaller also. Eventually, if the height intervals are made very small (0.1 in., for example), very few men will be found in any one interval. The plot of relative distribution versus height approaches zero and is of little value to us.

To do away with the dependence on the size of the interval Δx chosen for the height distribution, we represent the height distribution, at a particular height and of interval Δx, by a rectangle of area n_x/n. The height of this rectangle will be $(1/\Delta x)(n_x/n)$, its width Δx. Such a relative-frequency curve is called a *histogram*. Two histograms for the height example used here are shown in Fig. 5-9*a* and *b*, one for $\Delta x = 2$ in.,

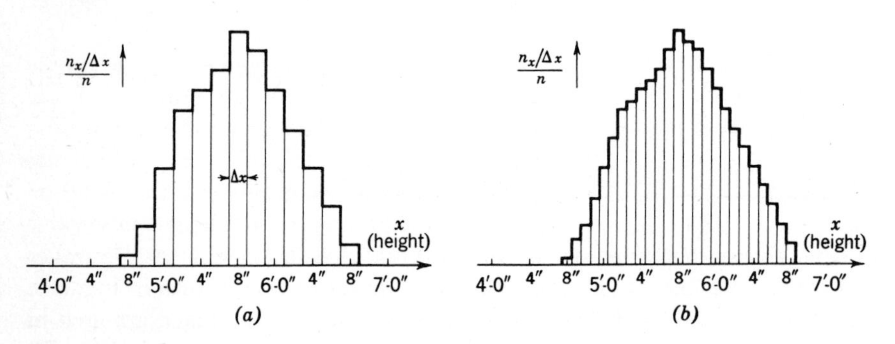

FIG. 5-9 Histograms of height distribution. (a) $\Delta x = 2$ in.; (b) $\Delta x = 1$ in.

the other for $\Delta x = 1$ in. For a large enough sample size the probability that a man's height will lie between two given heights, 5 ft 7 in. and 5 ft 9 in., for example, will now be the area of the histogram between these two limits.

If the histogram approaches a smooth curve as $\Delta x \to 0$, the ordinate takes on the form of a *probability-density function*, with the *area* under the curve between two points giving the probability that the height will be found between these two points. The probability-density function as the limiting case of the two histograms of Fig. 5-9 is shown in Fig. 5-10. The area between the two points, 5 ft 6 in. and 5 ft 8 in., shown crosshatched in the figure, represents the probability that a man's height will be found in that range.

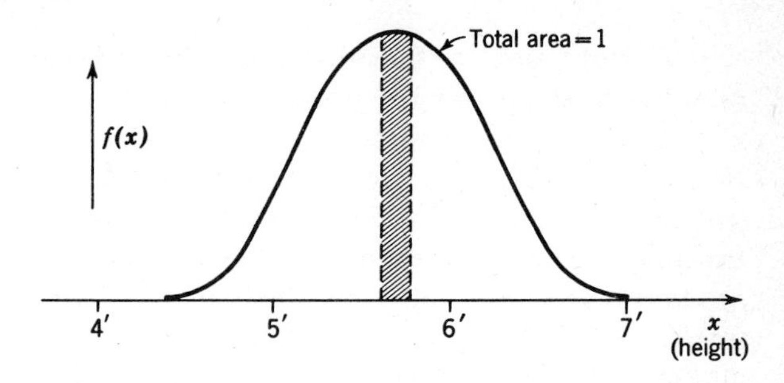

FIG. 5-10 Probability-density function corresponding to height histogram.

The assumption of a smooth curve corresponds mathematically to assuming that the probability density is continuous. Under this assumption we can formally define the probability density $f(x)$ by the limit

$$f(x) = \lim_{\substack{\Delta x \to 0 \\ n \to \infty}} \frac{n_x/\Delta x}{n} \qquad [5\text{-}25]$$

where n_x represents the number of samples of the total n found in the range $x - (\Delta x/2)$ to $x + (\Delta x/2)$. The probability that the variable x will lie in the range x_1 to x_2 is then the area under the $f(x)$ curve, or

$$\text{Prob}\,(x_1 < x < x_2) = \int_{x_1}^{x_2} f(x)\,dx \qquad [5\text{-}26]$$

In particular the probability that x will lie somewhere in its allowable range of variation must be 1. In general x can range from $-\infty$ to ∞ (the voltage at the output of an amplifier is an example). In special cases it may range only between 0 and ∞ (the variation of height, for example) or between 0 and 2π (for example, the rotation of a pointer on a wheel). For the general case of x ranging between $-\infty$ and ∞

$$\int_{-\infty}^{\infty} f(x)\,dx = 1 \qquad [5\text{-}27]$$

The histogram, or probability-density curves, must be normalized to have unity area.

From the definition of probability density $f(x)$ must be a function which is always positive. Thus

$$f(x) \geq 0 \qquad [5\text{-}28]$$

Equations [5-27] and [5-28] represent two conditions that must be satisfied by any probability-density function, as, for example, the one of Fig. 5-10.

To emphasize the fact that $f(x)$ represents only a probability-*density* function, with $f(x)\,dx$ the probability that x will be found in the range $x \pm dx/2$, we define another important function, the *cumulative-distribution function* $F(x)$. (For conciseness this will often be referred to as the distribution function.) This is defined to be the probability that the variable will be less than or equal to some value x. Since all values of x are mutually exclusive (a noise voltage can have only one value at any instant; a pointer may stop only at one point), $F(x)$ must be the sum of all the probabilities from $-\infty$ to x. This is just the area under the $f(x)$ curve from $-\infty$ to x. $F(x)$ is thus given by

$$F(x) = \int_{-\infty}^{x} f(x)\,dx \qquad [5\text{-}29]$$

Equation [5-29] is the indefinite integral of $f(x)$. If $F(x)$ possesses a first derivative, we have

$$f(x) = \frac{dF(x)}{dx} \qquad [5\text{-}30]$$

Equations [5-29] and [5-30] relate the probability-density function and the distribution function.

Since $f(x) \geq 0$ and $\int_{-\infty}^{\infty} f(x)\,dx = 1$, $F(x)$ must satisfy the inequality

$$0 \leq F(x) \leq 1 \qquad [5\text{-}31]$$

$F(x)$ is a continuously or monotonically increasing function, going from 0 to 1. A typical density function and its corresponding distribution function are shown in Fig. 5-11. $F(x_1)$ corresponds to the crosshatched area of Fig. 5-11*a*. It is left to the reader to show very simply that the probability that x will lie somewhere between x_1 and x_2 can be found

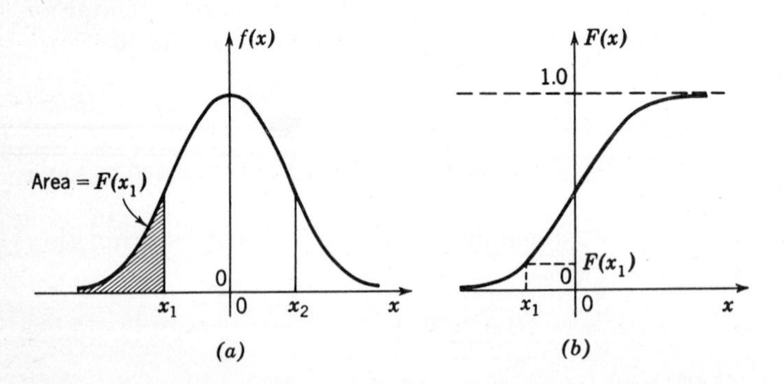

FIG. 5-11 Probability-density and probability-distribution functions. (a) $f(x)$ = density function; (b) $F(x)$ = distribution function.

from the distribution function by the relation

$$\text{Prob}\,(x_1 < x \leq x_2) = \int_{x_1}^{x_2} f(x)\,dx = F(x_2) - F(x_1) \qquad [5\text{-}32]$$

The rotating pointer on a wheel serves as a good example of these different relations. Intuitively we feel that a freely rotating pointer has an equally likely chance of stopping anywhere on the wheel. The probability that it will stop at any one angle θ is zero since there is an infinity of points, but the probability of its stopping within some angular range $d\theta$ is proportional to $d\theta$ and independent of the particular value of θ (Fig. 5-12). Since this probability must be $f(\theta)\,d\theta$, with $f(\theta)$ the density function, $f(\theta)$ must be a constant K for this example. The constant is found by invoking the specification that the area under the $f(\theta)$ curve must be 1. (The pointer will obviously stop somewhere on the wheel rim.) Thus

$$\int_0^{2\pi} f(\theta)\,d\theta = 1 = K \int_0^{2\pi} d\theta = 2\pi K \qquad [5\text{-}33]$$

and

$$f(\theta) = 1/2\pi \qquad [5\text{-}34]$$

This serves to normalize $f(\theta)$ properly. $f(\theta)$ in this case is one example of the rectangular density function to which we shall have occasion to refer in more detail later.

The cumulative-distribution function $F(\theta)$ is the probability that the variable will be less than or equal to θ. In this case the lower limit of θ is 0, and

$$F(\theta) = \int_0^{\theta} f(\theta)\,d\theta = \frac{\theta}{2\pi} \qquad [5\text{-}35]$$

This agrees with our intuitive feeling that all values of θ are equally likely. Both $f(\theta)$ and $F(\theta)$ are shown sketched in Fig. 5-12. The probability that θ will be less than $\pi/4$, for example, is $\frac{1}{8}$. The probability that it will be less than π is $\frac{1}{2}$.

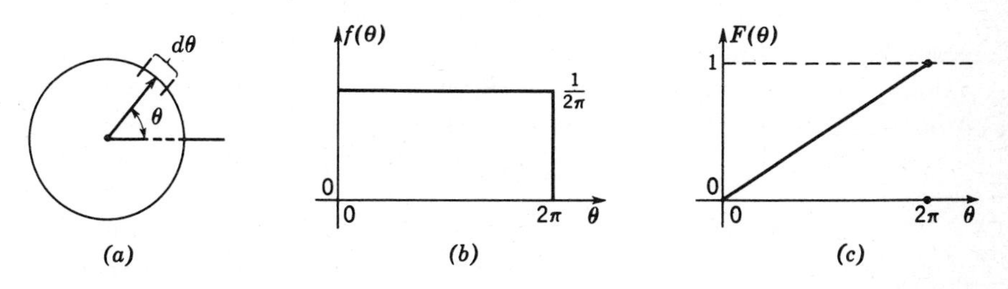

FIG. 5-12 Rotating pointer and probability functions. (a) Rotating pointer; (b) probability-density function; (c) distribution function.

The probability that θ will lie between $\pi/2$ and π is

$$F(\pi) - F\left(\frac{\pi}{2}\right) = \frac{\pi}{4\pi} = \frac{1}{4}$$

as is to be expected.

We have been discussing the probability-density function $f(x)$ for continuous variables. Can we also define such a function for discrete variables? This will be of particular interest in problems where a variable may have both discrete and continuous ranges. We shall now show that the impulse, or delta, function serves to connect the discrete and continuous cases.

We can develop the required relationship quite simply from the cumulative-distribution function. Assume a discrete random variable x with values $x_1, x_2, \ldots, x_n$. Corresponding to each value of x is a probability $P_1, P_2, \ldots, P_n$. An example is depicted in Fig. 5-13a. Since

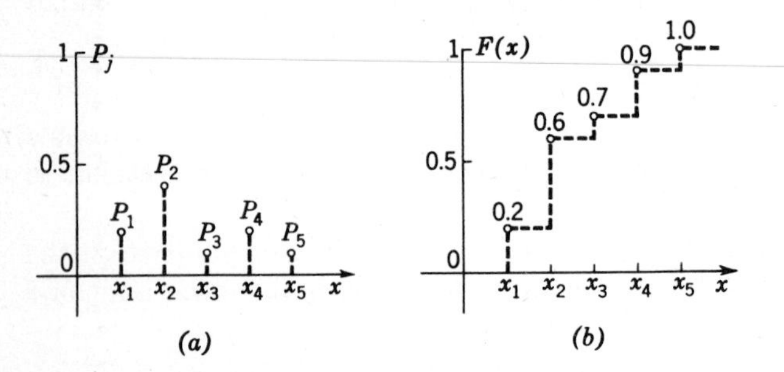

FIG. 5-13 Probability-distribution function for discrete variable. (a) Individual probabilities; (b) cumulative-distributive function.

the values of x are mutually exclusive and some one value of x must occur,

$$P_1 + P_2 + \cdots + P_n = 1 \qquad [5\text{-}36]$$

The probability that x will be less than or equal to some value x_j can again be defined as a cumulative-distribution function $F(x)$. Since all values of x are mutually exclusive, $F(x_j)$ will be equal to the sum of the probabilities $P_1 + P_2 + \cdots + P_j$. Thus

$$F(x_j) = P_1 + P_2 + \cdots + P_j \qquad [5\text{-}37]$$

For example, assume that x represents the number given by the toss of two dice. The numbers possible are $x = 2, 3, \ldots, 8, \ldots, 12$, and the corresponding probabilities are $\frac{1}{36}, \frac{2}{36}, \ldots, \frac{5}{36}, \ldots, \frac{1}{36}$, respectively. (The reader can check this for himself.) The probability that a number less than 5 will come up is then the sum of the probabilities of the numbers 2, 3, 4, or $\frac{1}{36} + \frac{2}{36} + \frac{3}{36} = \frac{1}{6}$.

Since the individual probabilities P_1, P_2, etc., are all greater than or equal to 0, the distribution function $F(x)$ must monotonically increase with x, just as in the case of the continuous random variable. Although it is defined only at the points $x_1, x_2, \ldots, x_n$, we may arbitrarily draw it as a series of ascending steps, as in Fig. 5-13*b*. The jump at step x_2 is just $P_2 = 0.4$, for example.

Although the derivative of $F(x)$ for the discrete variable does not exist in the usual sense (limit of ratio $\Delta F/\Delta x$ as $\Delta x \rightarrow 0$), we may define the derivative in terms of impulse functions. Thus, at x_j the derivative of $F(x)$ is $P_j\,\delta(x - x_j)$. This we *define* to be the probability-density function $f(x)$ at $x = x_j$.

$$f(x_j) \equiv P_j\,\delta(x - x_j) \tag{5-38}$$

The area under the impulse function is thus the actual probability at the point in question. In terms of this impulse-type density function the sum representation of the distribution function $F(x)$ given by Eq. [5-37] becomes the integral

$$F(x) = \int_{-\infty}^{x} f(x)\,dx \tag{5-39}$$

where

$$f(x) = \sum_{j=1}^{n} P_j\,\delta(x - x_j) \tag{5-40}$$

We thus generalize our concept of probability density to include discrete random variables as well as continuous random variables by utilizing the delta function.

MASS ANALOGY

It is instructive to relate the concepts of density and distribution functions to the analogous and familiar relations involving the distribution of mass. (Another analogy is that involving charge distributions.)

Consider first a weightless bar as shown in Fig. 5-14 with masses $M_1, M_2, \ldots, M_n$ suspended at distances $x_1, x_2, \ldots, x_n$ from the left end. The weights are normalized to a total weight of 1 lb, so that

$$M_1 + M_2 + \cdots + M_n = 1$$

The masses are then analogous to the probabilities at different values of the discrete variable and could be plotted as in Fig. 5-13*a*.

The cumulative mass distribution $M(x_j)$ can be defined to be the total

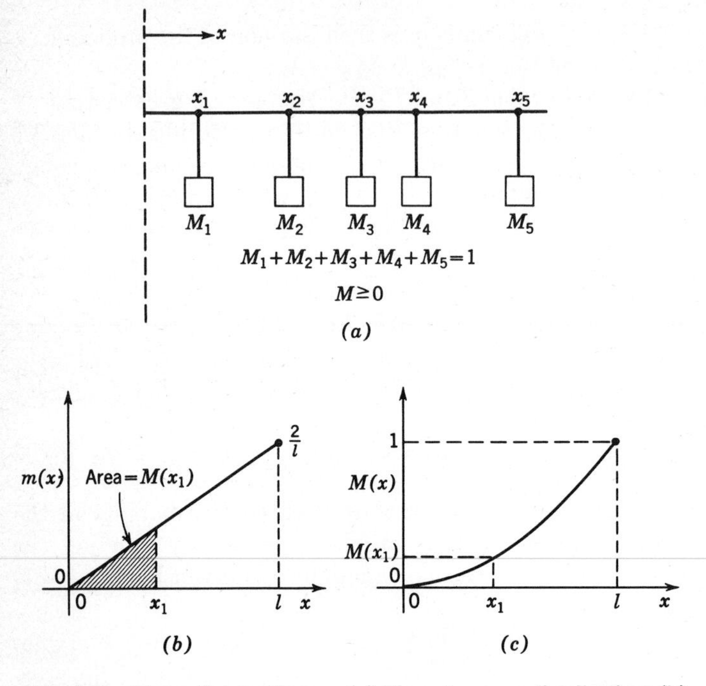

FIG. 5-14 Mass distributions. (a) Discrete mass distribution; (b) density function; (c) distribution function.

mass to the left of and including x_j and is obviously

$$M(x_j) = M_1 + M_2 + \cdots + M_j \leq 1$$

$M(x_j)$ would plot in the manner of $F(x_j)$ in Fig. 5-13.

Now consider a second bar with no hanging weights, but with a mass density $m(x)$, or (mass)/(unit length), varying in some manner along the bar. The *total* mass is still normalized to be 1. If the total length of the bar is l ft, $m(x)$ must satisfy the relation

$$\int_0^l m(x)\, dx = 1$$

Between any two infinitesimally separated points the mass included is $m\, dx$ lb. For example, if the mass is uniformly distributed, $m(x)$ is a constant equal to $1/l$ lb/ft. A sketch of $m(x)$ would be similar to the $f(\theta)$ function shown in Fig. 5-12*b*.

The cumulative-distribution function $M(x)$, or the mass of the bar included at all points less than x, is

$$M(x) = \int_0^x m(x)\, dx = \frac{x}{l} \qquad \text{lb}$$

for the uniformly distributed case. This corresponds to the sketch of $F(\theta)$ in Fig. 5-12*c*.

In all cases we must have $m(x) \geq 0$ and $0 \leq M(x) \leq 1$ is a monotonically increasing function. The mass included between two points x_1 and x_2 is given by

$$\int_{x_1}^{x_2} m(x)\, dx = M(x_2) - M(x_1)$$

All these relations are obviously identical with those set down for the probability-density and distribution functions. Just as in the probability case we must remember that $m(x)$ is a mass *density*, that only $m(x)\, dx$ represents mass (in particular, the mass between $x - dx/2$ and $x + dx/2$), and that the mass at any *one* point is zero if the mass is distributed smoothly throughout the bar.

Another example of a mass-density function might be one in which the density progressively increases along the bar. For example, let $m(x) = Kx$ lb/ft. With the total weight maintained at 1 lb the constant K is given by

$$\int_0^l m(x)\, dx = 1 = K \int_0^l x\, dx = \frac{Kl^2}{2}$$

Then $K = 2/l^2$, and $m(x) = 2x/l^2$ lb/ft. The mass-distribution function is now

$$M(x) = \int_0^x m(x)\, dx = \frac{x^2}{l^2}$$

Both $m(x)$ and $M(x)$ for this example are shown sketched in Fig. 5-14*b* and *c*.

We can of course extend this analogy, just as in the probability case, to include a bar with both discrete weight and distributed mass. The mass-density function would now include delta functions corresponding to the discrete masses.

GENERALIZED APPROACH TO DISTRIBUTION FUNCTIONS

The concepts of probability-distribution function and probability-density function may be developed in a more formal way, without resource to the relative frequency of occurrence of events and histograms, by expanding on the set-theory approach of the last section.[1] This again is a much more powerful approach because of its generality. Here one first defines a random variable $\underline{x}$ by associating every possible outcome in the set S with a real number x. The real number x can, in general, range between

[1] A. Papoulis, *op. cit.*, chap. 4.

$-\infty$ and $+\infty$. For any real number x the subset $\{\underline{x} \leq x\}$ of S then corresponds to all outcomes with numbers less than or equal to x. We may then define the probability-distribution function $F(x)$ as the probability that the random variable $\underline{x} \leq x$:

$$F(x) \equiv P(\underline{x} \leq x) \tag{5-41}$$

Two conditions must be appended here:

$$P(\underline{x} = +\infty) = P(\underline{x} = -\infty) = 0$$

It is then easy to demonstrate that $F(x)$ has the following properties:

1. $F(+\infty) = P(\underline{x} \leq \infty) = 1$. [This is the probability $P(S)$ and corresponds to all possible outcomes in the set S.]
2. $F(-\infty) = P(\underline{x} = -\infty) = 0$.
3. If $x_1 < x_2$, $F(x_1) \leq F(x_2)$. (The proof here consists of associating $\{\underline{x} \leq x_1\}$ with the event A, $\{\underline{x} \leq x_2\}$ with B, and using the definitions of probability of the last section.) The cumulative-distribution function is thus a monotonically increasing function, starting at 0 and increasing to 1 as x increases.

Note that both continuous and discrete random variables are contained within this development of the distribution function. If the random variable x is continuous, $F(x)$ may be defined as the integral of a density function $f(x)$:

$$F(x) = \int_{-\infty}^{x} f(x)\,dx \qquad f(x) = \frac{dF(x)}{dx} \tag{5-42}$$

It is then apparent that the density function $f(x) \geq 0$, and that $\int f(x)\,dx = F(+\infty) = 1$. [If x has discrete values only, $F(x)$ will be stepwise continuous and $f(x)$ is definable in terms of impulse functions, as previously.]

The basic difference between this approach and that beginning with the histogram and the density function is that the emphasis here is placed on the *distribution function* $F(x)$. Since this function represents the probability that the random variable $\underline{x} \leq x$, it may always be found once the possible outcomes of an experiment are associated with the real numbers x. The density function $f(x)$ is then defined in terms of this fundamental function, rather than the other way around.

As an example of this approach in first defining a random variable, and then its distribution function, consider a digital system emitting a train of binary symbols. In any one binary interval we have either one of two binary symbols. This is then a particularly simple set with only two possible (nonoverlapping) events. Associate the real number 0 with one event, 1 with the other. The random variable $\underline{x}$ then has only two values here, 0 and 1. (Although in some cases a 0 and 1 may actually be transmitted, more often the binary symbols will be two different and distinguishable pulse shapes. The random variable $\underline{x}$ does not necessarily have to coincide with the numerical values of the events, if such exist. We could equally well let the real numbers be -10 and -5, -1 and 0, or any other pair of numbers. Obviously, 0 and 1 serve as a particularly good representation for binary numbers.) Assume we know beforehand (a priori) that the symbol represented by the 1 has a probability p of being transmitted, while that represented by the 0 has probability $q = 1 - p$. (As noted in the last section one finds p by a judicious guess, by making measurements on the system, or by knowledge of the signal source and its statistics.) It is apparent that the distribution function here is particularly simple:

$$\begin{aligned} F(0^-) &= P(\underline{x} < 0) = 0 \\ F(0) &= P(\underline{x} \le 0) = q \\ F(1) &= P(\underline{x} \le 1) = q + p = 1 \end{aligned}$$

This is sketched in Fig. 5-15*a*. Note that $F(x)$ has the monotonic property indicated previously, rising from 0 to a maximum of 1. In this binary case it has discrete jumps of q and p at the values of $x = 0$ and 1, respectively. The density function is shown in Fig. 5-15*b*.

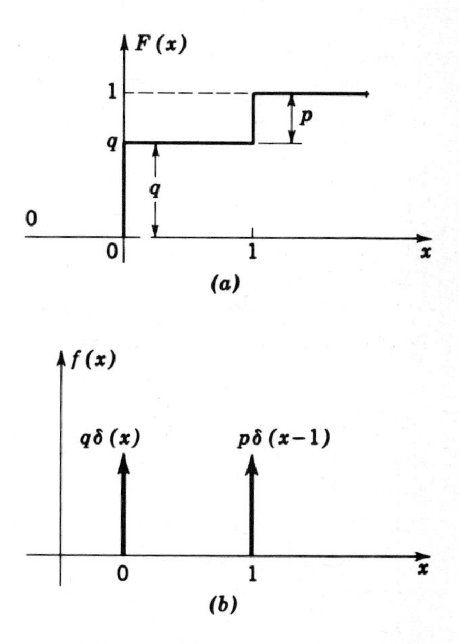

FIG. 5-15 Distribution function, binary random variable. (a) Distribution function; (b) density function.

Now consider n successive bits to be transmitted. Each bit is assumed independent of all others. We then ask for the distribution of 0's and 1's among the bits. In particular, what is the probability of obtaining no 1's (and hence n 0's), one 1 (and $n - 1$ 0's), etc.? In this case let the space S include all possible combinations of 1's and 0's. For simplicity's sake let the random variable $\underline{x}$ here take on the numbers corresponding to the possible numbers of 1's transmitted. Then $\underline{x} = k$, $k = 0, 1, \ldots, n$. Since the successive bits are assumed independent, the probability of transmitting a particular string of k 1's and $(n - k)$ 0's must be $p^k q^{n-k}$. But there are $\binom{n}{k} = n!/(n-k)!k!$ possible ways in which such a string could be transmitted. Since these different occurrences are mutually exclusive, the probability of transmitting exactly k 1's and $(n - k)$ 0's is just the sum of the probabilities for each possible occurrence, or

$$P(\underline{x} = k) = \binom{n}{k} p^k q^{n-k} \qquad [5\text{-}43]$$

The cumulative distribution $F(m)$, or the probability that $\underline{x} \leq m$, is then just the sum of the probabilities:

$$F(m) = \sum_{k=0}^{m} \binom{n}{k} p^k q^{n-k} \qquad [5\text{-}44]$$

Note that this too is a monotonically increasing function, changing with steps of $P(\underline{x} = k)$ at $\underline{x} = k$ (Fig. 5-16). As a check, let $m = n$. Then

$$F(n) = \sum_{k=0}^{n} \binom{n}{k} p^k q^{n-k} = (p + q)^n = 1 \qquad [5\text{-}45]$$

by the binomial expansion, and invoking the identity $p + q = 1$.

This distribution corresponding to the probability of k occurrences in n tries of a binary variable is called the *binomial distribution*. It is

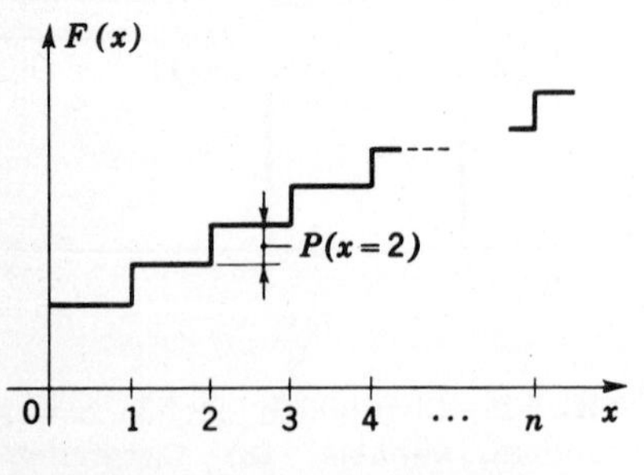

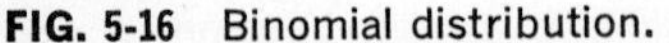
FIG. 5-16 Binomial distribution.

probably the most common discrete distribution and occurs quite often in science and engineering. The probability of k heads in n tosses of a coin is of course given by the binomial distribution. (If $p = q = \frac{1}{2}$, the coin is unbiased.) If in a particular electronic system successive voltage samples are divided into two groups—those above a certain level, those below—the distribution of the number exceeding the specified level is, of course, binomial. The binomial distribution will be utilized in Chap. 7 in studying noise in semiconductors and lasers, for example.

Note that the density function for the binomial distribution is again made up of a sequence of delta functions. In particular, we have, from Eq. [5-44],

$$f(x) = \sum_{k=0}^{n} P(\underline{x} = k)\delta(x - k) \qquad [5\text{-}46]$$

5-5 STATISTICAL AVERAGES AND EXAMPLES OF DENSITY FUNCTIONS

The probability-density functions discussed in the previous section (including impulse functions for discrete variables) provide us with information as to the probability or chance that a random variable will occupy a specified portion of its range. As such, they may also be interpreted as weighting functions, analogous to the weighting functions discussed in Chap. 2. Because of this weighting property, the density function also provides us with information as to the average value of a given random variable.

To demonstrate this, assume first that we have a discrete variable x, whose average value we would like to calculate. x can take on the values $x_1, x_2, \ldots, x_n$. We again perform repetitions of a hypothetical experiment, measuring the number of times N_1 that x_1 appears, the number of times N_2 that x_2 appears, etc. The total number of trials N is

$$N_1 + N_2 + \cdots + N_n = N \qquad [5\text{-}47]$$

The arithmetic or sample average of x is then found in the usual way by *weighting* each value of x by the number of times it appears, summing the weighted values of x, and dividing by the total number of trials N. Thus, av x, the average value of x, is simply

$$\text{av}\, x = \frac{x_1N_1 + x_2N_2 + \cdots + x_nN_n}{N} \qquad [5\text{-}48]$$

For a large enough number of trials, however, N_1/N is just the relative frequency of occurrence of x, or its probability P_1. (The expression "large enough number of trials" is taken to mean in an intuitive sense the number of trials beyond which we can detect no significant difference in the calculation of N_1/N and of av x.) We can thus write

$$\text{av } x = \sum_{j=1}^{n} P_j x_j \qquad \sum_{j=1}^{n} P_j = 1 \tag{5-49}$$

The average value of x in terms of its probability of occurrence, as given by Eq. [5-49], is taken as the *definition* of statistical average. Although it is apparent that the sample average of Eq. [5-48] should presumably approach the statistical average in some sense as the number of samples N becomes very large,[1] we shall henceforth distinguish between the two by using the symbol E to represent statistical average. This symbol stands for the phrase *expected value*, which is commonly used as a synonym for average value. Thus we define the statistical average or expected value of x to be

$$E(x) = \sum_{j=1}^{n} P_j x_j \qquad \sum_{j=1}^{n} P_j = 1 \tag{5-49a}$$

The term *mean* value is also commonly used to denote the statistical average.

In a similar way we could find the average of the square of x, the average of the square root of x, etc. In general, if we are interested in the average value of some function $g(x)$, we determine it in the same manner and write

$$E[g(x)] = \sum_{j=1}^{n} P_j g(x_j) \tag{5-50}$$

Note that the individual probabilities P_j behave like weighting parameters; those values of x most likely to occur (i.e., those with the highest probabilities) are weighted most heavily in determining the average value. As a special case, if x_5 has a probability $P_5 = 1$ of occurring and all other values of x have zero probability, $E[g(x)] = g(x_j)$, as expected.

As an example consider the binary variable introduced in the previous section. The average value of x in any one binary interval is just

$$E(x) = q \cdot 0 + p \cdot 1 = p$$

as might be expected. The expected value of the binomial distribution is more interesting. Recall that we asked for the probability of k 1's occurring in an n-bit binary sequence. If n becomes very large, we might expect from relative-frequency considerations to have roughly np 1's and nq 0's occurring in the sequence. (See Chap. 1 for a previous discussion of

[1] A. Papoulis, *op. cit.*, pp. 245, 246.

this point.) We now show in fact that the *average* number of 1's, $E(k)$, is just np!

To demonstrate this, we use Eqs. [5-49a] and [5-43] and write

$$E(k) = \sum_{k=0}^{n} kP(\underline{x} = k)$$

$$= \sum_{k=0}^{n} k \binom{n}{k} p^k q^{n-k} \qquad [5\text{-}51]$$

To evaluate this sum we use a simple trick. Note that

$$k\binom{n}{k} = \frac{n!}{(n-k)!(k-1)!} = n\binom{n-1}{k-1} = n\binom{n-1}{j}$$

with $j = k - 1$, a newly-defined index. Rewriting Eq. [5-51] in terms of j, we get

$$E(k) = np \sum_{j=0}^{n-1} \binom{n-1}{j} p^j q^{(n-1)-j} = np \qquad [5\text{-}51a]$$

(The sum over j is just 1 from the binomial theorem.) On the average, then, one expects to have np 1's and nq 0's appearing in a sequence of n binary symbols. The qualification *average* is highly important here. For the actual number of 1's in n bits is random, varying statistically (and unpredictably) from one n sequence to another. Thus k may generally take on any value from 0 to n. But over many repetitions of this experiment one would expect to find the average becomes np. Alternately, as n gets very large (effectively equivalent to many independent repeats of the experiment), the number of 1's approaches np. We shall demonstrate this somewhat more quantitatively later on after introducing the concept of variance.

The calculation of the average value of x or of some function $g(x)$ is carried out in the same way for continuous variables. Assume that we perform a hypothetical series of measurements on x. We choose x at Δx intervals, just as in the previous section, in determining the histogram of x. From these measurements we calculate $E(x)$ or $E[g(x)]$ as a straight arithmetic average. Alternatively we can utilize the information that $f(x)\,\Delta x$ is the probability that x will be found in the range $x \pm \Delta x/2$. From Eq. [5-50], then,

$$E[g(x)] = \sum_j g(x_j)f(x_j)\,\Delta x$$

In the limit, as $\Delta x \to 0$,

$$E[g(x)] = \int_{-\infty}^{\infty} g(x)f(x)\,dx \qquad [5\text{-}52]$$

As a special case the average value of x is

$$E(x) = \int_{-\infty}^{\infty} xf(x)\,dx \qquad [5\text{-}53]$$

in agreement also with Eq. [5-49*a*]. Equations [5-52] and [5-53] are taken as *definitions* in the set theoretic approach to probability.

The average value of x, $E(x)$, is frequently also called the first moment m_1 of x by analogy with the concept of moments in mechanics. In mechanics, the first moment of a group of masses is just the average location of the masses, or their center of gravity. For example, the center of gravity of the masses $M_1, M_2, \ldots$ in Fig. 5-14*a* can be found simply by taking moments about $x = 0$. The average value of x, or the center of gravity, is

$$\text{av } x = \frac{M_1x_1 + M_2x_2 + \cdots}{M_1 + M_2 + \cdots}$$

identical with the form of Eq. [5-48]. If the mass is not concentrated at discrete points but is smeared out over the bar, we can calculate the center of gravity by considering a differential mass $m(x)\,dx$ [with $m(x)$ the linear mass density] located x units from the origin. The center of gravity is then found by summing all the mass contributions. This sum again becomes an integral, and we have

$$\text{av } x = \int_0^l xm(x)\,dx$$

just as in Eq. [5-53].

The second moment in mechanics is just the moment of inertia of a mass or the turning moment of a torque about a specified point. By analogy with mechanics the second moment of a random variable is the average value of the square of the variable. For discrete variables this is given by

$$m_2 = E(x^2) = \sum_{j=1}^{n} P_jx_j^2 \qquad [5\text{-}54]$$

as a special case of Eq. [5-50]. For a continuous variable we have

$$m_2 = E(x^2) = \int_{-\infty}^{\infty} x^2f(x)\,dx \qquad [5\text{-}55]$$

as a special case of Eq. [5-52].

We can continue defining higher moments if we wish. In general the nth moment of x, or av x^n, is given by

$$m_n = E(x^n) = \int_{-\infty}^{\infty} x^nf(x)\,dx \qquad [5\text{-}56]$$

The density function $f(x)$ plays the role of a weighting function throughout. If some particular range of x, say, in the vicinity of x_j, occurs most frequently, the probability-density function will be highly peaked about that point. The nth moment will then be very nearly x_j^n. In particular, if

$$f(x) = \delta(x - x_j) \qquad [5\text{-}57]$$

$$m_n = \int_{-\infty}^{\infty} x^n \, \delta(x - x_j) \, dx = x_j^n \qquad [5\text{-}58]$$

Here x_j is the only point weighted, to the exclusion of all other points.

The delta function is one example of a probability-density function. It enables us to include the case of a variable defined to have one value only. If x is a discrete variable, the density function may be written as a sum of weighted impulse functions as in Eq. [5-40]. Thus

$$f(x) = \sum_{j=1}^{m} P_j \, \delta(x - x_j) \qquad [5\text{-}59]$$

and

$$m_n = E(x^n) = \int_{-\infty}^{\infty} x^n \left[\sum_{j=1}^{m} P_j \, \delta(x - x_j) \right] dx$$

$$= \sum_{j=1}^{m} P_j x_j^n \qquad [5\text{-}60]$$

checking Eq. [5-50].

What is the significance of these moments?

We shall see later on in signal and noise problems that m_1 gives just the dc voltage or current. m_2 will be found to give the mean-squared voltage (current) or the mean power. These quantities can easily be measured with meters. Since m_1 and m_2 can be determined from the probability-density function $f(x)$, it should be possible in turn to say something about $f(x)$, given the measured values of m_1 and m_2.

Consider, for example, the two possible probability-density functions shown in Fig. 5-17. What are the distinguishing characteristics of these

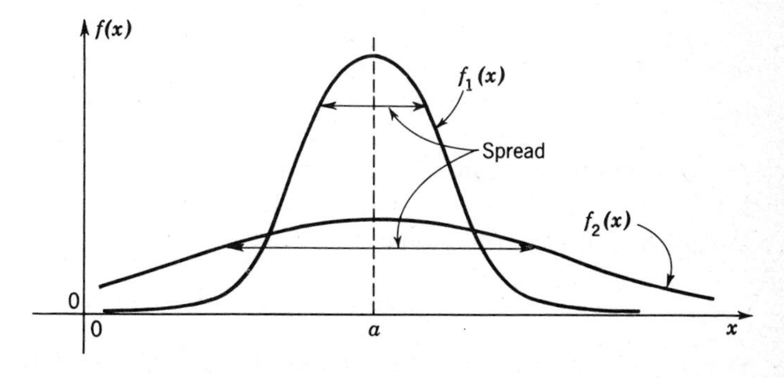

FIG. 5-17 Two possible density functions.

two density functions? Obviously $f_1(x)$ is more highly peaked about point a. $f_2(x)$ is more squat and spread out over a larger range of x. A wider range of values of x will thus appear in the case of $f_2(x)$, while in the case of $f_1(x)$ values close to a will appear much more frequently than those somewhat removed. We would expect that the average value of x would be close to, or equal to, a in both cases.

In this particular case the curves happen to have two distinguishing parameters: the location of the peak (a in this case), and the width, or spread, of the curves. We would probably agree intuitively that the first moment m_1 serves as some measure of the location of the peak in this case. (If the curve is symmetrical and unimodal, i.e., one peak, the peak will occur at m_1.) In general m_1 is one possible measure of the location of the range of most probable values of x. Other measures used and, in general, providing results not too different from m_1 are the median, or point at which the cumulative distribution $F(x)$ is 0.5, and the mode, or actual value at which the peak occurs if a single peak exists. Figure 5-18 shows one example of an unsymmetrical density function that we shall encounter later. Here $m_1 >$ modal point.

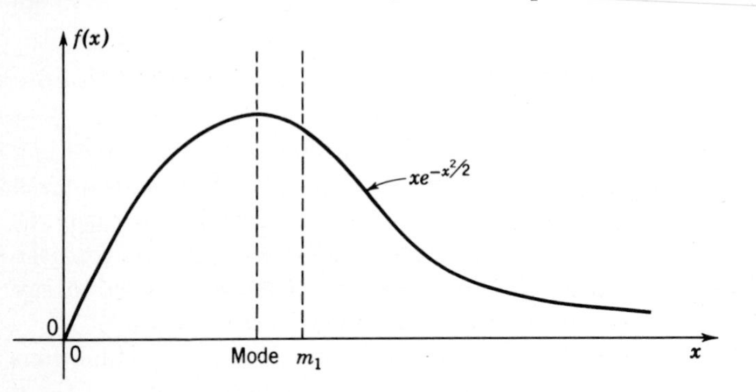

FIG. 5-18 An unsymmetrical density function.

How do we set up a measure of the width of the curves? We could use the 0.707 or 3-db points as in Chap. 2 for the amplitude-frequency-response curves. We could also use a specified percentage of the total area about point a as a measure of this spread. In general the measure of the width or criterion used to define a spread parameter is completely arbitrary. One possible measure of the width of the curve about a is the mean-squared variation about a:

$$s = \text{spread} \equiv \int_{-\infty}^{\infty} (x - a)^2 f(x)\, dx \qquad [5\text{-}61]$$

Because of the squared term, values of x to either side of a are equally significant in measuring variations away from a. If $f(x)$ happens to be a broad function such as $f_2(x)$ in Fig. 5-17 values of x far removed from a will still have sufficiently large values of $f_2(x)$ to be weighted strongly and provide a large value of spread, as desired. If $f(x)$ is a narrow function, such as $f_1(x)$, only those values of x close to a will be weighted significantly, and the spread will be correspondingly smaller.

In general this measure of the width of the curve $f(x)$ will vary with the value of a chosen. Although a was picked as the peak location in Fig. 5-17 it could have been chosen anywhere else, with Eq. [5-61] defined to be the spread about that point. The value of a for which the spread is a minimum is of interest. To find this point, differentiate Eq. [5-61] with respect to a. This gives

$$\begin{aligned}\frac{ds}{da} &= 0 = -2\int_{-\infty}^{\infty}(x-a)f(x)\,dx \\ &= 2a\int_{-\infty}^{\infty}f(x)\,dx - 2\int_{-\infty}^{\infty}xf(x)\,dx \end{aligned} \qquad [5\text{-}62]$$

But $\int_{-\infty}^{\infty} f(x)\,dx = 1$ and $m_1 = \int_{-\infty}^{\infty} xf(x)\,dx$ by definition. Then

$$a = m_1 = \int_{-\infty}^{\infty} xf(x)\,dx$$

or the point about which this mean-squared measure of the curve width is a minimum is just the average value, or first moment. In the symmetrical curves of Fig. 5-17 this happens to coincide with the peak. In the unsymmetrical curve of Fig. 5-18 it is to the right of the peak. The fact that $a = m_1$ gives a minimum and not a maximum is shown very simply by taking the second derivative of Eq. [5-61].

The spread about m_1, or the mean-squared variation about m_1, is called the *variance* or second central moment μ_2.

$$\mu_2 \equiv \int_{-\infty}^{\infty}(x-m_1)^2 f(x)\,dx = E[(x-m_1)^2] \qquad [5\text{-}63]$$

The square root of this term is called the *standard deviation* and is frequently given the symbol σ. We shall use σ as the measure of the spread of the density function about m_1. Thus

$$\mu_2 = \sigma^2$$

Multiplying out the squared term in the integral of Eq. [5-63] and integrating term by term, we get

$$\begin{aligned}\mu_2 = \sigma^2 &= \int_{-\infty}^{\infty}(x-m_1)^2 f(x)\,dx \\ &= \int_{-\infty}^{\infty}x^2 f(x)\,dx - 2m_1\int_{-\infty}^{\infty}xf(x)\,dx + m_1^2\int_{-\infty}^{\infty}f(x)\,dx \\ &= m_2 - m_1^2 \end{aligned} \qquad [5\text{-}64]$$

from Eq. [5-53].

The variance is thus the second moment less the square of the first moment. m_2 represents the spread of the curve about $x = 0$, μ_2 about $x = m_1$. Equation [5-64] is analogous to the parallel-axis theorem of mechanics, with m_2 the moment of inertia about the origin, μ_2 the moment of inertia about the center of gravity. In terms of mean-squared voltages and mean power, μ_2 or σ^2 will be considered later on to be the mean ac power and σ the rms ac voltage.

We now consider three typical distributions as illustrative examples of the above ideas. The first is the binomial distribution again, as an example of discrete random variables; the other two refer to continuous random variables.

Binomial distribution We found earlier that the expected value or first moment in this case was $E(k) = np$ (Eq. [5-51*a*]). We also indicated that as n increased, one would expect the number of 1's in a sequence of n binary symbols to approach the expected value np. In fact, from relative-frequency considerations one would expect p, the a priori probability of a 1 occurring, to be given "very closely" by k/n, as n becomes very large. The number k here represents the actual number of 1's counted in a sequence n bits long. All this intuitive reasoning implies that the variance or spread about p becomes small as n gets large. We now show by calculating the variance that these qualitative arguments are in fact true. This will then also serve as a quantitative justification of the relative frequency argument.

From Eq. [5-64], $\sigma^2 = m_2 - m_1^2$. (Although derived for continuous variables, this is easily proved as well for discrete variables. The proof is left as an exercise for the reader.) Since we already know

$$E(k) = m_1 = np$$

for the binomial distribution, we must now find m_2 to determine the variance σ^2. Specifically, then,

$$m_2 = E(k^2) = \sum_{k=0}^{n} k^2 \binom{n}{k} p^k q^{n-k} \qquad [5\text{-}65]$$

We use a trick similar to the one used to find the first moment $E(k)$. Subtracting and adding $E(k)$ to the right-hand side of Eq. [5-65], we have

$$\begin{aligned} m_2 &= \sum_{k=2}^{n} k(k-1) \binom{n}{k} p^k q^{n-k} + np \\ &= n(n-1)p^2 \sum_{j=0}^{n-2} \frac{(n-2)!}{j!(n-2-j)!} p^j q^{(n-2)-j} + np \end{aligned}$$

(Here we have introduced the dummy index $j = k - 2$.) But the sum is again just 1, by the binomial theorem. Then

$$m_2 = n(n-1)p^2 + np \qquad [5\text{-}65a]$$

and the variance σ^2 is

$$\sigma^2 = m_2 - m_1^2 = np - np^2 = npq \qquad [5\text{-}66]$$

with $q = 1 - p$. The standard deviation or spread about $E(k) = np$ is thus

$$\sigma = \sqrt{npq} \qquad [5\text{-}67]$$

As an example let $n = 9$, $p = q = \frac{1}{2}$. Then $E(k) = 4.5$, and $\sigma = 1.5$. The sketch of the density function in Fig. 5-19 shows the clustering in

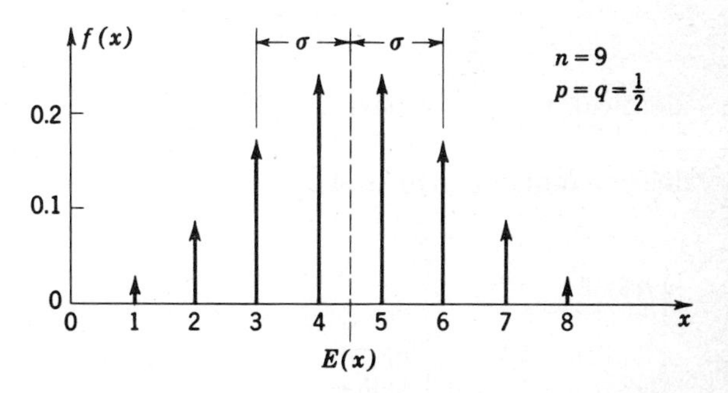

FIG. 5-19 Binomial distribution: density function. (After A. Papoulis, "Probability, Random Variables, and Stochastic Processes," fig. 4.9, p. 102, McGraw-Hill, New York, 1965.)

the region $E(k) \pm \sigma$ or 3 to 6. Note that as n increases, the *relative* spread about $E(k)$, $\sigma/E(k)$, approaches zero:

$$\frac{\sigma}{E(k)} = \frac{1}{\sqrt{n}}\sqrt{\frac{q}{p}} \qquad [5\text{-}68]$$

This indicates that for n "large enough," the number of 1's in an n-bit sequence is highly concentrated about the average value np, the concentration packing in more and more as n increases. Alternately, assume we take a long string of 1's and 0's and measure the number k of 1's occurring. We say this is a "good" measure or *estimate* of the (unknown) a priori probability. Call this estimate $\hat{p}$. Thus

$$\hat{p} = \frac{k}{n} \qquad [5\text{-}69]$$

Since k is a random number, $\hat{p}$ is a random variable as well. What are its properties? We have as its average value

$$E(\hat{p}) = \frac{E(k)}{n} = p \qquad [5\text{-}70]$$

so that *on the average* our estimate gives us p. (Such an estimate is said to be *unbiased.*) The variance of this estimate is then seen to be

$$\hat{\sigma}^2 = E(\hat{p} - p)^2 = \frac{\sigma^2}{n} = \frac{pq}{n} \qquad [5\text{-}71]$$

and the relative spread about the expected value p of the estimate is

$$\frac{\hat{\sigma}}{p} = \frac{\sigma}{E(k)} = \frac{1}{\sqrt{n}}\sqrt{\frac{q}{p}} \qquad [5\text{-}68a]$$

identical to the relative spread about $E(k)$. (This is obvious, since $\hat{p} = k/n$, and we are simply normalizing, dividing through by n.) The density function $f(\hat{p})$ is shown sketched in Fig. 5-20 for large n. As an

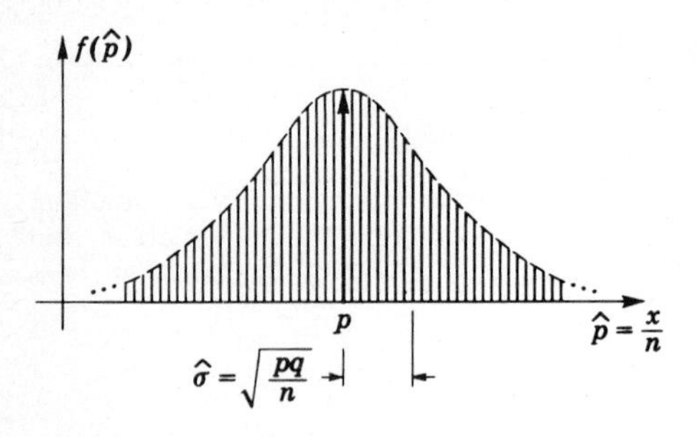

FIG. 5-20 Estimation of p, binomial distribution.

example, say $p = 10^{-2}$, $q = 0.99 \approx 1$, $n = 10^5$. Then one would expect the number of 1's in a binary string 10^5 bits long to cluster about an average value of 10^3. The spread about 10^3 is $\sigma = \sqrt{npq} \doteq \sqrt{1{,}000} \doteq 33$. The relative spread is $\sigma/E(k) = 0.033$. If n is now increased to 10^7, $E(k) = 10^5$, $\sigma \doteq \sqrt{10^5} \doteq 330$, and $\sigma/E(k) = 0.0033$, indicating the relative reduction in the spread by a factor of 10, as n increases by 100.

Rectangular distribution Assume that the random variable x is continuous and uniformly distributed with density function $f(x) = K$ between $x = a$ and $x = b$ (Fig. 5-21). The pointer problem of the pre-

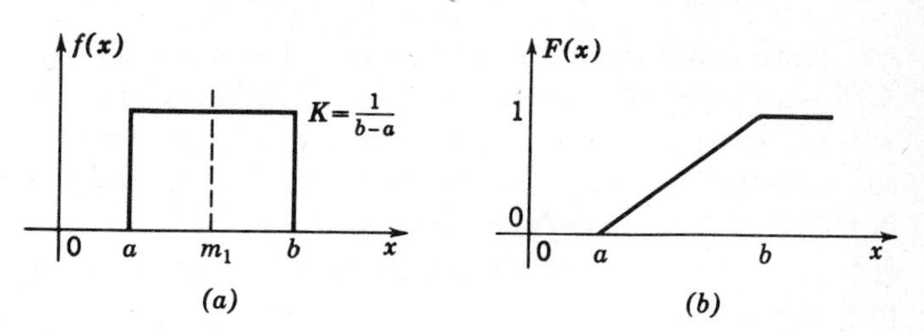

FIG. 5-21 Rectangular distribution. (a) Probability density; (b) distribution function.

vious section is one example of this density function. (There we had $a = 0$, $b = 2\pi$.) The PCM quantization error discussed in Chap. 3 was assumed to be uniformly distributed and hence has this form of density function. The variable x has no values less than a and greater than b and is equally likely to be found anywhere from a to b.

Judging from the curve of Fig. 5-21, we should expect the average value to be halfway between a and b, or at $m_1 = (a + b)/2$. The width of the curve should obviously be related to $b - a$.

Since $f(x)$ must be normalized to have unity area [the probability that x lies somewhere between a and b is $\int_a^b f(x)\,dx = 1$], the constant K must be $1/(b - a)$. m_1 is given by

$$m_1 = \int_a^b xf(x)\,dx = \frac{1}{b-a}\int_a^b x\,dx = \frac{a+b}{2} \qquad [5\text{-}72]$$

as expected. The second moment m_2 is given by

$$\begin{aligned} m_2 &= \int_a^b x^2 f(x)\,dx = \frac{1}{b-a}\int_a^b x^2\,dx = \frac{b^3 - a^3}{3(b-a)} \\ &= \frac{b^2 + ab + a^2}{3} \end{aligned} \qquad [5\text{-}73]$$

The variance is then

$$\sigma^2 = m_2 - m_1{}^2 = \frac{(b-a)^2}{12} \qquad [5\text{-}74]$$

This result could have been obtained just as well by finding σ^2 directly from Eq. [5-63]. The standard deviation σ is then

$$\sigma = \frac{b-a}{2\sqrt{3}} \qquad [5\text{-}75]$$

It thus appears in terms of $b - a$, as expected.

Note that if the width of the curve, $b - a$, is made smaller, the height $1/(b - a)$ increases correspondingly. The variable x is constrained to lie within a narrower range of variation and has a greater probability of doing so. Ultimately, as $b - a \to 0$, by letting b approach a, the height $\to \infty$, but with the area under the curve always equal to 1. In the limit $f(x) \to \delta(x - a)$, and the variable has only one possible value, $x = a$, with a probability of 1.

As an example of the use of the rectangular distribution, we can let the variable x be the PCM quantization error ϵ of Chap. 3. This error was the error due to quantization of continuous voltages measured within a specified range about a particular voltage level. If all voltages about this level are assumed equally likely to occur, ϵ will be uniformly distributed with zero average value. The voltage separation between adjacent levels is just the $b - a$ variation here. If this is set equal to a volts, as in Chap. 3 the rms quantization error is the standard deviation σ, or $a/(2\sqrt{3})$ volts. This was exactly the result obtained in Chap. 3.

As pointed out by Bennett,[1] these results are applicable in general to quantizing noise arising when analog signals are converted to their quantized digital form and are not limited to the PCM problem alone.

The cumulative-distribution function, or the probability that x will be less than some specified value, is found to be

$$
\begin{aligned}
F(x) &= 0 && x < a \\
F(x) &= \frac{x - a}{b - a} && a < x < b \\
F(x) &= 1 && x \geq b
\end{aligned}
\tag{5-76}
$$

Gaussian or normal distribution The gaussian density function is one of the most important density functions in probability theory and statistics. We shall find it recurring over and over again in our work. The gaussian function for one variable is given by the equation

$$f(x) = \frac{e^{-(x-a)^2/2\sigma^2}}{\sqrt{2\pi\sigma^2}} \tag{5-77}$$

When plotted, it has the characteristic bell-shaped curve of Fig. 5-22. The curve is symmetrical about the point $x = a$ and has a width proportional to σ. This is apparent from Eq. [5-77], for if we pick the point $x - a = \sqrt{2}\,\sigma$ at which the exponential is unity and let σ increase, $x - a$ increases correspondingly. Intuitively we would expect a to be just the

[1] W. R. Bennett, Methods of Solving Noise Problems, *Proc. IRE*, vol. 44, no. 5, pp. 609–638, May, 1956.

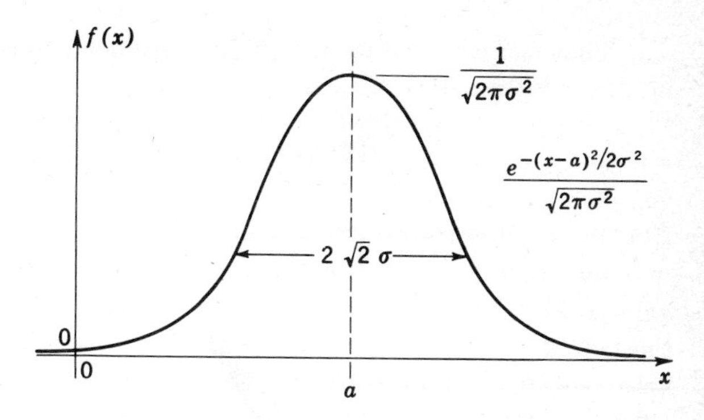

FIG. 5-22 Density function, gaussian distribution.

first moment, or average value of x, for this distribution. We can check this by actually performing the required integration. Thus

$$m_1 = \int_{-\infty}^{\infty} xf(x)\,dx = \int_{-\infty}^{\infty} \frac{xe^{-(x-a)^2/2\sigma^2}\,dx}{\sqrt{2\pi\sigma^2}} \qquad [5\text{-}78]$$

Upon changing variables by letting $y = (x - a)/\sqrt{2\sigma^2}$, $dx = \sqrt{2}\,\sigma\,dy$. The limits of integration remain $-\infty$ and ∞, so that we get

$$m_1 = \int_{-\infty}^{\infty} \frac{(\sqrt{2}\,\sigma y + a)e^{-y^2}}{\sqrt{\pi}}\,dy$$

But

$$\int_{-\infty}^{\infty} ye^{-y^2}\,dy = 0$$

(either by direct integration, or by noting that the function is odd), and

$$\int_{-\infty}^{\infty} e^{-y^2}\,dy = \sqrt{\pi}$$

from any table of definite integrals. This gives us immediately

$$m_1 = a \qquad [5\text{-}79]$$

for this particular function.

In a similar manner we can show by integration that σ^2 in Eq. [5-77] is the second central moment or variance μ_2 of this distribution.

$$\mu_2 = \int_{-\infty}^{\infty} \frac{(x - a)^2 e^{-(x-a)^2/2\sigma^2}\,dx}{\sqrt{2\pi\sigma^2}} = \sigma^2 \qquad [5\text{-}80]$$

The actual calculation is left as an exercise for the reader.

Since σ in Eq. [5-77] has been shown to be a measure of the width of the gaussian curve, this result agrees with our interpretation of the standard deviation as a measure of the spread or width of a probability-density curve.

That $f(x)$ as given by Eq. [5-77] is properly normalized can be shown by direct integration. Thus

$$\int_{-\infty}^{\infty} f(x)\,dx = \int_{-\infty}^{\infty} \frac{e^{-(a-x)^2/2\sigma^2}}{\sqrt{2\pi\sigma^2}}\,dx = 1 \qquad [5\text{-}81]$$

This integration is also left as a simple exercise for the reader.

This gaussian curve weights values of x near a most heavily. The value of $f(x)$ at the peak is $1/\sqrt{2\pi\sigma^2}$, so that, as the width σ decreases, the height of the curve in the vicinity of $x = a$ increases. Ultimately, for $\sigma \to 0$, this curve approaches the delta function $\delta(x - a)$, and the variable x becomes a constant a with a probability of 1. Note that the sketch of Fig. 5-22 is very similar to that of Fig. 5-20 for the binomial distribution. It may in fact be shown under some simple conditions that the binomial distribution approaches the gaussian for n large enough.[1]

The cumulative-distribution function, or the probability that the variable will be less than some value x, is

$$F(x) = \int_{-\infty}^{x} \frac{e^{-(x-a)^2/2\sigma^2}}{\sqrt{2\pi\sigma^2}}\,dx \qquad [5\text{-}82]$$

Since the $f(x)$ curve is symmetrical about $x = a$, half the area is included from $-\infty$ to a. The probability that $x \leq a$ is thus 0.5, or

$$F(a) = 0.5 \qquad [5\text{-}83]$$

As mentioned before, the 0.5 probability point is called the median of a statistical distribution. For the gaussian function the median, the average value, and the modal point [peak of $f(x)$] all coincide.

$F(x)$ is shown plotted in Fig. 5-23. The curve is symmetrical about the point $x = a$.

[1] A. Papoulis, *op. cit.*, p. 269.

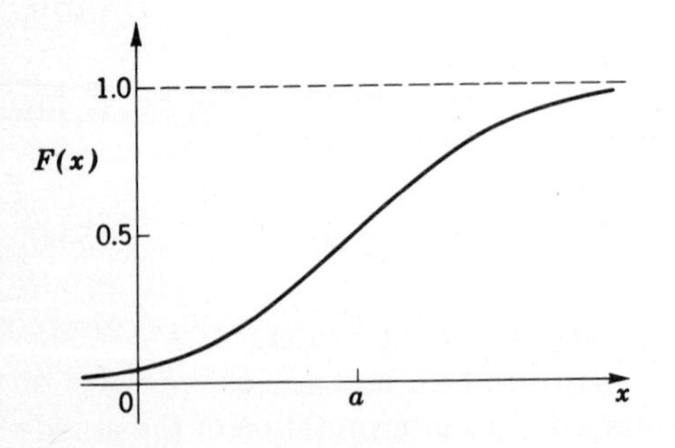

FIG. 5-23 Cumulative-distribution function, gaussian distribution.

The probability distribution of fluctuation noise is of the gaussian form, as we shall note in the next section. The average value of the noise is 0 volts, however, so that the curve is symmetrical about the origin. σ then represents the rms value of the noise voltage. A question frequently asked is: What is the probability that the noise voltage will be less than some prescribed $K\sigma$ (K a constant)? (Positive and negative voltages are included.) Letting x represent the instantaneous noise voltage, we can answer the question by writing

$$\text{Prob}\,(-K\sigma < x < K\sigma) = \int_{-K\sigma}^{K\sigma} f(x)\,dx = \int_{-K\sigma}^{K\sigma} \frac{e^{-x^2/2\sigma^2}}{\sqrt{2\pi\sigma^2}}\,dx \qquad [5\text{-}84]$$

This integral cannot be evaluated in closed form. Instead the integrand must be expanded in a power series and the resultant integral evaluated term by term. It can be put into a form more convenient for tabulation by letting $y = x/\sqrt{2}\,\sigma$. Doing this, and utilizing the symmetry of $f(x)$, we get

$$\text{Prob}\,(-K\sigma < x < K\sigma) = \frac{2}{\sqrt{\pi}} \int_0^{K/\sqrt{2}} e^{-y^2}\,dy \qquad [5\text{-}85]$$

This integral is frequently called the *error function* and is abbreviated erf $(K/\sqrt{2})$. In general,

$$\text{erf}\; x \equiv \frac{2}{\sqrt{\pi}} \int_0^x e^{-y^2}\,dy \qquad [5\text{-}86]$$

and

$$\text{Prob}\,(|x| < K\sigma) = \text{erf}\,\frac{K}{\sqrt{2}} \qquad [5\text{-}87]$$

The error function is tabulated in various books of mathematical tables[1] and in books on probability and statistics. Using these tables, we find that, for $K = 1$,

$$\text{erf}\,\frac{1}{\sqrt{2}} = 0.683$$

and, for $K = 2$,

$$\text{erf}\,\frac{2}{\sqrt{2}} = 0.955$$

The probability that the noise voltage will be less than σ volts in magnitude is thus 0.68. The probability that the voltage will be less than twice the rms noise voltage (2σ) is 0.95.

[1] See, for example, B. O. Peirce, "A Short Table of Integrals," Ginn, Boston, 1929.

Although we have used fluctuation noise as an example here, these results are more general. Thus it is easy to show that for any variable which has a gaussian probability-density function, the probability that the variable will deviate from the average value by less than σ is 0.68, while the chance of a deviation greater than 2σ is $1 - 0.95 = 0.05$. For example, assume that 100,000 resistors are to be manufactured with 100 kilohms nominal resistance. Owing to variations in the raw material and in the manufacturing process used the resistors will actually vary about the 100-kilohm nominal value. Assume that the variations away from this value follow a gaussian curve with a standard deviation of 10 percent of the average value. σ is then 10 kilohms, and 68 percent of the group of 100,000, or 68,000 resistors, should on the average have resistances within ± 10 kilohms of the 100-kilohm rated value. On the average 95 percent, or 95,000, should lie within ± 20 kilohms of the rated value.

The cumulative probability distribution of Eq. [5-82] can be related quite simply to the error function defined by Eq. [5-86]. It is left as an exercise to the reader to show that

$$F(x) = \frac{1}{2}\left(1 + \operatorname{erf}\frac{x - a}{\sqrt{2}\sigma}\right) \qquad [5\text{-}88]$$

5-6 APPLICATION: ERROR RATES IN PCM

With the discussion of probability density as a background we are now in a position to tackle an extremely important problem in digital data transmission: the calculation of probability of error due to noise introduced on the channel.

We have already stressed several times in this book that noise plays a significant role in determining the performance of communication systems. In fact, as first pointed out in Chap. 1 and again in Chap. 3, the presence of unwanted random disturbances limits the communication capacity, or rate at which we can communicate over a given channel. We shall have a great deal to say about sources of noise as well as the statistical properties of noise in the following two chapters. At this point, we simply state that noise has been added to the signals transmitted so that at the receiver the voltage measured consists of the algebraic sum of the two.

A typical oscillogram or pen recording of the noise voltage $n(t)$ might appear as in Fig. 5-24 (see Fig. 5-2*b* as well). Although the noise is assumed random so that we cannot specify in advance particular voltage values as a function of time, we assume we know the noise statistics. In particular, we assume first that the noise is zero-mean gaussian; i.e., it has

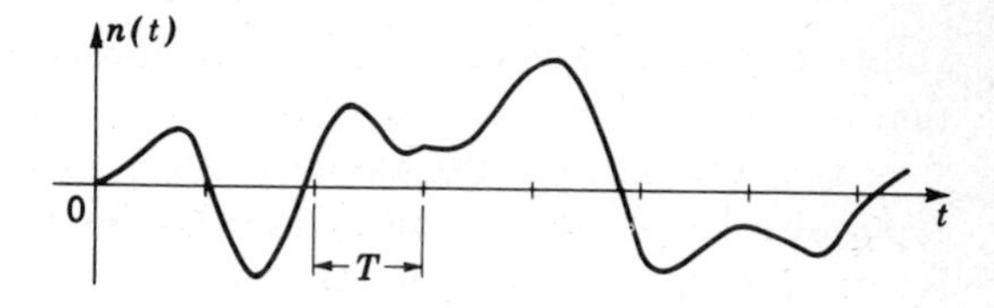

FIG. 5-24 Typical oscillogram, noise voltage.

a gaussian probability-density function, with $E(n) = 0$. Specifically, if we sample the noise at any arbitrary time t_1 the probability that the measured sample $n(t_1)$ will fall in the range n to $n + dn$ is given by $f(n)\, dn$, with

$$f(n) = \frac{e^{-n^2/2\sigma^2}}{\sqrt{2\pi\sigma^2}} \qquad [5\text{-}89]$$

This is, as we shall point out at length in the next chapter, the most commonly used statistical model for additive noise in communications, and is in most applications a valid representation for actual noise present. We assume the noise variance σ^2 is known. (As mentioned earlier and as will be demonstrated in the next chapter σ^2 may be either measured digitally or by a long time constant true power meter.)

Now assume we are receiving binary pulses in a binary PCM system. The noise $n(t)$ is added to the incoming group of pulses in the receiver, and there is a possibility that the noise will cause an error in the decoding of the signal. Specifically, if the system is of the on-off type, in which pulses represent 1's (or marks) in the binary code, the absence of pulses representing 0's or spaces (see Chap. 3), the error will occur if noise in the absence of a signal happens to have an instantaneous amplitude comparable with that of a pulse when present or if noise in the presence of a signal happens to have a large enough negative amplitude to destroy the pulse. In the first case the noise alone will be mistaken for a pulse signal, and a 0 will be converted to a 1; in the second case the 1 actually transmitted will appear as a 0 at the decoder output.

How often will such errors occur on the average? Is it possible to decrease the rate of errors below a tolerable number by increasing the pulse amplitude? If so, how much increase is necessary? What is the effect on the error rate of reducing the noise? All these questions are readily solved using the theory of the previous sections if we know the noise statistics or have a reasonably good model for these. We shall demonstrate this using the gaussian noise model of Eq. [5-89]. (Note that this model seems intuitively right in many situations, since it indicates the noise is equally likely to be positive and negative, occurs with higher probability at smaller values, and has a rapidly decreasing probability of attaining values in excess of several times the standard deviation σ.)

Assume that the pulse amplitudes are all A volts, as in Chap. 3. The composite sequence of binary symbols plus noise are sampled once every binary interval and a decision is made as to whether a 1 or a 0 is present. One particularly simple way of making the decision is to decide on a 1 if the composite voltage sample exceeds $A/2$ volts, and a 0 if the sample is less than $A/2$ volts. Errors will then occur if, with a pulse present, the composite voltage sample is less than $A/2$, or, with a pulse present, if the noise above exceeds $A/2$.

An example of a possible signal sequence, indicating the two possible types of error, is shown in Fig. 5-25. The signal pulses are shown trian-

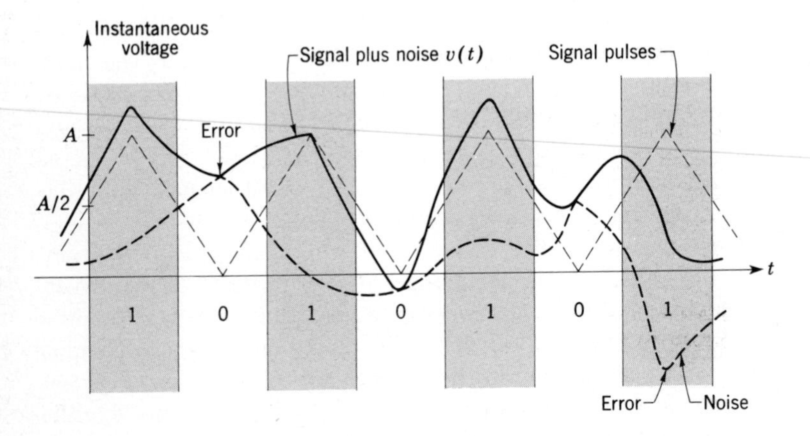

FIG. 5-25 Effect of noise in binary pulse transmission.

gular for simplicity's sake. The pulses and noise are shown dotted, while the resultant or composite voltage $v(t)$ is represented by the solid line. In this case samples are taken at the *center* of each binary interval, the system decoder then responding to the amplitude of these samples.

To determine the probability of error *quantitatively* we consider the two possible types of error separately. Assume first that a *zero* is sent, so that no pulse is present at the time of decoding. The probability of error in this case is just the probability that noise will exceed $A/2$ volts in amplitude and be mistaken for a pulse or a 1 in the binary code. Alternately, since $v(t) = n(t)$ if a 0 is present, v is a random variable with the same statistics as the noise. The probability of error is then just the probability that the sampled value of v will appear somewhere between

$A/2$ and ∞. Thus the density function for v, assuming a zero present, is just

$$f_0(v) = \frac{1}{\sqrt{2\pi\sigma^2}} e^{-v^2/2\sigma^2} \qquad [5\text{-}90]$$

The subscript 0 denotes the presence of a 0 symbol and the probability of error P_{e0}, in this case is just the area under the $f_0(v)$ curve from $A/2$ to ∞.

$$P_{e0} = \text{Prob}\left(v > \frac{A}{2}\right) = \int_{A/2}^{\infty} f_0(v)\, dv \qquad [5\text{-}91]$$

The density function $f_0(v)$ is shown sketched in Fig. 5-26a, with the probability of error indicated by the shaded area.

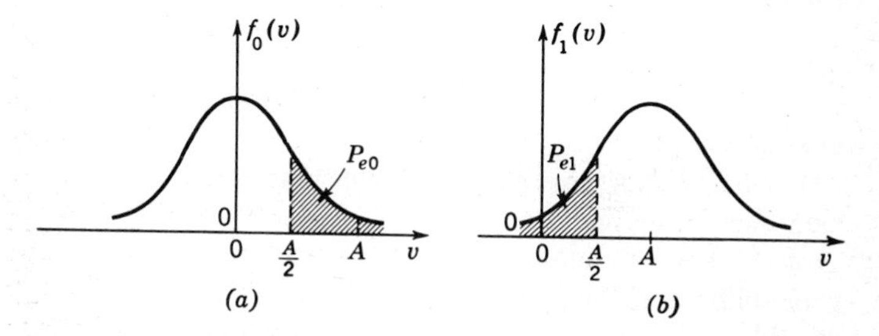

FIG. 5-26 Probability densities in binary pulse transmission. (a) Noise only (0 transmitted); (b) pulse plus noise (1 transmitted).

Assume now that a 1 is transmitted by the system encoder. This appears at the decoder as a pulse of amplitude A volts plus superimposed noise. A sample $v(t)$ of the composite voltage taken at time t is now the random variable $A + n(t)$. The fixed quantity A serves to shift the noise level from an average of 0 volts to an average of A volts. The random variable v has the same statistics as n, fluctuating about A, however, rather than 0. Its density function is the same gaussian function, with the same variance, but with an average value of A. We have

$$f_1(v) = \frac{1}{\sqrt{2\pi\sigma^2}} e^{-(v-A)^2/2\sigma^2} \qquad [5\text{-}92]$$

This is shown sketched in Fig. 5-26b. The probability of error now corresponds to the chance that the sample v of signal plus noise will drop below $A/2$ volts and be mistaken for noise only (or be judged, incorrectly, a 0). This is just the area under the $f_1(v)$ curve from $-\infty$ to $A/2$ and is given by

$$P_{e1} = \text{Prob}\left(v < \frac{A}{2}\right) = \int_{-\infty}^{A/2} f_1(v)\, dv \qquad [5\text{-}93]$$

This probability of error is indicated by the shaded area in Fig. 5-26*b*.

How do we now find the probability of error of the *system?* Note that the two possible types of error considered belong to *mutually exclusive* events; the 0 precludes a 1 appearing and vice versa. Probabilities can thus be added. However, in this case, it is apparent that P_{e0} and P_{e1} are both *conditional* probabilities, the first assuming a 0 present, the second a 1 present. To remove the conditioning we must multiply each by its appropriate a priori probability of occurrence. Thus, assuming the probability of transmitting a 0 is *known* to be P_0, while the probability of transmitting a 1 is *known* to be P_1, ($P_0 + P_1 = 1$), we have as the total error of the system

$$P_e = P_0 P_{e0} + P_1 P_{e1} \qquad [5\text{-}94]$$

It is apparent from Fig. 5-26 and from the symmetry of the gaussian curves that the two conditional probabilities P_{e0} and P_{e1} are equal in this example. (This may also be shown mathematically by a linear translation of coordinates, letting $x = v - A$ in Eq. [5-93], or by noting that the probability that noise alone will be less than $-A/2$ is the same as the probability that noise will exceed $A/2$.) If we also make the rather reasonable assumption that the two binary signals 0 and 1 are equally likely to occur, $P_0 = P_1 = \frac{1}{2}$, and we are left with the result that the total probability of error P_e is the same as P_{e0} or P_{e1}. It is left to the reader as an exercise to show that the probability of error is then simply given in terms of the error function as follows:

$$P_e = \frac{1}{2}\left(1 - \text{erf}\,\frac{A}{2\sqrt{2}\,\sigma}\right) \qquad \text{erf}\, x = \frac{2}{\sqrt{\pi}} \int_0^x e^{-y^2}\, dy \qquad [5\text{-}95]$$

With the 1's and 0's assumed equally likely in a long message, Eq. [5-95] gives the probability of an error in the decoding of any digit. Note that P_e depends solely on A/σ, the ratio of signal amplitude to the noise standard deviation. This latter quantity σ is commonly referred to as the *rms noise* (as noted earlier and as will be shown in the next chapter). The ratio A/σ is then the peak *signal-to-rms-noise ratio.*

The probability of error is shown plotted versus A/σ, in decibels, in Fig. 5-27. Note that for $A/\sigma = 7.4$ (17.4 db), P_e is 10^{-4}. This means that on the average 1 bit in 10^4 transmitted will be judged incorrectly. If 10^5 bits/sec are being transmitted, this means a mistake every 0.1 sec, on the average, which may not be satisfactory. However, if the signal is increased to $A/\sigma = 11.2$ (21 db), a change of 3.6 db, P_e decreases to

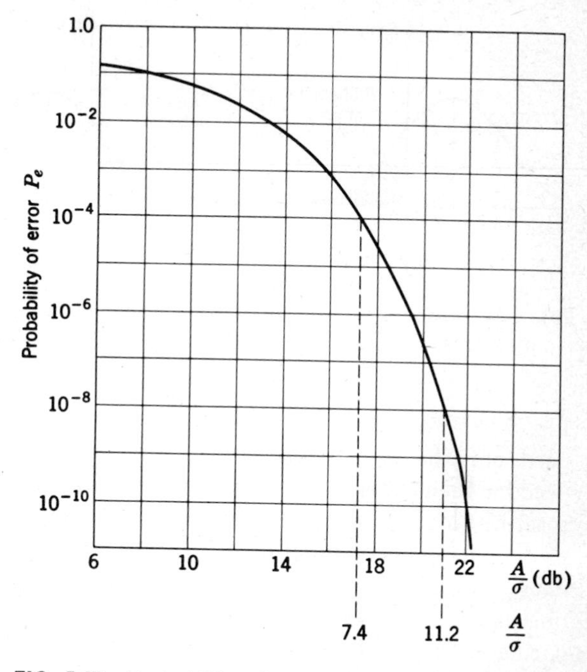

FIG. 5-27 Probability of error for binary detection in gaussian noise.

10^{-8}. For 10^5 bits/sec this means a mistake every 1,000 sec or 15 min on the average, which is much more likely to be tolerable. (In practice designers often use $P_e = 10^{-5}$ as a design goal for binary communication systems.)

Above $A/\sigma = 6$ or 16 db (approximately), the probability of error decreases very rapidly with small changes in signal. In the example just cited, increasing the signal by a factor of 3.6 db ($A/\sigma = 7.4$ to $A/\sigma = 11.2$) reduces the error rate by 10^4. The existence of a narrow range of signal-to-noise ratios above which the error rate is tolerable, below which errors occur quite frequently, is termed a *threshold effect*. The signal-to-noise ratio at which this effect takes place is called the threshold level. For the transmission of binary digits the threshold level may be chosen somewhere between $A/\sigma = 6$ to $A/\sigma = 8$ (16 to 18 db). Note that this does not imply complete noise suppression above the threshold level. It merely indicates that for pulse amplitudes greater than ten times the rms noise, say, errors in the transmission of binary digits will occur at a tolerable rate.

That the above error analysis for the transmission of on-off binary pulses holds true for bipolar pulses (positive and negative) is shown by

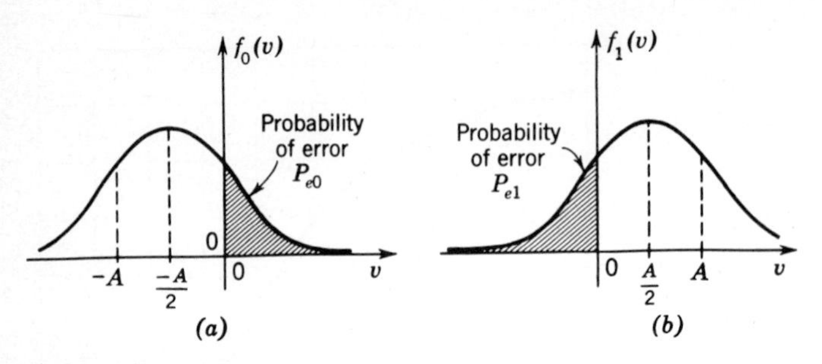

FIG. 5-28 Probability densities in the transmission of bipolar binary pulses. (a) Negative pulse transmitted; (b) positive pulse.

Fig. 5-28. We recall from Chap. 3 that the positive and negative pulses need only be transmitted at $A/2$ or $-A/2$ volts, respectively. The decoder then determines which pulse is present from the polarity of the total instantaneous voltage (signal plus noise). Figure 5-28*a* shows the probability-density function for the negative pulse of $-A/2$ volts plus noise. Figure 5-28*b* shows the corresponding curve for the positive pulse plus noise. The error probability is in each case the same, and, comparing with Fig. 5-26, the same as for the case of on-off pulses. The error curve of Fig. 5-27 thus applies to either type of binary-digit transmission. The bipolar signal requires only half the signal amplitude of the on-off signal, however (this was first noted in Sec. 3-7), or one-fourth of the peak power. (The *average* power is one-half that of the on-off signal, since that signal sequence is zero half the time, on the average.) It is thus apparent that a bipolar signal is preferable where possible.

Figure 5-27 was utilized in the discussion of PCM channel capacity in Sec. 3-7.

OPTIMUM DECISION LEVELS

In this discussion of the probability of error in binary transmission we have so far relied on more-or-less intuitive judgments in developing the expression for the probability of error. Thus we have assumed the signals to be equally likely to be transmitted, we have arbitrarily chosen as our decision level the value $A/2$ for the on-off pulse sequence, or 0 in the case of the bipolar sequence, etc. One may readily ask how significant are these assumptions? Is it possible to *decrease* the probability of error by another choice of decision level? Is there a *minimum* P_e one can attain? We shall consider these questions and others related to them ("best" choice of binary waveshapes, the effect of multiple sampling, extension to M-ary transmission, etc.) in a quite general way in Chap. 8.

At this point, however, we can say something about the possibility of decreasing P_e for this particular problem of binary transmission by appropriate choice of decision level. In particular, assume a bipolar sequence of pulses transmitted with gaussian noise added, so that we have a composite sample $v(t)$ of signal plus noise, as previously. How shall we make the decision about whether a 1 or a 0 was transmitted? In this communications problem it is apparent that a reasonable design criterion is that of minimizing the probability of error P_e. A system with minimum P_e is then optimum from our point of view.

Since the decoder can only base its decision on the voltage amplitude of the sample $v(t)$ taken, it is apparent that the only possible way to adjust P_e or to optimize the system is to vary the amplitude level at which the decision is made. Call this decision level d. It is apparent from Fig. 5-29, in which this level is shown superimposed on the prob-

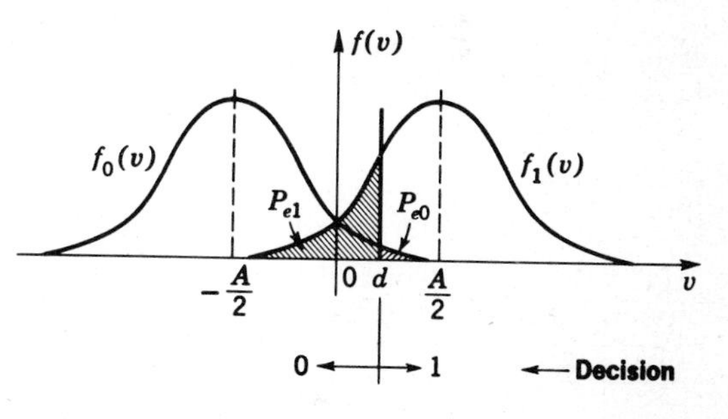

FIG. 5-29 Choice of decision level in binary transmission.

ability-density plots for bipolar transmission, that increasing d positively decreases the chance of mistaking a 0 for a 1 (P_{e0}) but at the cost of increasing P_{e1}. It is also apparent from the symmetry of the figure and the form of the gaussian functions shown that an optimum solution is $d = 0$, just our previous intuitive guess, *assuming 0's and 1's equally likely*. It is equally apparent that if for some reason 0's occur more often on the average ($P_0 > P_1$), one would tend to shift d positively. If on the other hand 1's tend to occur more often ($P_1 > P_0$), one would shift d negatively. The optimum choice of d thus depends on the a priori probabilities P_0 and P_1.

To make this discussion more quantitative we must return to our original formulation of the probability of error. From Eqs. [5-91], [5-93], and [5-94], we have, with d as an arbitrary decision level,

$$P_e = \underbrace{P_0 \int_d^{\infty} f_0(v)\,dv}_{P_{e0}} + \underbrace{P_1 \int_{-\infty}^{d} f_1(v)\,dv}_{P_{e1}} \qquad [5\text{-}96]$$

(Recall that the previous equations were written for the on-off signal case. Had they been written for the bipolar case, 0 would have been used in place of A/2.)

An optimum choice of d corresponds to minimum P_e, according to our criterion. Since P_e is a function of d in Eq. [5-96], we simply differentiate with respect to d to find the optimum level. In particular, we then have

$$\frac{\partial P_e}{\partial d} = 0 = -P_0 f_0(d) + P_1 f_1(d)$$

or

$$\frac{f_1(d)}{f_0(d)} = \frac{P_0}{P_1} \qquad [5\text{-}97]$$

invoking the usual rules of differentiation with respect to integrals.

It is thus apparent that the optimum d (in the sense of minimum error probability) depends on the form of the two conditional density functions [$f_0(v)$ and $f_1(v)$], as well as the a priori probabilities P_0 and P_1. If $P_0 = P_1 = \frac{1}{2}$, the optimum value of d is given by the point at which the two density functions intersect. For bipolar signals in gaussian noise this is just the point $d = 0$ (Fig. 5-29). For $P_0 \neq P_1$, the level shifts, as expected. Specifically, it is left for the reader to show that for the case of bipolar signals in additive gaussian noise (Fig. 5-29), the solution to Eq. [5-97] is given by

$$d_{\text{opt}} = \frac{\sigma^2}{A} \ln \frac{P_0}{P_1} \qquad [5\text{-}98]$$

As expected, d increases positively if $P_0 > P_1$, and negatively, if $P_1 > P_0$. The actual shift depends on the signal amplitude, noise variance, and P_0/P_1.

In practice this choice of optimum d is not very critical, and in the case of bipolar signals in additive gaussian noise one would normally pick $d = 0$ as the decision level. For one generally does not know the a priori probabilities accurately, and even if one did, the signal-to-noise ratio required for effective binary communication makes d_{opt} from Eq. [5-98] very close to zero. Specifically, if $A/\sigma = 8$, and $P_0/P_1 = 3$ (highly unlikely, since this requires $P_0 = \frac{3}{4}$, $P_1 = \frac{1}{4}$), then $d_{\text{opt}} = \sigma/8$. The optimum shift away from a $d = 0$ decision level is thus a fraction of the rms noise, an insignificant change. (Some thought would indicate that this corresponds approximately to changing the signal amplitude by the same amount, with the level held fixed. With the signal-to-noise ratio initially at 18 db, this represents a shift of approximately 0.13 db away! From Fig. 5-27 the change in P_e is not noticeable.)

If the optimization is inconsequential in this case, why discuss it? There are several reasons.

1. We at least know that our first, intuitive, guess was a valid one. The optimization procedure tells us that *in this case* there is no sense looking for an improvement in system performance by varying the threshold level. This is very often exactly the reason for attempting to carry out system optimizations in more sophisticated situations.
2. If the received signal amplitude (or noise variance for that matter) is not accurately known, or varies due to disturbances along the transmission channel, the probability of error will change as well. The sensitivity of P_e to changes in the amplitude A is obviously related to the sensitivity due to changes in d. (Alternately, one would approach the problem of sensitivity to changes in A in a way similar to that done here.)
3. If the statistics of the received signals plus noise are *not* gaussian, and do not have the nice symmetry of the density functions of Fig. 5-29, intuition fails in determining the desired decision level. Examples of such situations will be encountered later in this book in dealing with the noncoherent detection of binary signals.[1] One must thus resort to the solution of Eq. [5-97] or to the solution of equations like it. One simple pictorial example of two conditional density functions obtained under certain conditions in noncoherent detection is shown in Fig. 5-30. Here $P_0 = P_1 = \frac{1}{2}$ is assumed, so that the optimum decision level occurs at the intersection of the two conditional density functions $f_1(v)$ and $f_0(v)$.

[1] See M. Schwartz, W. R. Bennett, and S. Stein, "Communication Systems and Techniques," figs. 7-4-2 and 7-4-3, McGraw-Hill, New York, 1966, for the effect of threshold variation in on-off-keyed (OOK) signals with noncoherent detection.

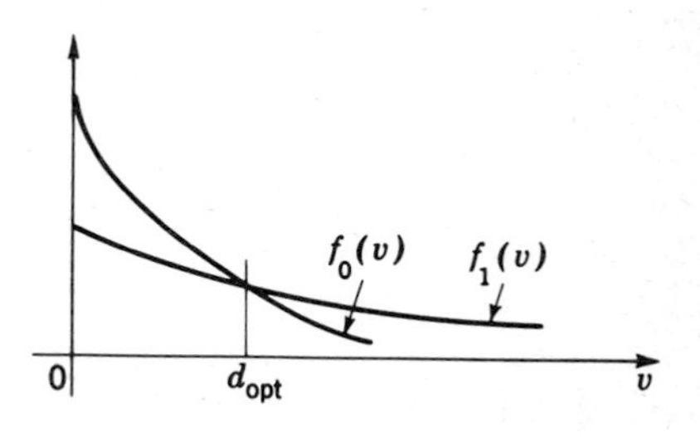

FIG. 5-30 Example of determination of optimum decision level ($P_0 = P_1 = \frac{1}{2}$).

4. This first discussion of optimum decision levels serves as a good introduction to the more general optimization problems to be considered in Chap. 8.

5-7 FUNCTIONAL TRANSFORMATIONS OF RANDOM VARIABLES

Up to this point in our discussion of random variables we have been assuming the probability density known. In the case of random noise this has been assumed to be the gaussian distribution.

We know from our discussion of modulation systems in Chaps. 3 and 4 that many operations, both linear and nonlinear, may be performed on a signal. Noise passing through a system is also operated on by modulators, rectifiers, amplifiers, filters, etc. It is of interest to investigate the effect of the system operation on the probability distribution. We shall restrict ourselves in this section primarily to nonlinear operations, with the output voltage y of a device related to the input v by the relation

$$y = G(v)$$

If v is a random variable with known probability density $f_v(v)$ (fluctuation noise of zero dc level, for example), what is the probability density $f_y(y)$ at the output?

As an example, assume that random noise is passed through a piecewise-linear rectifier of the type analyzed as a modulator and demodulator in Chap. 4. We assume that the characteristics are given by

$$\begin{aligned} y &= v \qquad v \geq 0 \\ y &= 0 \qquad v < 0 \end{aligned} \qquad [5\text{-}99]$$

What is the probability density $f_y(y)$ of the noise at the output (see Fig. 5-31a)?

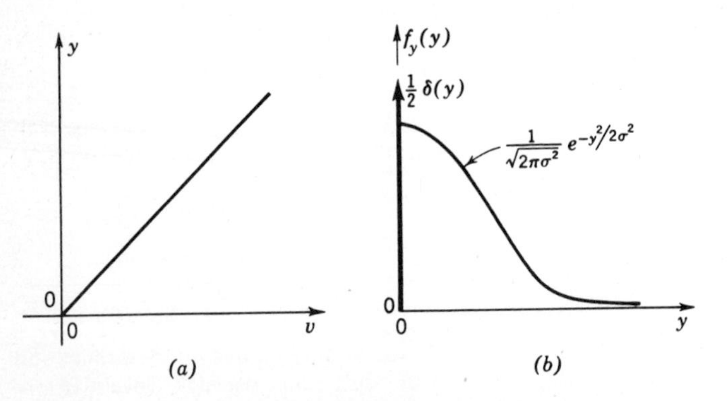

FIG. 5-31 Piecewise-linear rectification of noise. (a) Piecewise-linear rectifier; (b) probability density, output noise.

If the input noise happens to be of positive amplitude, it will be passed undisturbed through the rectifier. The probability that $y > 0$ must be the same as the probability that $v > 0$. For $v > 0$, then,

$$f_y(y) = f_v(v) = \frac{1}{\sqrt{2\pi\sigma^2}} e^{-y^2/2\sigma^2} \qquad v > 0 \qquad [5\text{-}100]$$

But for $v < 0$ the rectifier produces zero output ($y = 0$). The probability that $v < 0$ is 0.5, and this must correspond to the probability that $y = 0$. Thus

$$\text{Prob } (v < 0) = 0.5 = \text{Prob } (y = 0) \qquad [5\text{-}101]$$

The probability-density function corresponding to this must be a delta function of area ½ centered at $y = 0$. The total probability density for y is then

$$f_y(y) = \tfrac{1}{2}\delta(y) + \frac{1}{\sqrt{2\pi\sigma^2}} e^{-y^2/2\sigma^2} \qquad y \geq 0 \qquad [5\text{-}102]$$

Note that y has both a discrete and a continuous part. $f_y(y)$ is properly normalized, as can be seen by integrating $f_y(y)$ over all values of y (Fig. 5-31*b*).

We can use $f_y(y)$ to calculate the expected value or dc component of the noise at the output (now nonzero), the rms voltage, etc.[1] The dc voltage, for example, is given by

$$m_1 = E(y) = \int_0^\infty \frac{ye^{-y^2/2\sigma^2}}{\sqrt{2\pi\sigma^2}}\, dy + \int_{0-}^\infty y\,\delta(y)\, dy$$

$$= \frac{\sigma}{\sqrt{2\pi}} \qquad [5\text{-}103]$$

where σ is the rms voltage of the input noise.

If the transformation $y = G(v)$ can be expressed algebraically and the inverse $v = g(y)$ readily found, a simple relation between $f_y(y)$ and $f_v(v)$ can be developed.

Let v vary between v_1 and v_2. y varies correspondingly between $y_1 = G(v_1)$ and $y_2 = G(v_2)$. The probability that v will range between v_1 and v_2 is $\int_{v_1}^{v_2} f_v(v)\, dv$, and this must be just the probability that y will vary between y_1 and y_2. Thus

$$\int_{v_1}^{v_2} f_v(v)\, dv = \int_{y_1}^{y_2} f_y(y)\, dy \qquad [5\text{-}104]$$

[1] As indicated earlier we shall show in the next chapter that the first moment and the dc voltage are the same under some simple conditions. Similarly, the standard deviation and the rms voltage will be shown to be the same.

Now let $v = g(y)$ in the left-hand integral. Assuming that $g(y)$ is a *single-valued function* of y, we get, after transforming the integrand and changing the limits of integration,

$$\int_{v_1}^{v_2} f_v(v)\, dv = \int_{y_1}^{y_2} f_v[g(y)]g'(y)\, dy = \int_{y_1}^{y_2} f_y(y)\, dy \qquad [5\text{-}105]$$

Comparing the two integrals on the right, we get the relation

$$f_y(y) = f_v[g(y)]g'(y) \qquad [5\text{-}106]$$

Geometrically Eq. [5-104] corresponds to equating the two areas shown in Fig. 5-32. In particular, as $v_1 \to v_2$, we must have

$$f_y(y)\, dy = f_v(v)\, dv$$

which is just the relation given by Eq. [5-106].

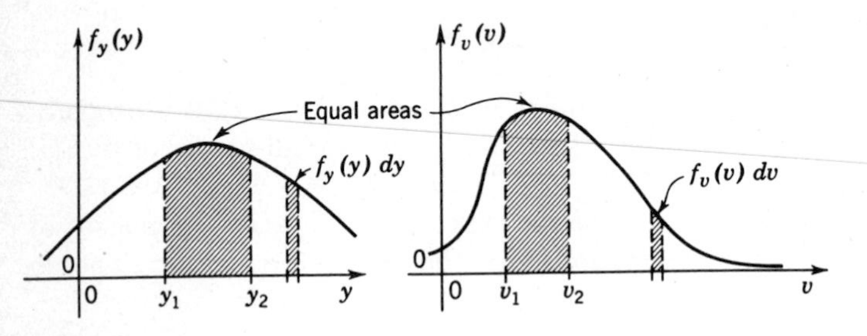

FIG. 5-32 Transformation of a random variable.

As an example, assume that v is uniformly distributed in the range $0 \leq v \leq 1$, with $f_v(v) = 1$. What is the distribution at the output of a square-law detector with the characteristic $y = v^2$?

Here $g(y) = \sqrt{y}$ (since v is positive, the positive square root must be used); $g'(y) = \frac{1}{2}y^{-\frac{1}{2}}$, and the density function at the output is

$$f_y(y) = \frac{y^{-\frac{1}{2}}}{2} \qquad 0 \leq y \leq 1 \qquad [5\text{-}107]$$

This is shown sketched in Fig. 5-33*c*. We can check to see whether or not $f_y(y)$ is properly normalized by calculating the area under the curve.

$$\int_0^1 f_y(y)\, dy = \frac{1}{2}\int_0^1 y^{-\frac{1}{2}}\, dy = 1$$

as expected.

The cumulative-distribution function is given by

$$F_y(y) = \int_0^y f_y(y)\, dy = y^{\frac{1}{2}} \qquad [5\text{-}108]$$

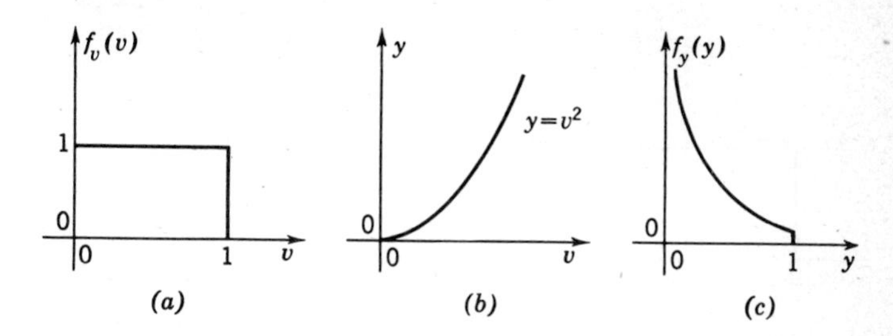

FIG. 5-33 Example of a transformation. (a) Input variable; (b) characteristic; (c) output variable.

In particular the probability that the output will be less than ½ is $F_y(\frac{1}{2}) = 0.707$.

The square-law transformation is also interesting because it gives us the density function of the instantaneous *power* if the input function v happens to be a voltage or a current applied to a 1-ohm resistor. For example, in the example just given, the probability that the voltage will be less than 0.707 is just 0.707 (uniform or rectangular distribution). This must correspond to the probability that the power will be less than ½, checking the answer obtained.

As another simple example of the transformation, assume $f_v(v) = e^{-v}$, $0 \leq v < \infty$. (The properties of this density function are left to the reader as an exercise.) What is the density function of $y = v^2$? (This again corresponds to passing the variable v through a square-law detector, or to finding the probability distribution of the power.)

Here $v = g(y) = y^{\frac{1}{2}}$ again, $f_v[g(y)] = e^{-\sqrt{y}}$, $g'(y) = \frac{1}{2}y^{-\frac{1}{2}}$, and

$$f_y(y) = \frac{e^{-\sqrt{y}}}{2\sqrt{y}} \qquad 0 \leq y < \infty \tag{5-109}$$

As a third example, assume that an angular variable θ is uniformly distributed between the values $-\pi/2 \leq \theta \leq \pi/2$. Then $f_\theta(\theta) = 1/\pi$ in that range. We would like to find the distribution of $y = a \sin \theta$. (This corresponds to finding the density function of a sine wave whose angle is uniformly distributed.)

Here $g(y) = \sin^{-1}(y/a)$, $g'(y) = 1/\sqrt{a^2 - y^2}$, and

$$f_y(y) = \frac{1}{\pi\sqrt{a^2 - y^2}} \qquad -a \leq y \leq a \tag{5-110}$$

$f_y(y)$ is shown sketched in Fig. 5-34*b*. This result is as might be intuitively expected, for the sine function spends most of its time in the region of $0 = \pi/2$ or $-\pi/2$ (where its derivative or rate of change has the smallest

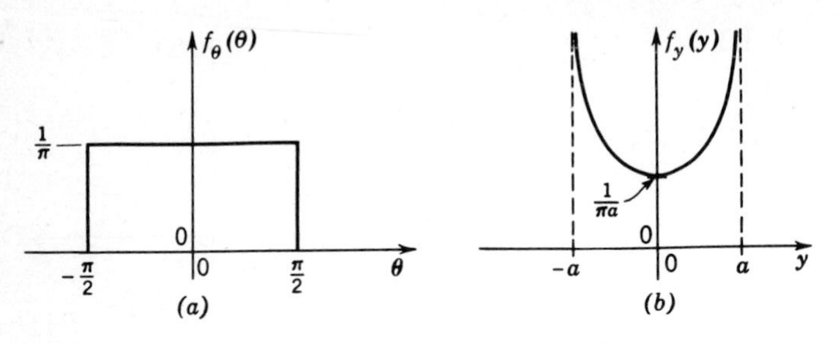

FIG. 5-34 Phase uniformly distributed. (a) $f_\theta(\theta)$; (b) $f_y(y)$: $y = a \sin \theta$.

value). y would thus be expected to have the largest probability density in the vicinity of a and $-a$.

In all three examples cited thus far $v = g(y)$ has been single-valued because of the range of v chosen.

As a final example, consider the case of gaussian-distributed noise passed through a square-law detector. Alternatively we may ask for the density function of the instantaneous power corresponding to a gaussian-distributed noise voltage. Thus with

$$f_v(v) = \frac{e^{-v^2/2\sigma^2}}{\sqrt{2\pi\sigma^2}} \qquad -\infty < v < \infty$$

and

$$y = v^2$$

find $f_y(y)$.

Here we have to be very careful. Both positive and negative v contribute to y, and $v = g(y) = \pm\sqrt{y}$ is no longer single-valued. To treat this case, assume that we are interested in the probability that y lies between two values y_1 and y_2. Since *two* sets of values of v give rise to these values of y and the values of v are *mutually exclusive*, we must sum the probabilities that v lies in the two different regions, to find

$$\text{Prob}\,(y_1 \le y \le y_2)$$

Calling the two sets of values of v, v_1, v_2 and v_1', v_2',

$$\text{Prob}\,(y_1 \le y \le y_2) = \text{Prob}\,(v_1 \le v \le v_2) + \text{Prob}\,(v_1' \le v \le v_2')$$

In terms of the density functions we get

$$\int_{y_1}^{y_2} f_y(y)\,dy = \int_{v_1}^{v_2} f_v(v)\,dv + \int_{v_1'}^{v_2'} f_v(v)\,dv \qquad [5\text{-}111]$$

for a double-valued function.

The gaussian function happens to be symmetrical about $v = 0$, so that $f_v(v)$ is the same for the positive and negative values of v that produce

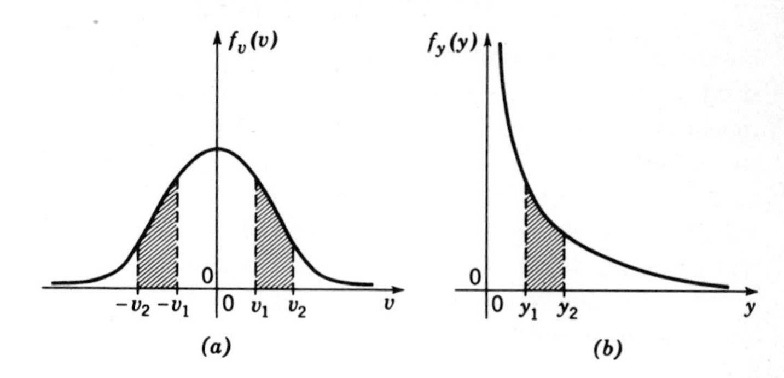

FIG. 5-35 Gaussian noise at output of square-law device. (a) Input distribution; (b) output distribution.

the same y. This is shown in Fig. 5-35*a*. In this case we get

$$f_y(y) = 2f_v[g(y)]g'(y) \tag{5-112}$$

from Eq. [5-111].

In particular, for $y = v^2$, $g'(y) = 1/2\sqrt{y}$, and

$$f_y(y) = \frac{e^{-y/2\sigma^2}}{\sqrt{2\pi\sigma^2 y}} \qquad 0 \leq y < \infty \tag{5-113}$$

is the distribution of the noise power, or the distribution at the output of a square-law device.

As a check, it is left to the reader to show that the average value of y is $m_1 = \sigma^2$. Since the input noise voltage has zero average value, its second moment, or total mean power, is just σ^2. (For the voltage, $m_2 = m_1 + \sigma^2 = \sigma^2$ in this case.) The second moment of the voltage is thus just the first moment of the power, as might be expected: the average value of the instantaneous power corresponds to the mean-squared voltage across a 1-ohm resistor.

5-8 DISTRIBUTION OF n INDEPENDENT VARIABLES

Up to this point we have been considering probability-density functions and cumulative distributions for one variable [for example, $f_x(x)$ and $F_x(x)$]. Just as in the case of discrete variables it is of interest to extend the notions of continuous probability functions to include the probability of joint occurrence of two or more continuous independent random variables.

This will enable us to then discuss the probability distribution of the sum of many independent random variables. Summation occurs commonly in many communication systems so that this problem is of substantial interest in its own right. As an example signal pulses may be repeated several times to enable signals to be better detected in the presence of noise. (A specific application of this technique will be considered in the next chapter in discussing radar systems. Multiple signal samples will also be discussed in Chap. 8.) The summation of many noise samples will be encountered in discussing PCM repeaters in the next section.

In addition we shall show by example that the sum of many independent random variables approaches a gaussian-distributed variable under some rather simple conditions. This is the so-called *central-limit theorem* of probability theory. It provides the motivation for spending so much time on gaussian statistics, as well as the justification for assuming noise statistics to be gaussian. This theorem will in fact be utilized in a later section in discussing signal-fading phenomena and transmission through random channels.

TWO INDEPENDENT VARIABLES

We recall that in Sec. 5-2 we introduced the concept of joint probability for discrete variables by defining a probability $P(AB)$ that both event A *and* event B occur. For example, considering playing cards, if A represents a heart, B a king, $P(AB)$ represents the probability of the king of hearts being drawn. If A represents a spade on the first of two card draws and B a spade on the second draw, $P(AB)$ is the probability of drawing two spades in succession.

If the event B is dependent on the event A, as is true in the two-spade case,

$$P(AB) = P(A)P(B|A)$$

where $P(B|A)$ is the conditional probability that event B will occur, it being known that A has occurred.

In the two-spade case, for example,

$$P(A) = 13/52 = 1/4 \qquad P(B|A) = 12/51 \qquad P(AB) = 1/17$$

We also showed, in Sec. 5-2, that if B is independent of event A, and vice versa,

$$\begin{aligned} P(B|A) &= P(B) \\ P(A|B) &= P(A) \\ P(AB) &= P(A)P(B) \end{aligned}$$

In the example given this would correspond to replacing the first card once drawn.

By extension to the continuous-probability case we can express the probability that two variables x and y will jointly take on values between

x_1 and x_2, y_1 and y_2, respectively, by the expression

$$\text{Prob}\,(x_1 \leq x \leq x_2, y_1 \leq y \leq y_2) = \int_{x_1}^{x_2} \int_{y_1}^{y_2} f_{xy}(x,y)\,dx\,dy \quad [5\text{-}114]$$

$f_{xy}(x,y)$ is a *two-dimensional* probability density, and the joint probability is the *volume* enclosed under the $f_{xy}(x,y)$ curve in the region bounded by x_1, x_2, y_1, y_2. This compares with the one-dimensional density function $f_x(x)$ for a single continuous variable, with the probability given by an area under the curve. $f_{xy}(x,y)\,dx\,dy$ represents the probability that x and y will lie jointly in the region

$$x \leq x \leq x + dx$$
$$y \leq y \leq y + dy$$

As an example of the need for the joint-probability formulation occurring in our work, we may ask for the probability that a noise voltage $n(t)$ will at time t appear between n_1 and $n_1 + dn$ volts and, τ sec later, between n_2 and $n_2 + dn$ volts. $n(t)$ then corresponds to x, $n(t + \tau)$ to y (Fig. 5-36). We shall discuss this specific problem in a later section.

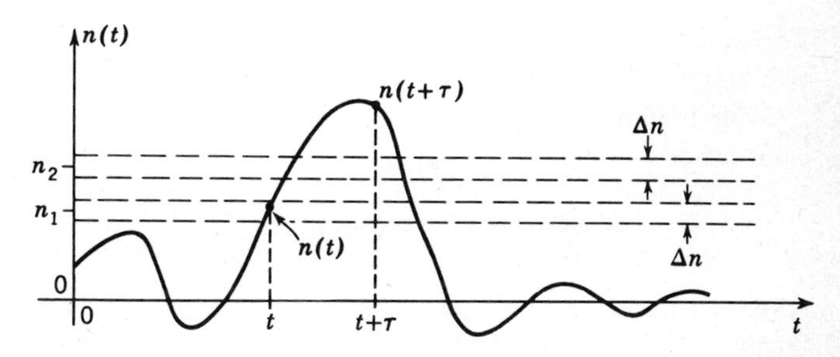

FIG. 5-36 Joint probability, noise voltage.

The two-dimensional cumulative distribution also becomes a volume integral and is given by

$$F_{xy}(x_1,y_1) = \int_{-\infty}^{x_1} \int_{-\infty}^{y_1} f_{xy}(x,y)\,dx\,dy \quad [5\text{-}115]$$

This represents the probability that x will be less than x_1, y less than y_1.

f_{xy} can be found from the limiting case of a relative-frequency expression, just as in the case of a single variable. For example, we could cut up the noise-voltage curve of Fig. 5-36 into many strips T sec long ($T \gg 1/\text{bandwidth}$), divide the voltage scale into many levels Δn volts wide, and sample the voltage at fixed times t sec and $(t + \tau)$ sec from the beginning of each strip. The fraction of times that the voltages

$n_1 \pm \Delta n/2$ at t and $n_2 \pm \Delta n/2$ at $t + \tau$ appeared would give us an expression approximating $f_{n_1 n_2}\, dn_1\, dn_2$. This procedure is of course identical with the method used in Sec. 5-2 to find joint-probability expressions from relative-frequency counts and is just an extension of the histogram idea to two variables.

Consider now the case where the two variables x and y are independent. (As will be shown in the next chapter, in the case of Fig. 5-36 this corresponds to $\tau \gg 1/\text{bandwidth}$.) By direct extension of our results for the discrete case we define the condition of independence to be

$$f_{xy}(x,y) = f_x(x) f_y(x) \qquad [5\text{-}116]$$

The joint-probability density of two independent variables is then just the product of the individual probability-density functions, and the individual probabilities multiply. The probability that x will range between x_1 and x_2 and y between y_1 and y_2 is

$$\begin{aligned}\text{Prob}\,(x_1 \le x \le x_2, y_1 \le y \le y_2) &= \int_{x_1}^{x_2} \int_{y_1}^{y_2} f_x(x) f_y(y)\, dx\, dy \\ &= \int_{x_1}^{x_2} f_x(x)\, dx \int_{y_1}^{y_2} f_y(y)\, dy \qquad [5\text{-}117]\end{aligned}$$

For example, assume that the instantaneous noise voltage across each of two resistors R_A and R_B is measured. Assuming the voltage across each resistor is gaussian-distributed and the two voltages are independent of each other (see Chap. 7),

$$f_{A,B}(n_A,n_B) = f_A(n_A) f_B(n_B) = \frac{e^{-(n_A{}^2/2\sigma_A{}^2 + n_B{}^2/2\sigma_B{}^2)}}{\sqrt{2\pi\sigma_A{}^2 \times 2\pi\sigma_B{}^2}} \qquad [5\text{-}118]$$

where σ_A and σ_B are the rms noise voltages of resistors A and B, respectively. The probability that voltmeter A across R_A will read between 2 and 3 volts and voltmeter B across R_B between 1 and 3 volts, for example, is found by integrating Eq. [5-118] between these sets of limits or by finding the individual probabilities and multiplying them together.

SUM OF INDEPENDENT VARIABLES: CHARACTERISTIC FUNCTIONS

As a special case of the handling of many independent random variables we now consider the situation where two or more variables are summed to find a new random variable. This, as noted earlier, has applicability to many communication problems. The techniques to be discussed also lead to the central-limit theorem mentioned earlier that the distribution of the sum of n independent variables approaches the gaussian (normal) distribution.

We start first by considering two random variables as in the previous paragraph. We then generalize to n independent variables. Consider then a variable $z = x + y$. With $f_x(x)$ and $f_y(y)$ given and x and y independent how do we find $f_z(z)$?

The most direct approach is that of performing a transformation of variables just as in the previous section. We therefore define two new variables given by

$$z = x + y \qquad [5\text{-}119]$$

and

$$\zeta = x \qquad [5\text{-}120]$$

The probability that x and y lie within a specified region must equal the probability that z and ζ lie within the corresponding region found from Eqs. [5-119] and [5-120]. Thus

$$\int_z \int_\zeta f_{z\zeta}(z,\zeta)\, dz\, d\zeta = \int_x \int_y f_x(x) f_y(y)\, dx\, dy \qquad [5\text{-}121]$$

Substituting Eqs. [5-119] and [5-120] into the right-hand side of Eq. [5-121], we obtain

$$\int_z \int_\zeta f_{z\zeta}(z,\zeta)\, dz\, d\zeta = \int_z \int_\zeta f_x(\zeta) f_y(z - \zeta)\, dz\, d\zeta \qquad [5\text{-}122]$$

and

$$f_{z\zeta}(z,\zeta) = f_x(\zeta) f_y(z - \zeta) \qquad [5\text{-}123]$$

The distribution for $f_z(z)$ is then found by integrating $f_{z\zeta}(z,\zeta)$ over all values of ζ. Since ζ is exactly x,

$$f_z(z) = \int_{-\infty}^{\infty} f_x(x) f_y(z - x)\, dx \qquad [5\text{-}124]$$

Note that Eq. [5-124] appears in the form of the convolution integral of Chap. 2. This implies that if we arbitrarily take the Fourier transforms of $f_x(x)$ and $f_y(y)$, the Fourier transform of $f_z(z)$ must be given by the product of the two transforms. This then gives us another (and frequently simpler) method of finding $f_z(z)$. Calling $G_x(t)$ the Fourier transform of $f_x(x)$ (t a new parameter), and $G_y(t)$ the corresponding transform of $f_y(y)$, we have

$$G_x(t) = \int_{-\infty}^{\infty} e^{jtx} f_x(x)\, dx \qquad [5\text{-}125]$$

$$G_y(t) = \int_{-\infty}^{\infty} e^{jty} f_y(y)\, dy \qquad [5\text{-}126]$$

(We use e^{+jtx} rather than e^{-jtx}, as in Chap. 2, to conform with the notation of probability theory.) These new functions, Fourier transforms of the probability-density functions, are called *characteristic functions*. The characteristic function $G_z(t)$ of the variable z is then

$$G_z(t) = G_x(t) G_y(t) \qquad [5\text{-}127]$$

and

$$f_z(z) = \frac{1}{2\pi} \int_{-\infty}^{\infty} e^{-jtz} G_z(t)\, dt \qquad [5\text{-}128]$$

from Fourier-integral theory. (Note that the $1/2\pi$ now appears in front of the e^{-jtz} term rather than the e^{jtz} term as in Chap. 2.)

Recalling the definition of the average of a function $g(y)$ of a random variable y,

$$E[g(y)] = \int_{-\infty}^{\infty} g(y) f_y(y)\, dy \qquad [5\text{-}129]$$

we see that the characteristic function is also the average of e^{jty}.

$$G_y(t) = E(e^{jty}) \qquad [5\text{-}130]$$

We can now extend this procedure to the sum of n independent variables. For $z = x_1 + x_2 + \cdots + x_n$ we have

$$\begin{aligned} G_z(t) &= E(e^{jtz}) = E(e^{jt(x_1+x_2+\cdots+x_n)}) \\ &= E(e^{jtx_1})E(e^{jtx_2}) \cdots E(e^{jtx_n}) \\ &= G_{x1}(t)G_{x2}(t) \cdots G_{xn}(t) \end{aligned} \qquad [5\text{-}131]$$

(The average of the product of n independent quantities is the product of the averages.) The probability density of z is then

$$f_z(z) = \frac{1}{2\pi}\int_{-\infty}^{\infty} e^{-jzt}G_z(t)\, dt \qquad [5\text{-}132]$$

with $G_z(t)$ given by Eq. [5-131].

Some examples will demonstrate the utility of the characteristic-function (c-f) method. They will also serve to illustrate some special cases of the central-limit theorem.

Sum of n gaussian-distributed variables

Here $z = \sum_{i=1}^{n} x_i$,

$$f_{x_i}(x_i) = \frac{1}{\sqrt{2\pi\sigma_i^2}} e^{-(x_i-a_i)^2/2\sigma_i^2} \qquad [5\text{-}133]$$

The Fourier transform of this gaussian function is itself a gaussian function. In particular it is left for the reader to show that

$$G_i(t) = e^{ja_it}e^{-\sigma_i^2t^2/2} \qquad [5\text{-}134]$$

Then

$$\begin{aligned} G_z(t) &= \prod_{i=1}^{n} G_i(t) \\ &= e^{jat}e^{-\sigma^2t^2/2} \end{aligned} \qquad [5\text{-}135]$$

where

$$a = \sum_{i=1}^{n} a_i$$

and

$$\sigma^2 = \sum_{i=1}^{n} \sigma_i^2$$

Comparing Eqs. [5-135] and [5-134],

$$f_z(z) = \frac{e^{-(z-a)^2/2\sigma^2}}{\sqrt{2\pi\sigma^2}} \qquad [5\text{-}136]$$

The distribution of the sum of n gaussian-distributed variables is thus also gaussian, with an average value given by the sum of the individual average values and a variance given by the sum of the variances.

As a matter of fact it can be shown that the average value and variance of a random variable

$$z = \sum_{i=1}^{n} x_i$$

with the x_i's independent and of any distribution whatsoever, are given by the sum of the individual average values and variances, respectively.[1] The average fluctuation powers of n independent variables thus add directly. (Here fluctuation refers to variations about the average values.)

Uniform distribution (Fig. 5-21)[2] For simplicity's sake we take the special case of a symmetrical uniform distribution. Letting $a = -b = -x_0/2$ in Fig. 5-21,

$$\begin{aligned} f_{x_i}(x_i) &= \frac{1}{x_0} \qquad \frac{-x_0}{2} < x_i < \frac{x_0}{2} \\ f_{x_i}(x_i) &= 0 \qquad \text{elsewhere} \end{aligned} \qquad [5\text{-}137]$$

This is just the rectangular pulse of Chap. 2, and its Fourier transform is the familiar $(\sin x)/x$ function.

$$G_i(t) = \frac{\sin (tx_0/2)}{tx_0/2} \qquad [5\text{-}138]$$

For the sum of n such variables, each assumed uniformly distributed over the *same range*,

$$f_z(z) = \frac{1}{2\pi} \int_{-\infty}^{\infty} \left[\frac{\sin (tx_0/2)}{tx_0/2} \right]^n e^{-jtz}\, dt \qquad [5\text{-}139]$$

[1] This is readily shown by taking the expectation of both left- and right-hand sides of $z = \sum_{i=1}^{n} x_i$. Then $E(z) = \sum_i E(x_i)$. Similarly, writing $[z - E(z)]^2 = \left\{\sum_i [x_i - E(x_i)]\right\}^2$, taking the expectation, and recalling that the x_i's are independent, one finds $\sigma_z^2 = \sum_i \sigma_{x_i}^2$. The details are left for the reader to work out.

[2] W. R. Bennett, Methods of Solving Noise Problems, *Proc. IRE*, vol. 44, no. 5, pp. 609–638, May, 1956.

Consider now the characteristic function

$$G_z(t) = \left[\frac{\sin (tx_0/2)}{tx_0/2}\right]^n$$

This has the value unity at $t = 0$ and damps out rapidly away from the origin for large n. Most of the contribution to the integral thus comes from small values of t. We can expand this function in a power series about $t = 0$ and get

$$\begin{aligned} G_z(t) &= \left[\frac{\sin (tx_0/2)}{tx_0/2}\right]^n = \frac{[tx_0/2 - (tx_0)^3/48 + \cdots]^n}{(tx_0/2)^n} \\ &= \left[1 - \frac{(tx_0)^2}{24} + \cdots\right]^n \\ &= 1 - \frac{n}{24}(tx_0)^2 + \cdots \end{aligned} \tag{5-140}$$

The first two terms in this power series are the same as those in the series for the exponential

$$e^{-(n/24)(tx_0)^2} = 1 - \frac{n}{24}(tx_0)^2 + \cdots$$

Using the exponential as an approximation to the $[(\sin x)/x]^n$ function in the vicinity of the origin, we have

$$G_z(t) = e^{-(n/24)(tx_0)^2} \qquad n \text{ large, } t \text{ small} \tag{5-141}$$

But this is just a gaussian-type function, and its Fourier transform must also be gaussian. In particular, comparing with Eqs. [5-135] and [5-136], we must have

$$f_z(z) \doteq \frac{1}{\sqrt{2\pi\sigma^2}} e^{-z^2/2\sigma^2} \tag{5-142}$$

where

$$\sigma^2 = \frac{n(x_0)^2}{12} \tag{5-143}$$

We showed previously that the standard deviation of the rectangular distribution was $(b - a)/\sqrt{12} = x_0/\sqrt{12}$ in the present notation. σ^2 here is then just the *sum* of the individual variances as noted in the previous example.

The gaussian approximation is plainly incorrect for large values of z. z can have no values greater than $nx_0/2$; yet the gaussian distribution predicts a finite probability of such values being attained. For large n, however, these values move out to the far tail of the gaussian curve, and the gaussian approximation becomes valid over most of the range of z. In general, just as in the cases of the distribution of the envelope or the phase of signal plus noise, the gaussian approximation is valid only in the vicinity of the peak ($z = 0$ here) and definitely not valid for values of z close to $nx_0/2$.

As noted by Bennett[1] this result for the sum of n uniformly distributed variables may be applied to the case of determining the quantizing noise resulting when a number of alternate analog-to-digital and digital-to-analog signal conversions are performed.

Both this example and the previous one are special cases of the central-limit theorem.

Random-phase distributions The final illustrative example concerns the sum of n independent sine waves of random phase. Thus, given

$$x_i = a_i \sin \theta_i$$

with θ_i uniformly distributed over 2π radians and a_i an arbitrary constant, we would like to find $f_z(z)$ for

$$z = \sum_{i=1}^{n} x_i$$

(This is sometimes used as a model for random noise.)

We recall that as an example of the calculation of the probability density of a transformed variable, we showed that the density function of x_i was

$$f_{x_i}(x_i) = \frac{1}{\pi} \frac{1}{\sqrt{a_i - x_i^2}} \qquad |x_i| < a_i \qquad [5\text{-}144]$$

The characteristic function $G_i(t)$ is just

$$G_i(t) = \int_{-a_i}^{a_i} \frac{e^{jtx_i}}{\sqrt{a_i^2 - x_i^2}} \frac{dx_i}{\pi}$$
$$= J_0(a_i t) \qquad [5\text{-}145]$$

with $J_0(a_i t)$ the Bessel function of the first kind and zeroth order. This may be shown by the simple change of variables $x_i = a_i \sin \theta$, from which we obtain the integral definition of the Bessel function given as Eq. [4-66].

Alternatively, and much more simply in this case, we can use the fact that $G_i(t) = E(e^{jtx_i})$ directly. Instead of averaging over x_i we can also write $G_i(t) = E(e^{jta_i \sin \theta_i})$ and average over all values of θ_i. Since θ_i is uniformly distributed, this gives

$$G_i(t) = \frac{1}{2\pi} \int_{-\pi}^{\pi} e^{jta_i \sin \theta_i} \, d\theta_i = J_0(a_i t) \qquad [5\text{-}146]$$

[1] Bennett, *op. cit.*

The characteristic function $G_z(t)$, for the sum of the n variables, is given by

$$G_z(t) = \prod_{i=1}^{n} J_0(a_i t) \qquad [5\text{-}147]$$

Now assume, as in the previous example, that the amplitudes a_i are all very small and t not too large. For $a_i t \ll 1$ we had, in Chap. 4,

$$J_0(a_i t) \doteq 1 - \tfrac{1}{4}(a_i t)^2 = e^{-\frac{1}{4}a_i^2 t^2} \qquad [5\text{-}148]$$

using the exponential approximation again. $G_z(t)$ now becomes, from Eq. [5-147],

$$G_z(t) \doteq e^{-\frac{1}{2}\sigma^2 t^2} \qquad [5\text{-}149]$$

where $$\sigma^2 = \tfrac{1}{2}(a_1^2 + a_2^2 + \cdots + a_n^2) \qquad [5\text{-}150]$$

Equation [5-149] represents the characteristic function of a normal distribution with zero average value and variance σ^2.

$$f_z(z) = \frac{1}{\sqrt{2\pi\sigma^2}} e^{-z^2/2\sigma^2} \qquad [5\text{-}151]$$

where $$\sigma^2 = \tfrac{1}{2}(a_1^2 + a_2^2 + \cdots + a_n^2)$$

Note that the term $a_i^2/2$ represents the average power in a sine wave of amplitude a_i. σ^2 is thus the total mean power in the sum of n sine waves of incommensurable frequencies.

Equation [5-151] is again an example of the central-limit theorem. Again care must be exercised in applying Eq. [5-151] to the calculation of probabilities. It is valid only for large n and in the vicinity of $z = 0$. It is again obviously incorrect for

$$z > \sum_{i=1}^{n} a_i$$

since the probability of this happening should be zero, while Eq. [5-151] predicts a finite probability. However, for n large, $f_z(z)$ falls off extremely rapidly at large values of z, and any errors become negligible if we stay away from the tail of the $f_z(z)$ curve.

5-9 APPLICATION TO PCM REPEATERS

The discussion of the sum of independent random variables, together with the material in Sec. 5-11 covering dependent random variables, concludes our formal presentation in this chapter of aspects of probability theory. The interested reader is referred to the references cited at the beginning of this chapter for more detailed treatments of probability and random variables. In this section and the one following, we apply some of the results obtained thus far to problems of interest to communications. In this section we extend the PCM error-probability analysis to the very practical problem of PCM repeaters. In the next section we discuss so-called fading channels and *multiplicative noise* effects obtained on these.

In Chap. 6 we continue the discussion by extending the concept of random variables to random or stochastic processes. This will then enable us to discuss the properties of random signals or noise passed through communication systems.

In discussing PCM systems in Chap. 3 we noted several times in passing that one significant reason for using all-digital signal transmission in communications is that the digital signals lend themselves so nicely to periodic conditioning and reshaping. Thus all communication channels serve to attenuate and distort signals passing through them. To ensure satisfactory reception at the final destination repeaters are commonly provided at appropriate spacings along the signal transmission route. These provide the necessary amplification and correction of distortion (by appropriately designed filters) to enable signal recognition to be satisfactorily carried out at the receiver.

Such repeaters are commonly used in cable and wire transmission, underwater cable transmission, microwave communication links, satellite communications, etc.[1] The number of repeaters and the spacing between them depends of course on the type of transmission medium used, its attenuation and phase distortion per unit length, the total transmission path length, carrier frequency used, etc. Typical attenuations range from a few tenths of a decibel per mile for wire lines and loaded cables in the voice frequency range to several decibels per mile at higher frequencies.[2] Phase shifts encountered also depend on the particular medium and frequency of transmission and vary roughly from a few tenths of a radian per mile to several radians per mile.

Although repeaters are used with all types of signals—analog and digital—we shall concentrate here on binary transmission only. We shall also ignore the modulation and demodulation processes necessary to get the signals on line (see Chap. 4), and will assume baseband transmission throughout.

Consider then a sequence of binary pulses transmitted along a line. The effect of the line may be characterized as equivalent to passing the signals through a linear network with frequency transfer function $H(\omega)$.

[1] See, e.g., the references to repeaters in D. H. Hamsher (ed.), "Communication System Engineering Handbook," McGraw-Hill, New York, 1967.

[2] *Ibid.*

The signals thus undergo amplitude and phase distortion, and filter networks, commonly called *equalizers*, must be introduced to compensate for the spectral distortion. In addition, noise and interference are introduced during transmission.

It is the introduction of the noise that poses a significant problem. For although the signal distortion may, at least in theory, be equalized to as fine a degree as required, the noise ultimately provides a limit on signal detectability. For over and above the relative amplitude-frequency distortion introduced by the line there is an overall attenuation of the signal level. If this level is allowed to drop too low, errors due to noise begin to limit the system transmission capability.

One may of course provide the proper amplification to bring the signal amplitude back up to the desired level, but this results in noise amplification as well. The only solution is to keep the signal level from dropping too low with respect to the noise. This therefore implies repeatedly conditioning the signal, at intervals along the line small enough to keep the attenuated signal well above the noise level. This provides the justification for the repeaters. In addition, the frequency distortion introduced by a short section of line is obviously less than that produced by a long line, so that amplitude and phase equalization is more readily carried out at each repeater, rather than at the ultimate destination.

Although repeaters are used to equalize and retime digital pulses, as well as to keep noise from accumulating during transmission, the discussion here will focus on the effects of noise only. This is admittedly an artificial situation, but the emphasis here is on the application of probability concepts to a communication problem, rather than on a discussion of repeaters per se.[1]

A little thought will indicate that two different classes of repeaters may be used. One employs straight amplification plus associated filtering to recoup the signal attenuation and compensate for the distortion. This is obviously the type of repeater that would be used for analog signal transmission. The second type, particularly appropriate for digital signal transmission, in essence makes a decision as to the binary symbol being transmitted (or decides on the appropriate level in the case of multilevel digital transmission), and then sends out a new, clean, noise-free binary symbol for further transmission down the line.

[1] As noted in Sec. 3-9 the intersymbol interference introduced by line distortion often poses more of a problem than noise. This is particularly true of many telephone systems. The reader is referred to W. R. Bennett and J. R. Davey, "Data Transmission," McGraw-Hill, New York, 1965, and R. W. Lucky, J. Salz, and E. J. Weldon, "Principles of Data Communication," McGraw-Hill, New York, 1968, for detailed discussions of telephone circuit equalization.

Which scheme is to be preferred for digital transmission? In the first scheme, involving amplifiers regularly spaced along the complete circuit, noise as well as signal is transmitted along the line, and in fact additional noise is introduced on each line section. A multilink line incorporating amplifiers is sketched in Fig. 5-37*a*. In the case of repeater signal

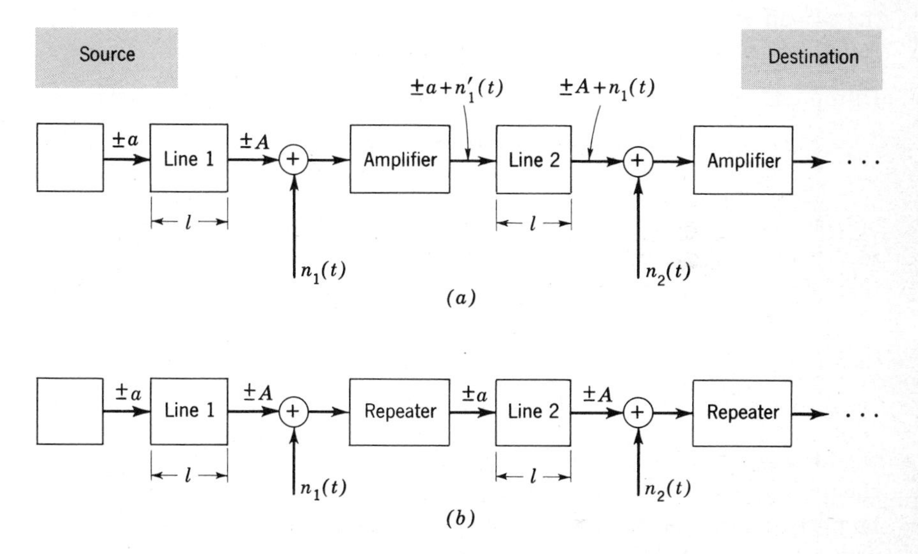

FIG. 5-37 Repeaters for PCM transmission: $A^2 = a^2\epsilon^{-\alpha l}$. (a) Repeater amplifiers; (b) repeater conditioners.

conditioners, or *regenerative* repeaters as they are called (Fig. 5-37*b*), noise is effectively eliminated at each repeater, and is not carried along from repeater to repeater. But in the process of making a decision at each repeater, errors may be made, and these may be propagated along the entire line. We shall now show with some simple calculations that for relatively low-noise (small probability of error) systems this latter scheme is nonetheless to be preferred in detecting binary signals in noise.

Assume the binary signals are bipolar, with transmitted amplitude $\pm a$ or transmitted power a^2. The individual lines, between repeaters, are each assumed to introduce a power attenuation α nepers/unit length. For a link length l, then, the signal power at the link output is $a^2e^{-\alpha l}$. The corresponding (attenuated) signal amplitude is A.

Assume further that gaussian noise, with variance σ^2, is introduced on each line link, the different noises being independent of one another. At the input to the first repeater, then, the composite signal plus noise[1] is

$$\pm A + n_1(t)$$

In the case of the amplifier scheme, both signal and noise power are amplified by $e^{\alpha l}$ to recoup the desired signal level. The signal and noise now propagate down the second link, the power again being attenuated by $e^{-\alpha l}$. An additional noise term $n_2(t)$ is now added, so that the input to the second amplifier is

$$\pm A + n_1(t) + n_2(t)$$

Ultimately, after m line links, the detector at the final destination receives

$$v(t) = \pm A + n_1(t) + \cdots + n_m(t) \tag{5-152}$$

From the results of the previous section we know that the sum of m independent gaussian variables is again gaussian, with a variance equal to the sum of the variances, and the means summed as well. In this case the individual noise terms are assumed to have zero mean. Since all the variances are assumed equal, the composite signal plus noise $v(t)$ at the destination may be written simply as

$$v(t) = \pm A + n(t) \tag{5-152a}$$

with the equivalent noise $n(t)$ gaussian of variance $m\sigma^2$. As in Sec. 5-6, the probability of an error is the probability that $v(t)$, when sampled, will be mistaken for the wrong binary symbol. Since the statistics are again gaussian, the probability of error is exactly the same as that calculated in Sec. 5-6, Eq. [5-95], with σ^2 replaced by $m\sigma^2$. (As in that section A remains the *received* signal amplitude.) The effect of m amplifiers in tandem is thus to increase the *rms* noise level by $\sqrt{m}$. Specifically, the probability of error in the m-link amplifier scheme is now given by

$$\begin{aligned} P_{amp}(e) &= \frac{1}{\sqrt{2\pi}} \int_{A/\sqrt{m}\sigma}^{\infty} e^{-y^2/2}\, dy \\ &= \frac{1}{2}\left[1 - \operatorname{erf} \frac{A}{\sqrt{2m}\,\sigma}\right] \equiv \frac{1}{2} \operatorname{erfc} \frac{A}{\sqrt{2m}\,\sigma} \end{aligned} \tag{5-153}$$

again using the previous definition of the error function, and defining erfc x, the *complementary* error function, as $1 - \operatorname{erf} x$.

Now consider a sequence of $(m - 1)$ repeaters which condition the signal and emit a clean noise-free pulse, as noted earlier. There are m sequential decisions that will be made on the one signal, including the

[1] The noise is assumed primarily introduced at the input to the repeater.

final decision at the ultimate destination. At the first repeater input the composite signal consists of $\pm A + n_1(t)$. The output of this repeater is simply $\pm a$ (the amplified and conditioned signal with the noise missing), or, if a mistake has been made in deciding on the correct signal, $\mp a$. Again the signal is attenuated to an amplitude level A. At the input to the second repeater the composite signal consists of $\pm A + n_2(t)$ if the first repeater decides correctly. This process is repeated on down the line.

It is apparent that errors initially made somewhere along the line may be corrected if an additional repeater further down the line makes a compensating error, or, in general, if there are an *even* number of incorrect decisions. A final error is made if an *odd* number of incorrect decisions is made along the line. If we assume as above that all the noises are independent and gaussian, with the same variance σ^2, it is apparent that the probability of an error at any one of the repeaters is

$$p = \frac{1}{\sqrt{2\pi}} \int_{A/\sigma}^{\infty} e^{-y^2/2}\, dy = \frac{1}{2} \operatorname{erfc} \frac{A}{\sqrt{2}\,\sigma} \qquad [5\text{-}154]$$

The probability of making k errors at the m decision points is then

$$P(k \text{ errors}) = \binom{m}{k} p^k (1 - p)^{m-k} \qquad [5\text{-}155]$$

with $\binom{m}{k} = m!/(m - k)!k!$ again the number of possible ways in which a sequence of k incorrect and $(m - k)$ correct decisions can be made. This is of course just the binomial distribution discussed earlier. The probability of error, $P_{\text{rep}}(e)$, on the entire m-link system, is then obtained by summing over all odd values of k, and is given by

$$\begin{aligned} P_{\text{rep}}(e) &= mp\,(1 - p)^{m-1} + \frac{m(m-1)(m-2)}{3!} p^3(1 - p)^{m-3} + \cdots \\ &= \sum_{\substack{k=1 \\ (k \text{ odd})}}^{m} \binom{m}{k} p^k (1 - p)^{m-k} \end{aligned} \qquad [5\text{-}156]$$

In particular, if $mp \ll 1$,

$$P_{\text{rep}}(e) \doteq mp \qquad (mp \ll 1) \qquad [5\text{-}157]$$

In this case the probability of a single incorrect decision is small enough that the probability of multiple errors is negligible. But the use of m decision points makes an individual error m times as likely.

The relative effectiveness of regenerative repeaters over amplifiers may now be found by comparing Eqs. [5-153] and [5-157.] Three sample calculations follow:[1]

1. $m = 100 \qquad \dfrac{A}{\sigma} = 6.36 \text{ (16 db)} \qquad p = 10^{-10}$

$$\text{Then} \qquad P_{\text{rep}}(e) \doteq 10^{-8} \qquad P_{\text{amp}}(e) = \frac{1}{2} \operatorname{erfc}\left(\frac{6.36}{10\sqrt{2}}\right) > 0.1$$

Note that here, with many repeaters and relatively high signal-to-noise ratio, the repeaters with conditioning circuitry far outperform the multiple amplifier scheme.

2. $m = 5 \qquad \dfrac{A}{\sigma} = \sqrt{10} \text{ (10 db)} \qquad p = 8 \times 10^{-4}$

$$P_{\text{rep}}(e) \doteq 4 \times 10^{-3} \qquad P_{\text{amp}}(e) = \frac{1}{2} \operatorname{erfc}(1) = 0.08$$

Again the effect of the repeaters is to increase the overall probability of error, but with the conditioning scheme far outperforming the amplifier scheme. This is again due to the fact that the probability of more than one error is negligible in the repeater scheme, while the rms noise level is effectively increased by $\sqrt{5}$ in the amplifier scheme.

3. $m = 3 \qquad \dfrac{A}{\sigma} = 1 \text{ (0 db)} \qquad p = 0.159$

$$P_{\text{rep}}(e) \doteq 0.34 \qquad P_{\text{amp}}(e) \doteq 0.28$$

Here with low signal-to-noise ratio and small numbers of repeaters, the two techniques are roughly comparable.

An interesting problem is to find the optimum number of repeaters for a fixed total line length and fixed *transmitter* power. As the number of repeaters is decreased, the overall probability of an error decreases as well, since there are fewer chances of making errors. The length of line between repeaters increases, however, so that the *received* signal decreases, leading to a higher probability of error at any one repeater. Specifically, let the *total* length of line be L, and let a^2 be the transmitter power. Then for small probability of error we have, from Eq. [5-157],

$$P_{\text{rep}}(e) = \frac{m}{2} \operatorname{erfc}\left(\frac{a}{\sqrt{2}\sigma} e^{-\alpha L/2m}\right) \qquad [5\text{-}158]$$

[1] Figure 5-27 may be used in determining the necessary probabilities. Since that figure refers to on-off binary transmission, while we are assuming bipolar transmission here, 6 db must be subtracted from the abscissa scale.

A trial-and-error minimization of this expression by varying m is left as an exercise for the reader.

The T1 carrier system operated by the Bell System is a 24-channel PCM voice-transmission system in which transistorized regenerative repeaters are spaced 6,000 ft apart. The transmission medium in this case consists of 19- or 22-gauge cable pairs. The output pulse rate in this system is 1.544 megapulses/sec.[1]

5-10 RAYLEIGH FADING AND MULTIPLICATIVE NOISE

We now consider another application of probability theory to a problem of great interest in communications. This concerns the propagation of signals through so-called random or fading media. Examples of such random channels include the ionosphere from which shortwave (h-f) signals are reflected back to the earth providing long-range radio transmission, troposcatter, underwater communications, seismic signal propagation through the earth, etc.

Detailed discussions of such channels and their influence on communications will not be attempted here; the reader is instead referred to the literature.[2] Basically, however, all such channels involve physical mechanisms in which an incident signal beam is scattered into a multitude of closely spaced beams propagating along different paths with correspondingly different attenuation and phase.

An extremely simple model of such a channel is shown in Fig. 5-38. Various mechanisms may be responsible for the multipath signal propagation: multiple reflections from inhomogeneities along the path of propagation, reflections from discontinuities in boundaries, presence of many physical scatterers, variations in dielectric constant, etc. From a terminal point of view, however, the different mechanisms produce the same result—a single beam is converted into a multitude of densely spaced beams. As we shall note below the random amplitude and phase introduced over the multiple paths leads to randomly varying (fading) signals.

To gain some insight into the effect of this scattering process, assume that a high-frequency carrier is incident on the medium producing the multiple beams.

[1] H. Cravis and T. V. Crater, Engineering of T1 Carrier System Repeatered Lines, *Bell System Tech. J.*, vol. 42, pp. 431–486, March, 1963.

[2] See, e.g., Schwartz, Bennett, and Stein, *op. cit.*, part III, as well as the references provided there.

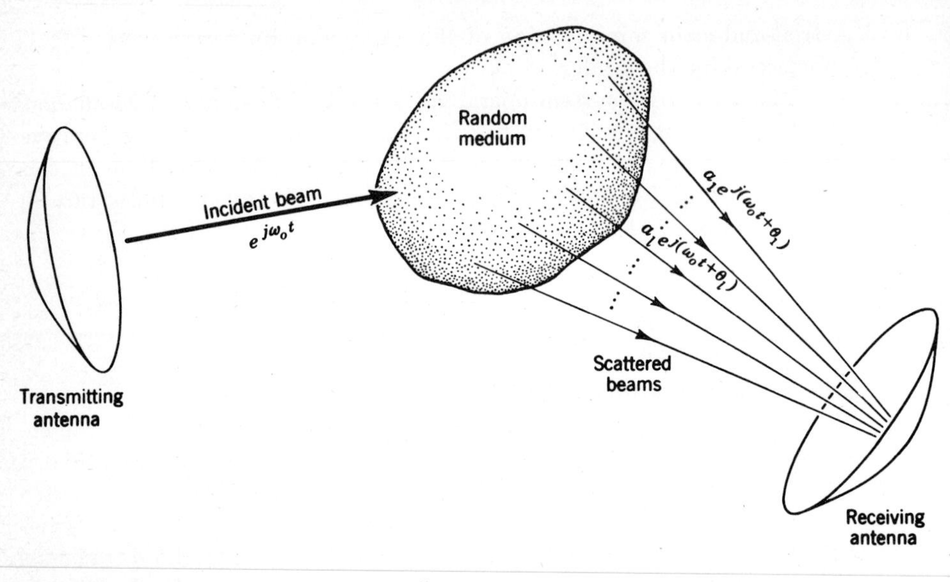

FIG. 5-38 Static model of random medium.

Using the exponential form for simplicity's sake, the incident signal is given by

$$s_i(t) = e^{j\omega_0 t} \tag{5-159}$$

(We can then take the real or imaginary part to determine the real signal. We also ignore here signal velocity, since *all* beams are assumed to travel at the same average velocity. This then provides a so-called static model of the channel.) The scattering mechanism produces a multitude of scattered beams. Each beam differs in amplitude and phase from the incident beam. (As a reasonable assumption the medium is assumed linear.) The lth scattered beam (Fig. 5-38) is for example given by $a_l e^{j(\omega_0 t+\theta_l)}$.

The received signal is the sum of all these beams and is given by

$$s_r(t) = \sum_l a_l e^{j(\omega_0 t+\theta_l)}$$

$$= r e^{j(\omega_0 t+\theta)} \tag{5-160}$$

where

$$r e^{j\theta} = \sum_l a_l e^{j\theta_l} \tag{5-161}$$

is the resultant complex phasor representing the additive effect of the beams. The complex addition is shown sketched in Fig. 5-39 for four scattered beams.

Aside from an average time delay involved in the propagation from transmitter to receiver the received signal has an amplitude r and phase

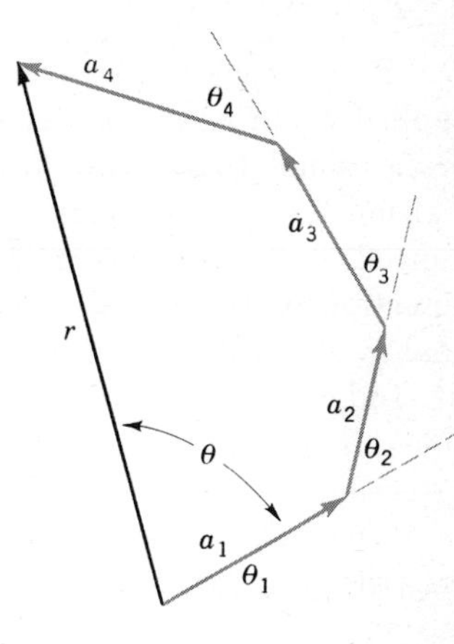

FIG. 5-39 Phasor sum of scattered beams.

angle θ relative to the incident wave. If the scatterers (or other mechanism responsible for the multipath effect) move slowly and randomly in the medium, the different amplitudes and phases a_l and θ_l may be expected to vary slowly and randomly with time. The resultant phasor will then have a randomly time-varying amplitude r and phase θ. Thus at some instant of time the individual beam phasors may be expected to line up, producing a large r; at other times they may interfere destructively with one another, producing a much smaller resultant amplitude. A typical sketch of the resultant received signal is shown in Fig. 5-40. Notice that the received signal resembles an AM signal. The incident signal with constant amplitude has been converted into one with randomly varying amplitude. At points such as a and a' the signal is seen to fade below the average received amplitude. Since the random term appears *multiplying* the carrier, this type of fluctuation is often referred to as *multiplicative noise.*

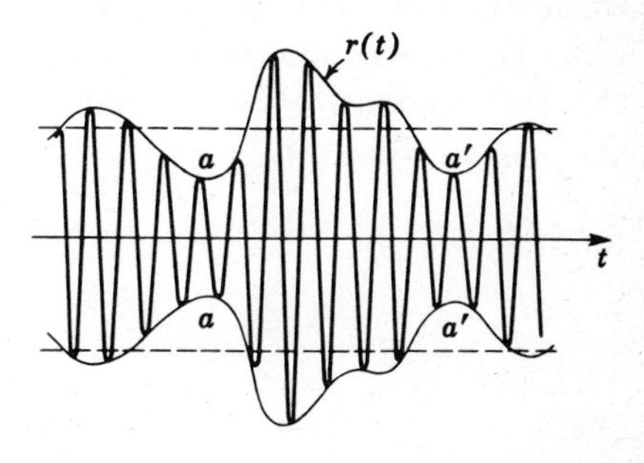

FIG. 5-40 Resultant received signal, unmodulated carrier input.

It is apparent from our discussion in Chap. 4 that such a signal applied to an envelope detector would produce at the output a voltage proportional to the amplitude term $r(t)$. If information is now being transmitted via the carrier, it is likewise apparent that fluctuations in r, as well as fluctuations in θ, the received phase, will interfere with the detection of the signal. This is true for all forms of signal, digital or analog, and all forms of modulation, AM, FM, etc. It is thus of interest to study the statistics of the randomly varying received signal in an attempt to determine quantitatively the effect of fading on signal detection.

To do this consider the complex phasor $re^{j\theta}$ of Eq. [5-161]. This may be written in terms of real and imaginary parts as

$$re^{j\theta} = x + jy \qquad [5\text{-}162]$$

where

$$x = \sum_l a_l \cos \theta_l \qquad [5\text{-}163]$$

and

$$y = \sum_l a_l \sin \theta_l \qquad [5\text{-}164]$$

If we now make the reasonable assumption that the θ_l's are all independent of one another and are *uniformly distributed* random variables (i.e., no particular phase angle is preferred), the central-limit theorem may be invoked to show that x and y are both *gaussian* random variables, of zero average value, and variance $\sigma_x^2 = \sigma_y^2 \equiv \sigma^2$. Although we shall not prove the applicability of the central-limit theorem here, it can be shown through arguments similar to those used in Sec. 5-8. Thus if all the a_l's were the same value and known, both x and y would be of the form considered in deriving the resultant density function for the sum of n independent sine waves of random phase. Here with the a_l's varying randomly it is sufficient to ensure that no one a_l is substantially larger than any other for the central-limit theorem to be applicable.

As indicated in Eq. [5-151] the sum of n random sine waves is gaussian in the limit of very large n. Also, as noted in Sec. 5-8, the variance of the sum is just the sum of the variances when the random variables added are independent. We thus have

$$f_x(x) = \frac{1}{\sqrt{2\pi\sigma_x^2}} e^{-x^2/2\sigma_x^2} \qquad [5\text{-}165a]$$

$$f_y(y) = \frac{1}{\sqrt{2\pi\sigma_y^2}} e^{-y^2/2\sigma_y^2} \qquad [5\text{-}165b]$$

with

$$\sigma_x^2 = \sigma_y^2 = \sum_l \frac{\sigma_{a_l}^2}{2} \equiv \sigma^2 \qquad [5\text{-}166]$$

Here $\sigma_{a_l}^2$ is the variance of the random variable a_l.

Although we shall not prove it here, it can be shown that x and y, although both gaussian with the same variance σ^2, are independent random variables.[1] Thus,

$$f_{xy}(x,y) = f_x(x)f_y(y) \qquad [5\text{-}167]$$

(See Eq. [5-116].)

Recapitulating, we have now shown that with the transmitted signal an r-f carrier, $s_i(t) = e^{j\omega_0 t}$, the received signal $s_r(t)$ is of the form

$$s_r(t) = re^{j(\omega_0 t + \theta)} = (x + jy)e^{j\omega_0 t} \qquad [5\text{-}160a]$$

with x and y independent *gaussian* random variables, and of equal variance. Assume now, specifically, that the incident or transmitted signal is

$$s_i(t) = \cos \omega_0 t = \text{Re } (e^{j\omega_0 t})$$

Then the received signal is just the real part of the expression in Eq. [5-160a], and is given by

$$s_r(t) = r \cos (\omega_0 t + \theta) = x \cos \omega_0 t - y \sin \omega_0 t \qquad [5\text{-}160b]$$

Note that the x term appears in phase with the transmitted carrier, the y term 90° out of phase. The two terms are commonly called the *inphase* and *quadrature* terms, respectively.

With x and y independent and gaussian what are the statistics of the random envelope r and the phase angle θ? It is apparent from the definition of x and y (Eq. [5-162]) that $r^2 = x^2 + y^2$ and $\theta = \tan^{-1} (y/x)$. Alternately we may write $x = r \cos \theta$, $y = r \sin \theta$. The variables x and y correspond to two-dimensional rectangular coordinates, the variables r and θ, to the equivalent polar coordinates. There is thus a unique transformation from one set of variables to the other. We use the known transformations to find the probability-density functions for r and θ. This is thus the extension to two dimensions (two random variables) of the one-dimensional transformation of Sec. 5-7.

In the one-variable case we equated areas under the probability-density curves (Fig. 5-32). Here too we equate probabilities:

$$\text{Prob } (x_1 < x < x_2, y_1 < y < y_2) = \text{Prob } (r_1 < r < r_2, \theta_1 < \theta < \theta_2) \qquad [5\text{-}168]$$

This corresponds, however, to equating volumes under the joint probability-density curves. As noted above we are effectively converting from rectangular (x,y) coordinates to polar (r,θ) coordinates in this case. With $f_{r\theta}(r,\theta)$ the probability-density function for the polar coordinates,

[1] Schwartz, Bennett, and Stein, *op. cit.*, chap. 1.

we must have

$$f_{xy}(x,y)\,dx\,dy = f_{r\theta}(r,\theta)\,dr\,d\theta \qquad [5\text{-}169]$$

From Eq. [5-167]

$$f_{xy}(x,y) = f_x(x)f_y(y) = \frac{e^{-(x^2+y^2)/2\sigma^2}}{2\pi\sigma^2} = \frac{e^{-r^2/2\sigma^2}}{2\pi\sigma^2} \qquad [5\text{-}170]$$

using $\sigma_x{}^2 = \sigma_y{}^2 = \sigma^2$ and $r^2 = x^2 + y^2$. Transforming differential areas, we have

$$dx\,dy = r\,dr\,d\theta \qquad [5\text{-}171]$$

(See Fig. 5-41.)

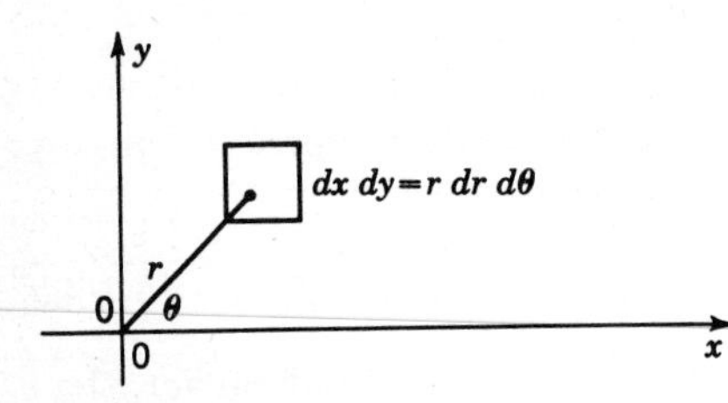

FIG. 5-41 Rectangular and polar coordinates.

From Eq. [5-169], then, with Eqs. [5-170] and [5-171],

$$f_{r\theta}(r,\theta)\,dr\,d\theta = \frac{re^{-r^2/2\sigma^2}}{2\pi\sigma^2}\,dr\,d\theta \qquad [5\text{-}172]$$

and

$$f_{r\theta}(r,\theta) = \frac{re^{-r^2/2\sigma^2}}{2\pi\sigma^2} \qquad [5\text{-}173]$$

To find the density function $f_r(r)$ for the envelope alone, we simply average Eq. [5-173] over all possible phases. Since the phase angle θ varies between θ and 2π, we get

$$f_r(r) = \int_0^{2\pi} f_{r\theta}(r,\theta)\,d\theta = \frac{re^{-r^2/2\sigma^2}}{\sigma^2} \qquad [5\text{-}174]$$

This is called the *Rayleigh distribution* and is shown sketched in Fig. 5-42. The peak of this distribution occurs at $r = \sigma$ and is equal to $e^{-1/2}/\sigma$. As σ (the standard deviation of the gaussian variables x and y) increases, the distribution flattens out, the peak decreasing and moving to the right. It is easily seen that $f_r(r)$ is properly normalized, so that $\int_0^\infty f_r(r)\,dr = 1$. Note that the normalization is from 0 to ∞ here, instead of from $-\infty$ to

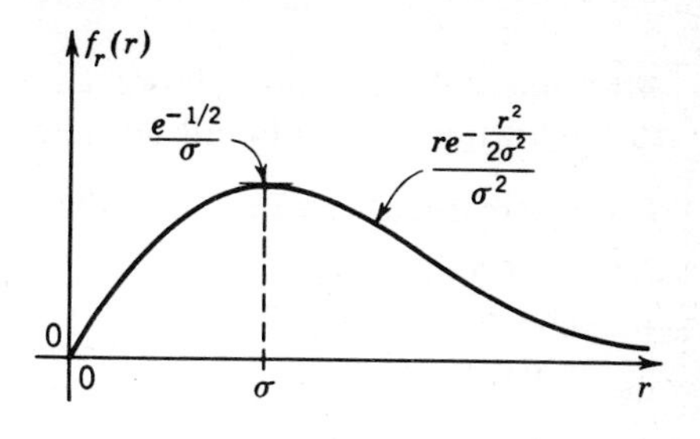

FIG. 5-42 Rayleigh distribution.

∞ for the gaussian distribution. The envelope can have only *positive* values.

The Rayleigh distribution appears in many other applications of statistics. We shall encounter it again in the next chapter in studying noise at the output of an envelope detector. Another simple example involves the firing of bullets at a target. Assume that the distribution of the bullets hitting the target is gaussian along the horizontal, or x, axis of the target and also gaussian along the vertical, or y, axis (i.e., a two-dimensional gaussian distribution) (Fig. 5-43). The average location of

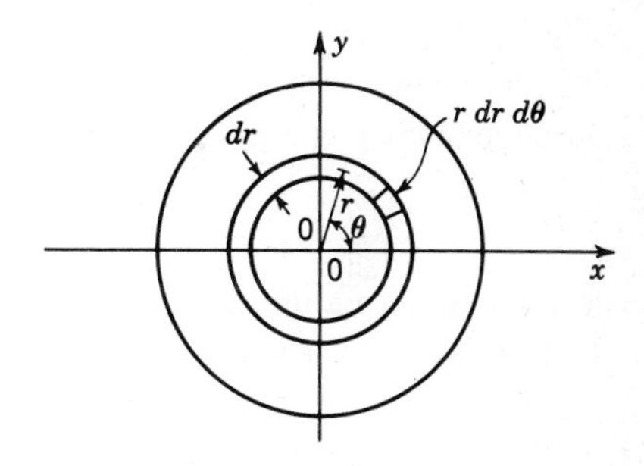

FIG. 5-43 A target.

the hits is at the origin, and the standard deviation in each direction is σ. The probability that the hits will lie within an annular ring dr units wide and r units from the origin is just $f_r(r)\,dr$, with $f_r(r)$ the Rayleigh distribution. {Note from Fig. 5-43 that, if $f_{r\theta}(r,\theta)\,dr\,d\theta$ represents the probability that the bullets will lodge within the differential area $r\,dr\,d\theta$, the probability that the bullets will lodge within the concentric annular ring is found by integrating over all values of θ. This is exactly what was done in Eq. [5-174].}

We have shown that the envelope of a constant amplitude sinusoidal signal passed through a so-called random channel follows a Rayleigh distribution. It is readily shown that this distribution is characterized by the following constants, all normalized to the given parameter σ:

1. Most probable value or mode $= \sigma$ (Fig. 5-42).
2. Expected value of the envelope, $E(r) = \sqrt{\pi/2}\,\sigma$.
3. The median or 50 percent cumulative distribution point, found by writing

$$0.5 = \int_r^\infty f_r(r)\,dr = e^{-r^2/2\sigma^2},$$

   is $r = 1.185\sigma$. (50 percent of the time, on the average, the signal is above this value, 50 percent of the time it is below it.)
4. The second moment is $E(r^2) = 2\sigma^2$.
5. The variance is $(2 - \pi/2)\sigma^2$.

Using the Rayleigh distribution, some simple questions concerning the expected variation of received amplitudes may be answered. For example, what is the probability that the signal will fade below the most probable value of σ? This is just

$$\int_0^\sigma \frac{re^{-r^2/2\sigma^2}}{\sigma^2}\,dr = 1 - e^{-\frac{1}{2}} = 0.39$$

Thus 39 percent of the time the signal would be expected to lie below σ. Again, what is the probability that the received signal level will exceed some arbitrary value $k\sigma$? This is just

$$\int_{k\sigma}^\infty f_r(r)\,dr = e^{-k^2/2}$$

Some typical values are given in the accompanying table, with the signal level relative to σ given in decibels.

Relative signal level, db	Probability of exceeding level
−20	0.995
−10	0.95
−6	0.88
−3	0.78
0	0.61
+3	0.37
+6	0.035

Information such as this is obviously useful in designing systems for transmission via Rayleigh fading channels. (The effect of Rayleigh fading on binary signal transmission will be considered in the next chapter.)

The Rayleigh fading model has been found to fit the experimental data quite well in many random channels.[1] Deviations from measured values are sometimes found at the higher signal levels, but this is to be expected since the central-limit theorem no longer applies in this region. (Recall that with a *finite* number of random variables added the gaussian distribution deviates from the true distribution at the tails.)

We have thus far discussed the fading envelope statistics only. What about the statistics of the phase angle θ? To find this we average $f_{r\theta}(r,\theta)$ over all values of r. This gives us

$$f_\theta(\theta) = \int_0^\infty f_{r\theta}(r,\theta)\,dr = \frac{1}{2\pi} \qquad [5\text{-}175]$$

The phase is thus uniformly distributed over 2π radians, as might be expected. Since $f_{r\theta}(r,\theta)$ can be written as $f_r(r)f_\theta(\theta)$ in this case, r and θ are independent random variables, just as were x and y.

RICIAN FADING

The Rayleigh fading model is due to the assumption of many densely spaced signal paths, no one of which predominates. In some fading situations, however, it is found that one particular signal path does dominate. This might be the case in which a line of sight path were superimposed on signals obtained from a scattering medium, or, in the case of ionospheric-skywave propagation, this might be due to the presence of one major strong stable path with a number of weak paths surrounding it. In this case the received signal with a steady sine wave transmitted will consist of a steady sine wave plus randomly fading signals superimposed. The received steady sine wave is often called the *specular component*. The envelope and phase distributions at the receiver differ from the Rayleigh and uniform distributions, respectively, in this case.

The determination of the statistics in this case is again straightforward, although a little more tedious. Thus, assume the received signal has a specular component $A_c \cos \omega_0 t$ added to the previous sum of random terms. The total received signal is thus written

$$s_r(t) = (x + A_c) \cos \omega_0 t - y \sin \omega_0 t \qquad [5\text{-}176]$$

[1] Schwartz, Bennett, and Stein, *op. cit.*, chap. 9.

where x and y are the previous gaussian-distributed terms, with

$$\sigma_x^2 = \sigma_y^2 \equiv \sigma^2$$

Considering the term $x + A_c$ alone, we note that the sum represents a gaussian variable with A_c the average value and σ^2 still the variance. Calling the sum a new parameter x',

$$x' \equiv x + A_c \qquad [5\text{-}177]$$

we have[1]

$$f(x') = \frac{e^{-(x'-A_c)^2/2\sigma^2}}{\sqrt{2\pi\sigma^2}} \qquad [5\text{-}178]$$

The envelope of the received signal $s_r(t)$ is now given by

$$r^2 = x'^2 + y^2 = (x + A_c)^2 + y^2 \qquad [5\text{-}179]$$

and the phase is

$$\theta = \tan^{-1}\frac{y}{x'} = \tan^{-1}\frac{y}{x + A_c} \qquad [5\text{-}180]$$

Again we have a received carrier term which is both amplitude and phase modulated by the randomly varying terms. We can again find the probability distributions for both the amplitude (envelope) and phase by a transformation to polar coordinates. This will give us the probability distribution at the output of an envelope detector, as well as the distribution at the output of a phase detector.

With x' and y independent random variables related to r and θ by the transformations $x' = r\cos\theta$, $y = r\sin\theta$, we have

$$f(r,\theta)\,dr\,d\theta = f(x',y)\,dx'\,dy = \frac{e^{-[(x'-A_c)^2+y^2]/2\sigma^2}}{2\pi\sigma^2}\,dx'\,dy$$

$$= \frac{e^{-A_c^2/2\sigma^2}re^{-(r^2-2rA_c\cos\theta)/2\sigma^2}}{2\pi\sigma^2}\,dr\,d\theta \qquad [5\text{-}181]$$

Note that we cannot write $f(r,\theta)$ as a product $f(r)f(\theta)$, since a term in the equation appears with both variables multiplied together as $r\cos\theta$. This indicates that r and θ are *dependent* variables. They are connected together in this case by the term $rA_c\cos\theta$. This is apparent from Eqs. [5-179] and [5-180] as well as Eq. [5-181]. If $A_c \to 0$, the two variables again become independent and $f(r,\theta)$ reduces to the product $f(r)f(\theta)$ found for the zero-signal case.

[1] We henceforth drop the subscript on the density functions for ease in printing.

We can find $f(r)$ again by integrating over all values of θ. This gives us

$$f(r) = \frac{e^{-A_c^2/2\sigma^2} re^{-r^2/2\sigma^2}}{2\pi\sigma^2} \int_0^{2\pi} e^{(rA_c \cos\theta)/\sigma^2} d\theta \qquad [5\text{-}182]$$

The integral in Eq. [5-182] cannot be evaluated in terms of elementary functions. Note, however, its similarity to the defining integral for the Bessel function of the first kind and zero order given by Eq. [4-66]. It is in fact related to the Bessel function of the first kind. In particular,

$$I_0(z) \equiv \frac{1}{2\pi} \int_0^{2\pi} e^{z\cos\theta} d\theta \qquad [5\text{-}183]$$

is called the *modified* Bessel function of the first kind and zero order. In terms of $I_0(z)$, Eq. [5-182] becomes

$$f(r) = \frac{r}{\sigma^2} e^{-(r^2+A_c^2)/2\sigma^2} I_0\left(\frac{rA_c}{\sigma^2}\right) \qquad [5\text{-}184]$$

The modified Bessel function can be written as an infinite series, just as in the case of the Bessel function of the first kind. This series can be shown to be

$$I_0(z) = \sum_{n=0}^{\infty} \frac{z^{2n}}{2^{2n}(n!)^2} \qquad [5\text{-}185]$$

For $z \ll 1$,

$$I_0(z) \doteq 1 + \frac{z^2}{4} + \cdots \doteq e^{z^2/4} \qquad [5\text{-}186]$$

Letting $A_c \to 0$ in Eq. [5-184] we get the Rayleigh distribution again, checking our previous result for the zero-signal case.

The envelope distribution of Eq. [5-184] is often called the *Rician* distribution in honor of S. O. Rice of Bell Telephone Laboratories, who developed and discussed the properties of this distribution in a pioneering series of papers on random noise.[1] (Some of this material will be treated later in our discussion of additive noise in systems.)

Equation [5-184] contains the term $A_c^2/2\sigma^2$. This is just the ratio of the average power in the specular component to the average power (variance) in either the inphase x or quadrature y terms. It is thus a measure of the fading statistics. As $A_c^2/2\sigma^2$ becomes larger, the fading or multiplicative noise becomes relatively less important. The distribution of

[1] S. O. Rice, Mathematical Analysis of Random Noise, *Bell System Tech. J.*, vol. 23, pp. 282–333, July, 1944; vol. 24, pp. 96–157, January, 1945 (reprinted in N. Wax, "Selected Papers on Noise and Stochastic Processes," Dover, New York, 1954).

received signal level becomes concentrated about the value of the steady (specular) component; the major remaining perturbations in that case are evidenced as phase fluctuations. Contrarywise, as $A_c^2/2\sigma^2$ becomes smaller, the specular component becomes indistinguishable, appearing as just one of the multipath components giving rise to the Rayleigh statistics.

We have indicated that for $A_c^2/2\sigma^2 \to 0$, the envelope of the received signal follows the Rayleigh distribution. For large $A_c^2/2\sigma^2$, however, it again approaches the original gaussian distribution of the inphase (x) signal term. This is apparent from Eq. [5-176] or Eq. [5-179]. For as A_c increases relative to σ, the inphase term dominates, the quadrature term contribution to the envelope becomes negligible, and the envelope becomes just

$$r \doteq A_c + x \qquad A_c \gg \sigma \tag{5-187}$$

This is just a gaussian random variable with average value A_c.

This can also be demonstrated directly from the Rician density function (Eq. [5-184]) itself. To show this we make use of a known property of the modified Bessel function—that is, that it approaches asymptotically, for large values of the argument, an exponential function. Thus for $z \gg 1$,

$$I_0(z) \doteq \frac{e^z}{\sqrt{2\pi z}} \tag{5-188}$$

Letting $rA_c \gg \sigma$ in Eq. [5-184], we make use of Eq. [5-188] to put $f(r)$ in the form

$$f(r) \doteq \frac{r}{\sqrt{2\pi r A_c \sigma^2}} e^{-(r-A_c)^2/2\sigma^2} \tag{5-189}$$

This function peaks sharply about the point $r = A_c$, dropping off rapidly as we move away from this point. Most of the contribution to the area under the $f(r)$ curve (or the largest values of the probability of a range of r occurring) comes from points in the vicinity of $r = A_c$. In this range of r, then, we can let $r = A_c$ in the nonexponential (and slowly varying) portions of $f(r)$ and get

$$f(r) = \frac{1}{\sqrt{2\pi\sigma^2}} e^{-(r-A_c)^2/2\sigma^2} \qquad rA_c \gg \sigma^2 \tag{5-190}$$

In the vicinity of the point $r = A_c$, then, the distribution approximates the normal (gaussian) distribution as noted above.

The Rician distribution is shown plotted in Fig. 5-44 for different values of $A_c^2/2\sigma^2$. Its application to fading phenomena and experiments verifying its validity in fading models are discussed further in the refer-

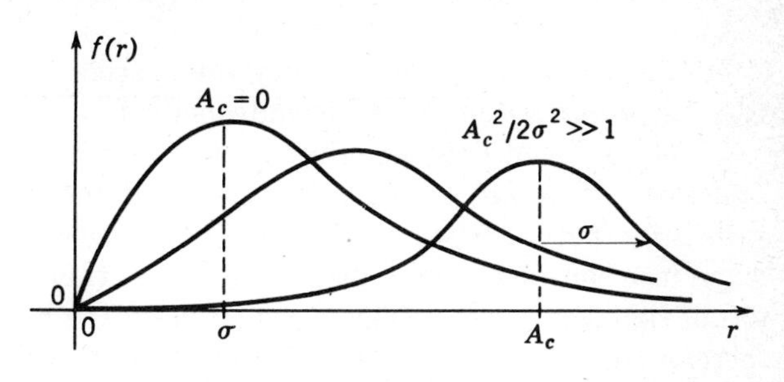

FIG. 5-44 Rician distribution.

ence cited.[1] We shall also have occasion to refer to it again in our discussion of noise in systems in Chap. 6.

5-11 DEPENDENT RANDOM VARIABLES

We have devoted considerable space in the past few sections to a discussion of independent random variables. It is of interest to extend our discussion of probability theory to include the case of dependent variables.

Dependent discrete variables were considered in Sec. 5-2. It was shown there, by example, that the study of dependent variables was important in studying the statistical properties of a language. (This was also alluded to in Chap. 1.) In general, as noted in Chap. 1, there is usually some dependence between successive signals of a message being transmitted. This reduces the rate of transmission of information.

In any study of communication in the presence of noise this dependence of successive voltages on one another must be taken into account. Consider, for example, the noise-voltage plot of Fig. 5-45. We have previously discussed the chance of noise at any one instant of time exceeding an arbitrary voltage level and related this to probabilities of error in PCM systems. We have thus far not discussed the joint statistics of two variables, such as n_1 and n_2 in Fig. 5-45, separated by a specified time

[1] Schwartz, Bennett, and Stein, *op. cit.*, pp. 372–374, 381.

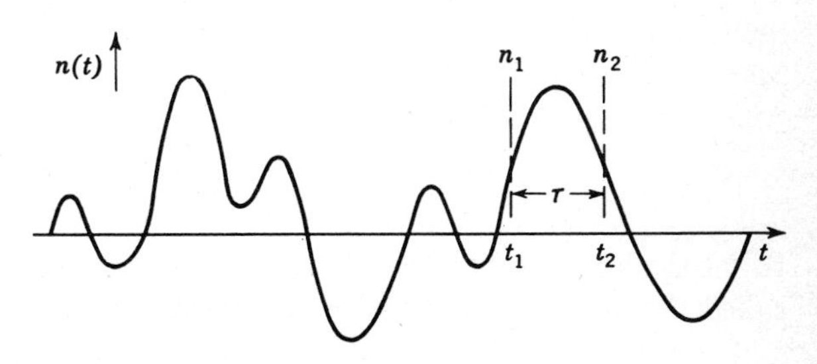

FIG. 5-45 Noise voltage.

interval, except to say that if they are spaced "far enough apart" they may be considered unrelated or independent.

Yet the band-limiting effect of any system through which noise or signal plus noise are passed will ensure that there is some connection or correlation between two closely spaced voltages. This in turn must be taken into consideration in discussing in any thorough manner the statistics of signals in noise. This we shall do in the next chapter. We shall therefore discuss to some extent the concept of joint statistics of two continuous dependent variables and a particular measure—the correlation function—of this dependency. These concepts can be extended to include dependency between many random variables, as was done with the sum of n independent variables in previous sections. We shall restrict ourselves to two dependent variables, however, in this book.

Consider two random variables x and y. By performing repeated measurements on these two variables jointly, as in the case of the discrete variables, we can develop a set of histogram curves relating the frequency of occurrence of different pairs of x and y. (In the example of Fig. 5-45, with $x = n_1$ and $y = n_2$ this would consist of measurements of the different combinations of pairs of voltages occurring.) From such a set of measurements, taken with x broken up into levels Δx units apart and y into levels Δy apart, a joint probability-density function $f(x,y)$ can be constructed. This is the same function introduced in Sec. 5-8 for the discussion of two independent variables.

$f(x,y)\,dx\,dy$ represents the probability that x and y will jointly be found in the ranges $x \pm dx/2$ and $y \pm dy/2$, respectively. Since some range of x and y must always occur jointly, we must have

$$\iint_{-\infty}^{\infty} f(x,y)\,dx\,dy = 1 \qquad [5\text{-}191]$$

The joint density function $f(x,y)$ is thus normalized so that the volume under the curve is 1. As noted in Sec. 5-8, $f(x,y)$ may be written as the product $f_x(x)f_x(y)$ if x and y are independent.

We can define a cumulative-probability function as in the case of one random variable, if we wish, and averages can also be computed. For example, the average of $G(x,y)$ is

$$E[G(x,y)] = \iint_{-\infty}^{\infty} G(x,y)f(x,y)\,dx\,dy \qquad [5\text{-}192]$$

In particular the averages of x and y raised to integer powers are again called the moments of x and y and are given by

$$m_{ij} = E(x^i y^j) = \iint_{-\infty}^{\infty} x^i y^j f(x,y)\,dx\,dy \qquad [5\text{-}193]$$

These moments play an important role in two-variable statistical theory, just as do the moments in the one-variable case. As special cases, we have

$$m_{10} = x_0 = E(x) = \iint_{-\infty}^{\infty} xf(x,y)\,dx\,dy$$

$$= \int_{-\infty}^{\infty} xf_x(x)\,dx \qquad f_x(x) = \int_{-\infty}^{\infty} f(x,y)\,dy \qquad [5\text{-}194]$$

$$m_{01} = y_0 = E(y) = \iint_{-\infty}^{\infty} yf(x,y)\,dx\,dy$$

$$= \int_{-\infty}^{\infty} yf_y(y)\,dy \qquad f_y(y) = \int_{-\infty}^{\infty} f(x,y)\,dx \qquad [5\text{-}195]$$

$$m_{11} = E(x,y) = \iint_{-\infty}^{\infty} xyf(x,y)\,dx\,dy \qquad [5\text{-}196]$$

$$m_{20} = E(x^2) = \int_{-\infty}^{\infty} x^2 f_x(x)\,dx \qquad [5\text{-}197]$$

$$m_{02} = E(y^2) = \int_{-\infty}^{\infty} y^2 f_y(y)\,dy \qquad [5\text{-}198]$$

Just as in the one-variable case we can also define central moments or moments about the average values. For example, the central moments of the second order, comparable with μ_2, are

$$\mu_{20} = E[(x - x_0)^2] = \int_{-\infty}^{\infty} (x - x_0)^2 f_x(x)\,dx$$

$$= m_{20} - m_{10}^2 = \sigma_x^2 \qquad [5\text{-}199]$$

$$\mu_{02} = E[(y - y_0)^2] = \int_{-\infty}^{\infty} (y - y_0)^2 f_y(y)\,dy$$

$$= m_{02} - m_{01}^2 = \sigma_y^2 \qquad [5\text{-}200]$$

and

$$\mu_{11} = E[(x - x_0)(y - y_0)] = E(xy) - E(x)E(y)$$

$$= m_{11} - m_{01}m_{10} \qquad [5\text{-}201]$$

as may be verified by the reader.

m_{11} and μ_{11} are the two second-order moments which serve as a measure of the dependence of two variables. For, as noted previously, two *independent* variables will have as their joint distribution function the *product* of the individual distribution function, or

$$f(x,y) = f_x(x)f_y(y) \qquad [5\text{-}202]$$

For this case m_{11} and μ_{11} become, respectively,

$$m_{11} = E(x)E(y) = m_{10}m_{01} \qquad [5\text{-}203]$$

and

$$\mu_{11} = 0 \qquad [5\text{-}204]$$

μ_{11} is called the *covariance* of the two variables. σ_1^2 and σ_2^2 are again called *variances*. If the two variables are independent, the covariance is zero, and m_{11}, the average of the product, becomes the product of the individual averages. (The converse of this statement is not true in general but does hold for the joint gaussian distribution.)

We may define a normalized quantity called the *normalized correlation coefficient* which serves as a numerical measure of the dependence between two variables. Using the symbol ρ for this quantity, we define

$$\rho \equiv \frac{\mu_{11}}{\sqrt{\mu_{20}\mu_{02}}} = \frac{\mu_{11}}{\sigma_1\sigma_2} \tag{5-205}$$

The correlation coefficient can be shown to be less than 1 in magnitude: $-1 \leq \rho \leq 1$. In particular assume first that y is completely determined by x and that a linear relation of the form $y = \alpha x + b$ exists between the two. Then $E(y) = \alpha E(x) + b$, or $y_0 = \alpha x_0 + b$. μ_{11} becomes

$$E[(x - x_0)(y - y_0)] = \alpha E[(x - x_0)]^2 = \alpha\mu_{20}$$

Also,

$$\mu_{02} = E[(y - y_0)^2] = \alpha^2 E[(x - x_0)^2] = \alpha^2\mu_{20}$$

Then $\rho = \pm 1$, depending on the sign of α. On the other hand, if x and y are independent, $\mu_{11} = 0$ and $\rho = 0$. So values of ρ close to 1 indicate high correlation; values close to 0 indicate low correlation.

The two-variable gaussian distribution serves as an example of the joint distribution function. It is given by the expression[1]

$$f(x,y) = \frac{1}{2\pi M} \exp\left[-\frac{1}{2M^2}(\mu_{02}x^2 - 2\mu_{11}xy + \mu_{20}y^2)\right]$$

$$= \frac{1}{2\pi M} \exp\left[\frac{-1}{2(1 - \rho^2)}\left(\frac{x^2}{\sigma_x^2} - \frac{2\rho xy}{\sigma_x\sigma_y} + \frac{y^2}{\sigma_y^2}\right)\right] \tag{5-206}$$

for zero average values ($x_0 = y_0 = 0$). Here

$$M^2 = \mu_{20}\mu_{02} - \mu_{11}^2 = \sigma_x^2\sigma_y^2(1 - \rho^2)$$

Note that if $\rho = 0$ ($\mu_{11} = 0$), $f(x,y)$ becomes the product of two single-variable distribution functions, indicating in this case that x and y are independent. For nonzero x_0 and y_0, x and y are replaced by $x - x_0$ and $y - y_0$, respectively. The two-dimensional gaussian density function is shown sketched in Fig. 5-46.[2]

The probability that x and y will be found in the ranges x_1 to x_2, y_1 to y_2, respectively, is just the volume under the curve of Fig. 5-46, enclosed by these points. For such a calculation the variances σ_1^2, σ_2^2

[1] A. Papoulis, "Probability, Random Variables, and Stochastic Processes," McGraw-Hill, New York, 1965.

[2] W. R. Bennett, Methods of Solving Noise Problems, *Proc. IRE*, vol. 44, no. 5, fig. 3, pp. 609–638, May, 1956.

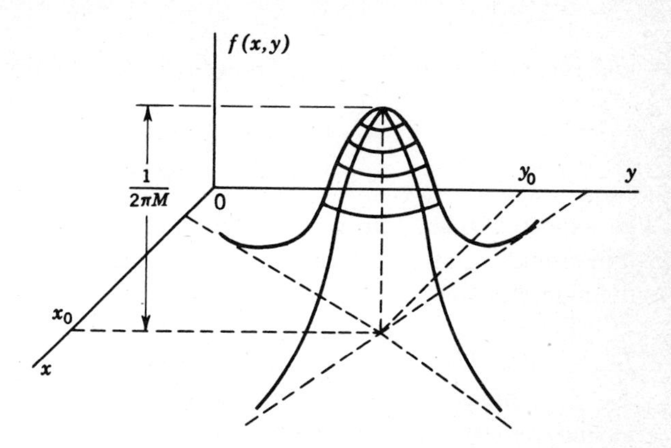

FIG. 5-46 Two-dimensional gaussian probability-density function.

and the correlation coefficient ρ must obviously be known from previous information.

That the parameters in Eq. [5-206] satisfy the defining relations for the moments expressed by Eqs. [5-194] to [5-201] may be verified by evaluating the appropriate integral. For example, assume that $E(x)$ or m_{10} and μ_{20} or σ_x^2 are to be found for the guassian distribution of Eq. [5-206]. Since we are interested only in average values of x, independent of variations in y, we first integrate over all values of y to find the distribution in x alone. This is given by

$$f_x(x) = \int_{-\infty}^{\infty} f(x,y)\,dy = \frac{\exp\{-x^2/[2(1-\rho^2)\sigma_x^2]\}}{2\pi M} \times \int_{-\infty}^{\infty} \exp\left[\frac{-1}{2(1-\rho^2)}\left(\frac{y^2}{\sigma_y^2} - \frac{2\rho xy}{\sigma_x\sigma_y}\right)\right] dy \quad [5\text{-}207]$$

The integral may be evaluated by first completing the square in the exponent. Thus

$$\frac{y^2}{\sigma_y^2} - \frac{2\rho xy}{\sigma_x\sigma_y} = \left(\frac{y}{\sigma_y} - \frac{\rho x}{\sigma_x}\right)^2 - \frac{\rho^2 x^2}{\sigma_x^2}$$

$f_x(x)$ now becomes

$$f_x(x) = \frac{\exp(-x^2/2\sigma_x^2)}{2\pi M} \int_{-\infty}^{\infty} \exp\left[-\frac{(y/\sigma_y - \rho x/\sigma_x)^2}{2(1-\rho^2)}\right] dy$$

Using the transformation of variables,

$$z^2 = \frac{(y/\sigma_y - \rho x/\sigma_x)^2}{2(1-\rho^2)}$$

and recalling that $\int_{-\infty}^{\infty} e^{-z^2}\, dz = \sqrt{\pi}$, we get finally

$$f_x(x) = \frac{e^{-x^2/2\sigma_x^2}}{\sqrt{2\pi\sigma_x^2}} \qquad [5\text{-}208]$$

as expected from our previous discussion of the one-variable or one-dimensional gaussian function. (Use has also been made here of the defining relation for M.)

From our knowledge of the gaussian function we recognize that $E(x) = 0$ and $\mu_{20} = \sigma_x^2$ for this example.

In a similar manner, $f_y(y)$, μ_{02}, $E(y)$, etc., may be found. These are left as exercises for the reader.

5-12 SUMMARY

As pointed out in the introduction to this chapter, any thorough study of the effects of noise in communication systems, as well as the transmission of random signals, must of necessity rely on the elements of probability theory. To make this book as self-contained as possible we have therefore included in this chapter an introduction to probability theory.

To develop some feel for the concepts involved we introduced the basic ideas via a relative-frequency approach. We then reintroduced them via the more generalized axiomatic approach using the elements of set theory. After defining probability we discussed in this chapter the ideas of statistical independence, conditional probability, and joint probability. We then went on to consider random continuous variables and the probability-density and probability-distribution functions utilized in discussing these quantities.

We introduced the gaussian (normal) density function as one of particular significance in probability and statistics, and we shall have occasion to refer to this distribution repeatedly in the chapters to follow. We also discussed in some detail the discrete binomial distribution. The discussion of the measurement or estimation of an unknown probability by counting the occurrence of 1's in a long string of 0's and 1's, showing the variation about the average estimate decreasing as the number of samples considered increased, is often used to justify the relative-frequency approach to probability introduced earlier.

Assuming that noise added during binary transmission obeyed gaussian statistics, we were able to apply the concept of probability to the calculation of error rates in binary PCM. A rather narrow range of signal-to-noise ratio was found to exist above which the probability of error dropped rapidly to negligible values, below which the chance of error was high. This was defined to be the *threshold* range of the signal-to-noise ratio. We shall return to the binary PCM problem in the next chapter, in comparing the error performance of FSK, PSK, and OOK transmission systems.

We also considered as an interesting problem that of choosing optimum binary decision levels for minimum probability of error. This important problem will be reconsidered in Chap. 8 as part of the overall problem of designing optimum signal waveshapes and receiver structures to minimize the probability of error of binary signals in noise.

Various types of nonlinear operations on signals are normally carried out in the process of transmitting signals through complex communication systems. Examples include the modulation and demodulation processes described in Chap. 4, frequency conversion, envelope detection, etc. Methods for determining the statistics of random signals and noise at the outputs of nonlinear devices in terms of the statistics at the input are therefore useful. Nonlinear transformations of random variables were accordingly considered in Sec. 5-7. The applications chosen also serve to further consolidate and solidify the reader's knowledge of probability theory.

Proceeding to the distribution of n independent random variables, we introduced the characteristic function, the Fourier transform of the probability-density function, as a tool in determining the probability distribution for the sum of many independent random variables. Using the characteristic function, we were able to demonstrate, in several cases, that the distribution of the sum approaches a gaussian distribution. These examples were all special cases of the more general central-limit theorem of probability theory.

As an application of these ideas in binary transmission we considered the question of PCM repeaters. We showed that reshaping the transmitted binary pulse stream periodically over the transmission channel was generally to be preferred to periodically amplifying the signal plus any noise added. In the first case the error rate may be kept within tolerable limits by spacing repeaters close enough together so that the signal never has a chance to attenuate sufficiently in the presence of added noise to result in significant numbers of errors. In the second case noise added at each amplifier continues along with the signal and previously added noise, resulting in a progressively deteriorating signal-to-noise ratio.

As a further application of probability theory to problems in communications we then considered the transmission of signals through so-called random fading media. Here, due to various physical effects, the signal propagates via a multitude of independent paths. A high-frequency carrier then arrives as the sum of many carrier terms (all at the same frequency for *linear* random channels) of differing amplitudes and phases.

As the transmission paths change slowly and unpredictably, the resultant phasor sum varies in amplitude and phase. The received carrier correspondingly shows a fluctuation (fading) in its amplitude. (This is well-known, for example, to anyone listening to short-wave broadcasts transmitted via the ionosphere.) For many independent paths of propagation the central-limit theorem may be invoked to show that the inphase and quadrature terms of the received carrier are both gaussian. It is then found that the fading amplitude or *envelope* of the carrier obeys *Rayleigh* statistics. If one propagation path dominates, the envelope is found to follow *Rician* statistics.

A knowledge of these statistics enables designers to more appropriately predict system performance and compare different types of systems. We shall apply these results to the transmission of high-frequency binary signals over fading media in the next chapter. The envelope statistics found here will also be found occurring in discussing noise, as well as signal plus noise, at the output of AM-type envelope detectors.

The final section of the chapter dealt with the joint probability distribution of two dependent random variables. In particular we introduced the covariance as one measure of the dependence of two random variables. This measure of statistical dependence will be elaborated on in detail in the next chapter when we discuss random processes (time functions) and explore the dependence of successive samples in time on one another.

PROBLEMS

5-1 An urn contains two white balls and six red ones. What is the probability that one ball drawn at random will be white?

5-2 An analysis of a long message transmitted in binary digits shows 3,000 zeros and 7,000 ones. What is the probability that any one digit in the message is a zero?

5-3 A selected list of words in the English language is to be transmitted by means of a binary code, with each word represented by 12 binary digits or fewer. How many words can there be on the list?

5-4 The numbers 1 through 10 are selected at random.

(*a*) What is the probability that the numbers are selected in the order 1, 2, . . . , 10?
(*b*) What is the probability of selecting the number 2 right after the number 1?

5-5 (*a*) Find the probability of getting a 7 in the toss of two dice.
(*b*) Find the probability of throwing a 6, 7, or 8 with two dice.

5-6 What is the probability that 4 cards drawn in succession from a deck of 52 will be aces?

5-7 A box contains five white balls, three red balls, and two black ones. What is the probability that two balls drawn from the box will both be red?

5-8 What is the probability of obtaining four tails if four coins are tossed? What is the probability that at least three heads will appear?

5-9 Calculate the conditional and joint probabilities for the typical message of Shannon's three-letter alphabet at the end of Sec. 5-2. Check with the probability tables shown there.

5-10 Refer to Fig. 5-5 in the text. Express the subsets AC and ABC in words. Show they are in fact given by the regions shown in Fig. 5-5*d*.

5-11 Using a Venn diagram prove the following set identities:

$$\begin{aligned} A\bar{B} + AB &= A \\ A(B + C) &= AB + AC \\ \overline{A + B} &= \bar{A}\bar{B} \\ \overline{AB} &= \bar{A} + \bar{B} \end{aligned}$$

5-12 Let x represent the age of husbands, y the age of wives. The set S consists of all possible elements corresponding to $\{15 < x,y < 80\}$. Let A represent the subset "husbands between the ages of 20 and 40," B, the subset "wives between 20 and 40," and C, the subset "husbands older than wives."

(*a*) Indicate S, A, B, C on a rectangular xy plot.
(*b*) Describe the following subsets in words, sketching them on the xy plot as well: AB, $\overline{AB}$, $\bar{A}$, $\bar{B}$, $\bar{A} + \bar{B}$, AC, $A + C$, ABC.

5-13 Two coins are tossed with the four possible outcomes HT, HH, TT, TH (H stands for head, T for tails). Each outcome is assumed equally likely. Let set S consist of these four elements. Indicate the probability of occurrence of, and elements contained within, the following subsets:

A: coin 1 comes up head
B: coin 2 comes up head
C: coin 1 is a tail
$A + C$, AB, $A + B$

Express the latter three subsets in words.

5-14 Consider a binary PCM system transmitting 1's with a probability $P_1 = 0.6$ and 0's with a probability $P_0 = 0.4$. The receiver recognizes 0's, 1's, and a third symbol E, called an erasure symbol. (A system like this is mentioned in Chap. 8.) There is a probability $P(0|1) = 0.1$ that the 1's will be received (mistakenly) as 0's, $P(E|1) = 0.1$ that they will be received as E's, and $P(1|1) = 0.8$ that they will be received (correctly) as 1's. Similarly, with 0 assumed transmitted, the appropriate probabilities of events at the receiver are given by $P(1|0) = 0.1$, $P(E|0) = 0.1$, $P(0|0) = 0.8$.

(*a*) Sketch a diagram indicating two transmit and three receive levels, show the appropriate transitions between them, and indicate the appropriate probabilities. *Note:* in the symbolism used above all conditioning refers to the *transmitter*.
(*b*) Calculate the probability of receiving a 0, a 1, and an E, respectively. Show these sum to 1, as required.
(*c*) Show the probability of an error is 0.1, the probability of a correct decision at the receiver is 0.8, and the probability of an erasure is 0.1.
(*d*) Repeat (*b*) and (*c*) if the transition probabilities, with a 0 transmitted, are changed to $P(1|0) = 0.05$, $P(E|0) = 0.05$, $P(0|0) = 0.9$.

5-15 Refer to Prob. 5-14. The symbol 1 is received. What is the probability it came from a 0? From a 1? Repeat for the symbols 0, and E, as received. (It may pay to adopt new symbols such as T_0, T_1, and R_0, R_1, R_E, or A_1, A_2, and B_1, B_2, B_3 to keep the appropriate conditional probabilities straight.) Check your results by summing appropriate probabilities.

5-16 Show that the probability of finding a continuous random variable x somewhere in the range x_1 to x_2 is given by $F(x_2) - F(x_1)$, with $F(x)$ the cumulative-distribution function (see Eq. [5-32]).

5-17 Tabulate the probabilities of getting the numbers 2 through 12 on the toss of two dice. Plot these to scale.

5-18 Consider a discrete variable with the probabilities $P_1, P_2, \ldots, P_n$ corresponding to its n possible values. Show that the relation between the variance, or second central moment μ_2 and the first and second moments m_1 and m_2, respectively, is given by $\mu_2 = m_2 - m_1^2$.

5-19 A given time interval T sec long is divided into $n = T/\Delta T$ time slots each ΔT sec long. Assume the chance of a phone call occurring in any one ΔT sec subinterval is $p \ll 1$, and that calls in adjacent subintervals occur independently.

(*a*) What is the probability of k phone calls occurring in T sec? At least k calls in T sec? Relate this problem to the binomial distribution discussed at the end of Sec. 5-4 in connection with binary transmission.
(*b*) What is the average number of phone calls per second, and the standard deviation about this? (See Sec. 5-5.)

5-20 *Electron emission.* A heated cathode has a probability $p \ll 1$ of emitting one electron in ΔT sec.

(*a*) What is the probability of emitting k electrons in n such ΔT-sec time intervals if electron emissions are independent of one another? Compare with Prob. 5-19. (Examples like this one are discussed in more detail in Chap. 7 in connection with shot noise generation.)
(*b*) What is the average number of electrons emitted per second? If each electron carries e coulombs, what is the average (dc) emission current?

5-21 (*a*) Consider a binomial distribution with $n = 4$ and $p = 0.4$. Plot the probability $P(k)$ versus k, as well as the cumulative distribution $F(m) = \text{Prob}\,[k \leq m]$.
(*b*) Repeat for $n = 10$ and $p = 0.4$.

5-22 *Poisson distribution.* A discrete random variable x takes on positive integer values 0, 1, 2, . . . , only. The probability that $x = k$ is then $P(x = k) = a^k e^{-a}/k!$, $k = 0, 1, 2, \ldots$.

(*a*) $a = 0.5$. Plot $P(x = k)$ and the cumulative-distribution function $F(m) \equiv P(x \leq m)$.
(*b*) Repeat for $a = 2$.
(*c*) Show $F(\infty) = 1$, independent of a.
(*d*) Show $E(k) = a$, $\sigma^2 = a$, $\sigma/E(k) = 1/\sqrt{a}$.

(The Poisson distribution is discussed in detail in Chap. 7.)

5-23 Photons in a laser communication system are emitted with a Poisson distribution. This means that if on the average a is emitted in an arbitrary interval, the probability of k being emitted in the same interval is $a^k e^{-a}/k!$ As noted in the book the ratio $\sigma/E(k)$ is a measure of the relative fluctuation or spread about the average number emitted. Calculate this ratio for $a = 10^6, 10^{10}, 10^{14}$. (See Prob. 5-22*d*.)

5-24 (*a*) Show that the distribution of the number of heads appearing in the tossing of a fair coin n times is given by the binomial distribution with $p = q = \frac{1}{2}$.
(*b*) What is the average number of heads expected in 1,000 tosses, in 10^6 tosses? What is the relative spread, $\sigma/E(k)$, about the average number expected in the two cases? What does this imply about the relative occurrence of heads as the number of tosses gets larger and larger?

5-25 n fair coins are tossed simultaneously.

(*a*) What is the probability of k heads appearing?
(*b*) What is the average number of heads expected? Compare with Prob. 5-24.

5-26 Consider the function $f(x) = kxe^{-x^2/2N}$, with k and N positive constants.

(*a*) To what range of values must x be restricted to ensure that the function represents a possible probability-density function?
(*b*) Determine k such that $f(x)$ is properly normalized. The resultant function is called the Rayleigh distribution and is discussed in Sec. 5-10. Plot $f(x)$ versus $x/\sqrt{N}$.
(*c*) What is the probability that a variable x obeying the Rayleigh distribution will have values between $x = \sqrt{2N}$ and $x = 2\sqrt{N}$?
(*d*) Sketch the cumulative distribution $F(x)$.

5-27 Calculate the mean value m_1 and the standard deviation σ of the Rayleigh distribution of the previous problem (see Sec. 5-10).

5-28 Consider the triangular probability-density function of Fig. P 5-28.

(*a*) Find b in terms of a so that the function is properly normalized.
(*b*) Calculate the mean value and standard deviation of the distribution.
(*c*) Plot the cumulative-distribution function $F(x)$.
(*d*) What is the probability that x will be greater than $a/2$?

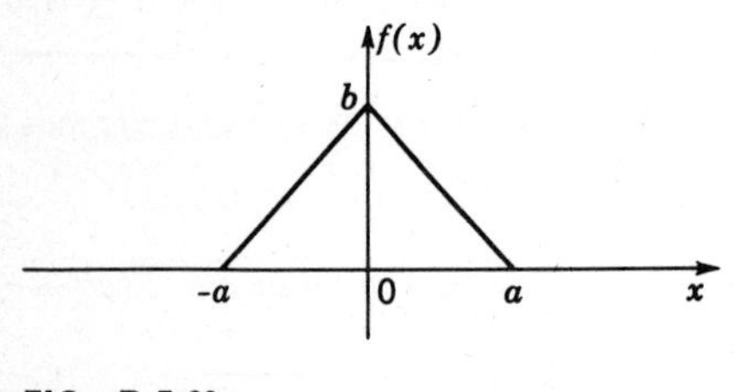

FIG. P 5-28

5-29 The Laplacian distribution is given by $f(x) = ke^{-|x|/c}$ (Fig. P 5-29).

(*a*) Determine k such that $f(x)$ is properly normalized.
(*b*) Show the standard deviation $\sigma = \sqrt{2}\,c$.
(*c*) Sketch the cumulative distribution $F(x)$.

Note: This distribution is sometimes used to model the amplitude distribution of so-called burst-type or impulse noise, occurring in high-frequency digital communications.

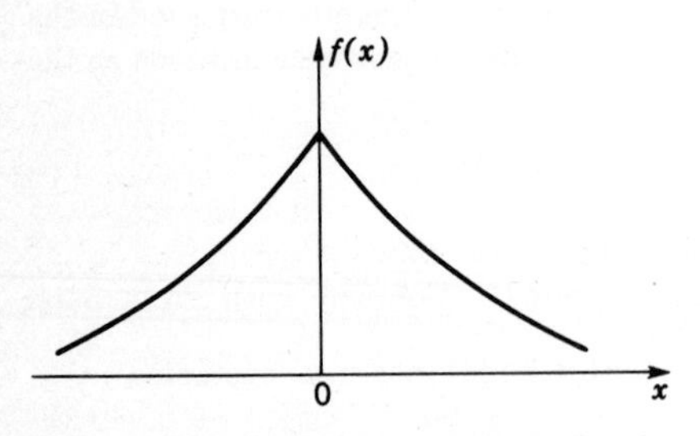

FIG. P 5-29 Laplacian density function.

5-30 Find the variance of the gaussian, or normal, distribution by evaluating the integral of Eq. [5-80].

5-31 Show that the gaussian function of Eq. [5-77] is properly normalized.

5-32 Show that the cumulative probability distribution for the gaussian function is given in terms of the error function by Eq. [5-88]. Plot $F(x)$, using any available table of the error function.

5-33 In the manufacture of resistors nominally rated at 100 ohms it is found that the probability distribution of actual resistance values is very closely given by a normal distribution with standard deviation $\sigma = 5$ ohms. What percentage of the resistors manufactured lie within the range 90 to 110 ohms?

5-34 Two voltmeters are available for a given experiment, and either one may be chosen at random. Depending upon temperature, humidity, etc., the readings of the voltmeters tend to deviate from the true voltage values. (The additional effect of fluctuation noise is assumed negligible here.) The distribution of the readings about each scale value follows a normal error curve. For one voltmeter the standard deviation is 0.1 volt, for the other 0.2 volt. What is the probability that any voltage read will be within 0.2 volt of its true value?

Note: In Probs. 5-35 to 5-38 that follow the words *rms noise* may be considered synonymous with standard deviation. This is explained fully in Chap. 6.

5-35 The output rms noise voltage of a given linear system is found to be 2 mv. The noise is gaussian fluctuation noise. What is the probability that the instantaneous noise voltage at the output of the system lies between -4 and $+4$ mv?

5-36 Repeat Prob. 5-35 if a dc voltage of 2 mv is added to the output noise.

5-37 Show that the probability of error in mistaking a binary pulse in noise for noise alone or of mistaking noise for a binary pulse is given by Eq. [5-95]. Calculate this error for various values of pulse height to rms noise, and check the curve of Fig. 5-27.

5-38 A binary transmission system transmits 50,000 digits per second. Fluctuation noise is added to the signal in the process of transmission, so that at the decoder, where the digits are converted back to a desired output form, the signal pulses are 1 volt in amplitude, with the rms noise voltage 0.2 volt. What is the average time between mistakes of this system? How is this average time changed if the signal pulses are doubled in amplitude?

5-39 Show the optimum decision level for bipolar signals in additive gaussian noise is given by Eq. [5-98] in the text.

5-40 A bipolar signal sequence of amplitude ± 1 volt has added to it noise with the triangular density function of Prob. 5-28, with parameter $a = 2$. Calculate the probability of error if 0 is chosen as the decision level for one sample of signal plus noise.

5-41 Noise having the Laplacian distribution of Prob. 5-29 is added to a bipolar signal sequence of amplitude $\pm A$. Find the probability of error in terms of A/σ (σ the noise standard deviation or rms value), if the decision level for one signal-plus-noise sample is set at zero. Plot the error curve and compare with Fig. 5-27 for gaussian noise.

5-42 A particular probability-density function is given by $f(x) = e^{-x}$, $0 \leq x < \infty$. Sketch both the density and cumulative-distribution functions, and show that $F(\infty) = 1$ (that is, the density function is properly normalized). Find the first and second moments and the variance of this distribution.

5-43 A density function is given by $f(x) = 1$, $0 \leq x \leq 1$. Show that the density function of $y = -\ln x$ is given by the function of Prob. 5-42.

5-44 Show that the density function of Eq. [5-109] is properly normalized.

5-45 Show that the average value of the distribution of Eq. [5-113] is σ^2.

5-46 The random variable x follows the Rayleigh distribution of Prob. 5-26. Find the distribution of the square of x.

5-47 *Computer generation of gaussian random variables from uniformly distributed random numbers.*

(*a*) Assume that random numbers x_i, uniformly distributed from 0 to 1, are available. Let $y = (b/n) \sum_{i=1}^{n} (x_i - \frac{1}{2})$, n a fixed number. Show y approximates a gaussian random variable of zero average value, and variance $\sigma^2 = b^2/12n$.

(*b*) The gaussian approximation of (*a*) is poor on the tails of the distribution. Why? A better approximation, using two independent uniform random numbers x and y, is obtained as follows:

(1) Let $r = \sqrt{-2\sigma^2 \log_e x}$. Show

$$f(r) = \frac{re^{-r^2/2\sigma^2}}{\sigma^2}$$

i.e., the Rayleigh distribution of Prob. 5-26 and Eq. [5-174] of the text.

(2) Show $z = r \cos 2\pi y$ is zero-mean gaussian, with variance σ^2. *Hint:* refer to the derivation of the Rayleigh density function in Sec. 5-10 (see Fig. 5-41).

5-48 Show that the Fourier transform of a gaussian function is itself a gaussian function. In particular, show that if a probability-density function is the gaussian function of Eq. [5-133], its characteristic function is given by Eq. [5-134].

5-49 A signal pulse has a gaussian-function shape. Find the relationship between the pulse width τ and the spread constant σ in the expression for the gaussian function if τ is the width measured between the half-power points in time. Find the frequency spectrum of the gaussian pulse, and relate the half-power frequency bandwidth to the pulse width τ. Does this result agree with the results of Chap. 2?

5-50 Find the characteristic function for the probability-density function of Prob. 5-42. Using this result, find the probability density for the sum of n such independent random variables. *Hint:* Note that $f(x)$ in Prob. 5-42 corresponds in form to the impulse response of a low-pass *RC* network. The probability density of the sum of the n variables should correspond to the impulse response of n isolated and cascaded *RC* networks.

5-51 Let

$$z = \sum_{i=1}^{n} a_i x_i$$

with the a_i's constants, the x_i's random variables.

(*a*) Show

$$E(z) = \sum_{i=1}^{n} a_i E(x_i)$$

(*b*) The x_i's are independent random variables. Show

$$\sigma_z^2 = \sum_{i=1}^{n} a_i^2 \sigma_{x_i}^2$$

Note: Actually, as shown in Prob. 5-59 below, the condition that the x_i's are *uncorrelated* suffices.

5-52 The input to a binary communications channel consists of 0's with a priori probability $P_0 = 0.8$, and 1's with a priori probability $P_1 = 0.2$. The transition

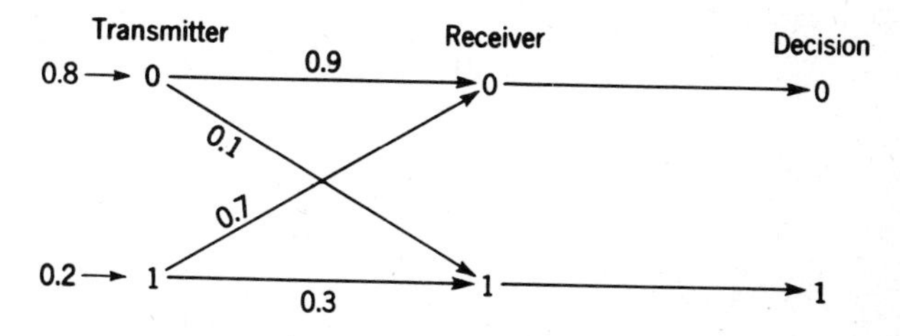

FIG. P 5-52 Transition probabilities.

probabilities on the channel, i.e., the probabilities during transmission that 1's will be received as 1's, 1's received as 0's, etc., are indicated in the figure.

(*a*) Find the probability of error P_e.
(*b*) Find P_e if the decisions are *reversed*, i.e., received 1's are called 0's and vice versa.
(*c*) The decision is made to *always* call a received signal a 0. Calculate the probability of error. Which is the best *decision rule* of the three described in (*a*), (*b*), (*c*)?
(*d*) Repeat (*a*), (*b*), (*c*) if $P_0 = P_1 = 0.5$.

5-53 A random variable x represents the input voltage, at a specified time, to each of the four devices shown in Fig. P 5-53. y is the corresponding random output voltage. Find and sketch the probability-density function of y for each of the following input-density functions:

(*a*) $f(x) = 1,\ 0 < x < 1.$
(*b*) $f(x) = e^{-x},\ x > 0.$
(*c*) $f(x) = (1/\sqrt{2\pi})e^{-x^2/2}.$

Check by calculating the area under each of the output-density functions.

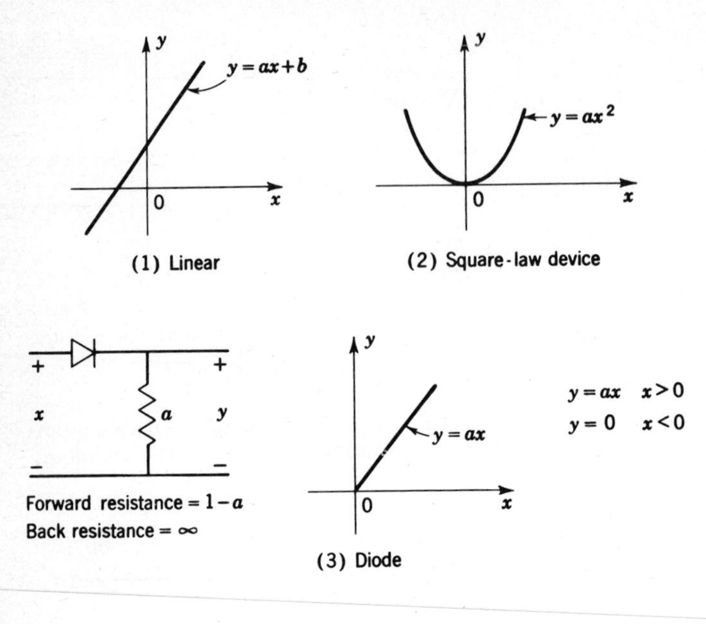

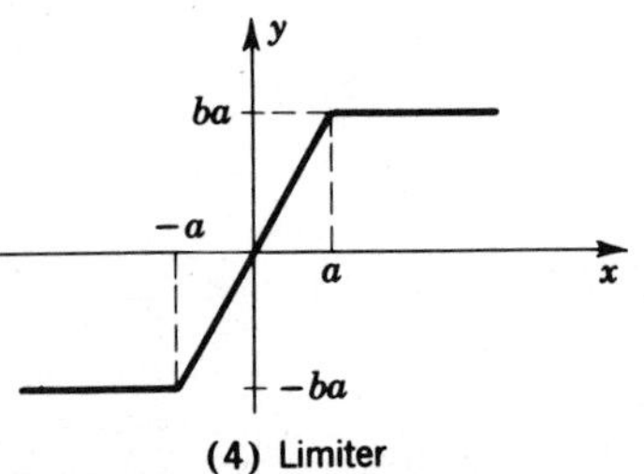

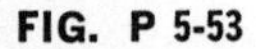
FIG. P 5-53

5-54 Bipolar signals are transmitted over a long line with 99 repeaters used. Find the overall probability of error at the output of the line if the probability of error at any repeater is $p = 10^{-7}$. Compare with the probability of error if amplifiers are used instead. (Additive gaussian noise is assumed.)

5-55 As will be shown in Chap. 6 gaussian noise applied at the input to a narrowband filter centered at f_c Hz appears at the output in the form

$$n(t) = x(t) \cos \omega_c t - y(t) \sin \omega_c t$$

with x and y slowly varying independent zero-mean gaussian voltages with the same variance σ_n^2. Show that $n(t)$ may also be written

$$n(t) = r(t) \cos (\omega_c t + \theta)$$

with r Rayleigh-distributed, and θ uniformly distributed between 0 and 2π.

5-56 A constant-amplitude sine wave of frequency f_c is transmitted through a fading channel, arriving at the receiver with a Rayleigh envelope distribution, with the parameter σ_s^2. Gaussian noise is added to the signal, and signal plus noise are both passed through narrowband r-f circuitry centered at f_c. Using the results of Prob. 5-55, show the envelope of the *sum* of signal plus noise at the narrowband circuitry output is also Rayleigh distributed, with the parameter $\sigma_n^2 + \sigma_s^2$. Calculate the probability that the envelope will exceed the rms noise voltage σ_n for several representative values of *signal-to-noise ratio* σ_s/σ_n.

5-57 A sinusoidal signal $A \cos \omega_c t$ plus gaussian noise appear at the output of a narrowband filter centered at f_c Hz. Using the result of Prob. 5-55, show that the envelope of the signal plus noise is *Rician-distributed* (Eq. [5-184] in the text).

5-58 A radar transmitter sends out high-frequency sinusoidal pulses. These are scattered back from a large complex target made up of many independent scatterers, no one of which predominates. Show the amplitudes of the received target echoes are Rayleigh-distributed.

5-59 n random variables, x_i, $i = 1, \ldots, n$, are pairwise uncorrelated so that $E[x_i x_j] = E(x_i)E(x_j)$, $i \neq j$. Show that the variance of the sum

$$z = \sum_{i=1}^{n} a_i x_i$$

is given by

$$\sigma_z^2 = \sum_{i=1}^{n} a_i^2 \sigma_{x_i}^2.$$

5-60 A fading received signal is of the form $s(t) = r \cos (\omega_c t + \theta)$, with r and θ relatively slowly varying, and f_c a high-frequency carrier.

(*a*) Show that the average power dissipated in a 1-ohm resistor is very nearly $P = r^2/2$ if a time average is taken over many periods of the carrier, during which time r and θ may be assumed to be nearly constant.

(*b*) Show the probability-density function of the average power P is given by $f(P) = e^{-P/\sigma^2}/\sigma^2$ if the envelope r of the signal is Rayleigh-distributed with parameter σ^2.

5-61 The two-variable (or two-dimensional) gaussian density function is given by Eq. [5-206]. Find the one-dimensional density function $f_y(y)$, $E(y)$, and $\mu_{02} = E[y - E(y)]^2$, and show they agree with previous (one-dimensional) results. Using the definition of the covariance μ_{11} given in Eq. [5-201], verify by actual integration that μ_{11} in Eq. [5-206] is appropriately labeled as such.

CHAPTER SIX

RANDOM SIGNALS AND NOISE

In the previous chapter we introduced the elements of probability theory and applied them to some important problems in communications—the effect of additive noise on digital transmission, the statistics of fading channels, etc. Some highly significant questions were left unanswered, however. Thus, although we were able in the case of digital transmission to quantitatively find the probability of error if the additive noise statistics were known, the final results depended on a knowledge of the noise variance σ^2. How does one physically measure this parameter, or calculate it if possible in physical systems? Since $n(t)$ is itself a time-varying wave, just as any signal it must be affected by the system through which it passes. How does one quantitatively determine the effect of systems on noise?

More important, if we now return to the premise of Chap. 1, that real *signals* must be time-varying and unpredictable (otherwise why transmit them?), it is apparent that the analysis we shall outline in the section that follows is not only of importance in studying noise, but is exactly the analysis we must use in studying real time-varying signals in systems.

In studying the properties of random signals and noise transmitted through systems we shall again find frequency or spectral analysis of signals and systems playing a dominant role. The random character of the signals and noise will, however, dictate a somewhat different approach than the Fourier analysis of deterministic signals, as in previous chapters. Interestingly, in extending the frequency analysis of deterministic (hence artificial) signals to the more realistic random-signal case, we shall find many of the results obtained in our study of deterministic signals still valid. These include such things as bandwidths, the filtering effects of linear systems, modulation and demodulation, etc. This indicates the usefulness of the approach we have adopted of studying deterministic signals first.

Using the spectral analysis of random signals and noise, we shall find ourselves in a position to compare various modulation and demodulation techniques for both digital and analog signals, not only on the basis of bandwidths (as done in previous chapters), but also on the basis of their efficiency in the presence of noise. Specifically, how do AM and FM systems compare insofar as their signal in noise performance is concerned? Why is phase-shift keying (PSK) as a modulation method to be preferred over frequency-shift keying (FSK) and on-off keying (OOK) when noise is present? What is the signal-to-noise advantage of synchronous detection over envelope detection? These and other questions will be answered in this chapter.

As in the previous chapter, in discussing the effect of noise in digital transmission, we shall assume noise somehow generated in the system. Questions regarding the mechanism of the generation of noise and the types of noise typically encountered in systems will be left for Chap. 7.

6-1 TIME AND ENSEMBLE AVERAGES

As noted in the previous chapter a typical oscillogram of noise would appear as in Fig. 6-1 (see also Figs. 5-2 and 5-24). Random-signal

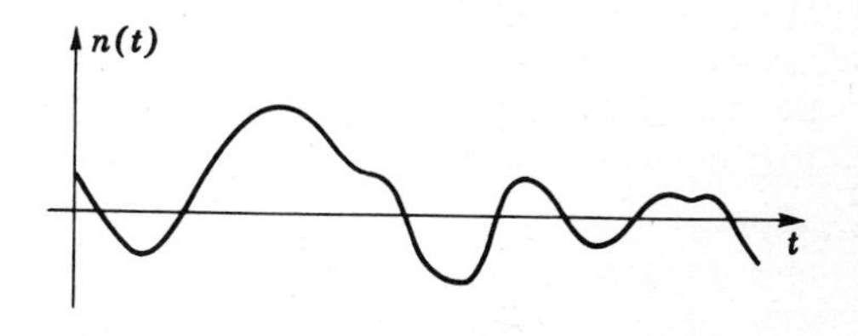

FIG. 6-1 Random process.

waveshapes would also have a similar irregular, unpredictable appearance. We call this random time-varying function a *random process* $n(t)$. [Although we shall for simplicity's sake use the notation $n(t)$ and frequently refer to the random wave as noise, random signals are included as well.] We pointed out in the last chapter that a sample of $n(t)$ taken at an arbitrary time t could be considered a random variable with some probability-density function $f_n(n)$. One technique indicated earlier for finding $f_n(n)$ was to perform a histogram analysis, sampling $n(t)$ at intervals "far enough apart" to ensure the statistical independence of the samples, setting levels at n and $n + \Delta n$, and counting the number of times the samples fell in this range of n.

But what do we mean by "independent samples"? What constitutes "far enough apart"? To answer these questions we shall have to discuss the effect of passing noise through linear systems, and how the wavelike structure of $n(t)$ is related to the linear system at whose output it appears. Although spectral analysis is needed to answer these questions, we first focus attention on the measurement of some simple statistical parameters such as $E(n)$, the variance σ_n^2, etc. In considering how one measures these quantities used so often in Chap. 5, we shall be led into the more general discussion of spectral analysis.

Although the average value and variance of n can be obtained from the same histogram analysis of $n(t)$ used to find $f_n(n)$, simply by averaging measured samples appropriately, it is apparent intuitively that one should be able to make the same measurements much more simply using time-averaging meters. For example, if we were to feed the wave $n(t)$ into a dc meter, we would intuitively expect to get a measure of the expected value $E(n)$. In this case we are implicitly comparing an *expected* or *statistical* average with a *time* average, as carried out by a dc meter. Specifically, if the meter has an effective time constant T, its reading should give a number

$$\bar{n} = \frac{1}{T}\int_0^T n(t)\,dt \qquad [6\text{-}1]$$

It is apparent that as $n(t)$ varies randomly, so will $\bar{n}$. Depending on *when* we perform the indicated average, we will get different numbers $\bar{n}$. So $\bar{n}$ is a random variable, with its own expected value, variance, etc. But we would still expect to find $\bar{n}$ some measure of $E(n)$, the statistical average of $n(t)$. To indicate the connection, we take the expected value of $\bar{n}$ itself; i.e., we visualize many meter readings over different sections of $n(t)$, each T sec long, the expected value of $\bar{n}$ then being the statistical average of these. Then

$$E(\bar{n}) = E\left[\frac{1}{T}\int_0^T n(t)\,dt\right] \qquad [6\text{-}2]$$

It may be shown that the operations of expectation and integration are interchangeable.[1] We then write

$$E(\bar{n}) = \frac{1}{T}\int_0^T E[n(t)]\,dt \qquad [6\text{-}3]$$

We now assume that the expected value $E[n(t)]$ is independent of time t. This is reasonable, for if the expected value *were* varying with time, one wouldn't expect the dc meter reading to be a fixed number anyway. Alternately, if we were to visualize many such strips of $n(t)$, each T sec long, placed one above the other [either obtained from the same record by

[1] A. Papoulis, "Probability, Random Variables, and Stochastic Processes," chap. 9, McGraw-Hill, New York, 1965.

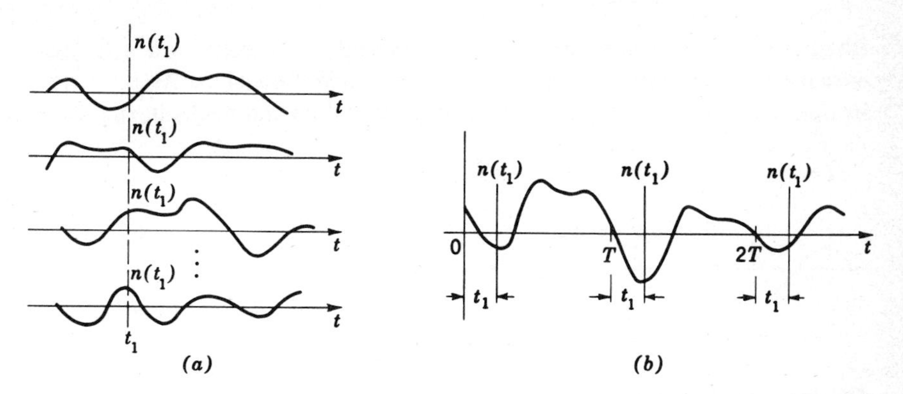

FIG. 6-2 Ensemble averages. (a) Many identical sources; (b) by cutting one record.

cutting the one strip at T-sec intervals, or by visualizing many identical sources providing independent outputs $n(t)$] (see Fig. 6-2), we could perform an ensemble average of the random variable $n(t_1)$ to actually find a close approximation to $E[n(t_1)]$. One would then assume that $E[n(t_2)]$, or $E[n(t)]$ at any *other* value of t, was the same. (This would be verified by averaging at each time interval, if so desired.)

With this assumption that $E[n(t)]$ is a *constant*, independent of time, we find, from Eq. [6-3], that

$$E(\bar{n}) = E(n) \tag{6-4}$$

So our intuition is justified here, indicating that at least in an *average* sense the time average $\bar{n}$ does provide a measure of $E(n)$. But we would also expect that the time interval T should play a role here. By making T longer, or by averaging over longer sections of $n(t)$, we would expect to find $\bar{n}$ approaching $E(n)$ more closely. In fact this is easily demonstrated by calculating the variance var$(\bar{n})$ of the random variable $\bar{n}$. Thus, by definition of the variance,

$$\text{var}\,(\bar{n}) = E[\bar{n} - E(\bar{n})]^2 = E\left\{\frac{1}{T}\int_0^T [n(t) - E(n)]^2\,dt\right\} \tag{6-5}$$

Although we shall not perform the calculation here, it is readily shown[1] that

$$\frac{\text{var}\,(\bar{n})}{E^2(n)} \to \frac{C}{BT} \qquad BT \gg 1 \tag{6-6}$$

where C is a fixed constant the order of 1, and B is the bandwidth of the process $n(t)$. [We shall define B precisely in the material to follow, but

[1] Papoulis, *ibid.*, pp. 324–328. This assumes of course that $E(n) \neq 0$.

roughly speaking it is a measure of the rapidity of variation of $n(t)$, just as with the deterministic signals encountered previously. As indicated in Fig. 6-3 the time spread between successive dips and peaks in the wave

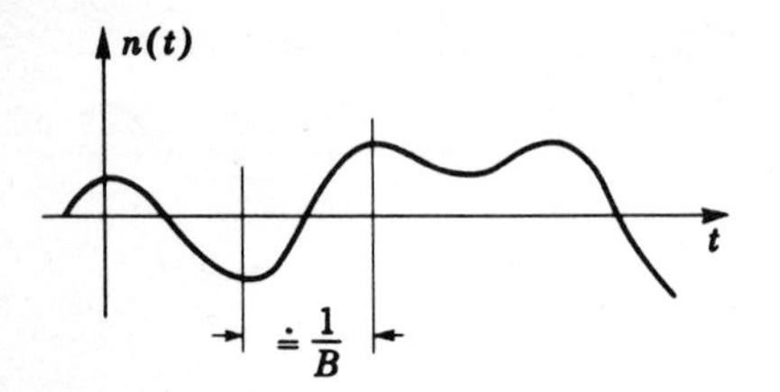

FIG. 6-3 Significance of bandwidth B.

is approximately the reciprocal of the bandwidth.] As an example, if $B = 100$ kHz and $T = 1$ msec (recall this is the meter averaging time), $1/BT = 0.01$. The spread about the expected value is thus 0.1 of the expected value.

As the dc meter averages over longer and longer time intervals, its reading $\bar{n}$ approaches more and more closely the parameter $E(n)$. (In practical situations it is usually sufficient to have $T \gg 1/B$.) For the variance of the reading $\bar{n}$ goes to zero as $1/T$, indicating that the variations about $E(n)$ decrease in the same manner. Thus, in the previous example if the meter time constant is increased to 100 msec, $1/BT = 10^{-4}$. The spread about the expected value is the square root of this, or 10^{-2}. Increasing the integration time by a factor of 100 has narrowed the deviation about $E(n)$ by a factor of 10, on the average. In the limit, as $T \to \infty$, we must have

$$\lim_{T\to\infty} \bar{n} = \lim_{T\to\infty} \frac{1}{T}\int_0^T n(t)\,dt = E(n) \qquad [6\text{-}7]$$

A random waveshape or process $n(t)$ for which Eq. [6-7] is true is said to be an *ergodic process;* i.e., time and ensemble (statistical) averages may be equated. Although there are processes for which this is not true,[1] we shall henceforth assume that $\bar{n}$ and $E(n)$ may be equated.

Since we have shown that one may use a dc meter to measure $E(n)$, one may reasonably ask if it is similarly possible to measure $\sigma_n{}^2$, the vari-

[1] A trivial example of a nonergodic process is that consisting of an ensemble of constant-voltage sources. Each source maintains its output absolutely constant with time. Each source provides, however, a different output. Choosing one source at random and averaging its output with time, we of course measure the particular source voltage. Measuring all sources simultaneously, however, and averaging these (the *ensemble* average), we get a different result than the time average, the ensemble average depending on the distribution of the source output.

ance of the noise. The answer is of course "yes," and in fact one uses a *power meter* for this purpose. Specifically, if one defines the average power P_{av} over an interval T sec long just as in previous chapters:

$$P_{av} \equiv \frac{1}{T}\int_0^T n^2(t)\,dt \qquad [6\text{-}8]$$

one shows, again by interchanging the order of ensemble averaging and integration, that

$$E(P_{av}) = E(n^2) = \sigma_n^2 + E^2(n) \qquad [6\text{-}9]$$

or

$$\sigma_n^2 = E(P_{av}) - E^2(n) \qquad [6\text{-}10]$$

if $E(n^2)$ is invariant with time. [Recall that $\sigma^2 = m_2 - m_1^2 = E(n^2) - E^2(n)$ for a random variable.]

Again P_{av}, as read by the power meter of time constant T, is a random variable, but *on the average* the readings will provide a measure of the second moment $E(n^2)$. Since $E^2(n)$ is very nearly the square of the dc value, it is apparent that the variance σ_n^2 must provide a measure of the fluctuating or non-dc power. To emphasize the fact that the variance on an ensemble average basis is the same as the time-averaged fluctuation power, we shall henceforth use the symbol N for the latter. One often calls this the ac power, as measured by true rms meters. N would then be the *square* of the rms meter reading.

One may again show[1] that

$$\frac{\text{var}\,(P_{av})}{E^2(P_{av})} = \frac{C'}{BT} \qquad BT \gg 1,\ C' \text{ a constant} \qquad [6\text{-}11]$$

Thus, as $T \to \infty$ (in practice $T \gg 1/B$ again suffices), the reading P_{av} approaches $E(n^2)$ with probability of 1, and we have

$$\lim_{T\to\infty} (P_{av} - \bar{n}^2) = \lim_{T\to\infty} \frac{1}{T}\int_0^T [n(t) - \bar{n}]^2\,dt = N \qquad [6\text{-}12]$$

A process $n(t)$ for which time and ensemble averaging are equal, in the sense of Eq. [6-12], is again spoken of as an ergodic process. We assume henceforth that one may interchange these two averages, although we shall encounter at least one example later in which this is not valid.

6-2 AUTOCORRELATION FUNCTION AND SPECTRAL ANALYSIS

We have just shown how one relates time and ensemble averages for a random wave or process $n(t)$. In particular, one may use a dc meter to

[1] Papoulis, *op. cit.* Here one must generally assume, however, that the statistics of $n(t)$ are *gaussian;* i.e., at any instant of time $f(n)$ is gaussian.

measure $E(n)$, and, if the dc term is blocked or absent $[E(n) = 0]$, one may use a true rms meter to measure N, the noise power or variance. These assume, however, that the meter time constant $T \gg 1/B$, with B the "bandwidth" of the process. How does one determine B? How is B related to the actual variations in time of $n(t)$? Is it possible to calculate N, and, if so, how does this depend on the physical systems through which $n(t)$ propagates? It is also apparent, extrapolating from our discussions of deterministic signals in previous chapters, that the bandwidth B must somehow relate to the physical system in which $n(t)$ is generated or through which it propagates. For example, one would not expect to find noise with a bandwidth of 1 MHz at the output of a system whose bandwidth is 100 Hz. Any noise terms varying this rapidly just would not appear at the system output.

Basically we require some measure of how the noise process may vary in a given time interval. Specifically, if we consider the noise wave $n(t)$ of Fig. 6-4 we note that as $t_2 \to t_1$, $n(t_2)$ as a random variable becomes more "closely related to" (or "predictable by") $n(t_1)$. As $(t_2 - t_1)$

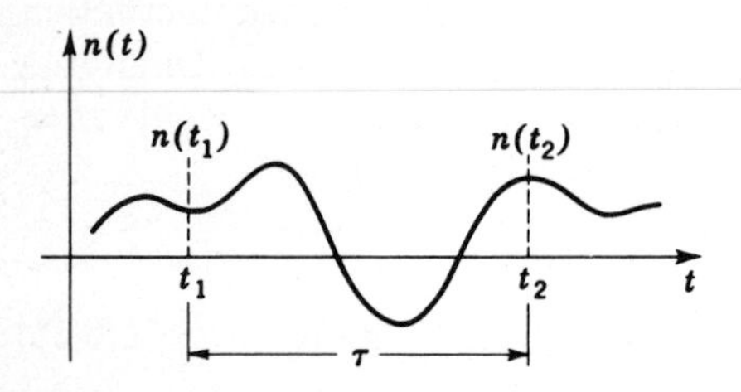

FIG. 6-4 Autocorrelation definition.

increases, we expect less dependence of one upon the other. We make this concept more precise by defining the autocorrelation function $R_n(t_1,t_2)$:

$$R_n(t_1,t_2) \equiv E[n(t_1)n(t_2)] \qquad [6\text{-}13]$$

Although many different definitions of "dependence" of one random variable on one another are possible, the autocorrelation function is probably the simplest and has many desirable properties that we shall explore later. Note that it is the extension to the *same* random variable (hence the prefix *auto*) of the definition for the *covariance* of two random variables introduced in Sec. 5-11. It is apparent that if $t_2 \to t_1$, $R_n \to E(n^2)$, or just the statistical second moment. If at some spacing $(t_2 - t_1)$, $n(t_2)$ and $n(t_1)$ tend to become statistically independent (one would expect this to occur at intervals greater than $1/B$), $R_n \to E^2(n)$, or 0, if $E(n) = 0$. $R_n(t_1,t_2)$ thus provides one possible measure of the dependence of $n(t_2)$ and $n(t_1)$.

To simplify the discussion we shall assume that $R_n(t_1,t_2)$ depends only on the interval $(t_2 - t_1) \equiv \tau$, and not on the time origin t_1. [This is similar to our assumption, made previously, that $E(n)$ is independent of time.] Then we can write

$$R_n(\tau) \equiv E[n(t)n(t + \tau)] \qquad [6\text{-}13a]$$

A process for which this is true, and for which $E(n)$ is independent of time, is called a *stationary process*.[1]

Now how would we actually measure $R_n(\tau)$? As we did previously we set up a time integral with the property that its expected value equals $R_n(\tau)$. Consider the integral

$$\bar{R}_n(\tau) \equiv \frac{1}{T} \int_0^T n(t)n(t + \tau)\, dt \qquad [6\text{-}14]$$

Note that this provides some measure of the dependence with time τ of $n(t)$ and $n(t + \tau)$. For as $\tau \to 0$, $\bar{R}_n(0) = P_{av}$; as τ increases and $n(t)$ and $n(t + \tau)$ vary relatively independently of one another, one would expect that the product of the two would be negative as often as positive, approaching zero if $E(n) = 0$.

Again $\bar{R}_n(\tau)$ is a random variable, depending on the interval T sec long over which evaluated. If we now ensemble average over all possible values of this variable, we get, from Eqs. [6-14] and [6-13*a*],

$$E[\bar{R}_n(\tau)] = R_n(\tau) \qquad [6\text{-}15]$$

(Again ensemble averaging and integration are interchanged.) So in an average sense the time average of Eq. [6-14] and the ensemble average of Eq. [6-13*a*] are the same.

As previously, one may calculate the variance of $\bar{R}_n(\tau)$ and show it goes to zero as $1/BT$, for large T.[2] We then have, as in the previous cases,

$$\lim_{T \to \infty} \bar{R}_n(\tau) = R_n(\tau) \qquad [6\text{-}16]$$

{Note from Eqs. [6-13*a*] and [6-14] that included as a special case here is the result $\lim_{T \to \infty} P_{av} = N + E^2(n) = E(n^2)$, previously shown as Eq. [6-12].}

We are now in a position to actually relate $\bar{R}_n(\tau)$ [or $R_n(\tau)$] to a spectral analysis of $n(t)$, and thus to a defined bandwidth. To do this

[1] Strictly speaking such a process is usually called a *wide-sense* stationary process. The term stationary process is then reserved for a more general case in which distribution functions are invariant with time. See Papoulis, *op. cit.*, chap. 9.

[2] One must assume $n(t)$ is gaussian to actually carry out the averaging necessary.

assume temporarily that $n(t)$ is deterministic. We take a section of $n(t)$ T sec long and expand it in a Fourier series. This will obviously consist of harmonics of the fundamental frequency $1/T$ (see Fig. 6-5). Thus,

$$n(t) = \frac{1}{T} \sum_{m=-\infty}^{\infty} c_m e^{j\omega_m t} \qquad \omega_m = \frac{2\pi m}{T} \tag{6-17}$$

Here

$$c_m = \int_{-T/2}^{T/2} n(t) e^{-j\omega_m t}\, dt \tag{6-18}$$

as in previous chapters. Note that c_m is a random variable because $n(t)$ is random. In fact, if $E(n) = 0$, it is easily shown that $E[c_m] = 0$. So we are in the peculiar situation of having a Fourier-series representation

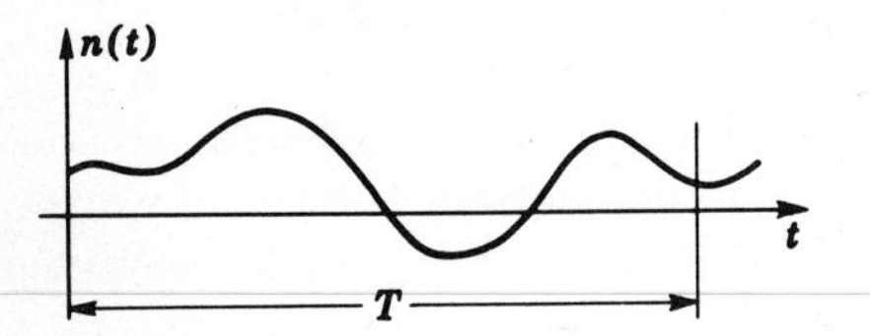

FIG. 6-5 Fourier analysis of $n(t)$.

of $n(t)$ which is valid for the *particular time interval T* over which determined, but which varies statistically each time we take another strip of $n(t)$ T sec long. Is it possible, however, to develop a spectral analysis of $n(t)$ that "settles down" to a fixed frequency spectrum without statistical fluctuations?

Recall that we had the ergodic relation of Eq. [6-16],

$$\lim_{T\to\infty} \bar{R}_n(\tau) = R_n(\tau) \tag{6-16}$$

This gives us a hint as to how to proceed. Let

$$n(t) = \frac{1}{T} \sum_{m=-\infty}^{\infty} c_m e^{j\omega_m t} = \frac{1}{T} \sum_{l=-\infty}^{\infty} c_l^* e^{-j\omega_l t} \qquad \omega_m = \frac{2\pi m}{T},\ \omega_l = \frac{2\pi l}{T} \tag{6-19}$$

[Since $n(t)$ is real, $n^*(t) = n(t)$, giving the last term on the right.] Then

$$\bar{R}_n(\tau) = \frac{1}{T}\int_0^T n(t)n(t+\tau)\,dt = \frac{1}{T}\int_0^T \left(\frac{1}{T}\sum_m c_m e^{j\omega_m t}\right)\left[\frac{1}{T}\sum_l c_l^* e^{-j\omega_l (t+\tau)}\right] dt$$

$$= \frac{1}{T}\sum_m \sum_l \left(\frac{1}{T^2}\right) c_l^* c_m e^{-j\omega_l \tau} \int_0^T e^{j(\omega_m - \omega_l)t}\,dt \tag{6-20}$$

But it is easy to show that

$$\begin{Bmatrix} \int_0^T e^{j\omega_m t}\,dt = 0 & \omega_m \neq 0 \\ = T & \omega_m = 0 \end{Bmatrix}$$

(The complex exponentials are examples of *orthogonal* functions, as might be expected since they are linear combinations of sines and cosines.)

Hence

$$\bar{R}_n(\tau) = \frac{1}{T^2} \sum_{l=-\infty}^{\infty} |c_l|^2 e^{-j\omega_l \tau} \tag{6-20a}$$

Just as in the Fourier-series→Fourier-transform transition of Chap. 2, the spacing between frequencies is $\Delta f = 1/T$. We may then write

$$\bar{R}_n(\tau) = \sum_{l=-\infty}^{\infty} \left(\frac{1}{T} |c_l|^2 \right) e^{-j\omega_l \tau}\, \Delta f \tag{6-20b}$$

and, on taking statistical averages,

$$E[\bar{R}_n(\tau)] = R_n(\tau) = \sum_{l=-\infty}^{\infty} E\left(\frac{1}{T} |c_l|^2 \right) e^{-j\omega_l \tau}\, \Delta f \tag{6-21}$$

But with $\tau = 0$, we also have

$$\bar{R}_n(0) = \frac{1}{T} \int_0^T n^2(t)\, dt = P_{\text{av}} = \sum_{l=-\infty}^{\infty} \left(\frac{1}{T} |c_l|^2 \right) \Delta f \tag{6-22}$$

and

$$E[\bar{R}_n(0)] = E(n^2) = \sum_{l=-\infty}^{\infty} E\left(\frac{1}{T} |c_l|^2 \right) \Delta f \tag{6-23}$$

Since $\bar{R}_n(0)$ represents the total power in the wave $n(t)$ (as measured by a true power meter), and, as we have indicated, $E[\bar{R}_n(0)] = R_n(0)$ for large enough T, the term $(1/T)E(|c_l|^2)$ has the dimensions of power/frequency range Δf, or power/Hz. Calling this term $G_n(f_l)$, we have, formally,

$$R_n(0) = E(P_{\text{av}}) = E(n^2) = \sum_{l=-\infty}^{\infty} G_n(f_l)\, \Delta f \tag{6-24}$$

$G_n(f_l)$ is thus a *power-density* term, providing a measure of the power in $n(t)$ [actually mean-squared $n(t)$] concentrated at each frequency range Δf wide.

In the limit, as $T \to \infty$, $\Delta f \to 0$ and we have

$$E(n^2) = \int_{-\infty}^{\infty} G_n(f)\, df \tag{6-25}$$

In the special case where the noise $n(t)$ is zero mean, $E(n) = 0$, and

$$N = \int_{-\infty}^{\infty} G_n(f)\,df \qquad E(n) = 0 \tag{6-26}$$

The noise power or variance N, the parameter on which the probability of error in the detection of pulses in noise was found to depend (Sec. 5-6), is thus directly related to the spectral density $G_n(f)$. It is the sum of the noise-power contributions at all frequencies. In the case of $T \to \infty$, we find here, just as in Chap. 2, that the frequency spectrum becomes *continuous*. $G_n(f)$ is referred to as the *power spectral density* of $n(t)$, or, frequently, just the *power spectrum*.

The difference in spectral analysis between deterministic and random signals (or noise) now becomes apparent. In the deterministic case we can expand the time function in its Fourier series or transform, resulting in an *amplitude* (and *phase*) *spectrum*. In the case of random signals it is the *power density* at each frequency that plays the equivalent role.

How does one actually determine the spectral density $G_n(f)$? Note from Eq. [6-21] that we can also write, in the limit of large T,

$$R_n(\tau) = \int_{-\infty}^{\infty} G_n(f)e^{-j\omega\tau}\,df \qquad \omega = 2\pi f \tag{6-27}$$

The power spectral density is thus formally the Fourier transform of $R_n(\tau)$, and may therefore be written

$$G_n(f) = \int_{-\infty}^{\infty} R_n(\tau)e^{j\omega\tau}\,d\tau \qquad [6\text{-}28]^1$$

Very often the autocorrelation function $R_n(\tau)$ of a process $n(t)$ may be measured, calculated, or even estimated on the basis of known data. For example, one might measure $\bar{R}(\tau)$ with analog devices using the time-average definition of Eq. [6-14]. With digital devices, one could sample $n(t)$ at discrete time intervals, and perform the equivalent digital calculation,

$$\bar{R}_n(\tau) = \frac{1}{M}\sum_{j=1}^{M} n(j\Delta)n(j\Delta + \tau) \qquad M\Delta = T \tag{6-14a}$$

to determine the autocorrelation function. (Δ is the spacing between samples, and we must have $\tau = m\Delta$, m an integer.)

[1] This is often used as the *definition* of $G_n(f)$, avoiding the need for a Fourier expansion, as developed here. See, for example, Papoulis, *op. cit.*, chap. 10. The Fourier integral relations of Eqs. [6-27] and [6-28] are *only* valid for *stationary processes*. In those cases where the process $n(t)$ is nonstationary, one may still define the autocorrelation function $R_n(t_1,t_2) \equiv E[n(t_1)n(t_2)]$, but a spectral density can no longer be defined.

The spectral density $G_n(f)$ is then found by taking the Fourier transform of $R_n(\tau)$.

Alternately, one might consider using a given strip T sec long of noise $n(t)$ to find the spectral density directly. Since we had

$$G_n(f_l) = E\left(\frac{1}{T}|c_l|^2\right) \qquad [6\text{-}29]$$

we might take as a possible estimate (approximation) of $G_n(f_l)$ the expression

$$\hat{G}_n(f_l) = \frac{1}{T}|c_l|^2 = \frac{1}{T}\left|\int_{-T/2}^{T/2} n(t)e^{-j\omega_l t}\,dt\right|^2 \qquad [6\text{-}30]$$

using the T-sec strip. This is a very useful approximation to $G_n(f_l)$ because with modern high-speed digital computers and the fast Fourier-transform method of finding Fourier coefficients extremely rapidly with such computers,[1] one may quite easily find $G_n(f_l)$ given a typical record of $n(t)$. There is a possible difficulty, however, of which one must be aware. Although we have, by definition, $G_n(f_l) = E[(1/T)|c_l|^2]$, it turns out that, unlike the previous cases considered, the *variance* of the random variable $\hat{G}_n(f_l)$ does *not* decrease with increasing T.[2] $\hat{G}_n(f_l)$ therefore does *not* approach $G_n(f_l)$ in a statistical sense; the approximation to the spectral density given by Eq. [6-30] does not settle down to the desired $G_n(f_l)$ but continues to oscillate randomly about that quantity no matter how long a time strip is taken. One way of avoiding this somewhat unpleasant dilemma is to take many (say K) time strips of the process $n(t)$, each T sec long ($BT \gg 1$), calculate $\hat{G}_n(f_l)$ for *each*, and then *average* the resultant calculated spectral densities to obtain an approximation to $G_n(f_l)$. If each calculated (and random) $\hat{G}_n(f_l)$ is statistically independent of the others, it is readily shown that the variance of the averaged calculated spectral density decreases as $1/K$, justifying the use of this technique.[3]

Typical examples of spectral-density and autocorrelation-function transform pairs are tabulated in Fig. 6-6. Note that in all three cases shown the autocorrelation function eventually goes to zero for "τ large enough." Since $R_n(\tau) = E[n(t)n(t + \tau)]$, and we have been assuming $E(n) = 0$, it is apparent that $n(t)$ and $n(t + \tau)$ (Fig. 6-7) eventually

[1] See, for example, the excellent paper, What Is the Fast Fourier Transform?, by W. T. Cochran et al., *Proc. IEEE*, vol. 55, no. 10, pp. 1664–1674, October, 1967.

[2] W. B. Davenport and W. L. Root, "Introduction to Random Signals and Noise," McGraw-Hill, New York, 1958.

[3] P. D. Welch, The Use of Fast Fourier Transforms for the Estimation of Power Spectra: A Method Based on Time Averaging Over Short, Modified Periodograms, *IEEE Trans. Audio and Electroacoustics*, vol. AU-15, pp. 70-73, June, 1967.

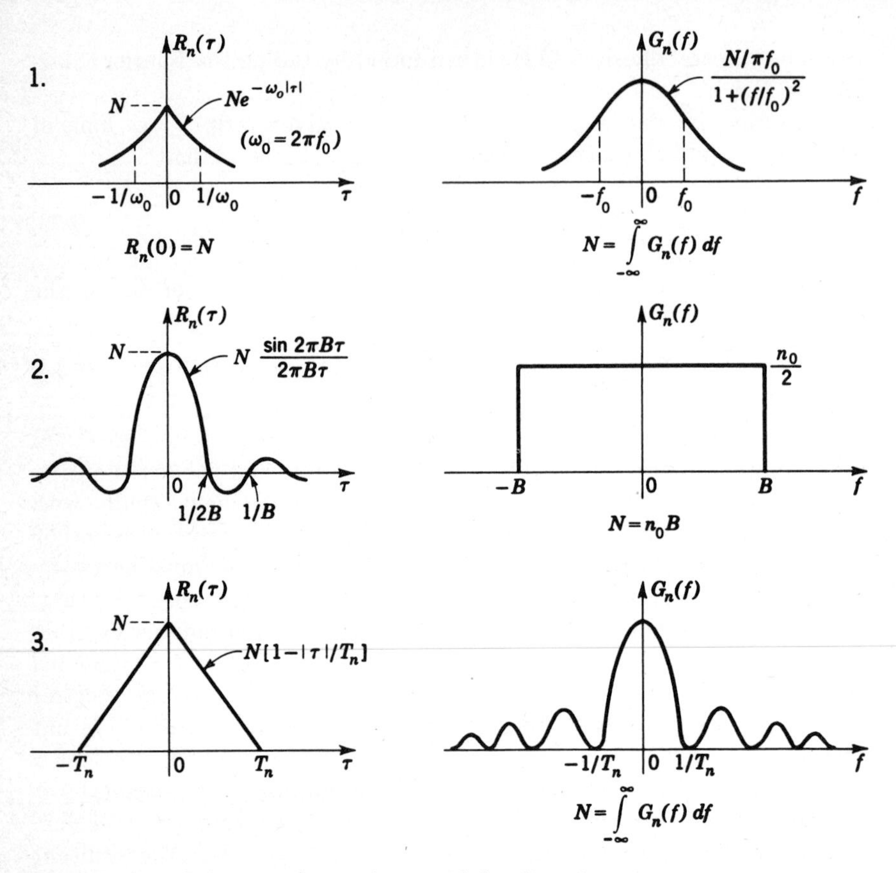

FIG. 6-6 Correlation function and spectral-density pairs.

become uncorrelated. If one defines an arbitrary bandwidth B, just as was done in Chap. 2, as some measure of the width of the spectral-density curve, $G_n(f)$, it is apparent from these curves and from the transform relations between $R_n(\tau)$ and $G_n(f)$ that $1/B$ plays an important role in determining the measure of correlation between a sample of $n(t)$ at time t,

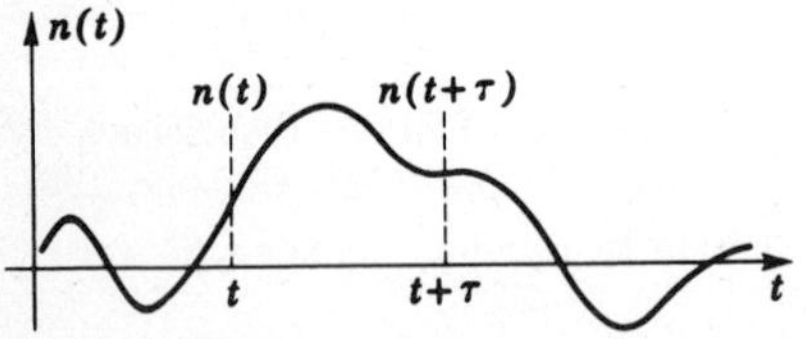

FIG. 6-7 Spacing between noise samples.

and a sample τ sec later. Thus, in example 1, with f_0 the 3-db bandwidth, $R_n(\pm 1/2\pi f_0) = e^{-1}R_n(0)$. [$G_n(f_0) = \frac{1}{2}G_n(0)$; this is *not* the same as half-power bandwidth, in which case $\int_{-B_{1/2}}^{B_{1/2}} G_n(f)\, df = N/2$.] In example 2, the noise $n(t)$ is truly *band-limited* to B Hz. It is apparent that at $\tau = \pm 1/2B$, and integral multiples thereof, $n(t)$ and $n(t + \tau)$ are *always* uncorrelated. This is also the case in example 3 for $|\tau| \geq T_n$, and we note that $1/T_n$, the first zero crossing of $G_n(f)$, is a measure of the bandwidth of $n(t)$.

Specifically, if the bandwidth B is 1 MHz, samples spaced more than 1 μsec apart are essentially uncorrelated in all three examples of Fig. 6-6. (In the case of example 3, if $T_n = 1\ \mu$sec, the samples *are* uncorrelated for all $\tau \geq 1\ \mu$sec.) Recall from Sec. 5-11 that two uncorrelated gaussian variables are *independent* as well. If the random wave in Fig. 6-7 is then gaussian, $n(t)$ and $n(t + \tau)$ are essentially independent if $\tau > 1/B$ sec.

Note that in examples 1 and 3 of Fig. 6-6 most of the noise power appears concentrated about the origin (dc), the bandwidths f_0 and $1/T_n$, respectively, then serving as measures of this concentration. The fact that relatively little noise power appears at the high frequencies thus indicates that the noise $n(t)$ rarely fluctuates at these rates, justifying our intuitive statements previously that 1/bandwidth is, roughly speaking, a measure of the time between significant changes in $n(t)$. This is of course also shown by the plots of the autocorrelation function $R_n(\tau)$; the value of τ for which $R_n(\tau)$ begins to decrease significantly is also a measure of the time between significant changes in $n(t)$ (this is exactly one possible interpretation of correlation), and is of course just 1/bandwidth, defined in some arbitrary sense. In example 2 *all* the power is assumed concentrated in the range 0 to B Hz. This case is of course the random signal equivalent of the band-limited deterministic signals of Chap. 2. $R_n(\tau)$ here is equivalent to $f(t)$ there; $G_n(f)$ is equivalent to $F(\omega)$.

One particular example of a spectral density that plays an extremely important role in communications and signal-processing analyses is that in which the spectral density $G_n(f)$ is flat or constant, say equal to $n_0/2$, over *all* frequencies:

$$G_n(f) = \frac{n_0}{2} \qquad (\text{all } f) \tag{6-31}$$

Although this is, strictly speaking, physically inadmissible, since it implies infinite noise power [$N = \int_{-\infty}^{\infty} G_n(f)\, df$], it is a good model for many typical situations in which the noise bandwidth is so large as to be out of the range of our measuring instruments (or frequencies of interest to us). We shall encounter some typical examples of noise of this type in Chap. 7 in our discussion of the origin and sources of noise.

Noise $n(t)$ with a flat spectral density $n_0/2$, as in Eq. [6-31], is called *white noise* because of its "equal jumbling" of all frequencies (compare with the common appellation "white light").

Note that in the band-limited noise case of example 2 of Fig. 6-6, one may obtain *white noise* by letting $B \to \infty$. The noise of example 2 is therefore often called *band-limited white noise.* It is apparent from the transform pair relations that the autocorrelation function for white noise is just an impulse or delta function centered at the origin:

$$G_n(f) = \frac{n_0}{2} \qquad R_n(\tau) = \frac{n_0}{2}\,\delta(\tau) \tag{6-31a}$$

The spectral-density and autocorrelation functions for white noise, as well as band-limited white noise, are shown in Fig. 6-8.

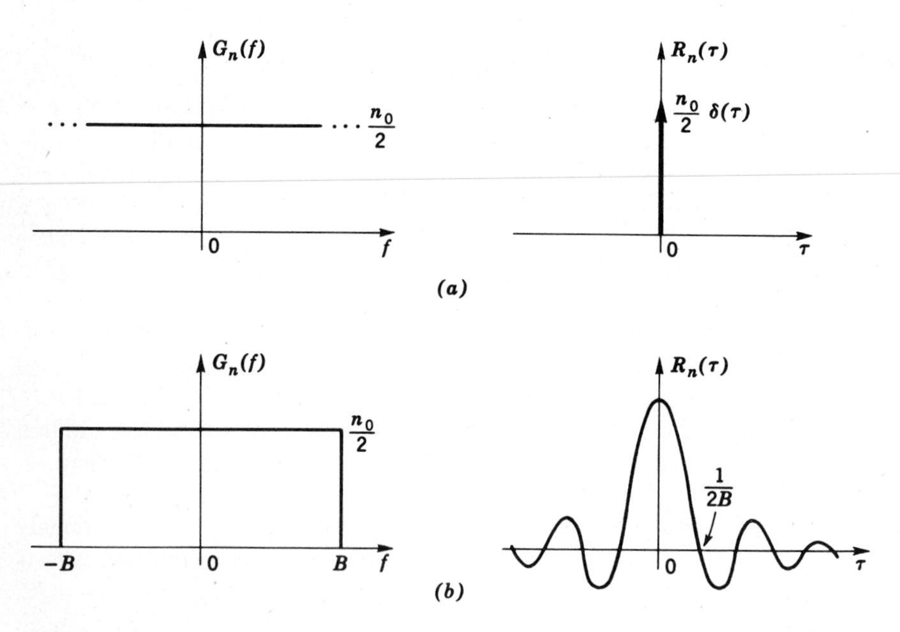

FIG. 6-8 White-noise power spectrum and autocorrelation function. (a) White noise; (b) band-limited white noise.

Since the autocorrelation function is an impulse in the case of white noise, this indicates that $n(t)$ is *always* uncorrelated with $n(t + \tau)$, no matter how small τ may be. The implication then is that $n(t)$ may vary infinitely rapidly, since it contains power at all frequencies. In practice of course, as just noted, this simply means that the high-frequency variations are beyond the capabilities of our instruments in a particular measurement we may be making. So although the white-noise model may appear to be physically inadmissible, we could never measure the rapid variations anyway. As an example, if our measuring devices have a time response $\gg 1/B$, with B the noise bandwidth of an actual physical noise process, the noise looks to us for all practical purposes like white noise. Band-limited white noise, with $B \gg$ significant frequencies in the frequency response of our measuring devices, thus appears to us as white noise. To an oscilloscope of bandwidth 50 MHz, input noise with bandwidth 500 MHz would obviously appear like white noise.

6-3 RANDOM SIGNALS AND NOISE THROUGH LINEAR SYSTEMS

Using the concept of power spectral density, as developed for random signals and noise, we are now in a position to consider the effect of linear filtering on these nondeterministic signals. This then parallels and extends our discussion of deterministic signals through linear systems in Chap. 2.

Consider then noise $n_i(t)$, with prescribed spectral density $G_{n_i}(f)$ and hence noise power N_i, as well as autocorrelation function $R_{n_i}(\tau)$, passed through a linear system with frequency transfer function $H(\omega)$, as shown in Fig. 6-9. What are the properties of the noise $n_o(t)$, at the output?

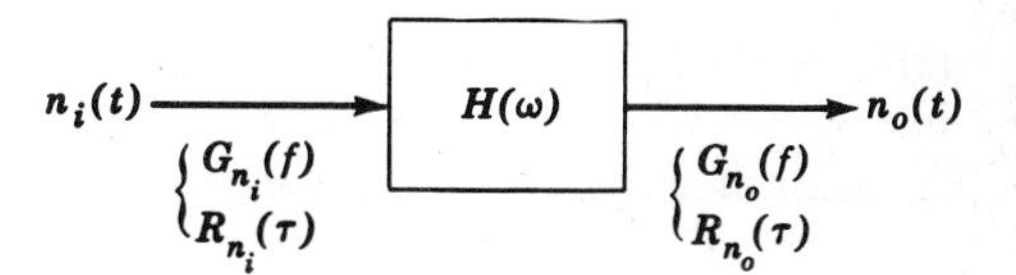

FIG. 6-9 Noise through a linear system.

That is, what are its spectral density $G_{n_o}(f)$, autocorrelation function $R_{n_o}(\tau)$, and output noise power N_o?

One would expect to find results similar to those obtained in Chap. 2. If the input noise $n_i(t)$ is varying roughly at a rate defined by its bandwidth B which is *slow* compared to the system bandwidth B_{sys}, the output noise $n_o(t)$ differs very little from $n_i(t)$. If on the other hand $B \gg B_{sys}$, the rapid fluctuations of $n_i(t)$ cannot get through (the system will not respond rapidly enough), and one would expect to find $n_o(t)$ varying at roughly the rate B_{sys}. We shall show that these intuitive arguments are of course valid, and that the effect of the system on the noise is given, quite simply, by the following relation between input and output spectral densities:

$$G_{n_o}(f) = |H(\omega)|^2 G_{n_i}(f) \qquad [6\text{-}32]$$

The two extreme cases noted above ($B \ll B_{sys}$, $B \gg B_{sys}$) are summarized qualitatively in the picture of Fig. 6-10.

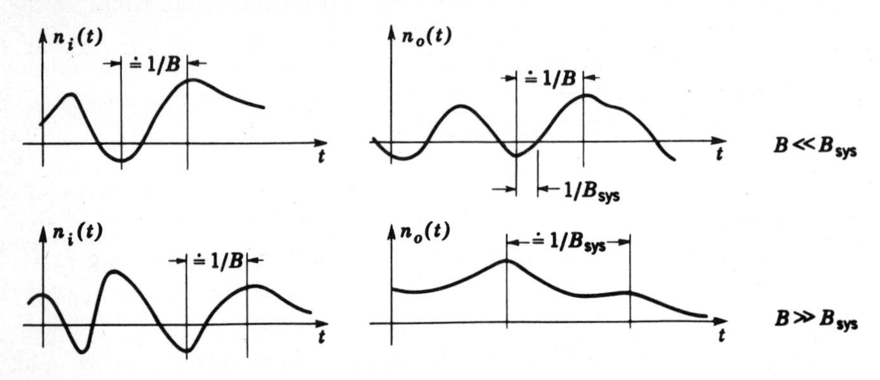

FIG. 6-10 System response to input noise.

We can demonstrate Eq. [6-32] very simply by referring back to the Fourier-series representation of a noise wave $n(t)$ as given by Eq. [6-17]:

$$n(t) = \frac{1}{T} \sum_{m=-\infty}^{\infty} c_m e^{j\omega_m t} \qquad \omega_m = \frac{2\pi m}{T} \qquad [6\text{-}17]$$

[Here we expand an arbitrary strip of $n(t)$ T sec long in a series involving harmonics of $1/T$.] We showed, in Eq. [6-30], that an estimate of the spectral density $G_n(f_l)$ is given, in terms of the Fourier coefficients c_l, by

$$\hat{G}_n(f_l) = \frac{1}{T} |c_l|^2$$

Now assuming the input-noise process $n_i(t)$ expanded into its (random) Fourier series, it is apparent that superposition applies, and each Fourier term appears at the output multiplied by $H(\omega_l)$:

$$\begin{aligned} n_i(t) &= \frac{1}{T} \sum_{l=-\infty}^{\infty} c_l e^{j\omega_l t} \qquad \omega_l = \frac{2\pi l}{T} \\ n_o(t) &= \frac{1}{T} \sum_{l} c_l H(\omega_l) e^{j\omega_l t} \end{aligned} \qquad [6\text{-}33]$$

The estimate of the input power spectral density is just

$$\hat{G}_{n_i}(f_l) = \frac{1}{T} |c_l|^2$$

and that of the output spectral density is therefore

$$\hat{G}_{n_o}(f_l) = \frac{1}{T}|c_l H(\omega_l)|^2 = |H(\omega_l)|^2 \hat{G}_{n_i}(f_l) \qquad [6\text{-}34]$$

It is thus apparent that if we now repeat this estimation of $G_{n_o}(f_l)$ many times, averaging as noted previously, that one then obtains the general statistical relation, valid at *all* frequencies,

$$G_{n_o}(f) = |H(\omega)|^2 G_{n_i}(f)$$

As an example assume we have white noise of spectral density $n_0/2$ applied at the input of an ideal low-pass filter of bandwidth B Hz (Fig. 6-11). The output noise is exactly the band-limited white noise of

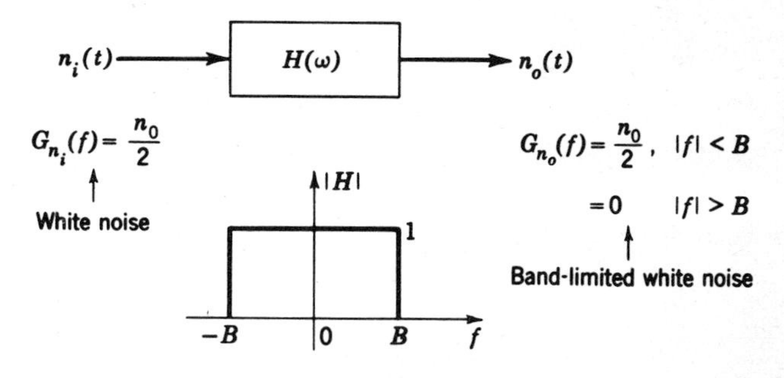

FIG. 6-11 White noise through an ideal filter.

spectral density $G_n(f) = n_0/2$, $|f| \leq B$, $G_n(f) = 0$, $|f| > B$, mentioned previously. The output noise power is then just

$$N_o = n_0 B$$

By *increasing* the filter bandwidth B, we *increase* the output *noise* power. We also increase the rate of variation of noise at the output, or decrease the correlation between $n(t)$ and $n(t + \tau)$, for a fixed τ.

As a second simple example consider the case of white noise applied to the input of an RC filter, as shown in Fig. 6-12. Since

$$|H(\omega)|^2 = \frac{1}{1 + (\omega RC)^2} = \frac{1}{1 + (2\pi RCf)^2}$$

for this filter, we have

$$G_{n_o}(f) = \frac{n_0/2}{1 + (2\pi RCf)^2} = \frac{n_0/2}{1 + (f/f_0)^2} \qquad [6\text{-}35]$$

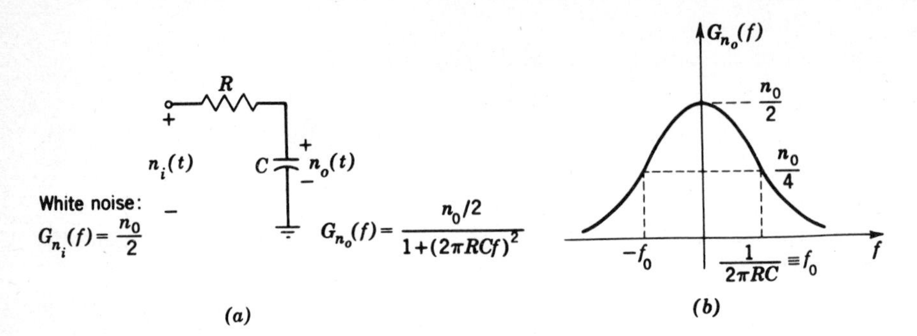

FIG. 6-12 *RC* filtering of white noise. (a) Filter; (b) output spectral density.

with $f_0 \equiv 1/2\pi RC$, just as in the first example of Fig. 6-6. The bandwidth of the output noise is thus inversely proportional to the filter time constant RC, just as in the case of the deterministic signals of Chap. 2. The correlation time is also of the order of RC. For we have, as in Fig. 6-6,

$$R_{n_o}(\tau) = N_o e^{-|\tau|/RC} \tag{6-36}$$

as the Fourier transform of this spectral density. The average noise power N_o may be readily found by integrating $G_{n_o}(f)$:

$$N_o = \int_{-\infty}^{\infty} G_{n_o}(f)\, df = \int_{-\infty}^{\infty} \frac{n_0/2}{1 + (f/f_0)^2}\, df = \frac{n_0 \pi}{2} f_0 \tag{6-37}$$

We now note an interesting fact shown in both of these examples—the low-pass filter and RC filter. The output *noise power* N_o is in both cases *proportional* to the system *bandwidth*. Thus in the first case we had $N_o = n_0 B$; in the second case we have $N_o = (n_0\pi/2)f_0$, with f_0 a measure of the system bandwidth. This thus indicates that as we *widen* the system bandwidth, we *increase* the *noise power* at the output. Similarly, we may *reduce* the noise at the output by *narrowing* the bandwidth.

We can generalize this statement to include all kinds of low-pass systems by defining a *noise equivalent bandwidth*. Assume white noise of spectral density $n_0/2$ applied at the input of an arbitrary linear system transfer function $H(\omega)$. The output noise power is then given by

$$N_o = \frac{n_0}{2} \int_{-\infty}^{\infty} |H(\omega)|^2\, df \tag{6-38}$$

since $G_{n_o}(f) = (n_0/2)|H(\omega)|^2$. Assume this same noise comes from an ideal low-pass filter of bandwidth B and amplitude $H(0)$, that is, the magnitude of the arbitrary filter transfer function at zero frequency. To have the same noise output we must then have

$$N_o = n_0 H^2(0) B = n_0 \int_0^{\infty} |H(\omega)|^2\, df \tag{6-39}$$

since $|H(\omega)|$ has even symmetry about the origin. Then we have the *noise equivalent bandwidth B* defined as

$$B \equiv \frac{1}{H^2(0)} \int_0^\infty |H(\omega)|^2 \, df \qquad [6\text{-}40]$$

This procedure essentially corresponds to replacing the arbitrary filter $H(\omega)$ by an equivalent ideal low-pass filter of bandwidth B, as shown in Fig. 6-13. One may also do this for bandpass filters, using in place of

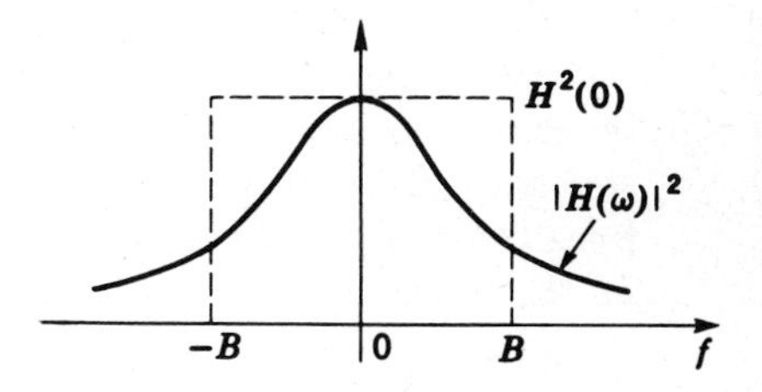

FIG. 6-13 Noise equivalent bandwidth.

$H(0)$ the value of $H(\omega)$ at the center frequency. As an example, for the RC filter, we have, from Eq. [6-37], $B = (\pi/2)f_0 = 1/4RC$.

To show the numerical quantities involved, assume that the white-noise spectral density at the input to a high-gain–low-pass amplifier is $n_0/2 = 10^{-14}$ volts²/Hz. (Detailed numerical calculations, applying these noise-spectrum concepts to some actual systems, will be considered in the next chapter.) The amplifier has a voltage gain of 10^3, and a high-frequency cutoff of 10 MHz. From Eq. [6-39], then, the mean-squared noise voltage at the amplifier output is $N_o = 2 \times 10^{-14} \times 10^6 \times 10^7 = 0.2$ volt².

In terms of the noise equivalent bandwidth we have then, as in Eq. [6-39], the general relation

$$N_o = n_0 H^2(0) B \propto B \qquad [6\text{-}41]$$

This indicates that the output noise power N_o is *proportional* to the *bandwidth*. As an interesting rule of thumb, then, one reduces noise in systems by *decreasing* the bandwidth, and increases the noise by widening the bandwidth. We include here, in the word "systems," measuring instruments, receivers, signal processors, etc.

If one attempts to detect or otherwise measure signals in the presence of noise, one thus tries to use as narrow a system bandwidth as possible—assuming the input noise is originally much wider in bandwidth than the system bandwidth. Of course one then has a limitation on the rate of variation of the signals themselves. For given classes of signals, one cannot narrow the bandwidth too far down to the point where the signal itself is adversely affected or disturbed in an undesirable manner. We shall have more to say about this tradeoff between signal detectability and noise power increase with increasing bandwidth shortly.

We can now summarize our study of noise thus far with two simple statements that serve as useful rules of thumb:

1. Increasing the system bandwidth increases the rate of fluctuation of the output noise (assuming the system bandwidth is initially much less than the input noise bandwidth).
2. Increasing the system bandwidth increases the output noise power. Since the output noise power N_o is a measure of the mean-squared statistical fluctuations about the average value (assumed zero here) [recall that N was the noise variance and hence a measure of the width of the noise probability-density function $f(n)$ about $E(n)$], larger instantaneous values of $n(t)$ thus become more probable as well.

These two rules appear expressed pictorially in the curves of Fig. 6-14. As the bandwidth B increases, the random process $n(t)$ is expected to deviate more often and more violently (in terms of peak excursions) away from its expected or average value.

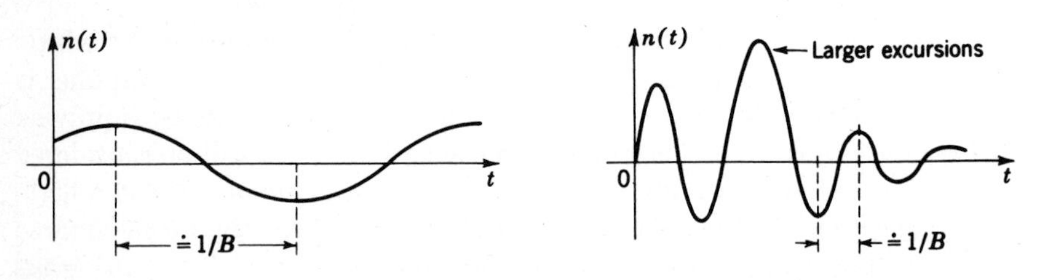

FIG. 6-14 Effect of increased noise bandwidth.

6-4 APPLICATION TO MATCHED-FILTER DETECTION

In Sec. 5-6 we calculated the probability of detection in a binary PCM system subject to additive gaussian noise. We implicitly assumed there that the intersymbol interference discussed in Chap. 3 (Sec. 3-9) was no problem. (In data transmission over telephone channels the reverse is usually true, as noted earlier; additive noise generally poses no problem, intersymbol interference does.)

We found there that the probability of error ultimately depended on the peak signal amplitude to rms noise ratio, A/σ or $A/\sqrt{N}$, using the present noise terminology (see Figs. 5-25 to 5-27, Chap. 5). An interesting and very practical question that we might pose would be: Is it possible to design the system up to the detector in order to maximize the ratio of peak signal amplitude to rms noise when sampled at the detector? Thus we envision white noise added to the sequence of binary pulses at the receiver input, with the composite sum passed through a linear filter (representing the system), then sampled and a decision made at the detector. Is it then possible to design the linear filter prior to detection to maximize the ratio $A/\sigma = A/\sqrt{N}$ at its output? This is shown pictorially in Fig. 6-15.

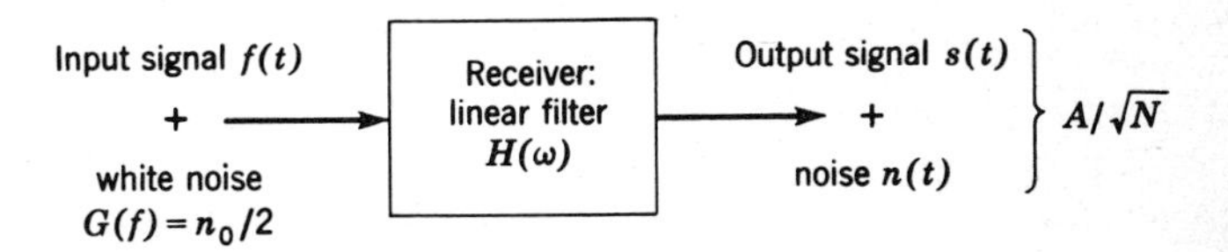

FIG. 6-15 Matched-filter problem.

(Although we assume throughout this section that the signal is a video pulse, the results obtained apply also to the case of binary carrier transmission. As discussed in Chap. 3 the high-frequency pulses may be synchronously detected, reducing them to the video pulse problem under consideration. Alternately, they may be envelope detected, if of the OOK or FSK type. This is shown in the sections that follow.)

We shall find that this problem is easily handled as an application of the power-spectrum concepts just introduced.

The results of the analysis to follow are useful in many other applications aside from binary pulse-code-modulation (PCM) transmission. They apply as well to the design of pulse-amplitude-modulation (PAM) (nonquantized) systems. There the question posed would be: Given pulsed samples of the signal to be transmitted, how should these be filtered so as to have maximum signal-to-rms-noise ratio at the output of the system? We might alternatively state this: How should we shape our pulses (by filtering them) so that the signal-to-noise ratio will be maximized?

A similar problem is encountered in radar systems, where it is required to detect the presence of a signal echo embedded in fluctuation noise. The amplitude of the signal relative to the noise should thus be maximized if possible.

In all these examples we are not specifically interested in maintaining fidelity of pulse shape. We are primarily interested in improving our ability to "see" or recognize a pulse signal in the presence of noise. This ability to see the pulse is assumed to be related to the ratio of peak signal to rms noise. In pulse-position-modulation (PPM) systems the time of occurrence of the pulse is of prime significance. (This is also true in many tracking radars.) Fidelity, or sharpness, of pulse rise time is thus of much greater significance there, and the concept of designing the system frequency characteristic to maximize peak signal-to-noise ratio has no meaning in that case.

This is an important point that bears stressing. The analysis to follow is of significance only in those communication systems in which the peak signal amplitude is of primary interest and is to be maximized relative to the noise. The signal shape is of secondary interest. An example of such a pulse signal embedded in noise, which is characteristic of the PCM, PAM, or radar problem, is shown in Fig. 6-16.

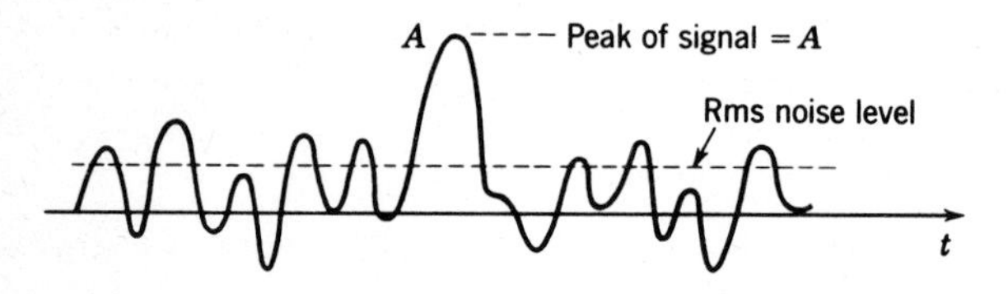

FIG. 6-16 Pulse signal embedded in noise.

How do we know that the peak signal-to-noise ratio can be maximized by properly choosing the filter characteristic? This is simply answered from our discussions of Chap. 2 and the previous section.

Assume, for simplicity's sake, that the input signal $f(t)$ is a rectangular pulse of width τ sec. The system is assumed to have the idealized filter characteristics of Sec. 2-5, with a variable bandwidth B Hz.

For very small bandwidth ($B \ll 1/\tau$) the peak of the output signal is small and increases with bandwidth. This is shown by the curves of Fig. 2-23. It is also indicated by the curve of Fig. 6-17a, showing the familiar $(\sin x)/x$ spectrum with the filter cutoff frequency (B) superimposed.

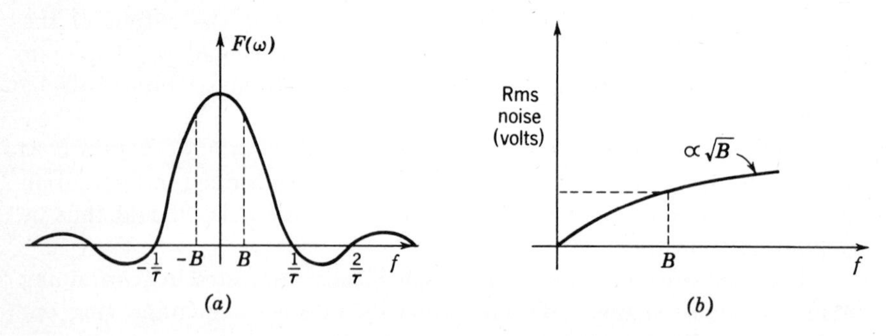

FIG. 6-17 Rectangular pulse and noise passed through ideal filter. (a) Signal pulse spectrum, with filter cutoff superimposed; (b) rms noise output.

The output-signal amplitude is proportional to the area under the curve. For $B \ll 1/\tau$ the frequency spectrum is flat, so that the output signal increases linearly with B. (This is also apparent from the sine integral of Eq. [2-66].)

The rms noise is proportional to $\sqrt{B}$ and so increases at a smaller rate than the signal for small bandwidth.

As the bandwidth increases, approaching $B = 1/\tau$, the signal amplitude begins to increase less rapidly with B. For $B > 1/\tau$ the signal remains approximately at the same amplitude as at $B = 1/\tau$ (Fig. 2-23). (Recall from Chap. 2 that increasing the bandwidth beyond $B = 1/\tau$ just served to fill out the fine details of the pulse. For a recognizable pulse a bandwidth $B = 1/\tau$ was all that was necessary.) The rms noise keeps increasing with bandwidth, however, so that the ratio of peak signal to rms noise begins to decrease inversely as $\sqrt{B}$. We would thus expect an optimum ratio at about $B = 1/\tau$. This will be borne out in the analysis to follow.

Note that this result for optimum B agrees with the time-bandwidth relation ($B\tau = 1$) demonstrated in Chap. 2.

In general, not only the filter bandwidth but the shape of the filter characteristic can be adjusted to optimize the peak signal-to-noise ratio. To show this, consider $f(t)$ impressed across a linear filter with frequency transfer function $H(\omega)$. Defining $F(\omega)$ to be the Fourier transform of $f(t)$,

$$F(\omega) = \int_{-\infty}^{\infty} f(t)e^{-j\omega t}\,dt \tag{6-42}$$

The output signal $s(t)$ is given by

$$\begin{aligned} s(t) &= \frac{1}{2\pi}\int_{-\infty}^{\infty} F(\omega)H(\omega)e^{j\omega t}\,d\omega \\ &= \int_{-\infty}^{\infty} F(\omega)H(\omega)e^{j\omega t}\,df \qquad \omega = 2\pi f \end{aligned} \tag{6-43}$$

The magnitude of $s(t)$ at the sampling time t_0 is just the desired output-signal amplitude A. Thus,

$$A = |s(t_0)| = \left|\int_{-\infty}^{\infty} F(\omega)H(\omega)e^{j\omega t_0}\,df\right|$$

The power spectrum of the white noise at the filter input is taken as

$$G(f) = \frac{n_0}{2} \tag{6-44}$$

as in the previous section. The power spectrum at the filter output is then

$$G_n(f) = \frac{n_0}{2}|H(\omega)|^2 \tag{6-45}$$

and the average output noise power (or mean-squared noise voltage across a 1-ohm resistor) is

$$N = \frac{n_0}{2} \int_{-\infty}^{\infty} |H(\omega)|^2 \, df \qquad [6\text{-}46]$$

$\sqrt{N}$ is the rms output noise in the absence of a signal.

We would now like to choose $H(\omega)$ such that the ratio $A/\sqrt{N}$ is maximized. This is the same as maximizing the square of the ratio, or A^2/N. This squared ratio is just the ratio of instantaneous peak signal power at $t = t_0$ to mean noise power, and will henceforth be referred to as the peak power signal-to-noise ratio, or peak SNR.

Since the input signal $f(t)$ is assumed given, its energy content $\int_{-\infty}^{\infty} f^2(t)\, dt$ is a constant. Calling the energy E, it is left to the reader as an exercise to show the following Fourier-transform identity is a valid one:[1]

$$E = \int_{-\infty}^{\infty} f^2(t)\, dt = \int_{-\infty}^{\infty} |F(\omega)|^2 \, df \qquad [6\text{-}47]$$

Dividing the peak signal power to mean noise power by the constant E will obviously not affect the determination of the maximum ratio. So we can take as our problem that of maximizing the ratio

$$\frac{A^2}{EN} = \frac{\left| \int_{-\infty}^{\infty} F(\omega) H(\omega) e^{j\omega t_0} \, df \right|^2}{(n_0/2) \int_{-\infty}^{\infty} |F(\omega)|^2 \, df \int_{-\infty}^{\infty} |H(\omega)|^2 \, df} \qquad [6\text{-}48]$$

This is readily done by means of *Schwarz's inequality*, relating the integral of products of complex functions:

$$\left| \int_{-\infty}^{\infty} X(\omega) Y(\omega) \, d\omega \right|^2 \leq \int_{-\infty}^{\infty} |X(\omega)|^2 \, d\omega \int_{-\infty}^{\infty} |Y(\omega)|^2 \, d\omega \qquad [6\text{-}49]$$

Schwarz's inequality for integrals of complex functions is just an extension of an inequality for real integrals, given by

$$\left[\int_{-\infty}^{\infty} f(t) g(t) \, dt \right]^2 \leq \int_{-\infty}^{\infty} g^2(t) \, dt \int_{-\infty}^{\infty} f^2(t) \, dt \qquad [6\text{-}50]$$

It might be termed a generalization of the familiar distance relation among vectors that the magnitude of the sum of two vectors is less than

[1] This relation is a basic theorem in Fourier-integral theory and is frequently called Parseval's theorem. It may be easily derived from Eq. [2-19b] in Chap. 2 by taking $(-T/2, T/2)$ as limits in the integral and letting $T \to \infty$. Alternately, it may be proved by applying the convolution theorem developed in Chap. 2. [Assume a function $f(t)$ passed through a linear filter with transfer function $H(\omega) = F^*(\omega)$.]

or equal to the sum of the magnitudes of the two vectors:

$$|\bar{a} + \bar{b}| \leq |\bar{a}| + |\bar{b}| \qquad [6\text{-}51]$$

In the vector case the equality is satisfied if $\bar{a} = K\bar{b}$, or $\bar{a}$ and $\bar{b}$ are collinear. Similarly in Eq. [6-50] the equality is satisfied if $f(t) = Kg(t)$. In the case of complex functions the equality is satisfied if

$$Y(\omega) = KX^*(\omega) \qquad [6\text{-}52]$$

K is a real number.

Schwarz's inequality is readily proved by considering the integral

$$I = \int_{-\infty}^{\infty} d\omega \int_{-\infty}^{\infty} |X(\omega)Y(y) - Y(\omega)X(y)|^2 \, dy \geq 0 \qquad [6\text{-}53]$$

where X and Y are complex functions. I is necessarily positive and real because it is the integral of the absolute value of a function. The quantity inside the vertical bars may be multiplied out and expanded to give four terms. Thus

$$\begin{aligned} |X(\omega)Y(y) - Y(\omega)X(y)|^2 &= |X(\omega)|^2|Y(y)|^2 + |Y(\omega)|^2|X(y)|^2 \\ &\quad - Y(\omega)Y^*(y)X^*(\omega)X(y) - Y^*(\omega)Y(y)X(\omega)X^*(y) \end{aligned}$$

Integrating term by term, the integrals of the first two terms are identical, as are the last two integrals. (ω and y are both dummy variables in any one integral.) Upon replacing y with ω in the first two terms and interchanging ω and y in the last term, the overall integral becomes

$$\begin{aligned} I = 2\int_{-\infty}^{\infty} |X(\omega)|^2 \, d\omega \int_{-\infty}^{\infty} |Y(\omega)|^2 \, d\omega \\ - 2\int_{-\infty}^{\infty} Y(\omega)X^*(\omega) \, d\omega \int_{-\infty}^{\infty} Y^*(y)X(y) \, dy \end{aligned}$$

But $\left|\int_{-\infty}^{\infty} X(\omega)Y(\omega)\, d\omega\right|^2 = \int_{-\infty}^{\infty} X(\omega)Y(\omega)\, d\omega \int_{-\infty}^{\infty} X^*(y)Y^*(y)\, dy$

ω being replaced by y in the last integral. Equation [6-53] finally becomes

$$\frac{I}{2} = \int_{-\infty}^{\infty} |X(\omega)|^2 \, d\omega \int_{-\infty}^{\infty} |Y(\omega)|^2 \, d\omega - \left|\int_{-\infty}^{\infty} X(\omega)Y(\omega)\, d\omega\right|^2 \geq 0 \qquad [6\text{-}54]$$

This proves the inequality stated as Eq. [6-49].

How do we apply Schwarz's inequality to our problem of maximizing peak signal-to-noise ratio?

Note that the ratio of Eq. [6-48] contains exactly the integrals of Eq. [6-49] if we let

$$X(\omega) = F(\omega)e^{j\omega t_0} \qquad Y(\omega) = H(\omega)$$

The ratio $n_0A^2/2EN$ must then be less than or equal to 1, and

$$\left|\int_{-\infty}^{\infty} F(\omega)H(\omega)e^{j\omega t_0}\, df\right|^2 \leq \int_{-\infty}^{\infty} |F(\omega)|^2 \, df \int_{-\infty}^{\infty} |H(\omega)|^2 \, df \qquad [6\text{-}55]$$

In particular, the ratio is a *maximum* when the equality holds, or

$$H(\omega) = K[F(\omega)e^{j\omega t_0}]^* = KF^*(\omega)e^{-j\omega t_0} \qquad [6\text{-}56]$$

As an example, if $f(t)$ is the rectangular pulse of width τ and height V, $F(\omega) = V\tau\{[\sin(\omega\tau/2)]/(\omega\tau/2)\}$ and

$$H(\omega) = K \frac{\sin(\omega\tau/2)}{\omega\tau/2} e^{-j\omega t_0}$$

for maximum ratio of peak signal to rms noise. The linear holding circuit of Figs. 3-16 to 3-18 has a transfer function of the $(\sin x)/x$ type.

Filters possessing the characteristic of Eq. [6-56] are said to be *matched filters*. The response at the output of such a filter, to $f(t)$ applied at the input, is

$$\begin{aligned} s(t) &= \int_{-\infty}^{\infty} F(\omega)H(\omega)e^{j\omega t}\,df \\ &= K\int_{-\infty}^{\infty} |F(\omega)|^2 e^{j\omega(t-t_0)}\,df \end{aligned} \qquad [6\text{-}57]$$

In particular $s(t)$ will have amplitude A when

$t = t_0$: $$|s(t_0)| = K\int_{-\infty}^{\infty} |F(\omega)|^2\,df = A$$

Note that the amplitude A is thus proportional to the signal energy E.

An interesting relation for the matched-filter output signal-to-noise ratio $A/\sqrt{N}$ may be derived by applying the matched-filter condition of Eq. [6-56] to Eq. [6-48]. Specifically, it is apparent that at the output of the matched filter the peak power SNR is

$$\frac{A^2}{N} = \frac{2E}{n_0}$$

or

$$\frac{A}{\sqrt{N}} = \sqrt{\frac{2E}{n_0}} \qquad [6\text{-}58]$$

The signal-to-noise ratio is thus a function solely of the energy in the signal and the white-noise spectral density. *The dependence on the signal input waveshape $f(t)$ has been obliterated by use of the matched filter.* Two different signal waveshapes will provide the same probability of error in the presence of additive white noise, providing they contain the same energy and are filtered by the appropriate matched filter in each case. It is the *energy* in the signal that provides its ultimate detectability in noise. This point will be pursued further in Chap. 8.

If the input time function $f(t)$ is symmetrical in time $[f(t) = f(-t)]$, $F(\omega)$ will be a real function of frequency (see Sec. 2-6). From Eq. [6-56] $H(\omega) = KF(\omega)e^{-j\omega t_0}$ for this case. This means that the impulse response $h(t)$ is $h(t) = Kf(t - t_0)$. The impulse response of a filter matched to a symmetrical input is a delayed replica of such an input.

If $f(t)$ is not symmetrical, $F(\omega)$ is complex. By using the Fourier-integral relations it may be shown that

$$h(t) = Kf[-(t - t_0)] \qquad [6\text{-}59]$$

in general. The proof is left to the reader as an exercise since it is identical to the procedure used to prove Parseval's theorem above. Since $f(t)$ is normally defined for positive t, $h(t)$ in the general case will be defined for negative t. As pointed out in Chap. 2, such a filter is physically not realizable. In the general case, then, the matched filter is *not realizable.*

Of what value then is this entire analysis leading to the matched-filter result? Just as in the case of the ideal filter of Chap. 2, we can approximate the matched-filter characteristics by those of an actual filter. We can also compare practical filters with the matched filter so far as the ratio of peak output signal to noise is concerned and can optimize their shape and bandwidth as far as practicable. This is the procedure we shall follow in the remainder of this section.

The input time function $f(t)$ is assumed to be a rectangular pulse of width τ sec and height V. We assume $V\tau = 1$. With $f(t)$ symmetrically located about $t = 0$, $F(\omega) = [\sin(\omega\tau/2)]/(\omega\tau/2)$, a real function as noted above. The impulse response of the matched filter is then also a rectangular pulse of τ sec duration. [$H(\omega) = F(\omega)$ here.] The response of this matched filter to the rectangular pulse will then be the convolution of two rectangular pulses. This was worked out in Chap. 2 as an example of the convolution integral and, as was shown there, gave a triangular-pulse output. We can check this by noting that for the matched filter $F(\omega)H(\omega) = \{[\sin(\omega\tau/2)]/(\omega\tau/2)\}^2 e^{-j\omega t_0}$ ($V\tau = 1$). This is just the Fourier transform of a triangular pulse of width 2τ sec, as shown in Chap. 2. This output pulse is τ sec wide at the half-amplitude points.

Note that such a triangular pulse is not too different in shape from the output of the ideal low-pass filter of Sec. 2-5, with $B = 1/\tau$ (see Fig. 2-23). In fact the ideal-filter output for $B = 1/\tau$ could very well have been approximated by such a triangle. If the bandwidth of the matched filter is assumed to be the frequency of the first zero in its amplitude charactersitic, the bandwidth is also just $1/\tau$ ($\sin\omega\tau/2 = 0$; $\omega\tau/2 = \pi$; $f = 1/\tau$).

This is an interesting point, for it agrees with our previous results that for producing a recognizable pulse a filter bandwidth of the order of $B = 1/\tau$ should be used. We actually assumed bandwidths of the order of $1/\tau$ in our discussion of PAM in Chap. 3. We shall see below that for an ideal low-pass filter $B\tau = 0.7$ actually gives maximum signal-to-noise ratio.

Just how significant the shape of the filter characteristic and bandwidth are in determining the output peak signal-to-noise ratio for a rectangular-pulse input can be found by applying Eq. [6-48] to various filters. The resulting signal-to-noise ratio can then be compared with that for the optimum matched filter. We shall actually calculate $n_0A^2/2EN$ so that the optimum value is normalized to 1. In all cases the rectangular-pulse input is assumed to have unit area ($V\tau = 1$), so that

$$F(\omega) = \frac{\sin(\omega\tau/2)}{\omega\tau/2}$$

Ideal low-pass filter, variable bandwidth Here

$$\begin{aligned} H(\omega) &= e^{-j\omega t_0} \qquad |\omega| \le 2\pi B \\ H(\omega) &= 0 \qquad |\omega| > 2\pi B \end{aligned} \tag{6-60}$$

The peak power SNR for this case, as obtained from Eq. [6-48], becomes

$$\frac{\left|\int_{-2\pi B}^{2\pi B} \frac{\sin(\omega\tau/2)}{\omega\tau/2}\, d\omega\right|^2}{\int_{-\infty}^{\infty}\left[\frac{\sin(\omega\tau/2)}{\omega\tau/2}\right]^2 d\omega \int_{-2\pi B}^{2\pi B} d\omega} = \frac{\left[\left(\frac{2}{\tau}\right)\int_{-a}^{a}\frac{\sin x}{x}\,dx\right]^2}{\frac{2}{\tau}\int_{-\infty}^{\infty}\left(\frac{\sin x}{x}\right)^2 dx\, 4\pi B}$$

with $x \equiv \omega\tau/2$ and $a = \pi B\tau$. Using the relation $\int_{-\infty}^{\infty} [(\sin x)/x]^2\, dx = \pi$, and recalling the sine integral definition, $\mathrm{Si}\, a = \int_0^a [(\sin x)/x]\, dx$, the ratio squared becomes

$$\frac{2}{\pi a}(\mathrm{Si}\, a)^2$$

This expression may be plotted by using tables of the sine integral[1] and is found to have a maximum at $a = \pi B\tau = 2.2$. This corresponds to $B\tau = 2.2/\pi = 0.7$. At this bandwidth the peak power SNR is found to be 0.83, as compared with 1 for the optimum filter. This corresponds to a relative deterioration of 0.8 db.

The ideal-low-pass-filter case is plotted in Fig. 6-18 as a function of $B\tau$. The decibel scale used is relative to the 0-db case for the optimum matched filter. Although the maximum ratio is found for $B\tau = 0.7$, the maximum is very broad and varies less than 1 db from $B\tau = 0.4$ to $B\tau = 1$. For a pulsed carrier or OOK signal the bandwidth would be twice the bandwidth shown here, so that $B\tau = 1.5$ would be optimum for a rectangular filter.

[1] S. Goldman, "Frequency Analysis, Modulation, and Noise," McGraw-Hill, New York, 1948; E. Jahnke and F. Emde, "Tables of Functions," Dover, New York, 1945.

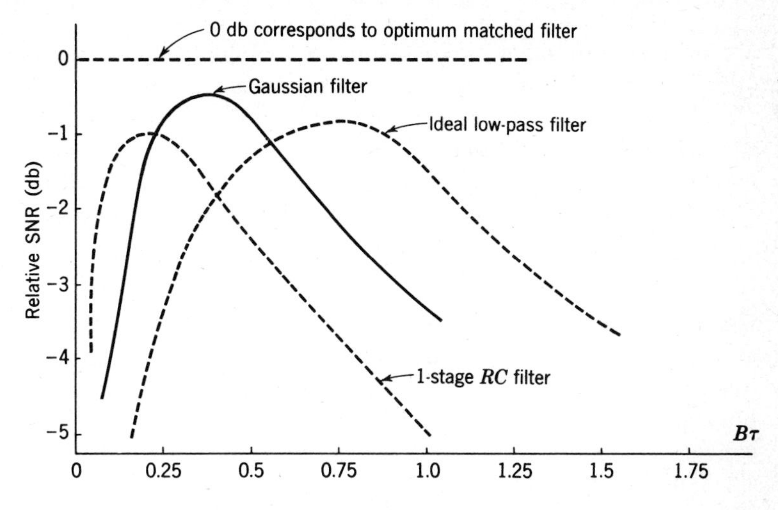

FIG. 6-18 Peak SNR for various filters, compared with matched filter (rectangular-pulse input).

One-stage RC filter, variable bandwidth

$$H(\omega) = \frac{1}{1 + j\omega RC} \tag{6-61}$$

with $2\pi B = 1/RC$ the 3-db radian bandwidth. The response of this filter to a rectangular-pulse input is plotted in Fig. 2-11. The peak output occurs at $t = \tau$ and is given by

$$s(\tau) = 1 - e^{-\tau/RC}$$

(This procedure is much simpler than the frequency-response method of Eq. [6-43] in this case.) The normalized signal-to-noise peak power ratio becomes for this case, after some manipulations,

$$\frac{(1 - e^{-2a})^2}{a} \qquad a = \frac{\tau}{2RC} = \pi B\tau$$

The details of this calculation are left to the reader as an exercise.

The signal-to-noise ratio for this case has also been plotted in Fig. 6-18 and shows a maximum value at $B\tau = 0.2$ ($B = 1/2\pi RC$). At this bandwidth the filter output is only 1 db worse than that for the matched-filter case. For $B\tau = 0.5$ the S/N ratio is 2.3 db worse than the matched-filter case so that the variation with bandwidth is again small.

Multistage RC filters A signal would normally be amplified by several stages of amplifiers, with filtering included in each amplifier. It is thus of interest to compare the matched-filter signal-to-noise output with that of a multistage amplifier. We assume a simple RC filter in each stage. (For a pulsed carrier we would use a single-tuned circuit for each stage.)

The peak SNR could of course be found by determining the overall frequency response of the multistage amplifiers. This becomes unwieldy mathematically. It can be shown,[1] however, that the transfer function of a large number of isolated RC sections approaches the form

$$H(\omega) = e^{-0.35(f/B)^2} e^{-jt_0\omega} \qquad [6\text{-}62]$$

where B is the 3-db bandwidth of the overall filter.

Note that this is the same mathematical form as the gaussian density function discussed in Chap. 5, and first included as an example among the filter shapes shown in Chap. 2. This gaussian filter is often used as a convenient mathematical model for systems analyses.

The gaussian filter response curve has a characteristic "bell" shape and is symmetrical about $f = 0$. It is shown sketched in Fig. 6-19.

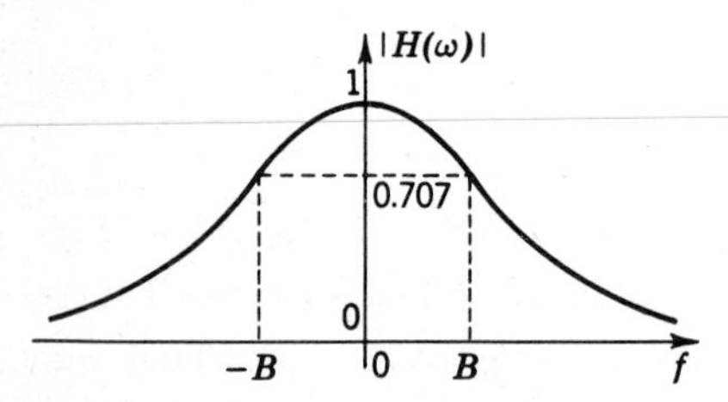

FIG. 6-19 Frequency response, multistage amplifier (gaussian error curve).

We can again use Eq. [6-48] to calculate the peak power SNR for a rectangular pulse applied to a network with this amplitude characteristic. The result will then approximate the output of a multistage RC amplifier. The analysis is identical to that carried out for the ideal filter. The details will not be presented here, but the results are plotted in Fig. 6-18, again compared with the optimum matched-filter case.

Actual calculations for two and three stages of amplification with RC filtering included produce results which do not differ substantially from those for the gaussian filter, so that the gaussian-error-curve analysis will

[1] C. Cherry, "Pulses and Transients in Communication Circuits," p. 311, Chapman and Hall, London, 1949; Dover, New York, 1950. The proof is similar to that discussed in connection with the central-limit theorem in Chap. 5.

be a good approximation to a multistage RC amplifier for two or more stages.

Note that the peak SNR occurs in this case for $B\tau = 0.4$. (For a two-stage amplifier the peak occurs at $B\tau = 0.3$.) The maximum is quite broad, however, varying by no more than 1 db from $B\tau = 0.2$ to $B\tau = 0.7$.

At the maximum the ratio is only 0.5 db less than that for the optimum matched filter. This would be quite negligible in most practical applications. The overall filter characteristic of a multistage amplifier will thus be close to optimum for transmission of rectangular pulses, with the overall bandwidth chosen as $B = 0.5/\tau$.

The filters considered here are all low-pass, and the signal and noise are both considered to be present at baseband. In practice, of course, sine-wave carrier transmission is used with all pulse systems—PAM, PCM, radar, etc. As shown in Chap. 4 the high-frequency circuits equivalent to the low-pass circuits here generally require twice the low-pass bandwidth ($\pm B$ about the carrier frequency). If the matched filters discussed here are included in the i-f section of a receiver, for example, all bandwidths shown in Fig. 6-18 must be multiplied by two. A multistage i-f amplifier would thus be designed to have a bandwidth of $2 \times 0.5/\tau$, or $1/\tau$, Hz, to optimize pulsed signal detection in noise.

Such a bandwidth choice is common practice in the design of pulse radars. It is also the optimum choice in those PAM or PCM systems in which noise is a more crucial factor than intersymbol interference. If intersymbol interference is a problem as well, the system bandwidth may be widened somewhat to narrow the pulses transmitted and hence decrease their overlap. The broad maxima in Fig. 6-18 indicate that the optimum bandwidth choice is not critical in combating noise, and hence widening the bandwidths somewhat will not deteriorate the noise performance too much.[1]

The design of filters to minimize both intersymbol interference and noise will not be considered here. The subject has received increasing attention in recent years and the interested reader is referred to the literature for theoretical investigations of this problem.[2]

The question of i-f matched filtering as contrasted with baseband (video) filtering will be further pursued in the sections following after a discussion of noise representations at high frequencies.

[1] M. Schwartz, W. R. Bennett, and S. Stein, "Communication Systems and Techniques," pp. 310–313, McGraw-Hill, New York, 1966.

[2] See, e.g., R. W. Lucky, F. Salz, and E. J. Weldon, Jr., "Principles of Data Communication," chap. 5, McGraw-Hill, New York, 1968, for a discussion and list of references.

6-5 NARROWBAND NOISE REPRESENTATION

In the discussion of noise through linear systems in Sec. 6-3 we stressed noise passed through low-pass devices. We assumed these low-pass structures to have an effective bandwidth B Hz, centered at 0 Hz (dc), and showed, for example, that the output noise power with white noise applied at the input was proportional to B. In the last section, in discussing the maximization of SNR by matched filtering, we also stressed, for simplicity's sake, low-pass filtering.

We did point out that the results obtained were applicable at carrier frequencies as well, with the equivalent bandpass filters having twice the low-pass bandwidth. As a matter of fact, the discussion in Sec. 6-3 of noise (and random signals) through linear systems is general enough to enable us to handle high-frequency carrier transmission and the various narrowband bandpass circuits encountered in practice. Thus, the transfer function $H(\omega)$ that appears in the spectral-density relation for input-output noise,

$$G_{n_o}(f) = |H(\omega)|^2 G_{n_i}(f) \tag{6-32}$$

is, in the case of high-frequency transmission, simply that of a filter centered at the desired center frequency.

We shall find it useful to develop a representation of noise particularly appropriate to narrowband transmission, however. This will enable us to realistically discuss the problem of detecting, at a receiver, high-frequency signals in the presence of noise. In particular we shall use this representation for narrowband noise to discuss, in a comparative way, the detection process in various types of digital and analog systems.

Thus we shall answer questions related to binary carrier transmission and reception that were first raised earlier in connection with our discussion of digital carrier systems: What *are* the reasons for selecting between OOK, PSK, and FSK transmission in a particular situation? What *are* the quantitative differences between synchronous (coherent) and envelope (noncoherent) detection of binary signals in noise?

In particular we shall find that PSK systems with synchronous detection offer a distinct improvement in either probability of error or signal-to-noise ratio over the other schemes and are therefore favored *if* phase coherence may be maintained. If envelope detection *must* be used (phase coherence is either not available or the cost does not justify the additional circuitry required to maintain it), FSK is found to be superior to OOK, but again with the requirement of somewhat more complex circuitry.

Using the narrowband representation of noise we shall also discuss briefly AM and FM detection in the presence of noise, obtaining the well-known SNR improvement of wideband FM over AM (this above the so-called FM threshold). We shall also study briefly the problem of noise through nonlinear devices, in this case the envelope detectors required in both AM and envelope detection of binary signals.

Consider then noise $n(t)$ at the output of a narrowband filter. Its spectral density $G_n(f)$ is centered about f_0 as in Fig. 6-20. For simplic-

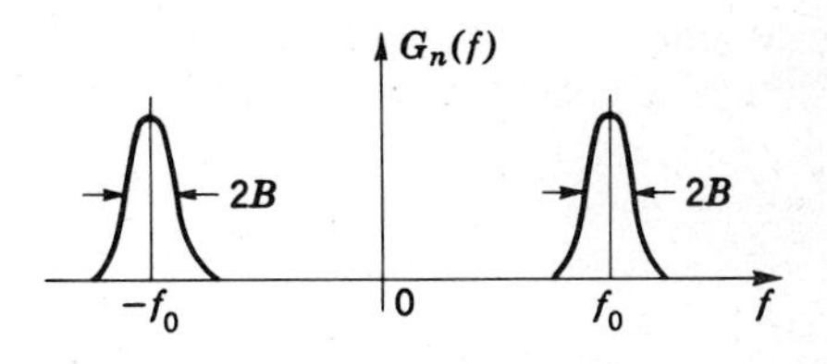

FIG. 6-20 Spectral density, narrowband noise.

ity's sake we shall assume $G_n(f)$ symmetrical about the frequency f_0, with bandwidth $2B \ll f_0$. (The noise may be assumed to have been generated by white noise passed through a narrowband filter, as shown in Fig. 6-21. These assumptions are not necessary in a more general

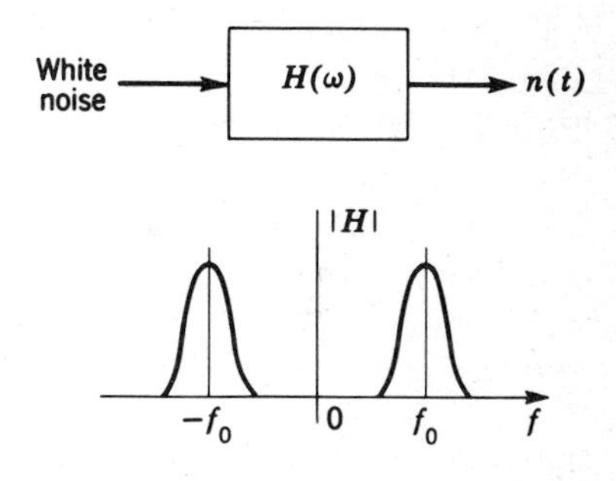

FIG. 6-21 Generation of narrowband noise.

approach to narrowband noise representation, but are used here to simplify the discussion.[1] They are, of course, frequently encountered in practice.) It is then apparent that the noise $n(t)$, although random, will be oscillating, on the average, at frequency f_0. (As the bandwidth $2B$ is

[1] See Schwartz, Bennett, and Stein, *op. cit.*, pp. 35–45, for a more systematic and general approach.

made smaller and smaller, the output should approach more and more that of a pure sine wave at frequency f_0.) We indicate this by writing $n(t)$ in the narrowband form

$$n(t) = r(t) \cos [\omega_0 t + \theta(t)] \qquad [6\text{-}63]$$

We would expect $r(t)$ and $\theta(t)$ to be varying, in a random fashion, roughly at the rate of B Hz, representing, respectively, the "envelope" and "phase" of the noise. This is indicated in Fig. 6-22.

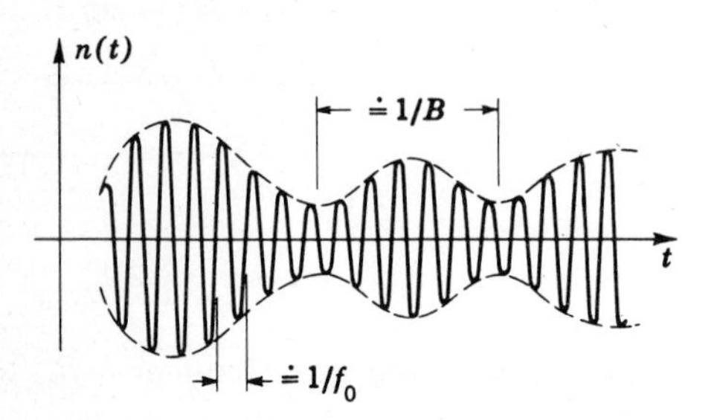

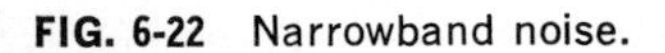

FIG. 6-22 Narrowband noise.

Note that in this form $n(t)$ is reminiscent of the randomly varying signal discussed in Sec. 5-10. We shall in fact show that r and θ are Rayleigh-distributed and uniformly distributed, respectively, just as in the fading channel discussion!

To actually develop $n(t)$ in the form of Eq. [6-63] with $r(t)$ and $\theta(t)$ given by these specific distributions, we use a simple artifice. We visualize the noise to be represented as the sum of many closely spaced sine waves, the spacing $\Delta f \ll B$. Note that this is then formally the same as the Fourier-series representation of noise used earlier in developing its spectral properties. Thus, let

$$n(t) = \sum_{l=-\infty}^{\infty} a_l \cos [(\omega_0 + l\Delta\omega)t + \theta_l] \qquad [6\text{-}64]$$

The variation about the center frequency f_0 is indicated by measuring the frequency of the different sine waves with respect to f_0. One would thus expect the coefficients a_l to be large in the vicinity of f_0 (i.e., small l), and small elsewhere.

Since we are not interested here in a unique representation of noise, but rather a model that will be useful in analysis, we now assume the θ_l's to be independent, uniformly distributed random variables. By the central-limit theorem, $n(t)$ *then has gaussian statistics*, just the property we assumed in the error calculations of Chap. 5. It is also apparent that $E(n) = 0$, averaging statistically over the random θ_l's. To find the coefficients a_l we now note that the sine-wave expansion of Eq. [6-64] is equivalent to assuming the continuous power spectral distribution of Fig. 6-23*a* to be replaced by a discrete spectrum of the same shape and power. This is shown in Fig. 6-23*b*. {We have concentrated on positive

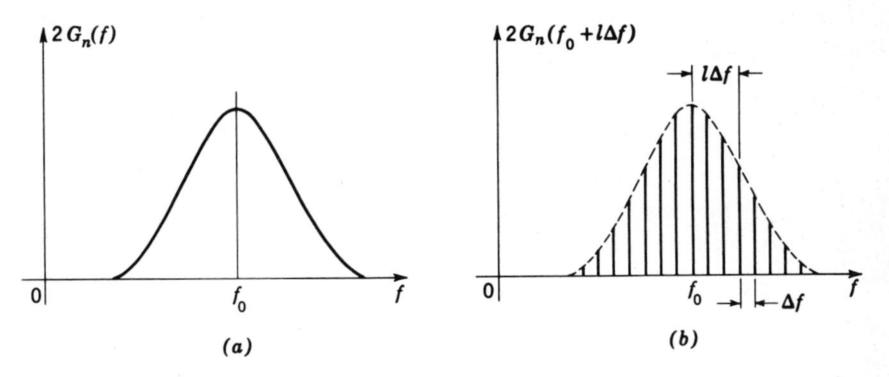

FIG. 6-23 Discrete representation of noise spectral density. (a) One-sided (positive-frequency) spectrum; (b) discrete equivalent.

frequencies only, as shown, because of the form of Eq. [6-64]. We then simply double the power spectral density at each frequency, as shown in the figure. The resultant spectral density, defined for positive frequencies only, is often called the one-sided spectral density, as contrasted to the *two-sided* $G_n(f)$, symmetrical in positive and negative frequencies.}

It is apparent from both figures that the total noise power N must be given by

$$N = 2\int_0^\infty G_n(f)\,df = \sum_{l=-\infty}^{\infty} 2G_n(f_0 + l\,\Delta f)\,\Delta f \qquad [6\text{-}65]$$

Again $l\,\Delta f$ represents the variation away from the center frequency f_0. Although $l\,\Delta f$ is indicated as ranging between $-\infty$ and ∞, it is of course only significant within the range $\pm B$ about f_0.

Now note that the average power in the representation $n(t)$ of Eq. [6-64] must be given by

$$N = E(n^2) = \sum_{l=-\infty}^{\infty} \frac{a_l^2}{2} \qquad [6\text{-}66]$$

a familiar result. {One may check this by multiplying together two series representations for $n(t)$ to get $n^2(t)$. Then ensemble averaging over the random variables θ_l, and noting that $E[\cos (l + m) \Delta t + \theta_l + \theta_m] = 0$, one gets Eq. [6-66].} Comparing with Eq. [6-65], we must have

$$\begin{aligned} a_l^2 &= 4G_n(f_0 + l\,\Delta f)\,\Delta f \\ a_l &= \sqrt{4G_n(f_0 + l\,\Delta f)\,\Delta f} \end{aligned} \qquad [6\text{-}67]$$

The a_l coefficients are thus uniquely known, once the θ_l's are assumed independent and uniformly distributed.

To get Eq. [6-64] more specifically in the narrowband form of Eq. [6-63], we now expand a typical sine-wave term about f_0 in the following manner:

$$\cos [\omega_0 t + (l\,\Delta\omega t + \theta_l)] = \cos (l\,\Delta\omega t + \theta_l) \cos \omega_0 t - \sin (l\,\Delta\omega t + \theta_l) \sin \omega_0 t \qquad [6\text{-}68]$$

Grouping the *low-frequency* terms $\cos (l\,\Delta\omega t + \theta_l)$ and $\sin (l\,\Delta\omega t + \theta_l)$ together, we then get

$$n(t) = \Big[\sum_l a_l \cos (l\,\Delta\omega t + \theta_l)\Big] \cos \omega_0 t - \Big[\sum_l a_l \sin (l\,\Delta\omega t + \theta_l)\Big] \sin \omega_0 t \qquad [6\text{-}69]$$

The resulting *low-frequency sums* shown in brackets we denote as $x(t)$ and $y(t)$, respectively:

$$\begin{aligned} x(t) &= \sum_l a_l \cos (l\,\Delta\omega t + \theta_l) \\ y(t) &= \sum_l a_l \sin (l\,\Delta\omega t + \theta_l) \end{aligned} \qquad [6\text{-}70]$$

We then have, finally,

$$\begin{aligned} n(t) &= x(t) \cos \omega_0 t - y(t) \sin \omega_0 t \\ &= r(t) \cos [\omega_0 t + \theta(t)] \end{aligned} \qquad [6\text{-}71]$$

as in Eq. [6-63], with

$$r^2 = x^2 + y^2 \qquad \theta = \tan^{-1} \frac{y}{x} \qquad [6\text{-}72]$$

Invoking the central-limit theorem it is apparent from Eq. [6-70] that both $x(t)$ and $y(t)$ are gaussian random processes. This is exactly the point we made in discussing signal models for fading channels in Sec. 5-10. As a matter of fact, it is readily shown that by writing $x^2(t)$ and $y^2(t)$ as double sums and statistically averaging over the θ_l's,

$$E(x^2) = E(y^2) = E(n^2) = N \qquad [6\text{-}73]$$

{The details are left to the reader, but note, as a hint, that

$$E[\cos (l\,\Delta\omega t + \theta_l) \cos (m\,\Delta\omega t + \theta_m)] = 0$$

with θ_l and θ_m *independent*, unless $l = m$, in which case $E(\quad) = \frac{1}{2}$.}

Both the inphase noise term $x(t)$ and the quadratic term $y(t)$ individually have the same variance or power as the original noise $n(t)$. [It is apparent by appropriate averaging that $E(x) = E(y) = 0$.]

In addition, it is readily shown by calculating $E(xy)$, using the same series representations of Eq. [6-70], that x and y are uncorrelated and, being gaussian, *independent*. (Recall the discussion of dependent random variables in Chap. 5.) Transforming the gaussian independent variables x and y to polar coordinates, as was done in Sec. 5-10, we find

$$f(r) = \frac{re^{-r^2/2N}}{N} \qquad f(\theta) = \frac{1}{2\pi} \tag{6-74}$$

The noise envelope r is thus *Rayleigh-distributed*, and the phase angle θ uniformly distributed. We shall use this extremely important result later in discussing envelope detection.

It is of interest to discuss the power spectral densities of $x(t)$ and $y(t)$. Comparing Eqs. [6-64] and [6-70], it is apparent that the noise terms $x(t)$ and $y(t)$ may be visualized as the original noise $n(t)$ shifted down to zero frequency. With $n(t)$ a random noise wave, oscillating, with a bandwidth B Hz, about the center frequency f_0, the inphase and quadrature terms $x(t)$ and $y(t)$ are both noise processes centered at dc; hence they are slowly varying at the bandwidth B. The calculation of the spectral densities in fact verifies this. It is readily shown that $G_x(f)$ and $G_y(f)$ are equal, and correspond to the original noise spectral density $G_n(f)$ shifted down to dc (zero frequency). In particular they are found to be given by

$$\begin{aligned} G_x(f) = G_y(f) &= G_n(f + f_0) + G_n(f - f_0) \qquad && -f_0 < f < f_0 \\ &= 0 && \text{elsewhere} \end{aligned} \qquad [6\text{-}75]^1$$

A sketch of a typical high-frequency noise spectral density and the low-frequency $G_x(f)$ are shown in Fig. 6-24.

If, as a special and very common case, $G_n(f)$ is symmetrical about the carrier frequency f_0, the positive- and negative-frequency contributions of $G_n(f)$ may be simply shifted down to zero frequency and added to give

$$G_x(f) = G_y(f) = 2G_n(f + f_0) \tag{6-76}$$

(This would be the case where the i-f filtering is symmetrical about f_0.) As an example, if $G_n(f)$ is gaussian shaped and given by

$$G_n(f) = \frac{N/2}{\sqrt{2\pi\sigma^2}} e^{-(f-f_0)^2/2\sigma^2} + \frac{N/2}{\sqrt{2\pi\sigma^2}} e^{-(f+f_0)^2/2\sigma^2} \tag{6-77}$$

[1] Schwartz, Bennett, and Stein, *op. cit.*, pp. 39, 40.

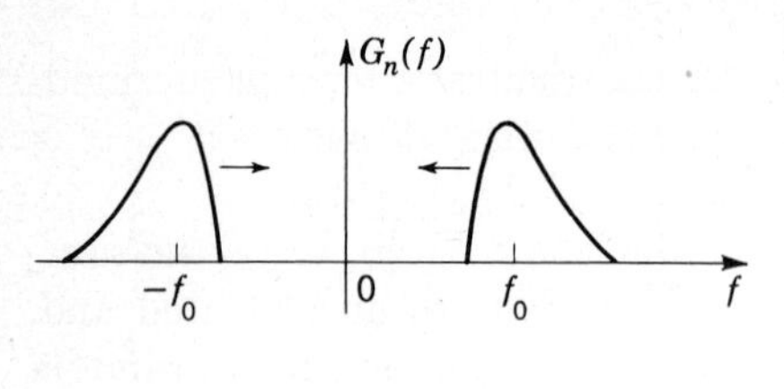

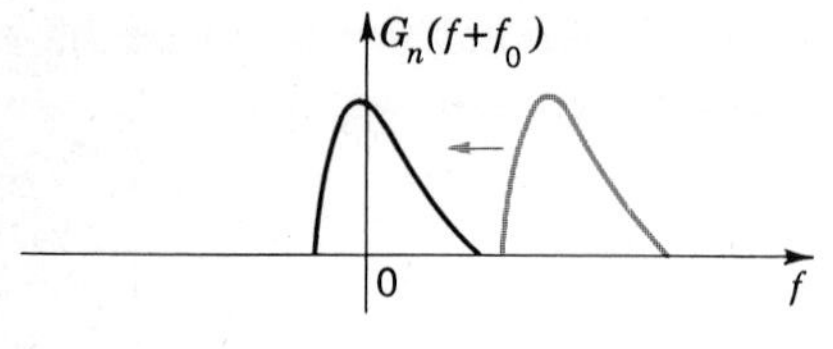

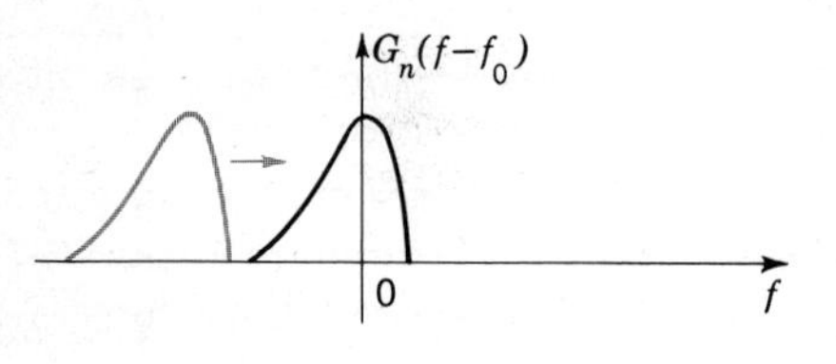

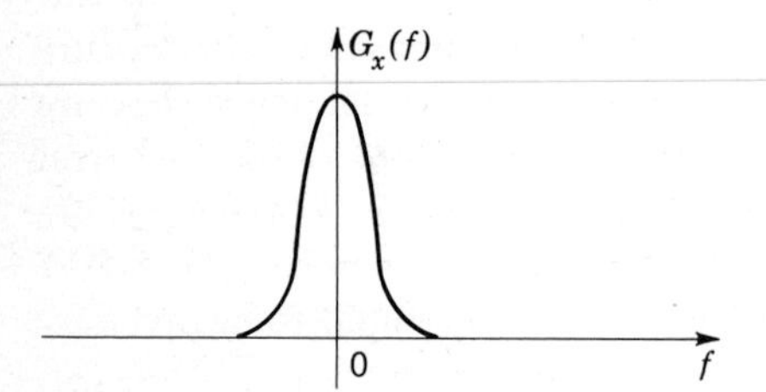

FIG. 6-24 Inphase (low-pass) noise spectral density (asymmetric noise spectrum).

with $\sigma \ll f_0$, we have

$$G_x(f) = G_y(f) = \frac{N}{\sqrt{2\pi\sigma^2}} e^{-f^2/2\sigma^2} \qquad [6\text{-}78]$$

The result is sketched in Fig. 6-25. [Note that σ may be defined as an rms bandwidth. For from the properties of gaussian functions discussed in Chap. 5, it is apparent that

$$\sigma^2 = \frac{\int_{-\infty}^{\infty} f^2 G_x(f)\, df}{\int_{-\infty}^{\infty} G_x(f)\, df} \qquad [6\text{-}79]$$

This is often used as the definition of bandwidth in spectral analysis, and extends the various possible definitions considered in Chap. 2.]

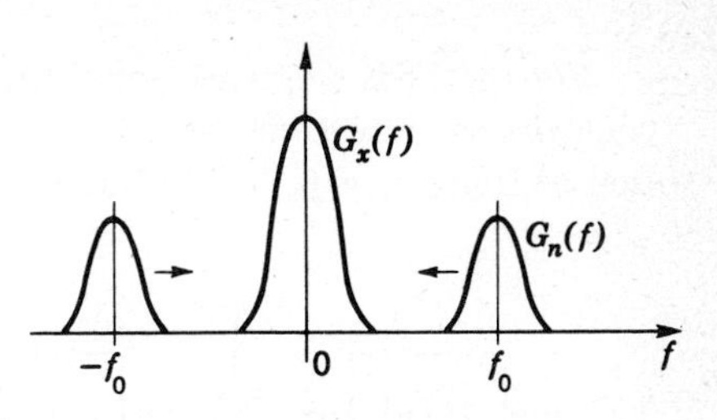

FIG. 6-25 Low-pass noise spectral density, symmetrical passband.

6-6 BINARY r-f TRANSMISSION COMPARED: SYNCHRONOUS DETECTION OF BINARY SIGNALS IN NOISE

Using the narrowband representation of noise, we are now in a position to study the comparative SNR properties of PSK, OOK, and FSK techniques (see Chap. 4), as well as others if so desired. To do this we shall assume gaussian noise added during transmission and assume decisions to be made after synchronous detection at the receiver. We shall then show, using the narrowband representation of noise, that probabilities of error may be written down directly using the PCM error calculations of Chap. 5. In the next section we extend this analysis to those cases where envelope detection is used instead.

Recall from Chap. 4 that synchronous detection requires carrier phase coherence to be maintained. The process of synchronous detection consists of multiplication of a received carrier signal by a locally generated sine wave of the same frequency and phase, the resultant product term then passed through a low-pass filter to eliminate second harmonic terms.

For a PSK binary sequence of the form $\pm A \cos \omega_0 t$, or for an OOK sequence consisting of either $A \cos \omega_0 t$ or 0, we simply multiply by $\cos \omega_0 t$ and filter. A synchronous detector for these signals is shown in Fig. 6-26.

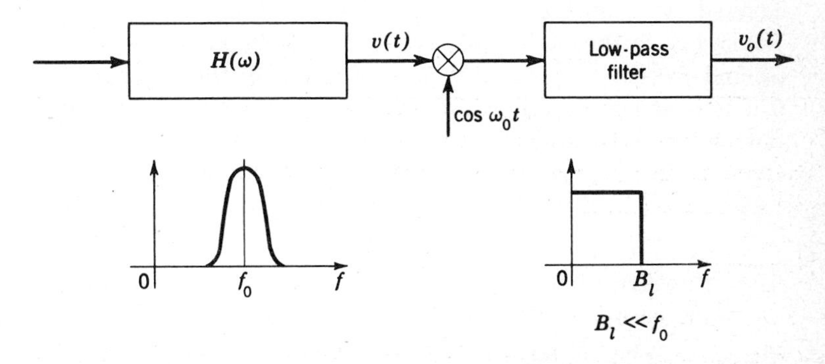

FIG. 6-26 Synchronous detector for PSK and OOK signals.

For the FSK sequence consisting of $A \cos \omega_1 t$ or $A \cos \omega_2 t$, *two* sets of synchronous detectors are needed, one operating at frequency f_1, the other at frequency f_2. The resultant FSK detector is shown in Fig. 6-27.

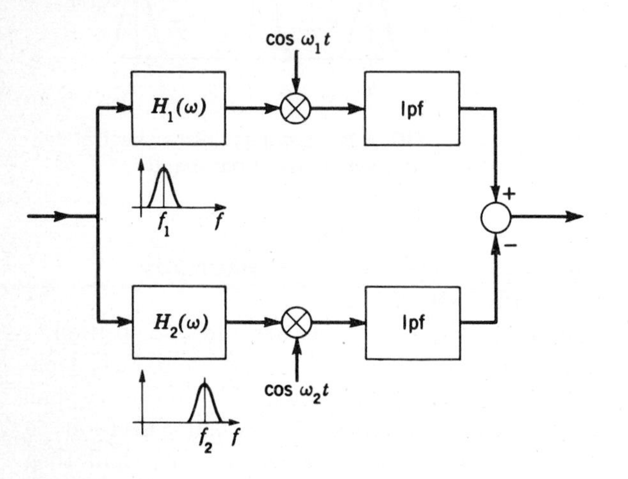

FIG. 6-27 Synchronous detection of FSK signals.

Note that in both figures predetection (i-f) filters are shown. These are narrowband filters with bandwidths chosen wide enough to pass the respective carrier signals (generally $2B_l$ Hz, if the low-pass filter bandwidths are B_l). Actually we shall show quite simply that they should be matched filters or reasonable approximations to these if additive noise is the primary source of error in detection.

Now assume that signal plus noise appear at the output of a particular narrowband filter. [This is labeled $v(t)$ in Fig. 6-26.] It is apparent, as pointed out in the discussion of Chap. 4, that all three binary cases are covered by assuming a signal of the form $f(t) \cos \omega_0 t$. [In the PSK case $f(t)$ is $\pm A$, in the OOK case, $+A$ or 0. In the FSK case ω_0 is either ω_1 or ω_2, and $f(t)$ is A if a signal is present in one of the two parallel channels, 0 if it is absent.] The composite signal plus noise at the input to the detector may thus be written

$$\begin{aligned} v(t) &= f(t) \cos \omega_0 t + n(t) \\ &= [f(t) + x(t)] \cos \omega_0 t - y(t) \sin \omega_0 t \end{aligned} \qquad [6\text{-}80]$$

using the narrowband noise representation of the last section.

The noise $n(t)$ is narrowband in form, its spectral density dependent on the i-f filters $H(\omega)$ shown in Figs. 6-26 and 6-27. The low-pass noise terms $x(t)$ and $y(t)$ thus have half the i-f bandwidth, assumed small compared to the center frequency f_0. Multiplication of $v(t)$ by the locally generated $\cos \omega_0 t$ therefore results in a low-frequency term $[f(t) + x]/2$ passed by the low-pass filter, as well as a double carrier, $2f_0$, term rejected by the filter.

The multiplicative factor of $\frac{1}{2}$ may be assumed absorbed in any gain factors. (For simplicity we have been ignoring these, since signal and noise are then both multiplied by the same constants.) The detector output $v_o(t)$ is then simply

$$v_o(t) = f(t) + x(t) \qquad [6\text{-}81]$$

Decision circuits then sample the low-pass $v_o(t)$ and decide on the binary symbol transmitted. We showed in the last section, however, that $x(t)$ has the same gaussian statistics, as well as the same variance (power) N as the noise $n(t)$. From the form of Eq. [6-81], it is then apparent that the binary decision problem here is *identical* with that first considered in Sec. 5-6. The $x(t)$ term here is the same as the low-pass $n(t)$ included there!

The process of synchronous detection performs the same operation on noise as on signal. It merely serves to translate the center frequency down, from f_0 to 0 frequency.

In particular, in the OOK case, the detector output is just

$$v_{o,\text{OOK}}(t) = \begin{matrix} A \\ \text{or} \\ 0 \end{matrix} + x(t) \qquad [6\text{-}82]$$

Since this output is identical with that discussed in Sec. 5-6, it is apparent that the probability of error is just

$$P_{e,\text{OOK}} = \frac{1}{2}\left(1 - \text{erf}\,\frac{A}{2\sqrt{2N}}\right) = \frac{1}{2}\,\text{erfc}\left(\frac{A}{2\sqrt{2N}}\right) \qquad [6\text{-}83]$$

with $\text{erf}\,x$ the error function, and $\text{erfc}\,x$, the so-called complementary error function, given by

$$\text{erfc}\,x = 1 - \text{erf}\,x$$
$$\text{erf}\,x \equiv \frac{2}{\sqrt{\pi}}\int_0^x e^{-y^2}\,dy \qquad [6\text{-}84]$$

(See Eq. [5-95]. We assume here, as in deriving that equation, that 1's and 0's in the binary sequence are equally likely, and that the decision level has been set at $A/2$.)

Figure 5-27, showing the probability of error as a function of A/σ, or $A/\sqrt{N}$ in this case, may thus be used directly in determining the performance of the OOK system in additive gaussian noise. The amplitude

A here is just the amplitude of the sine wave representing the 1 (or mark) symbol. Since minimum probability of error corresponds to maximizing $A/\sqrt{N}$, it is apparent that the $H(\omega)$ filter preceding the synchronous detector should be a matched filter (or a good approximation to this), matched to the pulsed sine wave $A \cos \omega_0 t$. Some thought indicates that this is just the low-pass matched filter translated up to the center frequency f_0. This assumes white noise at the input to the i-f filter $H(\omega)$.

Alternately, a little thought will indicate that the $H(\omega)$ filter may be widened considerably if desired, and the matched filtering performed in the low-pass filter following the multiplier. The synchronous detector merely serves to translate frequencies down, so that the *overall* filtering is effectively due to the cascaded effect of $H(\omega)$ and the low-pass filter. It is this overall filter that should be matched to the signal. The low-pass filter is often called a *post-detection* filter.

In the PSK case the synchronous detector output consists of a bipolar signal $\pm A$ plus noise. This thus corresponds exactly to the bipolar signal analysis in Sec. 5-6. Here, however, we have $\pm A$ as the signal, rather than $\pm A/2$, as assumed there. Again choosing 0 as the decision level ($v_o > 0$ is called a 1, $v_o < 0$ a 0 signal), and assuming equally likely binary symbols, the probability of error is just

$$P_{e,\text{PSK}} = \frac{1}{2} \operatorname{erfc} \left(\frac{A}{\sqrt{2N}} \right) \qquad [6\text{-}85]$$

As noted in Sec. 5-6, and as is apparent by comparing Eqs. [6-83] and [6-85], the PSK system requires only half the signal amplitude that the OOK system does, for the same probability of error. There is thus a 6-db peak SNR improvement. On an average power basis, however, the improvement is only 3 db because the OOK system is off half the time, on the average, requiring only half as much power.

In the case of the FSK system the outputs of two detectors are compared. At any one time one detector has signal plus noise, the other noise only. Calling the noise output of one channel x_1, that of the other x_2, we have, on subtracting the two channel outputs, the FSK output given by

$$v_{o,\text{FSK}} = \begin{matrix} +A \\ \text{or} \\ -A \end{matrix} + (x_1 - x_2) \qquad [6\text{-}86]$$

The output signal is again bipolar: $+A$ appears if a 1 has been transmitted, $-A$ for a 0 transmitted. The total noise output is, however, $x_1 - x_2$. If the noises in the two channels are *independent* [true, if the system input noise is white, and the two bandpass filters $H_1(\omega)$ and $H_2(\omega)$

do not overlap,[1] the usual case], the variances *add.*[2] We have effectively *doubled the noise* by subtracting the two outputs! However, since the output signal is bipolar, the effective signal excursion, as in the PSK case, is twice that of the OOK case. The FSK system thus provides results intermediate between the OOK and PSK cases:

$$P_{e,\text{FSK}} = \frac{1}{2}\,\text{erfc}\left(\frac{A}{2\sqrt{N}}\right) \qquad [6\text{-}87]$$

For a specified probability of error the FSK system requires 3 db more signal power than the equivalent PSK system with the same noise power, but is 3 db better than the OOK system on a peak power basis. (Recall from Chap. 4 that FSK requires wider transmission or channel bandwidths than either of the other two systems. The channel bandwidth is measured prior to the H_1 and H_2 filters of Fig. 6-27.)

Again a minimum probability of error requires maximization of the ratio $A/\sqrt{N}$ at the input to the synchronous detector. For this purpose the two filters H_1 and H_2 in Fig. 6-27 should be matched as closely as possible to their respective signal inputs.

It is apparent from this simple analysis of binary signals in noise that PSK transmission is to be preferred if *phase coherence* is available. As noted in Sec. 4-2, some of the Pioneer deep-space probes used PSK modulation successfully in their telemetry systems. Rather sophisticated techniques were of course required to establish and maintain the necessary phase synchronism. More commonly, FSK transmission with envelope detection at the receiver has been used in commercial data (teletype) transmission. The analysis of high-frequency binary transmission with envelope detection will consequently be considered in the next section. We shall find there, as expected, that envelope detection results in a somewhat higher probability of error, or a corresponding loss in SNR.

6-7 NONCOHERENT BINARY TRANSMISSION

If phase coherence cannot be maintained, or if it is just uneconomical to incorporate phase-control circuits in the receiver, one generally resorts to

[1] Schwartz, Bennett, and Stein, *op. cit.*, pp. 44, 45.

[2] If $y = x_1 - x_2$, and $E(x_1) = E(x_2) = 0$, as here,

$$\text{var}\,(y) = E(y^2) = E(x_1 - x_2)^2 = E(x_1^2) - 2E(x_1x_2) + E(x_1^2) = E(x_1^2) + E(x_2^2)$$

since x_1 and x_2 are independent and hence uncorrelated. More generally, the reader is asked to show for himself that if $y = a_1x_1 + a_2x_2$, x_1 and x_2 independent, var $(y) = a_1^2$ var $(x_1^2) + a_2^2$ var (x_2^2), *independent* of the first moments or mean values.

envelope detection of high-frequency carriers. The resultant system is generally referred to as noncoherent. The envelope-detection process has been discussed in previous chapters. We limit ourselves here to recalling that if we have a high-frequency sine wave of the form $r(t)(\cos \omega_0 t + \theta)$, with $r(t)$ a positive quantity, the envelope detector provides $r(t)$ at its output. In practice a nonlinear device plus low-pass filtering is needed to recover $r(t)$.

It is apparent that PSK signals require phase coherence to be demodulated. We therefore consider here only OOK or FSK signals, detected with envelope detectors. As in Fig. 6-27 the FSK receiver consists of *two* channels, one tuned to frequency f_1, the other to frequency f_2. Each synchronous detector in Fig. 6-27 is replaced by an envelope detector. The outputs of the two detectors are then compared to determine whether one binary symbol or the other was transmitted. The FSK envelope-detection scheme is shown in Fig. 6-28*b*. The OOK receiver consists of one channel, tuned to the carrier frequency of f_0, with one envelope detector providing the desired output. The OOK receiver is shown in Fig. 6-28*a*. A decision level on the output then decides whether a 1 or a 0 was transmitted.

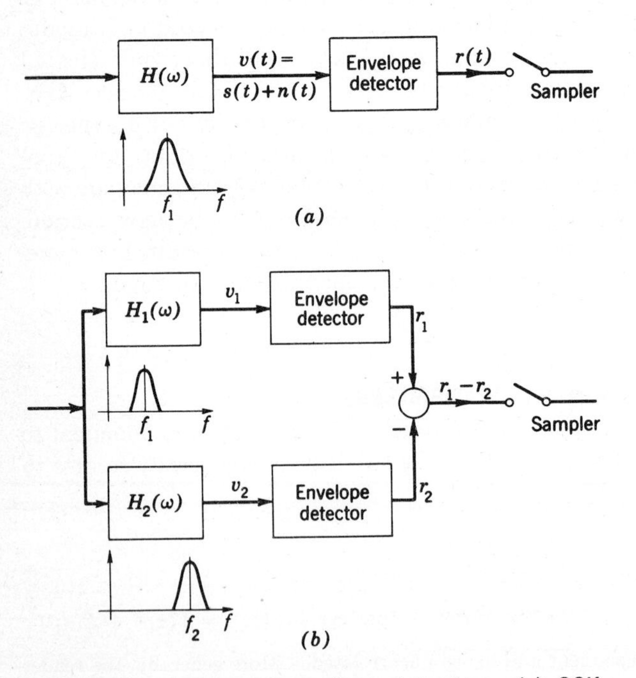

FIG. 6-28 Noncoherent envelope detection. (a) OOK receiver; (b) noncoherent FSK detection.

For simplicity's sake we shall discuss here only the simple OOK system. Results for the FSK case are found in a similar manner, by an extension of the analysis. If we assume an OOK sequence of symbols, the output $v(t)$ of a narrowband filter centered at f_0 is again, adding signal and noise,

$$v(t) = [f(t) + x(t)] \cos \omega_0 t - y(t) \sin \omega_0 t \qquad [6\text{-}88]$$

Here $f(t) = A$ or 0. Rewriting this in the equivalent form,

$$v(t) = r(t) \cos [\omega_0 t + \theta(t)] \qquad [6\text{-}89]$$

with $\quad r = \sqrt{(f + x)^2 + y^2} \quad$ and $\quad \theta = \tan^{-1} \dfrac{y}{f + x}$

it is apparent that an envelope detector will produce $r(t)$ at its output. Sampling $r(t)$ once each binary period, we then make a decision about whether a 1 or a 0 is present by noting whether the sampled $r(t) > b$, or $r(t) < b$, respectively, b a specified decision level.

The probability of error depends on the statistics of r in the two cases, $f = A$ or 0. Recall, however, from Sec. 5-10 that we have already solved this problem. For we showed there in the context of the statistics of a fading channel that the random variable r was Rayleigh-distributed with $A = 0$, and Rician-distributed with $A > 0$. There the random variables x and y were respectively the inphase and quadrature terms obtained by summing the contributions of the signal scattered over many independent paths. Here x and y represent additive noise terms. By the central-limit theorem, the random variables in both cases are gaussian-distributed. The two-dimensional transformation employed in Sec. 5-10 to find the statistics of r and θ from x and y applies here as well, and so we write directly:

1. Noise-only case, $A = 0$:

$$f_n(r) = \frac{re^{-r^2/2N}}{N} \qquad r \geq 0 \qquad [6\text{-}90]$$

2. Signal-plus-noise case, $A > 0$:

$$f_s(r) = \frac{re^{-r^2/2N}}{N} e^{-A^2/2N} I_0\left(\frac{rA}{N}\right) \qquad r \geq 0 \qquad [6\text{-}91]$$

(The subscript n stands for noise, s for signal plus noise.) Recall again that $I_0(x)$ is the modified Bessel function. N is of course the noise variance or the mean noise power at the output of the narrowband filter $H(\omega)$ in Fig. 6-28a. Both density functions are sketched in Fig. 6-29. Note again, as shown in Sec. 5-10, that the Rician distribution approaches

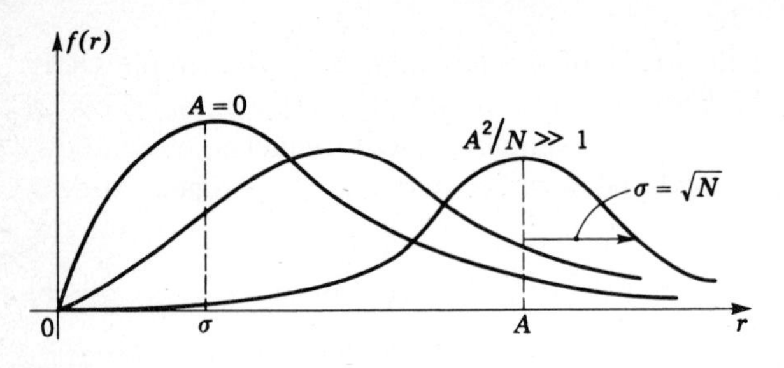

FIG. 6-29 Envelope distribution for signal plus noise.

the gaussian for $A^2/N \gg 1$. (Here $A^2/2$ is the average signal power with a 1 transmitted. $A^2/2N$ is thus a power signal-to-noise ratio.)

A typical sketch of noise at the output of the narrowband filter is shown in Fig. 6-30. Note that it looks exactly like the fading sine-wave

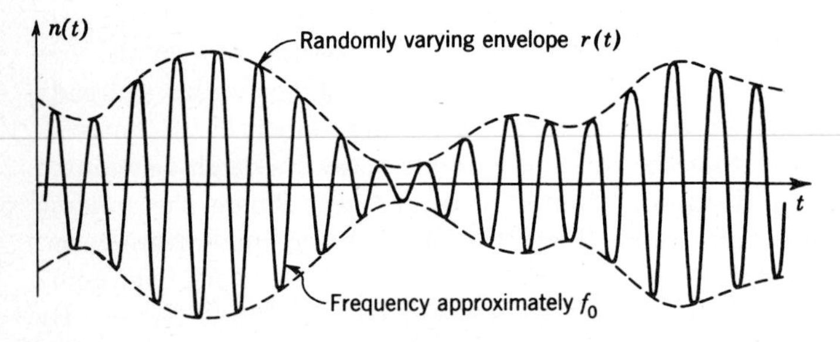

FIG. 6-30 Noise at output of narrowband filter.

carrier discussed in Sec. 5-10; it resembles a sine wave at frequency f_0, slowly varying in amplitude and phase at a rate determined by the filter bandwidth. The envelope $r(t)$ shown dashed in Fig. 6-30 has the Rayleigh statistics of Eq. [6-90]. The functional form of the noise shown is of course given by Eq. [6-89], with $A = 0$. The random phase angle $\theta(t)$ is uniformly distributed between 0 and 2π, again as shown in Sec. 5-10. Further properties of narrowband noise and the statistics of noise at the envelope-detector output will be considered in the sections following.

Using the two density functions, that of Eq. [6-90] for noise only (a 0 or *space* transmitted), and that of Eq. [6-91] for signal plus noise (a 1 or *mark* transmitted), we are now in a position to calculate the probability of error for OOK signaling, with envelope detection. The analysis is similar to that of the error analysis of Sec. 5-6, the only difference being that we must now consider envelope statistics rather than the gaussian statistics assumed there.

Here we decide on a 0 transmitted if $r < b$, a 1 transmitted if $r > b$, as noted earlier. The decision level b thus corresponds to the decision level d of Sec. 5-6. Although d could take on any value, positive or negative, b is of course restricted to positive values only because of the envelope characteristics. Figure 6-31 shows the two decision regions introduced by defining the decision level b.

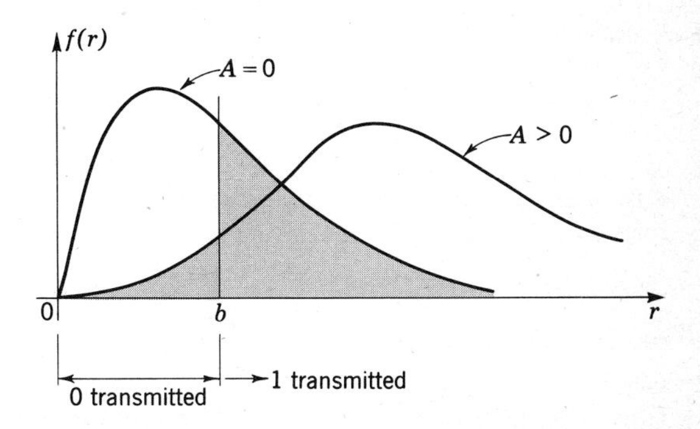

FIG. 6-31 Decision regions with envelope-detected OOK signals.

Assume, as in Sec. 5-6, that the a priori probabilities of transmitting a 0 and a 1 are, respectively, P_0 and $P_1 = 1 - P_0$. The overall probability of error is then given by

$$P_e = P_0 \int_b^\infty f_n(r)\,dr + P_1 \int_0^b f_s(r)\,dr \qquad [6\text{-}92]$$

with the Rayleigh-density function of Eq. [6-90] used in the first integral and the Rician-density function of Eq. [6-91] used in the second. The two integrals are indicated by the crosshatching in Fig. 6-31.

Although the first integral is directly evaluable, giving $e^{-b^2/2N}$, the second integral cannot be evaluated in closed form. It has, however, been numerically evaluated and tabulated by numerous investigators.[1] Carrying out the indicated integrations of Eq. [6-92] in the special case where $P_0 = P_1 = \frac{1}{2}$, and the two integrals are the same (i.e., the two crosshatched areas of Fig. 6-31 are the same), one obtains the curve of Fig. 6-32. Also shown is the comparable curve as obtained earlier for coherent (synchronous) detection. Note that the envelope-detection process is somewhat inferior to coherent detection. For the same probability of error P_e it requires somewhat more signal power (higher $A/\sqrt{N}$),

[1] See Schwartz, Bennett, and Stein, *op. cit.*, pp. 27, 28; 289, 290; appendix A.

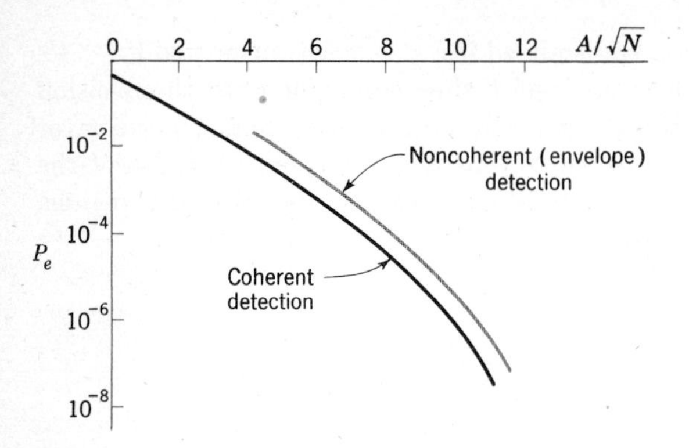

FIG. 6-32 Binary error probabilities, OOK transmission.

or for the same SNR $A/\sqrt{N}$, P_e is somewhat larger. This is as might be expected since we are essentially throwing away useful information by ignoring the phase in the envelope-detection case.

Actually one can again optimize the decision level b, choosing it to minimize the probability of error P_e from Eq. [6-92], as was done in Sec. 5-6. The result is not too different from that of Fig. 6-32. (As will be shown in Chap. 8 the optimum corresponds to the point at which the two density functions in Fig. 6-31 intersect. For $A^2/N \gg 1$, which is necessary to ensure low probability of error, that optimum point does not differ substantially from the values of b used in calculating Fig. 6-32. A discussion of the optimum b, and the effect on the probability of error in deviations from it appears in the reference.[1])

Note one very important point, however. Again *the probability of error depends uniquely on the ratio* $A/\sqrt{N}$, the ratio of *peak signal* at the *narrowband filter output* to the *rms noise* measured *at the same point.* The only way to decrease the probability of error is to increase $A/\sqrt{N}$. How does one do this, aside from the obvious ways of increasing transmitter power and trying to decrease the additive noise? It is apparent that a matched filter is again called for. The *narrowband filter* $[H(\omega)]$ should thus be *matched to the signal* representing the "1" symbol or mark, in this case a sinusoidal burst at frequency f_0, lasting the binary interval.

A word of caution should again be injected here, however. As noted earlier, in first discussing matched filters, we are neglecting intersymbol interference and are assuming that additive noise is the prime culprit in causing errors in detection. The point made in first introducing matched

[1] Schwartz, Bennett, and Stein, *op. cit.*, pp. 291, 292.

filters can now be repeated. In all digital transmission systems with additive gaussian noise the prime cause of detection errors, matched filters or reasonable approximations to them will normally be used to minimize the probability of error. This includes baseband digital systems, coherent or synchronous carrier systems (as studied in the previous section), the FSK system to be discussed briefly in the paragraph following, search radar systems to be discussed in the next section, etc. If individual pulse decisions may be considered, independent of streams of data preceding or following the pulse in question, a matched filter is usually called for. If intersymbol interference is the major problem, however, the filter design of Chap. 3, or that presented in the references, should be considered instead.[1]

In practice, of course, the matched-filter condition is usually met by designing the filter to have the appropriate bandwidth. This is usually the reciprocal of the signal pulse width (or the reciprocal of the binary interval), as noted earlier in discussing matched filters. Again, as noted earlier, some widening of the bandwidth may be tolerated and in fact is often employed to decrease pulse overlap (intersymbol interference) without affecting the SNR critically.

How does one now determine the probability of error of the noncoherent, or envelope-detected, FSK system of Fig. 6-28*b*? Assume here that frequency f_1 corresponds to a 1 (mark) transmitted, f_2 to a 0 (space). If a mark is actually transmitted, signal plus noise will appear on channel 1, noise alone on channel 2. The sampled value of $(r_1 - r_2)$ should be positive for a correct decision to be made. Similarly, $(r_1 - r_2)$ should be negative if a space is transmitted. (Note that although phase synchronism is no longer assumed between transmitter and receiver, binary-interval synchronism must be maintained. This is true as well with the OOK receiver of Fig. 6-28*a*.)

An error will obviously be made if $(r_1 - r_2)$ is negative, with a mark transmitted, or positive with a space transmitted. If 1's and 0's are assumed equally likely to be transmitted, the probability of either type of error is the same by the symmetry of the system of Fig. 6-28*b*. Assuming as an example a mark transmitted, the probability of error is just the probability that the noise causing r_2 will exceed the signal-plus-noise envelope signal r_1. This is given by

$$P_e = \int_{r_1=0}^{\infty} f_s(r_1) \left[\int_{r_2=r_1}^{\infty} f_n(r_2)\, dr_2 \right] dr_1 \qquad [6\text{-}93]$$

with the density functions of Eqs. [6-90] and [6-91] used where indicated. (The inner integral provides the probability of error for a *fixed* value of r_1.

[1] Lucky, Salz, and Weldon, *op. cit.*

Averaging over all possible values of r_1, one then obtains Eq. [6-93].) Since $f_n(r)$ is just the Rayleigh density, the inner integral integrates readily to $e^{-r_1^2/2N}$. The expression for the probability of error is then just

$$P_e = \int_0^\infty \frac{r_1}{N} e^{-r_1^2/N} e^{-A^2/2N} I_0\left(\frac{r_1 A}{N}\right) dr_1 \qquad [6\text{-}93a]$$

To integrate this expression we use a simple trick. We define a new (dummy) variable $x = \sqrt{2}\, r_1$. Equation [6-93a] then becomes, with a little manipulation

$$P_e = \tfrac{1}{2} e^{-A^2/4N} \int_0^\infty \frac{x}{N} e^{-x^2/2N} e^{-A^2/4N} I_0\left(\frac{xA}{\sqrt{2}\, N}\right) dx \qquad [6\text{-}93b]$$

But the integrand is exactly in the form of the Rician-density function of Eq. [6-91] if A there is replaced by $A/\sqrt{2}$ here. The integral must then just equal 1, and we have, quite simply,

$$P_e = \tfrac{1}{2} e^{-A^2/4N} \qquad [6\text{-}94]$$

The probability of error of the noncoherent FSK system thus decreases exponentially with the power SNR $A^2/2N$. The error curve corresponding to Eq. [6-94] is plotted in Fig. 6-33, together with the corresponding curves for synchronous (coherent) PSK and FSK transmission, using the results of the previous section and Sec. 5-6. Note that just as in the case of noncoherent versus coherent OOK (Fig. 6-32), there is a penalty paid for using envelope rather than synchronous detection. The noncoherent systems require slightly more signal power for the same probability of error.

This loss in SNR due to envelope detection becomes negligible at high SNR, however, as is apparent from Fig. 6-33, and as may be shown from the synchronous-detection results of the last section. For it is readily shown[1] that the asymptotic ($x \gg 1$) form for the complementary error function is given by

$$\operatorname{erfc} x \doteq \frac{e^{-x^2}}{x\sqrt{\pi}} \qquad x \gg 1 \qquad [6\text{-}95]$$

The coherent FSK error probability derived in the last section and indicated in Fig. 6-33 is then just

$$P_e = \frac{1}{2} \operatorname{erfc} \frac{A}{2\sqrt{N}} \doteq \frac{1}{\sqrt{\pi}\, A/2\sqrt{N}} \frac{e^{-A^2/4N}}{2} \qquad [6\text{-}96]$$

[1] See, for example, B. O. Peirce, "A Short Table of Integrals," p. 29, Ginn, Boston, 1929.

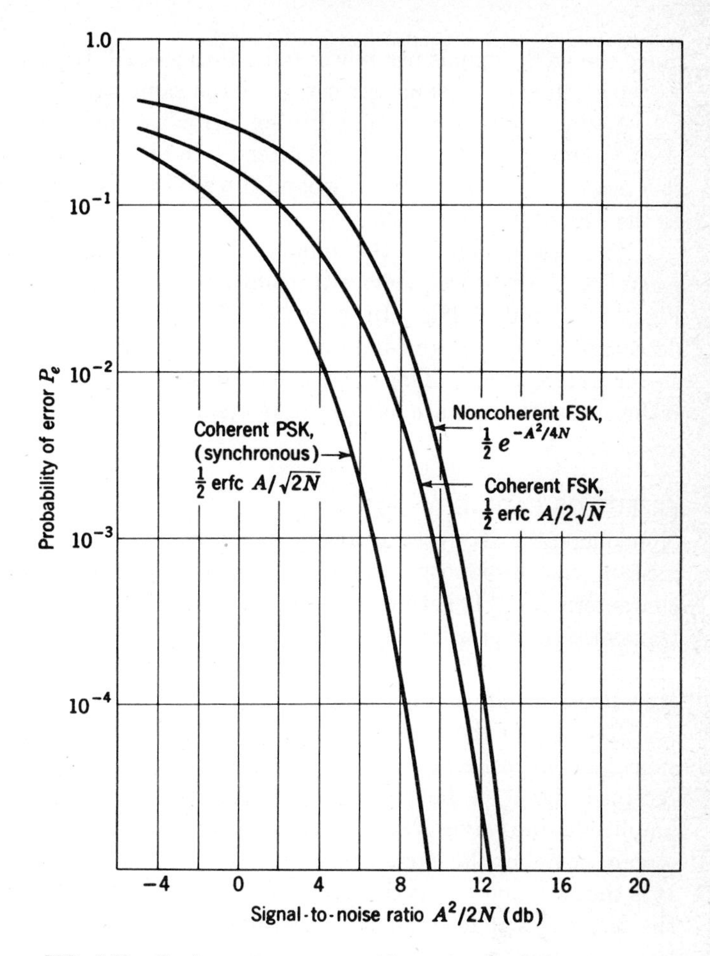

FIG. 6-33 Performance curves, binary transmission.

Note that the dominating exponential behavior is just that of the noncoherent FSK expression of Eq. [6-94]. At high $A^2/2N$ the other terms in the denominator rapidly become negligible and noncoherent and coherent FSK approach one another.

It is similarly possible to show that the asymptotic (high SNR) error probability for both coherent and noncoherent OOK is given by[1]

$$P_e \doteq \tfrac{1}{2}e^{-A^2/8N} \qquad [6\text{-}97]$$

Comparing this with Eqs. [6-94] and [6-96], it is apparent that the OOK systems require twice the SNR ($A^2/2N$) that the FSK systems require.

[1] Schwartz, Bennett, and Stein, *op. cit.*, p. 292.

Since the OOK signals use power only during mark transmission, the two systems achieve the same error rate at the same *average signal power*.

Again, minimum probability of error is ensured by maximizing $A^2/2N$, and hence using matched filters. In the case of noncoherent FSK this means that the two narrowband filters (Fig. 6-28*b*) are *each* matched to their respective channel.

The stress here in the binary FSK case has been on two-filter receivers; the two frequencies transmitted are separately detected, as in Figs. 6-27 and 6-28*b*. In practice, discriminators and zero-crossing detectors (Chap. 4) are commonly used for binary FM detection, as well as for analog FM. The error calculations for these detectors become rather involved and are left to the references.[1]

EFFECT OF FADING

Noncoherent FSK systems are quite commonly used for h-f data transmission via the ionosphere as well as other channels possessing fading characteristics. It is thus of interest to consider briefly the effect of data transmission over a fading channel.

Assume the channel has the Rayleigh fading described in Sec. 5-10. How does this affect the data error rates derived in this section?

Recall that the basis of the Rayleigh fading model was the transmission of a constant-sine-wave carrier. We then found the amplitude of the received signal to be Rayleigh-distributed. In this section, constant-amplitude sine waves were also assumed transmitted (although the binary stream switches the carrier on or off, or from one frequency to another). It is thus apparent that the effect of a Rayleigh fading channel is to cause the received signal amplitude A to fluctuate randomly following a Rayleigh distribution. With A *fixed* we get the error rate calculated here. With A *fluctuating* we must in turn *average* over the statistics of A.

In particular, if we assume noncoherent FSK transmission through a Rayleigh fading channel, with gaussian noise added at the receiver, the *average* probability of error $E(P_e)$ must be given by

$$\begin{aligned} E(P_e) &= \int_0^\infty P_e(A) \frac{Ae^{-A^2/2\sigma^2}}{\sigma^2} dA \\ &= \int_0^\infty \tfrac{1}{2} e^{-A^2/4N} \frac{Ae^{-A^2/2\sigma^2}}{\sigma^2} dA \\ &= \frac{1}{2 + (\sigma^2/N)} \end{aligned} \qquad [6\text{-}98]$$

[1] W. R. Bennett and J. R. Davey, "Data Transmission," chap. 9, McGraw-Hill, New York, 1965; Lucky, Salz, and Weldon, *op. cit.*, chap. 8.

carrying out the simple integration indicated. Here σ^2 is the mean signal power over the fading path, and N is the mean added noise power.

The ratio σ^2/N is thus an average SNR, averaged essentially over the signal fading statistics. It is apparent that the effect of fading is to increase the probability of error drastically, for the probability of error now decreases *inversely* as the average SNR, rather than *exponentially*, as found previously. This inverse dependence on average SNR is found for OOK transmission and other forms of binary communication as well.[1]

The effect of fading channels is thus to require substantially increased transmitter power for reliable communication. Are there other ways of compensating for the deterioration due to fading? Most commonly, various forms of *diversity* communications are used. Multiple receivers are used to detect the same data stream picked up over two or more *independently* fading channels. The hope here is that if one channel has a signal fading substantially below the median, the other (or others), may have a signal considerably larger in amplitude. Various types of diversity systems are used in practice: space diversity, with two or more separate transmitting or receiving antennas (or both) used, physically spaced several wavelengths apart so that the received signals fade independently; angle diversity, with two receiving antennas pointed in somewhat different directions; polarization diversity; frequency diversity; time diversity; etc. Space diversity is the technique most commonly used in practice. Time diversity corresponds to the repetition of signals at specified time intervals. This particular method of enhancing signal detectability in noise is commonly used in radar, sonar, and other signaling systems as well. There the problem is not necessarily that of detecting fading signals, but of detecting small-amplitude signals. This application of signal repetition will be considered quantitatively in the next section, in discussing radar signal detection.

In addition to the different diversity methods used for achieving independently fading signals, there are various ways of combining the independently fluctuating signals. One may select the largest signals, simply add all the outputs, or add them after weighting each according to the average SNR measured on each diversity channel. This last technique, called *maximal-ratio* combining, may be shown to be optimum in the sense of providing the largest effective SNR or minimum probability of error. Details of diversity techniques, including extensive analyses and performance curves, appear in the current literature, and are summarized in the reference indicated.[2]

[1] Schwartz, Bennett, and Stein, *op. cit.*, pp. 402–409.

[2] *Ibid.*, chaps. 10 and 11.

(Although the discussion of fading and diversity techniques was motivated here for the case of data transmission, fading is of course a problem with analog data transmission as well, and diversity combining is commonly used to combat it there also. This is discussed in detail in the same reference.)

6-8 RADAR SIGNAL DETECTION AND PHASE DETECTION

As an additional application of the techniques we have developed in this chapter for handling random signals and noise we now consider a simple radar system. Envelope detection of narrowband pulses is commonly encountered here as well, so that much of the discussion of the last section will be applicable in calculating the appropriate error probabilities for radar. Envelope detection is used in many standard radar systems, because of the difficulty in maintaining signal phase coherence. Radar return signals usually vary randomly in phase because of the nature of the complex targets from which they are reflected. Synchronous or coherent carrier injection at the receiver is thus normally not possible.

A simple radar is shown pictorially in Fig. 6-34. As indicated by the block diagram of Fig. 6-35 the radar transmitter provides a periodic series

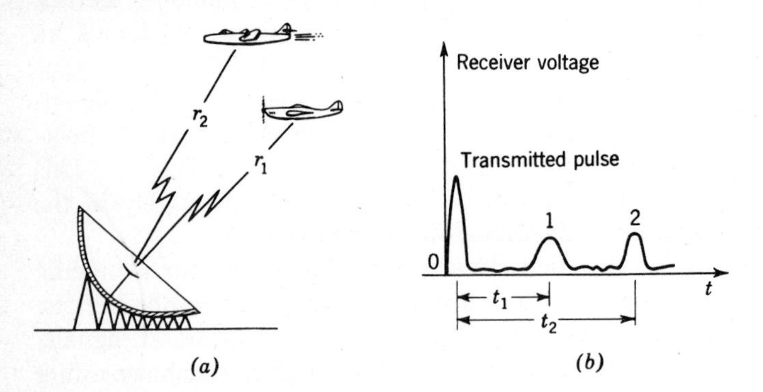

FIG. 6-34 Simple radar system. (a) Transmitter-receiver and targets; (b) transmitted and received signals.

of pulsed sine waves. These are normally generated at microwave frequencies in order to obtain a well-defined narrow beam of electromagnetic energy with antennas of reasonable size. Targets in the path of the beam reflect the pulsed high-frequency energy. A portion of this reflected energy is then picked up at the antenna, heterodyned down to an intermediate frequency, envelope-detected, and sent on to indicating circuits.

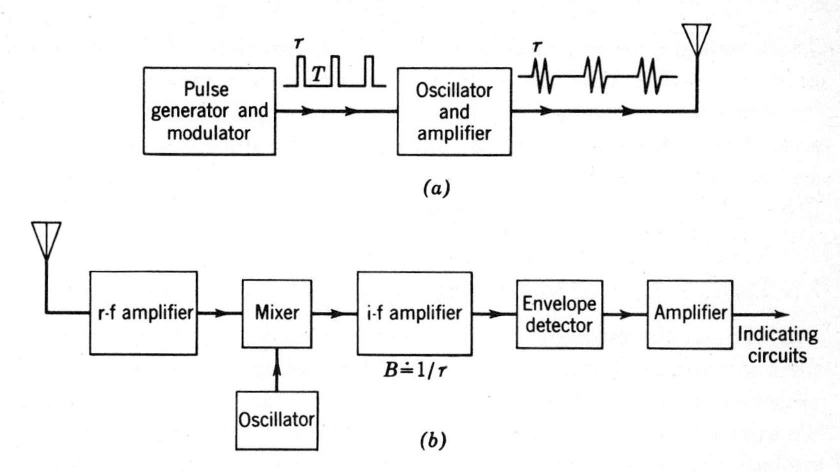

FIG. 6-35 Block diagram, simple search radar. (a) Transmitter; (b) receiver.

The component parts of the radar, excluding necessary synchronizing circuits, are thus essentially the same as those encountered in the usual AM system.

The information carried by the received pulses is primarily range (distance) information. Since electromagnetic energy travels at a speed of $c = 3 \times 10^8$ meters/sec in air (actually vacuum), the time taken for the transmitted pulse to travel a distance r_1 meters, be reflected, and return to the antenna is

$$t_1 = \frac{2r_1}{c} \qquad \text{sec}$$

An indicator presenting transmitted and received pulses in the sequence of time occurrence can thus be calibrated directly in range. Figure 6-34 shows two targets located a radial distance of r_1 and r_2 meters, respectively, from the radar. The return pulses appear t_1 and t_2 sec, respectively, after the transmitted pulse. Other information as to target location (azimuth and elevation angle, for example) is obtained from the known direction in which the antenna is pointing at the time of transmission and reception. This is the reason why narrow beams of energy are required.

Search radars, as implied by the name, "search" the skies for the possible appearance of a target. For such radars signal detection consists primarily in determining the presence or absence of a pulse. Pulse shape is secondary. For such radars, then, the system bandwidth is usually chosen as the reciprocal of the pulse width to maximize received signal-to-noise ratio. The matched-filter criterion thus applies in this case.

Tracking radars require information as to the precise time of reception of a pulse, however, and the actual pulse shape used is important.

We shall concentrate on the search radar only. Since a small target or one at a relatively far radial distance will produce only a small pulse signal at the receiver, it may be difficult to distinguish between signal and noise. In particular, two types of error may occur:

1. Noise may be mistaken for signal.
2. There is a chance that the signal will be lost in the noise.

We can obviously decrease the errors by increasing the signal level or, for a given target and given distance, increasing the transmitter power. But what SNR at the receiver is required for given (tolerable) error rates? To answer this question quantitatively, we shall calculate the two error probabilities mentioned above. (In actual radar practice the chance of error is reduced and the chance of detecting the signal increased by hitting the target with a repeated series of pulses. For a fixed target distance the received signal pulses will coincide at the same point on a viewing screen. If the screen is of a long-persistence type, these repeated pulses will add up, eventually rising to a detectable level above the surrounding noise.)

For calculation purposes we assume that the detection process consists in calling a voltage a signal if it exceeds a certain predetermined threshold level.

In the case of binary data transmission the threshold was chosen to minimize probability of error. For coherent on-off transmission the resultant threshold was shown to be at one-half the signal amplitude. For noncoherent OOK transmission, the threshold was very nearly one-half the signal amplitude for small probability of error. For bipolar (or coherent PSK transmission) the optimum threshold was just zero. Here there is no one probability of error to be minimized. The two types of error noted above cannot be combined into one error probability, as in the binary data case, because a priori transmission probabilities are not defined in the radar case; that is, one never knows beforehand what the chance of detecting a target is. In fact, most of the time targets are absent and one sees noise only in the system. The threshold level here is then chosen to reduce to a tolerable level the chance of noise, in the absence of signal, being mistaken for a signal. Once this level is fixed, the other error probability, that of losing the signal in the noise, will depend only on the SNR and can only be maintained at a tolerable level by keeping the SNR to within a specified minimum. Thus with the level chosen there is a chance that a signal, when it does appear, will encounter an instantaneously negative noise voltage. The resulting signal plus noise voltage may drop below the threshold level so that the signal will be lost. To keep the probability of detecting a signal, once it does appear, to as close to a value of 1 as desired will then require a specified minimum signal-to-noise ratio at the input to the envelope detector. With the receiver noise and target characteristics known the required transmitter power for a given probability of detection and probability of mistaking noise for signal can be calculated. The probability of detection of a signal is just 1 − (probability of losing a signal).

A typical radar time strip, with threshold level indicated, is shown in Fig. 6-36.

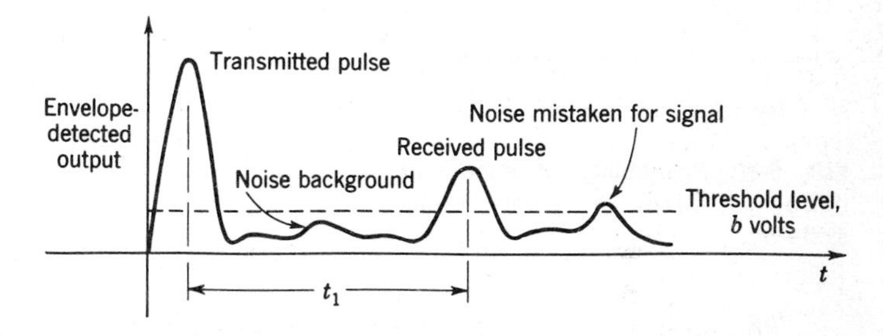

FIG. 6-36 Threshold-level signal detection.

The two probabilities to be calculated depend on the probability distributions found for the envelope of narrowband noise and narrowband signal plus noise. The chance that noise will exceed the level set at b volts and be erroneously labeled a signal is

$$P_{nb} = \int_b^\infty f_n(r)\,dr \qquad [6\text{-}99]$$

where

$$f_n(r) = \frac{re^{-r^2/2N}}{N} \qquad [6\text{-}100]$$

the Rayleigh distribution discussed in the previous section. N is the total average power at the input to the detector. The subscript n is used here to distinguish this envelope distribution from that for signal plus noise.

Integrating Eq. [6-99] directly, we get

$$P_{nb} = e^{-b^2/2N} \qquad [6\text{-}101]$$

This equation relates the probability of noise being mistaken for signal to the threshold level chosen. A curve of P_{nb} versus $b/\sqrt{N}$ is shown plotted in Fig. 6-37. P_{nb} is indicated pictorially by the crosshatched area under the $f_n(r)$ curve in Fig. 6-38a.

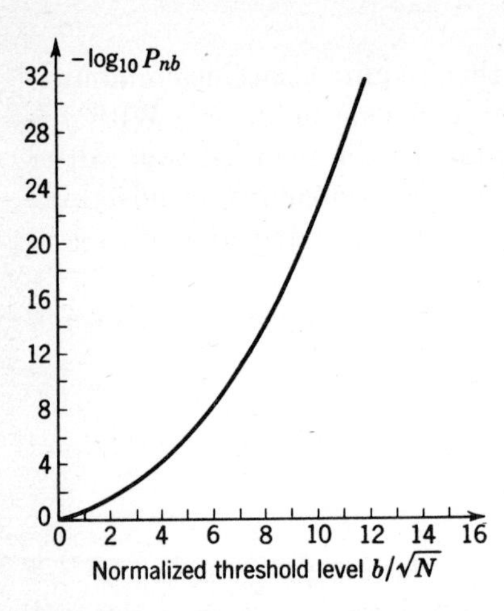

FIG. 6-37 Probability of mistaking noise for signal in simple radar system.

The probability of detecting a signal in the presence of noise is just

$$P_{sb} = \int_b^\infty f_s(r)\,dr \qquad [6\text{-}102]$$

where

$$f_s(r) = \frac{e^{-s^2} r e^{-r^2/2N}}{N} I_0\left(rs\sqrt{\frac{2}{N}}\right) \qquad [6\text{-}103]$$

is the Rician function for signal plus noise (Eq. [6-91]) introduced previously. We have assumed that the received signal pulse has an amplitude A as previously, and have defined $s^2 = A^2/2N$ as the mean power SNR at the detector input. Then

$$1 - P_{sb} = \int_0^b f_s(r)\,dr \qquad [6\text{-}104]$$

can be alternately defined and is just the probability of losing a signal. This is of course just the integral obtained in evaluating the performance

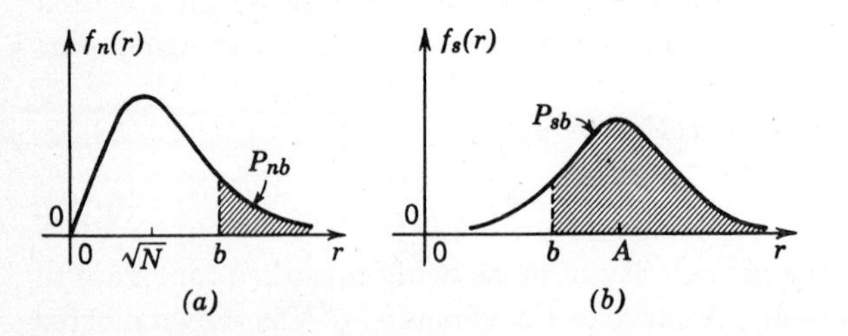

FIG. 6-38 Calculation of error and detection probabilities. (a) Noise only; (b) signal plus noise ($s^2 \gg 1$).

of the noncoherent OOK receiver (Eq. [6-92] and Fig. 6-31). As noted there, tables are available for its evaluation.

Alternatively we can approximate $f_s(r)$ by the gaussian distribution valid for large s^2. This gives

$$P_{sb} \doteq \int_b^\infty \frac{e^{-(r-A)^2/2N}}{\sqrt{2\pi N}}\,dr \qquad b \gg \sqrt{N},\ s^2 \gg 1 \qquad [6\text{-}105]$$

Both $f_s(r)$ and P_{sb} are indicated graphically in Fig. 6-38*b*.

Defining a new variable, $y = (r - A)/\sqrt{2N}$, in Eq. [6-105] and transforming the integrand and the limits accordingly, we get

$$P_{sb} \doteq \frac{1}{\sqrt{\pi}} \int_{b/\sqrt{2N}-s}^\infty e^{-y^2}\,dy \qquad b \gg \sqrt{N},\ s^2 \gg 1,\ s^2 \equiv \frac{A^2}{2N} \qquad [6\text{-}106]$$

Equation [6-106] can now be put into the error-function form for calculation purposes. Recalling from our previous work that

$$\frac{1}{\sqrt{\pi}} \int_0^\infty e^{-y^2}\,dy = \frac{1}{2}$$

and that $\operatorname{erf}(x) \equiv (2/\sqrt{\pi}) \int_0^x e^{-y^2}\,dy$, we get

$$P_{sb} \doteq \frac{1}{2} \operatorname{erfc}\left(\frac{b}{\sqrt{2N}} - s\right) \qquad b \gg \sqrt{N},\ s^2 \gg 1 \qquad [6\text{-}107]$$

For small threshold level and small signal-to-noise ratio s^2 we must evaluate P_{sb} using the actual density function for $f_s(r)$ given by Eq. [6-103]. For large s^2, however, Eq. [6-107] approximates the actual result very well. As an example, let $b/\sqrt{N} = \sqrt{2}\,s = 8$ ($s^2 = 32$). Then $P_{sb} \doteq 0.5$, since $\operatorname{erf} 0 = 0$. But $b/\sqrt{N} = 8$ corresponds to $P_{nb} = 10^{-14}$, from Fig. 6-37. So for a noise-error probability of 10^{-14} a power SNR of 32, or 15 db, is required to have signals detected 50 percent of the time on the average.

The probability of detecting signals may be increased by decreasing the threshold level b. This increases the noise-error probability, however. Alternatively, with b fixed and hence P_{nb} determined (Fig. 6-37), we may increase P_{sb} by increasing s^2. Equations [6-101] and [6-107] (or Eq. [6-102] in the exact case) may be combined to give a set of radar performance curves relating the two probabilities P_{nb} and P_{sb} to the input-power SNR s^2. Such a set of curves is shown plotted in Fig. 6-39. We must emphasize here that these correspond to the simplest type of radar system, in which *individual* signal pulses are required to exceed a threshold level. (In actual systems a group of repetitive pulses are

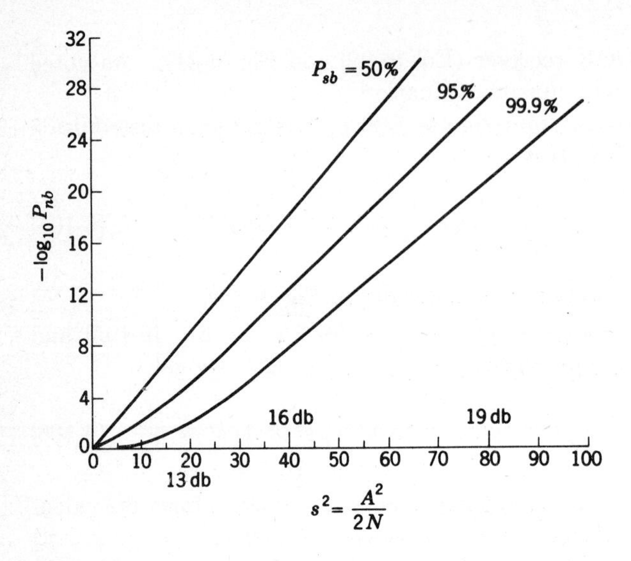

FIG. 6-39 Radar performance curves (single pulses only).

commonly summed or integrated. This is usually done automatically by scanning across a long-persistence viewing screen, as noted previously. The analysis for this case is discussed briefly in the next paragraph.)

How do we utilize the curves of Fig. 6-39? P_{nb} represents the probability that noise, in the absence of signal, will be mistaken for a signal. For a narrowband i-f section, as assumed here, the envelope of the i-f noise voltage will appear roughly as a sequence of pulses of random height. The width of these pulses will be approximately $1/W$, with W the i-f bandwidth. This means that approximately W noise pulses per second will appear on the average. If the radar searches for signals at *all* ranges and azimuths, with no particular sections of the volume scanned gated out, each of the W noise pulses per second has a chance of being mistaken for signal. On the average, however, only the fraction $P_{nb}W$ will manage to exceed the threshold level and be erroneously labeled signal. The number of errors per second, or false alarms, is thus $P_{nb}W$. It is this number (or its reciprocal, $1/P_{nb}W$, the interval between errors on the average) that must be fixed at a tolerable value.

With $P_{nb}W$ given and the system bandwidth known P_{nb} is of course known. With the required probability of detection of signals, P_{sb}, specified we find the minimum SNR required from Fig. 6-39. With the receiver noise power given we can work backward to the input to the receiver to find the signal power required at that point. For a known target and target distance and given antenna gains we can calculate the signal power that the radar must transmit in order to detect the given target with the specified values of P_{sb} and noise-error probability P_{nb}.

As a simple example of the calculation of s^2, assume a system bandwidth of 1 MHz, a required probability of detection of 95 percent, and an average interval of 3 hr (10^4 sec) between false alarms. Then $P_{nb}W = 10^4$, or $P_{nb} = 10^{-10}$. For $P_{sb} = 95$ percent Fig. 6-39 indicates that a minimum SNR ratio of 35 (15.5 db) is required at the detector input. From this required value of s^2 we can work backward to find the transmitted power necessary. An example of such a calculation is included among the problems at the end of the chapter. If P_{sb} is increased to 99.9 percent, s^2 increases to 50 (17 db).

IMPROVED PERFORMANCE USING SIGNAL INTEGRATION

The discussion thus far has concentrated on the simplest type of system, in which individual signal pulses are required to exceed a specified threshold level. In actual systems, however, as noted previously, there will be a number of pulses, spaced at intervals of the repetition period, reflected from a target. Either these are summed automatically as they are presented on a long-persistence screen, or summing devices (delay lines, storage tubes, etc.) may actually be incorporated in the system. We can assume, as one possible method of signal detection, that the sum of these returned pulses, each embedded in noise, is required to exceed a threshold level to be detected.[1]

Normally the individual returned pulses are envelope-detected before being summed. The envelope of each signal pulse plus its accompanying noise will be statistically distributed according to the Rician signal-plus-noise envelope distribution discussed previously. The statistical distribution of the sum of n such pulses can be found by using the techniques of Sec. 5-8. The probability of detection of a target P_{sb} will then be given by the area under the sum distribution curve to the right of the threshold level.

Noise alone, in the absence of a signal, also adds up on the viewing screen or on a summing device purposely incorporated in the system and may be falsely detected as signal. Since the noise samples added are also spaced a repetition period apart, they may be considered independent in their statistical characteristics. The probability P_{nb} that the sum of n successive noise samples will exceed the threshold level and be falsely labeled noise can thus be found by first determining the distribution of the sum of n Rayleigh-distributed independent variables.

[1] Note that this is analogous to time-diversity combining, mentioned in the previous section.

The details of the calculation of P_{sb} and P_{nb} for the n-pulse radar system will not be presented here because of the rather tedious computations involved.[1] Instead we include one typical set of radar performance curves found from such an analysis. These are shown in Fig. 6-40 for

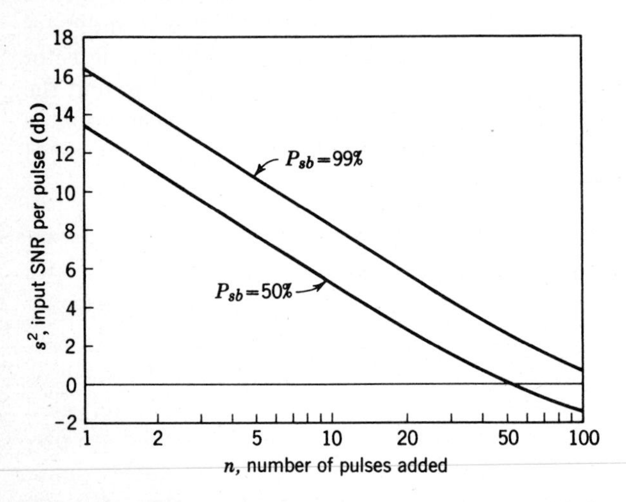

FIG. 6-40 Radar performance curves, n pulses added. $P_{nb} = 10^{-10}$.

the one case of $P_{nb} = 10^{-10}$. They assume that the successive signal amplitudes are all of the same known value (this depends on the target characteristics). The SNR s^2 is the ratio of mean signal power to mean noise power *per pulse* at the input to the envelope detector. (For a signal amplitude of A volts and mean-squared noise voltage N this is

$$s^2 = \frac{A^2}{2N}$$

as in Fig. 6-39.) The block diagram of the radar receiver assumed is shown in Fig. 6-41. In an actual system the summing, threshold, and indicating devices might all be incorporated in one viewing screen.

Although the curves of Fig. 6-40 are for $P_{nb} = 10^{-10}$ only, the variation with P_{nb} is rather small. The values of s^2 and P_{sb} required for a given number n of pulses added do not differ substantially over the range $P_{nb} = 10^{-8}$ to 10^{-12}.

[1] M. Schwartz, A Statistical Approach to the Automatic Search Problem, Ph.D. dissertion, Harvard, Cambridge, Mass., May, 1951.

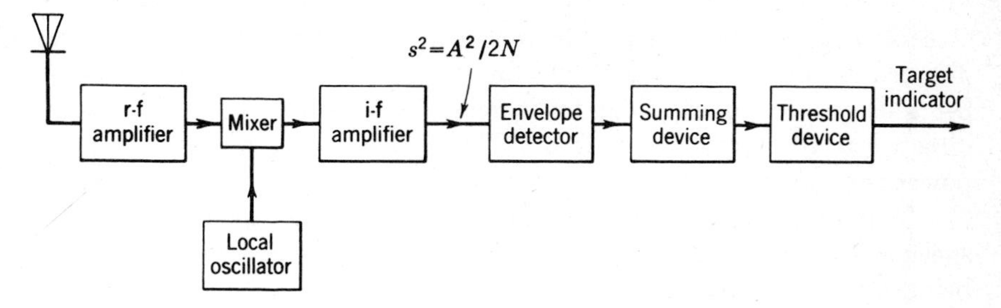

FIG. 6-41 n-pulse radar receiver (triggering circuits not shown).

Note that, as the number of pulses, n, added increases, the signal-to-noise ratio required to maintain a given probability of detection, P_{sb}, and noise-error probability P_{nb} decreases. Initially the decrease is at the rate of about 2.5 db/octave, but this drops to about 2 db/octave for n close to 100. These curves may be used in the same manner as those of Fig. 6-39. For a given P_{sb} and P_{nb} we find the signal-to-noise ratio required at the narrowband i-f section output (detector input). From the system noise figure, target distance and characteristics, etc., we then calculate the transmitter power necessary.

As an example of the application of the curves of Fig. 6-40, consider the radar example previously cited. We required an average interval between false alarms (noise mistaken for signal) of 3 hr (10^4 sec) and assumed a system bandwidth of 1 MHz. (The system pulse widths are then approximately 1 μsec.) This gave a required noise-error probability $P_{nb} = 10^{-10}$. For $P_{sb} = 95$ percent s^2 was then required to be at least 15.5 db (35 numerically), while for $P_{sb} = 99.9$ percent s^2 was at least 17 db (50). The point $n = 1$ and $P_{sb} = 99$ percent in Fig. 6-40 gives $s^2 = 16.3$ db, checking the previous numbers.

What is now the effect of adding together a group of pulses (signal plus noise, or noise alone) before requiring them to exceed a given threshold level and be labeled (perhaps erroneously) signal? For $P_{nb} = 10^{-10}$ again and $P_{sb} = 99$ percent, 10 pulses summed require only 8 db SNR per pulse rather than the 16.3-db figure found for the signal-pulse case. The case of 100 pulses received from a target requires only 0.5 db per pulse for the same noise-error probability and signal-detection probability.

The *peak power* of each pulse is thus *reduced* considerably by requiring a group of pulses to be summed. The *average energy* required to detect a given target *increases*, however. For, by directing 10 pulses of fixed pulse width at a given target, ten times as much energy is required if the pulse amplitudes remain unchanged. The energy required would remain the same if the peak power per pulse could be reduced by a factor of 10 db. Since the curves of Fig. 6-40 indicate a peak power reduction of 8 db the net result is a 2-db increase in the energy required to be directed at the target. Similarly, for 100 pulses added, the peak power per pulse would be expected to decrease by 20 db to keep the energy directed at the target unchanged. Figure 6-40 indicates that the peak power is reduced by about 16 db, however.

This apparent increase in total energy over the group of pulses is due specifically to the fact that the pulses were each envelope-detected before being summed. If the signal could be assumed coherent in phase, either the summing of the signals could be carried out in the i-f stages of the receiver, or an inphase local carrier could be injected to heterodyne the signal pulses down to video frequencies. The phase-coherent signals would then add linearly in voltage, the random noise linearly in power. The resultant SNR would decrease linearly with increasing n, and the total energy over the group of pulses would remain fixed.

Return signal-phase coherence is not in general possible because of the complex types of radar targets encountered, atmospheric changes along the path of propagation, etc. This lack of knowledge of the signal phase leads to the requirement of envelope detection of the individual signals, with a resulting loss in the total energy required for detection. This loss in detectability due to noncoherent (envelope) detection is similar to that noted in the last section, in comparing probabilities of error for coherent and noncoherent binary systems.

Experimental investigations of the effect of signal-pulse addition on signal detectability, for existing systems with observers locating targets on indicator screens, have shown that the signal-to-noise ratio decreases approximately as $1/\sqrt{n}$, with n the number of pulses added. This is somewhat worse than the results of the threshold analysis. Signal detection by human observers is a very complex process, taking into account many more factors than the simple threshold-level procedure utilized in this book. The threshold analysis serves as a crude first approximation to the process of human signal detection. It applies specifically to an automatic detection system, labeling any voltage above a specified level a signal.

PROBABILITY DISTRIBUTION OF PHASE

We have discussed at length in this section and in the one before the probability distribution of the envelope of a narrowband signal embedded in noise. The probability distribution of the phase of the signal plus noise is frequently also of interest.

We have given some examples, in the chapters on modulation systems, of systems where the phase of a sinusoidal signal carries significant information. Other examples include phase-sensitive navigational systems, such as cycle-matching loran, where the relative phase lag between two sinusoidal signal pulses or the difference in time of arrival of two signals carries positional information.

In all these examples we compare the phase of a received signal to that of a reference signal. (The reference signal may also be received from another transmitting station.) What is the error in determining the desired phase angle because of the presence of fluctuation noise? (Although we have been emphasizing noise generated in the receiver, it could just as well be due to additive noise along the path of signal transmission.)

Assuming a narrowband receiver again, the joint probability density of the envelope and phase of a sinusoidal signal in gaussian noise is the same as that found in Chap. 5 when discussing fading (Eq. [5-181]). (The actual phase angle of the signal is assumed to be 0° for convenience.) Repeating that equation here, with $s^2 \equiv A^2/2N$,

$$f(r,\theta) = \frac{e^{-s^2}}{2\pi N} re^{-(r^2-2Ar\cos\theta)/2N} \tag{6-108}$$

$f(\theta)$, the probability-density function of the phase, is found by integrating over all values of the envelope r.

$$\begin{aligned} f(\theta) &= \int_0^\infty f(r,\theta)\,dr \\ &= \frac{e^{-s^2}}{2\pi N}\int_0^\infty re^{-(r^2-2Ar\cos\theta)/2N}\,dr \qquad s^2 = \frac{A^2}{2N} \end{aligned} \tag{6-109}$$

The integral can be evaluated rather simply by completing the square in the exponent. Doing this, and using the symbol I to denote the integral, we get

$$I = e^{s^2\cos^2\theta}\int_0^\infty re^{-(r-A\cos\theta)^2/2N}\,dr \tag{6-110}$$

A transformation of variables, $y = (r - A\cos\theta)/\sqrt{2N}$, then gives, after some manipulation,

$$I = N + A\sqrt{\frac{N\pi}{2}}\cos\theta e^{s^2\cos^2\theta}(1 + \operatorname{erf} s\cos\theta) \qquad \operatorname{erf} x \equiv \frac{2}{\sqrt{\pi}}\int_0^x e^{-y^2}\,dy \tag{6-111}$$

The probability-density function for the phase thus becomes the rather formidable-looking expression

$$f(\theta) = \frac{e^{-s^2}}{2\pi} + \frac{1}{2}\sqrt{\frac{s^2}{\pi}}\cos\theta e^{-s^2\sin^2\theta}(1 + \operatorname{erf} s\cos\theta) \tag{6-112}$$

For $s^2 = 0$ (no signal) the equation reduces to $f(\theta) = 1/2\pi$, as expected.

$f(\theta)$ is shown sketched in Fig. 6-42. Note that the curve is symmetrical about $\theta = 0$, the assumed phase angle of the signal. (For any other phase angle, say, $\theta = \alpha$, the curve is symmetrical about that angle.)

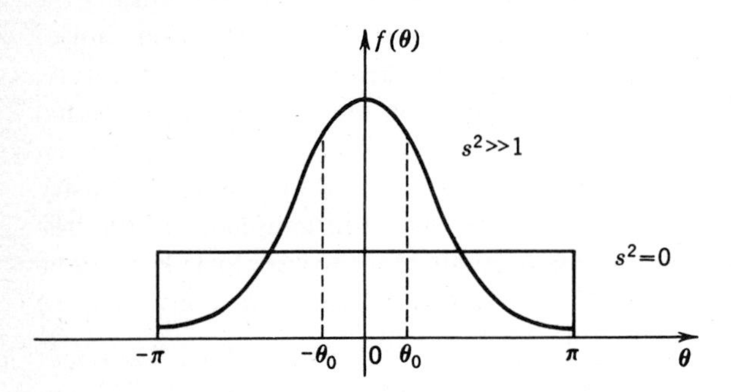

FIG. 6-42 Probability-density function of phase (signal plus noise).

For large signal-to-noise ratio s^2 the curve peaks markedly in the vicinity of $\theta = 0$, with the major contribution to the probability (or area under the curve) coming from that vicinity. As s^2 increases, then, the probability that θ will lie within specified limits about $\theta = 0$ increases as well. For large s^2 and in the vicinity of $\theta = 0$ the curve may be approximated by a gaussian (normal) distribution. Thus, for $s^2 \gg 1$ and $|\theta| < 5°$, $\sin \theta \doteq \theta$, $\cos \theta \doteq 1$, and $\operatorname{erf} s \cos \theta \doteq 1$. With these approximations we get

$$f(\theta) \doteq \frac{e^{-s^2\theta^2}}{\sqrt{\pi/s^2}} \qquad s^2 \ll 1,\ \theta < 5° \tag{6-113}$$

a gaussian distribution with zero mean value and variance $\sigma^2 = 1/2s^2$. The width of the curve thus decreases inversely as s, the voltage signal-to-noise ratio, approaching an impulse function centered about $\theta = 0$ for $s^2 \to \infty$.

The probability that θ will lie within a range $\pm\theta_0$ radians from 0 radians is given by

$$\begin{aligned} \text{Prob}\,(|\theta| < |\theta_0|) &= \int_{-\theta_0}^{\theta_0} f(\theta)\, d\theta \\ &\doteq \frac{2}{\sqrt{\pi/s^2}} \int_0^{\theta_0} e^{-s^2\theta^2}\, d\theta \end{aligned} \tag{6-114}$$

where $s^2 \gg 1$ and $\theta_0 < 0.1$ radian. Upon using the change of variables, $x = s\theta$, this becomes

$$\text{Prob}\,(|\theta| < |\theta_0|) \doteq \text{erf}\, s\theta_0 \qquad [6\text{-}115]$$

As an example, assume that the power signal-to-noise ratio is 100 at the output of the narrowband i-f section. The probability that the instantaneous phase angle of the signal pulse will not deviate more than $2\pi/100$ radians (3.6°) from the true phase angle is

$$\text{erf}\,\frac{2\pi}{10} = 0.625$$

The probability that the phase angle will be in error by no more than 0.02 cycle (7.2°) is

$$\text{erf}\, 1.356 = 0.945$$

(In this last calculation we have exceeded the limits of the gaussian approximation somewhat, but the probability found is accurate enough, since the error in the approximation occurs at the tail of the curve, where the contribution to the probability is very small.)

The "jitter" in the signal phase angle due to noise fluctuations will thus be within 3.6° 62.5 percent of the time and within 7.2° 94.5 percent of the time on the average. This is for a voltage SNR of 10. If the signal happens to be a timing signal, the signal phase error will constitute a timing error. For example, a 3.6° phase error in a 100-kHz wave (as in cycle-matching loran, for example), corresponds to a 0.1-μsec error in timing. The 7.2° phase error corresponds to a 0.2-μsec timing error.

6-9 SIGNAL-TO-NOISE RATIOS IN FM AND AM

We alluded in Chap. 4 to the fact that wideband FM gives a significant improvement in noise or interference rejection over AM. Similarly, wideband PPM was stated to provide noise-rejection improvement over PAM. We would now like to demonstrate these statements and see just how well, and under what conditions, FM provides an improvement over AM, and PPM over PAM.

In previous sections we were able to determine system performance quite uniquely by calculating probabilities of errors. An analogous approach is more difficult to adopt here because of the *continuous* nature of the signals with which we deal. (The calculation of probabilities of errors requires discrete levels of transmission.) Instead we shall use here a less satisfactory, although still useful, approach: continuous-wave (c-w) systems such as AM and FM will be compared on the basis of SNR at the receiver input and output.[1] This will essentially hinge on a comparative discussion of the detection process: the envelope detector in the AM case, and the discriminator in the FM case. By comparing SNR at the input and output of the detector in the two cases, we shall find that the wideband FM system produces the well-known SNR improvement with bandwidth, while in the AM system the input and output SNR can at best be the same.[2] Similar results will be shown to hold for PPM versus PAM.

This exchange of bandwidth for SNR in the two cases of FM and PPM, with SNR improvement obtained at the expense of increased transmission bandwidth, is not as efficient, however, as the optimum exponential exchange predicted by the Shannon capacity expression (Chap. 3). Since the form of the information-capacity expression for PCM (Sec. 3-7) is similar to that of the Shannon optimum, PCM systems provide a SNR bandwidth exchange much more efficient than that for FM.

We shall first compare FM and AM; we then take up, very briefly, PPM and PAM. For the purpose of comparison of FM and AM we shall assume the same carrier power and noise spectral density at the input to each system. We shall calculate the SNR at the system outputs and compare.

AMPLITUDE MODULATION

A typical AM receiver to be analyzed is shown in Fig. 6-43. The amplitude-modulated carrier at the input to the envelope detector has the form

$$v(t) = A_c[1 + mf(t)] \cos \omega_0 t \qquad |mf(t)| \leq 1 \tag{6-116}$$

[1] Other measures of c-w system performance, including, for example, mean-squared error between a random signal input and the noisy output of a receiver, are considered in J. M. Wozencraft and I. M. Jacobs, "Principles of Communication Engineering," chap. 8, Wiley, New York, 1965; see also D. J. Sakrison, "Communication Theory: Transmission of Waveforms and Digital Information," chaps. 7 and 9, Wiley, New York, 1968, and A. J. Viterbi, "Principles of Coherent Communication," part II, McGraw-Hill, New York, 1966.

[2] D. Middleton, "Introduction to Statistical Communication Theory," McGraw-Hill, New York, 1960, and J. L. Lawson and G. E. Uhlenbeck, "Threshold Signals," McGraw-Hill, New York, 1950, contain additional extensive discussions of both AM and FM. See also Schwartz, Bennett, and Stein, *op. cit.*, chap. 3.

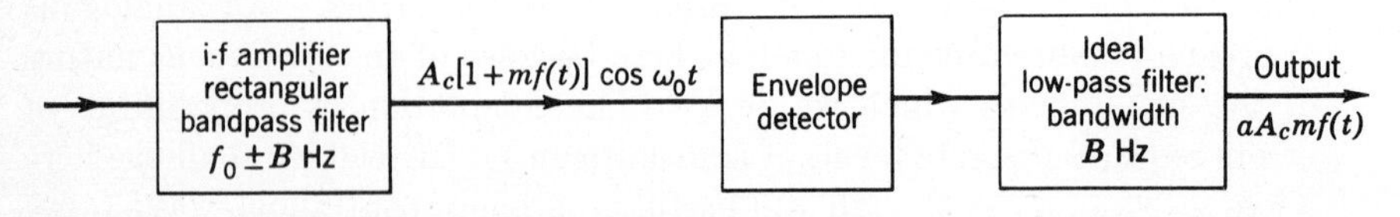

FIG. 6-43 Idealized AM receiver.

if noise is assumed absent. A_c is the unmodulated carrier amplitude measured at the same point.

If the modulating signal $f(t)$ is band-limited to B Hz, the intermediate-frequency (i-f) amplifier preceding the detector should have a bandwidth of at least $2B$ Hz, centered about f_0 Hz. The amplifier is assumed to have the characteristics of an ideal rectangular filter.

This AM signal is now envelope-detected and passed through an ideal filter B Hz wide. As shown in Chap. 4 the output of the envelope detector $f_d(t)$ will be proportional to $f(t)$, or

$$f_d(t) = aA_cmf(t) \tag{6-117}$$

with a a constant of proportionality of the detector. (This constant will henceforth be set equal to 1, since amplifiers can always be introduced to change the gain arbitrarily. In addition we shall be interested in *ratios* of signal to noise, and such constants cancel out anyway.)

The actual envelope-detection process was shown in Chap. 4 to be that of a nonlinear operation on the input AM signal followed by a low-pass filter as in Fig. 6-43. Two types of nonlinearity were examined in Chap. 4: a piecewise-nonlinear characteristic and one possessing smooth curvature. Both were shown to provide the desired envelope-detected output term. It turns out that the signal-to-noise analysis is most readily carried out assuming a detector with quadratic curvature. We shall for this reason concentrate on such a *quadratic envelope detector* here. Analyses for other types of detectors lead to similar results and so will not be considered in detail here.[1]

Our study of the signal-to-noise properties of AM detection is simplified still further by considering the rather artificial case of an *unmodulated* carrier in the presence of additive noise. This zero-modulating-signal case is a common artifice (we shall use the same approach in discussing FM noise), and is useful because the major results found apply in the modulated carrier case as well.[2] Interestingly, this implies modifying the low-pass filter somewhat. The envelope output of Eq. [6-117] assumes zero dc transmission at the filter, blocking the unmodulated carrier portion of the AM wave of Eq. [6-116]. For the unmodulated carrier model we shall be considering, however, it is just this blocked term that will represent the output signal. But this apparent inconsistency notwithstanding, the approach does provide an understanding of, and useful

[1] Middleton, *op. cit.;* W. B. Davenport, Jr., and W. L. Root, "Introduction to Random Signals and Noise," McGraw-Hill, New York, 1958.

[2] Middleton, *op. cit.;* Davenport and Root, *op. cit.;* Schwartz, Bennett, and Stein, *op. cit.*

quantitative information about, the signal-to-noise properties of AM detection.

Assume accordingly for a signal-to-noise analysis that the instantaneous voltage $v(t)$ at the detector input (i-f output) consists of the unmodulated carrier portions of Eq. [6-116] plus gaussian noise $n(t)$. Then, as in previous sections, we may write $v(t)$ in the form

$$\begin{aligned} v(t) &= A_c \cos \omega_0 t + n(t) \\ &= (x + A_c) \cos \omega_0 t - y(t) \sin \omega_0 t \\ &= r(t) \cos [\omega_0 t + \theta(t)] \end{aligned} \qquad [6\text{-}118]$$

Here $$r^2 = (x + A_c)^2 + y^2$$

and $$\theta = \tan^{-1} \frac{y}{x + A_c}$$

With the noise assumed white at the input to the narrowband rectangular-shaped i-f filter, the noise spectral density $G_n(f)$ at the filter output must have the rectangular shape shown in Fig. 6-44. The mean

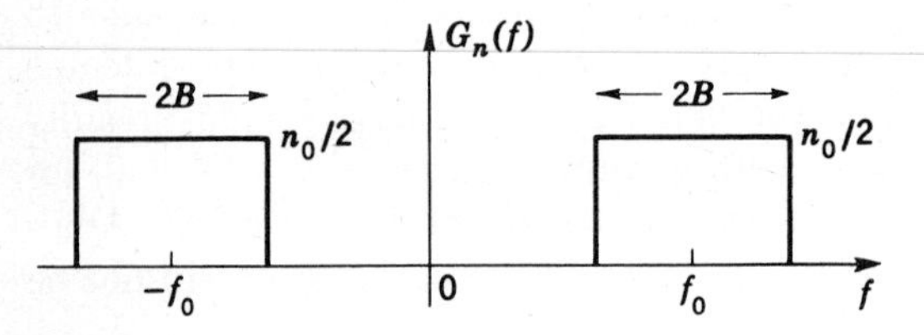

FIG. 6-44 Noise spectral density at i-f amplifier output.

power N is thus given by

$$E(n^2) = N = \int_{-\infty}^{\infty} G_n(f)\, df = 2n_0 B$$

We now assume $v(t)$ passed through a quadratic envelope detector, and ask for the SNR at the low-pass filter output (Fig. 6-45). Recalling

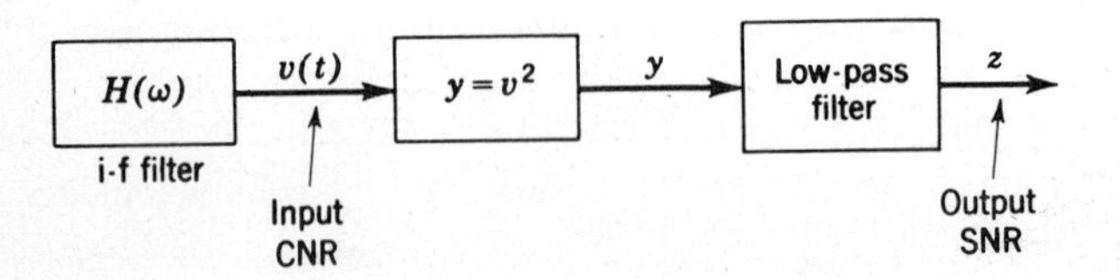

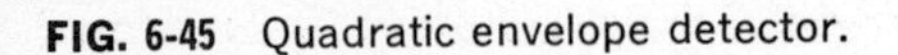
FIG. 6-45 Quadratic envelope detector.

from Chap. 4 that the quadratic envelope detector squares the input $v(t)$ and then passes low-frequency components only ($f < f_0$), we have, at the low-pass filter output,

$$z = v_{\text{lowpass}}^2 = \frac{r^2}{2} = \frac{(x + A_c)^2 + y^2}{2} \qquad [6\text{-}119]$$

from Eq. [6-118]. (Recall that x, y, and r are *slowly varying* random functions, with bandwidth $B \ll f_0$.) We have again ignored a detector constant with dimensions of volts/volts², since we shall shortly take the ratio of signal to noise powers, the constant then dropping out anyway. We shall in fact henceforth ignore the ½ factor appearing in front of r^2, since it is, similarly, an immaterial constant for our purposes.

It is apparent that the SNR analysis of the AM detector must be tied up with the statistics of the envelope $r(t)$. The properties of the Rician distribution discussed earlier can in fact be utilized in determining the output SNR. The use of the quadratic detector, however, enables us to find the output SNR much more simply in terms of the statistics of the gaussian-distributed x and y.

Specifically, we note the randomly varying output voltage z has a signal-dependent part (providing the output signal term), as well as a noise-dependent part. We now *define* the output SNR S_o/N_o to be the ratio of the output signal power in the absence of noise to the mean noise power at the output.[1] Since z is a *voltage*, z^2 must be used to find the powers.

Setting $n(t)$ in Eq. [6-118], or, equivalently, x and y in Eq. [6-119], equal to zero, it is apparent the output signal power in the absence of noise is

$$S_o = A_c^4 \qquad [6\text{-}120]$$

(Recall that we are neglecting the immaterial constant ½.)

The mean noise power at the output must then be the average power or second moment of the random variable z less the signal term:

$$N_o = E(z^2) - A_c^4 \qquad [6\text{-}121]$$

From Eq. [6-119], with the ½ factor again dropped,

$$E(z^2) = E(r^4) = E[(x + A_c)^2 + y^2]^2 \qquad [6\text{-}122]$$

[1] This is unfortunately not a unique definition. This is one of the difficulties with using a SNR formulation at the output of nonlinear devices. However, other possible definitions provide similar results so that the specific definition of SNR to be used is not critical. One simply selects a definition simple enough to be evaluable and yet meaningful at the same time. See Schwartz, Bennett, and Stein, *op. cit.*, pp. 102–120, for various approaches.

It is here that the simplicity of the quadratic detector becomes apparent. For the desired second moment of the output z is for this detector type just the fourth moment of the envelope r, found either by appropriate integration of the Rician density function, or, more simply, by expanding the right-hand side of Eq. [6-122] and taking the indicated average of the *gaussian* variables term by term. In this latter case we make use of the following identities:

1. The mean noise power at the i-f output, as shown previously, is

$$E(x^2) = E(y^2) = N$$

2. A known property of gaussian functions (this is easily checked either by direct integration, or by the moment-generating property of characteristic functions[1]) is

$$E(x^4) = E(y^4) = 3N^2$$

3. By assumption

$$E(x) = E(y) = 0$$

4. Since x and y are *independent* gaussian variables,

$$E(x^2y^2) = E(x^2)E(y^2) = N^2$$

Carrying out the indicated averaging of Eq. [6-122] then (the details are left as an exercise for the reader), we get the very simple expression

$$E(z^2) = E(r^4) = 8N^2 + 8NA_c^2 + A_c^4 \qquad [6\text{-}123]$$

The mean noise power at the envelope-detector output is thus

$$N_o = \underbrace{8N^2}_{n \times n} + \underbrace{8NA_c^2}_{s \times n} \qquad [6\text{-}124]$$

Note the two terms appearing. The first is often called the $n \times n$ term, which is due, as we shall see shortly, to the detector input noise nonlinearly beating with itself. The second is the so-called $s \times n$ term, which is due to the noise nonlinearly mixing with the carrier (or with a modulating signal if present). This second term obviously disappears when the unmodulated carrier goes to zero. Interestingly, this also predicts an *increase* in output noise level when an unmodulated AM carrier is tuned in. This is in fact easily noticed on commercial AM receivers. An opposite effect, a noise *suppression* or quieting effect, will be found to occur in FM.

[1] A. Papoulis, "Probability, Random Variables, and Stochastic Process," *op. cit.*

Combining Eqs. [6-120] and [6-124], the output SNR is found to be given by

$$\frac{S_o}{N_o} = \frac{A_c^4}{8N^2 + 8NA_c^2} \tag{6-125}$$

This may be written in a more illuminating form by defining the carrier-to-noise ratio as

$$\text{CNR} \equiv \frac{A_c^2}{2N} \tag{6-126}$$

($A_c^2/2$ is of course the average power in the unmodulated sine-wave carrier.) Then the output SNR becomes

$$\frac{S_o}{N_o} = \frac{1}{2}\frac{(\text{CNR})^2}{1 + 2\,\text{CNR}} \tag{6-127}$$

Note that for the CNR $\ll 1$ (0 db), the output SNR drops as the *square* of the carrier-to-noise ratio. This is the so-called *suppression characteristic* of envelope detection. The output SNR deteriorates rapidly as the carrier-to-noise ratio drops below 0 db. This quadratic SNR behavior below 0 db is characteristic of all envelope detectors, and is due specifically to the $n \times n$ noise term dominating at low CNR.

For high CNR (CNR $\gg 1$), on the other hand,

$$\frac{S_o}{N_o} \doteq \tfrac{1}{4}\,\text{CNR} \qquad \text{CNR} \gg 1 \tag{6-128}$$

The output SNR is then linearly dependent on the carrier-to-noise ratio, again a common characteristic of envelope detectors. (Here the $s \times n$ noise term dominates so far as the output noise is concerned.) Thus, *no SNR improvement is possible with AM systems.*

With FM, on the other hand, we shall find it possible to get significant improvement in output S_o/N_o by increasing the modulation index at the expense of course of increased transmission bandwidth. Here in the AM case an increase in the transmission bandwidth beyond the bandwidth $2B$ needed to pass the AM signals serves only to increase the noise N and hence decrease the output SNR.

The complete signal-to-noise characteristic for the envelope detector, showing both the asymptotic high and low CNR cases, is shown sketched in Fig. 6-46. Although derived here only for the quadratic detector case, similar characteristics may be derived for other types of nonlinearity or detector law. The exact intersection of the two asymptotic lines depends on the detector law assumed, but the slopes, or shape of the detector signal-to-noise characteristic, will be the same for all detector laws.

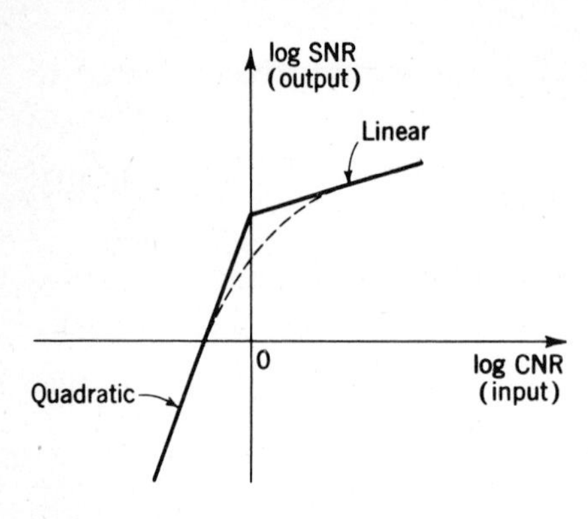

FIG. 6-46 Asymptotic signal-to-noise characteristics, envelope detector.

It is apparent from the discussion of previous sections that a synchronous detector does not produce the quadratic suppression characteristics, due to envelope detection, at small CNR. The inphase or coherent carrier injection provides the same signal and noise powers at the output as at the input. Output SNR is thus everywhere linearly proportional to the input CNR. The reader is asked to demonstrate this himself to his own satisfaction.

The change from the nonlinear (quadratic) to the linear portion of the detector characteristic of Fig. 6-46 may be accounted for in an alternate and rather instructive fashion by considering the envelope Rayleigh- and Rician-density functions discussed previously. Referring for example to Fig. 6-29, we note that for low $A/\sqrt{N}$ (low CNR) the envelope will not deviate too far on the average from the $r = 0$ axis. Since r can never become negative by definition of an envelope, we would expect a lopsided probability distribution with most of the contribution coming in the vicinity of $r = 0$. This is just the Rayleigh distribution. As the carrier A_c increases, however, we would eventually expect the envelope to be symmetrically distributed about A_c. Although variations of the envelope above A_c have theoretically no limit, while below A_c the envelope is constrained never to become negative, the $r = 0$ axis is so remote for $A_c \gg \sqrt{N}$ that this nonzero constraint becomes insignificant and the probability-density curve approaches just the symmetrical bell-shaped characteristic of the gaussian function.

So far as the signal plus noise at the output is concerned, then, the nonlinear operation of the envelope detector has no effect on the distribution for large CNR ratio. The same result is found to hold true for other nonlinear demodulators with high CNR ratio: the output distribution of signal plus noise remains the same as that of the input.

This is exactly the reason why we have been assuming the noise distribution to remain gaussian as the noise progresses from the r-f stages of a receiver down to the narrowband i-f output. Even with no transmitted signal present the local oscillator injects a large enough signal voltage at the nonlinear mixer to ensure a gaussian noise distribution at the mixer output. The mixer then serves only to translate r-f energy down to the i-f spectrum. The signal and noise properties remain relatively unchanged.

As noted earlier, the local carrier injection required for suppressed-carrier demodulation produces the same result; the noise distribution remains gaussian at the demodulator output.

6-10 AM DETECTOR SPECTRAL ANALYSIS

The envelope-detector signal-to-noise characteristic may be obtained quantitatively in a completely different fashion by focusing attention on the spectral aspects of the detection process. Specifically we determine the noise density at the output of the nonlinear device, and then investigate the filtering effect of the low-pass filter.

Not only is this approach valid in its own right, enabling us to specifically consider the low-pass filter design and its effect on the output noise, but it also enables us to introduce an extremely powerful tool in random signal and noise analysis—the use of correlation functions in determining the spectral properties of random signals passed through nonlinear devices or in calculating the spectra of various random signals.

We showed in Sec. 6-2 that the power spectral density and the autocorrelation function are Fourier transform pairs. We did not pursue this point further (until the present paragraph of course) because most of the emphasis in preceding sections has either been on noise passed through linear devices or on probability-of-error calculations. The spectral density at the output of linear filters is of course obtained by multiplying the input spectrum by the square of the magnitude of the transfer function. But how does one handle spectra at the output of nonlinear devices such as the envelope detector under consideration here? How does one calculate the spectrum of particular classes of random signals? Many techniques have been developed for handling these problems.[1] One of the most common and most useful is that of calculating the autocorrelation function at the nonlinear device output, and then taking its

[1] Middleton, *op. cit.;* Davenport and Root, *op. cit.;* J. L. Lawson and G. E. Uhlenbeck, "Threshold Signals," McGraw-Hill, New York, 1950. Schwartz, Bennett, and Stein, *op. cit.,* sec. 3-2.

Fourier transform to find the spectrum at the same point. We shall demonstrate this here (as well as in the last section of this chapter, which is devoted to random signals and the properties of correlation functions). Other applications, to FM spectral analysis for example, appear in the references just cited, as well as in the current periodical literature.[1]

The quadratic detector is again stressed here for simplicity's sake. The output voltage $z(t)$ will thus again contain the low-pass components of $v^2(t)$ (Fig. 6-45). We shall focus attention on the quadratic term $y = v^2$, however, from which $z(t)$ can be found by filtering.

Recall from Sec. 6-2 that the autocorrelation function of a random process $y(t)$ is simply given by taking the expected value of the product of $y(t)$ and the delayed term $y(t + \tau)$:

$$R_y(z) \equiv E[y(t)y(t + \tau)] \qquad [6\text{-}129]$$

Expressing $y(t)$ here in terms of the input $v(t)$, we have

$$R_y(\tau) = E[v^2(t)v^2(t + \tau)] \qquad [6\text{-}129a]$$

Evaluating this expression and formally taking its Fourier transform, we find the spectral density $G_y(f)$. Passing this through the low-pass filter $H_L(\omega)$, we then find of course that $G_z(f) = |H_L|^2G_y(f)$, the desired output spectral density. Here we shall simply say that the filter is an ideal low-pass filter of bandwidth $< f_0$, but high enough to pass all low-frequency components of $y(t)$. Instead of formally writing out $G_y(f)$, we shall then go directly to $G_z(f)$ by ignoring all high-frequency terms in $G_y(f)$.

Before formally evaluating Eq. [6-129a], however, we introduce one modification into the expression previously used for $v(t)$. Recall that the reason for using envelope detection in the first place was that we assumed lack of phase coherence. This implies that the received carrier signal must be of the form $A_c \cos(\omega_0 t + \phi)$, with ϕ a random, unknown phase. We have tacitly ignored this phase term so far, writing the unmodulated carrier signal as $A_c \cos \omega_0 t$, because it played no real role in the analysis, and could arbitrarily be set equal to zero without destroying the validity of the treatment. Here, however, where we are taking statistical averages to find autocorrelation functions, the fact that ϕ *is* random does affect the result. The indicated statistical averaging will thus be taken over the random carrier phase angle, as well as the noise. We shall assume, as previously, that ϕ is uniformly distributed over 2π radians, and statistically independent of the noise.

[1] See Schwartz, Bennett, and Stein, *op. cit.*, chap. 3, for a summary of applications to FM noise and signal bandwidth determination.

With this modification we now write the composite signal at the i-f filter output as

$$v(t) = A_c \cos(\omega_0 t + \phi) + n(t) \quad [6\text{-}130]$$

The expression for $v(t + \tau)$ is then of course[1]

$$v(t + \tau) = A_c \cos[\omega_0(t + \tau) + \phi] + n(t + \tau) \quad [6\text{-}131]$$

Squaring both $v(t)$ and $v(t + \tau)$ and multiplying them together, the autocorrelation function for $y(t)$ becomes

$$R_y(\tau) = E\{[A_c \cos(\omega_0 t + \phi) + n(t)]^2[A_c \cos[\omega_0(t + \tau) + \phi] + n(t+\tau)]^2\} \quad [6\text{-}132]$$

Expanding the terms in brackets, performing the indicated ensemble averages term-by-term over ϕ *and* n, and ignoring terms that will lead to high-frequency components of the spectral density, we get, as the autocorrelation function of the low-pass z,

$$R_z(\tau) = \underbrace{\frac{A_c^4}{4}}_{s\times s} + \underbrace{NA_c^2 + 2R_n(\tau)A_c^2 \cos \omega_0\tau}_{s\times n} + \underbrace{R_{n^2}(\tau)}_{n\times n} \quad [6\text{-}133]$$

Note again the various terms appearing: the signal component $A_c^4/4$ that we have labeled $s \times s$, the $s \times n$ term corresponding to signal beating (or mixing) with noise, and the $n \times n$ term $R_{n^2}(\tau)$ corresponding to noise mixing nonlinearly with itself. The $n \times n$ term is defined as

$$R_{n^2}(\tau) \equiv E[n^2(t)n^2(t + \tau)] \quad [6\text{-}134]$$

This particular autocorrelation term may be further simplified by again using a known property of two dependent gaussian variables [$n(t)$ and $n(t + \tau)$ here]: Calling them, for ease in writing, n_1 and n_2, it may be

[1] As a check the reader may find $R_v(\tau) = E[v(t)v(t + \tau)]$. Inserting Eqs. [6-130] and [6-131] into this expression, multiplying, performing the indicated ensemble average over ϕ and n and discarding terms centered at $2f_0$, it is easy to show that

$$R_v(\tau) = \frac{A_c^2}{2} \cos \omega_0\tau + R_n(\tau)$$

Taking Fourier transforms, we then get

$$G_v(f) = \frac{A_c^2}{4} [\delta(f - f_0) + \delta(f + f_0)] + G_n(f)$$

which is just the spectral density corresponding to the sine wave at frequency f_0 plus the additive noise.

shown[1] that

$$E(n_1^2 n_2^2) = N^2 + 2R_n^2(\tau) \qquad [6\text{-}135]$$

Then $R_z(\tau)$ simplifies to

$$R_z(\tau) = \left(\frac{A_c^2}{2} + N\right)^2 + \underbrace{2R_n(\tau)A_c^2 \cos \omega_0\tau}_{s \times n} + \underbrace{2R_n^2(\tau)}_{n \times n} \qquad [6\text{-}133a]$$

Taking the Fourier transform of this expression term-by-term to find the desired spectral density $G_z(f)$, we obtain the following interesting results.

1. The first term gives rise to an impulse at dc, of area $(A_c^2/2 + N)^2$:

$$G_1(f) = \left(\frac{A_c^2}{2} + N\right)^2 \delta(f) \qquad [6\text{-}136]$$

   This dc term is shown sketched in Fig. 6-47 for the rectangular-shaped $G_n(f)$ assumed here.

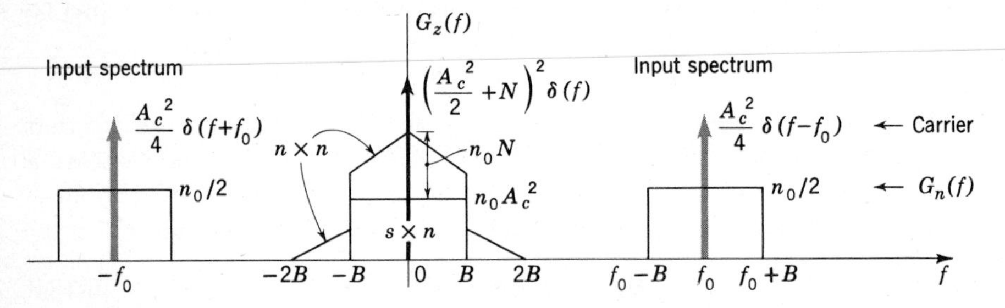

FIG. 6-47 Low-pass spectral density, quadratic envelope detector: band-limited white noise + carrier at input.

2. The second term results in a shift down to dc of the noise spectrum $G_n(f)$, originally centered at f_0:

$$\begin{aligned} G_2(f) &= 2A_c^2 \int_{-\infty}^{\infty} R_n(\tau) \cos \omega_0\tau e^{-j\omega\tau}\, d\tau \\ &= A_c^2 \int_{-\infty}^{\infty} R_n(\tau)[e^{-j(\omega-\omega_0)\tau} + e^{-j(\omega+\omega_0)\tau}]\, d\tau \\ &= A_c^2[G_n(f - f_0) + G_n(f + f_0)] \qquad (s \times n) \quad [6\text{-}137] \end{aligned}$$

[1] Schwartz, Bennett, and Stein, *op. cit.*, p. 118; Papoulis, "Probability, Random Variables, and Stochastic Processes," *op. cit.* More generally, for two dependent gaussian variables x_1 and x_2, $E(x_1^2x_2^2) = \mu_{11}\mu_{22} + 2\mu_{12}\mu_{21}$, with the μ's the central moments defined in Sec. 5-11.

Here use is made of the Fourier-transform relation between $G_n(f)$ and $R_n(\tau)$. (The shift indicated includes one up to $2f_0$ Hz as well, but we ignore this, concentrating on the low-pass expressions.)

This shift in frequency is as expected, since this $s \times n$ spectral term corresponds precisely to the spectrum of the term $n(t) \cos (\omega_0 t + \phi)$, obtained when squaring $v(t)$. Just as in previous chapters, multiplication by $\cos \omega_0 t$ corresponds to a shift up and down by f_0. Since $n(t)$ is itself centered at f_0, the resultant multiplication shifts $n(t)$ down to 0 Hz and up to $2f_0$ Hz, as shown here. This $s \times n$ contribution to the output spectral density is shown sketched in Fig. 6-47 as well.

3. The third term, $2R_n^2(\tau)$, results in a spectral contribution obtained by convolving the input noise spectral density with itself:

$$G_3(f) = 2 \int_{-\infty}^{\infty} G_n(f')G_n(f - f')\, df' \qquad (n \times n) \qquad [6\text{-}138]$$

This is easily demonstrated by recalling from Chap. 2 that the Fourier transform of a product is just the convolution of the two individual Fourier transforms. Thus, if $G(\omega) = F(\omega)H(\omega)$, we have

$$g(t) = \int_{-\infty}^{\infty} f(\tau)h(t - \tau)A\tau$$

Here the two functions corresponding to F and H are both $R_n(\tau)$, and we are going from the τ domain to the f domain, rather than from f to t, but by the symmetry of Fourier transforms it is apparent Eq. [6-138] is valid.

Specifically, for the rectangular $G_n(f)$ centered at f_0 (and $-f_0$ as well), convolution results in the triangular-shaped $n \times n$ term shown centered at 0 Hz in Fig. 6-47. (Another contribution at $2f_0$ is again not shown.) The triangular shape is of course exactly the result obtained in Chap. 2 in demonstrating graphically the convolution of two rectangles (Fig. 2-43).

Physically this $n \times n$ term, the power spectrum of $n^2(t)$, appears because of the multiplication of the input $n(t)$ by itself in forming v^2. The triangular spectrum may be verified qualitatively by visualizing $n(t)$ represented by a large number of equal-amplitude sine waves, all in the vicinity of f_0 Hz. Each one beating with itself, as well as with all the others, gives rise to difference frequencies extending from 0 to a maximum of $2B$ Hz. (This maximum contribution is due to the two extreme frequency terms located at $f_0 - B$ and $f_0 + B$, respectively.) Some thought will indicate that most contributions to the $n \times n$ spectrum come from sine waves closely spaced to one another. (There are proportionately more of these.) As the spacing between sine waves to be beat together increases, there are correspondingly fewer terms involved, and the overall contribution drops. This method of approximating the spectrum by discrete lines, and determining the beat frequencies and the number of contributions to each, is an alternate (albeit much more tedious) way of determining the noise spectra at the output of this quadratic envelope detector.[1]

The dc spectral contribution $G_1(f)$ may be checked quite easily by considering the output random process $z(t)$. We have

$$E(z) = E(v^2)_{\text{lowpass}} = \frac{A_c^2}{2} + N \qquad [6\text{-}139]$$

from Eq. [6-130] directly. $\{$Again $E(n) = 0$; $E[\cos^2(\omega_0 t + \phi)] = \frac{1}{2}$.$\}$ Then the *dc power* is just $(A_c^2/2 + N)^2$, as found here.

The total spectrum $G_z(f) = G_1(f) + G_2(f) + G_3(f)$. As first noted in Sec. 6-2 the total output power is then obtained by summing $G_z(f)$ over the entire frequency range. In this case the indicated integration can be done by inspection, using Fig. 6-47, and adding up the three contributions term-by-term. Thus

$$\begin{aligned} E(z^2) &= \int_{-\infty}^{\infty} G_z(f)\, df \\ &= \underbrace{\left(\frac{A_c^2}{2} + N\right)^2}_{\text{dc}} + \underbrace{2Bn_0A_c^2}_{s\times n} + \underbrace{2Bn_0N}_{n\times n} \\ &= \left(\frac{A_c^2}{2} + N\right)^2 + NA_c^2 + N^2 \end{aligned} \qquad [6\text{-}140]$$

(In the last line we have replaced $2Bn_0$ by its equivalent N.)

Expanding out and collecting like terms, we have, finally,

$$4E(z^2) = A_c^4 + 8NA_c^2 + 8N^2 \qquad [6\text{-}141]$$

just as in Eq. [6-123] previously! (The apparent factor of 4 difference is just due to the $\frac{1}{2}$ factor neglected in deriving Eq. [6-123].) The spectral approach thus provides us with the same result obtained previously. $\{$As another check, using Eq. [6-133*a*], $R_z(0)$ gives the same answer, noting that $R_n(0) = N$.$\}$

But not only do we have the total noise power at the output, we also have its spectral distribution as well. As pointed out earlier, this enables us to determine the effects of different filters on the result. For example, it is apparent that the output noise we have calculated—previously, and again just now using the spectral approach—is actually somewhat more

[1] Schwartz, Bennett, and Stein, *op. cit.*, pp. 108–117.

than one would get in practice using the quadratic detector and low-pass filter. For it includes noise spectral contributions out to $2B$ Hz (Fig. 6-47), whereas all that is required is a low-pass filter cutting off at B Hz. The power due to the $n \times n$ noise term is then reduced from N^2 to $0.75N^2$ (Fig. 6-47), increasing the output SNR somewhat. With a modulating signal introduced, the dc terms can also be eliminated and S_o/N_o (as redefined to include the modulated output) improved further.

A similar analysis for a piecewise-linear detector (often called a *linear envelope detector*) results in the low-frequency spectrum shown in Fig. 6-48.[1]

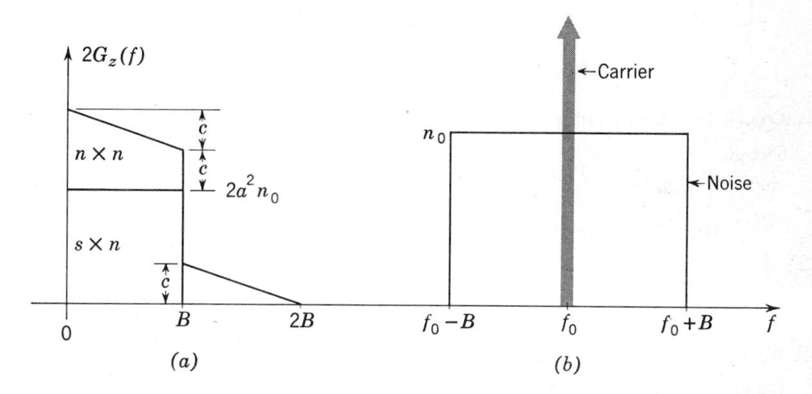

FIG. 6-48 Low-frequency power spectrum, noise plus carrier, at output of piecewise-linear detector. (a) Output spectrum; (b) input spectrum. (S. O. Rice, Mathematical Analysis of Random Noise, *Bell System Tech. J.*, vol. 24, pp. 46–156, fig. 9, part 4, January, 1945. Copyright, 1945, The American Telephone and Telegraph Co., reprinted by permission.)

(The negative-frequency terms have been folded over and combined with the positive ones.) Again the two types of noise, $s \times n$ and $n \times n$, appear. As shown in the figure the parameter c determines the $n \times n$ contribution, the parameter a the $s \times n$ contribution. The two parameters depend on CNR in a rather complicated way, as shown in Fig. 6-49. For CNR > 4, however,

$$c \doteq \frac{2(1/\pi)^2}{\text{CNR}} n_0$$

and

$$a \doteq \frac{1}{\pi}$$

[1] S. O. Rice, Mathematical Analysis of Random Noise, *Bell System Tech. J.*, vol. 24, pp. 46–156, part 4, January, 1945.

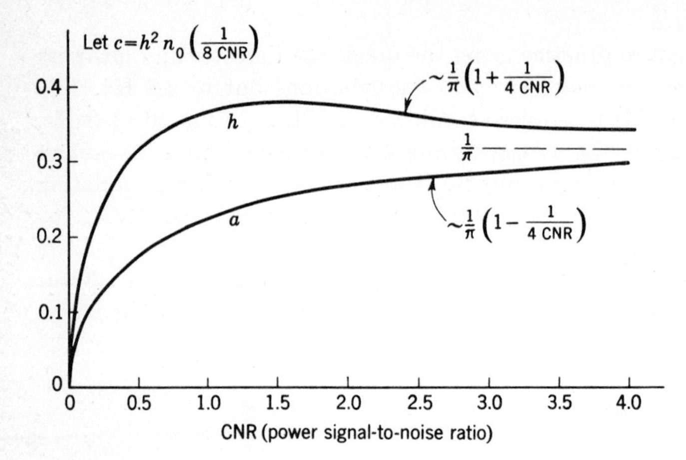

FIG. 6-49 Coefficients for linear detector output shown in Fig. 6-48. (S. O. Rice, Mathematical Analysis of Random Noise, *Bell System Tech. J.*, vol. 24, pp. 45–156, fig. 10, part 4, January, 1945. Copyright, 1945, The American Telephone and Telegraph Co., reprinted by permission.)

So even for CNR = 4, the $s \times n$ contribution predominates, as in the quadratic-detector case. For smaller values of CNR, the curves of Fig. 6-49 may be used to obtain c and a.

The dc contributions to the spectral density are not shown here. Note that the dimensions of the noise spectral components are not the same as those of Fig. 6-49 for the quadratic detector. But this is as it should be, since the output voltage in the linear detector case is proportional to r, the envelope itself, rather than r^2, as in the quadratic case. In particular, the output signal here is just $S_o = A_c^2/2$, rather than A_c^4, as previously. The ratio of S_o to N_o, however, has the same form in both cases.

6-11 FREQUENCY-MODULATION NOISE IMPROVEMENT

We now consider the case of frequency modulation. Again assuming the received signal to have gaussian noise added to it, we shall show that, contrary to the AM case, widening the transmission bandwidth (as is required for wideband FM) *does* improve the output SNR. As previously the SNR will be defined as the ratio of mean signal power with noise absent to the mean noise power in the presence of an unmodulated carrier.

For simplicity's sake the analysis will be confined to the case of large carrier-to-noise ratio (CNR).[1]

The idealized FM receiver to be discussed here is shown in Fig. 6-50. A frequency-modulated signal of transmission bandwidth B_T Hz is first

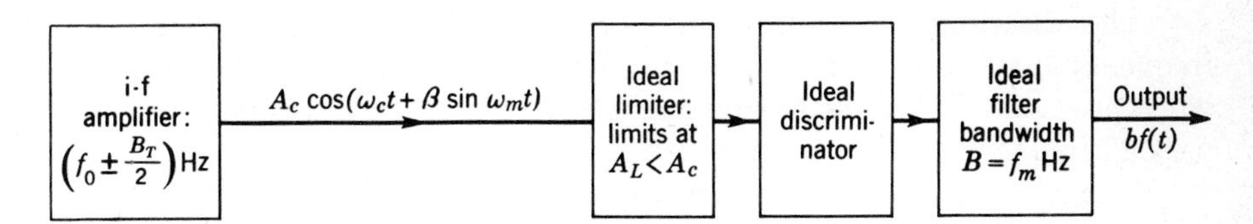

FIG. 6-50 Block diagram, idealized FM receiver.

passed through an ideal limiter which removes all amplitude variations. The limiter output, after filtering, goes to the discriminator, assumed to give an output directly proportional to the instantaneous frequency of the signal, and then to an ideal low-pass filter of bandwidth B Hz ($B < B_T/2$). B is the maximum bandwidth of the actual information signal $f(t)$ being transmitted.

Assuming, as in Chap. 4, a sine-wave signal of the form

$$f(t) = \Delta\omega \cos \omega_m t \tag{6-142}$$

the frequency-modulated carrier measured at the *output* of the *i-f amplifier* is given by

$$f_c(t) = A_c \cos (\omega_0 t + \beta \sin \omega_m t) \tag{6-143}$$

Noise is assumed absent here. $\beta = \Delta f/B$ is the modulation index and $\Delta f = \Delta\omega/2\pi$ the maximum-frequency deviation.

The average power of the FM wave is simply

$$S_c = \tfrac{1}{2}A_c^2 \tag{6-144}$$

independent of the modulating signal.

The instantaneous frequency is given by

$$\begin{aligned} \omega = \frac{d\theta}{dt} &= \omega_0 + \beta\omega_m \cos \omega_m t = \omega_0 + \Delta\omega \cos \omega_m t \\ &= \omega_0 + f(t) \end{aligned} \tag{6-145}$$

The discriminator output is proportional to the instantaneous frequency deviation away from ω_0, or $\omega - \omega_0$. This frequency deviation is

[1] See Schwartz, Bennett, and Stein, *op. cit.*, chap. 3, for a discussion of the complete FM noise problem, including references to the literature.

just $f(t)$. The output signal is then

$$f_d(t) = b\,\Delta\omega \cos \omega_m t = bf(t) \tag{6-146}$$

where b is a constant of the discriminator. As in the AM case this constant will be set equal to 1.

The discriminator output must be filtered to eliminate higher-frequency distortion terms. Filtering will also reduce the output noise when present. The filter bandwidth is chosen as $B = f_m$ Hz in order to pass all frequency components of $f(t)$.

From Eq. [6-146] the average output signal power is simply

$$S_o = \frac{(\Delta\omega)^2}{2} \qquad \text{watts} \tag{6-147}$$

if a 1-ohm normalized load resistor is assumed.

The case of noise plus unmodulated carrier (signal absent) can now be treated in a manner directly analogous to that for AM. We again assume fluctuation noise of spectral density $n_0/2$ watts/Hz uniformly distributed (i.e., band-limited white noise) at the output of the i-f amplifier. Here, as shown in Fig. 6-51, the noise is uniformly distributed over the

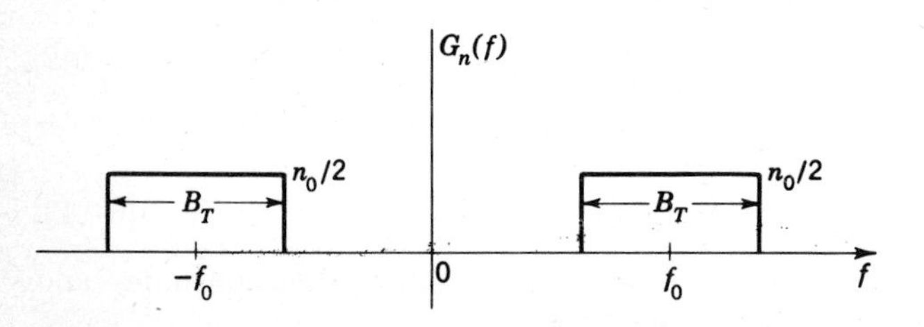

FIG. 6-51 Noise spectral density, FM receiver.

range $\pm B_T/2$ Hz about the carrier frequency f_0. (Compare with Fig. 6-44 for the AM case, where the transmission bandwidth $2B$ is generally less than the FM transmission bandwidth B_T.) The noise power at the i-f output (detector input) is thus

$$N = B_T n_0 \tag{6-148}$$

As previously, we may use the narrowband representation for the noise to write the unmodulated carrier plus noise in the form

$$\begin{aligned} v(t) &= A_c \cos \omega_0 t + n(t) \\ &= (A_c + x) \cos \omega_0 t - y(t) \sin \omega_0 t \\ &= r(t) \cos [\omega_0 t + \Phi(t)] \end{aligned} \tag{6-149}$$

In the AM case we focused attention on the properties of the envelope $r(t)$. It is apparent that in the FM case we focus attention on the phase term $\Phi(t)$. For in this unmodulated carrier case it is apparent that $\Phi(t)$ represents the noise at the discriminator output. In particular, since

$$\Phi = \tan^{-1} \frac{y}{x + A_c} \qquad [6\text{-}150]$$

the discriminator output is given by

$$\dot{\Phi} = \frac{(x + A_c)\dot{y} - y\dot{x}}{y^2 + (x + A_c)^2} \qquad [6\text{-}151]$$

This is a rather formidable-looking expression and is the reason why FM noise analysis, even in the unmodulated carrier case, is so difficult to carry out. [The straightforward, albeit mathematically difficult, way of determining the spectral density of the noise at the discriminator output is to find the autocorrelation function of the random noise process $\dot{\Phi}(t)$, relating it to the known input correlation function $R_n(\tau)$. One then takes the Fourier transform to find the output spectral density $G_{\dot{\Phi}}(f)$.†]

For large CNR this expression for the discriminator output, $\dot{\Phi}(t)$, simplifies considerably. Thus, recalling that $E(x^2) = E(y^2) = N$, and assuming that $A_c^2 \gg N$ (the CNR is actually $S_c/N = A_c^2/2N$), the discriminator output is given simply by

$$\dot{\Phi} \doteq \frac{1}{A_c} \dot{y} \qquad A_c^2 \gg N \qquad [6\text{-}152]$$

This is apparent from either Eq. [6-150] or Eq. [6-151]. The discriminator output noise is thus proportional to the time derivative of the quadrature noise term $y(t)$. (If the standard deviation $\sqrt{N}$ of x is small compared to A_c, it is apparent that the probability is small that the gaussian function x will be comparable to A_c in magnitude. It may then be neglected in comparison with A_c. Similarly, y/A_c is then small with a high probability, so that $\dot{\Phi} \doteq \dot{y}/A_c$.)

† See Schwartz, Bennett, and Stein, *op. cit.*, chap. 3, for a discussion of this approach. Detailed analyses also appear in S. O Rice, Statistical Properties of a Sine Wave Plus Random Noise, *Bell System Tech. J.*, vol. 27, secs. 7 and 8, pp. 138–151, January, 1948; J. L. Lawson and G. E. Uhlenbeck, "Threshold Signals," chap. 13, McGraw-Hill, New York, 1950; D. Middleton, "An Introduction to Statistical Communication Theory," *op. cit.*, chap. 15. More recently Rice has developed a simpler approach to FM noise analysis, using the so-called "clicks" phenomenon, that is readily extended to other types of FM receivers. See S. O. Rice, Noise in FM Receivers, in M. Rosenblatt (ed.), "Proceedings, Symposium of Time Series Analysis," chap. 25, pp. 375–424, Wiley, New York, 1963. This approach is also summarized in Schwartz, Bennett, and Stein, *op. cit.*, pp. 144–154.

Since the discriminator is followed by a low-pass filter (Fig. 6-50), we must now find the spectral density of the discriminator output noise $\dot{\Phi}$ in order to take into account the effect of the filter. To do this we simply note that differentiation is a linear operation. Hence, Eq. [6-152] indicates that $\dot{\Phi}$ may be considered the response at the output of a (linear) differentiator $H(\omega)$ with y applied at the input. From our discussion of random signals and noise we then write directly

$$G_{\dot{\Phi}}(f) = |H(\omega)|^2 G_y(f) \qquad [6\text{-}153]$$

But differentiation of a time function corresponds to multiplication of its Fourier transform by $j\omega$. Then $H(\omega) = j\omega/A_c$, $|H(\omega)|^2 = \omega^2/A_c^2$, and we have, very simply,

$$G_{\dot{\Phi}}(f) = \frac{\omega^2}{A_c^2} G_y(f) \qquad A_c^2 \gg N \qquad [6\text{-}154]$$

This is shown schematically in Fig. 6-52.

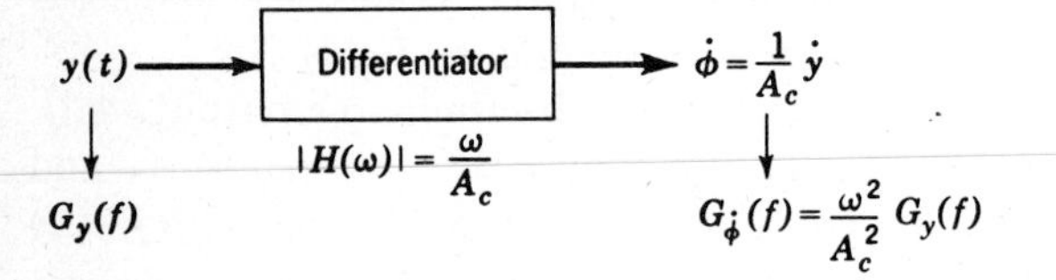

FIG. 6-52 Output-noise spectral density, FM receiver, high CNR.

Recall from our initial discussion of the narrowband noise representation, however, that for symmetrical bandpass filters,

$$G_x(f) = G_y(f) = 2G_n(f + f_0) \qquad [6\text{-}155]$$

Thus we simply find the spectral density of the quadrature (or inphase) noise component by shifting the noise spectral density $G_n(f)$ down to 0 frequency. Then we also have

$$G_{\dot{\Phi}}(f) = \frac{2\omega^2}{A_c^2} G_n(f + f_0) \qquad A_c^2 \gg N \qquad [6\text{-}156]$$

In particular, for the band-limited white-noise case assumed here (Fig. 6-51),

$$G_{\dot{\Phi}}(f) = \frac{\omega^2 n_0}{A_c^2} \qquad -\frac{B_T}{2} < f < \frac{B_T}{2}$$

$$= 0 \qquad \text{elsewhere} \qquad [6\text{-}157]$$

This is shown schematically in Fig. 6-53.

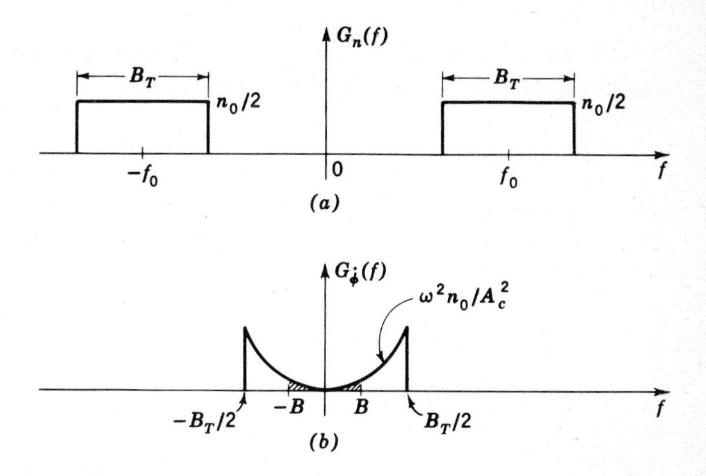

FIG. 6-53 FM noise spectrum (high CNR case). (a) Spectrum of i-f noise; (b) spectral density at FM detector output.

We are now in a position to readily determine the total noise at the FM receiver output. For assuming a low-pass filter of known spectral shape $H_L(\omega)$ following the discriminator, the noise spectral density at the low-pass filter output is simply $|H_L(\omega)|^2 G_{\dot\Phi}(f)$. We then integrate this spectral density over all frequencies to find the output noise N_o. Again for simplicity's sake assume this final filter to be an ideal low-pass one of bandwidth $B < B_T/2$. The bandwidth B should just be sufficient to pass all signal components, yet no larger to avoid increasing the noise passed. B is thus just the bandwidth of the original modulating signal (see Chap. 4). In the case of the sine-wave signal of frequency f_m, B is just f_m.

The output noise N_o is thus found by integrating $G_{\dot\Phi}(f)$ from $-B$ to B (Fig. 6-53).

$$
\begin{aligned}
N_o &= \int_{-B}^{B} G_{\dot\Phi}(f)\,df \\
&= \frac{2(2\pi)^2 n_0}{A_c^2} \int_0^B f^2\,df \\
&= \frac{(2\pi)^2 n_0}{3S_c} B^3
\end{aligned}
\qquad [6\text{-}158]
$$

Note that the output noise, although proportional to B^3, is *inversely* proportional to the carrier power $S_c = A_c^2/2$. As the carrier increases, therefore, the noise power drops. This *noise quieting* effect is of course well known in FM, and is just the opposite of the effect encountered in AM. (Recall that there the introduction of the carrier *increased* the noise.)

By using Eq. [6-147], the expression for the output signal power (the signal is assumed to be a cosine wave with noise absent), the mean SNR at the output becomes

$$\frac{S_o}{N_o} = 3\left(\frac{\Delta f}{B}\right)^2 \frac{S_c}{2n_0 B} \qquad [6\text{-}159]$$

But $2n_0B = N_c$, the average power in the AM sidebands, and $\beta = \Delta f/B$ is the modulation index. Then

$$\frac{S_o}{N_o} = 3\beta^2 \frac{S_c}{N_c} \qquad [6\text{-}160]$$

Note that S_c/N_c corresponds to the CNR of an *AM system* with the same carrier power and noise spectral density. The total noise power over the transmission bandwidth B_T of the FM system is greater than N_c since $B_T > 2B$.

For a specified carrier amplitude and noise spectral density at the i-f amplifier output Eq. [6-160] shows that the output SNR increases with the modulation index or the transmission bandwidth. This is in contrast to the AM case, where increasing the bandwidth beyond $2B$ was found only to deteriorate the output SNR.

We may specifically compare the FM and AM systems by assuming the same unmodulated carrier power and noise spectral density n_0 for both. (These quantities are measured here at the output of the i-f amplifier.) For a 100 percent modulated AM signal we have

$$\left(\frac{S_o}{N_o}\right)_{\text{AM}} = \frac{S_c}{N_c}$$

Equation [6-160] may thus be modified to read

$$\left(\frac{S_o}{N_o}\right)_{\text{FM}} = 3\beta^2 \left(\frac{S_o}{N_o}\right)_{\text{AM}} \qquad [6\text{-}161]$$

For large modulation index (this corresponds to a wide transmission bandwidth, for, with $\beta \gg 1$, the bandwidth approaches $2\,\Delta f$), we can presumably increase S_o/N_o significantly over the AM case. As an example, if $\beta = 5$, the FM output SNR is 75 times that of an equivalent AM system. Alternatively for the same SNR at the output in both receivers the power of the FM carrier may be reduced 75 times. But this requires increasing the transmission bandwidth from $2\,B$ (AM case) to $16\,B$ (FM case); see Fig. 4-41. Frequency modulation thus provides a substantial improvement in SNR, but at the expense of increased bandwidth. This is of course characteristic of all noise-improvement systems.

Can we keep increasing the output SNR indefinitely by increasing the frequency deviation and hence the bandwidth? If we keep the transmitter power fixed, S_c is fixed. With the noise power per unit bandwidth ($n_0/2$) fixed and the audio signal bandwidth B fixed N_c presumably remains constant, *but*, as the frequency deviation increases and the bandwidth with it, more noise must be accepted by the limiter. Eventually the noise power at the limiter becomes comparable with the signal power. The above simplified analysis, which assumes large carrier-to-noise power ratios, does not hold any more, and noise is found to "take over" the system.

This effect is found to depend very sharply upon FM carrier-to-noise ratio S_c/N and is called a *threshold* effect. For this ratio greater than a specified threshold value FM functions properly and shows the significant improvement in SNR predicted by Eq. [6-161]. For the ratio below this threshold level the noise improvement is found to deteriorate rapidly, and Eq. [6-161] no longer holds. The actual threshold level depends upon the FM carrier-to-noise ratio and upon β. For large β the level is usually taken as 10 db.

This threshold phenomenon is a characteristic of all wideband noise-improvement systems and was encountered previously in discussing PCM systems.

Two conditions must thus ordinarily be satisfied for an FM system to show significant noise-"quieting" properties:

1. Frequency-modulation carrier-to-noise ratio > 10 db to avoid the threshold effect.
2. With FM carrier-to-noise ratio > 10 db, $\beta > 1/\sqrt{3}$ if $S_o/N_o > S_c/N_c$ (see Eq. [6-161]).

But $\beta > 1\sqrt{3} \doteq 0.6$ corresponds to the transition between narrowband and wideband FM. *Narrowband FM thus provides no* SNR *improvement over AM.* This is of course as expected, for the improvement is specifically the result of restricting the noise phase deviations of the carrier to small values, while the signal variations are assumed to be large.

Experimental and theoretical studies of FM noise characteristics show the threshold phenomenon very strikingly and also bear out the validity of Eq. [6-161] above the threshold value. Figure 6-54 is taken from some experimental work of M. G. Crosby[1] and shows a comparison of AM and FM receivers for $\beta = 4$ and $\beta = 1$. Note that for $\beta = 4$ the FM signal-to-noise ratio deteriorates rapidly for $S_c/N_c < 13$ db. In fact, for $S_c/N_c < 8$ db, the AM system becomes superior. For $S_c/N_c >$ 15 db, however, the FM system shows an improvement of 14 db. For

[1] M. G. Crosby, Frequency Modulation Noise Characteristics, *Proc. IRE*, vol. 25, pp. 472–514, fig. 10, April, 1937.

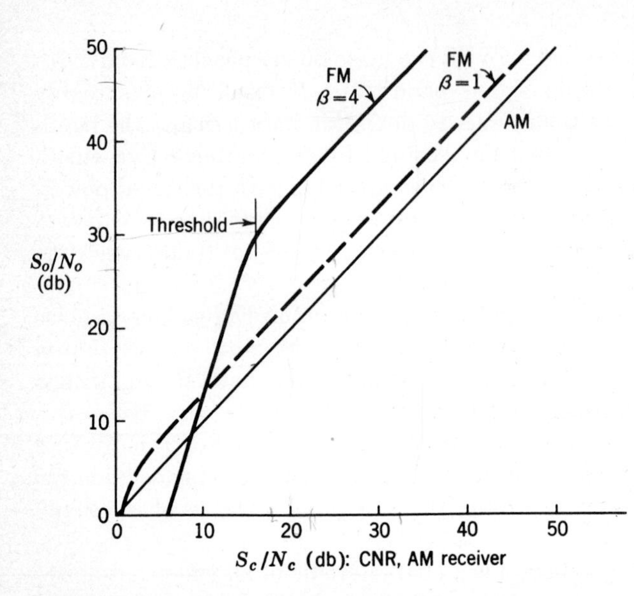

FIG. 6-54 Measured characteristics, FM and AM receivers. (Adapted from M. G. Crosby, Frequency Modulation Noise Characteristics, *Proc. IRE*, vol. 25, pp. 472–514, fig. 10, April, 1937, by permission.)

$\beta = 4$ we would expect the theoretical improvement to be $3\beta^2 = 48$, or 17 db. For $\beta = 1$ the threshold level is experimentally found to occur at 2 db. Above this value of AM carrier-to-noise ratio the FM improvement over AM is 3 db. The theoretical improvement would be expected to be $3\beta^2 = 3$, or 4.8 db.

These measured characteristics of M. G. Crosby are of historical significance because they were among the first obtained in quantitative studies of FM noise. They have of course since been reproduced countless times by many investigators. The detailed theoretical analyses noted earlier bear these results out as well.

Much engineering time has been devoted in recent years to the development of so-called threshold-improvement receivers designed to reduce threshold in FM receivers. These include the FM receiver with feedback (FMFB), the phase-locked loop, and the frequency-locked loop.[1]

[1] Schwartz, Bennett, and Stein, *op. cit.*, pp. 157–163; A. J. Viterbi, "Principles of Coherent Communications," McGraw-Hill, New York, 1966; K. K. Clarke and D. F. Hess, Frequency-locked Loop Demodulator, *IEEE Trans. Commun. Technol.*, pp. 518–524, August, 1967. The equivalence of all three devices under limiting conditions is discussed by D. T. Hess, Equivalence of FM Threshold Extension Receivers, *IEEE Trans. Commun. Technol.*, October, 1968.

SIGNAL-TO-NOISE IMPROVEMENT THROUGH DEEMPHASIS

We showed in Chap. 4 that the transmission bandwidth of an FM system is determined by the maximum-frequency deviation Δf produced by the highest modulating frequency f_m to be transmitted. In particular, for $f_m = 15$ kHz and a maximum-frequency deviation Δf of 75 kHz we found that the required bandwidth was 240 kHz.

In practice the higher-frequency components of the modulating signal rarely attain the amplitudes needed to produce a 75-kHz frequency deviation. Audio signals of speech and music are found to have most of their energy concentrated in the lower-frequency ranges. The instantaneous signal amplitude, limited to that required to give the 75-kHz deviation, is due most of the time to the lower-frequency components of the signal. The smaller-amplitude high-frequency (h-f) components will, on the average, provide a much smaller frequency deviation. The FM signal thus does not fully occupy the large bandwidth assigned to it.

The spectrum of the noise introduced at the receiver does, however, occupy the entire FM bandwidth. In fact, as we have just noted, the noise-power spectrum at the output of the discriminator is emphasized at the higher frequencies. (The spectrum is proportional to f^2 for large carrier-to-noise ratio.)

This gives us a clue as to a possible procedure for improving the SNR at the discriminator output: we can artificially *emphasize* the h-f components of our input audio signal at the transmitter, *before the noise is introduced*, to the point where they produce a 75-kHz deviation most of the time. This *equalizes* in a sense the l-f and h-f portions of the audio spectrum and enables the signal fully to occupy the bandwidth assigned.

Then, at the output of the receiver discriminator, we can perform the inverse operation, or *deemphasize* the higher-frequency components, to restore the original signal-power distribution. But in this deemphasis process we reduce the h-f components of the noise also and so effectively increase the SNR.

Such a preemphasis and deemphasis process is commonly used in FM transmission and reception and provides as we shall see, 13 to 16 db of noise improvement. Note that this procedure is a simple example of a signal-processing scheme which utilizes differences in the characteristics of the signal and the noise to process the signal more efficiently. The entire FM process is itself an example of a much more complex processing scheme in which use is made of the fact that random noise alters the instantaneous frequency of a carrier much less than it does the amplitude of the carrier (for large carrier-to-noise ratio). The noise-improvement properties of PCM and PPM (considered in the next section), are again due to differences in the characteristics of random noise and signal.

A simple frequency transfer function that emphasizes the high frequencies and has been found very effective in practice is given by

$$H(\omega) = 1 + j\frac{\omega}{\omega_1} \qquad [6\text{-}162]$$

An example of an RC network that approximates this response very closely is shown in Fig. 6-55*a*. The asymptotic logarithmic amplitude-frequency plot for this network is shown in Fig. 6-55*b*.

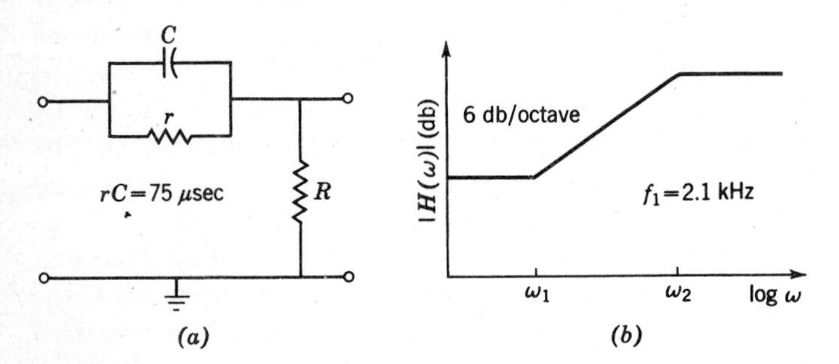

FIG. 6-55 Example of a preemphasis network. (a) Preemphasis network, $r \gg R$, $rC = 75$ μsec; (b) asymptotic response, $\omega_1 = 1/rC$, $\omega_2 \doteq 1/RC$.

With $r \gg R$ the amplitude response has two break frequencies given by $\omega_1 = 1/rC$ and $\omega_2 \doteq 1/RC$. Signals in the range between ω_1 and ω_2 are thus emphasized. (Actually the higher-frequency components are passed unaltered, and the lower-frequency components are attenuated. The attenuation can of course be made up by amplification.) The choice of $f_1 = \omega_1/2\pi$ is not critical, but 2.1 kHz is ordinarily used in practice ($rC = 75$ μsec). $f_2 = \omega_2/2\pi$ should lie above the highest audio frequency to be transmitted. $f_2 \geq 30$ kHz is a typical number. In the range between these two frequencies $|H(\omega)|^2 \doteq 1 + (f/f_1)^2$, and all audio frequencies above 2.1 kHz are increasingly emphasized.

The receiver deemphasis network, following the discriminator, must have the inverse characteristic given by

$$H(\omega) = \frac{1}{1 + jf/f_1} \qquad [6\text{-}163]$$

with $f_1 = 2.1$ kHz as before. This then serves to restore all signals to their original relative values. The simple RC network of Fig. 6-56 ($rC = 75$ μsec) provides this deemphasis characteristic. (Note incidentally that the two networks are identical with the lead and lag net-

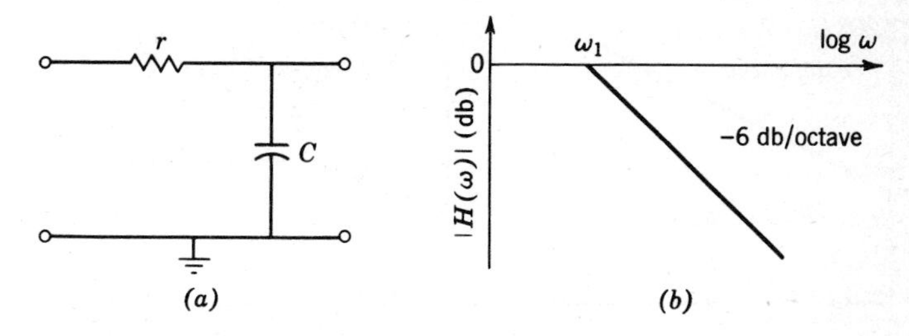

FIG. 6-56 Example of a deemphasis network. (a) Deemphasis network, $rC = 75\ \mu$sec; (b) asymptotic response, $f_1 = 2.1$ kHz.

works, respectively, of the servomechanisms and feedback amplifier fields.)

How much does the deemphasis network improve the SNR ratio at the discriminator output?

From Eq. [6-157] of the FM noise analysis, the noise spectral density at the discriminator output (for large carrier-to-noise ratio) is

$$G_{\dot{\Phi}}(f) = \frac{n_0\omega^2}{A_c^2} = \frac{n_0\omega^2}{2S_c} \qquad [6\text{-}164]$$

where n_0 is the input noise spectral density, A_c the carrier amplitude, and $S_c = A_c^2/2$ the mean carrier power at the discriminator input.

If this noise is now passed through the RC deemphasis network of Fig. 6-56, the modified spectral density at the network output is

$$G_H(f) = G_{\dot{\Phi}}(f)|H(\omega)|^2 = \frac{n_0\omega^2}{2S_c}\,\frac{1}{1 + (f/f_1)^2} \qquad [6\text{-}165]$$

The original noise-power spectrum $G_{\dot{\Phi}}(f)$ at the output of the discriminator and the modified spectrum $G_H(f)$ are shown sketched in the logarithmic plots of Fig. 6-57. (Only the one-sided spectra, defined for positive f only, are shown because of the logarithmic plots.)

Note that for $f > f_1$ the noise spectrum with deemphasis included becomes a uniform spectrum; the deemphasis network has canceled out the ω^2, increasing noise frequency, factor of Eq. [6-164].

The total mean noise power at the output of an ideal low-pass filter of bandwidth B Hz is given by

$$\begin{aligned} N_{oD} &= \int_{-B}^{B} G_H(f)\,df \\ &= \int_{-B}^{B} \frac{n_0\omega^2}{2S_c}\,\frac{df}{1 + (f/f_1)^2} \qquad [6\text{-}166] \end{aligned}$$

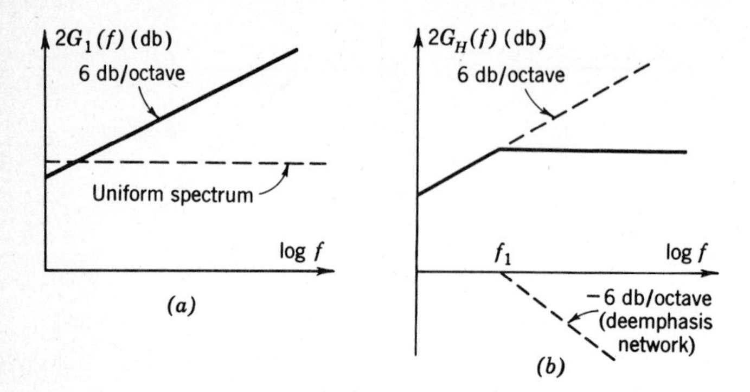

FIG. 6-57 Logarithmic FM noise-power spectrum: output of discriminator ($S_c \gg N_c$). (a) Spectrum without deemphasis; (b) spectrum with deemphasis.

where N_{oD} represents the mean noise power with the deemphasis network included, as compared with the symbol N_o used previously for the noise power with no deemphasis network.

Equation [6-166] is readily integrated to give

$$N_{oD} = \frac{n_0}{2\pi S_c} \omega_1{}^3 \left(\frac{B}{f_1} - \tan^{-1}\frac{B}{f_1}\right) \qquad [6\text{-}167]$$

Multiplying numerator and denominator by $3/B^3$, and recalling from Eq. [6-158] that $N_o = (2\pi)^2 n_0 B^3/3S_c$, the mean noise power can be written in the form

$$N_{oD} = N_o D \qquad [6\text{-}168]$$

where

$$D = 3\left(\frac{f_1}{B}\right)^3 \left(\frac{B}{f_1} - \tan^{-1}\frac{B}{f_1}\right) \qquad [6\text{-}169]$$

The parameter D represents the effect of the deemphasis network and is readily seen to be less than or equal to 1 in value. The deemphasis network thus reduces the output noise. For $B \ll f_1$ (the deemphasis effect is then introduced beyond the range of the low-pass filter) $D \to 1$, as is to be expected. For $B \gg f_1$, $\tan^{-1}(B/f_1) \to \pi/2$, and

$$D \to 3\left(\frac{f_1}{B}\right)^2 \qquad f_1 \ll B \qquad [6\text{-}170]$$

The output noise power thus decreases rapidly with increasing B.

Increasing B indefinitely provides no *absolute* improvement in the output noise, however, for N_o increases as B^3. The output filter bandwidth B should thus be restricted to just the bandwidth required to pass the highest audio frequency and no more.

With $f_1 = 2.1$ kHz, as noted previously, and $B = 15$ kHz, $D = \frac{1}{20}$. The noise is thus reduced by a factor of 20, or 13 db. If $B = 10f_1 = 21$ kHz, the improvement due to the deemphasis network is 16 db.

The signal power at the output of the discriminator is $S_o = (\Delta\omega)^2/2$ (Eq. [6-147]), for a single sine wave, independent of the preemphasis and deemphasis procedure. The SNR ratio with deemphasis is thus

$$\frac{S_o}{N_{oD}} = \frac{1}{D}\frac{S_o}{N_o} = \frac{3\beta^2}{D}\frac{S_c}{N_c} \qquad [6\text{-}171]$$

The signal-to-noise improvement in decibels is the same as the noise reduction: 13 db for $B = 15$ kHz, 16 db for $B = 21$ kHz.

Preemphasis and deemphasis techniques are obviously not restricted to FM only. They are possible because the audio signals to be transmitted in practice are concentrated at the low end of the spectrum. All modulation systems can thus use deemphasis techniques to improve the SNR at the receiver output.

As an example, consider AM. Here we showed that for large carrier-to-noise ratio (CNR $\gg 1$) the noise-power spectrum at the detector output was uniform. (See Figs. 6-47 and 6-48. Recall that for large CNR the $s \times n$ terms dominate.) In particular

$$G_z(f) = K \qquad [6\text{-}172]$$

at the detector output, where K is, for example, $n_0 A_c^2$ in the quadratic detector case and $a^2 n_0$ in the linear detector case.

Using the preemphasis and deemphasis networks of the FM analysis,

$$G_H(f) = G_z(f)|H(\omega)|^2 = \frac{K}{1 + (f/f_1)^2} \qquad [6\text{-}173]$$

is the spectral density at the output of the deemphasis network. Both $G_H(f)$ and $G_z(f)$ are sketched in Fig. 6-58. The total mean noise power at

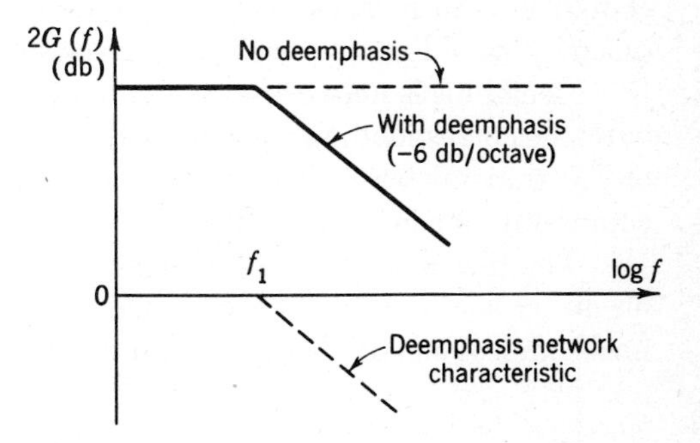

FIG. 6-58 Amplitude-modulation spectral density, detector output (CNR $\gg$ 1).

the output of an ideal low-pass filter of bandwidth B is

$$\begin{aligned} N_{oD} &= \int_{-B}^{B} G_H(f)\,df \\ &= 2Kf_1 \tan^{-1}\frac{B}{f_1} \\ &= N_o D' \end{aligned} \qquad [6\text{-}174]$$

where $N_o = 2BK$ is again the mean noise power with no deemphasis network and

$$D' = \frac{f_1}{B}\tan^{-1}\frac{B}{f_1} \qquad [6\text{-}175]$$

is the improvement factor due to deemphasis in the AM case.

For $B \ll f_1$, $D' \to 1$, so that there is no improvement in this case (the filter bandwidth is well within the deemphasis range). For $B/f_1 \gg 1$, $\tan^{-1}(B/f_1) \doteq \pi/2$, and

$$D' \doteq \frac{\pi}{2}\frac{f_1}{B} \qquad \frac{B}{f_1} \gg 1 \qquad [6\text{-}176]$$

The noise power thus decreases inversely with the filter bandwidth B. In the FM case the noise power decreased inversely as the square of the bandwidth (Eq. [6-170]), so that the RC deemphasis network provides relatively greater improvement for FM than for AM.

As an example, if $f_1 = 2.1$ kHz and $B = 15$ kHz, D' is 6 db for the AM case and D is 13 db for the FM case. With $B = 21$ kHz, D' is 8 db for AM, while D is 16 db for FM. For $B = 5$ kHz, $D' = 3$ db.

Although this deemphasis technique is not as efficient for AM as for FM, it still would provide some improvement. Why then has it not been adopted for AM systems? Several reasons can be listed:[1] Its use would require modification of all existing receivers. Emphasizing the higher-frequency sidebands would increase interference between adjacent AM channels. An increase in m, the modulation factor, at higher audio frequencies would put severe requirements on detector design.

Deemphasis networks are commonly used in FM receivers, however, with the corresponding preemphasis networks built into the audio section of FM transmitters. The same "equalization" principle is also used quite commonly in sound recording to reduce "scratch."

The discussion of SNR improvement by deemphasis has proceeded on an *ad hoc* basis. A specific deemphasis network was postulated and noise improvement demonstrated by calculation. The discussion can,

[1] L. B. Arguimbau and R. D. Stuart, "Frequency Modulation," p. 5, Methuen, London, 1956.

however, be handled on a more general level by postulating the use of preemphasis and deemphasis networks and then finding the networks that minimize the output noise. Specifically, we shall use an approach similar to that adopted previously in discussing matched filters. Here too the Schwarz inequality will be found to play a role.[1] The one distinction is that the signal as well as the noise is now assumed random.

Specifically, let the signal at the transmitter have a known spectral density $G_s(f)$. The signal is then passed through a preemphasis network $H(\omega)$, and the resultant signal actually transmitted. This output signal now has a spectral density $G_s(f)|H|^2$, and the average transmitter power is

$$P = \int_{-\infty}^{\infty} G_s(f)|H(\omega)|^2 \, df \qquad [6\text{-}177]$$

We assume, as is usually the case in practice, that P is *fixed* at some maximum level.

Noise of spectral density $G_n(f)$ is added during transmission, and signal plus noise passed through a deemphasis network with transfer function $H^{-1}(\omega)$, just the inverse of the preemphasis network. The output signal power is then just

$$S_o = \int_{-\infty}^{\infty} G_s(f) \, df \qquad [6\text{-}178]$$

a fixed quantity, and the output noise power is

$$N_o = \int_{-\infty}^{\infty} G_n(f)|H(\omega)|^{-2} \, df \qquad [6\text{-}179]$$

How do we now choose $|H(\omega)|$ to maximize S_o/N_o (or minimize N_o, since S_o is fixed), with P a known constant? (Note that the phase of the two correction networks is arbitrary, provided however, that the phase of the deemphasis network corrects for that of the preemphasis network to avoid signal distortion.)

Since P is constant, we can just as well minimize

$$PN_o = \int_{-\infty}^{\infty} G_s(f)|H|^2 \, df \int_{-\infty}^{\infty} G_n(f)|H|^{-2} \, df \qquad [6\text{-}180]$$

But recall that the Schwarz inequality for real integrals may be written

$$\int A^2(f) \, df \cdot \int B^2 \, df \geq \left[\int A(f)B(f) \, df\right]^2 \qquad [6\text{-}181]$$

If we let

$$G_s(f)|H|^2 \equiv A^2(f)$$

and

$$G_n(f)|H|^{-2} \equiv B^2(f)$$

[1] This approach to deemphasis network analysis was suggested to the author in a private communication by Dr. Robert Price.

we have immediately

$$PN_o \geq [\int \sqrt{G_s(f)G_n(f)}\, df]^2 \qquad [6\text{-}182]$$

with the minimum output noise attained when $A(f) = B(f)$, or

$$|H(\omega)|^2_{\text{opt}} = \sqrt{\frac{G_n(f)}{G_s(f)}} \qquad [6\text{-}183]$$

Providing that the signal and noise have *different* spectral shapes (an almost obvious consideration), preemphasis does pay off (i.e., optimum networks in the sense indicated here do exist).

As an example assume that $G_s(f)$ varies as $1/f^2$ (this is a particularly simple model for indicating relatively lower energy at higher frequencies) and $G_n(f)$ increases over a limited frequency range as f^2. This is then the high CNR FM case discussed earlier, where attention is focused just on the baseband portions of the system. Then the optimum $|H| \propto f$, just the preemphasis network assumed earlier (Fig. 6-55). (The nonlinear modulation portions of the FM system may be ignored here, since the discussion is based solely on the optimization of the baseband preemphasis and deemphasis networks, not on the entire system.)

For an equivalent AM system, with the output-noise spectral density flat, and $G_s(f)$ again assumed dropping off as $1/f^2$, at least over a limited range, the optimum $H(\omega) \propto \sqrt{f}$, somewhat different than the networks considered.

This optimization may be carried one step further without assuming any particular structure for the system (i.e., the preemphasis-deemphasis network pair assumed here), and complete receivers can be designed that demodulate analog signals in some optimum sense in the presence of noise. Such criteria as least mean-squared error between transmitted and received signals, maximum SNR, maximum a posteriori probability, etc., have been adopted, and optimum, albeit not necessarily realizable, structures obtained. But approaches such as these are beyond the scope of this book.[1]

6-12 SIGNAL-TO-NOISE RATIO IN PPM

The PPM system described qualitatively in Chap. 3 can be considered the pulse-modulation analog of a phase-modulation or frequency-modulation system. Here information is transmitted by displacement of the positions

[1] See H. Van Trees. "Detection, Estimation, and Modulation Theory: I," Wiley, New York, 1968; A. J. Viterbi, *op. cit.*, chap. 5.

of the train of pulses constituting the carrier. Since the pulses are maintained at fixed width and amplitude, the positional information is also carried by the location of the leading edge of a pulse or by the point of zero crossing of the leading edge.

A train of unmodulated trapezoidal pulses is shown in Fig. 6-59. A_c represents the fixed amplitude of these pulses. (This corresponds to the

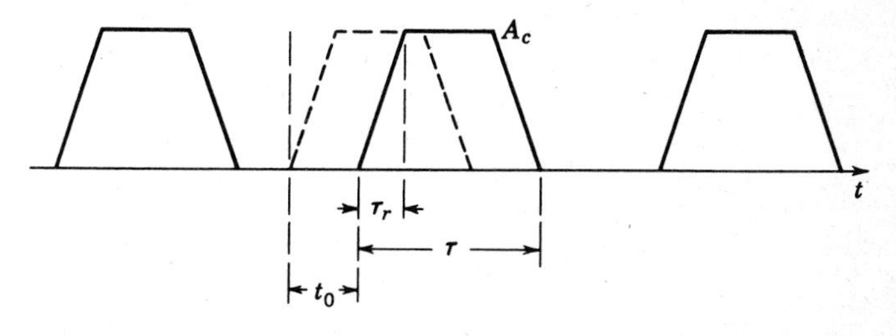

FIG. 6-59 Pulse-position-modulation pulse train.

constant amplitude of the carrier in FM.) τ is the pulse width, τ_r the pulse rise time. In the Federal Telephone Laboratories PPM system described in Chap. 3 a pulse width of 0.8 μsec and pulse rise time of 0.15 μsec are used. t_0 represents the maximum modulation displacement in any one direction away from the unmodulated position of the pulse. The maximum displacement in the Federal Telephone Laboratories system is 1 μsec. Thus, if the modulating signal is a sinusoidal test signal of frequency f_m and of maximum amplitude, the corresponding position of the pulse will vary sinusoidally in time about the unmodulated pulse location following the equation $t_0 \cos \omega_m t$. This corresponds of course to a sinusoidal phase modulation of the carrier in a phase-modulation system, or to the $\Delta\omega \cos \omega_m t$ frequency displacement in FM.

For such a sinusoidal test signal the rms time displacement of the pulse (or its leading edge) is $t_0/\sqrt{2}$ sec. An ideal converter in the output of a PPM receiver converts these pulse-position variations to voltage variations given by $Kt_0 \cos \omega_m t$, with K a constant of proportionality of the converter.

If noise is now superimposed on the pulse, the pulse amplitude and the point of zero crossing will both be perturbed (Fig. 6-60). This corresponds to the amplitude and frequency (or phase) modulation of a sinusoidal carrier by noise, discussed in the previous sections. An amplitude limiter can be used to eliminate the pulse-amplitude variations due to the noise, just as in the FM system. The uncertainty in the pulse position due to noise fluctuations remains, however, and gives rise to a noise voltage at the output of the converter.

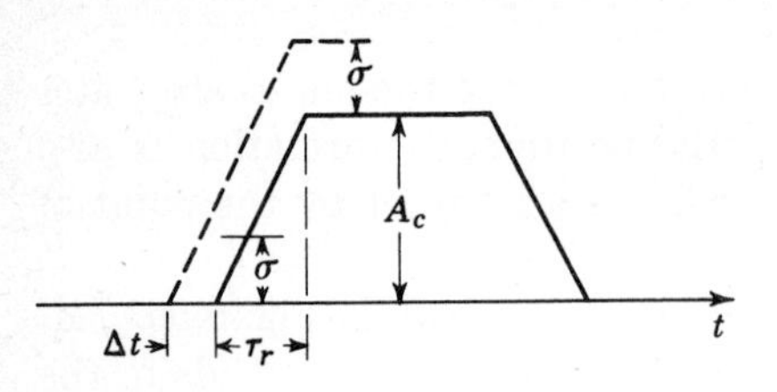

FIG. 6-60 Variation in pulse position due to noise or interference.

If the mean noise power is small compared with the average pulse power and the noise varies slowly over the rise-time portion of the pulse, the pulse perturbation may be represented as in Fig. 6-60. Since we are eventually interested in a mean signal-to-noise ratio at the converter output, we represent the amplitude of the noise voltage over the time interval shown by the rms noise voltage σ and assume this voltage constant over the pulse rise time. The corresponding rms error in determining the pulse location will be Δt sec.

Figure 6-60 indicates that the rms positional error Δt may be reduced by increasing the slope of the pulse leading edge. For a given mean noise power the noise effect may be minimized by using pulses with sharply defined leading edges. But this requires either decreasing the rise time, correspondingly *widening* the transmission *bandwidth*, or increasing the pulse amplitude A_c. We thus see that, just as in the FM case, an exchange of input signal to noise for bandwidth is possible: here decreasing the rise time (or increasing bandwidth) enables a smaller amplitude pulse to be used for a given rms positional error. This is obviously not possible in a PAM system, where the amplitude alone carries the signal information.

We can obtain a simple expression for output signal-to-noise ratio by assuming the same conditions as in the FM-AM analysis: the mean output noise power is evaluated in the absence of a signal and under the assumption of large carrier-to-noise ratio. The output signal power is calculated under noise-free conditions and for a sinusoidal test signal of maximum amplitude. The signal-to-noise ratio is then defined to be the ratio of these two quantities.

The rms output signal voltage is simply $Kt_0/\sqrt{2}$ in terms of the preceding notation. The rms output noise voltage due to the rms positional displacement Δt will similarly be $K\,\Delta t$ at the output of the converter. The SNR at the converter output is then

$$\frac{S_o}{N_o} = \left(\frac{t_0/\sqrt{2}}{\Delta t}\right)^2 \qquad [6\text{-}184]$$

From Fig. 6-60 we get the simple relation

$$\frac{\Delta t}{\tau_r} = \frac{\sigma}{A_c} \qquad [6\text{-}185]$$

valid for $A_c \gg \sigma$, and in the absence of a signal. Substituting for Δt in Eq. [6-184], the output SNR is given by

$$\frac{S_o}{N_o} = \frac{1}{2}\left(\frac{t_0}{\tau_r}\right)^2 \left(\frac{A_c}{\sigma}\right)^2 \qquad [6\text{-}186]$$

But $(A_c/\sigma)^2$ is the ratio of peak pulse power to mean noise power and is analogous to the carrier-to-noise power ratio of the previous section. Using the same symbols as used there,

$$\frac{S_c}{N_c} \equiv \left(\frac{A_c}{\sigma}\right)^2$$

and

$$\frac{S_o}{N_o} = \frac{1}{2}\left(\frac{t_0}{\tau_r}\right)^2 \frac{S_c}{N_c} \qquad [6\text{-}187]$$

The term S_c/N_c may also be considered the effective signal-to-noise ratio of a PAM system.

Equation [6-187] indicates very simply the improvement in S_o/N_o possible in a PPM system by decreasing the pulse rise time τ_r or correspondingly widening the transmission bandwidth.

If the relation $B = 1/\tau_r$,† as developed in Chap. 2, is used to represent the system bandwidth necessary to pass the trapezoidal pulses chosen, the output SNR becomes

$$\frac{S_o}{N_o} = \frac{1}{2} t_0^2 B^2 \frac{S_c}{N_c} \qquad [6\text{-}188]$$

The power ratio S_o/N_o is thus proportional to B^2, just as in the FM case (Eq. [6-161]). If $B = 1/2\tau_r$ is used instead,

$$\frac{S_o}{N_o} = 2 t_0^2 B^2 \frac{S_c}{N_c} \qquad [6\text{-}189]$$

As an example, we can evaluate the SNR improvement possible for the Federal Telephone Laboratories PPM system of Chap. 3 as compared with a PAM system. With $t_0 = 1$ μsec and $\tau_r = 0.15$ μsec, $S_o/N_o = 22 S_c/N_c$. With these numbers the SNR is improved by 15 db. The bandwidth required is correspondingly increased, however.

† This is the baseband bandwidth. A sinusoidal carrier amplitude modulated with these pulses would require twice this bandwidth.

A PAM system incorporating the same 24 channels, each sampled at the same 8-kHz rate, could conceivably utilize the full 5 μsec between samples for each channel pulse. The minimum bandwidth required for the system would then be approximately 200 kHz. For the PPM system 2.8 MHz is specified as the system bandwidth. The improvement in SNR has thus been obtained at the expense of increased bandwidth, just as in the FM case. (Note again that the bandwidths referred to are the bandwidths required to transmit the dc pulses. Intermediate-frequency amplifiers of twice these bandwidths would be required if the pulses amplitude modulated a sinusoidal carrier.)

Assuming the system bandwidth unlimited, can we improve S_o/N_o indefinitely? As previously, the noise power introduced at the receiver increases with bandwidth. Eventually the noise becomes comparable with the signal, the zero crossings become completely erratic, and the noise "takes over," or "captures," the system. A threshold level thus also exists, just as in the FM case. The carrier-to-noise ratio S_c/N_c must be much larger than the threshold value for the wideband improvement to work. This threshold level is usually taken as $A_c/\sigma = 2$, or $S_c/N_c = 4$ (6 db).

Again as in the FM case equalizing networks may be utilized to improve further the SNR properties of PPM systems designed to transmit speech and music.

6-13 AUTOCORRELATION FUNCTION AND POWER SPECTRUM OF RANDOM SIGNALS

We have stressed thus far in this chapter the utility of the power-spectrum concept in determining the output-noise properties of various systems. In most cases discussed we have relied on the extremely important relation,

$$G_o(f) = |H(\omega)|^2 G_i(f) \qquad [6\text{-}190]$$

connecting spectral densities at the input and output of a linear device, in carrying out the noise calculations. We have also indicated that the same results apply to random signals.

In the one nonlinear case considered, that of the envelope detector, we showed how the spectrum at the output of a nonlinear device may be found by first finding the autocorrelation function and then taking its Fourier transform to obtain the desired quantity. This technique has been utilized extensively in practice in determining power and spectra at the output of nonlinear devices. Examples other than the one calculated have already been noted in passing (the linear envelope detector and the general FM noise problem are two such examples), and many others may be found in the references cited.[1]

The intimate (Fourier-transform) relation between the power spectral density and autocorrelation function has also been used extensively in determining the power spectra of various types of random signals. This is obviously extremely useful information in system design, in determining bandwidths necessary to transmit the signals, in judging the effect of filtering and nonlinear operations on signals, etc. As pointed out earlier the random signal is often a much more realistic model for real signals than the deterministic pulse streams and sine waves considered in earlier chapters. (However, it is still important to stress that deterministic signal models provide useful information for system design.)

As has been noted repeatedly throughout this book, actual signals to be transmitted in practice are always random and unpredictable in form. Otherwise why transmit them? Although the use of a series of test pulses or a sine wave frequently provides us with useful information as to the operation of a particular information-transmission system, such waveforms are artificial representations of actual signals.

Once the power spectrum of a particular random signal is known, the filtering effect of various networks through which the signal passes may be determined. Knowing the power spectrum of both the signal and of any noise mixed in with it may enable us to develop schemes for eventually filtering out the signal from the noise. Or we may just as well find that a simple linear filter will not suffice to distinguish between signal and noise.

Knowledge of the signal-power spectrum is thus important in the design of information-transmission systems. The calculation of the power spectrum of a random type of signal is the first step in the more realistic approach of designing systems to handle prescribed random inputs rather than arbitrarily assumed pulse and sine-wave inputs.

We shall restrict ourselves in this book to discussing one typical example of such a random-signal spectral calculation. Others appear in the references cited.[2] In the next chapter we use an autocorrelation-function approach to calculate the statistics of the shot-noise process.

[1] See, e.g., D. Middleton, *op. cit.*, and Davenport and Root, *op. cit.*

[2] See, e.g., Lawson and Uhlenbeck, *op. cit.;* W. R. Bennett and J. R. Davey, "Data Transmission," McGraw-Hill, New York, 1965; and H. E. Rowe, "Signals and Noise in Communication Systems," Van Nostrand, Princeton, N.J., 1965, for calculations of the spectra of pulse systems. Rowe, *ibid.;* Schwartz, Bennett, and Stein, *op. cit.;* and D. Middleton, *op. cit.*, among others, consider random FM signal spectra. In some of these references the power spectrum is found directly by appropriate averaging over the signals, in others by first finding the autocorrelation function.

The particular random signal chosen here to illustrate the power spectral calculation via autocorrelation functions is the so-called "random telegraph signal." This signal is defined to be a voltage $s(t)$ of binary amplitudes $+a$ or $-a$, as shown in Fig. 6-61. The length of time for

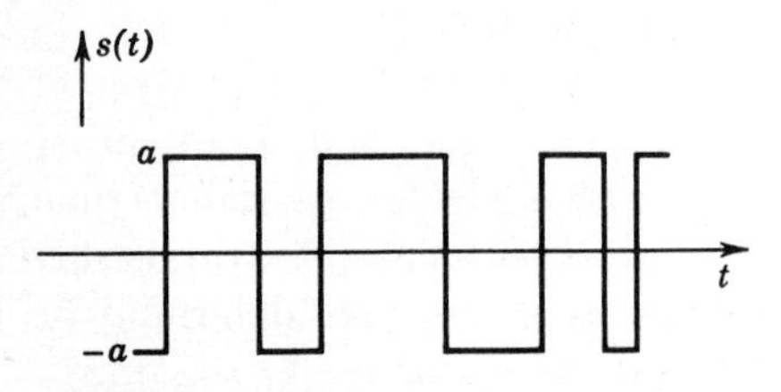

FIG. 6-61 Random telegraph signal.

which $s(t)$ remains in either position is random. In particular the number of times the function changes sign (shifting from either amplitude to the other) is given by the so-called "Poisson distribution"

$$P(K) = \frac{(\mu T)^K e^{-\mu T}}{K!} \qquad K \geq 0 \tag{6-191}$$

where $P(K)$ is the probability of exactly K sign changes in T sec and μ is the average number of sign changes (zero crossings) per second.

We shall first discuss briefly the particular statistical distribution, the Poisson distribution, assumed for the random telegraph signal. The calculation of the autocorrelation function, and from that the power spectrum, will then follow.

It is well to repeat at this point that the telegraph signal represents one of many examples of random signals with prescribed statistical properties. This signal happens to have a Poisson distribution. In other examples other distributions would occur. The method to be used for finding the power spectrum would be the same, however.

The Poisson distribution of Eq. [6-191] is an example of a discrete probability distribution, since it is defined only for discrete values of K. That $E(K)$ is μT may be shown readily from the definition of ensemble average. Thus

$$\begin{aligned} E(K) &= \sum_{K=0}^{\infty} KP(K) \\ &= e^{-\mu T} \sum_{K=0}^{\infty} \frac{K(\mu T)^K}{K!} \\ &= \mu T e^{-\mu T} \sum_{K=1}^{\infty} \frac{(\mu T)^{K-1}}{(K-1)!} \end{aligned} \tag{6-192}$$

since the $K = 0$ term is zero. But

$$\sum_{K=1}^{\infty} \frac{(\mu T)^{K-1}}{(K-1)!} = 1 + \mu T + \frac{(\mu T)^2}{2!} + \cdots$$

is just the series expansion of $e^{\mu T}$. From Eq. [6-192], then,

$$E(K) = \mu T \qquad [6\text{-}193]$$

and μ represents the average number of sign changes per second.

The Poisson distribution is obtained from the simple assumption for the random telegraph signal that the probability of a sign change in a differential interval t to $t + \Delta t$ is proportional to the length of the interval Δt, independent of occurrences outside this interval.

Interestingly the Poisson distribution will be derived as a limiting form of the binomial distribution first discussed in Chap. 5. Both distributions will again be encountered in the next chapter in discussing shot noise.

To derive the Poisson distribution, consider a large interval T sec long divided into M subintervals Δt sec long. Then $M = T/\Delta t \gg 1$ (Fig. 6-62). The probability of a zero crossing (or sign change) in any

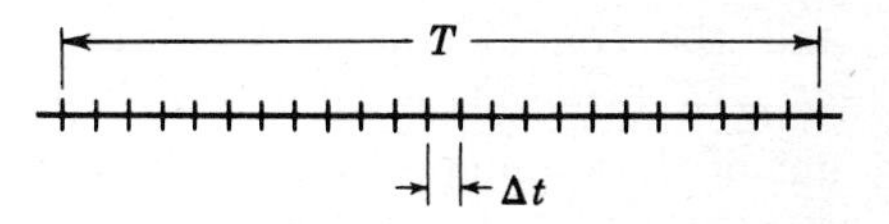

FIG. 6-62 Derivation of Poisson distribution.

one interval Δt sec long is $\mu\,\Delta t$ by hypothesis, with μ a proportionality factor ($\mu\,\Delta t \ll 1$). The probability of no zero crossing in this small interval must be $1 - \mu\,\Delta t$.

Under the assumption that the probability of an occurrence in one interval is independent of occurrences in the other intervals the probability that there will be *no* zero crossings in any of the M independent intervals is the product of the probabilities. This is just $P(0)$, the probability of no sign changes in T sec, and is given by

$$P(0) = (1 - \mu\,\Delta t)^M = (1 - \mu\,\Delta t)^{T/\Delta t} \qquad [6\text{-}194]$$

Upon letting $\Delta t \to 0$, $M \to \infty$, and $P(0)$ becomes

$$P(0) = e^{-\mu T} \qquad [6\text{-}195]$$

in the limit. [Recall that $\lim_{\epsilon \to 0} (1 - \epsilon)^{1/\epsilon} = 1/e$ and that

$$\lim_{\epsilon \to 0} (1 - n\epsilon)^{1/\epsilon} = e^{-n}]$$

Similarly the probability of exactly *one* sign change in T sec is

$$\begin{aligned} P(1) &= M(\mu\, \Delta t)(1 - \mu\, \Delta t)^{M-1} \\ &= (\mu T)(1 - \mu\, \Delta t)^{M-1} \end{aligned} \qquad [6\text{-}196]$$

since there are M possible intervals in which the one sign change can take place. Each of these possibilities is mutually exclusive, and so the probabilities add.

Again taking the limit $\Delta t \to 0$,

$$P(1) = \mu T e^{-\mu T} \qquad [6\text{-}197]$$

The probability of exactly K sign changes in the M intervals Δt sec long $(K < M)$ is

$$P(K) = \frac{M!}{K!(M - K)!} (\mu\, \Delta t)^K (1 - \mu\, \Delta t)^{M-K} \qquad [6\text{-}198]$$

For the probability of one sign change in each of K intervals is $(\mu\, \Delta t)^K$ and the concurrent probability of no sign changes in the remaining $M - K$ intervals is $(1 - \mu\, \Delta t)^{M-K}$. The overall probability is the product of these independent probabilities. But there are $M!/K!(M - K)!$ ways in which K sign changes can occur (the number of combinations of M things, K at a time). These are mutually exclusive, and the resulting probabilities add, giving Eq. [6-198].

The distribution of $P(K)$, given by Eq. [6-198], is of course just the binomial distribution. In the limit, as $\Delta t \to 0$, this distribution approaches the Poisson distribution of Eq. [6-191].

This is easily demonstrated by considering the limit as $\Delta t \to 0$ and $M \to \infty$. For large M, $M!$ may be approximated by Stirling's approximation.[1]

$$M! \doteq \sqrt{2\pi}\, e^{-M} M^{M+\frac{1}{2}} \qquad [6\text{-}199]$$

(This is even a good approximation for $M = 2$, as may be shown by putting in numbers.) With K fixed, $(M - K)!$ may be similarly represented by

$$(M - K)! \doteq \sqrt{2\pi}\, e^{-(M-K)} (M - K)^{M-K+\frac{1}{2}} \qquad [6\text{-}200]$$

[1] P. Franklin, "A Treatise on Advanced Calculus," Wiley, New York, 1940.

Substituting these approximations for $M!$ and $(M - K)!$ into Eq. [6-198], we get, for large M,

$$K!P(K) \doteq e^{-K}\left(1 - \frac{K}{M}\right)^{-(M+\frac{1}{2})}\left(1 - \frac{K}{M}\right)^{K}(M\mu\,\Delta t)^{K}(1 - \mu\,\Delta t)^{M-K} \qquad [6\text{-}201]$$

By keeping $M\,\Delta t = T$ fixed, as well as K fixed, and taking the limit as $M \to \infty$, Eq. [6-191] is obtained. Again use is made of the relation

$$\lim_{M\to\infty}\left(1 - \frac{K}{M}\right)^{-M} = e^{K}$$

as well as

$$\lim_{\Delta t\to 0}(1 - \mu\,\Delta t)^{T/\Delta t} = e^{-\mu T}$$

How do we now find the autocorrelation function $R_s(\tau)$, from the known distribution $P(K)$, for the number of sign changes in T sec?

Recall that $R_s(\tau)$ was defined as

$$R_s(\tau) = E[s(t)s(t + \tau)] \qquad [6\text{-}202]$$

where we take the statistical (ensemble) average over the product of the random signal $s(t)$ and its value τ sec later. To apply this definition to the case of the random telegraph signal, assume that a large number of identical sources of such a signal are available. One member of such an ensemble is shown in Fig. 6-63. Since the product $s(t)s(t + \tau)$ is either

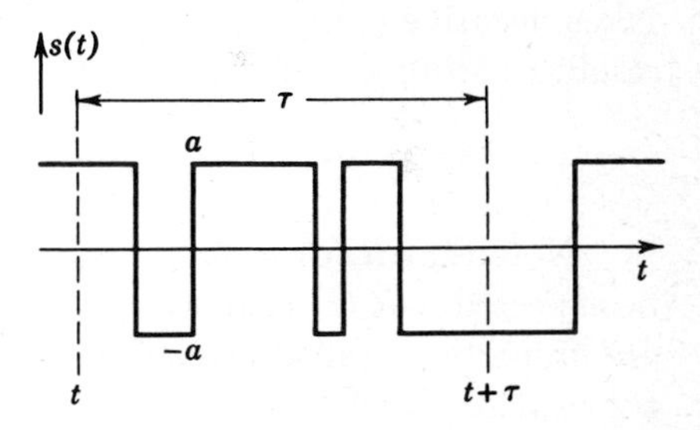

FIG. 6-63 Calculation of correlation function, random telegraph signal.

$+a^2$ or $-a^2$, the average of this product is simply $+a^2$ times its probability of occurrence plus $-a^2$ times its probability.

The desired probabilities may be calculated from the given Poisson distribution. For the products $+a^2$ will occur if the sign of $s(t)$ changes an even number of times in τ sec, while the product $-a^2$ will be obtained for an odd number of sign changes. In the sample wave of Fig. 6-63, for example, $s(t)s(t + \tau)$ is $-a^2$, since the wave has changed sign five times in the τ-sec interval chosen. The ensemble average $E[s(t)s(t + \tau)]$ is thus $+a^2$ times the probability of an even number of sign changes of $s(t)$ in τ sec, plus $-a^2$ times the probability of an odd number of sign changes.

The autocorrelation function is thus

$$\begin{aligned} R_s(\tau) &= a^2[P(0) + P(2) + P(4) + \cdots] \\ &\quad - a^2[P(1) + P(3) + P(5) + \cdots] \end{aligned} \qquad [6\text{-}203]$$

where $P(0)$ is the probability of no sign changes, $P(1)$ the probability of one sign change, etc. The appropriate values for the probabilities are obtained from the Poisson distribution of Eq. [6-191], with T replaced by τ.

$$P(K) = \frac{(\mu\tau)^K}{K!} e^{-\mu\tau} \qquad [6\text{-}204]$$

The autocorrelation becomes, after interleaving the positive and negative terms,

$$\begin{aligned} R_s(\tau) &= a^2 e^{-\mu\tau}\left[1 - \frac{\mu\tau}{1!} + \frac{(\mu\tau)^2}{2!} - \frac{(\mu\tau)^3}{3!} + \cdots\right] \\ &= a^2 e^{-\mu\tau} e^{-\mu\tau} \\ &= a^2 e^{-2\mu\tau} \end{aligned} \qquad [6\text{-}205]$$

since the sum in brackets is just the power-series expansion for the exponential. Although τ was assumed positive, it could just as well have been taken negative (with $t + \tau$ and t in Fig. 6-63 interchanged), with the same results. More generally then we have

$$R_s(\tau) = a^2 e^{-2\mu|\tau|} \qquad [6\text{-}206]$$

Note that $R_s(0) = a^2$ is just the average power in the signal, as is to be expected. $R_s(\tau)$ is an even function of τ, and $R_s(\tau) < R_s(0)$. (These are properties of autocorrelation functions in general, as will be shown in the next section.)

Equation [6-206] is shown sketched in Fig. 6-64*a*. An example of an autocorrelation function of this form has already been discussed. In particular the autocorrelation function of white noise through a simple *RC* network was shown earlier to be of exactly this form. The two spectral densities must thus also be of the same form. In particular it is left for the reader to show that the power spectrum of the random telegraph signal must be given by

$$G_s(f) = \frac{a^2/\mu}{1 + (\pi f/\mu)^2} \qquad [6\text{-}207]$$

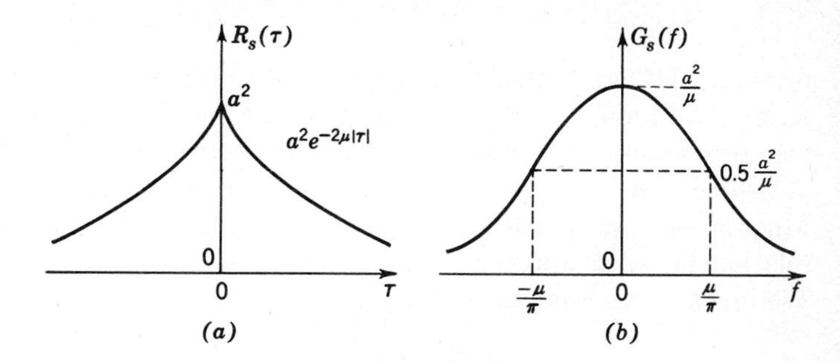

FIG. 6-64 Autocorrelation function and power spectrum, random telegraph signal. (a) Correlation function; (b) power spectrum.

The power spectrum of the random telegraph signal is thus identical with the spectrum of white noise passed through a single RC network with time constant $RC = 1/2\mu$. (μ represents the average number of sign changes of the signal per second.) The power spectrum is sketched in Fig. 6-64*b*.

This simple example of the calculation of the power spectrum of one type of random signal has also demonstrated the nonunique character of the power spectrum. The random telegraph signal and white noise at the output of an RC network are indistinguishable as far as their frequency spectral behavior is concerned. This is in general true of all power spectra or autocorrelation functions. A given spectrum may represent many different types of random inputs.

The reason for this is easily explained. We have chosen to emphasize here one particular statistical parameter, the correlation function. Although the correlation function is uniquely specified once the appropriate joint-probability-distribution function is given, the inverse is not at all true. Many different distribution functions can give rise to the same correlation function. To characterize a random process of some type more completely, be it noise or a signal to be transmitted, the joint-probability-density function must be known. In many practical cases of interest this is not possible. The correlation function or power spectrum, either calculated or measured experimentally, must thus suffice as a much more limited measure of the statistical properties of a given random function.

6-14 PROPERTIES OF THE AUTOCORRELATION FUNCTION

We conclude this chapter by tabulating some useful properties of the autocorrelation function. Some of these have already been noted in passing. Others, although not previously pointed out as general properties, may have already been noted by the reader in studying the examples throughout this chapter. These properties are frequently useful in obtaining some overall information quickly or in checking results obtained. Most of these properties are obtained by utilizing the Fourier-transform relation between autocorrelation function and power spectral density.[1] We include this relation for the sake of reference:

$$R_n(\tau) = \int_{-\infty}^{\infty} G_n(f)e^{-j2\pi f\tau}\, df \qquad [6\text{-}208]$$

1. $R_n(0) = N$, the total noise (or random signal) power. This has been discussed in detail earlier, and is apparent from Eq. [6-208]:

$$R_n(0) = \int_{-\infty}^{\infty} G_n(f)\, df = N$$

2. $R_n(0) \geq R_n(\tau)$. This is apparent from all the examples discussed; in all cases $R_n(\tau)$ decreases for $\tau > 0$. The proof depends on the fact that the phase factor $e^{j2\pi f\tau}$ can never increase the value of a function. Thus

$$|R_n(\tau)| = \left|\int_{-\infty}^{\infty} G_n(f)e^{-j2\pi f\tau}\, df\right| \leq \int_{-\infty}^{\infty} |G_n(f)|\, df = R_n(0) \qquad [6\text{-}209]$$

since $G_n(f)$ is always a positive number. (Why?)

3. $R_n(\tau) = R_n(-\tau)$ for a stationary random process. $R_n(\tau)$ is thus symmetrical in τ. This is of course demonstrated by the previous examples of correlation functions. The proof follows from the definition of the autocorrelation function.

$$\begin{aligned} R_n(\tau) &= E[n(t)n(t+\tau)] \\ &= E[n(t-\tau)n(t)] = R_n(-\tau) \end{aligned}$$

since the time origin can be shifted arbitrarily if the process is stationary.

From this and the Fourier-transform relationship follows the very important relation

$$G_n(f) = G_n(-f) \qquad [6\text{-}210]$$

Thus the spectral density is an even function of frequency. This has been noted time and again.

4. Correlation function of a random periodic function is also periodic. Thus, if

$$n(t) = A\cos(\omega_0 t + \theta) \qquad [6\text{-}211]$$

[1] See A. Papoulis, *op. cit.*, for a detailed discussion of both the correlation function and the power spectral density.

with θ uniformly distributed from 0 to 2π,

$$R_n(\tau) = \frac{A^2}{2} \cos \omega_0 \tau \qquad [6\text{-}212]$$

This was first noted here in connection with the quadratic envelope detector. The spectral density is then just a pair of impulse functions at frequencies f_0 and $-f_0$:

$$G_n(f) = \frac{A^2}{4} [\delta(f + f_0) + \delta(f - f_0)] \qquad [6\text{-}213]$$

Equation [6-212] is readily proved by writing

$$R_n(\tau) = E[n(t)n(t + \tau)] = A^2E\{\cos [\omega_0(t + \theta)] \cos [\omega_0(t + \tau) + \theta]\}$$

expanding into sum and difference terms, and noting that $E[\cos (b + \theta)] = 0$, with b a deterministic parameter. Equation [6-213] follows, of course, from Fourier-transform properties.

As a check, note that $R_n(0) = A^2/2$ is the mean-squared value of $n(t)$, $R_n(\tau) = R_n(-\tau)$, and $R_n(0) \geq R_n(\tau)$ in this case (see Fig. 6-65).

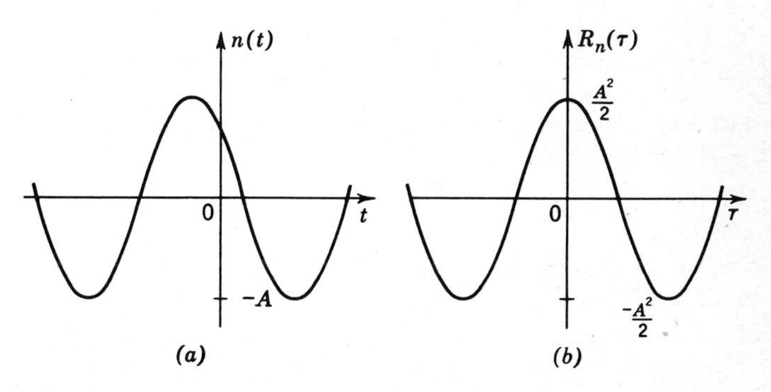

FIG. 6-65 Autocorrelation function of a random periodic function. (a) $n(t) = A \cos (\omega t + \theta)$; (b) $R_n(\tau) = (A^2/2) \cos \omega r$.

$R_n(\tau)$ retains the periodicity of $n(t)$ but is independent of the phase angle. [This is as expected because of the transform relation between $R(\tau)$ and the power spectral density $G_n(f)$.]

If $n(t)$ is a dc voltage $n(t) = A_0$, $R_n(\tau) = A_0{}^2$ from the definition of autocorrelation function. This of course agrees with the intuitive concept of correlation. The Fourier transform of a constant is just the impulse function at $f = 0$.

In general, if

$$n(t) = A_0 + \sum_{k=1}^{n} A_k \sin(\omega_k t + \theta_k) \qquad [6\text{-}214]$$

with all the frequencies incommensurable ($\omega_m \neq \omega_n$), and the phase angles θ_k independent and random,

$$R_n(\tau) = A_0^2 + \sum_{k=1}^{n} \frac{A_k^2}{2} \cos \omega_k \tau \qquad [6\text{-}215]$$

The corresponding spectral density is represented by a set of impulse functions at the frequencies of the sine waves. This of course agrees with intuition, since one would expect random signals at fixed frequencies to have their power concentrated at those frequencies only (Fig. 6-66).

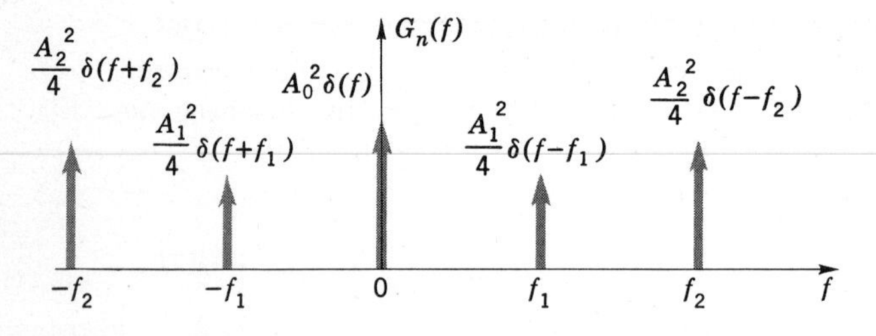

FIG. 6-66 Power spectral density, $n(t) = A_0 + A_1 \cos(\omega_1 t + \theta_1) + A_2 \cos(\omega_2 t + \theta_2)$.

6-15 SUMMARY

In this chapter we were finally able to consider the effect of passing random (and hence more realistic) signals through systems. The mechanism of doing this was to define the autocorrelation function and its Fourier transform, the power spectral density. These two functions are central to the study of random processes and recur over and over again in more advanced treatments than that given here.

The power spectral density represents, as implied by the name, the signal (or noise) power distribution over all frequencies. If the power is concentrated in a definable range of frequencies, we can talk of the signal (or noise) *bandwidth* B, just as in earlier chapters. By the usual Fourier relations, the width of the corresponding autocorrelation function is proportional to $1/B$. This in turn is, by definition of the autocorrelation function, a measure of the correlation between signal samples: for a time separation $\tau < 1/B$, the samples are essentially correlated; for $\tau > 1/B$, the correlation usually goes to zero. This in turn describes, in an intuitive way, the rate of change of the signal (noise) in time.

The effect of passing a random signal through a linear system $H(\omega)$ is to multiply the signal spectral density by $|H(\omega)|^2$. This simple relation then enabled us to determine both the band-limiting effect of systems on signals, and the actual signal (noise) power at the system output.

With these basic concepts of random processes as an introduction we were able to discuss in a systematic way the comparative signal-to-noise properties of a representative group of high-frequency binary transmission systems as well as more traditional AM and FM systems. Thus, after first introducing the *matched filter* both as an application of the noise spectral analysis and because of its importance in its own right, we went on to discuss the reception of high-frequency binary signals in noise. Both synchronous and noncoherent (envelope) detection were considered, the discussion including the PSK, FSK, and OOK signals first introduced in Chap. 4. The narrowband representation of noise enabled us to actually follow through the detection process. We found that matched filtering was appropriate to *all* binary receivers to minimize probability of error, that synchronous detection provides lower error rates (for the same SNR) than envelope detection, and that, ideally, coherent PSK was the system to be preferred.

A further application of our signal and noise analysis was to radar detection. Here instead of one unique probability of error we had to retain two types of error: false-alarm probability, and the probability of missing a desired signal (or, alternately, the probability of signal detection). We were able to show how one develops performance characteristics for the radar, relating the two probabilities and SNR, as well as the number of received signal pulses added up.

In the case of analog modulation systems we were able to show that FM and PPM, as examples of "wideband" modulation schemes, provide a SNR improvement over such systems as AM and PAM. This in turn provides an improvement in the ability to transmit information. In particular, in both FM and PPM systems, an increase in bandwidth provides a proportional increase in output SNR, effectively suppressing the noise more and more relative to the signal. This is only possible above the so-called *threshold region* of carrier-to-noise ratio, however. Thus, below a CNR of approximately 10 db the output SNR plunges rapidly.

This bandwidth-SNR exchange does not occur in AM-type systems. There, above a 0-db CNR, the output SNR is strictly proportional to the input SNR. (Below 0 db a suppression effect is also encountered in AM.)

The SNR-bandwidth exchange of FM systems is of course not as efficient as the exponential exchange encountered with coded PCM systems in Chap. 3. (The high-frequency binary systems discussed in this chapter are all examples of binary PCM systems, but using different modulation techniques.) We shall refer to this exponential exchange again in Chap. 8 in discussing systems designed to transmit as best as possible over band-limited channels that introduce gaussian noise.

As an interesting and important by-product of the spectral analysis of random processes introduced in this chapter, we showed how one could calculate the noise (or signal) spectral distribution at the output of a nonlinear device by first relating autocorrelation functions at the input and output, and then taking the Fourier transform of the output correlation function to find the required spectral density. This approach was carried out in detail for the quadratic or square-law envelope detector. We were similarly able to find the power spectrum of a typical random signal (the random telegraph wave in this case) by first finding its autocorrelation function by appropriate ensemble-averaging, and then taking the Fourier transform to find the desired spectrum. Another example will be encountered in the next chapter in analyzing shot-noise processes.

PROBLEMS

6-1 A random signal $s(t)$ of zero average value has the triangular spectral density of Fig. P 6-1.

(*a*) What is the average power (mean-squared value) $S \equiv E(s^2)$ of the signal?
(*b*) Show that its autocorrelation function is

$$R_s(\tau) = S\left(\frac{\sin \pi B\tau}{\pi B\tau}\right)^2$$

Hint: Refer back to Chap. 2 for the Fourier transform of a triangular pulse.
(*c*) $B = 1$ MHz, $K = 1\ \mu\text{v}^2/\text{Hz}$. Show that the rms value of the signal is $\sqrt{S} = 1$ mv and that samples of $s(t)$ spaced 1 μsec apart are uncorrelated.

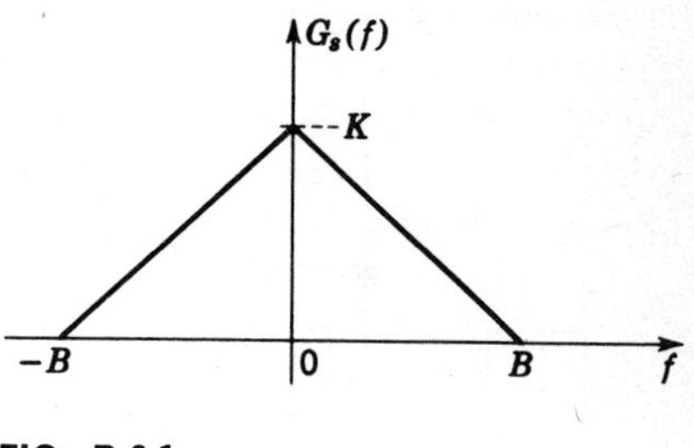

FIG. P 6-1

6-2 Band-limited white noise $n(t)$ has spectral density $G_n(f) = 10^{-6}$ volt²/Hz, over the frequency range from −100 to +100 kHz.

(*a*) Show the rms value of the noise is approximately 0.45 volt.
(*b*) Find $R_n(\tau)$. At what spacings are $n(t)$ and $n(t + \tau)$ uncorrelated?
(*c*) $n(t)$ is assumed gaussian. What is the probability at any time t that $n(t)$ will exceed 0.45 volt? 0.9 volt?
(*d*) The noise, again assumed gaussian, is added to bipolar signals of amplitude $\pm A$. What is the probability of error if the binary signals are equally likely, the decision level is taken as 0, and $A = 0.9$ volt? Repeat for $A = 1.8$ volts, and 4.5 volts.

6-3 $R_n(\tau) = N \cos \omega_0\tau$. Show $G_n(f) = (N/2)\delta(f - f_0) + (N/2)\delta(f + f_0)$. Sketch both $R_n(\tau)$ and $G_n(f)$.

6-4 Consider a random signal $s(t) = A \cos (\omega_0 t + \theta)$, with θ a uniformly distributed random variable. Show

$$E[s(t)s(t + \tau)] = \frac{A^2}{2} \cos \omega_0\tau$$

by averaging over the random variable θ, as indicated. Compare this result with that of Prob. 6-3. Can you explain this result?

6-5 Consider a random signal

$$s(t) = \sum_{i=1}^{n} a_i \cos (\omega_i t + \theta_i)$$

with the frequencies all distinct, and the θ_i's random and independent. Calculate the autocorrelation function of $s(t)$ by averaging $s(t)s(t+\tau)$ over the random θ_i's, and show

$$R_s(\tau) = \sum_{i=1}^{n} \frac{a_i^2}{2} \cos \omega_i \tau$$

Calculate the spectral density $G_s(f)$ and sketch. Compare this result with that of Probs. 6-3 and 6-4.

6-6 Consider the periodic rectangular pulse train of Fig. P 6-6. Timing jitter causes the entire train to be shifted relative to a fixed time origin by a random variable Δ as shown. Δ may be assumed uniformly distributed over the range 0 to T. Show by averaging $s(t)s(t+\tau)$ over Δ that the autocorrelation function is a periodic sequence of triangular pulses of peak amplitude A^2t_0/T, base width $2t_0$, and period T.

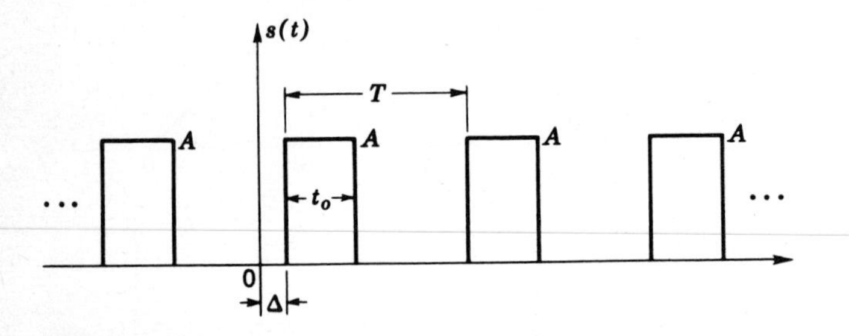

FIG. P 6-6

6-7 An autocorrelation function that frequently arises in practical problems is given by

$$R(\tau) = R(0)e^{-\alpha|\tau|} \cos \beta\tau$$

(*a*) Calculate the power spectrum $G(f)$.
(*b*) Typical values for α and β are $\alpha = 1$, $\beta = 0.6$. Plot $R(\tau)/R(0)$ and $G(f)$.
(*c*) Check the results of (*a*) by considering the two limiting cases (1) $\alpha = 0$, (2) $\beta = 0$.

6-8 White noise of spectral density $G_n(f) = n_0/2$ watts/Hz is applied to an ideal low-pass filter of bandwidth B Hz and transfer amplitude A. Find the correlation function of noise at the output. Calculate the total average power at the filter output from the spectral density directly, and compare with $R(0)$.

6-9 White noise of spectral density $G(f) = 10\ \mu v^2/\text{Hz}$ is passed through a noiseless narrowband amplifier centered at 400 Hz. The amplifier may be represented by an ideal bandpass filter of 50-Hz bandwidth about the 400-Hz center frequency and amplification factor of 1,000.

(*a*) Write an expression for the autocorrelation function $R(\tau)$ at the amplifier output.
(*b*) The output noise voltage is to be sampled at intervals far enough apart so as to ensure uncorrelated samples. How far apart should the samples be taken?

6-10 Gaussian noise $n(t)$ with correlation function $R_n(\tau)$ is sampled at time t and τ sec later, at $t + \tau$. Call the two samples n_1 and n_2, respectively.

(*a*) Using Eq. [5-206] in Chap. 5, write an expression for the two-dimensional density function $f(n_1,n_2)$, relating the samples taken. The moments and correlation coefficient appearing in Eq. [5-206] should all be written in terms of $R_n(\tau)$.
(*b*) Write the two-dimensional density function $f(n_1,n_2)$ specifically for the case of the noise in Prob. 6-2, assuming that the noise is gaussian. Take two cases: (1) $\tau = 2.5$ μsec, (2) $\tau = 5$ μsec. Compare with the expressions for the one-dimensional density functions, $f(n_1)$ and $f(n_2)$ in each case.

6-11 White noise of spectral density 10^{-6} volt²/Hz is applied at the input of the circuit in Fig. P 6-11, as shown.

(*a*) Find the noise spectral densities $G_1(f)$ and $G_2(f)$, at points 1 and 2, respectively, in terms of the bandwidth $B \equiv 1/2\pi RC$.
(*b*) Find the output rms noise voltage $\sqrt{N_2}$ in terms of the bandwidth B. $R = 10$ kilohms. What is $\sqrt{N_2}$ if $C = 0.0001$ μf? 0.01 μf?

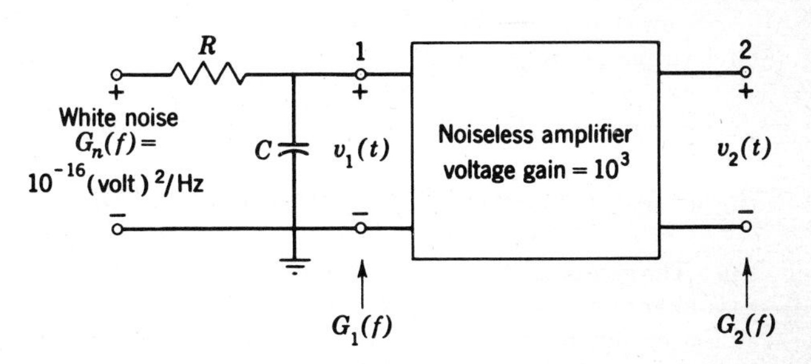

FIG. P 6-11

6-12 Consider the circuit of Fig. P 6-12.

(*a*) Show that the power spectral density of $v(t)$ at the amplifier input is $G_v(f) = 10^{-14}/[1 + (f/B)^2]$ volt²/Hz; $B = G/2\pi C = 1{,}600$ Hz.
(*b*) Show that the rms noise voltage at the output of the amplifier is 7 mv.

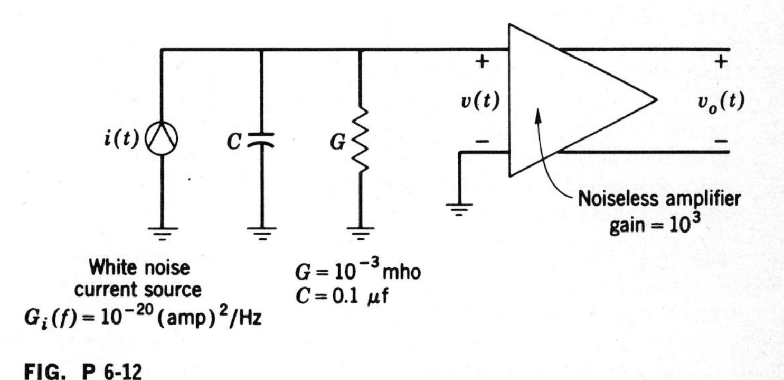

FIG. P 6-12

6-13 A rectangular pulse of amplitude V volts and width τ sec is applied to a matched filter. Show that the output is a triangular-shaped pulse. Find the peak value of this pulse. Calculate the total noise power at the output of the filter (assumed noiseless) if white noise of spectral density $n_0/2$ volts²/Hz is applied at the input. Calculate the output signal-to-noise ratio at the peak of the signal pulse if signal and noise appear together at the filter input.

6-14 (*a*) The signal pulse of Prob. 6-13 is applied to a single RC section, with $RC = 2\tau/3$. Sketch the output pulse, and compare with that of Prob. 6-13. Calculate the total output noise power if white noise of spectral density $n_0/2$ volts²/Hz is again added to the input signals. Calculate the peak SNR at the output, and compare with the result of Prob. 6-13, checking a point on the appropriate curve of Fig. 6-18 in the text.

(*b*) Let the pulse width τ vary, and repeat (*a*). Calculate the peak SNR at the output, and plot as a function of τ. Compare with the appropriate curve of Fig. 6-18.

6-15 A signal pulse of unit peak amplitude and half-power width τ is given mathematically by the so-called "gaussian" error curve $f(t) = e^{-0.35(2t/\tau)^2}$. Added to the signal pulse is white noise of spectral density $n_0/2$ volts²/Hz. The signal plus noise is applied to a filter.

Show that the optimum filter characteristic for maximizing the peak SNR at the filter output is given by a gaussian curve in frequency,

$$|H(\omega)| = \sqrt{\frac{\pi}{1.4}}\,\tau e^{-(\omega\tau)^2/5.6}$$

(Recall that $\int_{-\infty}^{\infty} e^{-x^2}\,dx = \sqrt{\pi}$.) Sketch $f(t)$ and $|H(\omega)|$.

6-16 The gaussian pulse and white noise of Prob. 6-15 are passed through an ideal low-pass filter of cutoff frequency $2\pi B$ radians/sec. Show that the maximum peak SNR at the output occurs at $2\pi B\tau = 2.4$. Show that this maximum value of the peak SNR is only 0.3 db less than that found by using the optimum filter characteristic of Prob. 6-15.

6-17 (*a*) Prove that the impulse response of a matched filter is given in general by Eq. [6-59].

(*b*) A signal pulse given by $f(t) = e^{-\alpha t}$, $t \geq 0$, is mixed with white noise of spectral density $n_0/2$ volts²/Hz. Find the impulse response of the matched filter, and compare with $f(t)$.

6-18 The i-f section of a radar receiver consists of four identical tuned amplifiers with an overall 3-db bandwidth of 1 MHz. The center frequency is 15 MHz. The amplifiers represent the bandpass equivalent of RC-coupled amplifiers. The input to the i-f strip consists of a 1-μsec rectangular signal pulse plus fluctuation noise generated primarily in the r-f stage and mixer of the receiver. Calculate the relative difference in decibels of the peak SNR at the receiver output for the amplifier given, as compared with an optimum matched filter. Repeat for overall amplifier bandwidths of 500 kHz and 2 MHz.

6-19 (*a*) Find and sketch the matched-filter output to the two waveforms shown in Fig. P 6-19.

(*b*) Compare the output of matched filter 2 at times $t = T$, $t = T - T/4n$, $t = T - T/2n$. How critical is the *phase* synchronization between transmitter and receiver in this case? Explain.

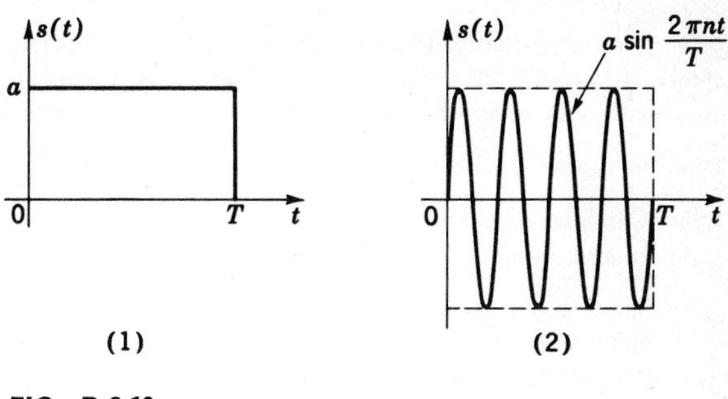

FIG. P 6-19

(*c*) The received signal $s(t)$ is a sinusoidal pulse at frequency $f_c = 1{,}000$ MHz, and pulse width $T = 1\ \mu$sec. The output of a filter matched to $s(t)$ is to be sampled precisely at the end of the pulse, $t = T = 1\ \mu$sec. Using (*b*) above, discuss the synchronization required.

6-20 (*a*) Prove the Fourier-transform relationship

$$E = \int_{-\infty}^{\infty} |f(t)|^2\, dt = \int_{-\infty}^{\infty} |F(\omega)|^2\, df$$

Hint: Consider the special case $f(t)$ real. Let $f(t)$ be the input to a matched filter. Find and equate the outputs using the convolution integral, and then Fourier transforms.

(*b*) Check this equality for $f(t)$ a rectangular pulse, a gaussian pulse (see Prob. 6-15), and $f(t) = e^{-a|t|}$.

6-21 (*a*) Using Eq. [6-70], show by averaging over θ_l that

$$E(x^2) = E(y^2) = N$$

(*b*) Show, similarly, that $E(x) = E(y) = 0$, and that $E(xy) = 0$, as well.

6-22 A noise process $n(t)$ has the spectral density $G_n(f)$ shown in Fig. P 6-22. $[E(n) = 0.]$

(*a*) Find the autocorrelation function and sketch for $f_0 \gg B$.
(*b*) Find the mean-squared value (power) of the process.
(*c*) $n(t)$ is written in the narrowband form $n(t) = x(t)\cos\omega_0 t - y(t)\sin\omega_0 t$. Sketch the spectral densities $G_x(f)$ and $G_y(f)$. What are $E(x^2)$ and $E(y^2)$?

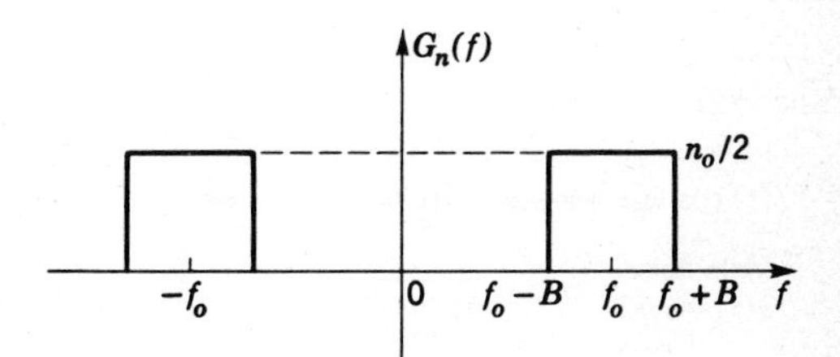

FIG. P 6-22

6-23 Find and sketch the spectral density of the derivative of the noise processes in Probs. 6-2 and 6-22. *Hint:* Differentiation is a linear operation and hence may be represented by a linear filter. What is $H(\omega)$ for this filter?

6-24 Compare the correlation functions for $n(t)$ and $x(t)$ in Prob. 6-22.

6-25 Consider the random process $z(t) = x \cos \omega_0 t - y \sin \omega_0 t$ where x and y are independent random gaussian variables with zero expected value and variance σ^2.

(*a*) Show that $z(t)$ is also normal (gaussian) with zero expected value and variance σ^2. Show the autocorrelation function $R_z(t, t + \tau) = \sigma^2 \cos \omega_0 \tau$, independent of t. (The process is therefore *wide-sense stationary.*) *Hint:* Set up formally the expression $E[z(t)z(t + \tau)]$, and then average over both x and y.
(*b*) Show that $z(t)$ can also be written as $r \cos (\omega_0 t + \theta)$. Find the joint density function of r and θ.
(*c*) Repeat (*a*) for $z(t) = x \cos (\omega_0 t + \varphi) - y \sin (\omega_0 t + \varphi)$ with φ a uniformly distributed $(0,2\pi)$ random variable independent of x and y.

6-26 Consider $z(t) = x(t) \cos \omega_0 t - y(t) \sin \omega_0 t$, where $x(t)$ and $y(t)$ are normal (gaussian), zero-mean, independent random *processes* with $R_x(\tau) = R_y(\tau)$. (They are therefore wide-sense stationary.)

(*a*) Show $R_z(\tau) = R_x(\tau) \cos \omega_0 \tau$. (It is therefore wide-sense stationary as well.) Distinguish between this problem and Prob. 6-25. *Hint:* Again average over both $x(t)$ and $y(t)$, as in Prob. 6-25.
(*b*) Find the power spectrum $G_z(f)$ in terms of the spectrum $G_x(f) = G_y(f)$.
(*c*) With $R_x(\tau) = \sigma^2 e^{-\alpha|\tau|}$ evaluate the spectra of $x(t)$ and $z(t)$. Sketch. Compare with Prob. 6-7.
(*d*) Show the density function of $z(t)$, at any time t, is zero-mean gaussian, with variance $R_x(0)$.

6-27 Consider the two random processes

$$s(t) = x(t) \cos \omega_0 t \qquad (1)$$
$$s(t) = x(t) \cos (\omega_0 t + \theta) \qquad (2)$$

$x(t)$ is a wide-sense stationary random process with $R_x(t, t + \tau) = R_x(\tau)$, and θ is a uniformly distributed $(-\pi,\pi)$ random variable independent of $x(t)$. Find the autocorrelation function for each process and show (1) is non-stationary (a function of t as well as τ), but that (2) is wide-sense stationary. *Hint:* In (2) one must average over θ as well as $x(t)$. Find and sketch the spectra for $x(t)$ and $s(t)$ in (2), if $R_x(\tau) = e^{-\alpha|\tau|}$.

6-28 A high-frequency PSK signal $f(t)\cos \omega_0 t$ plus gaussian white noise appear at the input to the receiver shown in Fig. P 6-28. The receiver consists of narrowband

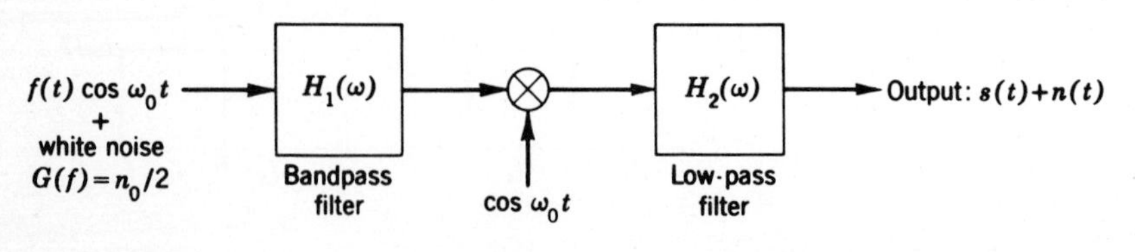

FIG. P 6-28

bandpass circuitry, a synchronous detector, and a low-pass filter before the decision circuitry at the output.

(*a*) The bandpass filter has a bandwidth large compared to the signal bandwidth (although small compared to the center frequency f_0). Characterize and sketch the amplitude characteristic $|H_2(\omega)|$ of the low-pass filter if the probability of error at the receiver output is to be minimized. [$f(t) = \pm a$: rectangular pulses T sec long.]

(*b*) The low-pass filter $H_2(\omega)$ has a bandwidth $\gg 1/T$. Characterize and sketch $H_1(\omega)$ if the probability of error at the receiver output is to be minimized.

(*c*) In both parts (*a*) and (*b*) above show the receiver output signal is $s(t) = \pm E$, while the output-noise power is $N = n_0E/2$ with $E = \int_0^T f^2(t)\,dt$ and $n_0/2$ the input-white-noise spectral density. Show that with equally likely binary signals the probability of error is given by

$$P_{e,\mathrm{PSK}} = \tfrac{1}{2}\,\mathrm{erfc}\sqrt{\frac{E}{n_0}}$$

This is then the matched-filter error probability for PSK signals (see Eq. [8-59] in Chap. 8).

(*d*) Filter $H_1(\omega)$ has a rectangular amplitude characteristic of bandwidth $\pm B$ about f_0. Filter $H_2(\omega)$ has a rectangular (ideal low-pass) amplitude characteristic of bandwidth B Hz. Find the peak SNR at the receiver output for $B = 0.5/T$, $0.75/T$, and $1/T$. *Hint:* Show that the ideal low-pass curve in Fig. 6-18 is applicable.

(*e*) $E/n_0 = 8$ db. Find the probability of error in part (*c*) and for the three bandwidths of part (*d*).

6-29 Repeat Prob. 6-28 for an OOK signal plus gaussian white noise at the input to the receiver of Fig. P 6-28. Show that the matched-filter error probability is in this case

$$P_{e,\mathrm{OOK}} = \tfrac{1}{2}\,\mathrm{erfc}\left(\frac{1}{2}\sqrt{\frac{E}{n_0}}\right)$$

Here $E = \int_0^T f^2(t)\,dt$ refers to the *on* signal.

6-30 Draw a block diagram for an FSK matched-filter receiver with synchronous detection. Show that with additive white gaussian noise at the receiver input and equally likely binary signals the probability of error for the system is

$$P_{e,\mathrm{FSK}} = \tfrac{1}{2}\,\mathrm{erfc}\sqrt{\frac{E}{2n_0}}$$

E and n_0 are the same as in Probs. 6-28 and 6-29. Calculate the probability of error for $E/n_0 = 8$ db and compare with the PSK and OOK results in Probs. 6-28 and 6-29 respectively. Replace the matched filter with the rectangular filters of Prob. 6-28*d* and evaluate the probability of error for the three bandwidths given there. Compare with the PSK and OOK results in Probs. 6-28 and 6-29.

6-31 An FSK system transmits 2×10^6 bits/sec. White gaussian noise is added during transmission. The amplitude of either signal at the receiver input is 0.45 μv, while the white-noise spectral density at the same point is $n_0/2 = \tfrac{1}{2} \times 10^{-20}$ volt^2/

Hz. Compare the probability of error for a receiver using synchronous detection with one using envelope detection. Assume matched-filter detection in both cases. Repeat for received signal amplitudes of 0.9 μv.

6-32 (*a*) Gaussian bandpass noise $n(t)$ with power spectral density $G_n(f) = n_0/2$, $f_0 - B < |f| < f_0 + B$, zero elsewhere, as in Fig. P 6-22, is applied to an envelope detector. Find the probability-density function of the detector output $r(t)$, as well as its dc and rms values.

(*b*) Repeat for another noise process $n(t)$ with spectral density n_0 over the range of frequencies $f_0 - 2B < |f| < f_0 + 2B$, zero elsewhere. Sketch the two probability-density functions on the same scale, and compare dc and rms values.

6-33 Zero-mean bandpass gaussian noise $n(t)$, as in Prob. 6-32*a* above, is applied to the *RC* filter shown in Fig. P 6-33. The output noise $n_o(t)$ is then gaussian as well. Find

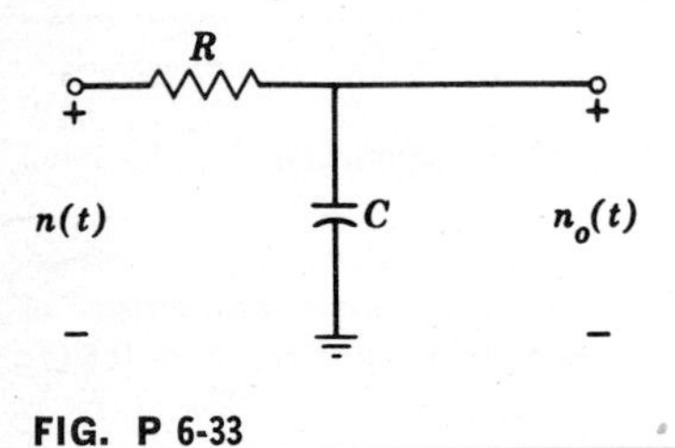

FIG. P 6-33

and sketch the power spectral density of $n_o(t)$ for various values of *RC*. Find the mean and mean-squared value of $n_o(t)$ as a function of *RC*. Write the probability-density function of $n_o(t)$ for some value of *RC*.

6-34 What is the probability that the envelope of narrowband gaussian noise will exceed three times its rms value?

6-35 Gaussian bandpass noise with mean-squared value N is detected by a quadratic (or square-law) envelope detector, whose output is proportional to the square of the instantaneous envelope voltage. Thus the detector output voltage $z(t)$ is cr^2, as shown in Fig. P 6-35, with c a constant of proportionality and r the envelope voltage.

(*a*) Show that the probability-density function of z at any time t is

$$f(z) = \frac{e^{-z/2cN}}{2cN} \qquad z \geq 0$$

(*b*) Calculate the voltages at the output of the detector that would be read by a long-time-constant dc meter, a long-time-constant rms meter, and a long-time-constant rms meter preceded by a blocking capacitor.

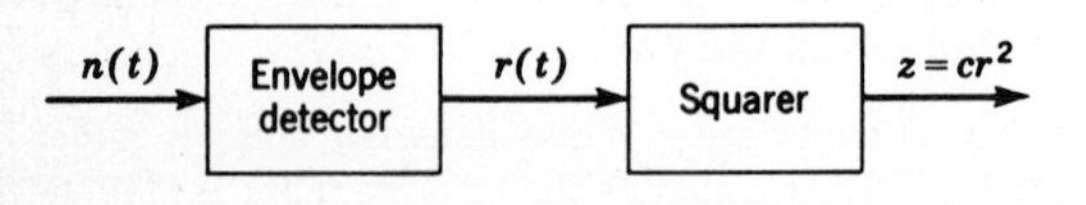

FIG. P 6-35

6-36 A non-phase-coherent communication receiver consists of a narrowband tuned amplifier and a quadratic envelope detector. The receiver accepts signal pulses of fixed amplitude. To distinguish between incoming signals and noise present in the receiver, a specific voltage level at the output of the detector is chosen such that all voltages exceeding this level are called signal.

(*a*) What is the probability that noise in the absence of signal will be mistaken for signal if the level is set at $2N$, with N the mean-squared noise voltage at the detector input? (Assume that the detector constant of proportionality is 1 volt/volt^2.)

(*b*) Plot the probability of error due to noise if the voltage level is varied from 0 to $20N$.

6-37 A received fading signal $s(t)$ may be written in the narrowband form of Chap. 5:

$$s(t) = A_x(t) \cos \omega_0 t - A_y(t) \sin \omega_0 t$$

where $A_x(t)$ and $A_y(t)$ are each slowly varying zero-mean gaussian processes with $E(A_x)^2 = E(A_y^2) = \sigma^2$. A_x and A_y may be assumed independent of one another. The signal has narrowband noise $n(t)$ added to it, so that the sum $v(t)$ is given by

$$v(t) = s(t) + n(t)$$

The mean power in $n(t)$ is N.

(*a*) Show that the probability distribution of the envelope of $v(t)$ is Rayleigh with parameter $\sigma^2 + N$.

(*b*) The fading signal $s(t)$ represents the signal reflected from a target in a simple radar system. Detection is accomplished by envelope detecting $v(t)$ and using a threshold level of b volts at the detector output. Indicate how you would determine this level and what mean SNR σ^2/N is required if the probability of mistaking noise alone for signal is to be 10^{-5} while the probability of detecting a signal is to be 0.99.

6-38 The probability of error for a noncoherent FSK system with Rayleigh fading, as given by Eq. [6-98], is to be verified by an alternative method. As shown in Prob. 6-37*a* the envelope of a Rayleigh-fading signal plus narrowband noise is itself Rayleigh. Using this result, calculate the probability that the envelope-detected output r_1 of the FSK channel containing the (fading) signal plus noise is less than the corresponding output r_2 of the channel containing noise alone. Show this result is exactly the probability of error given by Eq. [6-98].

6-39 A simple radar system is to be designed, using a threshold-detection criterion. The probability that noise alone will be mistaken for signal is specified at 10^{-10}. What must the average SNR at the input to the envelope detector be to keep the probability of missing the signal pulse, when it appears, to 0.1? Assume that the SNR and threshold level are high enough so that the gaussian-density-function approximation to the distribution of signal plus noise may be used. Verify this.

6-40 The receiver of the radar system of Prob. 6-39 has a noise spectral density of $n_0/2 = 2 \times 10^{-20}$ watts/Hz measured at the antenna input. The transmitted radar pulses are 1 μsec wide. (The pulse width is normally fixed by the required resolution of targets in space.) Calculate the peak signal power required at the antenna to attain the specifications of Prob. 6-39. (This may in turn be used to calculate the peak transmitter power required if the system antenna gain and the target range and characteristics are specified.) *Note:* Matched-filter detection may be assumed.

6-41 Consider the density function $f(z)$ of noise at the output of a quadratic envelope detector derived in Prob. 6-35*a*. Find the characteristic function for this density function. Using this result, find the density function of the sum of n such independent noise samples (see Prob. 5-50).

Use this density function in turn to find the probability that the sum of n independent samples of noise, each measured at the output of a quadratic envelope detector, will exceed kN volts. N is the mean-squared voltage of each of the noise samples measured at the input to the quadratic envelope detector. Assume that the detector constant is 1. (Note that the results of Probs. 6-35 and 6-36 are a special case of the results of this problem, with $n = 1$.)

Plot the probability versus the parameter k for $n = 1$, 2, and 5. This gives the probability of error due to noise in the n-pulse radar system using threshold-level detection that is discussed in Sec. 6-8. It applies to the special case of a quadratic envelope detector. Further analysis, including the case where signal plus noise is present, shows that the final radar performance curves of Fig. 6-40 are very nearly independent, however, of the type of envelope detector chosen.

6-42 Carry out the indicated statistical averaging in Eq. [6-122] to obtain Eq. [6-123].

6-43 Assume that an unmodulated carrier plus narrowband noise are synchronously detected. Show that the output SNR is proportional to the input CNR for *all* values of CNR.

6-44 (*a*) Gaussian bandpass noise $n(t)$ with power spectral density $G_n(f) = n_0/2$, $f_0 - B < |f| < f_0 + B$, zero elsewhere, is applied to the quadratic envelope detector of Fig. P 6-35. (Assume the detector constant $c = 1$.) Find the spectral density of the output noise $z(t)$ by first finding the correlation function $R_z(\tau)$ in terms of the correlation function of the input $x(t)$ and $y(t)$. *Hint:* $r^2 = x^2 + y^2$. Assume x and y independent. Recall that $R_x(\tau) = R_y(\tau)$. Use Eq. [6-135] for gaussian processes to relate $R_{x^2}(\tau)$ to $R_x(\tau)$. The spectral density of $G_x(f)$ is given in terms of the input noise spectral density by Eq. [6-76].

(*b*) Repeat part (*a*), using the procedure of Sec. 6-10. Thus assume quadratic envelope detection is equivalent to squaring the input $n(t)$ and passing $n^2(t)$ through a low-pass filter. Show that the resultant output spectrum is the same as that found in (*a*) above, except for the $\frac{1}{4}$ constant factor noted in Secs. 6-9 and 6-10.

6-45 An FM receiver consists of an ideal bandpass filter of 225-kHz bandwidth centered about the unmodulated carrier frequency, an ideal limiter and frequency discriminator, and an ideal low-pass filter of 10-kHz bandwidth in the output. The ratio of average carrier power to total average noise power at the input to the limiter is 40 db. The modulating signal is a 10-kHz sine wave that produces a frequency deviation Δf of 50 kHz.

(*a*) What is the signal-to-noise ratio S_o/N_o at the output of the low-pass filter?

(*b*) A deemphasis network with a time constant $rC = 75$ μsec is inserted just before the output filter. Calculate S_o/N_o again.

(*c*) Repeat (*a*) and (*b*) if the modulating signal is a 1-kHz sine wave of the same amplitude as the 10-kHz wave. Repeat with the amplitude reduced by a factor of 2. (The carrier amplitude and filter bandwidths are unchanged.)

6-46 An audio signal is to be transmitted by either AM or FM. The signal consists of either of two equal-amplitude sine waves: one at 50 Hz, the other at 10 kHz. The amplitude is such as to provide 100 percent carrier modulation in the AM case and a frequency deviation $\Delta f = 75$ kHz in the FM case.

The average carrier power and the noise-power density at the detector input are the same for the AM and FM systems.

(*a*) Calculate the transmission bandwidth and output filter bandwidth required in the AM case and the FM case. The carrier-to-noise ratio is 30 db for the AM system. Calculate the output S_o/N_o for each sine wave for each of the systems. (Preemphasis and deemphasis networks are not used.)
(*b*) Repeat (*a*) if the amplitude of the 10-kHz wave is reduced by a factor of 2.

6-47 The ratio of average carrier power to noise spectral density, S_c/n_0, at the detector input is 4×10^6 for both an FM and an AM receiver. The AM carrier is 100 percent modulated by a sine-wave signal, while a sine-wave signal produces a maximum deviation of 75 kHz of the FM carrier.

Calculate and compare the output signal-to-noise ratio for both the FM and the AM receivers if the bandwidth B of the low-pass filter in each receiver is successively 1, 10, and 100 kHz. Deemphasis networks are not used. (Assume that the r-f bandwidths are always at least twice the low-pass bandwidth.)

6-48 The low-pass filter bandwidth of an AM and an FM receiver is 1 kHz. The ratio of average carrier power to noise spectral density, S_c/n_0, is the same in both receivers and is kept constant. The AM carrier is 100 percent modulated.

Calculate the relative SNR improvement of the FM over the AM system if the frequency deviation Δf of the FM carrier is 100 Hz; 1 kHz; 10 kHz.

6-49 The time constant of an *RC* deemphasis network is chosen as 75 μsec. Plot the improvement in output SNR due to the deemphasis network for both FM and AM detectors as a function of the bandwidth of the output low-pass filter. (Assume that S_c/n_0 is the same and constant in both cases.) What is the minimum filter bandwidth to be used if audio signals from 0 to 10 kHz are to be passed?

6-50 The signal pulses in a PPM system have a peak amplitude of 1 volt. The pulse rise time is 0.15 μsec. The maximum modulation displacement is 1 μsec. Noise of value 0.1 volt rms is added to these pulses during transmission. Compare the SNR after demodulation with the SNR during transmission. What bandwidth is required for transmission?

6-51 The received signal $s(t) = m(t) \cos(\omega_c t + \theta) - \hat{m}(t) \sin(\omega_c t + \theta)$ is embedded in white gaussian noise with spectral density $n_0/2$. [$m(t)$ is the message; $\hat{m}(t)$ its Hilbert transform.]

(*a*) Show that $s(t)$ is a SSB signal.
(*b*) Assuming that θ is known, draw a block diagram of a synchronous demodulator, including appropriate low-pass filtering.
(*c*) Find the output SNR if $m(t)$ is random, $E(m) = 0$, $E(m^2) = S$.

6-52 The output of a "jittery" oscillator is described by $g(t) = \cos(\omega t + \theta)$, where ω and θ are independent random variables with the following probability-density functions:

1. θ uniform $(0,2\pi)$.
2. ω has a probability-density function $f_\omega(\omega)$.

Show the power spectral density of $g(t)$ is given by

$$G_g\left(\frac{\omega}{2\pi}\right) = \frac{\pi}{2} f_\omega(\omega)$$

Sketch for $f_\omega(\omega)$ gaussian, with expected value ω_0. *Hint:* Write the autocorrelation function for $g(t)$, averaging over *both* ω and θ. Show $R_g(\tau) = \frac{1}{2}E(\cos \omega\tau)$, referring to expectation over ω. Write out the form for this expectation in terms of $f_\omega(\omega)$, then compare with the expression for the Fourier transform of $G_g(f)$.

6-53 The output SNR and input SNR of an FM discriminator *above threshold* are related by the equation

$$\frac{S_o}{N_o} = 3\beta^2(\beta + 1)\frac{S_i}{N_i}$$

with β the modulation index. The complete SNR characteristic is shown in Fig. P 6-53. An FM signal at the discriminator input has a power of 55 mw. The input

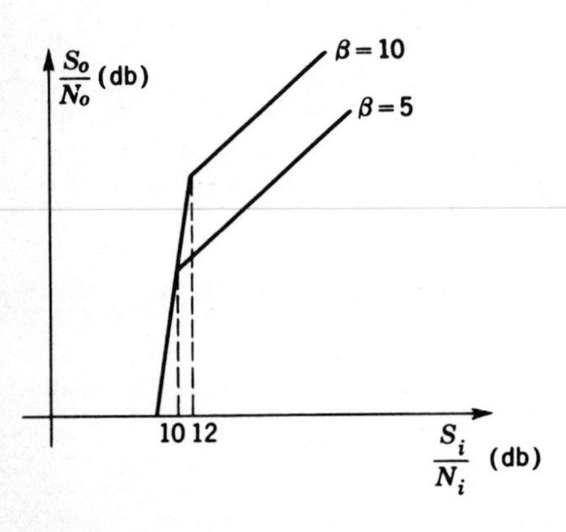

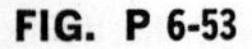
FIG. P 6-53

noise is white with spectral density $n_0/2 = 0.25 \times 10^{-10}$ watts/Hz. The maximum modulation frequency is 5 MHz. The frequency deviation is 25 MHz. If the bandwidth following the discriminator is B Hz, while that preceding the discriminator is $2(\beta + 1)B$, find

(*a*) S_i/N_i in decibels.
(*b*) S_o/N_o in decibels.
(*c*) Repeat if the deviation is increased to 50 MHz.

6-54 A communication system produces at its output an impulse term plus random noise given by

$$e_0(t) = 2\pi\delta(t - t_i) + n(t)$$

The random noise has power spectral density

$$G_n(f) = kf^2$$

up to very high positive and negative frequencies. (This is a model of impulse and random noise occurring at the output of FM receivers.)

This output voltage is to be passed through a linear filter $H(\omega)$ designed to optimize the recognition of the impulse term. Determine the expression for $H(\omega)$ which maximizes, at time $t_i + t_0$, the ratio of the output due to the impulse to the rms output due to $n(t)$. *Hint:* Use the Schwarz inequality here.

6-55 Band-limited white gaussian noise with spectral density $n_0/2$, $-B < f < B$, is passed through a square-law detector whose output is $n^2(t)$, and then through an ideal low-pass filter of bandwidth B.

Determine the power spectrum of $n^2(t)$, the spectrum at the low-pass filter output, and the rms value of the output noise.

6-56 Show that the spectral density in Eq. [6-207] is correct.

CHAPTER SEVEN

PHYSICAL SOURCES OF NOISE

In the previous chapter we postulated the existence of noise sources in communication systems. Focusing attention on the defined autocorrelation function as one measure of the degree of dependence of successive noise samples on each other, we were able to develop extremely useful information as to how a random noise wave is affected by linear systems through which it passes. In particular, defining the spectral density $G_n(f)$ as the Fourier transform of the autocorrelation function we found the noise spectral densities at the input and output of a linear circuit to be related by

$$G_o(f) = |H(\omega)|^2 G_i(f) \tag{7-1}$$

Here $H(\omega)$ is the frequency transfer function of the linear circuit.

In the special case where the input noise is white, with spectral density $G_i(f) = n_0/2$, we have

$$G_o(f) = \frac{n_0}{2} |H(\omega)|^2 \tag{7-2}$$

The total noise power at the output of the system is then

$$N = \int_{-\infty}^{\infty} G_o(f)\, df \tag{7-3}$$

It was this noise power N that was found to play the dominant role in signal-to-noise calculations as well as calculations of error probability in digital transmission.

In all this previous work of Chap. 6 the noise power N (or equivalently, the input-white-noise spectral density $n_0/2$) was assumed known. In many real situations specific values for N (or $n_0/2$) may actually be obtained by measurement. The discussion at the beginning of Chap. 6 indicates how one might use true power meters or digital processing techniques for this purpose.

It is, however, quite important to be able to make some estimate from first principles of the noise power in a system. Singling out the critical sources of noise, one can not only calculate their effect in a system but attempt to minimize this effect by proper system design. We therefore propose in this chapter to discuss actual sources of noise, obtaining quantitative measures for the noise they produce. This enables us in many cases to actually calculate or estimate the white-noise spectral density, $n_0/2$, at a system input and hence the total noise power N at the output.

In this chapter we focus attention on the two most common sources of noise, *thermal noise and shot noise.* We shall show that over most frequency ranges of interest these sources may be approximated quite well as white-noise sources.

Just one word of caution, however. The term *noise* is used generally to refer to any spurious or undesired disturbances that tend to observe or mark the signal to be transmitted. Various kinds of noise can be distinguished—the so-called "man-made" kind, the erratic disturbances normally beyond our control and occurring irregularly, and the type due to "spontaneous fluctuations." Although all three categories occur in real life and tend to limit system performance capabilities, we shall concentrate here only on the third category.

There are several reasons for this. First, the man-made type of disturbance can with appropriate case be minimized or even eliminated. Examples of that type of disturbance include electromagnetic pickup of another interfering signal; radiation from nearby electrical sources (electrical appliances, ignition radiation, etc.); pickup through a power supply; spurious voltages generated in a balanced modulator; mechanical vibrations converted to electrical disturbances (noisy relay contacts for example), etc. There are literally hundreds of examples of such spurious disturbances that could be listed. They all have the property, however, that their effects can be eliminated or at least minimized by relocating the communication system, by proper shielding, by filtering circuits, by improving the mechanical and electrical design, etc.

Second, both this "man-made" noise and the erratic type (due to electrical storms in the atmosphere, to sudden and unexpected voltage sources, etc.), by their very definitions, do not occur continuously. They may appear, disappear, and reappear. This makes it very difficult to specify the quantitative effects of this type of noise. In practice these categories of noise are subsumed under the titles impulse, or *burst noise.* They may be catastrophic in their effects on communication systems when they appear (this is particularly true of digital communications), but their erratic behavior makes them very difficult to analyze.[1]

[1] The interested reader is referred to the periodical literature for studies of impulse noise and its effect on systems. See, e.g., the *IEEE Transactions on Information Theory* and the *IEEE Transactions on Communication Technology,* for occasional discussions.

The type of noise we concentrate on here, that due to "spontaneous fluctuations" of voltage, current, or their analogs in linear systems, is *ever present* and so represents a basic limitation on the transmission of information. Spontaneous fluctuations occur generally throughout our physical world and are due to the noncontinuous, or granular, character of physical quantities. The thermal noise and shot noise mentioned above are the two most prominent classes of this type of noise.[1]

Thermal noise is associated with random motion of particles in a force-free environment. For example, the air pressure in a given room is due to the summed effect of countless air molecules moving chaotically in all directions. The molecules are in continuous turbulent motion, striking and rebounding from one to another. When we talk of the "pressure" at a point we refer in effect to the resultant force per unit area of all molecules striking a surface located at the point in question. This force will fluctuate or vary in time as fewer or more molecules strike the wall from time to time. Since the numbers of molecules involved are ordinarily tremendously large for normal-sized surfaces, the *average* force in time will remain constant so long as the average molecular energy remains constant. The instantaneous force as a function of time, however, will vary randomly about this average value. Increasing the temperature increases the molecular energy, and the average pressure goes up, as do the fluctuations about the average. Thermodynamics and the kinetic theory of heat applied to this problem indicate that for an ideal gas (one for which intermolecular forces may be neglected), the average kinetic energy of motion per molecule in any one direction is $kT/2$, with $k = 1.38 \times 10^{-23}$ joule/°K the Boltzmann constant, and T the absolute temperature in degrees Kelvin. We shall thus find the mean-squared thermal noise proportional to kT.

In electrical work we encounter similar fluctuations which are thermally induced. Conductors contain a large number of so-called "free" electrons and ions strongly bound by molecular forces. The ions vibrate randomly about their normal (average) positions, however, this vibration being a function of temperature. Collisions between the free electrons and the vibrating ions continuously take place. There is a continuous transfer of energy between electrons and ions. This is the source of the resistance in the conductor. The freely moving electrons constitute a current, which over a long period of time averages to zero, since as many electrons on the average move in one direction as another. (This is the

[1] Comprehensive references to the material of this chapter include the following books: W. R. Bennett, "Electrical Noise," McGraw-Hill, New York, 1960; D. A. Bell, "Electrical Noise, Fundamentals and Physical Mechanism," Van Nostrand, London, 1960; A. Van der Ziel, "Noise," Prentice-Hall, Englewood Cliffs, N.J., 1954.

analog of the pressure case noted above, in which the *average* molecular velocity is zero, although chaotic motion exists.) There are random fluctuations about this average, however, and, in fact, we shall see that the mean-squared fluctuations in current are proportional to kT also.

Both cases noted—pressure fluctuations and current fluctuations—deal with the chaotic motion of particles (molecules or electrons) possessing thermal energy. There are no forces present "organizing" this motion in preferred directions. Both cases may therefore be treated by equilibrium thermodynamics, with the mean-squared fluctuations found proportional to kT.

The second type of noise that we shall discuss, *shot noise*, is also due to the discrete nature of matter, but here we assume an average flow in some direction taking place: electrons flowing between cathode and anode in a cathode-ray oscilloscope or vacuum tube, electrons and holes flowing in semiconductors, photons emitted in some laser systems, photoelectrons emitted in photodiodes, fluid moving continuously under the action of a pressure gradient, etc. Although averaging over many particles we find the average flow, or average number moving per unit time, to be a constant, there will be fluctuations about this average. The mechanism of the fluctuations depends on the particular process; in the vacuum-tube case it is the random emission of the electrons from the cathode, in the semiconductor it is the randomness in the number of electrons that continually recombine with holes, or in the number that diffuse, etc. Thus the processes that give rise to an average flow have statistical variations built in, producing fluctuations about the average. We shall find that in all these cases the mean-squared fluctuations about the average value are proportional to the average value itself, so that *shot noise* is characterized by a dependence of the *noise* on the *average value*. (Because these processes involve forces producing the flow of particles, they are in nonequilibrium, thermodynamically speaking, so that they are difficult to treat using classical thermodynamics.)

We shall consider shot noise in more detail first, and then move to the case of thermal noise; finally, we shall consider situations (such as noise in semiconductors) in which *both* types of fluctuation occur simultaneously.

7-1 SHOT NOISE

We are basically interested here in developing an expression for the spectral density of a shot-noise source. As noted earlier, this type of noise involves random fluctuations about an average particle flow. Interestingly the expression we shall come up with is that the current spectral density in the case of electrical charge flow is given by the simple relation

$$G(f) = eI_{dc} \equiv \frac{n_0}{2} \qquad [7\text{-}4]$$

where e is the electron charge, and I_{dc} is the average or dc current flowing. (Analogous expressions arise when nonelectrical particle flow is considered.) The white-noise form of this expression is only an approximation, as was noted earlier.

To develop this simple equation relating mean-square current fluctuations about the dc value to the dc value two approaches will be used. The first uses time averaging to determine mean-squared fluctuations. The second uses statistical or ensemble averaging. Recall from the discussion in Chap. 6 that if the underlying process is ergodic, the two types of averages may be equated. (In the case of the time average we must assume the averaging time T becoming very large. In practice of course it generally suffices to have $BT \gg 1$, with B the bandwidth of the process under investigation.)

TIME-AVERAGE APPROACH

In this approach we shall calculate the mean-square fluctuations in a frequency band B Hz wide, show this is proportional to B, and use this result to obtain the white-noise form, Eq. [7-4], for the noise spectral density. For simplicity we shall focus attention at first on shot noise in a temperature-limited diode. Extensions can then be made to shot noise in semiconductors (this is done in the ensemble-average case of the next paragraph), photomultipliers, etc.

Consider then the simple diode circuit of Fig. 7-1. If we were to

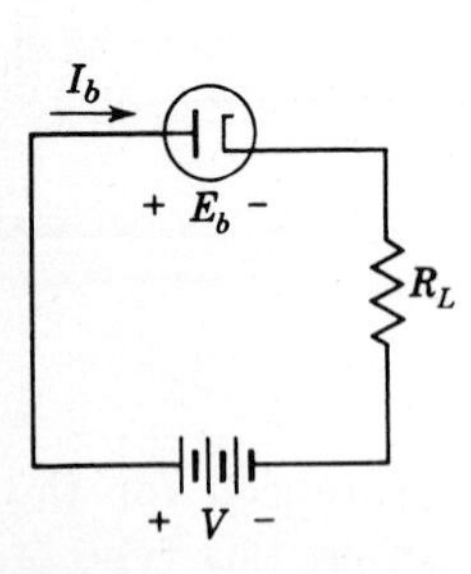

FIG. 7-1 Diode circuit.

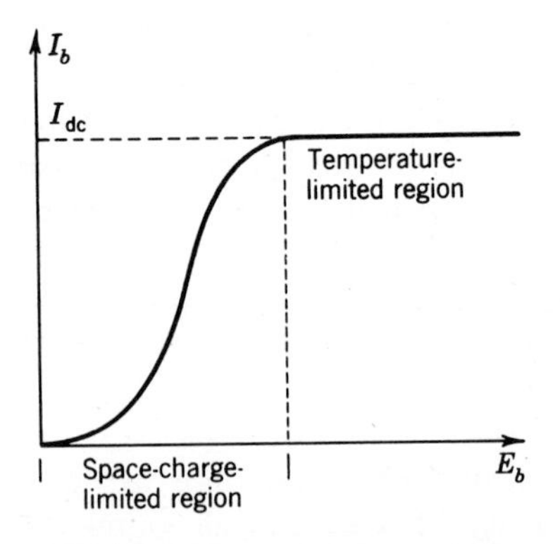

FIG. 7-2 Diode current-voltage characteristic.

vary the diode plate–cathode voltage E_b with the filament temperature fixed, we would obtain the usual diode characteristic of Fig. 7-2.

Heat supplied by the filament causes electrons to be boiled off, or emitted from, the cathode. If the plate voltage is high enough, all the electrons emitted are swept across to the plate and the plate current saturates at the value of I_{dc}, given by the average number of electrons emitted per second. This depends of course on the temperature at the cathode. For smaller values of E_b not all the electrons receive enough energy to reach the anode. They remain in the vicinity of the cathode and constitute a space-charge cloud, or "virtual cathode," which serves to limit the number of electrons reaching the plate. In this space-charge-limited region[1]

$$I_b \doteq CE_b^{3/2}$$

The current flow is always due to the emission of electrons from the cathode. Although we talk of a specified current I_b or I_{dc}, the emission of electrons is actually "erratic," or random in form—more electrons are emitted one instant, fewer the next, with the *average* number per unit time the same over a long period of time.

In the temperature-limited case this average number and the resulting average current I_{dc} are functions of the cathode temperature.

A very sensitive and fast-responding current meter placed in the diode circuit would trace out a current-time curve like that of Fig. 7-3 as

[1] We neglect, in all this section, second-order effects due to initial velocities, high-field emission, retarding fields, etc. The derivation of the three-halves-power law above can be found in many electronics books. See, e.g., T. S. Gray, "Applied Electronics," chap. 3, Wiley, New York, 1954; W. B. Davenport, Jr. and W. L. Root, "Introduction to Random Signals and Noise," chap. 9, McGraw-Hill, New York, 1958.

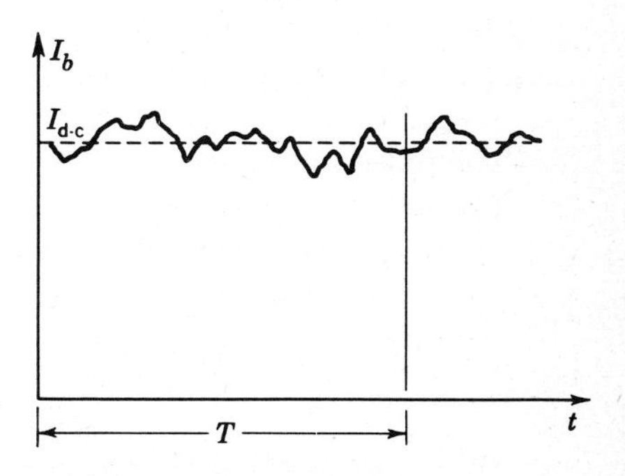

FIG. 7-3 Random current fluctuations in diode.

a function of time. Because of the random emission of the electrons the current fluctuates randomly about its average value I_{dc}.

Using a time average, as in Chap. 6, the desired mean-square variations about I_{dc}, $\overline{i_b^2}$, can be written

$$\overline{i_b^2} \equiv \overline{(I_b - I_{dc})^2} = \frac{1}{T}\int_0^T (I_b - I_{dc})^2\, dt \qquad [7\text{-}5]$$

The quantity $\overline{i_b^2}$ is the same as the term $\overline{n^2} = N$ used in Chap. 6. (The small letters on averaging will always imply variations about some constant quantity.)

The total *instantaneous* current I_b represents the superposition of the number of electrons leaving the cathode and hitting the plate over a given time interval, while I_{dc} is the average over a much larger interval. Each electron leaving the cathode thus contributes to the total current. We shall make the *basic assumption*, for temperature-limited conditions, that each electron contributes independently to the current. That this assumption is valid can be checked only by comparing the calculated results with experiment.

On this assumption we can calculate the power dissipated in the resistor by the current due to one electron leaving the cathode (if one can stretch the imagination that far!) and sum over all the electrons in a given time interval to get the total average power dissipated.[1] Part of this power will be the dc power $I_{dc}^2R_L$. The remainder will be just our desired $\overline{i_b^2}R_L$.

Consider, then, one electron leaving the cathode. As the electron moves from the cathode to the plate, it induces a current i_{b1} in the plate circuit. One possible representation for i_{b1} is the current pulse of Fig. 7-4*a*. τ is the transit time, or time taken for the electron to move from the cathode to the anode.

As we shall see very shortly, the actual shape of the current pulse is unimportant. Important are two factors:

1. The pulse lasts τ sec.
2. The area under the curve is e, the charge of a single electron, for e coulomb is ultimately deposited on the plate. (Another form for the pulse might have been a rectangular pulse of amplitude e/τ amp and lasting τ sec.)

[1] Recall from Chap. 5 that the variances of independent random variables add when the variables are summed. Expected, or dc, values also add.

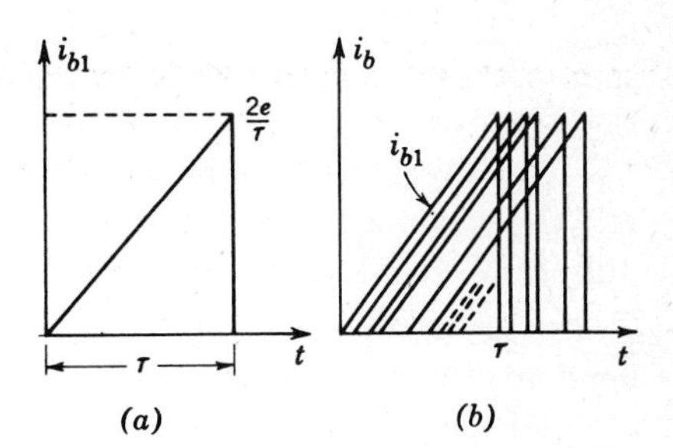

FIG. 7-4 Variations in plate current due to random occurrence of current pulses. (a) Plate-current pulse due to one electron (τ = transit time); (b) overlapping current pulses and their random occurrence.

If any initial velocity the electron might have is neglected, the transit time is simply given by[1]

$$\tau = \sqrt{\frac{2m}{eE_b}}\, d_s \qquad [7\text{-}6]$$

where E_b is the plate-cathode voltage, d_s the plate-cathode spacing (in meters), and m the mass of the electron (in kilograms). For example, taking d_s as 1 mm, E_b as 10 v, e/m as 1.96×10^{11} coulombs/kg,

$$\tau \doteq 10^{-9} \text{ sec}$$

But

$$I_{\text{dc}} = \frac{re}{T} \qquad [7\text{-}7]$$

if r electrons leave the cathode in T sec on the average. For $I_{\text{dc}} = 1$ ma, $r/T = 6 \times 10^{15}$ electrons per second, or in $\tau = 10^{-9}$ sec there are 6×10^6 electrons leaving the cathode.

There is thus a tremendous overlap of current pulses. In this simple example 6×10^6 current pulses would be initiated in the time taken for one to be completed. The tremendous numbers involved give us the right to use averaging and statistical concepts where needed. The overlap of current pulses is shown in Fig. 7-4*b*. The total plate current is the

[1] This assumes a uniform electric field. The equation can be derived by equating the kinetic energy of the electron at the plate to the decrease in potential energy in moving across the tube. The average speed of the electron, from which τ can be found, is just one-half the final speed for a uniform field.

sum of these randomly occurring pulses and thus varies randomly about I_{dc} (Fig. 7-3).

Our procedure will be simply to choose an interval $T \gg \tau$ and calculate the mean-squared current deviation about the average value due to one pulse. We shall then add the mean-square deviations due to r pulses in T sec to obtain $\overline{i_b^2}$. (Since the pulses are assumed statistically independent, their power contributions must add.)

To find the mean-squared current deviation due to one pulse, we consider the pulse repeated over T-sec intervals and expand in a Fourier series (Fig. 7-5). Thus

$$i_{b1}(t) = \frac{1}{T}\left[c_0 + 2\sum_{n=1}^{\infty} |c_n| \cos\left(\frac{2\pi nt}{T} + \theta_n\right)\right] \qquad [7\text{-}8]$$

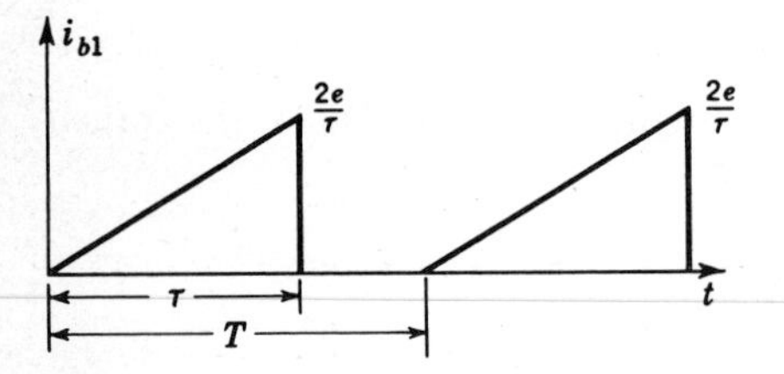

FIG. 7-5 Calculation of pulse Fourier series.

For example, if a triangular pulse,

$$|c_n| = e\left[\frac{\sin(n\pi d/2)}{n\pi d/2}\right]^2 \qquad d = \frac{\tau}{T}, \; n = 0, 1, 2, 3, \ldots$$

and if a rectangular pulse,

$$|c_n| = e\left|\frac{\sin n\pi d}{n\pi d}\right| \qquad \text{etc.}$$

But no matter what the shape of the pulse, the "bandwidth" of the pulse spectrum is approximately $1/\tau$ (see Chap. 2). For our previous example this would be 10^9 Hz. (Other choices of numbers might have given 10^8 Hz.) For $f \ll 1/\tau$, then, the spectrum amplitude $|c_n|$ is very nearly constant and equal to e,

$$|c_n| = e \qquad f \ll \frac{1}{\tau} \qquad [7\text{-}9]$$

This of course agrees with the discussion of the impulse, or delta, function in Chap. 2. We showed there that many types of pulses approached an impulse function for very small time duration. The Fourier transform, or frequency spectrum, of such pulses is just a constant equal to the area of the pulse. We are essentially representing the individual currents, due to individual electrons here, as impulses. The spectrum is thus independent of the actual pulse shape up to very high frequencies. If we include Fourier components within a frequency range $\Delta f \ll 1/\tau$,

$$i_{b1,\Delta f} = \frac{e}{T}\left[1 + 2\sum^{\Delta f} \cos\left(\frac{2\pi nt}{T} + \theta_n\right)\right] \tag{7-10}$$

In this paragraph we discuss low-frequency (l-f) shot-noise phenomena only, so that we need consider only $\Delta f \ll 1/T$. Higher-frequency shot-noise phenomena involving transit time effects may be incorporated by assuming the current impulses band-limited by some transfer function $H(\omega)$. The overall effect is to introduce an equivalent bandwidth $B \doteq 1/\text{transit time}$. This approach is considered in the next paragraph using the ensemble-average approach. These transit time effects are treated in detail in a book by A. Van der Ziel.[1]

On this assumption—that we restrict ourselves at this point to frequencies far below the reciprocal of the transit time—the mean-squared deviation about direct current due to one current pulse is simply the sum of all the mean-squared cosine terms of Eq. [7-10]. The number of terms to be included in the Δf interval is $\Delta f/(1/T)$ (Fig. 7-6). Thus

$$\begin{aligned} \overline{i_{b1}^2} &\equiv \overline{[i_{b1} - (\text{direct current})]^2} \\ &= \left(\frac{2e}{T}\right)^2 \frac{1}{2}\left(\frac{\Delta f}{1/T}\right) = 2\frac{e}{T}\,e\,\Delta f \end{aligned} \tag{7-11}$$

where the term ½ comes from averaging the square of the cosine term in Eq. [7-10]. Adding the mean-squared deviations due to r electrons

[1] A. Van der Ziel, "Noise," Prentice-Hall, Englewood Cliffs, N.J., 1954.

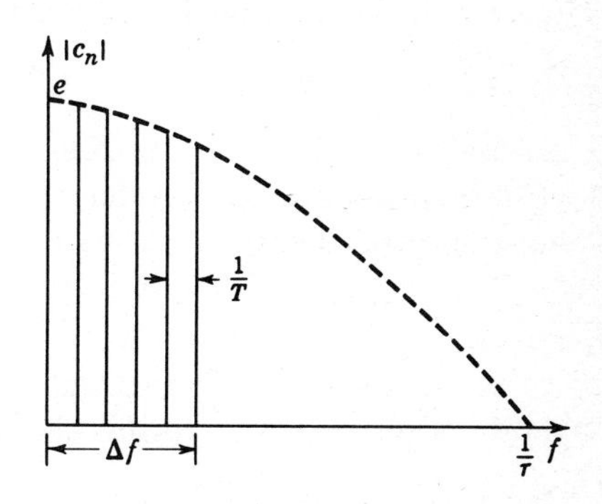

FIG. 7-6 Typical pulse spectrum.

in T sec,

$$\overline{i_b^2} \equiv \overline{(I_b - I_{dc})^2} = 2r\frac{e}{T}e\,\Delta f \tag{7-12}$$

But $$I_{dc} = \frac{re}{T}$$

Therefore $$\overline{i_b^2} = 2eI_{dc}\,\Delta f \tag{7-13}$$

Equation [7-13] is the so-called *shot-effect formula* for the diode mean-squared current fluctuations under *temperature-limited conditions.* Note again that this is an l-f equation only, valid for frequencies where transit-time effects are negligible. The mean-squared noise current is thus proportional to the plate current and to the frequency bandwidth. The rms noise current is proportional to the *square root* of current and bandwidth. *Widening the bandwidth in circuit applications increases the amount of noise power available.*

As an example of numbers involved, take I_{dc} as 1 ma, Δf as 5 kHz. Then

$$\overline{i_b^2} = 2(1.6 \times 10^{-19}) \times 10^{-3} \times (5 \times 10^3) = 1.6 \times 10^{-18}$$

The rms current generated is thus 1.26×10^{-9} amp rms. If this current flows through a 5-kilohm resistor, the rms noise voltage due to this current (we are neglecting at this point additional noise generated in the resistor) is 6.3 μv rms.

If the bandwidth is now quadrupled to 20 kHz the rms noise voltage doubles to 12.6 μv rms. Equation [7-13] has been checked experimentally many times and has even been used for precise measurements of the electron charge e.†

From the development of the expression for the mean-squared current fluctuation noise $\overline{i_b^2}$ it is apparent that the corresponding current spectral density is just $\overline{i_b^2}/2\,\Delta f$ (the factor of 2 is needed to ensure a spectral density defined over positive *and* negative frequencies). Thus,

$$G(f) = eI_{dc} \tag{7-4}$$

precisely the expression noted earlier. If shot noise is then the *only* form of noise generated in a system, this is just the expression for the white-noise spectral density $n_0/2$, used throughout Chap. 6.

$$n_0 = 2eI_{dc} \tag{7-14}$$

As was done previously in Chap. 6, we can now proceed to calculate the total noise power, due to this shot-noise term at the output of any linear

† E. B. Moullin, "Spontaneous Fluctuations of Voltage," Oxford, New York, 1938.

system, using the techniques of Chap. 6. For the example given above, with $I_{dc} = 1$ ma, $n_0 = 3.2 \times 10^{-22}$ amp²/Hz of bandwidth.

STATISTICAL AVERAGING: POISSON PROCESS

The simple expression obtained above for shot-noise spectral density may be derived as well, as noted earlier, using statistical (ensemble) averaging. There are several reasons for doing this:

1. The statistical approach is much more general, enabling one to apply and extend the results to many more physical situations than that of current emission in a vacuum diode,
2. It provides an interesting example of an ergodic process, in which time and statistical averaging may be interchanged,
3. The analysis depends, as we shall see, on evaluating autocorrelation functions and provides some further insight into the meaning of $R_n(\tau)$. We shall also encounter, along the way, a simple example of a non-stationary random process, in which case we shall have to evaluate $R_n(t_1,t_2)$.
4. Transit time effects come in quite naturally as a band-limiting effect. For variety's sake we shall refer here to charge flow and the resultant shot noise in a semiconductor slab. The reader will find it instructive to repeat the same arguments in the case of a photodiode or other device.

Consider then the following model for the generation of our noise: large numbers of electrons (or holes, or other particles in other physical situations) per unit time are moving in a semiconductor material. These constitute the steady-state (dc) current flow. From time to time, however, the number per unit time changes, providing current fluctuations about the average or dc value. (In the semiconductor case the dc current flow is determined by recombination and diffusion mechanisms. Both these mechanisms are statistical, and changes in the number of recombinations per unit time, or in the number of holes or electrons diffusing per unit time in a specified direction, produce the fluctuations in the current.)

As each electron (or hole) moves through the semiconductor it induces a current pulse of area e, just as in the vacuum-diode case of the previous section. The summed effect of *all* the pulses due to *all* electrons moving provides the total instantaneous current $i(t)$. Assuming the shape of an individual current pulse to be some arbitrary function $h(t)$, we then have

$$i(t) = e \sum_j h(t - t_j) \qquad [7\text{-}15]$$

[The area under each pulse must be unity:

$$\int h(t)\, dt = 1$$

since the current due to each electron is e.] As previously, the starting time t_j of each pulse is random. Simple assumptions about t_j will enable us to develop the shot-noise equation [7-14] through statistical averaging. An example of $h(t)$, and the superposition of many $h(t - t_j)$'s is shown in Fig. 7-7. It is the shape of $h(t)$ that incorporates the transit time effects mentioned earlier.

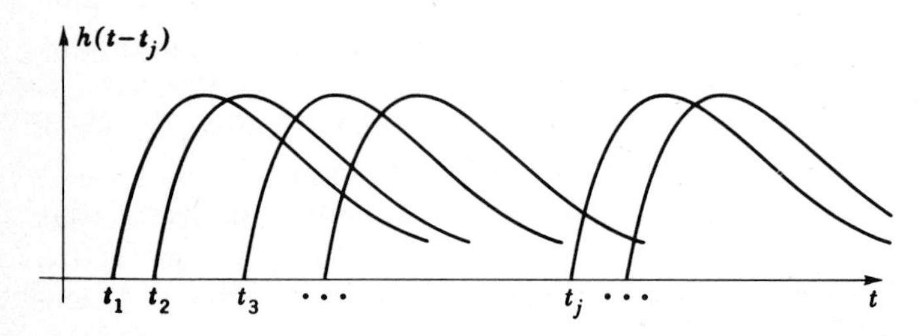

FIG. 7-7 Shot noise, random pulse starting times.

As noted earlier, even for small dc currents the number of current pulses (or starting times t_j) per unit time is astronomical, providing the numbers needed to ensure the validity of statistical averaging. Thus with $e = 1.6 \times 10^{-19}$ coul, a current I_{dc} of 1.6×10^{-9} amp corresponds to 10^{10} electrons per second on the average, while a dc current of 1.6 ma corresponds to 10^{16} per second.

We now make the following (*Poisson*) hypothesis concerning the starting times t_j. Pick a small time interval Δt. If Δt is "small enough," the *chance* of a current pulse beginning in this time slot is assumed to be *proportional* to Δt. We also assume that the probability of more than one pulse occurring is very unlikely and that there is no memory in the system; the probabilities in each adjacent interval Δt sec wide are independent. (The occurrence of a pulse in one interval has no effect on occurrences in other intervals.) See Fig. 7-8.

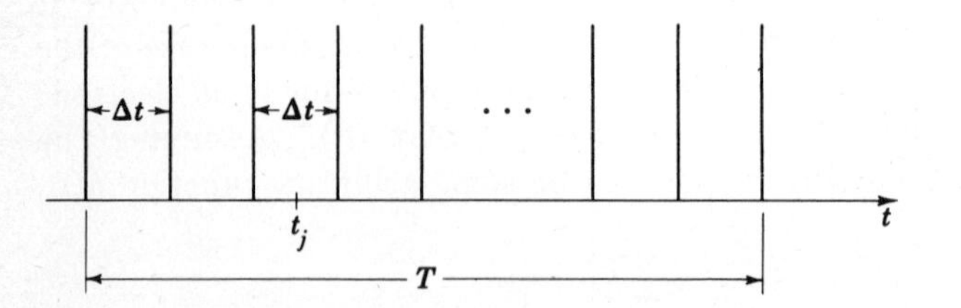

FIG. 7-8 Significant time intervals, Poisson process.

But this hypothesis is exactly that used in Chap. 6 in deriving the statistics of the random telegraph signal. We can therefore use the results desired there. Specifically, if we now pick a long time interval T, as in Fig. 7-8, and let $T/\Delta t = m \gg 1$, the probability of K pulses occurring in T sec is just the binomial distribution

$$P(K) = \binom{m}{K} p^K q^{m-K} \qquad [7\text{-}16]$$

Here

$$\binom{m}{K} \equiv \frac{m!}{K!(m-K)!}$$

is the number of combinations of m things, taken K at a time. The probability p, as in Chap. 6, is just the probability of a pulse occurring in any interval Δt sec long, and is, by hypothesis, given by

$$p = \mu \, \Delta t \qquad [7\text{-}17]$$

The parameter μ is the proportionality factor alluded to above. (As previously, of course, $q = 1 - p$.)

As we now let the interval $\Delta t \rightarrow 0$, with $m = T/\Delta t \rightarrow \infty$, the binomial distribution again approaches the *Poisson distribution* first encountered with the random-telegraph-signal problem in Chap. 6. The Poisson distribution is the one underlying all shot-noise phenomena; that is, we deal with many ($m \gg 1$) *independent events* in a specified time interval. As shown in Chap. 6 the probability of K occurrences in a T-sec interval for the Poisson distribution is just

$$P(K) = \frac{(\mu T)^K e^{-\mu T}}{K!} = \frac{a^K e^{-K}}{K!} \qquad [7\text{-}18]$$

where $a \equiv \mu T$, $K = 0, 1, 2, \ldots$.

As also shown in Chap. 6, the *average* or expected number of occurrences in the T-sec interval is just

$$E(K) = \sum_{K=0}^{\infty} K P(K) = \mu T = a \qquad [7\text{-}19]$$

The parameter μ thus corresponds to the number of pulses occurring per unit time. Since each pulse carries a charge e, $e\mu$ represents the charge flowing per unit time, or just the average current I_{dc}. We shall return to this point later.

The second moment and variance of the Poisson distribution are also readily found to be given, respectively, by

$$E(K^2) = \sum_{K=0}^{\infty} K^2 P(K) = a + a^2 \qquad [7\text{-}20]$$

and

$$\sigma_K{}^2 = E(K^2) - E^2(K) = a = \mu T \qquad [7\text{-}21]$$

Note then that the variance, the total fluctuation about the average (or dc current), is the same as the average value. This is precisely the reason why the shot-noise current is proportional to the dc current. This is a specific characteristic of shot noise and depends, of course, on the Poisson distribution assumptions.

Now returning to our current formulation of Eq. [7-15], we have again

$$i(t) = e \sum_j h(t - t_j) \qquad [7\text{-}15]$$

where by hypothesis we take the pulse starting times t_j to be *Poisson-distributed.* The probability of a pulse being initiated in an interval Δt is $\mu \, \Delta t$, while probabilities in adjacent time intervals are independent, so that one pulse triggered off does not affect the chance of another being triggered off. Then the probability of exactly K pulses initiated in T sec is given by Eq. [7-18].

Note from the form of Eq. [7-15] that we can equally well consider each of the pulses to correspond to the impulse response $h(t)$ of a fictitious linear network, with $i(t)$ then the response to a sequence of impulses initiated at the random times t_j. Thus if we let

$$z(t) = e \sum_j \delta(t - t_j) \qquad [7\text{-}22]$$

$i(t)$ is simply the output response of a linear filter of impulse response $h(t)$, excited by $z(t)$ (see Fig. 7-9). It thus suffices to find the spectral density

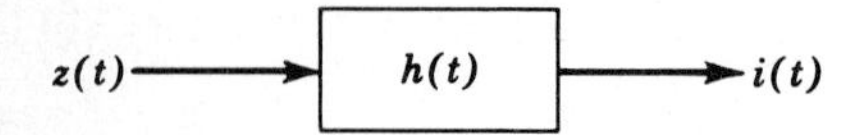

FIG. 7-9 Relation between $z(t)$ and $i(t)$.

$G_z(f)$ of the random (Poisson) process $z(t)$, for the spectral density of $i(t)$ is then given by

$$G_i(f) = |H(\omega)|^2 G_z(f) \qquad [7\text{-}23]$$

To find the statistics of $z(t)$ we use the following stratagem.[1] Let each current pulse, as initiated, give rise to a unit count, starting at time $t = 0$ say. Call the count at time t $x(t)$. An example is shown in Fig. 7-10. Assuming the times t_j ($j = 1, 2, \ldots$) are Poisson-distributed, $x(t)$

[1] A. Papoulis, "Probability, Random Variables, and Stochastic Processes," pp. 284–288, 357–360, McGraw-Hill, New York, 1965.

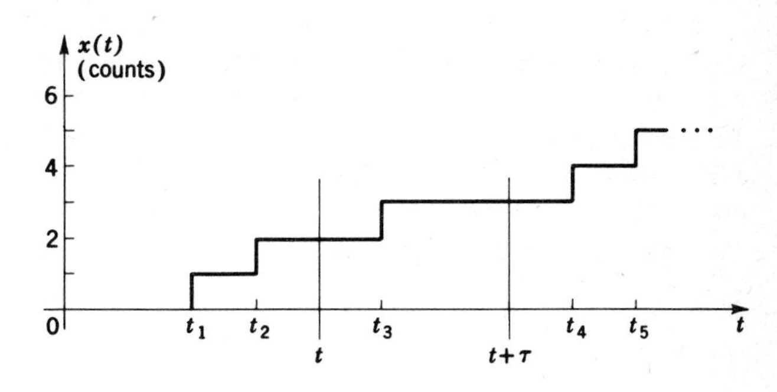

FIG. 7-10 Poisson process.

at any specified time t is a random variable with Poisson statistics, and the evolution of $x(t)$, as time progresses, is a random process. In particular, at time t, the probability that K counts have occurred is

$$P(x = K) = \frac{(\mu t)^K e^{-\mu t}}{K!} \tag{7-24}$$

The average or expected value of $x(t)$, at time t, is just

$$E[x(t)] = \mu t \tag{7-25}$$

and its variance is

$$\sigma_x^2 = \mu t \tag{7-26}$$

(Refer to Eqs. [7-19] and [7-21].) Note that $x(t)$ thus is an example of a *nonstationary process,* since its average value varies with time.

Since $x(t)$ changes by unit steps at the Poisson-distributed times t_j, it is also apparent that our desired process $z(t)$ is just given by

$$z(t) = e\frac{dx}{dt} \tag{7-27}$$

Thus by finding the appropriate statistics of $x(t)$, we shall find the equivalent (and desired) statistics of $z(t)$ from Eq. [7-27]. In particular we shall find the autocorrelation function of $x(t)$ quite readily. From this we shall find the autocorrelation function $R_z(\tau)$ of $z(t)$, and hence, by Fourier transform, $G_z(f)$.

Since $x(t)$ has already been shown to be nonstationary, we must use the general definition of the autocorrelation function here:

$$R_x(t, t + \tau) = E[x(t)x(t + \tau)] \tag{7-28}$$

We would expect this quantity to depend, in this case, on the initial value t, as well as on the difference τ, in times. Note that t and $t + \tau$ are any two *arbitrary* times, not specifically the switching times (see Fig. 7-10).

To calculate this autocorrelation function, we use a simple trick. Recalling from the Poisson hypothesis that counts in nonoverlapping time intervals are independent (the occurrence of a count at time t_j has no effect on the occurrence of a count at time t_{j+1}, or any other times), we write $x(t+\tau)$ as the sum of counts in the two nonoverlapping intervals $(0,t)$ and $(t, t+\tau)$:

$$x(t+\tau) = x(t) + [x(t+\tau) - x(t)] \qquad [7\text{-}29]$$

This decomposition of $x(t+\tau)$ is shown in Fig. 7-11. The two random variables $x(t)$ and $[x(t+\tau) - x(t)]$ are by hypothesis independent.

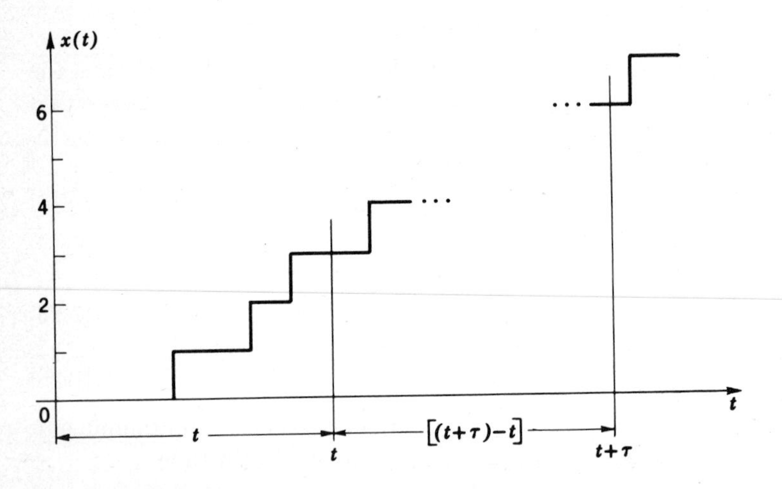

FIG. 7-11 Decomposition of $x(t+\tau)$ into independent quantities.

Recalling that $E(x_1x_2) = E(x_1)E(x_2)$ for two independent random variables x_1 and x_2, we then have, from Eq. [7-28],

$$\begin{aligned} R_x(t, t+\tau) &= E[x(t)\underbrace{\{x(t) + [x(t+\tau) - x(t)]\}}_{=x(t+\tau)}] \\ &= E[x^2(t)] + E[x(t)]E[x(t+\tau) - x(t)] \end{aligned} \qquad [7\text{-}30]$$

From Eqs. [7-25] and [7-26] we have $E[x(t+\tau) - x(t)] = \mu\tau$, $E[x(t)] = \mu t$, $E[x^2(t)] = \sigma_x^2 + E^2(x) = \mu t + (\mu t)^2$. We then get

$$R_x(t, t+\tau) = \mu^2 t(t+\tau) + \mu t \qquad \tau \geq 0 \qquad [7\text{-}31]$$

on substituting into Eq. [7-30] and combining terms. Note, as expected, that $R_x(t, t+\tau)$ depends on t as well as τ, the characteristic of a *nonstationary process.*

If we consider the case $\tau < 0$, the time $(t + \tau)$ is *less* than t. The autocorrelation function $R_x(t, t + \tau)$ is found in a similar manner for this case by simply interchanging t and $t + \tau$. We then get

$$R_x(t, t + \tau) = \mu^2 t(t + \tau) + \mu(t + \tau) \qquad \tau \leq 0 \tag{7-31a}$$

To find the autocorrelation function of $z(t)$, $z(t) = e(dx/dt)$, from which we may then find the desired $G_z(f)$, we simply write

$$\frac{z(t)}{e} = \lim_{\epsilon \to 0} y(t) \equiv \frac{x(t + \epsilon) - x(t)}{\epsilon} \tag{7-32}$$

We first find the autocorrelation function $R_y(t, t + \tau)$ of the defined random process $y(t)$, and from this, by taking limits, $R_z(\tau)$. (The validity of this limiting process for random processes is readily justified.[1]) Now note that with ϵ fixed, $y(t)$ is just $1/\epsilon$ the number of counts in the interval $(t, t + \epsilon)$. Thus the probability that y has the value K/ϵ, $K = 0, 1, 2, \ldots$, is given by the Poisson distribution

$$P\left(y = \frac{K}{\epsilon}\right) = \frac{(\mu\epsilon)^K e^{-\mu\epsilon}}{K!} \tag{7-33}$$

The expected value of y is

$$E(y) = \frac{1}{\epsilon} E(K) = \frac{\mu\epsilon}{\epsilon} = \mu \tag{7-34}$$

since the average number of counts in the interval ϵ sec wide is just $\mu\epsilon$. The expected value of y is thus a constant independent of time. We shall in fact see that $R_y(t_1,t_2)$ is dependent on $(t_2 - t_1)$ only, so that $y(t)$ is a *stationary* random process.

To find $R_y(t, t + \tau)$, we formally write

$$\begin{aligned} R_y(t, t + \tau) &\equiv E[y(t)y(t + \tau)] \\ &= \frac{1}{\epsilon^2} E\{[x(t + \epsilon) - x(t)][x(t + \tau + \epsilon) - x(t + \tau)]\} \end{aligned} \tag{7-35}$$

We now multiply out, as indicated, and take the appropriate expectation values in x. We use the same trick of rewriting the appropriate x's in a nonoverlapping form to make the desired calculations. Two separate cases must now be distinguished, however:

1: $\quad \tau < \epsilon \quad t < t + \tau < t + \epsilon$
2: $\quad \tau > \epsilon \quad t + \tau > t + \epsilon$

[1] Papoulis, *op. cit.*, pp. 314–318.

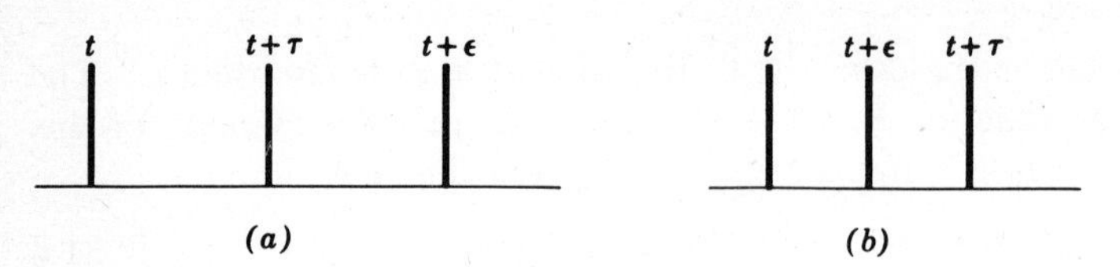

FIG. 7-12 Time intervals. (a) Overlapping intervals: $\tau < \epsilon$; (b) nonoverlapping intervals: $\tau > \epsilon$.

These are indicated in Fig. 7-12. In the second case, $\tau > \epsilon$, and $[x(t + t) - x(t)]$ and $[x(t + \tau + \epsilon) - x(t + \tau)]$ are already independent random variables, since the respective intervals are nonoverlapping. In this case it is then readily shown, from Eq. [7-35], that

$$R_y(t, t + \tau) = \frac{1}{\epsilon^2} (\mu\epsilon)(\mu\epsilon) = \mu^2 \qquad |\tau| \geq \epsilon \tag{7-36}$$

while in the other case, that of overlapping intervals $(\tau < t)$, a little calculation shows that

$$R_y(t, t + \tau) = \mu^2 + \frac{\mu}{\epsilon} - \frac{\mu|\tau|}{\epsilon^2} \qquad |\tau| < \epsilon \tag{7-37}$$

The resultant triangular-shaped autocorrelation function is shown in Fig. 7-13.

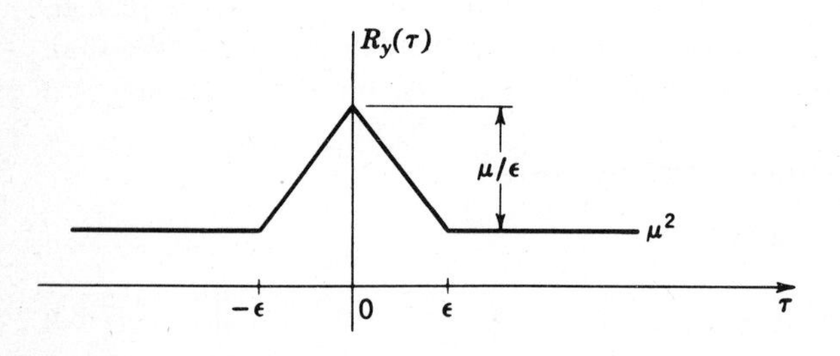

FIG. 7-13 Autocorrelation function, $R_y(\tau)$.

Since $R_y(\tau)$ depends on τ only, and *not* on the time t, $y(t)$ is an example of a *stationary random process.* {Contrast this with $R_x(t, t + \tau)$ in Eq. [7-31], which is nonstationary.} Its spectral density $G_y(f)$ is thus readily found as the Fourier transform of $R_y(\tau)$, and is in fact given by

$$G_y(f) = \mu^2\delta(f) + \mu \left(\frac{\sin \pi f\epsilon}{\pi f\epsilon}\right)^2 \tag{7-38}$$

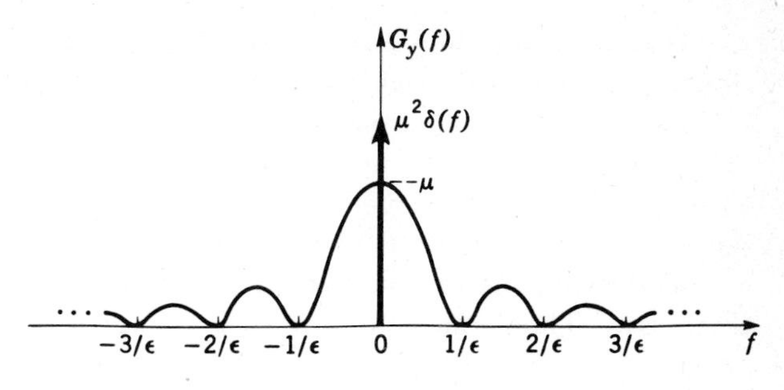

FIG. 7-14 Spectral density of $y(t)$.

(See Fig. 7-14.) The $\delta(f)$ component in $G_y(f)$ is of course easily explained; it represents the dc power in the random process $y(t)$: $E(y) = \mu$ (Eq. [7-34]), so that the dc power is μ^2. This nonzero average value accounts in the same way for the nonzero $R_y(\tau)$, $\tau \to \infty$. If we consider the random process $y'(t) = y(t) - \mu$, corresponding to blocking the dc (here the average number of counts per second), we get a process with triangular $R(\tau)$ and $[(\sin x)/x]^2$ $G(f)$. It is *this* process which represents fluctuations about the direct current that accounts for the shot noise we are considering.

In particular, as we let $\epsilon \to 0$ now, we get the desired spectrum of $z(t)$, the sequence of current impulses:

$$z(t) = \lim_{\epsilon \to 0} ey(t) \qquad [7\text{-}39]$$

$$R_z(\tau) = e^2\mu\delta(\tau) + \underbrace{e^2\mu^2}_{\text{dc term}} \qquad [7\text{-}40]$$

$$G_z(f) = \underbrace{e^2\mu}_{\text{white-term noise}} + \underbrace{e^2\mu^2\delta(f)}_{\text{dc term}} \qquad [7\text{-}41]$$

Thus, as $\epsilon \to 0$, the triangular component of $R_y(\tau)$ approaches an impulse of amplitude μ, corresponding to a white-noise spectral density of magnitude μ as well. This is shown in Fig. 7-15.

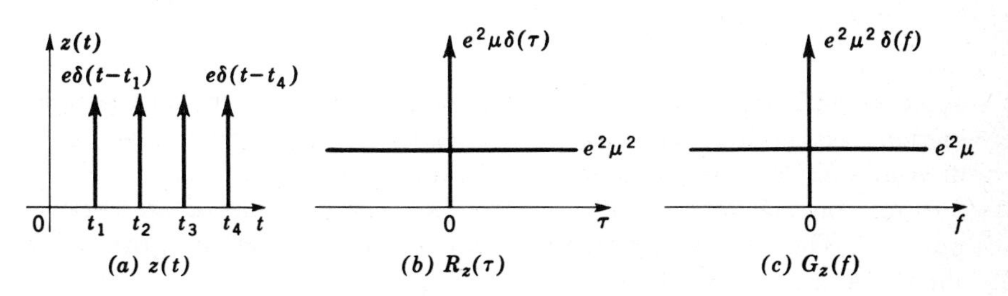

FIG. 7-15 White-noise process: Poisson impulses.

It is apparent that if the current pulses due to the liberation of an electron in a vacuum diode (or CRO), or to the interaction of holes and electrons in a semiconductor (as a result of recombination, diffusion, and other processes), may be treated as impulses, the resultant shot noise or variation about the dc level is truly white noise. More realistically, let the actual current flow $i(t)$ be represented by the Poisson impulses transmitted through the linear filter $h(t)$, as in Eqs. [7-15], [7-22], and [7-23] (Fig. 7-9). The shape factor $h(t)$, with $\int h(t)\,dt = 1$, represents the current pulse induced by the passage of a single electron in transit between cathode and plate, the current shape induced by an electron before liberation and recombination in a semiconductor, etc. Then we have

$$i(t) = \int_{-\infty}^{\infty} z(\tau)h(t-\tau)\,d\tau \qquad [7\text{-}42]$$

and

$$G_i(f) = |H(\omega)|^2 G_z(f) \qquad [7\text{-}43]$$

In particular, since $i(t)$ is a random process, we have for its expected value

$$E(i) = \int_{-\infty}^{\infty} E(z)h(t-\tau)\,d\tau = E(z) = \mu e \qquad [7\text{-}44]$$

since $\int h(t)\,dt = 1$ by assumption. But this expected value is just the dc current I_{dc}. It represents the average number μ of electrons emitted per second, each with charge e. It is also the current that would be read by a dc meter with time constant long compared to $1/B$, with bandwidth B just the width of $G_i(f)$.

From Eqs. [7-41] and [7-43] we also have

$$G_i(f) = e^2\mu^2 H^2(0)\delta(f) + \mu e^2|H(\omega)|^2 \qquad [7\text{-}45]$$

Note: with $\int_{-\infty}^{\infty} h(t)\,dt = 1$, we also have $H(0) = 1$. If we call $[i(t) - I_{dc}]$ the shot noise or fluctuation current $i_b(t)$ we have of course, with $I_{dc} = \mu e$,

$$G_{i_b}(f) = e^2|H(\omega)|^2 = eI_{dc}|H(\omega)|^2 \qquad [7\text{-}46]$$

This is the desired expression for the shot-noise spectral density. In particular, if our measuring instruments have bandwidths small compared to the width of $|H(\omega)|^2$, we have

$$G_{i_b}(f) = eI_{dc} \qquad f \text{ small} \qquad [7\text{-}47]$$

agreeing with the more intuitive, time-averaged, result of the previous section (Eq. [7-4]). This is just the white-noise spectral density $n_0/2$ discussed in Chap. 6, but now specifically evaluated.

As pointed out earlier, we thus have an example of an ergodic process. The time-averaged result of the previous section agrees with the statistical (ensemble) averaging of this section.

The white-noise-model result of Eq. [7-47] may be specified more precisely by defining a noise-equivalent bandwidth, as in previous sections. Thus, the total shot-noise power is, from Eq. [7-46],

$$\begin{aligned} \overline{i_b^2} &\equiv \int_{-\infty}^{\infty} G_{i_b}(f)\, df = 2eI_{dc} \int_0^{\infty} |H(\omega)|^2\, df \\ &= 2eI_{dc}H^2(0)B = 2eI_{dc}B \end{aligned} \qquad [7\text{-}48]$$

using the noise-equivalent bandwidth B, defined in Chap. 6. We also use the fact that $H(0) = \int h(t)\, dt = 1$ here. The noise power is thus found to be proportional to the bandwidth B, as expected, and Eq. [7-47] may now be modified to read

$$G(f) = eI_{dc} \qquad f \ll B \qquad [7\text{-}49]$$

As some examples of typical current pulse shapes $h(t)$, consider those shown in Fig. 7-16:

1. $h(t)$ the linearly increasing pulse of Fig. 7-16*a*. Here it is readily shown[1] that

$$|H(\omega)|^2 = \frac{4}{(\omega T_r)^4}[2 - 2\cos\omega T_r + (\omega T_r)^2 - 2(\omega T_r)\sin\omega T_r] \qquad [7\text{-}50]$$

with T_r the pulse duration (electron transit time). In particular, the noise-equivalent bandwidth is

$$B = \frac{1}{2}\int_{-\infty}^{\infty} |H(\omega)|^2\, df = \frac{1}{2}\int_{-\infty}^{\infty} h^2(t)\, dt = \frac{2}{3T_r} \doteq \frac{1}{2\pi}\frac{4}{T_r} \qquad [7\text{-}51]$$

This noise-equivalent bandwidth is indicated in the figure. As an example, if $T_r = 10^{-8}$ sec, $B \sim 10^8$ Hz. If $T_r = 10^{-9}$ sec $B \sim 10^9$ Hz $= 1$ GHz. So long as all subsequent filters, amplifiers, and measuring instruments have bandwidths less than B, the shot noise may be assumed effectively white.

2. $h(t)$ the triangular pulse of Fig. 7-16*b*. Here

$$|H(\omega)|^2 = \left[\frac{\sin(\pi f T_r/2)}{\pi f T_r/2}\right]^4 \qquad [7\text{-}52]$$

and

$$B = \int_0^{\infty} |H(\omega)|^2\, df = \frac{1}{2}\int_0^{\infty} h^2(t)\, dt = \frac{2}{3T_r} \qquad [7\text{-}53]$$

The noise-equivalent bandwidth is thus the same as in the case above.

3. $h(t)$ the sinusoidal pulse of Fig. 7-16*c*.

$$B = \frac{1}{2}\int_0^{T_r} h^2(t)\, dt = \frac{\pi^2}{16T_r} \doteq \frac{0.62}{T_r}$$

[1] Papoulis, *op. cit.*, pp. 359–360.

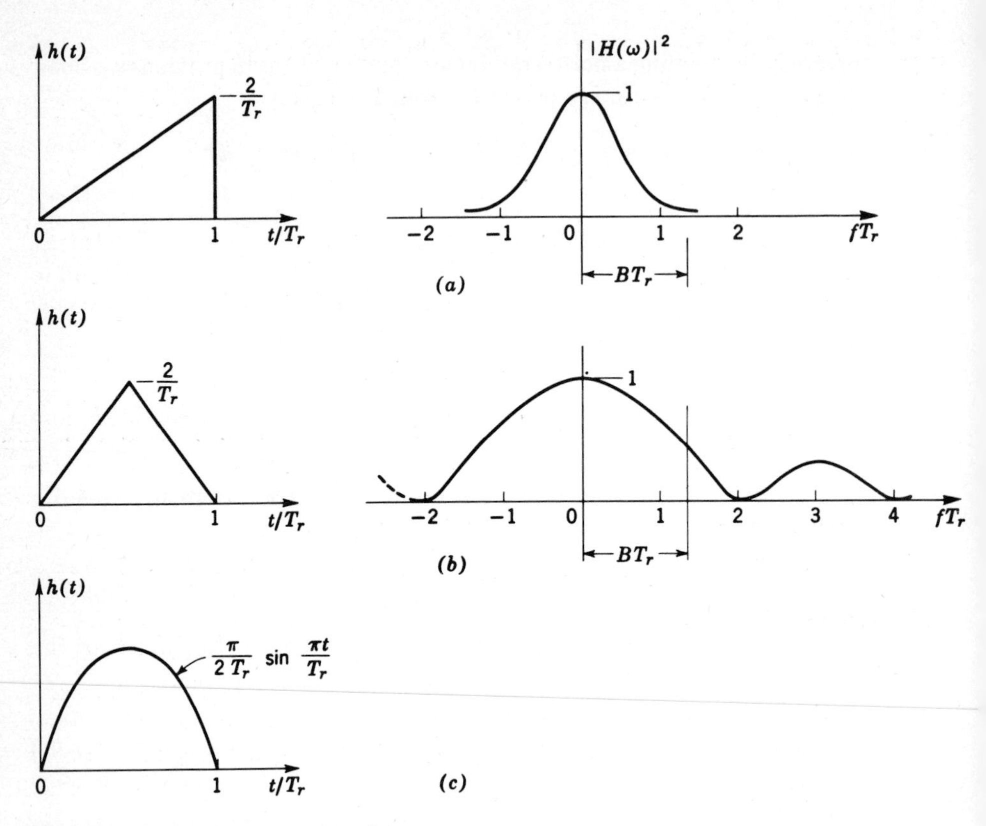

FIG. 7-16 Typical current pulse shapes.

Note that in all these examples the noise-equivalent bandwidth $B \sim 1/T_r$. If, as noted previously, $T_r = 10^{-9} - 10^{-8}$ sec, $B \sim 10^8 - 10^9$ Hz. Assuming the circuitry following operates at frequencies substantially below these values (or, if at a high frequency, has a relatively narrow bandwidth), the shot-noise spectrum is essentially flat. If we now represent the system following by an equivalent linear transfer function $H_o(\omega)$, we of course have, at the system output, a shot-noise component $n_o(t)$ with power spectral density

$$G_{n_o}(f) = |H_o(\omega)|^2 G(f) = eI_{\text{dc}}|H_o(\omega)|^2 \tag{7-54}$$

The total noise power at the system output, due to shot noise at the input is

$$N_o = \int_{-\infty}^{\infty} G_{n_o}(f) \doteq 2eI_{\text{dc}} \int_0^{\infty} |H_o(\omega)|^2 \, df$$
$$= 2eI_{\text{dc}} H_o^2(0) B_{\text{sys}} \tag{7-55}$$

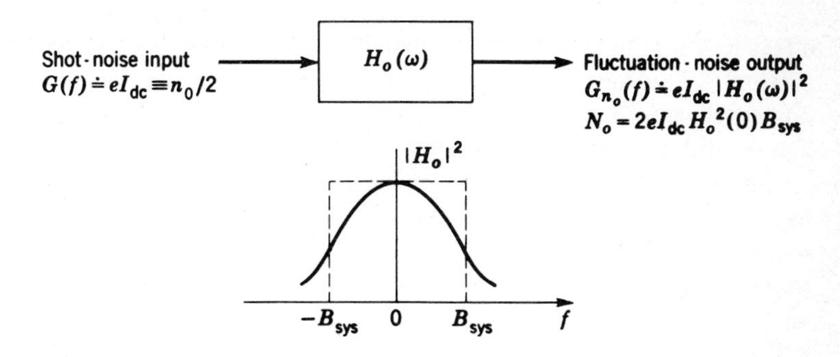

FIG. 7-17 Shot noise at circuit output.

with B_{sys} the system noise-equivalent bandwidth, and $H_o(0)$ the low-frequency gain of the system (Fig. 7-17). This agrees of course with Eq. [7-13], derived using time averaging.

It is apparent that the rms fluctuation current (or voltage) $\sqrt{N_o}$ at a system output is proportional to the gain $H_o(0)$. Amplification thus increases noise at the output, just as it does *any* signal—random or deterministic.

It is this high amplification of noise *and* signal both that makes it important to develop techniques for minimizing noise or for determining the deleterious effects of noise. With the basic shot-noise spectral density of Eq. [7-49] obtained, we now turn to the other major source of fluctuation noise in electronic (and other) circuits: *thermal noise.* We shall similarly find the spectral density for thermal noise to be essentially flat over very wide frequency ranges. Thermal noise will then also be approximated by a white-noise model. The approximation here will in fact be better than that for shot noise, since the bandwidth at which thermal noise begins to deviate from a flat spectrum will be roughly 10^{13} Hz at room temperature. For all normal circuit work we can forget this bandwidth limitation. At optical frequencies, however, so important with modern laser techniques, the white-noise assumption has to be reevaluated.

After obtaining the spectral density of thermal noise, we shall combine both shot noise and thermal noise and consider *both* going through circuits.

7-2 THERMAL NOISE

As was mentioned in the introduction to this chapter, two basic types of spontaneous fluctuation voltage are present in electrical circuits. The first, the so-called shot effect, is due to the random flow of particles. Shot effect in vacuum tubes and semiconductors was discussed in the previous section. The second type of spontaneous fluctuation, for example that due to thermal interaction between the free electrons and vibrating ions in a conduction medium (normally a resistor), is the subject of this section.

In these first sections shot noise and thermal noise are considered to be independently generated in different devices. However, modern semiconductors (semiconductor diodes, transistors, etc.) have been found to generate both types of noise, as well as other types due to other mechanisms. These combined noise effects in junction diodes and transistors will be discussed in a later section.

Thermal noise was first thoroughly studied experimentally by J. B. Johnson of Bell Laboratories in 1928. His experiments, together with the accompanying theoretical studies by H. Nyquist, demonstrated that a metallic resistor could be considered the source of spontaneous fluctuation voltages with mean-squared value

$$\overline{v^2} = 4kTRB \qquad [7\text{-}56]$$

where T is the temperature in degrees Kelvin of the resistor, R its resistance in ohms, k the Boltzmann constant already referred to (1.38×10^{-23} joule/°K), and B any arbitrary bandwidth. Johnson was able to measure the value of k fairly accurately using this equation and thus demonstrated its validity. He also showed that $\overline{v^2}$ was proportional to temperature.

This expression for the mean-squared thermal noise due to a resistor R again implies that the noise is white. We shall see shortly that it is valid up to extremely high frequencies of the order of 10^{13} Hz. At these high frequencies quantum-mechanical effects set in. (This assumes of course that R is independent of frequency over this tremendous range.) We shall develop a more general relation later which includes the quantum-mechanical effects at high frequencies. (As noted earlier the more general expression is necessary when considering communication at optical frequencies.)

From Eq. [7-56] the thermal-noise spectral density is given by the simple expression

$$G_v(f) = \frac{\overline{v^2}}{2B} = 2kTR \qquad [7\text{-}57]$$

We shall develop this white-noise spectral density as a special case of the more general relation to which reference has just been made.

Nyquist's original derivation of Eq. [7-56][1] was based on thermo-

[1] H. Nyquist, *Phys. Rev.*, no. 32, p. 110, 1928.

dynamic reasoning, assuming temperature equilibrium. The actual mechanism of thermal-noise generation—assumed to be due to the random interaction between the conduction electrons and ions in a metallic conductor—is not necessary for the derivation. Although this at first appears disconcerting, it is actually a blessing in disguise. For using the same thermodynamic reasoning it may be shown that *any linear passive* device, mechanical, electromechanical, microphones, antennas, etc., has associated with it thermal noise of one form or another. In some cases this may be due to random agitation of the air molecules, in others to random electrical effects in the ionosphere and atmosphere, etc. This is so because the $\frac{1}{2}kT$ term occurs generally in thermodynamics as the energy associated with any mode of oscillation (as temperature increases, gas molecules move more rapidly, ions vibrate more violently in a lattice structure, etc.). This was the basis of Nyquist's derivation. We shall use a somewhat similar approach here.

Because of this reasoning the *resistors in the electrical analog of a passive linear physical device* may be considered *sources of noise voltage* as given by Eq. [7-56]. This concept is commonly used in antenna work, acoustics, etc.

Equation [7-56] has also been derived from an assumed model for electrical conduction through a metal.[1] The derivation is somewhat similar to our time-averaging derivation of the temperature-limited shot effect, with a current wave due to one electron expanded into a Fourier series periodic over a long interval of time.

A voltage-model representation of Eq. [7-56] is shown in Fig. 7-18*a*. R is assumed noise-free, with the noise effect lumped into the noise-voltage source shown. An application of Norton's theorem gives the current-

[1] See J. L. Lawson and G. E. Uhlenbeck, "Threshold Signals," McGraw-Hill, New York, 1950.

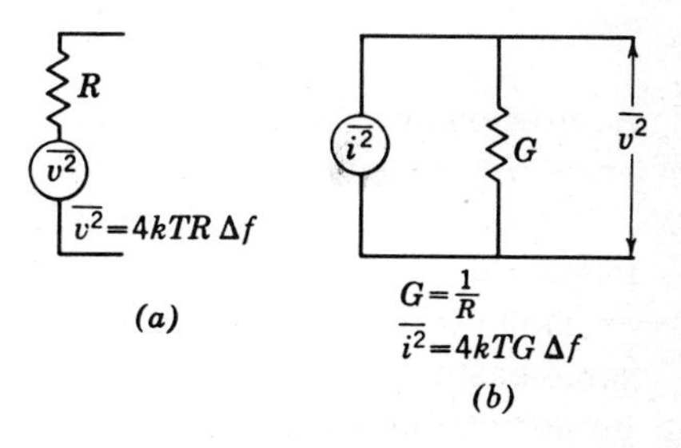

FIG. 7-18 Thermal-noise circuit models. (a) Voltage model; (b) current model.

source equivalent of Fig. 7-18*b*. (Since $i = v/R$, $\overline{i^2} = \overline{v^2}/R^2 = 4kTGB$; $G = 1/R$.)

Either model may be used, although we shall frequently find the current-source model more convenient, especially when calculating noise voltages across parallel elements. Some typical numbers are again of interest. Let B be 5 kHz, T equal 293°K (this is normal room temperature, or 20°C), and R equal 10 kilohms. Then $\overline{v^2} = 0.8 \times 10^{-12}$ volts², or

$$\sqrt{\overline{v^2}} = 0.90 \ \mu\text{v rms}$$

$\sqrt{\overline{i^2}} = 0.90 \times 10^{-10}$ amp rms. If the bandwidth is quadrupled to 20 kHz, the rms noise voltage and current are doubled to 1.8 μv and 1.8×10^{-10} amp, respectively. The rms noise voltage is proportional to the square root of the resistance and to the square root of the bandwidth.

If the temperature is increased, the resistance value used refers to the new temperature, as does T. Since the derivation of Eq. [7-56] depends on thermal equilibrium, the equation holds only after a steady-state temperature has been reached, not during the heating or cooling period.

Recapitulating, both theory and experiment indicate that a resistance R at temperature T is found to be the source of a fluctuation (noise) voltage with mean-squared value

$$\overline{v^2} = 4kTRB \qquad [7\text{-}56]$$

and spectral density

$$G_v(f) = 2kTR \qquad [7\text{-}57]$$

The corresponding mean-squared value and spectral density for the current-source model may be written, respectively,

$$\overline{i^2} = 4kTGB \qquad [7\text{-}58]$$

and

$$G_i(f) = 2kTG \qquad [7\text{-}59]$$

As previously, B represents the noise-equivalent bandwidth of the measuring instrument or circuit used to measure these quantities.

Using these spectral densities, one may, as previously, proceed to follow the noise (voltage or current) through any linear device. For example, if the transfer function between the thermal-noise voltage and some output voltage $n(t)$ (or current) is $H(\omega)$, we have for the spectral density at the output

$$\begin{aligned} G_n(f) &= G_v(f)|H(\omega)|^2 = 2kTR|H(\omega)|^2 \\ &= 2kTRH^2(0)B \end{aligned} \qquad [7\text{-}60]$$

in terms of the noise-equivalent bandwidth B of the linear system. This is indicated in Fig. 7-19. Note that $H(\omega)$ must include the resistance R as well as any other elements and networks between v and n.

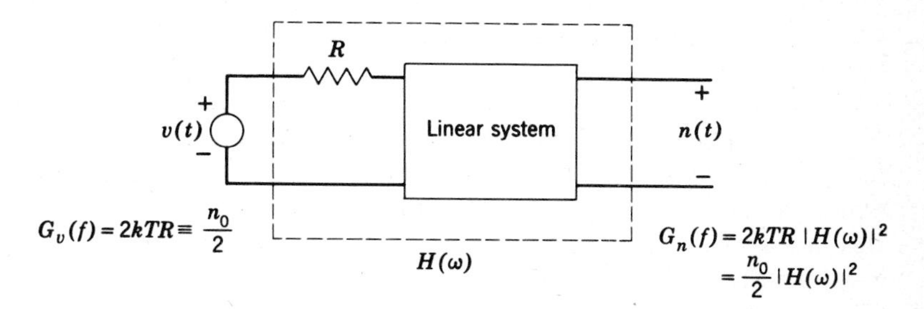

FIG. 7-19 Thermal noise through linear system.

Similar considerations hold for the thermal-noise current i. We shall consider several examples in the following pages.

Before proceeding with the actual thermal-noise calculation for some simple circuits, however, we digress to discuss the actual derivation of the thermal-noise spectral densities of Eqs. [7-57] and [7-59]. This will enable us to determine the validity of the white-noise approximation and the frequency (in the infrared range) at which the spectral density begins to vary with frequency. It will also enable us to readily consider extensions to masers and lasers—i.e., thermal-noise calculations at infrared and optical frequencies—as well as to thermal noise generated in space and received at an antenna.

Recall again that we assume we have a resistor R sitting at an equilibrium temperature of T°K. This resistor is the source of random current (and hence voltage) fluctuations produced by the heat energy assumed provided by its surroundings. As the temperature T increases, the electrons moving freely inside the conducting material are raised to higher energy levels and the mean-squared current flow increases. (The average current of course remains zero with no electric field applied, since currents in opposite directions cancel, on the average.) The actual physical mechanism of current flow in the conductor depends on the interaction of the electrons with the thermally vibrating ions in the metallic crystal. Since the noise derivation using an assumed model for electrical conduction through a metal is rather complicated,[1] we resort instead to the stratagem first used by H. Nyquist[2] in deriving the thermal-noise equation

[1] Lawson and Uhlenbeck, *ibid.*

[2] Nyquist, *Phys. Rev.*, *op. cit.*

for a resistor. This technique is based on simple equilibrium thermodynamics, avoids the problem of the specific physical mechanism involved in the noise generation, and has the advantage of easily being extended to thermal-noise calculations where physical resistors may not be present, e.g., blackbody radiation, thermal noise in microwave circuits, masers, lasers, etc.

The stratagem consists of assuming the resistor connected electrically to one or more energy-storage elements. The elements store up the thermal energy provided by the resistors. By knowing the energy stored and equating it to the thermal energy supplied by the resistor, one then finds the mean-squared thermal-noise current (or voltage) available at the resistor terminals.

Although various combinations of energy-storage elements may be used for this calculation, we shall select the oscillatory circuit consisting

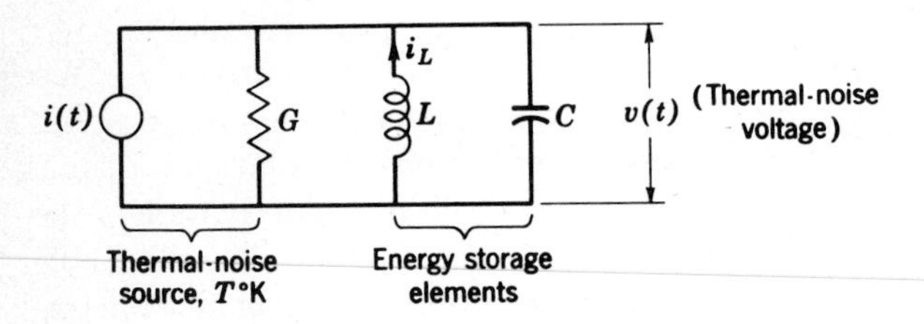

FIG. 7-20 Tuned circuit for thermal-noise calculations.

of a parallel inductor and capacitor, as shown in Fig. 7-20. We do this for two reasons:

1. The stored thermal energy of the tuned circuit is readily written down.
2. The approach used enables us to extend the results to many other situations: nonelectrical systems where G, L, C are the electrical analogs (for example, a mass-spring combination, with G representing the dissipative elements assumed at temperature T); microwave and optical circuits, with LC representing a resonant cavity, G its dissipation, etc.

This oscillatory circuit, oscillating at the resonant frequency $f_0 = 1/2\pi\sqrt{LC}$ is the prototype of the so-called *harmonic oscillator* of modern physics. In most books on modern physics and quantum mechanics[1-3] it is shown that the harmonic oscillator, resonant at fre-

[1] R. B. Leighton, "Principles of Modern Physics," McGraw-Hill, New York, 1959.
[2] R. L. Sproull, "Modern Physics," 2d ed., Wiley, New York, 1963.
[3] R. T. Weidner and R. L. Sells, "Elementary Modern Physics," pp. 433–435, 491–494, Allyn and Bacon, Boston, 1960.

quency f_0, can possess discrete stored energies only, its so-called discrete energy levels given by

$$E_n = (nh + \tfrac{1}{2})f_0 \qquad n = 0, 1, 2, \ldots \tag{7-61}$$

(Fig. 7-21).

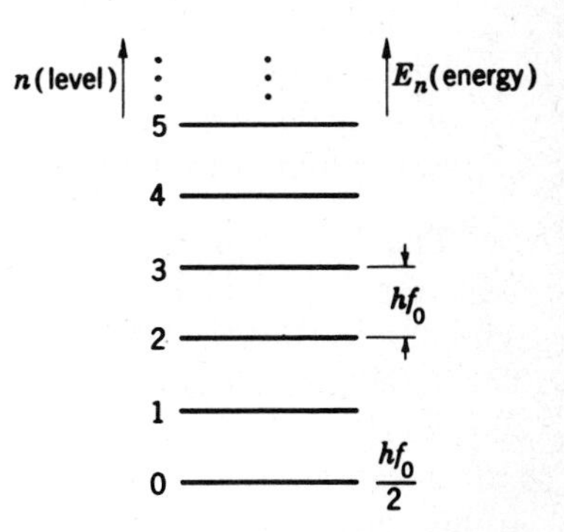

FIG. 7-21 Harmonic-oscillator energy levels.

The constant h is Planck's constant, $h = 6.6257 \times 10^{-34}$ joule/sec. If one now energizes this harmonic oscillator thermally at T°K, one finds that all energy levels may be occupied, but with differing probabilities. In particular the probability of exciting the nth energy level is found to exponentially decrease with n, and is proportional to $e^{-E_n/kT}$, k again the Boltzmann constant, $k = 1.38 \times 10^{-23}$ joule/°K. A simple statistical average to provide the average energy of the harmonic oscillator then gives

$$\bar{E} = \frac{\sum_{n=0}^{\infty} E_n e^{-E_n/kT}}{\sum_{n=0}^{\infty} e^{-E_n/kT}} = \frac{hf_0}{2} + \frac{hf_0}{e^{hf_0/kT} - 1} \tag{7-62}$$

after substituting in Eq. [7-61] and performing the indicated summations.

As an example, for very low temperatures, $kT \ll hf_0$, $\bar{E} = hf_0/2$, the so-called "zero-point energy" or lowest energy level. If the thermal energy provided is very small, the harmonic oscillator will on the average remain at its lowest energy level. If the temperature is now *high*, $kT \gg hf_0$, $e^{-hf_0/kT} = 1 - hf_0/kT$, and

$$\bar{E} = kT \qquad kT \gg f_0 \tag{7-63}$$

This is the so-called "classical" result, which says that the average energy of the oscillator is proportional to the absolute temperature. It is instructive to note that by average we again mean either of two possibilities—an ensemble or time average:

1. Many identical oscillators, oscillating independently of each other, are available. Each one will randomly occupy a particular energy level. The average energy of the *ensemble* is then given by Eq. [7-62].
2. Alternatively, *one* oscillator will occupy only one level at any one time. But over many observation times it will occupy different levels with different probabilities, the *average* being given by Eq. [7-62].

The LC circuit of Fig. 7-20 is just one of many possible "harmonic oscillators" to which Eq. [7-62] applies. (In most physics texts the symbol ν, rather than f_0, is used to denote the frequency of oscillation.) The "oscillator" may be made up of two bound atoms forming a molecule such as H_2. Heat applied to a gaseous system of such molecules tends to pull the atoms apart, restoring forces tend to keep them together, the two atoms then vibrating at a characteristic frequency determined by their masses and restoring force constants. $\bar{E}$ is then the so-called average vibrational energy of the molecules. The thermally vibrating ions in a crystal, bound to their specific locations by molecular forces but vibrating due to heat energy applied, may be modeled by harmonic oscillators. The application of Eq. [7-62] then enables one to quite accurately calculate the specific heat of a solid.[1,2] Thermal radiation produced by individual atoms radiating in a given substance is also calculable using Eq. [7-62]. Here one assumes the radiation "captured" by a resonant structure of arbitrary size. Standing electromagnetic waves are set up in this structure, at frequencies appropriate to its boundary conditions. Each frequency is then assumed to correspond to a fictitious harmonic oscillator at that frequency, and has average energy given by Eq. [7-62]. Summing over all possible frequencies, one finds the thermal energy distribution is found to be exactly that of so-called *blackbody radiation*[1,2]: the radiant energy emanated by many heated sources at T°K (a furnace, the sun, stars, the atmosphere, etc.). It is this blackbody radiation, produced by the sun, stars, sources of radiation in the earth's atmosphere, etc., that appears as additional thermal noise at an antenna input in any high-frequency radio receiver.

Consider now specifically the parallel G, L, C combination of Fig. 7-20. The resistor at temperature T°K may be an actual resistor connected across the LC circuit, or it could be the equivalent dissipation

[1] Leighton, *op. cit.*

[2] F. Reif, "Fundamentals of Statistical and Thermal Physics," McGraw-Hill, New York, 1965.

resistance of the LC circuit, assumed connected in parallel, or it could be the equivalent resistance of an antenna tuned to frequency $f_0 = 1/2\pi\sqrt{LC}$, etc. We visualize the resistor kept at temperature T°K by literally being "immersed in a heat bath" at that temperature; e.g., if the resistor is part of a circuit in a room, T is the normal room temperature. On being connected to the LC circuit the resistor provides the energy $\bar{E}$, given by Eq. [7-62], eventually stored in the L and C after the connection has been made for a while. The connection also enables a current to flow through the resistor, producing a power loss in that element. We relate all of these by saying that there is an average stored energy in the capacitor,

$$\bar{E}_1 = \tfrac{1}{2}C\overline{v^2}$$

equal to the stored energy $\bar{E}_2 = \tfrac{1}{2}L\overline{i_L^2}$ in the inductor. The total stored energy is then

$$\bar{E} = C\overline{v^2} = \frac{hf_0}{2} + \frac{hf_0}{e^{hf_0/kT} - 1} \tag{7-64}$$

from Eq. [7-62]. The quantity $\overline{v^2}$ is just the mean-squared fluctuation of the voltage $v(t)$ appearing across the tuned circuit, and thus represents the thermal noise measured at that point. If we visualize this noise term having a spectral density $G_v(f)$, we must have

$$\overline{v^2} = \int_{-\infty}^{\infty} G_v(f)\,df \tag{7-65}$$

Assuming now that the source of the thermal noise in the resistor is represented by the current source $i(t)$ in Fig. 7-20, we must have the following relation connecting the spectral densities of $v(t)$ and $i(t)$:

$$G_v(f) = |H(\omega)|^2 G_i(f) \tag{7-66}$$

From Eqs. [7-64] to [7-66], we then get the following equation, which enables us to find the spectral density $G_i(f)$ of the noise source in terms of the known stored thermal energy of the tuned circuit:

$$\overline{v^2} = \frac{\bar{E}}{C} = \int_{-\infty}^{\infty} |H(\omega)|^2 G_i(f)\,df \tag{7-67}$$

For simplicity's sake assume now that the resonant frequency f_0 is low enough to enable the stored energy $\bar{E}$ to be adequately approximated by kT:

$$f_0 \ll \frac{kT}{h}$$

For normal room temperature (290°K), this implies

$$f_0 \ll 10^{13}\ \text{Hz}$$

in the infrared region of the electromagnetic spectrum. It is apparent that for all normal electronic circuits (including those at microwave frequencies) this is a valid assumption. [Although it is not valid at infrared and optical frequencies, at which masers and lasers operate, we shall indicate shortly how one can still use the approach followed here, and obtain a more general expression for $G_i(f)$ appropriate at these frequencies.]

We thus have

$$\frac{kT}{C} = \int_{-\infty}^{\infty} |H(\omega)|^2 G_i(f)\, df \qquad f_0 \ll \frac{kT}{h} \tag{7-68}$$

Now assume further that the bandpass characteristic $H(\omega)$ is *narrow* compared to the noise spectral density $G_i(f)$. This immediately implies that we are modeling $G_i(f)$ as a *white-noise source* so far as the GLC circuit is concerned (Fig. 7-22). We then have

$$\frac{kT}{C} \doteq G_i(f_0) \int_{-\infty}^{\infty} |H(\omega)|^2\, df \tag{7-69}$$

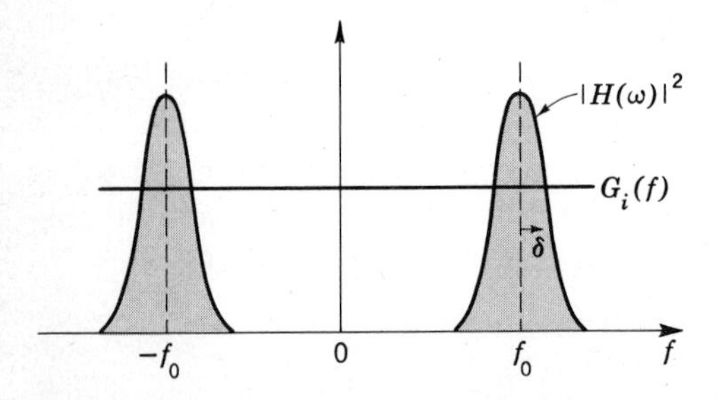

FIG. 7-22 White-noise approximation.

To evaluate the integral of Eq. [7-69], we first note that $|H(\omega)|$ may be written

$$|H(\omega)| = \left|\frac{V(\omega)}{I(\omega)}\right| = \frac{1}{|G + j\omega C + 1/j\omega L|}$$
$$= \frac{1/G}{\sqrt{1 + (\omega C - 1/\omega L)^2/G^2}} \doteq \frac{1/G}{\sqrt{1 + (\delta/\alpha)^2}} \tag{7-70}$$

where we have made the substitutions $\delta = \omega - \omega_0$, $\alpha = G/2C$, and have assumed that $|H(\omega)|^2$ is narrowband enough about ω_0 to have $\delta \ll \omega_0$.

Equation [7-70] then becomes

$$\frac{kT}{C} = \frac{2G_i(f_0)}{2\pi G^2} \int_0^\infty \frac{d\delta}{1 + (\delta/\alpha)^2}$$

$$= \frac{\alpha G_i(f_0)}{\pi G^2} \int_0^\infty \frac{dx}{1 + x^2} \qquad [7\text{-}71]$$

using the change of variables $x = \delta/\alpha$, after noting that $d\omega = d\delta$. The indicated integral is just $\pi/2$, so that one finally gets, after solving for $G_i(f)$,

$$G_i(f) = 2kTG \qquad f \ll \frac{kT}{h} \qquad [7\text{-}72]$$

This is of course the desired spectral density of the noise-current source of the resistor, given originally in Eq. [7-59].

More generally, if $G_i(f)$ varies with frequency, but *slowly enough* compared to the bandwidth of the tuned circuit, we can still pull $G_i(f_0)$ out of the integral of Eq. [7-68]. Then, using the more general form of Eq. [7-62] for the stored energy of the tuned circuit, we have

$$\frac{\bar{E}}{C} = \frac{1}{C}\left(\frac{hf_0}{2} + \frac{hf_0}{e^{hf_0/kT} - 1}\right) = \frac{\alpha G_i(f_0)}{\pi G^2} \int_0^\infty \frac{dx}{1 + x^2} \qquad [7\text{-}73]$$

It is then apparent, by comparison with Eqs. [7-71] and [7-72], that we have finally

$$G_i(f) = 2\left(\frac{hf}{2} + \frac{hf}{e^{hf/kT} - 1}\right)G \qquad [7\text{-}74]$$

as the spectral density of the resistive noise-current source. Here we have dropped the "0" subscript in the frequency term, since f_0 was an arbitrary frequency to which we assumed the oscillatory circuit tuned.

Equation [7-74] represents the more general spectral-density expression for the thermal-noise-current source associated with the resistor. It can of course be converted to a voltage equivalent by simply replacing G by R.

This more general form for spectral density now holds for *any* dissipative element at T°K, at any frequency, with an equivalent resistance R, or equivalent conductance $G = 1/R$.

The first term $hf/2$ in the expression for spectral density, due to the zero-point energy of the harmonic oscillator, is a strictly quantum-mechanical one. This quantum-mechanical noise is negligible at frequencies $f \ll kT/h \sim 10^{13}$ Hz. At infrared and optical frequencies, however, it begins to play a dominant role.

As an example of the applicability of Eq. [7-74], or its approximate, white-noise, form, Eq. [7-72], assume that the antenna of a radio receiver is connected directly to a tuned circuit at the desired r-f (carrier) frequency. In addition to the desired signals coming in at the antenna, thermal noise also appears. Assume that this noise is primarily due to blackbody radiation from space at an average temperature of $T_s = 1000°K$, say. It is well known[1] that the antenna may be assumed to have an effective radiation resistance R_r such that I^2R_r = power at antenna input, with I the rms current actually measured at the antenna terminals. This is shown in Fig. 7-23*a*.

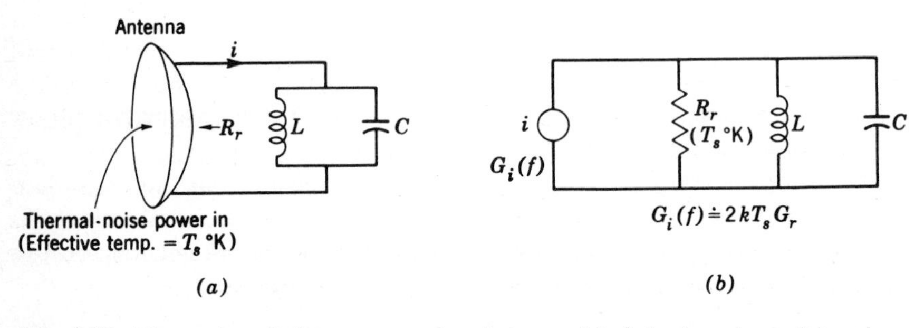

FIG. 7-23 Thermal radiation power at antenna. (a) Actual system; (b) noise model.

Using the same argument as previously, the tuned circuit must have an average stored energy $\bar{E} = kT_s$ (for frequency $f_0 \ll kT_s/h$), since it is the thermal or blackbody radiation from space, at effective temperature T_s °K, that provides the stored energy, by hypothesis. This is important to stress. It is *not* the temperature of the tuned circuit that determines the thermal-noise energy stored, but the temperature of whatever mechanism is responsible for providing the noise power.

Since the tuned circuit sees, effectively, the radiation resistance R_r in parallel with it, it is apparent that the noise source acts as if it were a resistance R_r at a temperature T_s °K. This results in the noise model shown in Fig. 7-23*b*.

How does one determine the effective temperature T_s? This depends of course on what the antenna "sees": it depends on the actual thermal radiators in the solid angle subtended by the antenna, their actual temperatures (these could be the earth's surface reradiating thermal energy, the atmosphere, radiation from the sun if the antenna is aimed in that direction, galactic radiation, etc.), the frequency of the antenna system

[1] E. C. Jordan, "Electromagnetic Waves and Radiating Systems," Prentice-Hall, Englewood Cliffs, N.J., 1950.

and the thermal noise received at this frequency, etc. All these must be measured. In practice one measures an effective rms noise current at the antenna terminals (or, perhaps, an rms noise voltage across the tuned circuit). If this is essentially due to incoming thermal radiation, and *not* to actual dissipation in the antenna itself, one must have

$$\overline{i^2} = 4kT_sG_rB \qquad [7\text{-}75]$$

with B the bandwidth of the measuring apparatus. With G_r, B, and $\overline{i^2}$ known (measured), this provides a measure of T_s.

How *does* one take into account actual antenna dissipation in the tuned circuit (including a resistor possibly put in parallel)? Let R represent the equivalent resistance corresponding to these quantities. Letting the actual temperature of the antenna system be $T°$ K, and the effective temperature of space T_s °K, *each* source must *independently* provide energy to the tuned circuit. Assuming the frequency f low enough so that $f \ll kT/h$ and $f \ll kT_s/h$, to simplify the analysis (otherwise we simply use the more complicated form, Eq. [7-74] for the spectral densities), the spectral density at the tuned circuit output (or *any* circuit output for that matter) is given by

$$G_v(f) = |H(\omega)|^2(2kT_sG_r + 2kTG) \qquad G = \frac{1}{R}, \; G_r = \frac{1}{R_r} \qquad [7\text{-}76]$$

One simply *adds* the independent noise contributions. (Recall from Chap. 5 that if independent or uncorrelated random variables are added, their variances add.) This is shown in Fig. 7-24. It is important to

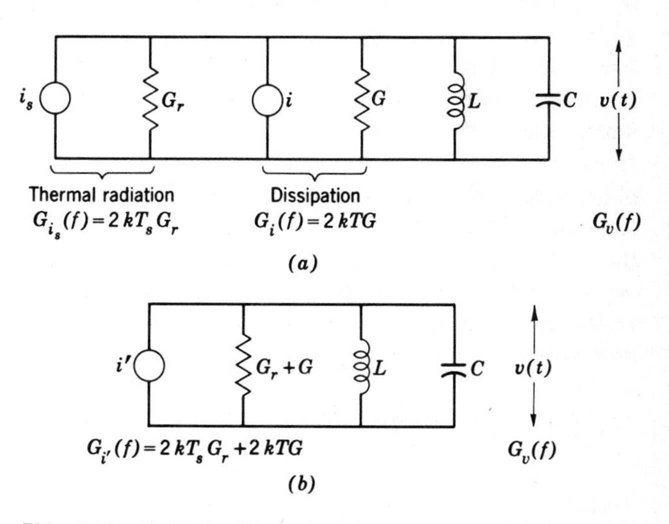

FIG. 7-24 Multiple thermal noise sources. (a) Complete model; (b) reduced model.

stress that the transfer function $H(\omega)$ includes the (fictitious) noise conductances G_r and G. As shown in Fig. 7-24b, $H(\omega)$ is the impedance $1/(G_r + G + j\omega C - 1/j\omega L)$ in this particular case.

As a special case, if $T_s = T$, so that both thermal radiation energy and locally generated thermal noise are at the same temperature, the overall current spectral density is simply the sum of the densities, a satisfying result. Thus, for two parallel resistors R_r and R at the same temperature T,

$$G_i(f) = 2kT(G_r + G) \tag{7-77}$$

Similarly, it is apparent that if two equi-temperature resistors R_1 and R_2 appear in series, the noise-voltage spectral density is given by

$$G_v(f) = 2kT(R_1 + R_2) \tag{7-78}$$

as would be expected.

If one is now interested in the *total* mean-squared noise, measured over all frequencies, one simply integrates the appropriate noise spectral density, whether Eq. [7-76], [7-77], or [7-78], over all frequencies. We have already done this previously in Chap. 6, as well as here, in the development leading to Eqs. [7-72] and [7-74]. (See Eqs. [7-67] to [7-72].)

It is apparent from the approach leading to Eqs. [7-76] to [7-78] that the noise power due to *independent* sources (actually *uncorrelated* sources suffice) is the sum of the individual noise powers. This is of course due to the fact that the variances of *independent* (or, more generally, *uncorrelated*[1]) random variables add. It is worthwhile stressing this point. We do *not* add rms noise currents or voltages. This would be the same as adding (incorrectly) the standard deviations of uncorrelated random variables. Mean-squared values only can be added. This point will occur again and again in the examples following on multiple noise sources. The stress on the condition of *uncorrelated noise sources* is also important. If, as an example, thermal noise due to a given resistor is amplified and the noise voltage then applied to two different circuits, the output voltages of the two circuits are generally correlated. If these voltages are now combined, the resulting voltage is obviously not found from the sum of the two mean-squared voltages. In this case one falls back on first principles. If the two output noise voltages are v_1 and v_2, we have as the average power due to the sum

$$N_o = E(v_1 + v_2)^2 = E(v_1^2) + E(v_2^2) + 2E(v_1 v_2) \tag{7-79}$$

[1] Recall that independence implies zero correlation, while uncorrelated random variables may not necessarily be independent. Only in the case of gaussian random variables does zero correlation imply independence.

(zero average values are assumed here for simplicity). The third term, involving the cross-correlation between v_1 and v_2, must be included in the power calculation.[1]

It is apparent by a simple extension of the arguments above that whenever resistors (or other thermal-noise sources) at the *same temperature* occur in parallel or in series, they may be combined and the mean-squared noise due to the equivalent resistance calculated. If the resistors are at different temperatures and their outputs uncorrelated, as in Eq. [7-76], the noise spectral densities must be individually calculated and then added.

7-3 TRANSISTOR NOISE MODEL: SHOT NOISE AND THERMAL NOISE COMBINED

The actual calculation of noise voltages in electronic systems requires the combination of noise voltages due to shot effect and thermal noise (both picked up at the antenna and generated in the circuits themselves). Total noise voltages (or currents) may be found simply by drawing the appropriate noise-source models and summing spectral densities or individual mean-squared voltages (currents).

A simple transistor amplifier serves as a good example of a circuit with shot and thermal noise both present. Consider therefore the grounded-emitter circuit of Fig. 7-25.

To calculate the noise output of this amplifier, we first have to replace the transistor by its noise equivalent circuit. Various circuits have been

[1] In the special case where the two outputs occupy separate and *nonoverlapping* frequency ranges, the cross-correlation term may be shown to be zero. See Schwartz, Bennett, and Stein, *op. cit.*, pp. 44–45.

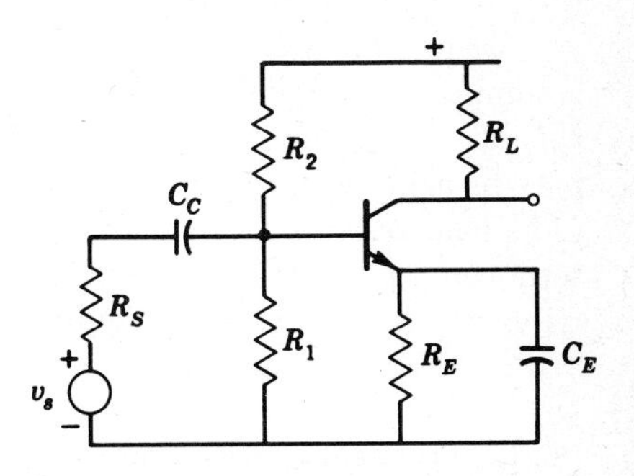

FIG. 7-25 Grounded-emitter transistor amplifier.

suggested for this purpose. One particularly useful one, that utilizes the hybrid-pi transistor model and that is consistent with both theory and experiment, is shown in Fig. 7-26.[1] Note the three noise sources shown.

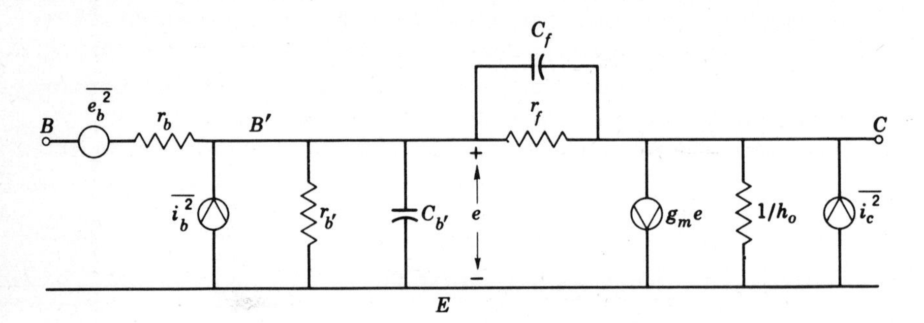

FIG. 7-26 Transistor noise model.

The term $\overline{e_b^2}$ represents the *thermal* noise generated in the base spreading resistance r_b, $\overline{i_b^2}$ is a *shot*-noise term due to current fluctuations in the base, and $\overline{i_c^2}$ is a *shot*-noise term due to fluctuations in the collector. The three terms have the spectral densities and mean-squared values in a frequency range B developed in the previous sections. Thus, they are, respectively,

$$\begin{aligned} \overline{e_b^2} &= 4kTr_bB & G_{e_b}(f) &= 2kTr_b \\ \overline{i_b^2} &= 2eI_bB & G_{i_b}(f) &= eI_b \\ \overline{i_c^2} &= 2eI_cB & G_{i_c}(f) &= eI_c \end{aligned} \qquad [7\text{-}80]$$

I_b is the dc base current and I_c the dc collector current. It is assumed in using this model that the transistor has high current gain and low leakage currents.[2] The three noise sources are assumed uncorrelated, so that their mean-squared contributions may be added.

Using this transistor noise model in the circuit of Fig. 7-25, and assuming for simplicity that we are operating in a frequency range where C_E and C_c provide perfect short circuits, and the coupling terms C_f and r_f may be neglected, we get the simplified amplifier noise model of Fig. 7-27. (The load resistor R_L is a source of thermal noise as well, and this noise term could be included if desired, but as will be shown later, its noise contribution is generally negligible for high-gain amplifiers.) The only

[1] E. R. Chenette, Low-noise Transistor Amplifiers, *Solid State Design,* February, 1964, pp. 27–30.

[2] *Ibid.*

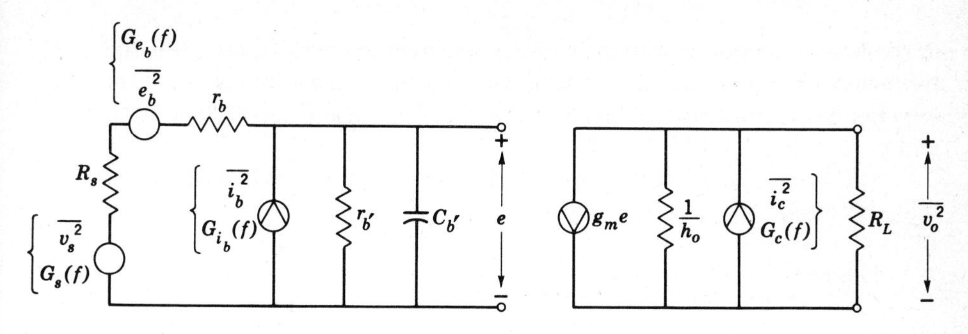

FIG. 7-27 Amplifier noise model.

additional noise term is the thermal noise $\overline{v_s^2}$, with spectral density $G_s(f) = 2kTR_s$ introduced by the source resistance R_s.

The object now is to find the mean-squared output noise $\overline{v_o^2}$. We may do this by working directly with the noise sources shown or by converting the sum of the source and base thermal noises to a current-source equivalent. In this latter case it is readily shown that the output-noise spectral density $G_o(f)$ is just

$$G_o(f) = \left(\frac{1}{h_0 + G_L}\right)^2 \left\{G_{i_c}(f) + \frac{g_m{}^2[G_i(f) + G_{i_b}(f)]}{(g_{b'} + g)^2 + (\omega C_{b'})^2}\right\} \qquad [7\text{-}81]$$

Here $g = 1/(R_s + r_b)$, $G_i(f) = 2kTg$, $g_{b'} = 1/r_{b'}$, $G_L = 1/R_L$. One may now integrate $G_o(f)$ over all frequencies to find $\overline{v_o^2}$. If the amplifier is in turn followed by a narrowband network of bandwidth B, the mean-squared output of this network is in turn $2G_o(f)B$.

Note that the overall noise spectral density of Eq. [7-81] has three terms, corresponding to each of the three noise sources. [The two thermal-noise terms have been combined, as noted, into the one term $G_i(f)$.] The terms $G_i(f)$ and $G_{i_b}(f)$ generated in the base circuit are effectively multiplied by the squared gain of the amplifier, before being added to the collector-noise term.

The relative size of the different terms in Eq. [7-81] is of interest in a typical case. Assume a source resistance $R_s = 1{,}000$ ohms, a base spreading resistance $r_b = 200$ ohms, and a collector current $I_c = 1$ ma. Assume also a dc current ratio $I_c/I_b = 100$. The hybrid-pi parameter g_m is just eI_c/kT for the transistor. At room temperature we have $e/kT = 40$, so that $g_m = 40I_c = 0.04$ mho here. The two other hybrid-pi parameters appearing are given by

$$r_{b'} = \frac{\beta}{g_m} \qquad \text{and} \qquad C_{b'} = \frac{g_m}{\omega_T}$$

β, the low-frequency ac current gain, is approximately I_c/I_b, and ω_T is the frequency at which $\beta = 1$. Taking $\beta = 100$, $r_{b'} = 2{,}500$ ohms.

For these numbers we then find $g = 8 \times 10^{-4}$, $(g_{b'} + g) = 1.2 \times 10^{-3}$, and $g_m/(g_{b'} + g) = 40/1.2 = 33$.

At low frequencies then, for which the frequency-dependent $\omega C_{b'}$ term may be neglected, the base-noise sources are effectively multiplied by $(33)^2 \doteq 1{,}000$ before being added to the collector-noise term.

The noise spectral densities are given respectively by

$$G_i(f) = 2kTg = 6.4 \times 10^{-24} \text{ amp}^2/\text{Hz}$$

$$G_{i_b}(f) = eI_b = \frac{eI_c}{100} = 1.6 \times 10^{-24} \text{ amp}^2/\text{Hz}$$

and

$$G_{i_c}(f) = eI_c = 1.6 \times 10^{-22} \text{ amp}^2/\text{Hz}$$

The low-frequency output-current spectral density

$$G_{i_c}(f) + \left(\frac{g_m}{g_{b'} + g}\right)^2 [G_i(f) + G_{i_b}(f)]$$

is then

$$1.6 \times 10^{-22} + (1{,}000)(6.4 \times 10^{-24} + 1.6 \times 10^{-24})$$
$$= 10^{-22}(1.6 + 64 + 16) \doteq 82 \times 10^{-22} \text{ amp}^2/\text{Hz}$$

Note that the collector-noise term, although larger than either of the two base terms, is overpowered by them because of the high gain of the amplifier. It is in fact easily shown that most of the noise is due to the thermal noise contributed by the source resistance! The transistor amplifier therefore introduces relatively little noise compared to the noise introduced at its input. (We shall see in a later calculation that the additional noise is 40 percent.) It is the noise at the input to a system that effectively determines the noisiness of the system, since noise terms introduced subsequently generally compete with the *amplified* input noise. We shall consider this further in the next section in discussing noise figure and noise temperature.

The output-current spectral density just calculated is effectively the short-circuited current density. It represents the noise current that would flow if the output were short-circuited. The rms short-circuit noise current in a 500-Hz bandwidth in the example given here is $\sqrt{82 \times 10^{-19}} \doteq 2.9 \times 10^{-9}$ amp. If $1/h_o \gg R_L$, the load resistance, and $R_L = 10$ kilohms, as an example, the rms output-noise voltage in a 500-Hz bandwidth is $2.9 \times 10^{-5} \doteq 30$ μv.

7-4 NOISE FIGURE AND NOISE TEMPERATURE

We have in the previous chapters stressed the deleterious effects of noise in representative communication systems. As a rough rule of thumb we found that signal-to-noise ratios of 10 to 20 db are necessary to ensure adequate signal transmission in both analog and digital systems. Using the noise calculations of this chapter, we are now in a position to actually specify signal powers required to ensure adequate signal transmission in both analog and digital systems.

It has already been noted that generally it is the noise introduced at the input to a system that is most critical in determining the resultant SNR. From our discussion of noise in both the previous chapter and this one, it is apparent that rms noise and signal terms are amplified (or attenuated) by the same amount in passing through successive stages of a system. [Recall that the output-noise *power* of a system with noise-equivalent bandwidth B and gain $H(0)$ is just $N_o = n_0|H(0)|^2B$. Here $n_0/2$ is again the input-white-noise spectral density. The rms noise $\sqrt{N_o}$ is thus proportional to the gain $H(0)$.]

Ideally we would like to deal with systems which introduce no noise additional to that present at the system input. This would imply a constant SNR throughout the system, since both signal and noise voltages are amplified or attenuated by the same amount. (We assume here linear systems or linear portions of systems. It is apparent from the previous chapters that nonlinear operations do not preserve signal-to-noise ratio.) In practice this constancy of SNR is impossible, since any additional resistors or transistors, among other devices, introduce additional noise. The SNR must thus continuously decrease throughout a system. (The decrease may be negligible, of course, if the amplification at the beginning of the system is made large enough so that the additional noise introduced later is negligible.)

A study of SNR at different points in a system enables us to pinpoint the significant contributors to the noise. It also enables us to design circuits and systems to minimize the noise. It is thus important to provide a measure of the "noisiness" of a system. One particularly useful measure is the so-called system *noise figure*. The noise figure F is defined as the ratio of the SNR S_S/N_S at the system input (source) to the SNR S_o/N_o at the system output. Specifically,

$$\frac{S_o}{N_o} \equiv \frac{1}{F}\frac{S_S}{N_S} \tag{7-82}$$

An ideal network is thus one whose noise figure is 1 (that is, no additional noise is introduced in the system). As F increases, the "noisiness" of the system increases.

Noise figures are frequently measured in decibels (since F is a ratio of SNR's in power), the conversion simply being $10 \log_{10} F$. Advances in h-f receiver circuits have resulted in spectacular decreases in receiver-noise figures. Radar receivers in the 1,000-MHz range, for example, had noise figures in the 1950s ranging from 10 to 15 db, or F ranging from 10 to 40. Most of the noise was thus developed in the system. The use of low-noise traveling-wave tubes at receiver front ends now makes noise figures of 2 to 3 possible at the 5-kMHz range. The advent of radio astronomy and the impetus it gave to the further development of low noise receivers has resulted in the development of parametric amplifiers and maser amplifiers with noise figures as low as 1.1. These devices require extremely low temperatures and hence elaborate cooling systems for their effective operation. In these latter cases the concept of noise figure no longer has a real significance—it is the external noise introduced at the antenna that provides the major problem, rather than internal receiver noise.[1,2] In these cases the concept of *noise temperature* is often more usefully introduced. We shall discuss this briefly at the end of this section.

The noise figure of a system, as defined by Eq. [7-82], requires the calculation of both signal and noise powers at two points in a system. An alternate definition, completely equivalent to this first one, involves taking the ratio of noise powers only and so is frequently much more readily carried out. To develop this alternate definition requires first the introduction of the concept of *available gain*.

An overall system commonly consists of a series of cascaded networks. Normally any computation of output signal becomes rather complicated because of the loading effect of each network on the one preceding. In calculating the signal-to-noise ratios to determine the noise figure, however, both signal and rms noise appear across the same impedance. The SNR is thus independent of load (the noise contribution of the next stage is taken into account at the output of that stage), and depends only on the output impedance. In calculating the signal voltage and, ultimately, SNR at a point in the system, then, we can assume any arbitrary load impedance—in particular, a load impedance that will simplify the calculations.

Since the SNR's are power ratios, or ratios of mean-squared voltages, the simplest choice of load impedance is the one maximizing the output power. We shall thus assume the load impedance matched to the output impedance. Generalized noise calculations are normally carried out on this basis.

The maximum power available at the output of a system, under matched conditions, is frequently called the *available power*. Thus, for

[1] J. V. Evans and T. Hagfors (eds.), "Radar Astronomy," McGraw-Hill, New York, 1968.

[2] A. E. Siegman, "Microwave Solid-state Masers," McGraw-Hill, New York, 1964.

a source represented by an rms signal voltage e_s and output resistance R_s, the available signal power is

$$S_s = \frac{e_s^2}{4R_s} \qquad [7\text{-}83]$$

(The load resistance is chosen equal to the output resistance.)

Assuming only thermal noise present, the mean-squared noise voltage is

$$\overline{e_n^2} = 4kTR_sB \qquad [7\text{-}84]$$

The available noise power is thus

$$N_s = \frac{\overline{e_n^2}}{4R_s} = kTB \qquad [7\text{-}85]$$

The SNR at the source is simply

$$\frac{S_s}{N_s} = \frac{e_s^2}{\overline{e_n^2}} = \frac{e_s^2}{4kTR_sB} = \frac{\text{available signal power}}{\text{available noise power}} \qquad [7\text{-}86]$$

from Eqs. [7-83] and [7-85].

This of course checks our statement that SNR's may be found from the available power and are independent of load impedances.

Now consider this source connected to an arbitrary linear network as shown in Fig. 7-28. e_o is the open-circuit signal output voltage and R_o the output impedance (assumed resistive here).

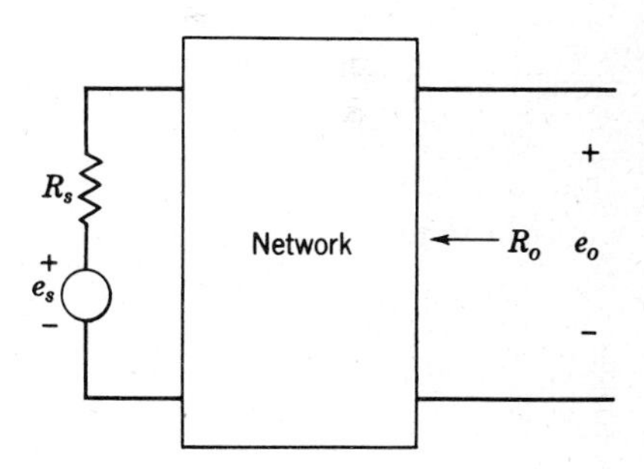

FIG. 7-28 Arbitrary linear network.

The *available power gain G* of this network is defined to be the ratio of available output signal power to available signal power. (If the network is noiseless, the same ratio will hold for available noise powers.) If the output power is *less* than the input, the gain will actually be a *loss*. Thus

$$S_o = \frac{e_o^2}{4R_o} \qquad S_s = \frac{e_s^2}{4R_s} \tag{7-87}$$

and

$$S_o \equiv GS_s \tag{7-88}$$

The output signal-to-noise ratio S_o/N_o is then given by

$$\frac{S_o}{N_o} = \frac{GS_s}{N_o} = \frac{1}{F}\frac{S_s}{N_s} \tag{7-89}$$

by definition of the network noise figure. The noise figure F can thus be written as

$$F = \frac{N_o}{GN_s} \tag{7-90}$$

Equation [7-90] represents an alternate form for the equation of noise figure (aside from the definition in terms of SNR) and is frequently given as the basic definition of F. Thus F may be defined as the ratio of actual noise power available from a network to that which would be available if the network were noiseless.

The available output-noise power N_o can be written in the form

$$N_o = GN_s + N_n \tag{7-91}$$

where GN_s is the available output noise due solely to the input and N_n any additional noise generated in the network itself. Then

$$F = 1 + \frac{N_n}{GN_s} = 1 + \frac{N_n}{GkTB} \tag{7-92}$$ [1]

from Eqs. [7-90], [7-91], and [7-85].

Equation [7-92] demonstrates simply the statements made previously that if there is enough amplification in a system the noises added will contribute negligibly to the overall noise. Here G represents the gain of the system. For $GN_s \gg N_n$, $F \doteq 1$, and the additional noise introduced (N_n) affects the system output SNR negligibly. A noiseless system would have $N_n = 0$ and $F = 1$ identically. The only noise at the output is then that due to the input and passed through the system. If the system does introduce additional noise and $G < 1$ (that is, the power gain is actually a loss), the noise figure of the system may be quite large. Here the incoming noise and signal are both deemphasized so that the noise added could be relatively large.

[1] B is, as previously, a noise bandwidth small enough that the noise spectral densities are effectively constant over this range of frequencies.

As an example of the applicability of this alternate definition of noise figure consider again the transistor amplifier of Fig. 7-25. What is the noise figure of this circuit?

We found previously that the output-current spectral density was 82×10^{-22} amp²/Hz. (Note that this is the same as the square of the *short-circuit* output current density.) If we now set the three noise sources $\overline{e_b^2}$, $\overline{i_b^2}$, and $\overline{i_c^2}$ in Fig. 7-27 equal to zero, the output-noise-current spectral density due to the source R_s only is simply

$$\left(g_m \frac{r_{b'}}{R_s + r_b + r_{b'}}\right)^2 2kTR_s = \left(40 \times 10^{-3} \times \frac{2.5}{3.7}\right)^2 (8 \times 10^{-21} \times 10^3)$$
$$= 58 \times 10^{-22} \text{ amp}^2/\text{Hz}$$

Since both of these output current terms appear across the same resistance, it is apparent that the noise figure is just the ratio of the two, or

$$F = 82/58 = 1.4$$

For these particular numbers, then, the amplifier itself contributes 40 percent more noise power to the output.[1]

This alternate definition of noise figure as the ratio of total noise out to that available with the system noiseless enables us to assess rather readily the relative contributions of successive cascaded networks to the overall noise figure. As an example, consider the two networks cascaded as shown in Fig. 7-29. How does the overall noise figure depend on the individual noise figures and available gains of the individual networks?

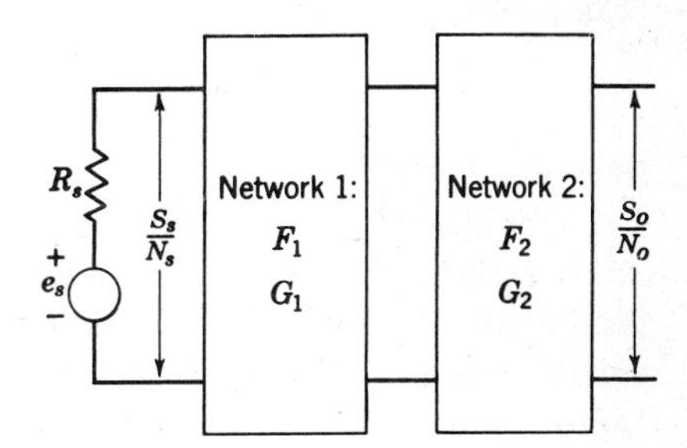

FIG. 7-29 Cascaded linear networks.

By definition of available power the available gains of the two networks multiply, so that

$$S_o = G_1G_2S_s \tag{7-93}$$

[1] In practice, measured noise figures may be considerably higher than this number due to other factors not taken into account.

The available noise power at the system output is now made up of three parts:

1. That due to the source noise and multiplied by the available gain of the overall network. This is simply $G_1G_2N_s$.
2. That introduced in network 1 and transmitted through network 2. This is simply G_2N_{n1}. But

$$F_1 = 1 + \frac{N_{n1}}{G_1N_s}$$

from Eq. [7-92] or

$$N_{n1} = (F_1 - 1)G_1N_s$$

This component of the output noise is thus

$$G_1G_2(F_1 - 1)N_s$$

3. The noise N_{n2} developed in network 2 directly. Again from Eq. [7-92], $N_{n2} = (F_2 - 1)G_2N_s$.

Upon adding together the three noise components the total available noise power at the output is

$$\begin{aligned} N_o &= G_1G_2N_s + G_1G_2(F_1 - 1)N_s + G_2(F_2 - 1)N_s \\ &= G_1G_2F_1N_s + G_2(F_2 - 1)N_s \end{aligned} \qquad [7\text{-}94]$$

But the noise figure F of the overall network is the ratio of total available noise power to that available if the networks were lossless (Eq. [7-90]). Thus

$$F = \frac{N_o}{GN_s} = \frac{N_o}{G_1G_2N_s} = F_1 + \frac{F_2 - 1}{G_1} \qquad [7\text{-}95]$$

from Eq. [7-94].

Obviously, then, in dealing with amplifiers (as in communication or radar receivers), where each stage may have appreciable amplification (and available gain), the first stages determine the SNR properties of the entire system. The noise introduced beyond this point is negligible in comparison with the amplified noise of the first stage or two. (In receivers with mixer circuits the mixer is the primary source of noise.) It is of prime importance then that the first stage of an amplifier be a relatively low-noise stage.

Equation [7-95] demonstrates very clearly the relative contribution of two cascaded networks to the overall noise figure. It can be interpreted exactly as was done for Eq. [7-92]. Thus, if the available gain of the first network, G_1, is large enough ($G_1 \gg F_2 - 1$), the second network contributes negligibly to the overall noise. The overall noise figure remains at F_1.

These results can be extended quite easily to more cascaded networks. Thus, for three networks in cascade,

$$F = F_1 + \frac{F_2 - 1}{G_1} + \frac{F_3 - 1}{G_1 G_2} \qquad [7\text{-}96]$$

In designing circuits therefore, one frequently tries for a design that minimizes the noise figure of the first stage.

The ability of a designer to control somewhat the noise figure of a circuit or a system is also demonstrated by using the transistor amplifier of Fig. 7-25 as an example. If we retain the symbols for the parameters rather than putting in specific numerical values, as we did previously, and disregard $C_{b'}$ to obtain a low-frequency, frequency-independent noise figure, we get as the noise figure of the circuit

$$\begin{aligned} F &= 1 + \frac{\overline{e_b^2}}{\overline{v_s^2}} + \frac{\overline{i_b^2}(R_s + r_b)^2}{\overline{v_s^2}} + \frac{\overline{i_c^2}(R_s + r_b + r_{b'})^2}{(g_m r_{b'})^2 \overline{v_s^2}} \\ &= 1 + \frac{r_b}{R_s} + \frac{I_b e}{2kTR_s}(R_s + r_b)^2 + \frac{eI_c(r_{b'} + R_s + r_b)^2}{2kTR_s(g_m r_{b'})^2} \end{aligned} \qquad [7\text{-}97]$$

Note that F depends on the source resistance R_s and the collector current I_c. It is presumably possible to minimize F with respect to either one of these. To indicate this more explicitly, we introduce the defining relations for g_m and $r_{b'}$ noted previously:

$$r_{b'} = \frac{\beta}{g_m} \qquad g_m = \frac{eI_c}{kT}$$

Also $I_b = I_c/h_f$, with h_f the dc current gain. Then if $h_f \gg 1$, $\beta^2 \gg h_f$, and $r_b \ll r_{b'}$ (usually valid assumptions), it is readily shown that the noise figure may be put in the following simplified form:

$$F \doteq 1 + \left(r_b + \frac{kT}{2eI_c}\right)\frac{1}{R_s} + \left(\frac{e}{2kT}\frac{I_c}{h_f}\right)R_s \qquad [7\text{-}98]$$

Note that a minimum would appear to exist as the source resistance R_s is varied, all other parameters as well as the operating point being held fixed. An optimum F also appears as the operating point (I_c) is varied if the dc current gain h_f may be assumed roughly independent of I_c.

Specifically, the optimum source resistance R_s, in terms of minimum F, is given by

$$R_{s,\text{opt}} = \frac{1}{g_m}\sqrt{h_f(1 + 2g_m r_b)} \qquad [7\text{-}99]$$

again introducing the defining relation $g_m = eI_c/kT$. For this optimum R_s the (minimum) noise figure becomes

$$F_{\min} = 1 + \frac{1}{\sqrt{h_f}} \sqrt{1 + 2g_m r_b} \qquad [7\text{-}100]$$

As an example, we use the same numbers noted previously. Assume that $h_f = 100$, $r_b = 200$ ohms, and $I_c = 1$ ma. Then, at room temperature,

$$g_m = 40I_c = 40 \times 10^{-3} \text{ mho}$$

and

$$F_{\min} = 1.3$$

$$R_{s,\text{opt}} = 750 \text{ ohms}$$

Note that our previous choice, $R_s = 1{,}000$ ohms, giving $F = 1.4$, was not too far from optimum.

If the operating point is varied, R_s remaining the same, a rough estimate of the optimum operating point is obtained by assuming the transistor parameters do not vary with operating point. On this basis, then, a minimization with respect to I_c gives

$$F_{\min} = 1 + \frac{r_b}{R_s} + \frac{1}{\sqrt{h_f}} \qquad [7\text{-}101]$$

with

$$I_{c,\text{opt}} = \frac{kT}{eR_s} \sqrt{h_f} = \frac{0.025}{R_s} \sqrt{h_f} \qquad [7\text{-}102]$$

at room temperature. Again using the numbers used previously, that is, $r_b = 200$ ohms, $R_s = 1{,}000$ ohms, $h_f = 100$, we have $F_{\min} = 1.3$ and $I_{c,\text{opt}} = 0.25$ ma.

We have stressed here the non-frequency-dependent behavior of the transistor-noise figure. The noise figure of a typical junction transistor as a function of frequency is sketched in Fig. 7-30.[1,2] The middle region corresponds to the discussion above. At high frequencies (ω approaching ω_T) the noise figure begins to decrease because of the $C_{b'}$ capacitive loading in the model of Fig. 7-26. At very low frequencies, typically below 1 kHz, the transistor-noise figure rises at a rate of 3 db per octave or functionally as $1/f$. The type of noise producing this differs from the shot and thermal noise previously considered, which have essentially a flat frequency spectrum. The same type of noise effect appears in vacuum-tube work and is called *flicker effect*. It is assumed due to

[1] Chenette, *op. cit.*

[2] E. G. Nielsen, Behavior of Noise Figure in Junction Transistors, *Proc. IRE*, vol. 45, no. 7, pp. 957–963, July, 1957.

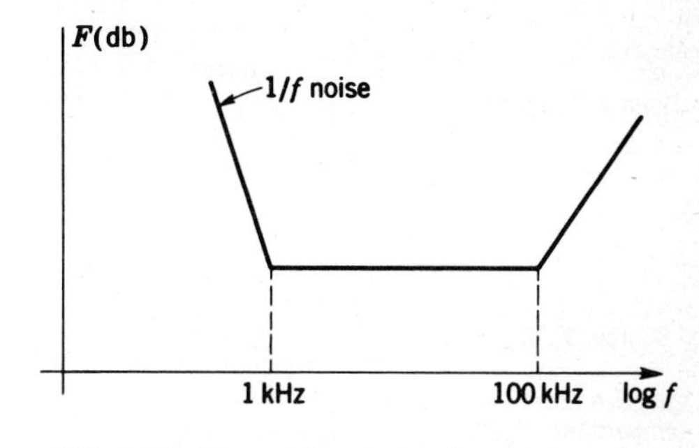

FIG. 7-30 Transistor-noise figure.

inhomogeneities in manufacture and cannot be predicted at present. It may be accounted for in the noise model of Fig. 7-26 by including an additional noise-current source between B' and E in Fig. 7-26, proportional to both I_b and $1/f$.[1]

NOISE TEMPERATURE

It was noted in the early part of this section that the noise-figure concept is not too useful in discussing low-noise devices such as masers and parametric amplifiers. The concept of an effective *noise temperature* of a system (or circuit) has gained favor instead. The noise temperature of any given system is simply defined to be the effective temperature of a thermal-noise source at the system input that would be required to produce the same noise power at the system output, the system then being considered noiseless. If this effective temperature T_e is much less than the equivalent temperature T_s of noise sources appearing at the system input, the system introduces effectively no noise of its own.

Specifically, consider the system of Fig. 7-31 with an equivalent thermal-noise source T_s °K, as shown, and an available gain G. The system output noise in a bandwidth B Hz wide is then

$$N_o = GN_S + N_n = GkT_sB + N_n \qquad [7\text{-}103]$$

with N_n the output noise due to internal noise sources in the system.[2] If we assume N_n due to a fictitious thermal-noise source of temperature T_e at the system input, we must have

$$N_n = GkT_eB \qquad [7\text{-}104]$$

[1] Chenette, *op. cit.*

[2] As previously, we assume B narrow enough so that possible variations with frequency are negligible.

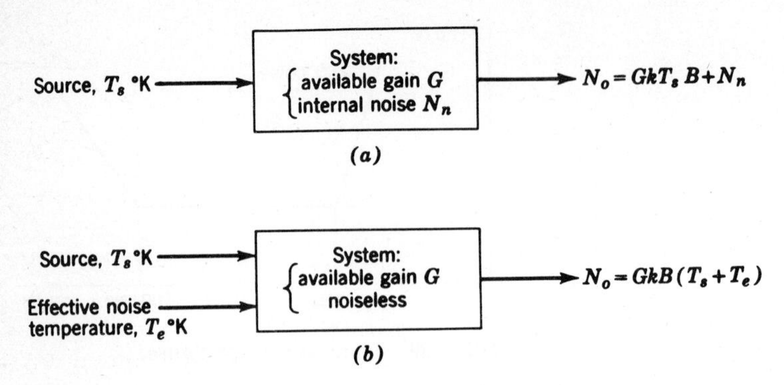

FIG. 7-31 Definition of effective noise temperature. (a) Actual system noise; (b) effective noise temperature.

This then defines the effective noise temperature T_e. Note, as mentioned above, that if $T_e \ll T_s$, the system itself introduces negligible noise.

The system noise figure F and effective noise temperature T_e are related, as is readily seen by writing for F

$$F = \frac{N_o}{GN_S} = 1 + \frac{N_n}{GN_S} = 1 + \frac{T_e}{T_s} \qquad [7\text{-}105]$$

The specific value for effective noise temperature depends on the equivalent source temperature T_s, but as an example, low-noise maser receivers may provide a T_e of 10 to 30°K. (If T_s is room temperature, 300°K, this corresponds to noise figures of 1.03 to 1.1.)

Further details of the noise-temperature approach and its applicability to specific devices and systems appear in some of the references cited in this chapter.[1]

7-5 SUMMARY

We have discussed in this chapter the sources of the fluctuation noise considered in depth in Chap. 6. Two major types of noise encountered in communication systems (as elsewhere) were considered: shot noise and thermal noise. The first, characterized by a current spectral density proportional to the average current flow, occurs in electron and hole flow in semiconductors, in electron flow in vacuum tubes, in light (photon) detection by photodetectors, etc. It appears in any particle flow process obeying Poisson statistics. The second arises in all dissipative processes

[1] See, e.g., Evans and Hagfors, *op. cit.;* Siegman, *op. cit.*

as well as in the random radiation of energy by many oscillators in thermodynamic equilibrium with their surroundings (blackbody radiation).

Both types of noise may be modeled by white-noise processes at sufficiently low frequencies (less than 10^{13} Hz in the case of thermal noise, the reciprocal of the transit time in the shot-noise case), or if followed by relatively narrowband systems and devices.

The total noise generated in, and transmitted through, a given system may be found by replacing active sources and dissipative elements by their appropriate noise models. The overall noise spectral density (or mean-squared noise) at a system output may then be found by appropriately superposing the noise output due to each source. The concept of noise figure was introduced as an aid in comparing the relative noisiness of different networks, as well as in determining the significant contributors to the noise in a complex system. The noise figure was defined essentially to be the ratio of mean noise power out of a system to mean noise power out if the system were noiseless.

We showed that in normal communication systems with several stages of amplification included, the noise generated in the first stage was the primary factor in determining the system noise figure.

Using the noise figure of a system (either measured or calculated), one is in a position to determine the total noise power, either at a specified point in a system or referred to the system input, that signals measured at the same point must contend with. It is apparent from our discussion in Chap. 6 (as well as in portions of Chap. 5) that the signal-to-noise ratios at various points in a system determine the system performance: whether the carrier-to-noise ratio at the input to an FM discriminator, the signal-to-noise ratio at the decision point in a binary PCM system, or the signal-to-noise ratio at the output of an envelope detector.

Knowing the actual noise power either at a specific point or at the system input, and the SNR required to attain a specified performance, we can determine the signal power required at the system input. (This is generally at the input to the antenna in radio-type systems.) This in turn enables us to specify the power required at the transmitter. Examples of such calculations appear among the problems for this chapter as well as in the final chapter that follows. (See Sec. 8-6 for a specific example.)

In modern low-noise receiver systems, particularly those designed for radio and radar astronomy applications, the noise figure is close to 1 and loses much of its utility as a measure of system noisiness. In these systems the concept of effective noise temperature has been found to be quite useful. This is measured by the temperature of an equivalent thermal-noise source at the input that would be required to produce the same noise at the output.

PROBLEMS

Note: All temperatures are assumed to be 20°C except where otherwise noted.

7-1 The diode of Fig. 7-1 is operated in the temperature-limited region. $I_b = 5$ ma, $R_L = 10$ kilohms.

(*a*) Find the rms noise voltage, due to the diode only, across R_L. Assume a bandwidth of 10 kHz.
(*b*) Find the rms noise voltage due to the resistive load in a 10-kHz bandwidth.
(*c*) Find the *total* rms noise voltage across R_L in a 10-kHz bandwidth.
(*d*) $I_b = 0.1$ ma, $R_L = 1$ kilohm, and the bandwidth is 10 kHz. Repeat (*c*).

7-2 Calculate the rms noise voltage in a 10-kHz bandwidth across:

(*a*) a 10-kilohm resistor.
(*b*) a 100-kilohm resistor.

Repeat for a 5-kHz bandwidth. Repeat for $T = 30$°K.

7-3 The resistive load of Prob. 7-1 is replaced by a resistor and capacitor in parallel. $R = 1$ kilohm, $C = 0.02$ μf, $I_b = 5$ ma.

(*a*) Find the total rms noise voltage across the *RC* combination, summing contributions at all frequency ranges.
(*b*) What portion of the total noise power lies below 30 kHz? 15 kHz?

7-4 Consider the circuit of Fig. P 7-4. The diode is temperature-limited, and its average current is $I_b = 1$ ma. $R = 100$ ohms. The $H(\omega)$ network shown is an amplifier with a noise-equivalent bandwidth of 5 kHz and a voltage amplification of 10^6. Assuming that the amplifier contributes negligible noise, find the rms output voltage.

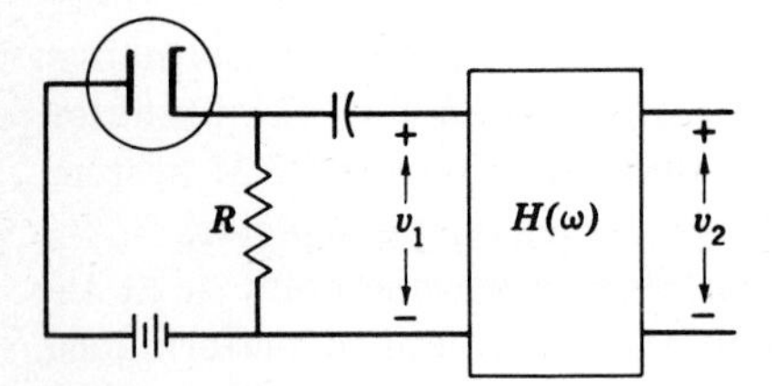

FIG. P 7-4

7-5 Consider a "noisy" resistor R at temperature T°K and an inductor L connected in series.

(*a*) Find the spectral density of the current through the inductor.
(*b*) Show that the average energy stored in the inductor is $\frac{1}{2}kT$.

7-6 A simplified model of an optical communications system is shown in Fig. P 7-6. The transmitter produces pulses of light which travel down the optical path to the photodetector. The photodetector current i may be written

$$i = \bar{I} + i_s + i_B$$

$\bar{I}$, the *signal*, is the average current due to the transmitted light. i_s is a shot-noise component with spectral density $G_s(f) = e\bar{I}$, due to the arrival of the photons at dis-

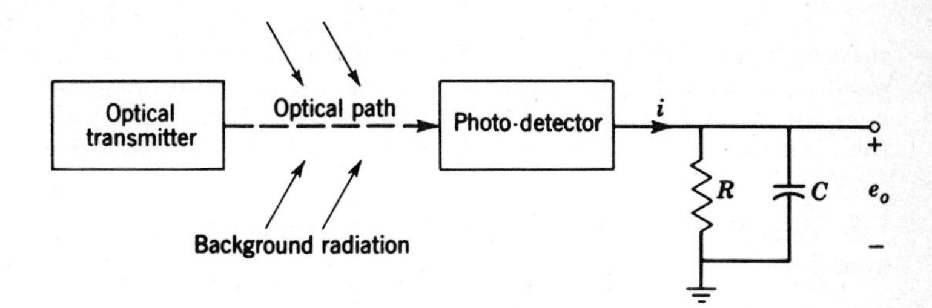

FIG. P 7-6 Optical communications system.

crete but random instants. i_B is a white-noise component with spectral density $G_B(f) = \eta$, due to background radiation.

(*a*) Derive an expression for the signal-to-noise power ratio at the output of the *RC* filter. Assume that the resistor *R* is noiseless.
(*b*) Repeat if *R* is a source of thermal noise.

7-7 Carry out the details of the calculation in deriving $R_y(\tau)$ of Eqs. [7-36] and [7-37] (Fig. 7-13) from Eq. [7-35].

7-8 Calculate the noise-equivalent bandwidths of the three pulses of Fig. 7-16.

7-9 Consider the random telegraph wave which switches between +1 and 0. The switching instants are described by the Poisson distribution, with μ the average number of switches per unit time.

(*a*) Find the autocorrelation function of this random signal.
(*b*) Find its power spectral density.
(*c*) Estimate the bandwidth of a communications channel which must transmit this wave. How is the bandwidth related to μ?

7-10 The random telegraph wave described in Prob. 7-9 above is applied to the input terminals of an ideal low-pass filter, with bandwidth (cutoff frequency) *B*.

(*a*) Find an expression for the mean-square value of the filter output.
(*b*) Repeat (*a*) if the low-pass filter is replaced by a simple *RC* circuit with time constant *T*. Compare with the answer to (*a*).

7-11 *Noise in junction diodes.* A semiconductor junction diode is connected in the circuit shown in Fig. P 7-11*a*. The noise model for the diode operating at a dc current

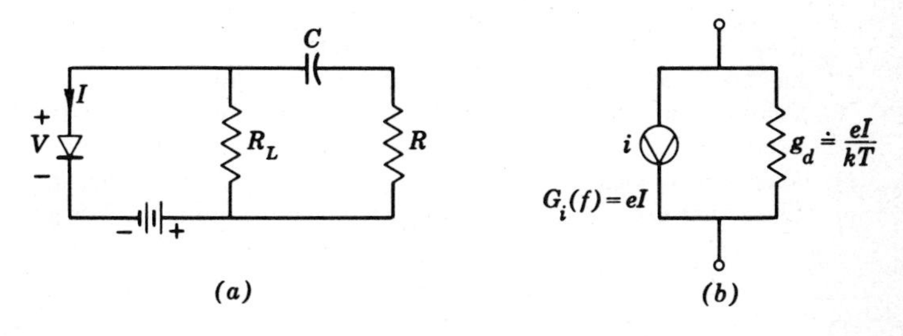

FIG. P 7-11 Junction-diode noise model. (a) Circuit; (b) junction-diode noise model.

of I amp consists of the shot-noise-current source in parallel with the diode incremental conductance $g_d = \partial I/\partial V \doteq eI/kT$, shown in Fig. P 7-11*b*. At room temperature, $g_d = 40I$, $R_L = 10$ kilohms, $R = 100$ kilohms, $I = 1$ ma. Find the total rms noise voltage across resistor R in a 5-kHz bandwidth. (The l-f cutoff frequency due to capacity C is 50 Hz.)

7-12 Using Eq. [7-61], show that the average energy of the harmonic oscillator is given by Eq. [7-62].

7-13 Carry out the details of the derivation in going from Eq. [7-69] to Eq. [7-72]. Derive Eq. [7-74] in detail.

7-14 A transistor-noise model appropriate for the common base configuration is shown in Fig. P 7-14. r_e is the incremental emitter resistance: $1/r_e = \partial I_e/\partial V_e \doteq eI_e/kT$.

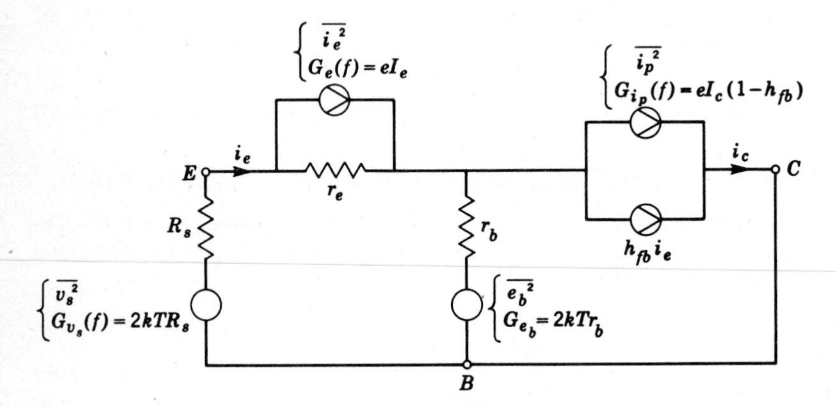

FIG. P 7-14 Common-base-transistor noise model.

$h_{fb} = i_c/i_e$ is the common base-current "gain" given by $h_f/(1 + h_f)$ in terms of the common emitter-current gain factor. R_s is the source resistance.

(*a*) Derive the noise figure for this circuit and show it may be written in the form

$$F = 1 + \frac{r_b + r_e/2}{R_s} + \frac{(r_b + r_e + R_s)^2}{2h_f r_e R_s} \doteq 1 + \frac{r_b(1 + r_b/2h_f r_e) + r_e/2}{R_s} + \frac{R_s}{2h_f r_e}$$

if $h_f \gg 1$, $r_b \ll h_f r_e \sim r_b'$, $r_e \ll r_b$. (Here the approximate dc relation $I_c = h_{fb}I_e$ has been used to eliminate I_c, and I_e has in turn been eliminated in terms of r_e.) Compare this result with that for the common emitter amplifier (Eq. [7-98] in the text).

(*b*) Calculate F with $r_b = 200$ ohms, $r_e = 20$ ohms, $h_{fb} = 0.98$, $R_s = 1{,}000$ ohms.

(*c*) Show that the optimum source resistance for minimum F is given by

$$R_{s,\text{opt}} \doteq \frac{1}{g_m}\sqrt{h_f(1 + 2g_m r_b) + g_m{}^2 r_b{}^2}$$

Compare this result with that for the common emitter amplifier (Eq. [7-99]). Calculate $R_{s,\text{opt}}$ for the numbers given in (*b*) above. Evaluate the noise figure F

for this value of R_s and show that F does not vary greatly with large changes in R_s.

7-15 Find the output-noise spectral density of the transistor amplifier of Figs. 7-25 and 7-27, verifying Eq. [7-81]. Do this two ways:

(*a*) By combining the three input-noise sources into one equivalent one.
(*b*) By finding the output noise due to each noise source separately, and then applying superposition.

7-16 (*a*) Derive the noise figure of the amplifier of Fig. 7-25, obtaining the result shown in Eq. [7-97].
(*b*) Show that the expression for the noise figure may be simplified to that given by Eq. [7-98].
(*c*) Verify the optimum source-resistance expression given by Eq. [7-99]. Show that the optimum operating point is given by Eq. [7-102].

7-17 A temperature-limited diode noise source is to be used to determine experimentally the noise figure of a given amplifier. The circuit used is shown in Fig. P 7-17.

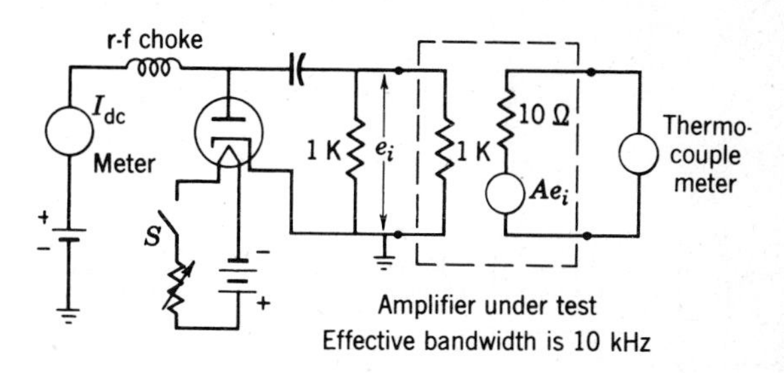

FIG. P 7-17 Network noise determination.

With switch S open, the output power is 1 mw. With S closed, the output power is 3 mw and I_{dc} = 0.3 ma.

(*a*) What is the amplifier noise figure?
(*b*) What is the decibel gain of the amplifier?
(*c*) What are the rms input and output noise voltages with S open?

7-18 Derive Eq. [7-96], the noise figure of three networks in cascade.

7-19 *Noise figure of lossy transmission line (cable).* Consider signals transmitted along a cable of length l with voltage attenuation constant α (Fig. P 7-19). The ratio of power out to power in is then $L = e^{-2\alpha l}$. It can then be shown[1] that the line introduces thermal noise due to the line losses given by $N_l = (1 - L)kTB$, $hf \ll kT$, with T the temperature of the line and B a specified bandwidth.

Show that the noise figure of the lossy cable at T°K is

$$F = 1 + \frac{1 - L}{L} = \frac{1}{L}$$

For a line with $L = \frac{1}{2}$, then, $F = 2$.

[1] Siegman, *op. cit.*, pp. 373–375.

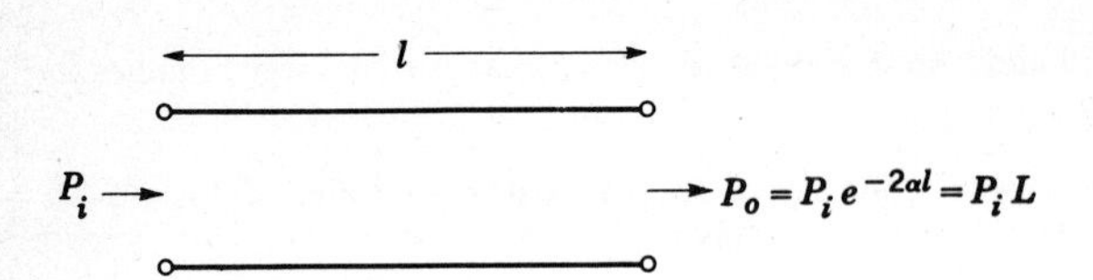

FIG. P 7-19 Lossy transmission line.

7-20 *Maser receiving systems.* Consider the low-noise maser receiver shown in Fig. P 7-20. Although the low-noise (low-temperature) maser amplifier is used to decrease the system noise as much as possible, it is found in practice that the line feed from antenna to the maser amplifier provides the limitation on the reduction of the system noise. To demonstrate this, consider the following examples:

(*a*) The line loss is $L \equiv -0.1$ db. The maser has an effective noise temperature of 3°K. Show that the effective noise temperature at the antenna input = 10°K. (The second-stage noise contribution may be neglected.)
(*b*) The maser amplifier noise is now 30°K. Show that the effective noise temperature at the antenna input is ≐ 37°K.

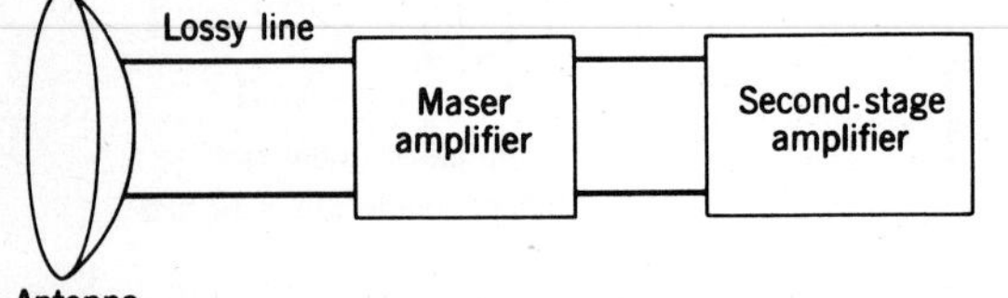

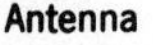

FIG. P 7-20 Maser receiver.

7-21 A high-frequency binary OOK communications system has a receiver with a noise figure of 3 db. The system transmits 10^6 bits/sec. Find the signal power required at the receiver input if an error probability of 10^{-5} is to be maintained. Repeat for a noise figure of 10 db.

7-22 (*a*) A single-pulse radar system is to be designed to have a noise-error probability of 10^{-10}, with a probability of detection of 50 percent. Find the received power required at the radar receiver input if the receiver noise figure is 3 db.
(*b*) Repeat if 100 return pulses are summed before a decision is made.

Note: Refer to Sec. 6-8 in working this problem.

7-23 *Measurement of effective noise temperature*

(*a*) Consider a system denoted by H in Fig. P 7-23 whose effective noise temperature T_e is to be measured. Two thermal-noise sources, one at temperature T_1 °K, the other at temperature T_2°K, as shown, are separately connected to the input of H. The output power, denoted respectively by P_1 and P_2, is measured in each

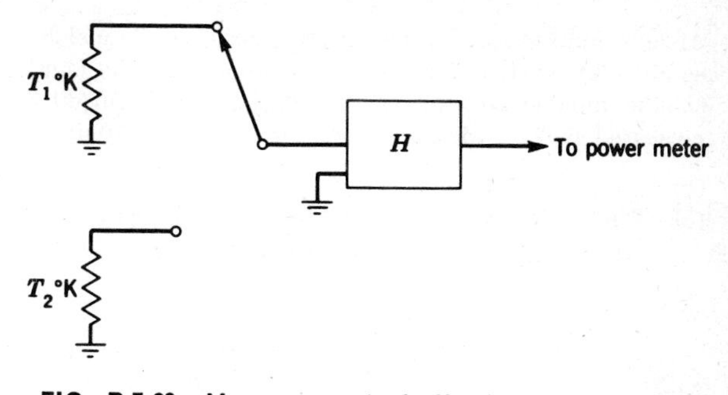

FIG. P 7-23 Measurement of effective noise temperature.

case. Show that the effective noise temperature is then given by

$$T_e = \frac{T_1 - yT_2}{y - 1}$$

with $y \equiv P_1/P_2$ the ratio of the two powers.

(*b*) Show how a calibrated attenuator plus some sort of power-indicating device at the output of H may be used to measure P_1/P_2. (Absolute power readings are then not necessary.)

7-24 *Noise-loading test for distortion measurement.* One hundred baseband channels each 5 kHz wide are frequency-multiplexed side by side as shown in Fig. P 7-24. The

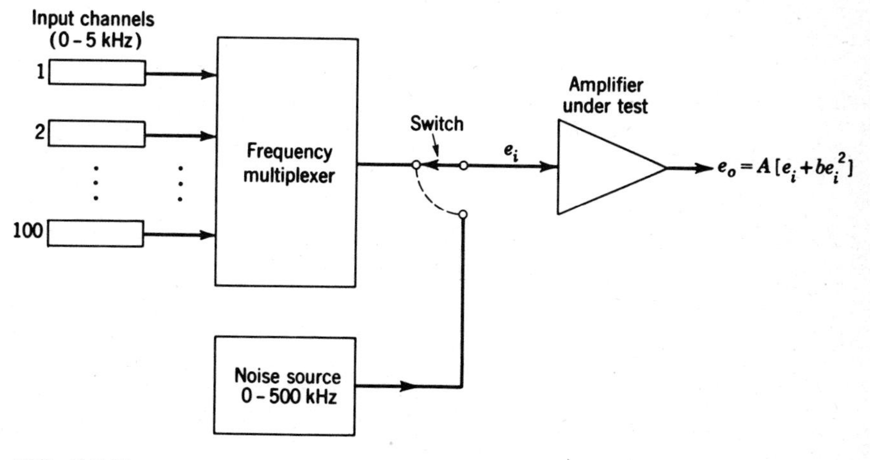

FIG. P 7-24

multiplexed output is passed through an amplifier with some quadratic distortion, as shown. A noise-loading test is performed as follows to determine the distortion coefficient b. (1) First the 100 channels are simulated by using in their place a spectrally flat gaussian noise source from 0 to 500 kHz, with an rms value of 1 volt.

At the amplifier output, the noise appearing in channel 20 (95 to 100 kHz) is measured as S_0. (2) Next, the same noise is applied to the amplifier, but with the input at channel number 20 notched out. The amplifier output in channel number 20 is now measured as N_0. Measurements show $S_0/N_0 = 40$ db.

(*a*) Show $b = 7.4 \times 10^{-3}$.
(*b*) Find S_0/N_0 when the same test is performed on channel number 60 (295 to 300 kHz). (*Ans.* -41 db.) Make all reasonable approximations.

Note: Recall from Chap. 6 that for gaussian stationary noise $n(t)$,

$$E[n^2(t)n^2(t+\tau)] = 2R_n{}^2(\tau) + R_n{}^2(0)$$

with

$$R_n(\tau) = E[n(t)n(t+\tau)]$$

Also,

$$E[n^2(t)n(t+\tau)] = E[n(t)n^2(t+\tau)] = 0$$

CHAPTER EIGHT

STATISTICAL COMMUNICATION THEORY AND DIGITAL COMMUNICATIONS

We have now come full circle in our discussion of information transmission. In Chap. 1 we pointed out qualitatively that the rate of transmission of information, in bits per second, was limited by two basic quantities, the time response or bandwidth of systems through which the information was to pass, and noise innately present in all systems. We then went on to discuss in detail system frequency response, noise, and their connection in various types of communication systems. An attempt was made to compare various communication systems on the basis of signal-to-noise and bandwidth exchange.

We have thus far concentrated on the analysis and comparison of the most commonly used communication schemes, whether for the transmission of digital or analog signals. We now return to the question of information transmission from a more fundamental viewpoint. We ask whether it is possible to *optimize* the design of systems in the sense of maximizing the information-transmission rate with prescribed constraints on error probability, signal power, noise, and bandwidth. This leads us into the realm of statistical communication theory.

We have already considered some aspects of system optimization at various points throughout this book. Thus, in discussing binary communications, we discussed the optimum setting of a decision threshold. The concept of matched filtering arose out of the discussion of maximizing SNR in binary transmission. We applied the Schwarz inequality in considering optimum emphasis networks for FM and AM analog transmission. We discussed the SNR-bandwidth exchange in pulse-code-modulation (PCM) systems, and indicated that it was similar to that found by Shannon for a hypothetical optimum digital transmission system.

These were essentially isolated cases, however, useful in developing familiarity with existing systems and their performance. We now attack the problem of optimum information transmission in a more systematic way. We use here as a tool the elements of statistical decision theory.

The emphasis throughout will be on *digital* communications, first because of its rapidly growing importance in modern technology and second because the optimization procedures, based primarily on the minimization of probability of error, are much simpler to carry out and interpret than for analog signal transmission.[1]

After a necessary introduction to statistical decision theory with specific reference to binary communication, we consider in detail the optimum design of binary communication systems, designed to perform with minimum probability of error in the presence of additive gaussian noise.[2] The specific questions to be answered are: (1) What is the optimum decision procedure at the receiver? and (2) Is it possible to optimally design signals at the transmitter?

Surprisingly, we shall find both questions answered simultaneously by our statistical decision approach. Bipolar signal transmission with matched-filter decision threshold detection will be found to be the optimum binary transmission scheme! This is, of course, exactly the system analyzed earlier—the system we found theoretically superior to both frequency-shift-keyed (FSK) and on-off-keyed (OOK) transmission.

We know from Chaps. 5 and 6, however, that even with optimum matched filtering the resultant probability of error attainable depends ultimately on the ratio of signal energy to noise spectral density. If this ratio is already as large as we can possibly make it, is it possible to further decrease the probability of error? The answer of course lies in the famous Shannon channel-capacity expression first introduced in Chap. 1, and then referred to again in Chap. 3 in studying PCM systems. This expression, to which we shall again refer, indicates that it is possible to transmit digital information with as low an error rate as desired. No recipe for finding specific systems that perform in this optimum manner is prescribed, but at least two specific schemes have been found in practice. One, requiring an exponentially increasing bandwidth as the error rate is decreased more and more, involves the transmission of one of M orthogonal signals. The other, which uses only finite bandwidth, requires the use of a feedback channel. Both techniques will be discussed here and provide further insight into the significance of the Shannon capacity expression.

We then conclude this chapter by considering briefly other ways of

[1] Optimum analog transmission relies on the concepts of statistical *estimation* theory. See A. J. Viterbi, "Principles of Coherent Communication," part 2, McGraw-Hill, New York, 1966, for an introduction and comprehensive bibliography.

[2] As noted earlier in the book, this model is particularly appropriate for space communications. We ignore intersymbol interference here, the major problem in telephone data transmission. See R. W. Lucky, J. Salz, and E. J. Weldon, Jr., "Principles of Data Communication," McGraw-Hill, New York, 1968, for a detailed treatment of both noise and intersymbol interference.

improving digital transmission performance. These involve error detection and error correction. Here the addition of extra (redundant) data symbols to the signal binary stream enables error detection and correction to be carried out at the cost, of course, of increased system complexity and reduction of data rate (or increased bandwidth).

8-1 STATISTICAL DECISION THEORY

It is apparent from all our discussions in this book that the problems of deciding between either of two signals transmitted in a binary communication case, or of appropriately processing analog signals in the AM or FM case, are essentially statistical in nature. The signals transmitted are, of course, random to begin with, the noise added enroute or at the receiver can generally only be described statistically, the fading and multiplicative noise possibly encountered enroute can likewise only be described statistically, etc. We must then look to the realm of statistics for techniques that may be directly carried over to the communication field to help develop schemes for optimumly transmitting and processing signals.

The fields of statistical decision and estimation theory have proven particularly fruitful in handling the problem of optimum information transmission. Statistical decision theory, as is apparent from the name, deals specifically with the problem of developing statistical tests for optimumly deciding between several possible hypotheses. One would expect techniques developed here to be particularly useful in digital communications, detection radar, and other systems where discrete *decisions* have to be made (which one of two signals transmitted, which one of M signals transmitted, is it a target or noise, etc.). Statistical estimation theory, on the other hand, deals with methods of estimating as "best" as possible continuous random parameters or time-varying random functions. It is thus applicable to problems of analog transmission. We concentrate here only on statistical decision theory and its application to digital communications.[1]

[1] Many fine books exist on statistical theory. Included are A. M. Mood and F. A. Graybill, "Introduction to the Theory of Statistics," 2d ed., McGraw-Hill, New York, 1963; H. Cramer, "Mathematical Methods of Statistics," Princeton, Princeton, N.J., 1946; M. Fisz, "Probability Theory and Mathematical Statistics," 3d ed., Wiley, New York, 1963. Books in communications that incorporate and expand on the material in this chapter include D. J. Sakrison, "Communication Theory: Transmission of Waveforms and Digital Information," chap. 8, Wiley, New York, 1968; J. M. Wozencraft and I. M. Jacobs, "Principles of Communication Engineering," Wiley, New York, 1965; Viterbi, *op. cit.*, part 3; C. W. Helstrom, "Statistical Theory of Signal Detection," Pergamon, New York, 1960; D. Middleton, "Introduction to Statistical Theory of Communication," McGraw-Hill, New York, 1960; W. B. Davenport and W. L. Root, "An Introduction to the Theory of Random Signals and Noise," McGraw-Hill, New York, 1958.

By optimum or best system we shall mean here for the most part one that minimizes the probability of error.[1]

To bring out the elements of statistical decision theory and its applicability to digital communication, consider the problem of distinguishing between either one of two possible signals received. Assume that s_1 or s_2 has been transmitted and a voltage v received. We sample v, and, based on the value of the sample measured, determine as "best" as possible which of the two signals was transmitted. (Later, we shall extend this to the case of many sequential samples.)

In statistical terminology, we are given the value of a statistical sample (or group of samples) and wish to select between either of two alternative hypotheses. Call these H_1 and H_2. Hypothesis H_1 corresponds to the decision s_1 transmitted; H_2 of course to the decision s_2 transmitted.

How does one establish a rule for deciding between the two hypotheses? Note that we can make two types of error. Assuming H_1 true, s_2 may very well have been transmitted, or alternately, assuming H_2 true, s_1 may have been transmitted. It is apparent that the total probability of error to be minimized is based on both these errors. It is also apparent that the rule to be chosen will consist of splitting up the one-dimensional space corresponding to all possible values of v into two nonoverlapping (mutually exclusive) regions V_1 and V_2 such that the overall error probability is minimized.[2] We assume that we know the a priori probabilities

[1] It is important to keep in mind that the word *optimum* is usually meant in a restricted sense; it refers to the "best" system according to the particular criterion adopted for evaluating the system performance.

[2] In most of the references cited, more general cost functions than simple error probability are considered. Generally, the resultant rule is a simple extension of the one developed here, however.

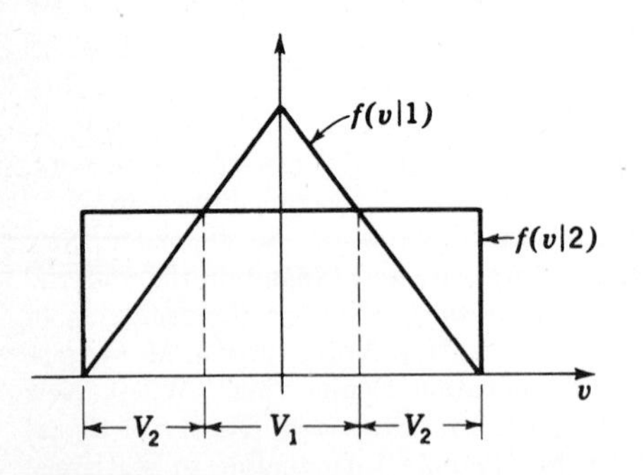

FIG. 8-1 Choice of binary decision regions.

P_1 and P_2 of transmitting s_1 and s_2, respectively ($P_1 + P_2 = 1$), as well as the *conditional* probability densities $f(v|1)$ and $f(v|2)$, corresponding respectively to the probability of receiving v given s_1 transmitted, and v given s_2 transmitted. A typical example is shown in Fig. 8-1.

We set up the expression for overall probability of error and then minimize it by adjusting V_1 and V_2. This then provides the desired rule.

Just as in the error-probability calculations of Chap. 5 we may find the probability of error by considering first the probability that v will fall in region V_2 even though a 1 has been transmitted. This is simply

$$\int_{V_2} f(v|1)\, dv$$

Similarly, the probability that v will fall in V_1 even though signal 2 has been transmitted is given by

$$\int_{V_1} f(v|2)\, dv$$

Again, as in Chap. 5 the overall probability P_e that is to be minimized by adjusting V_1 (or V_2) is found by weighting each of the integrals above by its respective a priori probability and then summing the two:

$$P_e = P_1 \int_{V_2} f(v|1)\, dv + P_2 \int_{V_1} f(v|2)\, dv \qquad [8\text{-}1]$$

We use a little trick now to perform the minimization quite directly. Since $V_1 + V_2$ covers all possible values of v, we have

$$\int_{V_1+V_2} f(v|1)\, dv = 1 = \int_{V_1} f(v|1)\, dv + \int_{V_2} f(v|1)\, dv \qquad [8\text{-}2]$$

We can then eliminate the integral over V_2 in Eq. [8-1], writing instead

$$P_e = P_1 + \int_{V_1} [P_2 f(v|2) - P_1 f(v|1)]\, dv \qquad [8\text{-}3]$$

Since P_1 is a specified number and assumed known, P_e is minimized by choosing the region V_1 appropriately. A little thought indicates that this is done by adjusting V_1 to have the integral term in Eq. [8-3] negative and as large numerically as possible. But we recall that probabilities and density functions are always positive. The solution then corresponds quite simply to picking V_1 as the regions of v corresponding to

$$P_1 f(v|1) > P_2 f(v|2) \qquad [8\text{-}4]$$

This is then the desired decision rule. As an example, let $P_1 = P_2 = \frac{1}{2}$. The region V_1 then corresponds to all values of v where $f(v|1) > f(\mathrm{v}|2)$. Three examples are shown in Fig. 8-2.

Note that the first example resembles that of choosing between two gaussian distributions, and shows the optimum decision level occurring precisely at the intersection of the two functions. This agrees of course

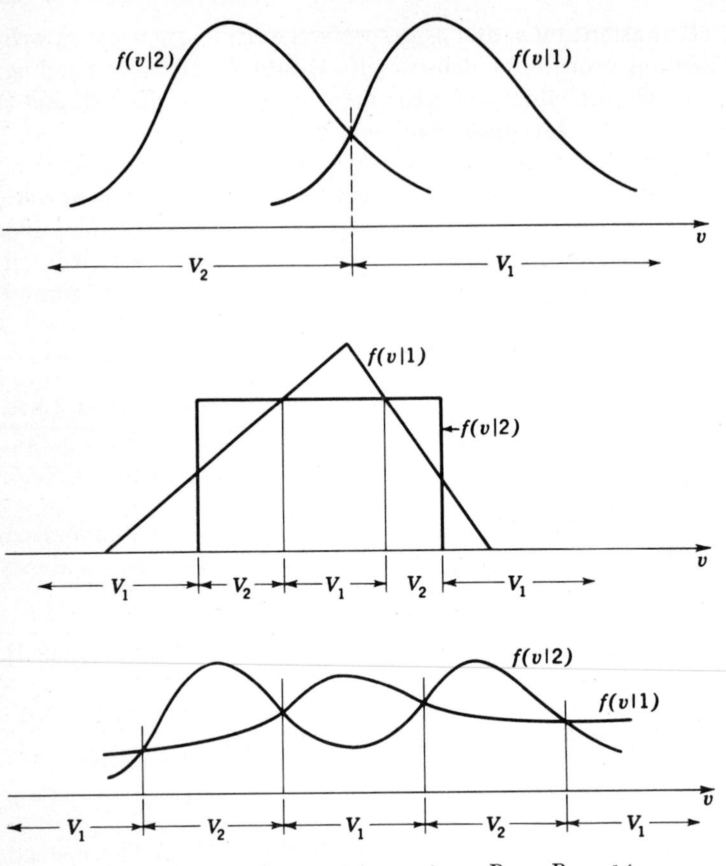

FIG. 8-2 Examples of binary decision regions, $P_1 = P_2 = \frac{1}{2}$.

with the optimum threshold location found in Chap. 5. Some simple mathematics will verify this result. Assume that the two signals transmitted are A and 0, with a priori probabilities P_1 and P_2, respectively. Gaussian noise of zero average value and variance N (mean noise power) is added on reception. As in Chap. 5 the two conditional density functions are, respectively,

$$f(v|1) = \frac{1}{\sqrt{2\pi N}} e^{-(v-A)^2/2N} \qquad [8\text{-}5]$$

and

$$f(v|2) = \frac{1}{\sqrt{2\pi N}} e^{-v^2/2N} \qquad [8\text{-}6]$$

The optimum threshold level is now found by substituting into Eq. [8-4].

The optimum decision rule often is written in the equivalent form

$$l \equiv \frac{f(v|1)}{f(v|2)} > \frac{P_2}{P_1} \tag{8-7}$$

The parameter l, a function of the sample value v, is called the *likelihood ratio*. The region V_1 corresponding to the decision that S_1 was transmitted then corresponds to those values of v for which $l > P_2/P_1$. More general approaches to optimum decision theory, in which specified costs rather than probability of error are to be minimized, lead to the likelihood ratio solution as well. The number P_2/P_1 is then replaced by suitable combinations of the costs and P_1 and P_2. This is discussed in some of the references cited earlier.

Since the logarithm of a function is monotonic with that function, it is apparent that the boundaries V_1 or V_2 will not change if we take the logarithm of both sides of Eq. [8-7]. This simplifies the algebra considerably, particularly in those cases where the density functions are exponential. Using the gaussian functions of Eqs. [8-5] and [8-6] as specific examples, it is apparent, after taking logarithms, that the region V_1 is given by

$$v^2 - (v - A)^2 > 2N \ln \frac{P_2}{P_1} \tag{8-8}$$

Expanding the term in parentheses and rewriting, we find as the desired region V_1 all those values of v corresponding to

$$v > \frac{A}{2} + \frac{N}{A} \ln \frac{P_2}{P_1} \tag{8-9}$$

Note that for equally likely signals, the optimum threshold is $A/2$, just the value found in Chap. 5. If signal s_2 is transmitted more often, the threshold moves to the right, indicating a greater willingness to choose s_2. This is shown in Fig. 8-3.

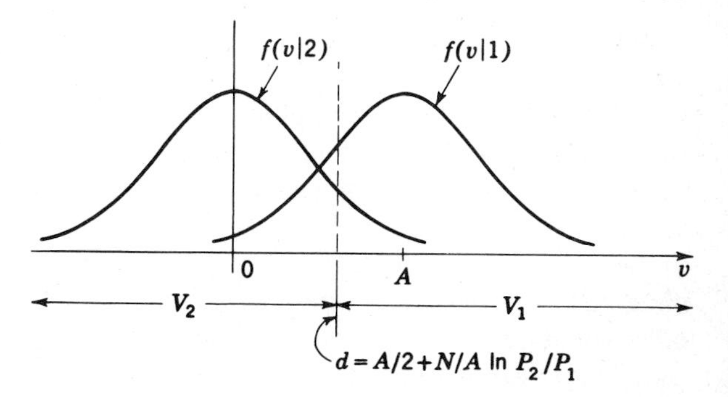

FIG. 8-3 On-off signals plus gaussian noise.

As another example assume OOK transmission, additive gaussian noise, and envelope detection at the receiver. From our discussion in Chap. 6 it is apparent that the detected envelope r is Rayleigh-distributed in the case of a zero transmitted, Rician-distributed if a 1 is transmitted. The two envelope-density functions are shown sketched in Fig. 8-4, with the optimum decision level shown at their intersection in the case of equally probable transmission.

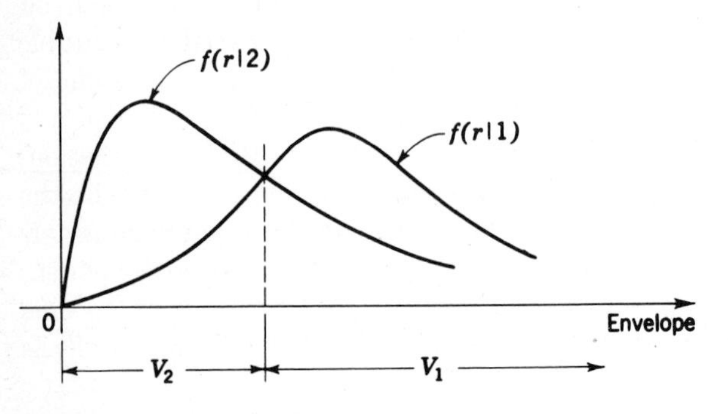

FIG. 8-4 Envelope-detected signals, optimum decision regions.

8-2 SIGNAL VECTORS—MULTIPLE-SAMPLE DETECTION

We have shown how one optimally decides between either of two signals received on the basis of one received sample. It is apparent that one should be able to improve the signal detectability by making available more received samples. The obvious questions then are:

1. How much improvement is to be expected?
2. What is the optimum decision rule?

In this latter case one can alternately ask: What is the optimum way of processing the samples?

We shall assume for simplicity that we have n *independent* samples on which to base the decision. (For the case of binary pulses in additive gaussian noise this corresponds to sampling at intervals $>1/\text{bandwidth}$ of the noise. As shown in Chap. 6 the correlation rapidly decreases to zero beyond this point.) These are generated by sequentially sampling the received waveform as in Fig. 8-5. Alternately, a particular pulse could be repeated n times, each one then being sampled once.

The n samples define n-fold density functions. In particular, if signal s_1 is transmitted, we get the n-dimensional conditional-density

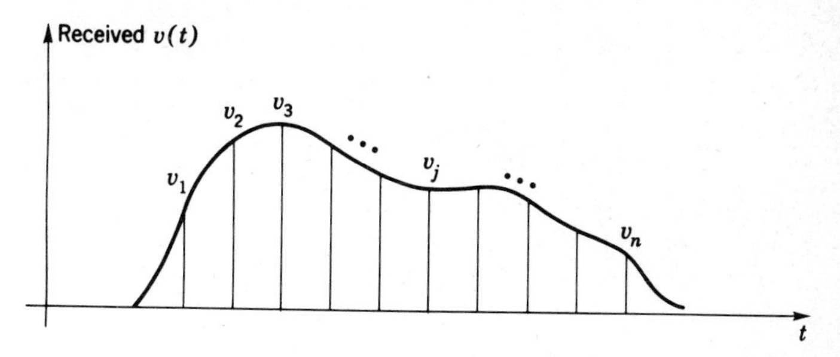

FIG. 8-5 Generation of n samples for processing.

function

$$f(v_1, v_2, \ldots, v_j, \ldots, v_n|1) = \prod_{j=1}^{n} f(v_j|1)$$

with successive samples assumed independent. Similarly, if s_2 is transmitted we may write the alternate n-fold conditional-density function

$$f(v_1, v_2, \ldots, v_j, \ldots, v_n|2) = \prod_{j=1}^{n} f(v_j|2)$$

As in the one-sample case we have to assume these functions known. These n-dimensional functions extend the two-dimensional functions discussed in Chap. 5. Note that the n samples may be visualized as describing an n-dimensional space, the possible range of values of each serving to define the range of that particular dimension. A little thought indicates that our statistical decision problem now boils down to dividing the n-dimensional space into two mutually exclusive regions V_1 and V_2 ($V_1 + V_2$ corresponding to the entire space of the n samples). We would like to choose the boundary between these two regimes such that the probability of error is minimized. If the n samples fall into region V_1, we declare signal s_1 present, if into region V_2, s_2 is declared present.

The composite samples $v_1, v_2, \ldots, v_j, \ldots, v_n$ now constitute an n-dimensional vector $\mathbf{v}$. This geometric and vector approach often simplifies quite considerably the analysis and optimum design of communication systems, and has been widely adopted by communication theorists. An example of the two-dimensional region corresponding to $n = 2$ samples is shown in Fig. 8-6. Received vector $\mathbf{v}$ is shown falling in region V_2, so that signal s_2 would be declared present.

How do we choose V_1 and V_2 optimumly now, in the sense of minimizing the error probability P_e? We again write the probability of error explicitly in terms of the two conditional-density functions. Since V_1 and V_2 are now n-dimensional, the integration must be carried out over

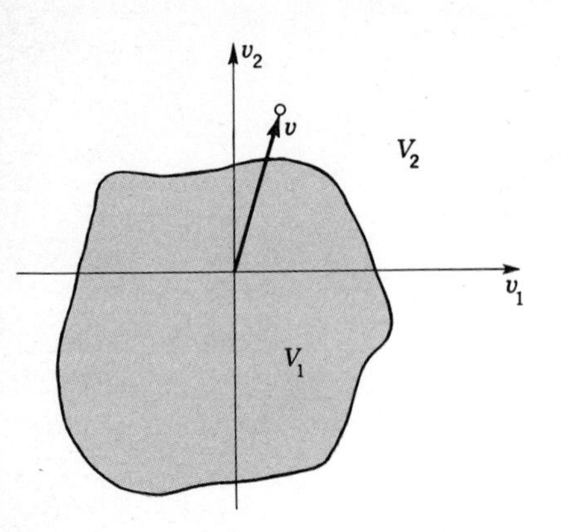

FIG. 8-6 Two-dimensional signal space.

the n-dimensional volumes:

$$P_e = P_1 \int_{V_2} f(\mathbf{v}|1)\, dv_1\, dv_2 \cdots dv_n + P_2 \int_{V_1} f(\mathbf{v}|2)\, dv_1\, dv_2 \cdots dv_n \qquad [8\text{-}10]$$

Here we have written the two conditional-density functions using the shorthand vector notation for $\mathbf{v}$.

But note now that aside from the extension to n dimensions the minimization process here is identical to that carried out earlier for one dimension. The optimum choice for the region V_1 is thus given by the likelihood ratio

$$l \equiv \frac{f(\mathbf{v}|1)}{f(\mathbf{v}|2)} > \frac{P_2}{P_1} \qquad [8\text{-}11]$$

As an example assume we again transmit an on-off signal of amplitude A or 0, and gaussian noise is added at the receiver. The received signal plus noise voltage $v(t)$ is sampled n times (alternately the signal may be assumed repeated n times) and the n samples $v_1 \cdots v_n$ used to make the decision. The conditional-density functions for the jth sample (assuming the samples independent) are respectively given by

$$f(v_j|1) = \frac{e^{-(v_j-A)^2/2N}}{\sqrt{2\pi N}} \qquad [8\text{-}12]$$

and

$$f(v_j|2) = \frac{e^{-v_j^2/2N}}{\sqrt{2\pi N}} \qquad [8\text{-}13]$$

The n-dimensional density functions are products of these individual sample functions. Again, taking logs to simplify, we find

$$\ln l = \frac{1}{2N}\left[\sum_{j=1}^{n} v_j^2 - \sum_{j=1}^{n} (v_j - A)^2\right] > \ln \frac{P_2}{P_1} \qquad [8\text{-}14]$$

In the special case where $P_2 = P_1 = \frac{1}{2}$, we have the region V_1 defined by

$$\sum_{j=1}^{n} v_j^2 > \sum_{j=1}^{n} (v_j - A)^2 \qquad [8\text{-}15]$$

Note, however, that not only can we define a vector $\mathbf{v} \equiv (v_1, v_2, \ldots, v_n)$, but we can talk of two-n-dimensional vectors $\mathbf{s}_1 \equiv (s_1^{(1)}, s_2^{(1)}, \ldots, s_n^{(1)})$ and $\mathbf{s}_2 \equiv (s_1^{(2)}, s_2^{(2)}, \ldots, s_n^{(2)})$, where the subscripts represent the sample number and the superscripts the particular signal, 1 or 2, under consideration. In this special case $\mathbf{s}_1 \equiv (A, A, A, \ldots, A)$ and $\mathbf{s}_2 \equiv (0, 0, 0, 0, \ldots, 0)$, since we have assumed a rectangular pulse of height A throughout the n-sample interval (Fig. 8-7).

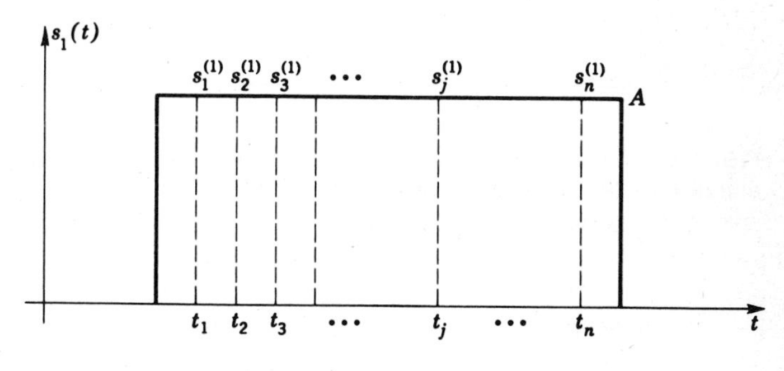

FIG. 8-7 n-dimensional signal s_1.

It is now apparent that the right-hand side of Eq. [8-15] represents the squared length of the vector $\mathbf{v} - \mathbf{s}_1$, and the left-hand side the squared length of the vector $\mathbf{v} - \mathbf{s}_2$.

In terms of the usual vector notation we have the squared length of a vector **a** given by the dot product $\mathbf{a} \cdot \mathbf{a}$, so that Eq. [8-15] may be equally well given by

$$(\mathbf{v} - \mathbf{s}_2) \cdot (\mathbf{v} - \mathbf{s}_2) > (\mathbf{v} - \mathbf{s}_1) \cdot (\mathbf{v} - \mathbf{s}_1) \qquad [8\text{-}16]$$

The n-dimensional vector notation and dot product is just the extension to n dimensions of the common three-dimensional vector notation. {The reader may find it instructive to use arbitrary but known pulse shapes $s_1(t)$ and $s_2(t)$ for the two signals. Vectors $\mathbf{s}_1$ and $\mathbf{s}_2$ can then again be defined as the composite of the n samples for each, and Eq. [8-16] obtained in this more general case.} The interpretation of the optimum decision rule in this case of two signals plus gaussian noise is now apparent from Eq. [8-16]. Pick signal $\mathbf{s}_1$ if the distance between the received vector $\mathbf{v}$ and the known vector $\mathbf{s}_1$ is less than the distance between $\mathbf{v}$ and $\mathbf{s}_2$. This is shown graphically in Fig. 8-8 for the special case of two dimensions.

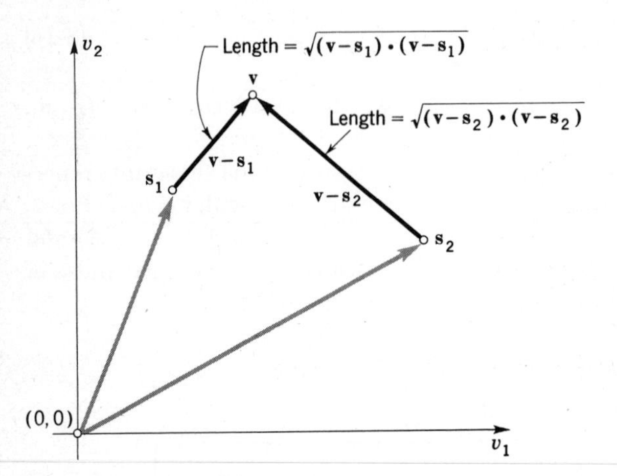

FIG. 8-8 Vector signals and decision rule in two dimensions (additive gaussian noise): received signal $\mathbf{v}$ is "closer to" signal $\mathbf{s}_1$.

Note that the receiver must have stored replicas of both vectors $\mathbf{s}_1$ and $\mathbf{s}_2$ [or, equivalently, the n sample values of $s_1(t)$ and of $s_2(t)$]. It measures the received vector $\mathbf{v}$, and then forms the necessary dot products and decides on $\mathbf{s}_1$ or $\mathbf{s}_2$, following the rule of Eq. [8-16].

The rule may be simplified in this case, however. For note that in both Eqs. [8-15] and [8-16] a common factor

$$\mathbf{v}\cdot\mathbf{v} = \sum_{j=1}^{n} v_j^2$$

may be canceled out on left- and right-hand sides. We then have remaining the rather simple expression

$$\sum_{j=1}^{n} v_j > \frac{nA}{2} \qquad [8\text{-}17]$$

for the special case of on-off signals. In words: Sum the n received samples and see if the sum exceeds the specified threshold $nA/2$. Note that this is just the extension to n dimensions of the one-dimensional result obtained earlier. The interesting point here is that the rule calls for the *sum* of the received samples. This is specifically due to the assumption of known binary signals in additive gaussian noise. In examples to follow we shall find different rules for different assumed statistics.

More generally, if instead of on-off binary signals we assume two *arbitrary* equally likely binary signals $s_1(t)$ and $s_2(t)$, the n samples of each define, respectively, the two vectors $\mathbf{s}_1$ and $\mathbf{s}_2$. In this case, the optimum decision rule from Eq. [8-16] becomes, after expanding the dot products and dropping the common term $\mathbf{v} \cdot \mathbf{v}$,

$$\mathbf{v} \cdot (\mathbf{s}_1 - \mathbf{s}_2) > \frac{\mathbf{s}_1 \cdot \mathbf{s}_1 - \mathbf{s}_2 \cdot \mathbf{s}_2}{2} \qquad [8\text{-}18]$$

In terms of the samples themselves, we have

$$\sum_{j=1}^{n} v_j[s_j^{(1)} - s_j^{(2)}] > \sum_{j=1}^{n} \frac{s_j^{(1)^2} - s_j^{(2)^2}}{2} \qquad [8\text{-}19]$$

The receiver now performs a weighted sum, weighting each received sample v_j with the stored samples $s_j^{(1)} - s_j^{(2)}$ before adding. The final sum is then compared to a known threshold.

The boundary between regions V_1 and V_2 is found by replacing the inequalities in the equations above by equal signs. In n dimensions the boundary between regions V_1 and V_2, given for additive gaussian noise by either Eq. [8-17], Eq. [8-18], or Eq. [8-19], is called a *hyperplane*. For the special case of on-off signals in two dimensions this degenerates into the line $v_1 + v_2 = A$, shown sketched in Fig. 8-9.

How much improvement does the use of the n samples provide over one sample? The answer of course depends on the particular signal shapes and noise statistics assumed. Consider, however, the on-off case as a specific example. If both signals are assumed equally likely for simplicity, either signal is equally likely as well to be mistaken for the other. The probability of an error is then the probability that vector $\mathbf{v}$, with $\mathbf{s}_1$ transmitted, appears in region V_2. From Eq. [8-17] this corresponds to the sum of the samples falling *below* the threshold level $nA/2$. Rather than calculate the probability of the n-dimensional gaussian vector $\mathbf{v}$ falling into the region V_2 we can deal with the much simpler sum of the v_j's. We note that with the v_j's each gaussian, the sum must be gaussian as well. In particular, from Chap. 5 we know that the variance of the sum is the sum of the variances, or just nN in this case, while the expected values add as well.

Letting

$$y \equiv \sum_{j=1}^{n} v_j$$

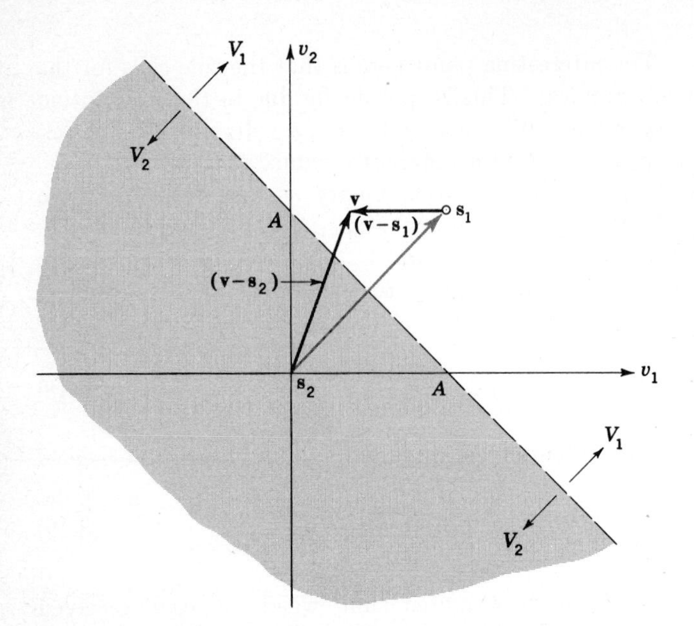

FIG. 8-9 Decision rule for on-off signals ($s_2 = 0$) in two dimensions (additive gaussian noise).

we then have the two conditional-density functions, one for signal s_1, the other for s_2,

$$f(y|1) = \frac{e^{-(y-nA)^2/2nN}}{\sqrt{2\pi nN}} \qquad [8\text{-}20]$$

and

$$f(y|2) = \frac{e^{-y^2/2nN}}{\sqrt{2\pi nN}} \qquad [8\text{-}21]$$

These are shown sketched in Fig. 8-10. The decision level $nA/2$ is of course just at the intersection of the two curves. The probability of error is then just that of the single sample case, but with the signal amplitude A replaced by nA, the noise power N replaced by nN! The improvement obtained by using the n samples optimumly (adding the received samples and requiring the sum to exceed a threshold level), corresponds to a net increase in the single sample signal-to-noise ratio $A/\sqrt{N}$ by a factor of $\sqrt{n}$. The n-sample system thus corresponds to a single-sample system with effective SNR given by $\sqrt{n}\, A/\sqrt{N}$. The SNR in power improves linearly with n. (The same result was mentioned briefly in Chap. 6 in discussing multipulse radar. There the point made was that the improvement with increasing numbers of samples or pulses

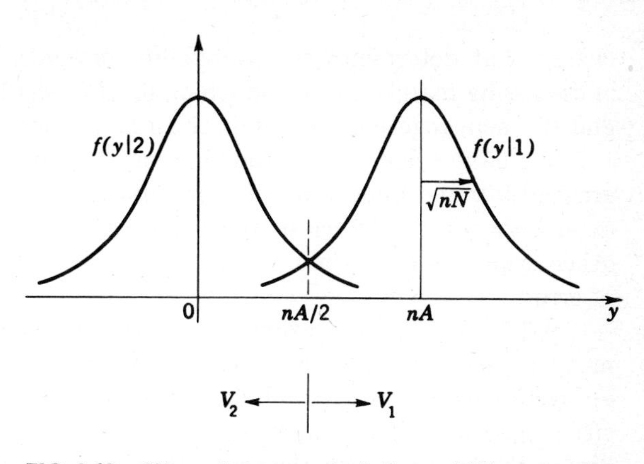

FIG. 8-10 Alternate error calculation, additive gaussian noise.

added is not as efficient as that found here, increasing less than linearly with n, because of the required envelope detection before pulse addition.)

Alternately, for the same probability of error the peak signal power A^2 may be reduced by a factor of n if n samples are added. Again there is an exchange of power and time (or bandwidth). For to obtain large numbers of independent samples, the signal duration must be increased accordingly. The bit rate must thus be decreased. (The same is true if instead of lengthening the signal duration the signal pulses are always repeated n times.) Note, however, that the signal *energy* $E = \int s^2(t)\,dt$ remains *fixed*. For if the basic bit interval is τ and the pulse height $\sqrt{n}\,A$, $E = nA^2\tau$. If pulses of height A and duration $n\tau$ are now transmitted, E remains the same (Fig. 8-11). Ultimately, then, it is the signal

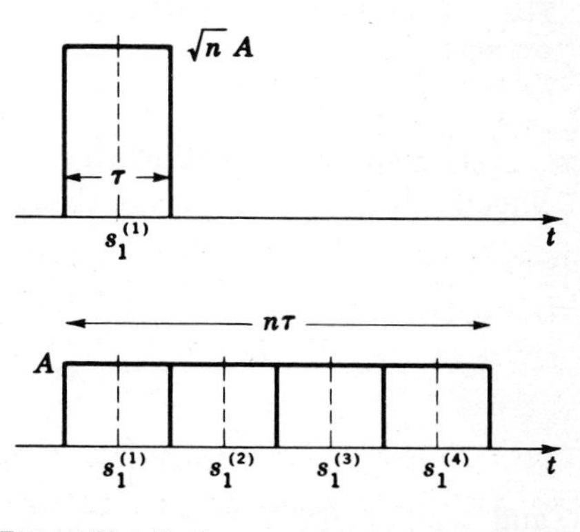

FIG. 8-11 Exchange of power for time.

energy that determines the probability of error. This was first noted in discussing matched filters in Chap. 6. We shall return to these points and tie them together once and for all in the next section.

We have spent substantial time on the gaussian case because of its great utility and importance in communication problems. We shall also extend the ideas further in the next section. We now consider some other examples of optimum signal processing for minimum probability of error.

Assume as the first example that the two signals s_1 and s_2 are zero-mean gaussian-distributed variables, but with differing variances σ_1^2 and σ_2^2 respectively. Again n independent samples of the received wave $v(t) = s_1(t)$ *or* $s_2(t)$ are taken, and we require the optimum way of combining the n samples in the sense of minimum error probability. Here we have as the conditional-density functions for the jth sample,

$$f(v_j|1) = \frac{e^{-v_j^2/2\sigma_1^2}}{\sqrt{2\pi\sigma_1^2}} \tag{8-22}$$

and

$$f(v_j|2) = \frac{e^{-v_j^2/2\sigma_2^2}}{\sqrt{2\pi\sigma_2^2}} \tag{8-23}$$

These are sketched in Fig. 8-12.

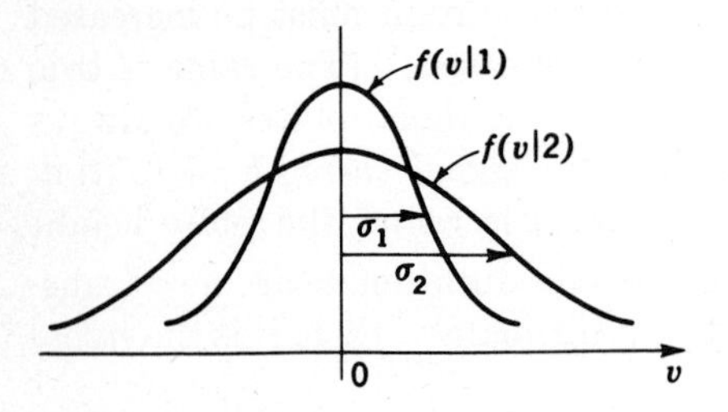

FIG. 8-12 Binary signals—another example.

The n-dimensional density functions needed for calculating the likelihood ratio of Eq. [8-11] are then, with independent samples assumed,

$$f(\mathbf{v}|1) = \frac{\exp\left[-(1/2\sigma_1^2)\sum_{j=1}^{n} v_j^2\right]}{(2\pi\sigma_1^2)^{n/2}} = \frac{e^{-(1/2\sigma_1^2)\mathbf{v}\cdot\mathbf{v}}}{(2\pi\sigma_1^2)^{n/2}} \tag{8-24}$$

and

$$f(\mathbf{v}|2) = \frac{\exp\left[-(1/2\sigma_2^2)\sum_{j=1}^{n} v_j^2\right]}{(2\pi\sigma_2^2)^{n/2}} = \frac{e^{-(1/2\sigma_2^2)\mathbf{v}\cdot\mathbf{v}}}{(2\pi\sigma_2^2)^{n/2}} \tag{8-25}$$

Here vector notation has again been introduced. Again taking the log of the likelihood ratio to simplify results, we now find as the rule for deciding signal s_1 present,

$$n \log \left(\frac{\sigma_2{}^2}{\sigma_1{}^2}\right) + \mathbf{v} \cdot \mathbf{v} \left(\frac{1}{\sigma_2{}^2} - \frac{1}{\sigma_1{}^2}\right) > 2 \log \frac{P_2}{P_1} \qquad [8\text{-}26]$$

If $\sigma_1{}^2 < \sigma_2{}^2$, as in Fig. 8-12, we get as the rule for deciding s_1 is present,

$$\mathbf{v} \cdot \mathbf{v} < \underbrace{\frac{n \log (\sigma_2{}^2/\sigma_1{}^2) + 2 \log (P_1/P_2)}{1/\sigma_1{}^2 - 1/\sigma_2{}^2}}_{d^2} \qquad [8\text{-}27]$$

The square of the length of vector $\mathbf{v}$ is to be compared to a threshold. If

$$\mathbf{v} \cdot \mathbf{v} \equiv \sum_{j=1}^{n} v_j{}^2 > d^2$$

declare s_2 present; if $<d^2$, declare s_1 present. The geometry here is again shown in the two-dimensional case of Fig. 8-13. In the n-dimensional

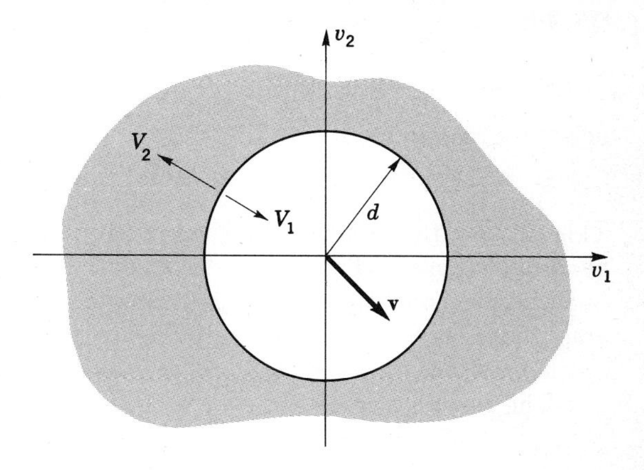

FIG. 8-13 Decision regions for case of Fig. 8-12.

case the boundary between V_1 and V_2 corresponds to a hypersphere concentric with the origin. Points inside the sphere correspond to $\mathbf{s}_1$, those outside to $\mathbf{s}_2$. This is consistent with our intuition, for with $\mathbf{s}_1$ and $\mathbf{s}_2$ the random gaussian variables of Fig. 8-12, it is apparent polarity should play no role in detection here. One would expect signal $\mathbf{s}_2$ to have larger amplitudes, on the average, than signal $\mathbf{s}_1$. Squaring the received samples eliminates polarity from consideration. We then compare the sum of the squared values with the threshold d^2.

As a second additional example we assume bipolar signals of amplitude $\pm A$ transmitted. Noise is again added, but this time the noise probability-density function is given by the Laplacian function

$$f(n) = \frac{1}{2c} e^{-|n|/c} \qquad [8\text{-}28]$$

This is sketched in Fig. 8-14. It is left to the reader to show that the function is properly normalized and that the standard deviation (rms

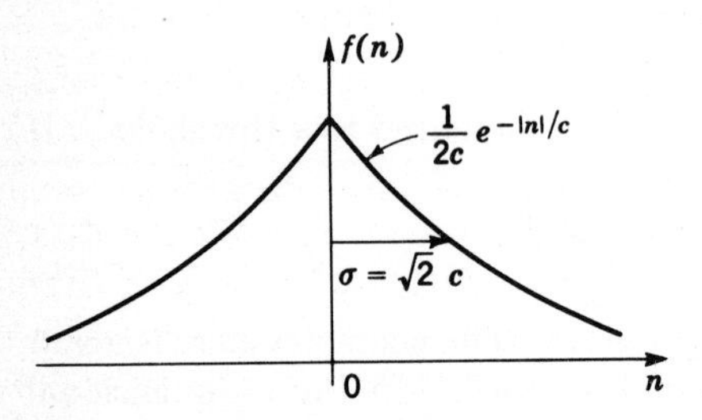

FIG. 8-14 Laplacian noise.

noise) is given by

$$\sigma = \sqrt{2}\, c \qquad [8\text{-}29]$$

This density function is sometimes used to model additive impulse or burst-type noise. The simple exponential behavior, rather than the quadratic exponential of the gaussian-density function, means higher amplitudes have correspondingly higher probabilities of appearing, a characteristic of this type of impulse noise.

The received signal sample v is then

$$v = \begin{matrix} s_1 \\ \text{or} \\ s_2 \end{matrix} + n \qquad [8\text{-}30]$$

The two conditional-density functions are given by

$$f(v|1) = \frac{e^{-|v-A|/c}}{2c} \qquad [8\text{-}31]$$

and

$$f(v|2) = \frac{e^{-|v+A|/c}}{2c} \qquad [8\text{-}32]$$

as sketched in Fig. 8-15.

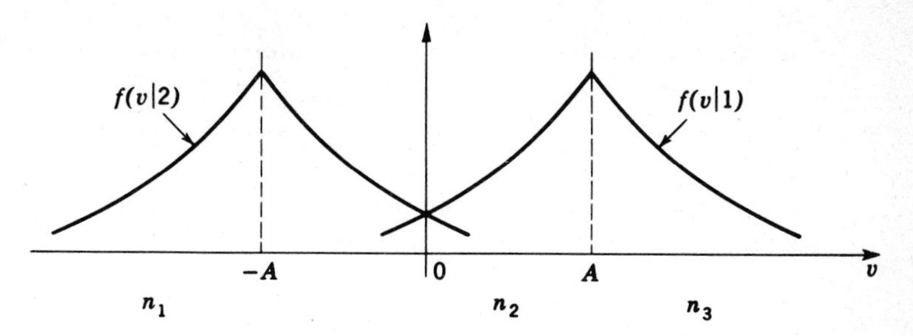

FIG. 8-15 Bipolar signals in Laplacian noise.

We now sample the received signal v n times and again ask for the optimum processing procedure. With independent samples assumed we again set up the likelihood ratio as in Eq. [8-11] and proceed to crank out the result. Here however we have to be somewhat careful because of the shape of $f(n)$. To take into account the abrupt change in the form of $f(v|1)$ and $f(v|2)$ at $v = \pm A$, we break the range of v into three regions and consider each separately. Consider a typical term in the likelihood ratio, $f(v_j|1)/f(v_j|2)$:

1. $v_j < -A$. Then

$$\frac{f(v_j|1)}{f(v_j|2)} = \frac{e^{(v_j-A)/c}}{e^{(v_j+A)/c}} = e^{-2A/c} \qquad [8\text{-}33]$$

Note that the dependence on v_j cancels out! The only knowledge retained is the fact that $v_j < -A$.

Assume now that n_1 of the n samples available fall in this range of v_j. There are then n_1 terms like Eq. [8-33] in the likelihood ratio, all to be multiplied together. Again taking the natural log of the likelihood ratio to simplify results, it is apparent that these n_1 terms in the likelihood ratio contribute

$$-\frac{2n_1A}{c}$$

to the total log l.

2. $v_j > A$. Here

$$\frac{f(v_j|1)}{f(v_j|2)} = \frac{e^{-(v_j-A)/c}}{e^{-(v_j+A)/c}} = e^{2A/c} \qquad [8\text{-}34]$$

Again the dependence on v_j cancels out, and each term in log l that corresponds to $v_j > A$ contributes $+2A/c$ to the total. Assuming that n_3 of n samples have $v_j > A$, we get as the contribution to log l,

$$\frac{+2n_3A}{c}$$

3. $-A < v_j < A$. Here

$$\frac{f(v_j|1)}{f(v_j|2)} = \frac{e^{(v_j-A)/c}}{e^{-(v_j+A)/c}} = e^{2v_j/c} \qquad [8\text{-}35]$$

Assume that n_2 of the n samples fall in this range. Taking logs, we find these n_2 terms in the likelihood ratio contribute the term

$$\frac{2}{c}\sum_{j=1}^{n_2} v_j$$

to the total.

The three numbers n_1, n_2, and n_3 are indicated in Fig. 8-15. It is apparent that we have the constraint

$$n_1 + n_2 + n_3 = n \qquad [8\text{-}36]$$

Combining all three terms, the rule for deciding on signal S_1 now becomes

$$(n_3 - n_1)A + \sum_{j=1}^{n_2} v_j > \frac{c}{2}\ln\frac{P_2}{P_1} \qquad [8\text{-}37]$$

The interpretation here is quite interesting. The optimum processor consists of three-level threshold circuitry followed by an appropriate counter and an adding device. If the threshold circuitry detects a sample in the $v_j > A$ region, a positive count of A is added to the counter. If a sample in the range $v_j < -A$ appears, A is *subtracted* from the counter. If $-A < v_j < A$, the sample value is stored in the adding circuit. At the end of the n samples, the counter and adder outputs are summed. If they exceed the decision level $(c/2)\ln(P_2/P_1)$, s_1 is declared present; otherwise s_2 is assumed present. Note that such a processor lends itself nicely to digital circuitry. Note also that if A/σ is small (small SNR), the number of samples n_2 falling in the central region would tend to be small, the adder output would become negligible, and the counter alone would suffice.

8-3 OPTIMUM BINARY TRANSMISSION

We now focus attention on one particular case of binary transmission: the transmission of binary symbols $s_1(t)$ or $s_2(t)$ in additive gaussian noise. We assume a fixed binary interval T sec long and ask for both the optimum receiver structure and optimum signal shapes $s_1(t)$ and $s_2(t)$. Part of the answer is already available to us. We showed in the previous section that with n samples of the received signal plus noise the optimum processing was that described by the vector Eq. [8-18], or its equivalent Eq. [8-19]. Thus choose $s_1(t)$ if

$$\mathbf{v} \cdot (\mathbf{s}_1 - \mathbf{s}_2) > \frac{\|\mathbf{s}_1\|^2 - \|\mathbf{s}_2\|^2}{2} \qquad [8\text{-}38]$$

(Here we use the symbol $\| \quad \|$ for magnitude of a vector. Also recall the assumption that the a priori probabilities P_1 and P_2 are equal.)

We now ask the obvious question: In a given interval T how many samples n should we use? To answer this we assume the noise is band-limited white noise, with spectral density $G_n(f) = n_0/2$, over the range of frequencies $\pm B$ Hz. The total noise power is then $N = n_0B$. We shall also assume the two signals $s_1(t)$ and $s_2(t)$ band-limited over the same band B.

For this model of noise we recall the autocorrelation function, the Fourier transform of the spectral density, is just

$$R_n(\tau) = N \frac{\sin 2\pi\tau B}{2\pi\tau B} \qquad [8\text{-}39]$$

Both $G_n(f)$ and $R_n(\tau)$ are shown sketched in Fig. 8-16. Note that there is zero correlation between samples spaced multiples of $1/2B$ sec apart. For gaussian noise this further indicates the samples are *independent*. There are thus $n = 2BT$ independent samples available to us. Recall also from the sampling theorem in Chap. 3 that samples spaced $1/2B$ sec

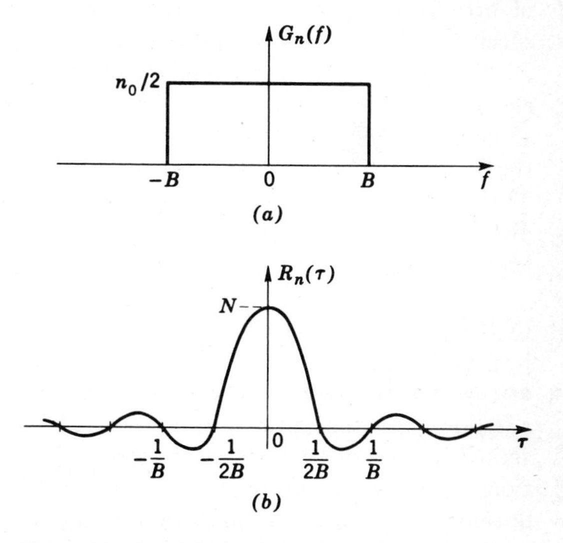

FIG. 8-16 Band-limited white noise. (a) Spectral density; (b) autocorrelation function.

apart suffice to uniquely characterize a band-limited signal. In fact we indicated that with the sample values $n(j/2B)$, $j = 0, \pm 1, \pm 2, \ldots,$ known, we could reproduce the original wave $n(t)$ as

$$n(t) = \sum_j n\left(\frac{j}{2B}\right) \frac{\sin 2\pi B(t - j/2B)}{2\pi B(t - j/2B)} \qquad [8\text{-}40]$$

The one catch here is that all samples $j = (-\infty, +\infty)$ have to be known, whereas we only have $n = 2BT$ of them. For $BT \gg 1$, however, Eq. [8-40] provides a good approximation to $n(t)$ in the binary interval T, and we shall make this assumption.

The $(\sin x)/x$ functions in Eq. [8-40] have an interesting property that proves extremely useful to us. They are examples of so-called orthogonal functions, with the property that

$$\int_{-\infty}^{\infty} \frac{\sin \pi(2Bt - k)}{\pi(2Bt - k)} \cdot \frac{\sin \pi(2Bt - m)}{\pi(2Bt - m)}\, dt = \frac{1}{2B}\, \delta_{km} \qquad [8\text{-}41]$$

Here δ_{km} is the Kronecker delta

$$\begin{aligned} \delta_{km} &= 1 \qquad k = m \\ &= 0 \qquad k \neq m \end{aligned}$$

The integral of the product of two $(\sin x/x)$ functions displaced in time is thus zero. The Fourier series of Chap. 2 is another example of a set of orthogonal functions, in which the integral of the product over a specified interval (finite or infinite) is zero.

The proof of the orthogonality of the $(\sin x)/x$ functions is left as an exercise for the reader. {As a hint, consider the integral of Eq. [8-41] to be a convolution integral. It is then equal to the integral containing the product of the Fourier transforms of the two $(\sin x)/x$ functions. These are just constants over the range $\pm B$, with appropriate exponential phase factors. It is then easy to show the integral of the resultant exponential terms goes to zero.}

If we now assume the two binary signals $s_1(t)$ and $s_2(t)$ band-limited to B Hz as well, with $BT \gg 1$, it is apparent that they can equally well be expressed as the series of Eq. [8-40] in $(\sin x)/x$, with the sample values of s_1 and s_2 appropriately inserted. The received signal $v(t) = [s_1(t)$ or $s_2(t)] + n(t)$ is thus also band-limited and equally expressible in the same series. The series expansions in terms of the orthogonal $(\sin x)/x$ functions have the same properties as those of the Fourier series of Chap. 2. Specifically, a generalized form of the Parseval theorem mentioned earlier is easily derived, relating sums of sample values to an integral of the analog time function. Thus, consider the integral $\int n^2(t)\, dt$.

[Although we should restrict the integration to the range $(-T/2, T/2)$, over which $n(t)$ is defined, with $BT \gg 1$ the integration may be taken over the infinite range $(-\infty, \infty)$.] Replacing each $n(t)$ by the sum of Eq. [8-40], interchanging the order of integration and summation, and noting the orthogonality relation of Eq. [8-41], we find

$$\int_0^T n^2(t)\,dt \doteq \frac{1}{2B}\sum_{j=1}^{2BT} n^2\left(\frac{j}{2B}\right) \qquad BT \gg 1 \tag{8-42}$$

This is the same as the Parseval relation found earlier, and a similar relation may readily be derived the same way for *all* orthogonal series expansions.

If we now consider two time functions $v(t)$ and $s(t)$, band-limited over $\pm B$, we readily demonstrate in the same way that

$$\int_0^T v(t)s(t)\,dt \doteq \frac{1}{2B}\sum_{j=1}^{2BT} v\left(\frac{j}{2B}\right) s\left(\frac{j}{2B}\right) \tag{8-43}$$

But this is quite interesting, for if we look at the optimum processor for binary signals in gaussian noise, Eq. [8-38], we note that the vector products shown are just in the form of the right-hand side of Eq. [8-43]. Specifically, then, applying Parseval's theorem to both sides of Eq. [8-38], we find the optimum processor to be equally well given by

$$\int_0^T v(t)[s_1(t) - s_2(t)]\,dt > \frac{1}{2}\int_0^T [s_1^2(t) - s_2^2(t)]\,dt \tag{8-44}$$

Alternately, note that the integrals on the right-hand side above are just the respective energies in the signals. The inequality can thus be written

$$\int_0^T v(t)[s_1(t) - s_2(t)]\,dt > \frac{E_1 - E_2}{2} \tag{8-44a}$$

with $\qquad E_1 \equiv \int_0^T s_1^2(t)\,dt \qquad E_2 \equiv \int_0^T s_2^2(t)\,dt$

The interpretation here is quite interesting. It says that instead of the digital operations on the samples indicated by Eq. [8-38], one may equally well take the incoming signal $v(t)$, multiply it by two stored replicas of $s_1(t)$ and $s_2(t)$, integrate, and sample the resultant output every T sec to see if it exceeds a specified threshold. Note that the $s_1(t)$ and $s_2(t)$ inserted at the receiver must be precisely in phase with the $s_1(t)$ or $s_2(t)$ portion of the received $v(t)$. The resultant operation is nothing more than our old friend coherent or synchronous detection! In the general form of Eq. [8-44*a*] it is also often called correlation detection.

A further alternate form may be obtained by separating the s_1 and s_2 terms:

$$\int_0^T v(t)s_1(t) - \frac{E_1}{2} > \int_0^T v(t)s_2(t)\,dt - \frac{E_2}{2} \qquad [8\text{-}44b]$$

The E_1 and E_2 thus serve as fixed-bias terms to equalize the detector outputs. The correlation detector and sampler of Eq. [8-44b] is shown sketched in Fig. 8-17.

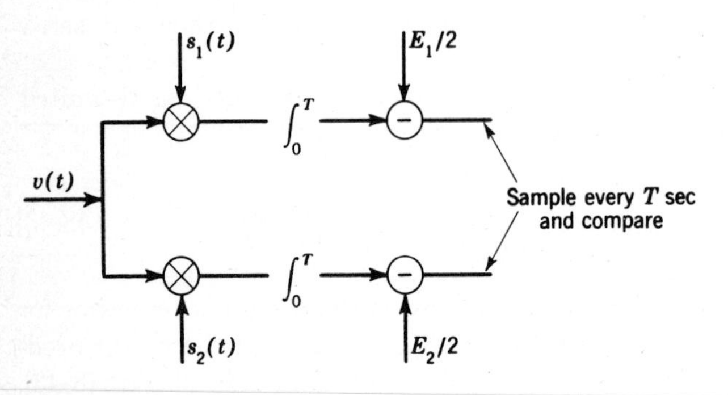

FIG. 8-17 Optimum binary processor—correlation detection.

As an example let $s_1(t) = \cos \omega_0 t$, $s_2(t) = -\cos \omega_0 t$, just the PSK signals of Chap. 6. Then $E_1 = E_2$, and the correlation detector of Fig. 8-17 becomes the synchronous detector discussed previously. This is shown in Fig. 8-18a. (The integrator of course provides the necessary low-pass filtering.) If $s_1(t) = A \cos \omega_0 t$, $s_2(t) = 0$, we have the OOK signal of Chap. 6. Then

$$E_1 = \int_0^T A^2 \cos^2 \omega_0 t \doteq \frac{A^2 T}{2}$$

if $\omega_0 T \gg 2\pi$, and the processing called for is just

$$\frac{2}{T}\int_0^T v(t) \cos \omega_0 t\,dt > \frac{A}{2}$$

This corresponds of course to synchronous detection, followed by the $A/2$ decision level (Fig. 8-18b). Finally, if $s_1(t) = \cos \omega_1 t$, $s_2(t) = \cos \omega_2 t$, $E_1 = E_2$, we get the FSK synchronous detector of Fig. 8-18c.

There is still a further interpretation of the optimum processor that leads directly to the matched filter of Chap. 6. The integral of Eq.

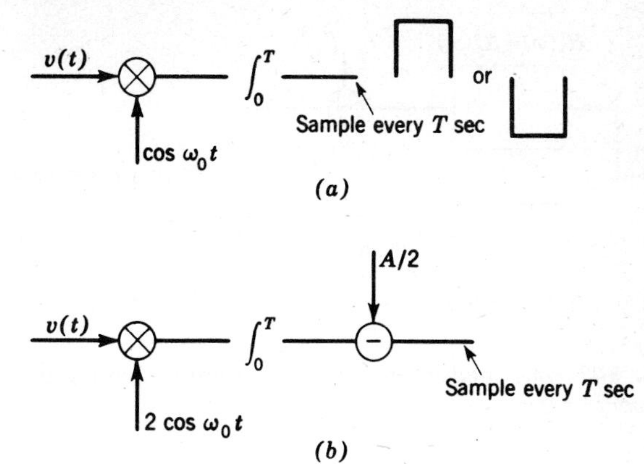

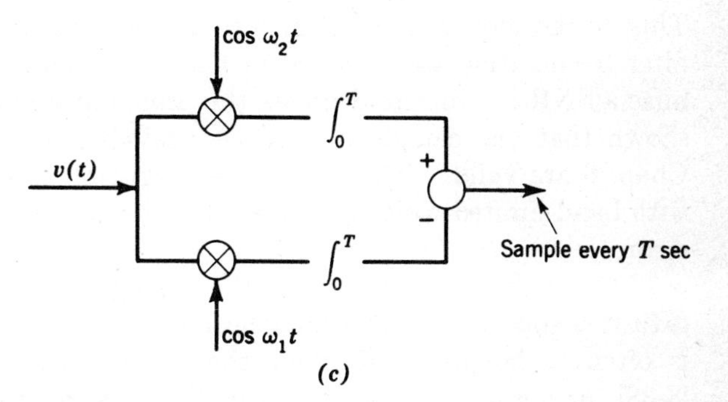

FIG. 8-18 Examples of Fig. 8-17. (a) PSK synchronous detector; (b) OOK synchronous detector; (c) FSK synchronous detector.

[8-44*a*] may be rewritten as the following convolution integral,

$$\int_0^T v(t)[h_1(T - t) - h_2(T - t)]\,dt > \frac{E_1 - E_2}{2} \qquad [8\text{-}44c]$$

where

$$h_1(T - t) \equiv s_1(t)$$
$$h_1(t) = s_1(T - t)$$

and

$$h_2(T - t) \equiv s_2(t)$$
$$h_2(t) = s_2(T - t) \qquad [8\text{-}45]$$

By taking Fourier transforms of both sides of Eq. [8-45], it is apparent that we must have

$$H_1(\omega) = e^{-j\omega T} S_1^*(\omega)$$

and

$$H_2(\omega) = e^{-j\omega T} S_2^*(\omega) \qquad [8\text{-}46]$$

just our earlier (Chap. 6) conditions for a matched filter. The optimum processor may thus be drawn in the matched-filter form of Fig. 8-19.

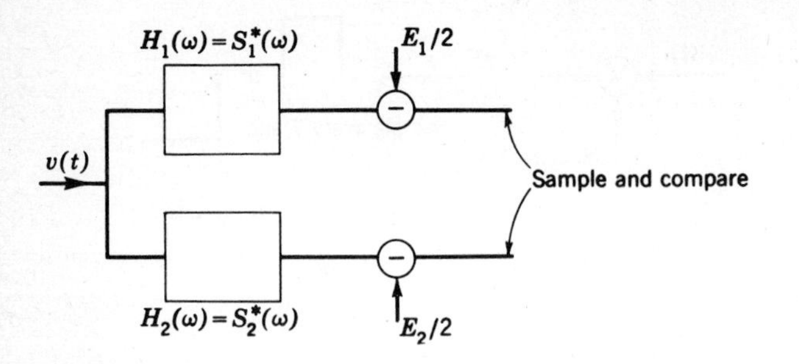

FIG. 8-19 Matched-filter form of optimum binary processor.

This, in the special cases referred to above, is identical with the matched-filter formulation we obtained in Chap. 6. There we wanted to maximize a SNR in order to decrease the probability of error. Here we have shown that one can do no better—providing the assumptions made in Chap. 6 are valid. Thus we assume negligible intersymbol interference, with band-limited white gaussian noise the only possible source of binary error.

Although we have stressed the analog processors of Figs. 8-17 and 8-19 it is apparent that with modern integrated circuits digital processing is often to be preferred. One then seeks ways of implementing Eq. [8-38], or digital equivalents, directly. One may consider carrying out the process

$$\frac{1}{2B} \sum_{j=1}^{n=2BT} v_j \left[s_1\left(\frac{j}{2B}\right) - s_2\left(\frac{j}{2B}\right) \right] > \frac{E_1 - E_2}{2} \qquad [8\text{-}47]$$

directly, using the incoming samples v_j and stored signal samples, or, alternately, pass the successive incoming samples v_j through a *digital* matched filter for processing. This is indicated in Fig. 8-20.

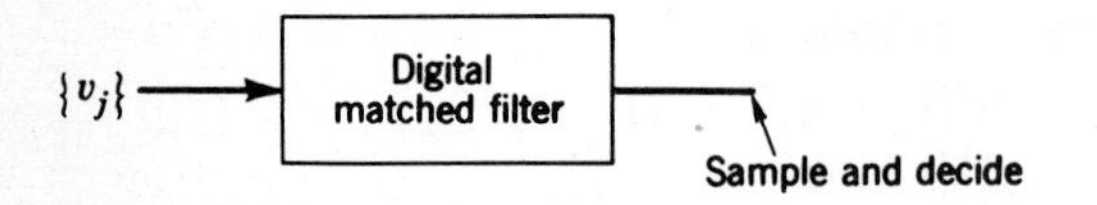

FIG. 8-20 Digital version, optimum processor.

OPTIMUM WAVESHAPES

It is apparent that the calculation of probability of error for the optimum binary processor must lead to the same erfc x curves obtained in Chap. 6, since we have just demonstrated the equivalence of the ad hoc detectors discussed there to the optimum processors studied here. It is still useful to repeat the calculation, however, for in the process we shall find a simple answer to a second question we raised earlier. What are the optimum waveshapes to be used for $s_1(t)$ and $s_2(t)$? We shall show the answer to be simply given by $s_1(t) = -s_2(t)$, just the condition satisfied by a bipolar or PSK sequence! So unless we find other waveshapes to be simpler to generate, it is apparent we need look no further for "better" waveshapes.

The probability of error is readily calculated using either Eq. [8-44*a*] or the digital equivalent Eq. [8-47]. Assume for example that we know $s_1(t)$ has been transmitted. The probability that Eqs. [8-44*a*] or [8-47] are not satisfied gives us the probability of error in this case. By symmetry it is also the same as the probability of mistaking $s_2(t)$ for $s_1(t)$, and therefore gives the total probability of error of the system P_e. (Recall again that we have assumed $P_1 = P_2 = \frac{1}{2}$ here for simplicity. The analysis is readily extended to the more general case $P_1 \neq P_2$ as well.)

With $s_1(t)$ transmitted, the received signal $v(t)$ is

$$v(t) = s_1(t) + n(t) \tag{8-48}$$

The probability of error is then simply the probability that

$$\int_0^T [n(t) + s_1(t)][s_1(t) - s_2(t)]\,dt < \frac{E_1 - E_2}{2}$$

or, alternately, using the Parseval identity,

$$\frac{1}{2B}\sum_{j=1}^{2BT} (n_j + s_{1_j})(s_{1_j} - s_{2_j}) < \frac{E_1 - E_2}{2}$$

[We have used the simpler notation s_{1_j} and s_{2_j} to represent the jth samples of $s_1(t)$ and $s_2(t)$, respectively.] Simplifying by leaving the $n(t)$ term only on the left-hand side, and recalling the definitions of E_1 and E_2,

$$\begin{aligned} E_1 &\equiv \int_0^T s_1^2(t)\,dt \doteq \frac{1}{2B}\sum_{j=1}^{2BT} s_{1_j}^2 \\ E_2 &\equiv \int_0^T s_2^2(t)\,dt \doteq \frac{1}{2B}\sum_{j=1}^{2BT} s_{2_j}^2 \end{aligned} \tag{8-49}$$

we get as the condition for an error to occur

$$y \equiv \int_0^T n(t)[s_1(t) - s_2(t)] < -b \equiv -\tfrac{1}{2}\int_0^T [s_1(t) - s_2(t)]^2\,dt \tag{8-50}$$

Equivalently, using the sample values, an error occurs if

$$y \equiv \frac{1}{2B} \sum_{j=1}^{2BT} n_j(s_{1_j} - s_{2_j}) < -b \qquad [8\text{-}51]$$

But the noise $n(t)$ is assumed gaussian. The random variable y is then gaussian as well. (This is apparent either from Eq. [8-51], where y is defined as the weighted sum of $2BT$ gaussian variables n_j, or equally well from Eq. [8-50], describing y as the output of a *linear* matched filter with $n(t)$ applied at the input.) The expected value and variance of y are given very simply by

$$E(y) = \frac{1}{2B} \sum_{j=1}^{2BT} E(n_j)(s_{1_j} - s_{2_j}) = 0 \qquad [8\text{-}52]$$

(we have assumed zero-mean noise), and

$$\begin{aligned}\sigma_y^2 &= E[y - E(y)]^2 \\ &= \left(\frac{1}{2B}\right)^2 E\left[\sum_i \sum_j n_i n_j (s_{1_j} - s_{2_j})(s_{1_i} - s_{2_i})\right] \\ &= \left(\frac{1}{2B}\right) \sum_i \sum_j E(n_i n_j)(s_{1_j} - s_{2_j})(s_{1_j} - s_{2_j}) \qquad [8\text{-}53]\end{aligned}$$

interchanging summation and expectation.

Recall, however, that for band-limited white noise, the samples spaced $1/2B$ sec apart are *uncorrelated*. Also, $E(n_i^2) = N$ (see Eq. [8-39]). Therefore, $E(n_i n_j) = N\delta_{ij}$. Equation [8-53] then simplifies to

$$\begin{aligned}\sigma_y^2 &= \frac{N}{(2B)^2} \sum_{j=1}^{2BT} (s_{1_j} - s_{2_j})^2 = \frac{N}{2B} \int_0^T [s_1(t) - s_2(t)]^2 \, dt \\ &= \frac{n_0}{2} \int_0^T [s_1(t) - s_2(t)]^2 \, dt \qquad [8\text{-}53a]\end{aligned}$$

again using the Parseval relation. Here we have also put $N = n_0 B$ for the band-limited white noise. (Recall that $n_0/2$ is the band-limited spectral density. See Fig. 8-16*a*.)

The probability density of the variable y defined by Eqs. [8-50] and [8-51] is then

$$f(y) = \frac{1}{\sqrt{2\pi\sigma_y^2}} e^{-y^2/2\sigma_y^2} \qquad [8\text{-}54]$$

and the probability of error is just

$$P_e = \int_{-\infty}^{-b} \frac{e^{-y^2/2\sigma_y^2}\,dy}{\sqrt{2\pi\sigma_y^2}}$$

$$= \int_{-\infty}^{-b/\sqrt{2}\sigma_y} \frac{e^{-x^2}\,dx}{\sqrt{\pi}}$$

$$= \tfrac{1}{2}\operatorname{erfc}\frac{b}{\sqrt{2}\,\sigma_y} \qquad [8\text{-}55]$$

The appropriate integration to obtain P_e is indicated by the shaded area in Fig. 8-21.

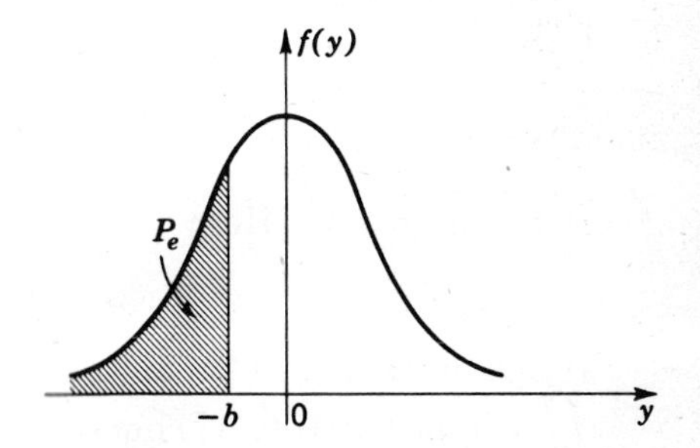

FIG. 8-21 Probability of error in binary waveshape transmission.

Using the definition of b in Eq. [8-50] and that of σ_y^2 in Eq. [8-53*a*], we have as the final result for P_e,

$$P_e = \tfrac{1}{2}\operatorname{erfc}\frac{a}{2\sqrt{2}} \qquad [8\text{-}55a]$$

where the new parameter a is defined by

$$a^2 \equiv \int_0^T \frac{(s_1 - s_2)^2\,dt}{n_0/2} \qquad [8\text{-}56]$$

Note how similar Eq. [8-55*a*] is to the form of Eqs. [6-83], [6-85], and [6-87]. As a matter of fact, those equations are all subsumed in Eq. [8-55*a*] if we assume matched filters used in those cases. To demonstrate this, recall that with matched filters, we have from Eq. [6-58], $A/\sqrt{N} = \sqrt{2E/n_0}$. The three cases considered in Chap. 6 and the equivalent result from Eq. [8-56] then provide the following results:

1. OOK:

$$P_e = \frac{1}{2} \operatorname{erfc}\left(\frac{1}{2}\sqrt{\frac{E}{n_0}}\right) \qquad [8\text{-}57]$$

from Eq. [6-83]. To show this is the same as Eq. [8-55a], let $s_2(t)$ in Eq. [8-56] $= 0$, and $s_1(t) = A \cos \omega_0 t$, just the OOK case. Then the parameter $a^2 = (2/n_0) \int_0^T A^2 \cos^2 \omega_0 t\, dt = 2E/n_0$, and

$$\frac{a}{2\sqrt{2}} = \frac{1}{2\sqrt{2}}\sqrt{\frac{2E}{n_0}} = \frac{1}{2}\sqrt{\frac{E}{n_0}}$$

just as in Eq. [8-57].

2. FSK:

$$P_e = \tfrac{1}{2} \operatorname{erfc}\sqrt{\frac{E}{2n_0}} \qquad [8\text{-}58]$$

from Eq. [6-87]. Here we have $s_1(t) = A \cos \omega_1 t$ and

$$s_2(t) = A \cos \omega_2 t$$

If $\omega_1 T \gg 2\pi$, $\omega_2 T \gg 2\pi$, and $(\omega_1 - \omega_2)T \gg 2\pi$, it is readily shown that $\int_0^T s_1(t)s_2(t)\, dt = 0$. [The signals $s_1(t)$ and $s_2(t)$ are examples of orthogonal signals.] Equation [8-56] then simplifies to

$$a^2 = \frac{2}{n_0} \int_0^T [s_1{}^2(t) + s_2{}^2(t)]\, dt = \frac{4E}{n_0}$$

since here $E_1 = E_2 = E$. Then we also have $a/2\sqrt{2} = \sqrt{E/2n_0}$, just as in Eq. [8-58].

3. PSK:

$$P_e = \tfrac{1}{2} \operatorname{erfc}\sqrt{\frac{E}{n_0}} \qquad [8\text{-}59]$$

from Eq. [6-85]. Here we have $s_1(t) = -s_2(t)$, and $a^2 = 8E/n_0$, from Eq. [8-56]. Then $a/2\sqrt{2} = \sqrt{E/n_0}$, agreeing of course with Eq. [8-59].

The optimum waveshapes to be used in binary signal transmission in additive gaussian white noise are also readily obtained from Eqs. [8-55a] and [8-56]. Since erfc $(a/2\sqrt{2})$ decreases with a, the larger a is, the smaller the probability of error P_e. A little thought will indicate that a is maximized and P_e minimized by setting $s_1(t) = -s_2(t)$. The PSK signals of course satisfy these conditions. Other pairs of bipolar signals would serve just as well. Since $a^2 = 8E/n_0$ for this case, with

$$E = \int_0^T s_1^2\,dt = \int_0^T s_2^2\,dt$$

it is again the ratio of average energy E to noise spectral density n_0 that determines the probability of error. This is the point we made in discussing matched filters in Chap. 6. To minimize error probability, one should use signals with as high an average energy content as allowable. In passing the signals through the matched filter, the specific dependence on the details of the waveshape disappears. The matched-filter signal-power output is proportional to the average energy E.

A universal probability-of-error curve for optimum binary transmission in additive white gaussian noise is shown in Fig. 8-22. This is

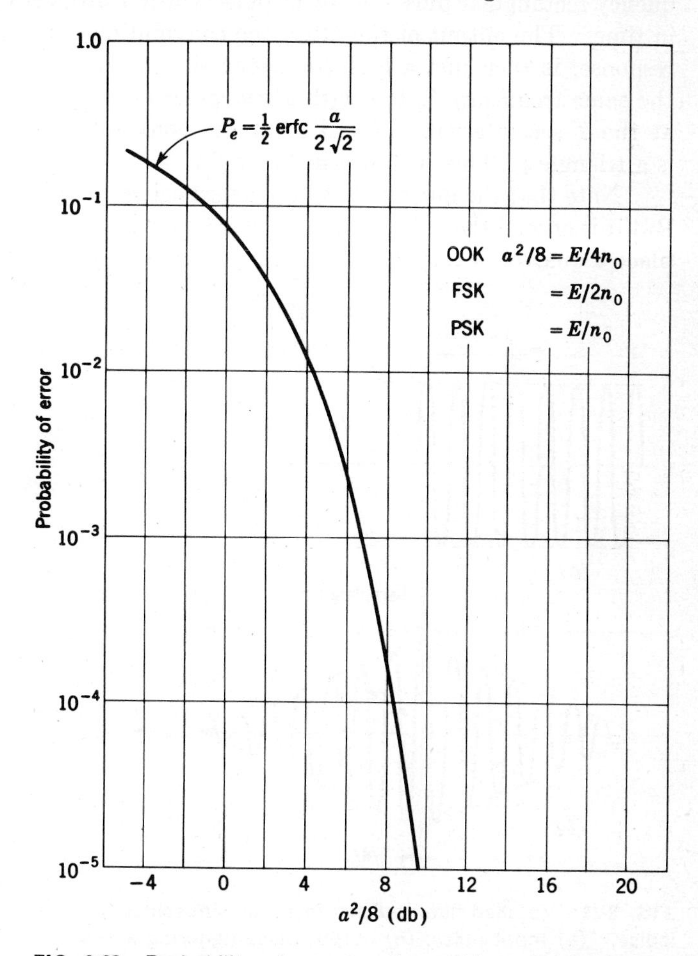

FIG. 8-22 Probability of error, optimum binary transmission.

just a plot of Eq. [8-55*a*]. As indicated above and again in the figure, the parameter $a^2/8$ on which the probability of error depends is a function only of E/n_0, the ratio of signal energy to noise spectral density.

In this discussion, synchronous or phase-coherent detection is again assumed available, as in Chap. 6. If it is not possible to maintain phase synchronism, one must again resort to envelope detection with its attendant SNR deterioration. The need for synchronous detection is implicit in the correlation or matched-filter detection circuits of Figs. 8-17 and 8-19. As an example assume the input to the matched filter is $A \cos \omega_0 t$, defined over the binary interval T as in the examples just cited. The matched filter must then be a bandpass filter centered at f_0, and of bandwidth $\sim 1/T$. Alternately, its impulse response is the same high-frequency rectangular pulse $\cos \omega_0 t$, of pulse width T although turned around in time. The output of the filter, the convolution of signal and impulse response, is then just a high-frequency sinusoidal pulse $2T$ sec long, at the same frequency f_0, but with a *triangular* envelope, reaching its peak at the T-sec interval. (Recall that the convolution of two rectangles is a triangle.) This is shown in Fig. 8-23.

Note that the output peaks up, as expected, at the sampling time T. But it is crucial that sampling take place *exactly* at T. If sampling takes place a quarter of a cycle, or $1/4f_0$ sec, early or late, the output drops to

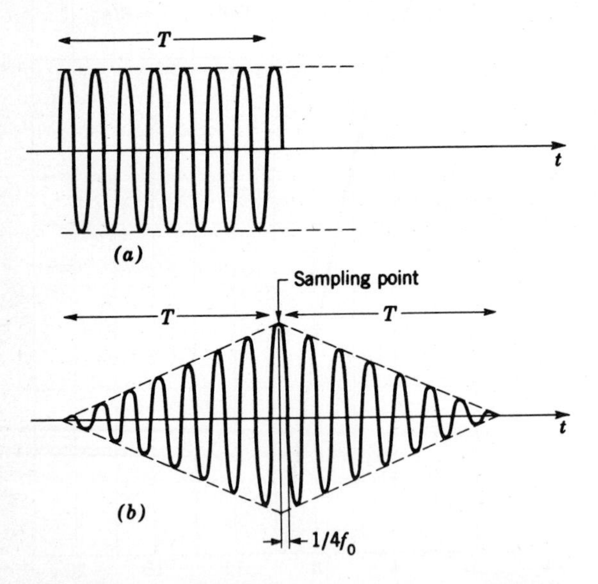

FIG. 8-23 Matched-filter output to input sinusoidal pulse. (a) Input pulse; (b) output pulse (ignoring a constant time delay).

zero! This synchronism must be maintained to within a fraction of $1/4f_0$ sec, which is not a mean task if the carrier frequency f_0 is in the VHF, UHF, or microwave range.

Actually in practice one would not allow the output pulse to extend $2T$ sec in length. For then successive binary pulses spaced $1/T$ sec apart would interfere with one another. One way of avoiding this problem is to use as the matched filter an *integrate and dump circuit.*[1] This consists of a resonant circuit tuned to frequency f_0, with provision for shorting out the tuned circuit every T sec, just before receiving a signal input, to *dump* energy stored from previous intervals.

8-4 SHANNON CAPACITY REEXAMINED

Figure 8-22 indicates the best performance possible using binary signals in the presence of additive gaussian white noise. As an example, to attain a probability of error of 10^{-5} with PSK transmission, the ratio of signal energy to noise spectral density required is $\doteq 10$ db. But what if the signal energy required is just not available? What does one do then? Does this indicate that relatively good (low-error-rate) data transmission is just not possible?

This question was essentially answered in Chap. 3 in discussing the Shannon capacity expression of Eq. [3-35]. We indicated there that Shannon had proven that virtually error-free digital transmission was possible over a channel with additive gaussian noise, providing one did not try to exceed the channel capacity in bits per second. Specifically, assume as in Chap. 3 an available transmission bandwidth of W Hz. Band-limited gaussian noise of spectral density $n_0/2$ is added during transmission. (This is just the assumption made in the optimum binary transmission case of the previous section.) Then Shannon was able to show that by appropriately *coding* a binary message sequence before transmission it should be possible to achieve as low an error rate as possible providing the channel capacity C was not exceeded. The capacity expression in this case of band-limited white noise was found by Shannon to be given by[2]

$$C = W \log_2 \left(1 + \frac{S}{N}\right) \qquad \text{bits/sec} \qquad [8\text{-}60]$$

[1] J. M. Wozencraft and I. M. Jacobs, "Principles of Communication Engineering," pp. 235, 236, Wiley, New York, 1965.

[2] C. E. Shannon, Communication in the Presence of Noise, *Proc. IRE*, vol. 37, pp. 10–21, January, 1949.

with S the average signal power, and $N = n_0 W$ the average noise power. (S/N is then the signal-to-noise ratio at the receiver.)

Thus, providing one did not attempt to transmit more than C bits/sec over such a channel, one could hope to attain tolerable error rates. Specifically, if the binary transmission rate is R bits/sec (the binary interval is then $1/R$ sec), and if $R < C$, it may be shown that the error probability is bounded by

$$P_e \leq 2^{-E(C,R)T} \qquad R < C \tag{8-61}$$

with $E(C,R)$ a positive function such as the one shown in Fig. 8-24. As the transmission R approaches C the probability of error $\rightarrow 1$. This is apparent from Eq. [8-61] and Fig. 8-24.

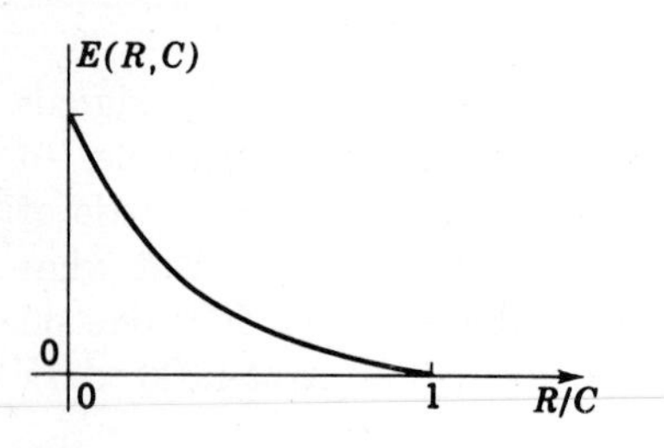

FIG. 8-24 Shannon error exponent.

The parameter T appearing in Eq. [8-61] indicates the time required to transmit the *encoded* signal. With binary transmission rate R and channel capacity C fixed, the probability of error may be reduced by increasing T. Although Shannon proved the capacity expression of Eq. [8-60] quite generally for this so-called *gaussian channel*, he did not provide a formula for actually designing the required encoder. His proof indicates that it is *possible* to transmit at as low an error rate as possible, but does not show how. A great deal of research activity in the years since Shannon developed his capacity expression has been devoted to the investigation of various types of encoders. In a later section we shall discuss a very specific communication system incorporating a feedback path that is found to obey Shannon's equation. This will provide further insight into the significance of Eqs. [8-60], [8-61], and the meaning of $R < C$.

Not only does the need to design encoders provide a drawback to the application of Shannon's theorem, but an additional complication enters into the picture. As one tries for lower and lower probabilities of error, the encoding time T becomes longer and longer. A delay of T sec in transmission is then incurred at the transmitter. A corresponding delay of T sec at the receiver is incurred in decoding the actual transmitted message, for a total delay of $2T$ sec. As usual then a price must be paid for the required decrease in SNR. The circuitry becomes much more complex and large time delays are incurred.

This tradeoff between probability of error and transmission time T is indicated more clearly by considering the system of Fig. 8-25. A binary

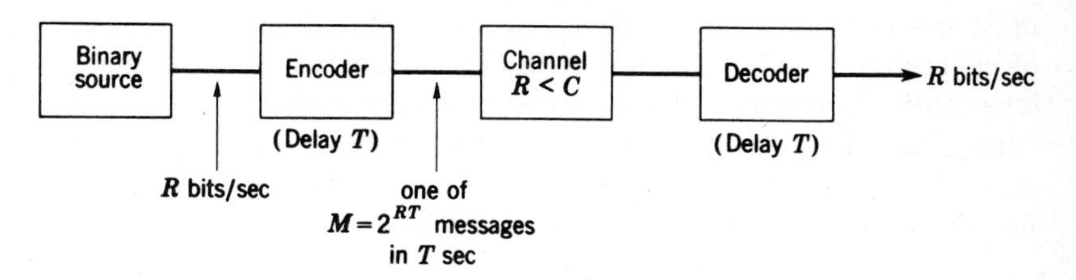

FIG. 8-25 Encoding for optimum transmission.

source spewing out, as an example, the bipolar pulses of Fig. 8-26, is followed by an encoder which examines a fixed number of successive binary pulses. For an encoder processing time of T sec, this number must be RT. (In the example of Fig. 8-26, $RT = 8$.) It is apparent

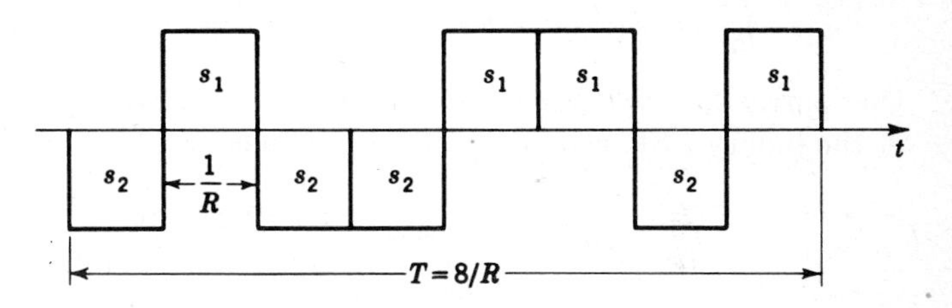

FIG. 8-26 Binary source output: relation between T and $1/R$. ($M = 2^{RT} = 2^8 = 256$.)

that 2^{RT} possible binary messages could have been transmitted in the interval T sec long. For every one of these possible inputs the encoder of Fig. 8-25 puts out a unique message T sec long. Every T sec, then, one of the $M = 2^{RT}$ possible encoded messages is transmitted by the encoder. It is this encoded message that is then corrupted by noise during transmission. The receiver decoder must now decide which one of the $M = 2^{RT}$ messages was actually transmitted. This requires storing the incoming message during its time of transmission T, and then sending out the corresponding sequence of binary symbols.

Note that in the encoding-decoding process the same bit rate of R bits/sec is always maintained. There is no increase or decrease in rate of transmission. A total delay of $2T$ sec is entailed, however, because of the encoding at the transmitter (time taken to examine RT bits) and the decoding at the receiver (time taken to examine the encoded message T sec long).

This encoding and decoding process is similar to the binary–M-ary conversion discussed in Chap. 3. There we talked solely of converting 1 of M possible amplitude levels to $\log_2 M$ binary samples; or, conversely, of collapsing n binary symbols into one of 2^n possible amplitude levels. The feedback system to be discussed in a later section actually uses this type of encoding. In the next section we discuss another form of encoding. In general, the form of the encoded messages is not specified and is one of the open questions in communications.

8-5 M-ARY ORTHOGONAL SIGNALS

One particular class of encoded signals that has been utilized successfully in practice in approaching the Shannon limit on capacity is the class of so-called M-ary orthogonal signals. By orthogonal signals $s_i(t)$ and $s_j(t)$ we mean, as noted in previous sections,

$$\int_0^T s_i(t)s_j(t)\,dt = E\delta_{ij} \qquad \delta_{ij} = 0,\ i \neq j;\ \delta_{ij} = 1,\ i = j \tag{8-62}$$

One particular example is the class of M-ary FSK signals, an extension of the binary FSK class considered previously. Thus let

$$s_i(t) = A\cos\omega_i t \qquad \begin{matrix} 0 \le t \le T \\ i = 1\ .\ .\ .\ ,\ M \end{matrix} \tag{8-63}$$

with

$$f_{i+1} - f_i = \frac{1}{T} \tag{8-64}$$

A little thought indicates that these signals do satisfy the orthogonality relation of Eq. [8-62]. These are equiamplitude sinusoidal pulses, each lasting T sec, of progressively increasing carrier frequency, with the spacing between adjacent frequencies just $1/T$ (Fig. 8-27). To show the binary–M-ary conversion of Fig. 8-25 in this case, take as an example $RT = 5$. The encoder then stores 5 successive bits, and, based on the particular 5-bit sequence read, sends out the appropriate 1 of

$$M = 2^{RT} = 32$$

possible frequencies (Fig. 8-27a).

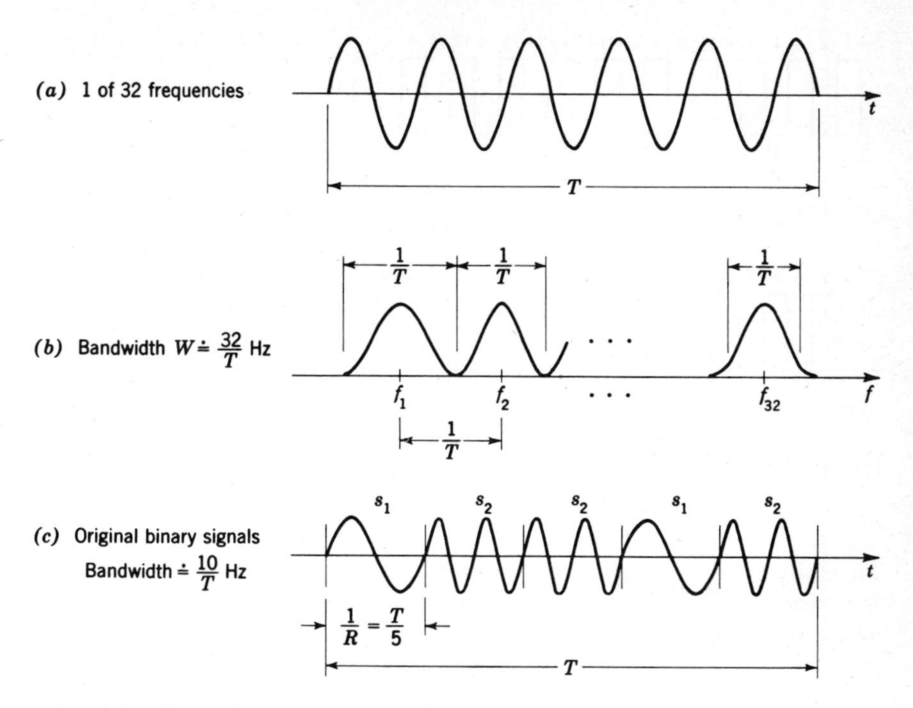

FIG. 8-27 M-ary FSK transmission, $M = 32$.

Note, however, that as the number M increases (by increasing the transmission time T), the bandwidth required to transmit the M possible signals goes up rapidly. For, as shown in Fig. 8-27*b*, the bandwidth required is

$$W \doteq \frac{M}{T} = \frac{2^{RT}}{T} \tag{8-65}$$

(Any one sine-wave pulse T sec long requires $\doteq 1/T$ Hz bandwidth.) We thus pay an additional price for the encoding operation. Since the bandwidth is increasing exponentially with T, low error rates may require extremely wide transmission bandwidths.

Another common example of an M-ary signal is the orthogonal pulse sequence shown in Fig. 8-28. There are 2^M possible sequences utilizing the on-off pulses of duration T/M shown. Of these a much smaller set of M orthogonal signals are selected as the encoded signals. An example of two such orthogonal signals is shown in Fig. 8-28. Note that these also require a (baseband) bandwidth of $M/T = 2^{RT}/T$ Hz. (If this sequence in turn amplitude modulates a high-frequency carrier, the transmission bandwidth required is of course $2M/T$.) Such orthogonal pulse sequences

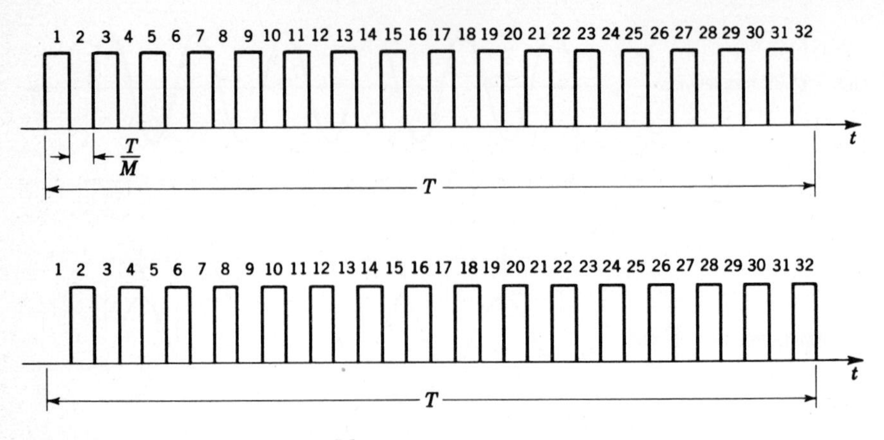

FIG. 8-28 Orthogonal signals, $M = 32$.

are easily generated using shift register techniques.[1] Another possibility is to use these elementary binary symbols to in turn modulate a PSK sequence so that $\pm \cos \omega_0 t$ pulses T/M sec long are transmitted in place of the 1's and 0's, respectively, in Fig. 8-28. The resultant sequences of $\pm \cos \omega_0 t$ pulses must of course be chosen to satisfy the orthogonality relation.

We have indicated that these M orthogonal signal sets may be shown to be examples of optimum encoded signals in the sense that the probability of error goes to zero as T (hence $M = 2^{RT}$) becomes very large. For M very large, the transmission bandwidth $W = 2^{RT}/T$ rapidly grows beyond bound, however. The Shannon capacity of Eq. [8-60] in turn approaches a limiting value C_∞ given by

$$C_\infty = \lim_{W \to \infty} C = \frac{S}{n_0 \log_e 2} \qquad \text{bits/sec} \tag{8-66}$$

[Recall that $\log_e (1 + \epsilon) \to \epsilon$, if $\epsilon \ll 1$.] But error-free transmission is only possible if $R < C$. In the limit of large T (and hence large W), we must then have

$$\frac{S}{n_0} > R \log_e 2 \tag{8-67}$$

or

$$\frac{S}{n_0 R} > \log_e 2 = 0.69 \tag{8-67a}$$

[1] S. W. Golomb (ed.), "Digital Communication with Space Applications," Prentice-Hall, Englewood Cliffs, N.J., 1964.

for low error probabilities to be attained. For a given signal power and noise spectral density, there is thus a maximum rate of transmission possible. We shall show that the M-ary orthogonal signal sets, $M \to \infty$, demonstrate exactly this type of behavior.

For finite M the error probability of the M orthogonal signals may be readily calculated, extending the approach used in Sec. 8-3 for binary signals.[1] Interestingly, one may find the optimum receiver for the M orthogonal signals in additive band-limited white gaussian noise, and just as in the previous section, it is given by a bank of M parallel matched filters, plus a sampling and decision circuit. Here samples are taken every T sec. The M filters are each matched to one of the $s_j(t)$,[2] $j = 1 \cdots M$.

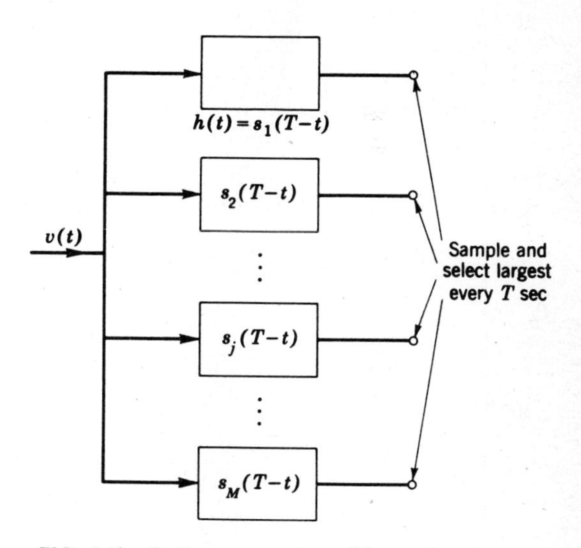

FIG. 8-29 Optimum receiver, M-ary signals.

A sketch of the optimum receiver is shown in Fig. 8-29. The largest sample of the M available every T sec provides the decision as to the $s_j(t)$ transmitted.

A. J. Viterbi has plotted the probability-of-error curves for the

[1] A. J. Viterbi, "Principles of Coherent Communication," chap. 8, McGraw-Hill, New York, 1966; also, M. Schwartz, W. R. Bennett, and S. Stein, "Communication Systems and Techniques," chap. 2, McGraw-Hill, New York, 1966.

[2] Schwartz, Bennett, and Stein, *ibid.*

optimum M-ary receiver.[1] These are reproduced here as Fig. 8-30. The parameter k indicated is $\log_2 M$. Note that the probability of error does decrease with M as expected and that a limit on S/n_0R exists for $M \to \infty$. For $S/n_0R > 0.69$, $P_e \to 0$; for $S/n_0R < 0.69$, $P_e \to 1$. This

[1] Viterbi, *op. cit.*, fig. 8.3, p. 222.

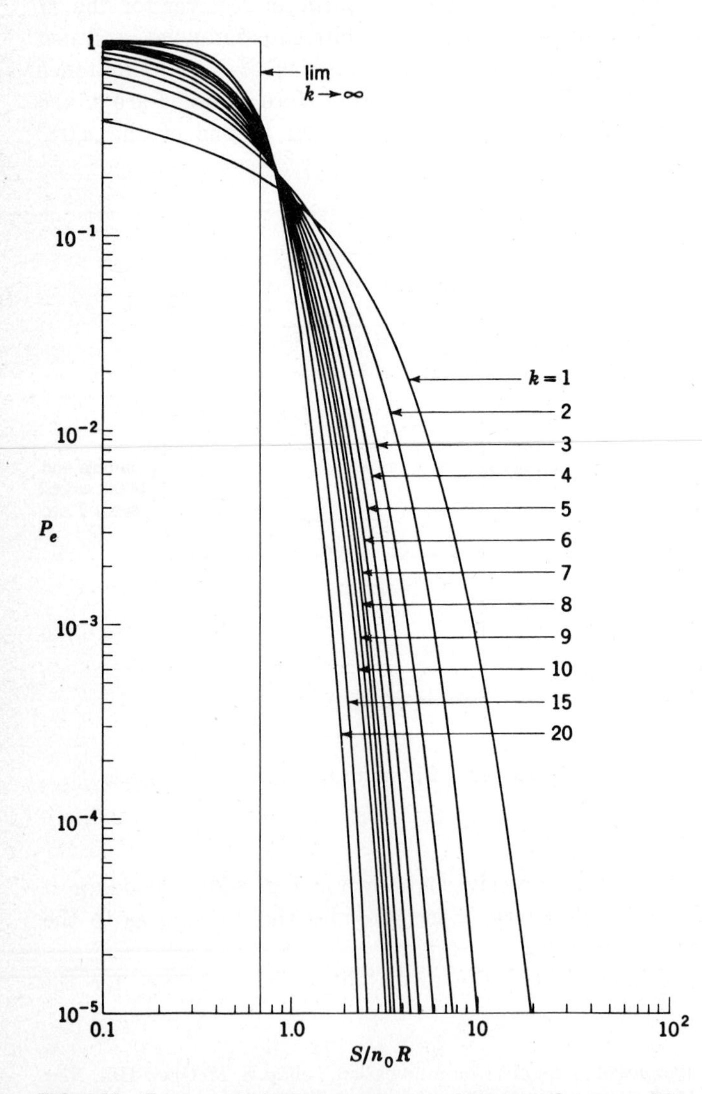

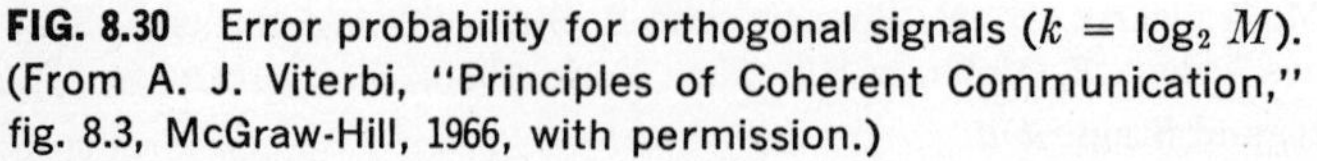
FIG. 8.30 Error probability for orthogonal signals ($k = \log_2 M$). (From A. J. Viterbi, "Principles of Coherent Communication," fig. 8.3, McGraw-Hill, 1966, with permission.)

is of course the same as obtained from the infinite-bandwidth Shannon capacity. The M-ary, orthogonal signal set, $M \to \infty$, is therefore an optimum code in the Shannon sense.

If the channel being used has no bandwidth constraint, these M orthogonal signals, $M \to \infty$, provide the best possible error performance for a channel with additive gaussian noise. The decrease in error probability is rather slow with M, however, as shown by Fig. 8-30.[1] (Note that the change in P_e in going from $M = 2^5$ to $M = 2^{10}$ is not very great, for example, yet the bandwidth increases 2^5 times, while the system complexity grows in a similar manner.) It is thus not practical to use extremely large M's. A telemetry system in use, called the *Digilock* system, uses $M = 32$ orthogonal signals.[2] The improvement possible in this special case of $M = 32$ over straight binary transmission is shown by the curve of Fig. 8-31. (The binary and M-ary signals compared are

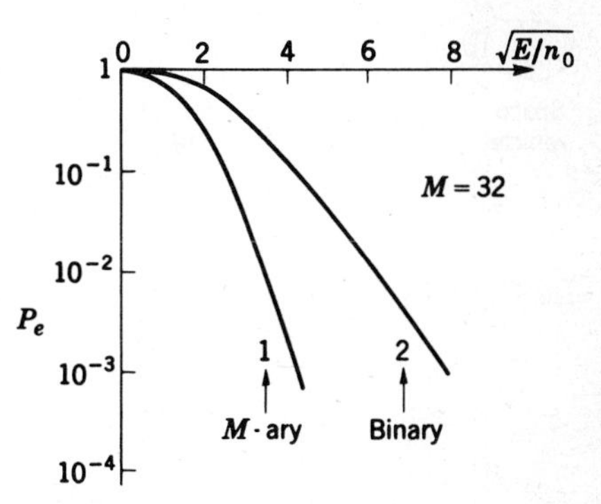

FIG. 8-31 Comparison of M-ary and binary transmission.

shown in Fig. 8-27.) The curves drawn here compare the character error probability for the same average signal energy in each of the two cases. Note that the M-ary scheme requires on the average about ⅓ the signal power at a cost of roughly three times the bandwidth ($32/T$ as compared to $10/T$ in the binary case—see Fig. 8-27*b* and *c*). There is

[1] These curves actually refer to *character* error probability, i.e., the probability that one of the M signals will be in error. This may readily be converted to bit error probability, if so desired. See Viterbi, *op. cit.*, fig. 8.6, p. 227. For $M \gg 1$, there is no essential difference between the two.

[2] R. M. Jaffee, Digilock Telemetry System for the Air Force Special Weapons Center's Blue Scout Jr., *IRE Trans. Space Electron. and Telemetry*, vol. SET-8, pp. 44–50, March, 1962.

of course additional cost in equipment as well as the $2T$ time delay involved.

8-6 A SPACE-COMMUNICATIONS EXAMPLE

The use of wideband M-ary transmission or other types of coded transmission systems is particularly important in long-distance space-communication systems. A typical example will indicate the problems involved. It also gives us an opportunity to consider some representative numbers, and clarifies further the significance of channel capacity.

Consider a digital-communications system aboard a space vehicle a distance d miles out in space (Fig. 8-32). The system operates at 500

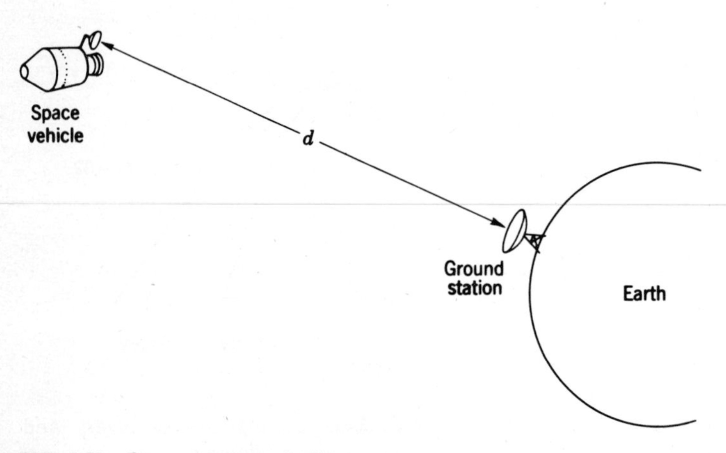

FIG. 8-32 Space communication.

MHz, and has an average transmitter-output power S_T at that frequency of 1 watt. The problem is to determine the SNR at the receiver on earth and see very specifically what the Shannon maximum-channel capacity means in this case.

If the power on the vehicle were radiated isotropically, the power density (watts/meter^2) at a distance d would be $S_T/4\pi d^2$. An antenna on the vehicle serves to concentrate or focus the electromagnetic energy transmitted into a beam. The larger the antenna size (in wavelengths) the narrower the beam and the greater the energy concentration. Specifically, this focusing effect is denoted by assigning to the antenna a gain G_T over isotropic radiation. The effective power density at a distance d from the transmitting antenna is then $(S_T/4\pi d^2)G_T$. At high frequencies aperture-type antennas (parabolas or lens antennas are examples) are commonly used, in which case it may be shown that the maximum gain attainable is proportional to the aperture area A_T (the solid angle subtended by the beam is inversely proportional to the area) and is given by $G_T = 4\pi A_T/\lambda^2$.† Here λ is the wavelength. At the receiver the receiving antenna of aperture area A_R ideally presents an effective area equal to A_R in picking up the received power. (In practice both the transmitting antenna gain and receiver antenna area are less than the quantities shown, reducing the received power typically by 3 or more db. We ignore these effects here.) The received power in watts is then

$$S_R = \frac{S_T}{4\pi d^2} G_T A_R$$

or, putting in the ideal expression for receiver antenna gain, $G_R = 4\pi A_R/\lambda^2$, we have, finally,

$$S_R = \frac{S_T}{4\pi d^2} G_R G_T \frac{\lambda^2}{4\pi} \qquad [8\text{-}68]$$

This simple expression relates received and transmitted powers. (This expression ignores transmission losses due to attenuation along the propagation path. It assumes free space transmission.) Quite commonly actual power calculations are carried out in decibels. This of course simplifies the calculation, and enables any additional loss or gain factors to be simply added or subtracted. (These would include antenna gain losses, propagation losses, etc.) As an example, at d = 1 million miles (1.6×10^9 meters), and a frequency of 500 MHz ($\lambda = 0.6$ meter), the "free space loss" $(\lambda/4\pi d)^2$, in decibels, is $\doteq -210$ db. A factor of 10 in distance changes this by 20 db.

Assume that the two antenna gains are, respectively,

$$G_T = 10 \equiv 10 \text{ db}$$

and

$$G_R = 10^4 \equiv 40 \text{ db}$$

(It is apparent that antennas at the ground station on earth can be much larger than those on the space vehicle, providing therefore considerably more gain.) At 500 MHz, $\lambda = 0.6$ meter, so that the ground antenna aperture is $A_R \doteq 300$ meters², or roughly 55 ft × 55 ft. Then, from Eq. [8-68].

$$S_{R,db} = S_{T,db} + 10 + 40 - 210 = S_{T,db} - 160$$

† D. J. Angelakos and T. E. Everhart, "Microwave Communications," sec. 5-7, McGraw-Hill, New York, 1968.

at $d = 10^6$ miles. For the 1-watt transmitter power assumed, then, $S_{R,db} = -160$ db, or 10^{-16} watt. At $d = 10$ million miles, the received power is 10^{-18} watt, while at 100,000 miles it is 10^{-14} watt.

If a receiver with a 3-db noise figure is assumed, the noise spectral density at 290°K is $n_0 = FkT \doteq 10^{-20}$ watt/Hz, or $\doteq -200$ db. The ratio of received power to noise spectral density is then

$$\frac{S_R}{n_0} = 10^4 \equiv 40 \text{ db}$$

Now assume that we want to use ordinary binary transmission for communication purposes. As noted earlier, the curve of Fig. 8-22 indicates that E/n_0 should be at least 10 to have a probability of error less than 10^{-5}. This is of course for optimum PSK transmission, with synchronous detection assumed available at the receiver. For an FSK system $E/n_0 \geq 20$ for the same error probability. But $E = S_R/R$, R the binary transmission rate in bits per second. Therefore, we must have a received power to noise spectral density ratio given by

$$\frac{S_R}{n_0} \geq 10R \qquad [8\text{-}69]$$

$P_e \leq 10^{-5}$. This indicates of course a maximum binary transmission rate of $S_R/10n_0$. In the example under consideration (with $d = 10^6$ miles), this is $R_{\max} = 1{,}000$ bits/sec. If the range increases to 10 million miles, $R_{\max} = 10$ bits/sec, a rather low data rate. (If orthogonal binary transmission or FSK is used the numbers are 500 bits/sec and 5 bits/sec, respectively.)

How much improvement do we get by going to M-ary techniques? Before considering this, what is the *maximum* rate at which we can transmit over this space channel? From Eq. [8-66] the maximum possible capacity (with an error probability approaching zero) at $d = 10^6$ miles is $C_\infty = 10^4/\log_e 2 = 14{,}500$ bits/sec. At $d = 10^7$ miles this becomes 145 bits/sec. So it *is* theoretically possible to transmit at more than ten times the binary rate, but at the expense of course of much larger bandwidths, additional circuitry, and time delay.

If we now consider a more practical M-ary communication system, with $M = 32$ orthogonal signals used, Fig. 8-30 indicates that for a *character* probability of error of at least 10^{-5}, $S_R/n_0R \geq 5$.† There is thus a factor of 2 improvement over optimal (PSK) binary transmission, or 4 over binary FSK. One could transmit up to 2,000 bits/sec at $d = 10^6$ miles, using the 32 orthogonal signals. At $d = 10^7$ miles, this again reduces by a factor of 100 to 20 bits/sec.

† The bit error probability is essentially the same. See Viterbi, *op. cit.*, pp. 226–227.

8-7 AN INFORMATION-FEEDBACK SCHEME FOR ACHIEVING BAND-LIMITED CHANNEL CAPACITY

We now turn to the discussion of an extremely ingenious scheme for M-ary communication that achieves the optimum *band-limited* capacity of Eq. [8-60]. Recall that Shannon's analysis nowhere came up with a prescription for a specific encoding scheme to achieve capacity. Instead random coding arguments have been advanced to show that there must exist a particular coding scheme that achieves capacity in the sense of Shannon.

The noise added is again assumed to be gaussian white noise, with a channel bandwidth limitation of W Hz.

The digital-communications system we shall discuss was suggested by J. P. M. Schalkwijk.[1] It requires the availability of a noiseless feedback channel over which to transmit information from the receiver back to the transmitter. Although strictly noiseless channels do not exist in real life, such channels may be readily approximated. One example is the path from ground to space vehicle in a space-communications system. As noted in the previous section the ground antenna as well as power available are considerably greater than the antenna and power available on the space vehicle. The SNR at the space vehicle due to transmission from ground is hence considerably greater than the corresponding SNR at the ground, due to transmission from space. It is thus reasonable to assume noise negligible on the ground-space channel, as compared to noise added on the space-ground path.

The use of a feedback path for information transmission results in a significant reduction in coding complexity, as contrasted to one-way encoding and decoding. The scheme is also of interest because the concept of channel capacity C in the sense of a probability of error approaching zero for binary rates $R < C$ arises quite naturally from the discussion. This then provides us with additional insight into the concept of capacity, binary to M-ary encoding, etc.

The specific scheme to be discussed is shown in Fig. 8-33. The encoder again stores the RT bits coming in, in a T-sec interval. It converts the particular bit sequence measured to the appropriate one of $M = 2^{RT}$ possible amplitude levels. (The encoder output is then quantized PAM, with 2^{RT} levels used.) These are shown in Fig. 8-34, with the maximum amplitude normalized to 1.

The particular amplitude $\theta_i = i/M$ $(i = 1, \ldots, M)$ to be transmitted, labeled θ here for simplicity, is then used to amplitude modulate a

[1] J. P. M. Schalkwijk and T. Kailath, A Coding Scheme for Additive Noise Channels with Feedback, part I, *IEEE Trans. Information Theory,* vol. IT-12, no. 2, April 1966; J. P. M. Schalkwijk, A Coding Scheme for Additive Noise Channels with Feedback, part II, *IEEE Trans. Information Theory,* vol. IT-12, no. 2, April, 1966.

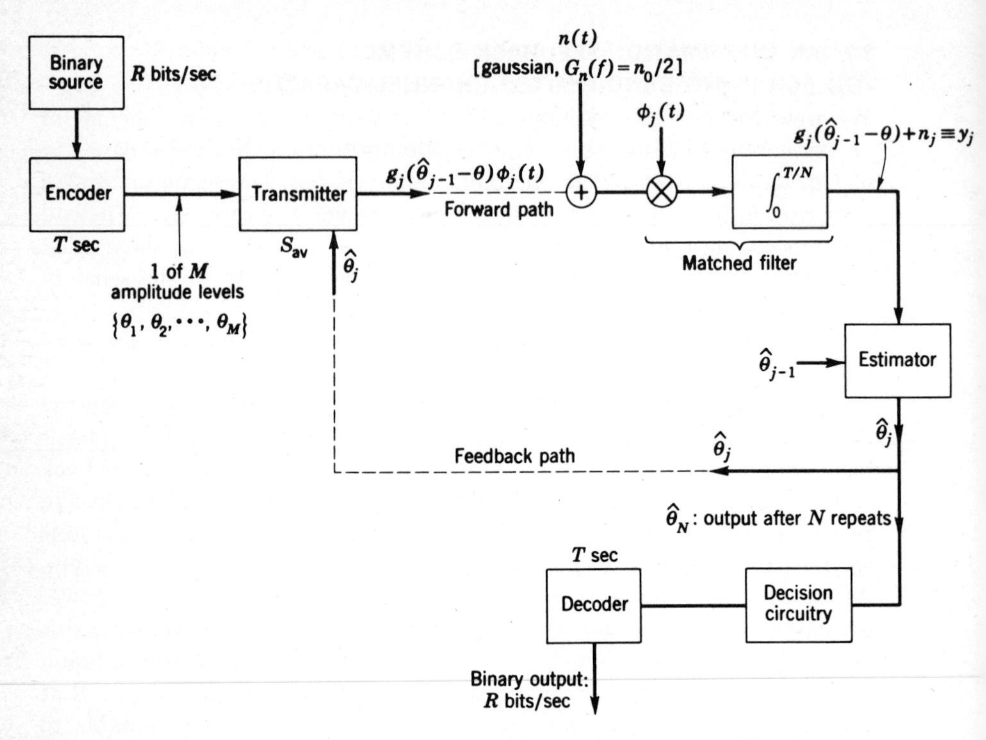

FIG. 8-33 Information-feedback transmission system, $j = 1, \ldots, N$.

transmitter as shown in Fig. 8-33. The total transmission time T required is here considered broken up into N time intervals, each T/N sec long. During each interval the transmitter transmits a burst of signal energy. This signal, plus noise added during transmission or at the receiver, are matched-filter detected for T/N sec at the receiver. At the end of the T/N sec transmission interval the receiver transmits an estimate of θ back to the transmitter. During the next time interval T/N sec long the transmitter uses the receiver estimate of θ to modify its amplitude-modulation signal. Again transmission takes place for T/N sec. After N repetitions for a total transmission time of T sec, the receiver makes its

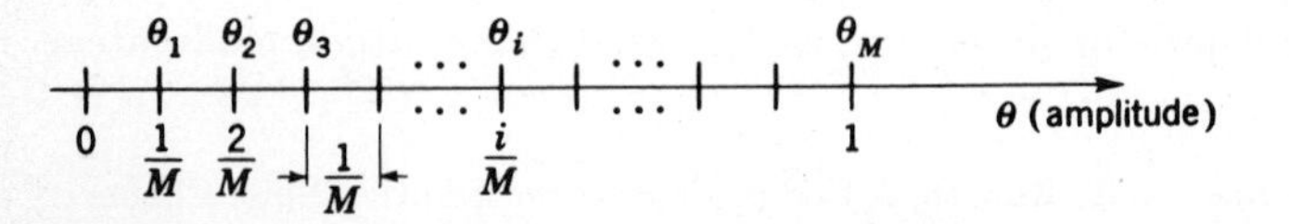

FIG. 8-34 Output amplitude levels, M-ary PAM.

final estimate $\hat{\theta}_N$ of the θ transmitted. This serves as the receiver output to the M-ary–binary decoder. The transmitter then begins the cycle again, using the next M-ary signal θ.

The transmitter carrier signal during the jth T/N sec interval is denoted by $\varphi_j(t)$ for simplicity (Fig. 8-35), while the amplitude to be

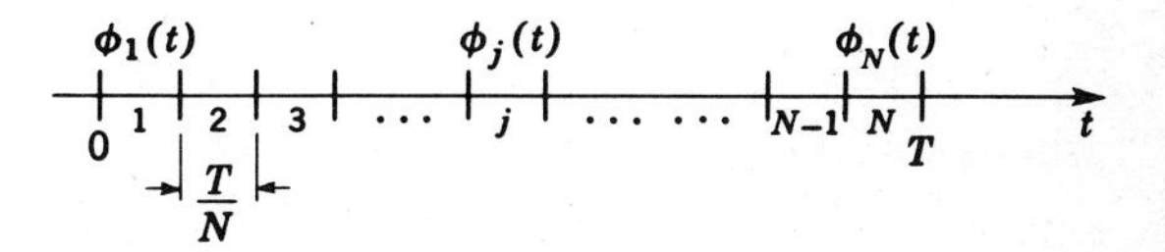

FIG. 8-35 Repetition intervals.

transmitted is given by $g_j(\hat{\theta}_{j-1} - \theta)$. g_j is a prescribed gain factor (to be discussed later), $\hat{\theta}_{j-1}$ denotes the receiver estimate of θ formed during the previous $(j - 1)$th interval, and θ (one of the M amplitudes of Fig. 8-34) is the signal to be transmitted.

For simplicity we assume the carrier $\varphi_j(t)$ to have unit energy:

$$\int_0^{T/N} \varphi_j^2(t)\, dt = 1$$

The actual amplitude is then subsumed in the gain factors g_j. [Strictly speaking we should integrate $\varphi_j^2(t)$ over the jth interval $(j - 1)(T/N)$ to $j(T/N)$, but we write the interval as $(0, T/N)$ for simplicity.]

With the transmission and reception strategy outlined, we now have to decide on the gain factors g_j and the method of finding the receiver estimates $\hat{\theta}_j$. Before doing this, we put in an additional important constraint: the transmitter is assumed average power-limited to S_{av} watts. (A peak power limitation may also be introduced, but this turns out not to be crucial to the development.) Because of the random nature of the transmission ($\hat{\theta}_{j-1}$ in the modulating signal term is random and depends on the noise added), we shall only be able to specify S_{av} in a statistical or *ensemble* average sense.

Specifically, the average power is just the total energy required over the full T-sec interval divided by T. On any one T-sec transmission the particular power required is

$$S = \frac{1}{T} \sum_{j=1}^{N} \int_0^{T/N} g_j^2(\hat{\theta}_{j-1} - \theta)^2 \varphi_j^2(t)\, dt$$

$$= \frac{1}{T} \sum_{j=1}^{N} g_j^2(\hat{\theta}_{j-1} - \theta)^2 \tag{8-70}$$

with

$$\int_0^{T/N} \varphi_j^2(t)\, dt = 1$$

as assumed. As the signal θ varies, while the estimate $\hat{\theta}_{j-1}$ varies due to noise, S varies as well. The ensemble average power is then

$$S_{\text{av}} = \frac{1}{T} \sum_{j=1}^{N} E[g_j^2(\theta_{j-1} - \theta)^2] \tag{8-71}$$

where we average over the random M-ary signal θ ($\theta = i/M$, $i = 1, \ldots, M$) and the random variables $\hat{\theta}_{j-1}$. Note that if after each T/N sec transmission the estimate $\hat{\theta}_{j-1}$ approaches θ more closely, as might be expected and as will in fact be shown, one would expect the gain g_j to *increase* with j. This is the only way to keep S_{av} fixed. We shall verify this later.

We now turn to a more detailed analysis of the system. As shown in Fig. 8-33 the matched-filter output at the end of the jth transmission is the random variable

$$\begin{aligned} y_j &= \int_0^{T/N} g_j(\hat{\theta}_{j-1} - \theta)\varphi_j^2(t)\, dt + \int_0^{T/N} n(t)\varphi_j(t)\, dt \\ &= g_j(\hat{\theta}_{j-1} - \theta) + n_j \end{aligned} \tag{8-72}$$

where

$$n_j = \int_0^{T/N} n(t)\varphi_j(t)\, dt \tag{8-73}$$

is the noise component of the matched-filter output due to $n(t)$ added during the T/N sec interval. Here we again make use of the fact that $\varphi_j(t)$ is assumed to have unit energy over the T/N sec interval. With $n(t)$ and hence n_j gaussian, y_j is a gaussian random variable as well.

The receiver must presumably use the number y_j measured to carry out its estimate of θ. How does it do this? We assume the previous estimate $\hat{\theta}_{j-1}$ stored and available for the new estimate as well. The new estimate $\hat{\theta}_j$ is then a function of y_j and $\hat{\theta}_{j-1}$:

$$\hat{\theta}_j = h(y_j, \hat{\theta}_{j-1}) \tag{8-74}$$

This is indicated by the box labeled *estimator* in Fig. 8-33. It is apparent that the "best" functional form to use is one that minimizes the probability of error, or, equivalently, the final mean-squared error $E(\hat{\theta}_N - \theta)^2$ between the receiver estimate and θ. ($\hat{\theta}_N$ is gaussian, with expected value θ, as will be shown shortly. This mean-squared error is then of course the variance of $\hat{\theta}_N$, and minimum variance corresponds to minimum probability of error here.) It may be shown[1] that the strategy

[1] Schalkwijk, *op. cit.*

guaranteed to minimize $E(\hat{\theta}_N - \theta)^2$ proceeds as follows:

1. From the incoming y_j and the stored $\hat{\theta}_{j-1}$, form the new gaussian random variable

$$z_j = \hat{\theta}_{j-1} - \frac{y_j}{g_j} = \theta - \frac{n_j}{g_j} \qquad [8\text{-}75]$$

(Note that if g_j does increase with j, as indicated earlier, the successive z_j's move in and more closely to θ, as desired.) This variable has the variance $\sigma_{z_j}{}^2 = (1/g_j{}^2)E(n_j{}^2) = \sigma^2/g_j{}^2$, if we use σ^2 to denote the noise variance.

2. Denote the variance of the random variable $\hat{\theta}_{j-1}$ by $\sigma^2_{\theta_{j-1}}$. Then the new estimate $\hat{\theta}_j$ is formed as follows:

$$\hat{\theta}_j = \frac{\sigma_{z_j}{}^2\hat{\theta}_{j-1} + \sigma^2_{\theta_{j-1}}z_j}{\sigma_{z_j}{}^2 + \sigma^2_{\theta_{j-1}}} \qquad [8\text{-}76]$$

This particular estimate has very interesting properties and in fact may be rewritten in a particularly simple form if we start with the first T/N sec transmission, $j = 1$, and eliminate $\hat{\theta}_{j-1}$ in terms of past values of z_l, $l = 1, \ldots, j-1$. Thus, on the first transmission we send $g_1(\hat{\theta}_0 - \theta)$ and receive $y_1 = g_1(\hat{\theta}_0 - \theta) + n_1$. [We can ignore the carrier terms $\varphi_j(t)$ since they obviously do not enter into the analysis.] Any arbitrary number may be used as the 0th estimate $\hat{\theta}_0$. We shall use 0.5 for simplicity. This is a known number, specified beforehand. The first estimate of θ is taken as

$$\hat{\theta}_1 = z_1 = 0.5 - \frac{y_1}{g_1} = \theta - \frac{n_1}{g_1} \qquad [8\text{-}77]$$

using the appropriate expression for y_1. Its variance is of course

$$\sigma_{z_1}{}^2 = \frac{\sigma^2}{g_1{}^2}$$

again using σ^2 to represent $E(n_1{}^2)$, the noise variance.

At the end of the second transmission the received signal is given by $y_2 = g_2(\hat{\theta}_1 - \theta) + n_2$. Forming $z_2 = \theta_1 - y_2/g_2 = \theta - n_2/g_2$, and using Eq. [8-76] to provide the estimate $\hat{\theta}_2$, we have

$$\hat{\theta}_2 = \frac{(\sigma^2/g_2{}^2)\hat{\theta}_1 + (\sigma^2/g_1{}^2)z_2}{\sigma^2/g_2{}^2 + \sigma^2/g_1{}^2} = \frac{g_1{}^2z_1 + g_2{}^2z_2}{g_1{}^2 + g_2{}^2} \qquad [8\text{-}78]$$

using $\hat{\theta}_1 = z_1$, as in Eq. [8-77].

Continuing this procedure step-by-step, it is left to the reader to show that at the end of j transmissions, the estimate $\hat{\theta}_j$ is given by

$$\hat{\theta}_j = \frac{\sum_{k=1}^{j} g_k^2 z_k}{\sum_{k=1}^{j} g_k^2} \qquad [8\text{-}79]$$

Note the particularly simple form of $\hat{\theta}_j$. Note also that by eliminating previous estimates we have managed to put $\hat{\theta}_j$ in a form depending only on the gaussian random variables z_j. These are all independent if successive noise samples are independent, as is true here. [$n(t)$ was assumed *white*, with spectral density $n_0/2$.] This may be simplified still further. Since z_k is gaussian, $\hat{\theta}_j$ is gaussian as well. In particular, we have from Eq. [8-79] and Eq. [8-75] describing z_j,

$$E(\hat{\theta}_j) = \theta \qquad [8\text{-}80]$$

and

$$\text{var}\,(\hat{\theta}_j) \equiv \sigma_j^2 = \frac{\sigma^2}{\sum_{k=1}^{j} g_k^2} \qquad [8\text{-}81]$$

As j increases, then, σ_j^2 decreases, providing the desired approach more and more closely to the transmitted signal θ.

After the specified N repeats the variance of the estimate $\hat{\theta}_N$, the actual receiver output to the decoder, is given by

$$\sigma_N^2 = \frac{\sigma^2}{\sum_{j=1}^{N} g_j^2} \qquad [8\text{-}82]$$

With this discussion as an introduction, we now ask: How well does this system perform? How does the channel-capacity concept arise?

We shall be in a position to answer these questions after determining the optimum set of gain coefficients g_j. These will be found on the basis of minimizing σ_N^2 (Eq. [8-82]), with the average transmitter power S_{av} (Eq. [8-71]) fixed. Interestingly, we shall find these coefficients increasing exponentially with j. The corresponding variance σ_N^2 is then found to decrease exponentially with the number of repetitions N. If the ratio T/N is held constant, increasing N corresponds to increasing T. So the variance σ_N^2 decreases rapidly with T, a desirable characteristic. The probability of error then presumably decreases as well.

To show this, consider the signal θ to be one of the M amplitude levels in Fig. 8-34. It is apparent that with all M levels equally likely to occur, an optimum decision procedure consists of setting decision levels $\pm 1/2M$ units away from each of the M levels (Fig. 8-36). So long as

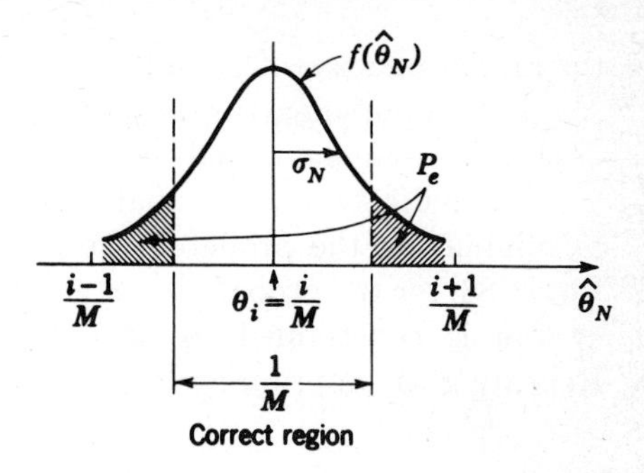

FIG. 8-36 Determination of probability of error.

the final estimate θ_N falls within this range $\pm 1/2M$ about the appropriate $\theta_i = i/M$, no error occurs. If the estimate falls outside this region, a mistake will be made. The evaluation of the probability of error is shown in Fig. 8-36. It is apparent that the probability of error is the same for all θ_i's transmitted, except possibly for the points near the ends of the θ interval (Fig. 8-34). For $M \gg 1$, however, the contributions due to these points may be neglected.

Since $\hat{\theta}_N$ is gaussian with expected value $\theta = \theta_i$ (the particular signal transmitted), and variance σ_N^2 given by Eq. [8-82], it is apparent from Fig. 8-36 that the probability of error is just

$$\begin{aligned} P_e &\doteq 2 \int_{\theta_i + \frac{1}{2M}}^{\infty} \frac{e^{-(\hat{\theta}_N - \theta_i^2)/2\sigma_N^2}}{\sqrt{2\pi\sigma_N^2}} \, d\hat{\theta}_N \\ &= 2 \int_{\frac{1}{2M\sigma_N}}^{\infty} \frac{e^{-x^2/2}}{\sqrt{2\pi}} \, dx \\ &= \operatorname{erfc}\left(\frac{1}{2\sqrt{2}\, M\sigma_N}\right) \end{aligned} \qquad [8\text{-}83]$$

after shifting the origin to θ_i, normalizing to σ_N, and using the appropriate definition of the complementary error function. (The upper limit in the integrals of Eq. [8-83] obviously cannot exceed 1, but if $M \gg 1$ as assumed, and $\sigma_N \ll 1$, the upper limit of ∞ used is quite accurate.) From the properties of the erfc function it is apparent that as σ_N decreases, P_e decreases as well. (This is also apparent from Fig. 8-36.)

But note from Eq. [8-83] that the error depends on $M\sigma_N$. If one attempts to increase M as σ_N decreases, it is apparent that the probability of error may *not* decrease. Since M is determined by the original binary transmission rate R, as well as by the transmission time T ($M = 2^{RT}$), increasing T or R could very well increase M. We shall see that limiting the rate of increase of M in fact brings the concept of capacity into play.

Before pursuing this argument further, we return to the evaluation of the optimum gain coefficients g_j.

We indicated earlier that minimizing the variance σ_N^2 corresponds to minimizing the probability of error P_e. This is also apparent from Eq. [8-83] for the probability of error. We also indicated earlier that the system is constrained to have fixed average transmitter power S_{av}. Repeating the appropriate expressions for σ_N^2 and S_{av} here, we have

$$\sigma_N^2 = \frac{\sigma^2}{\sum_{j=1}^{N} g_j^2} \qquad [8\text{-}82]$$

and

$$S_{av} = \frac{1}{T} \sum_{j=1}^{N} E[g_j^2(\hat{\theta}_{j-1} - \theta)^2] \qquad [8\text{-}71]$$

This latter equation for average power S_{av} may be simplified considerably by noting that $E[(\hat{\theta}_{j-1} - \theta)]^2 = \text{var}\,(\hat{\theta}_{j-1}) = \sigma^2/\left(\sum_{k=1}^{j-1} g_k^2\right)$, from Eq. [8-81]. Also, extracting the energy contribution due to the first transmission, we have $E[g_1^2(0.5 - \theta)^2] = g_1^2/12$. [All M values of θ are assumed equally likely, and for large M ($M \gg 1$), θ is very nearly continuously and uniformly distributed.] Finally, then,

$$S_{av} = \frac{g_1^2}{12T} + \frac{\sigma^2}{T} \sum_{j=2}^{N} \frac{g_j^2}{\sum_{k=1}^{j-1} g_k^2} \qquad [8\text{-}84]$$

The object then is to minimize σ_N^2 (Eq. [8-82]) with S_{av} (Eq. [8-84]) held fixed. It is easily shown that this minimization corresponds alternately to minimizing the power S_{av} with σ_N^2, and hence P_e, fixed.

The desired minimization is simplified considerably by first performing a change of variables. Let

$$z_j \equiv \frac{g_j^2}{\sum_{k=1}^{j-1} g_k^2} \qquad [8\text{-}85]$$

Also let $g_1^2 \equiv \alpha^2$ for simplicity in notation. Then

$$\begin{aligned} g_2^2 &= z_2\alpha^2 \\ g_3^2 &= z_3(g_1^2 + g_2^2) = z_3\alpha^2(1 + z_2) \\ g_4^2 &= z_4(g_1^2 + g_2^2 + g_3^2) = z_4\alpha^2(1 + z_3)(1 + z_4) \end{aligned}$$

using Eq. [8-85], and simplifying. In particular g_j is then given simply by

$$g_j^2 = z_j\alpha^2 \prod_{k=2}^{j-1} (1 + z_k) \tag{8-86}$$

Note that all the gain coefficients appear normalized to the first gain coefficient $g_1 \equiv \alpha$.

The minimization of σ_N^2 corresponds to the maximization of its denominator $\sum_{j=1}^{N} g_j^2$. Using Eqs. [8-85] and [8-86] to replace the g_j's by the corresponding z_j's, we have as the equivalent optimum solution that of *maximizing*

$$\sum_{j=1}^{N} g_j^2 = \frac{g_{N+1}^2}{z_{N+1}} = \alpha^2 \prod_{j=2}^{N} (z_j + 1) \tag{8-87}$$

with

$$S_{av} = \frac{\alpha^2}{12T} + \frac{\sigma^2}{T} \sum_{j=2}^{N} z_j \tag{8-88}$$

held fixed.

The solution to this problem turns out to be given by the extremely simple result

$$z_{j,\text{opt}} = \lambda \qquad j = 2, \ldots, N \tag{8-89}$$

with λ an arbitrary constant that can be found in terms of the specified S_{av} by substituting into Eq. [8-88].

This result is readily obtained through the use of *Lagrange multipliers:*[1] Since a logarithm of a function is monotonic with the function itself, we have the problem of maximizing

$$\ln \prod_{j=2}^{N} (z_j + 1) = \sum_{j=2}^{N} \ln (1 + z_j) \tag{8-90}$$

subject to $\sum_{j=2}^{N} z_j$ held fixed (Eq. [8-88]). This is equivalent to maximizing

$$f(z_2, \ldots, z_n) \equiv \sum_{j=2}^{N} \ln (1 + z_j) + \gamma \sum_{j=2}^{N} z_j \tag{8-91}$$

with γ the Lagrange multiplier. The solution is obviously given by differentiating $f(z_2, \ldots, z_n)$ with respect to each z_j, and setting each

[1] See, for example, P. Franklin, "Methods of Advanced Calculus," McGraw-Hill, New York, 1944.

derivative equal to zero. In particular,

$$\frac{df}{dz_j} = 0 = \frac{1}{1 + z_j} + \gamma \qquad [8\text{-}92]$$

and it is apparent that $z_j = \lambda$ (a constant) is the required solution.

Substituting this solution back into Eq. [8-88], we have immediately

$$S_{av} = \frac{\alpha^2}{12T} + \frac{\sigma^2}{T}(N - 1)\lambda \qquad [8\text{-}93]$$

This justifies our statement much earlier that as the estimate $\hat{\theta}_j$ approaches θ more closely (as j increases), the gain g_j must increase correspondingly. For z_j is proportional to the average energy on the jth transmission, $g_j^2 E(\hat{\theta}_{j-1} - \theta)^2 = g_j^2 \sigma_{j-1}^2$. (From Eqs. [8-85] and [8-81], $z_j = g_j^2 \sigma_{j-1}^2/\sigma^2$.) The solution $z_j = \lambda$ thus indicates that the optimum solution is the one for which the average energy on each transmission is fixed.

Now assume the number of repetitions N large enough so that the average energy expended on the first transmission, $\alpha^2/12$, is small compared to the energy on the $(N - 1)$ other transmission. Then, from Eq. [8-93],

$$S_{av} \doteq \frac{\sigma^2}{T}(N - 1)\lambda \qquad [8\text{-}93a]$$

and the constant λ is just

$$\lambda = \frac{S_{av}T}{(N - 1)\sigma^2} \doteq \frac{S_{av}T}{N\sigma^2} \qquad N \gg 1 \qquad [8\text{-}94]$$

From Eq. [8-86] the optimum gain coefficient on the jth repetition is

$$g_j^2 = \alpha^2 \lambda (1 + \lambda)^{j-2} \qquad j \geq 2 \qquad [8\text{-}95]$$

so g_j increases exponentially with j, as noted earlier. The variance of the final estimate of the signal θ, $\sigma_N^2 \equiv E(\hat{\theta}_n - \theta)^2$, is also given by

$$\sigma_N^2 = \frac{\sigma^2}{\alpha^2(1 + \lambda)^N} \qquad [8\text{-}96]$$

and decreases exponentially with N, as noted earlier.

With this minimum possible variance evaluated, we can now return to the discussion of the probability of error P_e. Substituting Eq. [8-96] for σ_N^2 into Eq. [8-83] for P_e, we have

$$P_e = \text{erfc}\left[\frac{\alpha}{2\sqrt{2}\,\sigma M (1 + \lambda)^{-N/2}}\right] \qquad [8\text{-}97]$$

Since we have assumed the channel band-limited to W Hz, it is apparent that in the T-sec transmission interval $N = 2WT$ independent signal samples are needed to completely specify the signal transmitted. [Alternately, since the $\varphi_j(t)$, $j = 1, \ldots, M$, each occupy T/N sec disjoint time intervals, they are automatically orthogonal:

$$\int_{(j-1)T/N}^{jT/N} \varphi_i(t)\varphi_j(t) = \delta_{ij}$$

They can thus be chosen to occupy the same interval W Hz wide in the frequency domain. Since they are pulsed carriers, lasting T/N sec, the minimum bandwidth required is also $1/(2T/N) = W$, or $N = 2WT$ again.] We can therefore replace N in Eq. [8-97] by $2WT$, indicating that as the transmission time T increases, N, the number of repetitions, does likewise. The time T/N for any one repetition is fixed and determined by the bandwidth W specified.

We can also rewrite the constant λ in terms of a new constant C by defining

$$(1 + \lambda)^{-N/2} = (1 + \lambda)^{-WT} \equiv 2^{-CT} \tag{8-98}$$

Then

$$C = W \log_2 (1 + \lambda) \tag{8-99}$$

Note that with this simple rewriting of a constant we have an expression remarkably similar to the Shannon capacity equation. As a matter of fact, replacing λ in Eq. [8-99] by its value in Eq. [8-94], we have, specifically,

$$\begin{aligned} C &= W \log_2 \left(1 + \frac{S_{\text{av}}T}{N\sigma^2}\right) \\ &= W \log_2 \left(1 + \frac{S_{\text{av}}}{2W\sigma^2}\right) \end{aligned} \tag{8-99a}$$

The quantity σ^2 is, as we recall, just the variance of the noise at the matched-filter output in Fig. 8-33. From Eq. [8-73] σ^2 is in fact given by

$$\begin{aligned} \sigma^2 = E[n_j^2] &= E \int_0^{T/N} n(t)\varphi_j(t)\,dt \int_0^{T/N} n(s)\varphi_j(s)\,ds \\ &= \int_0^{T/N}\int_0^{T/N} E[n(t)n(s)]\varphi_j(t)\varphi_j(s)\,dt\,ds \end{aligned} \tag{8-100}$$

But we assumed additive *white* noise, with spectral density $G_n(f) = n_0/2$. Then

$$R_n(\tau) = \frac{n_0}{2}\,\delta(\tau) \tag{8-101}$$

or,

$$\begin{aligned} \sigma^2 &= \frac{n_0}{2}\int_0^{T/N}\int_0^{T/N} \delta(t - s)\varphi_j(t)\varphi_j(s)\,dt\,ds \\ &= \frac{n_0}{2} \end{aligned} \tag{8-102}$$

using the known property of impulse functions, and the assumption that

$$\int_0^{T/N} \varphi_j^2(t)\, dt = 1$$

We get finally for C in Eq. [8-99a] just the Shannon capacity expression for a band-limited channel with additive gaussian white noise:

$$C = W \log_2 \left(1 + \frac{S_{av}}{n_0 W}\right) \qquad \text{[8-102}a\text{]}$$

To indicate its significance in the system under discussion, we note that M in Eq. [8-97] may be rewritten directly in terms of the bit rate R and the transmission time T, as $M = 2^{RT}$. Finally, then, the probability of error is given by

$$P_e = \text{erfc}\left[\frac{\alpha}{2\sqrt{2}\,\sigma} 2^{(C-R)T}\right] \qquad \text{[8-103]}$$

Note the significance of this result:

1. $R < C$: As $T \to \infty$, $P_e \to 0$.
2. $R > C$: As $T \to \infty$, $P_e \to 1$.

The system thus performs well only if $R < C$. The constant C defined above in terms of the other number λ thus displays the properties of a channel capacity: it represents the *maximum* rate of transmission (in bits per second) of information for this system. For rates R less than this number, effective transmission, in the sense of any probability of error approaching zero desired, may be attained by taking T large enough. This simply means letting the number of binary digits RT processed by the encoder of Fig. 8-33 increase to the appropriate amount. In terms of the error calculation of Fig. 8-36 this implies letting the correct decision region $1/M$ units wide decrease, but slowly compared to the corresponding decrease in σ_N, such that $M\sigma_N$ becomes progressively smaller with T. This is the meaning of binary–M-ary encoding, processing time T, and channel capacity in this case.

The probability-of-error expression, Eq. [8-103], has the as yet undetermined gain normalizing factor α in it. Although the choice of this factor is not critical, an optimum value does exist in the sense of minimizing P_e. To show this recall that α^2 appeared in the average power expression of Eq. [8-93] as the coefficient of the first term (due to the energy on the first T/M sec repetition). If we do not neglect this term, but carry it along, the capacity C in Eq. [8-103] is found to depend on it. Maximum $\alpha 2^{C(\alpha)T}$ in Eq. [8-103] then corresponds to minimum P_e. Maximizing with respect to α, we readily find that

$$\alpha_{\mathrm{opt}}^2 \doteq 12\sigma^2 \left(1 + \frac{S_{\mathrm{av}}}{n_0 W}\right) \doteq 12\sigma^2 \qquad N \gg 1 \qquad [8\text{-}104]$$

for wideband systems. The error probability is then given by

$$P_e = \mathrm{erfc}\left[\sqrt{\frac{3}{2}}\, 2^{(C-R)T}\right] \sim \exp\left[-\frac{3}{2}\, 2^{2(C-R)T}\right] \qquad [8\text{-}105]$$

for small error probabilities.

Note that P_e decreases with time T as a *double* exponential, rather than as the single exponential found by Shannon. Can this system then outperform Shannon's "optimum"? The apparent contradiction is resolved by recalling that the average power S_{av} here is a statistical average, rather than a true time-averaged power. There is therefore a small but significant probability that arbitrarily high signal powers may occur, the system then performing exceptionally well. If a *peak power limitation* is imposed on the system as well, however, the expression for error probability reduces to the same single exponential found by Shannon.[1]

For peak to average power ratios of 16 or more and transmission times T that are not too large ($P_e > 10^{-9}$ or so) it turns out that the peak power constraint is not too significant. The probability of error then does follow Eq. [8-105] rather closely, decreasing in this range much more rapidly with T than a corresponding nonfeedback (one-way) transmission scheme. So although in the limit of $T \to \infty$, $P_e \to 0$, this system does *not* outperform Shannon's optimum system, for finite T it may do considerably better than other systems.

As an example consider a very wideband situation, with $S_{\mathrm{av}}/n_0 W \ll 1$. How does the feedback scheme compare with the wideband M-ary orthogonal signaling scheme discussed earlier? To be specific, let $C_\infty = S/n_0 \ln 2$ (Eq. [8-66]) be 14,500 bits/sec as in the previous section. (Recall that this was the case for a space vehicle–earth distance of 10^6 miles, 1-watt transmitter power at $f = 500$ MHz, 3-db receiver noise figure, and the antenna gains discussed there.) Let $R/C_\infty = 0.8$, so that $R = 11{,}600$ bits/sec. For a probability of error of $P_e = 10^{-7}$, we find from Eq. [8-105] that a transmission time $T = 0.56$ msec is required. Then $M = 2^{RT} = 2^{6.5} \sim 2^7 = 128$ amplitude levels are required. The bandwidth W must satisfy $W \gg S/n_0 = 10^4$ bits/sec. These numbers do not seem unreasonable.

[1] A. D. Wyner, On the Schalkwijk-Kramer Coding Scheme with a Peak Energy Constraint, *IEEE Trans. Information Theory,* vol. IT-14, no. 1, pp. 129–134, January, 1968.

For M-ary orthogonal signaling, however, it may be shown that for the same numbers, that is, $R/C_\infty = 0.8$, $R = 11{,}600$ bits/sec, and $P_e = 10^{-7}$, the required transmission time is $T = 0.14$ sec, or 250 times as long as the time required with the feedback scheme! The number of orthogonal signals is $M = 2^{RT} = 2^{1,600}$, obviously an impossible number! [The bandwidth required is then $W = M/T = (2^{1,600}/0.14)$ Hz!] This is due to the *single* exponential decrease of P_e with T in this case.

M-ary orthogonal signaling with the same performance is thus impossible to attain. If instead we restrict M to be $32 = 2^5$, and ask for a more reasonable probability of error, say $P_e = 10^{-5}$, we find from Fig. 8-30 that $S/n_0R \doteq 5$. With $S/n_0 = 10^4$ as assumed here, $R = 2{,}000$ bits/sec. The transmission time T is $(\log_2 M)/R = 2.5$ msec, and the bandwidth required is $W = M/T = 12.8$ kHz. These numbers are of course not unreasonable, but note the deterioration in performance from the feedback scheme numbers. We can increase the bit transmission rate here somewhat by increasing the number of orthogonal signals. Thus, let $M = 2^7 = 128$ orthogonal signals. From Fig. 8-30, $S/n_0R \doteq 3.9$ for $P_e = 10^{-5}$. Then $R = 10^4/3.9 \doteq 2{,}600$ bits/sec, $T = 7/R = 2.7$ msec, and $W = M/T = 47.5$ kHz. Using four times as many signals (2^5 to 2^7), we increase the bit rate possible from 2,000 to 2,600 bits/sec, with almost a fourfold increase in bandwidth required.

It would thus appear that for the numbers considered here an information feedback scheme is highly deserving of consideration. It too has drawbacks that must be considered in an overall evaluation, however: a feedback channel must be available, the SNR on this channel must be very high (requiring high power and large antenna on the ground), there must be some sort of storage on the space vehicle to store the signal and successive receiver messages during the time for transmission and over many (N) repeats. This problem becomes even more significant if delay time, i.e., the time for the signal to travel from space vehicle to earth and back, is taken into account. (The apparent empty gaps in signal transmission during this relatively long delay time may be filled by time multiplexing other signals.)

8-8 ERROR DETECTION AND CORRECTION

We have indicated in the previous sections that it is possible to reduce error rates in data transmission to as small a level as desired by appropriate encoding and decoding techniques. The two examples discussed earlier involved coding into M-ary signals; one example used M-ary orthogonal signals; the other, M amplitude levels (M-ary PAM) plus feedback, to achieve lower error rates.

An alternate approach toward which much effort has been devoted in recent years is that of designing *binary codes* to detect or correct errors. Although many different techniques have been investigated, the most common procedure is that of adding an additional number r, say, of bits to every sequence of k bits originally generated, sufficient to correct one, two, or more errors if desired in the k-bit sequence. The additional r bits added are called *redundant* or *check* bits. We shall provide only a simple introduction to error correction coding here, referring the interested reader to the literature for a detailed treatment.

Before discussing error correction coding, we introduce the concept of detection and correction of errors by discussing a simple feedback system for error detection.

NULL-ZONE DETECTION

Recall in Chap. 5 and again in Chap. 6, in discussing the detection of binary signals in additive gaussian noise, that a simple binary decision level was used to decide on whether a 1 or 0 had been transmitted. In particular for equally likely bipolar pulses the decision level was just zero, a positive signal-plus-noise sample indicating a 1 transmitted, a negative sample, a 0. The probability of error was then given by the complementary error function, as plotted in Fig. 8-22 (see also Figs. 5-27 and 6-33). We indicated both in Chap. 5 and in this chapter that this was the optimum decision level, in the sense of minimizing the probability of error. Shannon's channel-capacity theorem was then invoked to indicate that one could reduce the error probability still further, for fixed binary transmission rate R bits/sec, by encoding large numbers of bits.

An alternate technique that does not require encoding and yet provides significant improvement in binary error rates involves the use of *two* decision thresholds for binary transmission, rather than the one considered previously. The technique is called *null-zone detection*[1] and is readily explained by referring to Fig. 8-37. It is apparent that if the received signal-plus-noise sample is of comparatively large amplitude, positive or negative, it is likely that the signal transmitted was of the same polarity. We know the noise has a very small probability of taking on amplitudes more than a few times its rms value $\sqrt{N}$. It is only when the received signal plus noise is in the vicinity of the origin (the noise is then of the opposite polarity to the signal, and roughly of the same magnitude) that errors may occur. To reduce the number of errors, we

[1] F. J. Bloom et al., Improvement of Binary Transmission by Null-zone Reception, *Proc. IRE*, vol. 45, no. 7, pp. 963–975, July, 1957.

then consider a scheme in which a signal decision is deferred if the received signal-plus-noise amplitude, of either polarity, is below some specified value. Specifically, assume that instead of the single decision threshold at 0, we put in *two* thresholds, $+d$ and $-d$, symmetric about the origin (Fig. 8-37). The region between these two levels is referred to as the

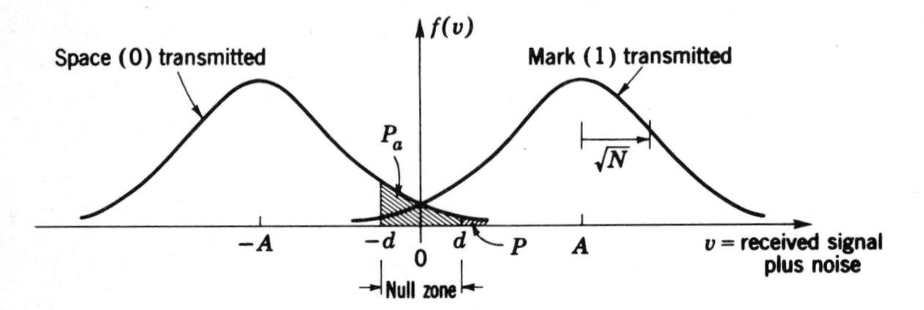

FIG. 8-37 Null-zone detection, bipolar transmission.

null-zone region. A decision is deferred if samples fall in this region. Only if the received sample v exceeds $+d$ or drops (negatively) below $-d$ is a decision made.[1]

In one application of a scheme of this type, used for error detection in a digital-computer binary stream, samples falling in the null-zone region are discarded, or erased. The scheme is then referred to as a null-zone *erasure* scheme. The output bit stream will then contain, in addition to 1's and 0's, blanks (or some other symbol), indicating no decision was made. It is apparent that here one trades binary errors for blanks. This is acceptable in some applications, although of course not in all.

Instead of erasing the null-zone samples one may ask the transmitter to simply repeat the particular bit. Decision is again deferred if the new received sample falls in the null zone. The whole procedure is repeated until a definite binary decision is made.[2] It is apparent that a feedback channel must be assumed available to signal the transmitter to repeat. Unlike the feedback channel in the previous section, however, the channel is only used for decision feedback (ask the transmitter to repeat, ask it to stop, etc.). This scheme is thus an example of a *decision-feedback* scheme, as contrasted with the *information-feedback* system of the previous section.

[1] It is apparent that matched filtering is still appropriate *before* the decision circuit, in order to maximize the signal-to-noise ratio (proportional to $A/\sqrt{N}$ in Fig. 8-37) at the decision point.

[2] Alternately, the transmitter may continue to repeat until told to stop by the receiver. Only then does it go on to the next bit.

Here too we assume noise added on the feedback channel to be negligible. This assumption may again be presumed valid for possible space-communication channels, in which the feedback channel is directed from earth to space. It is a reasonable assumption for many other situations as well, however, since the decision messages to be sent over the feedback channel are highly specialized and may therefore be readily coded to overcome any noise added.

There is a distinct difficulty with schemes of this type, however. Since the time taken to come to a decision is a random variable, the resultant information bit rate fluctuates randomly as well. A buffer is then required between the fixed rate source input and the variable rate output. (Alternately a buffer may be used to even out the varying bit rate.) Buffer design is a complex problem with as yet unanswered questions regarding the buffer storage required, the probability of overflow, effect on error rate, etc.[1]

How well does this null-zone scheme work? In particular, how much improvement does one get over the single-decision-level binary scheme? To answer this question quantitatively, assume 1's and 0's (marks and spaces) equally probable to be transmitted. As previously, then, the probability of error of this scheme may be calculated by simply assuming either symbol (0 or 1) transmitted. In particular, assume a 0 (space) transmitted. Referring to Fig. 8-37 let the probability of signal plus noise exceeding the upper decision level $+d$ be P. This obviously results in an error. (Note that P is much less than the previous error probability due to received signal plus noise going positive, or exceeding 0. This is the reason why this scheme is effective.) Also let the probability that the received signal plus falls in the null zone be labeled P_a, as indicated in Fig. 8-37. The overall probability of error, P_e, is then the sum of the probabilities that (1) the first sample received appears above $+d$; (2) the first sample falls in the null zone and the second, retransmitted, sample above $+d$; and (3) the original sample and the first retransmitted one fall in the null zone, the second retransmitted one then being detected incorrectly, etc. Assuming the possibility of indefinite cycling between receiver and transmitter, the probability of error is then

$$P_e = P + P_aP + P_a^2P + P_a^3P + \cdots = \frac{P}{1 - P_a} \qquad [8\text{-}106]$$

after summing the indicated geometric series.

[1] See T. W. Eddy, Some Results for Additive Noise Channels with Noiseless Information Feedback, Ph.D. (E.E.) dissertation, chap. 11, Polytechnic Institute of Brooklyn, New York, June, 1968.

The expression for the average number of transmitted samples, $E(n)$, per bit transmitted is also readily obtained. This is just the average number of transmissions needed to come to a decision (correct or incorrect), and is given by

$$E(n) = (1 - P_a)[1 + 2P_a + 3P_a^2 + \cdots + jP_a^{j-1} + \cdots] = \frac{1}{1 - P_a} \qquad [8\text{-}107]$$

(The probability of termination on any one transmission is just $1 - P_a$, as is apparent from Fig. 8-37. The probability of cycling to the jth transmission is P_a^{j-1}, the probability of falling into the null zone $j - 1$ times in succession. The sum in brackets is then easily evaluated by noting that it may be written

$$S = \sum_{j=1}^{\infty} jP_a^{j-1} = \sum_{j=1}^{\infty} \frac{dP_a^j}{dP_a} = \frac{d}{dP_a}\sum_{j=1}^{\infty} P_a^j = \frac{d}{dP_a}\left(\frac{P_a}{1 - P_a}\right) = \left(\frac{1}{1 - P_a}\right)^2 \qquad [8\text{-}108])$$

Note then that probability of error and average transmission time are related by the simple expression

$$P_e = E(n)P \qquad [8\text{-}109]$$

Although $E(n)$ increases as the null zone widens, the increase in $E(n)$ is slow compared to the corresponding decrease in P, the resultant error probability P_e decreasing rapidly.

As an example of the numbers involved, say the error probability with no null zone and feedback is $P_e = 10^{-5}$. From Fig. 8-22 (or Fig. 6-33), we then find $A/\sqrt{N} = 4.2$. (Recall that $A^2/N = 2E/n_0$, with matched filtering assumed.) Now let $d/\sqrt{N} = 1$, as an example. It is apparent for the numbers chosen $P \ll P_a$ (see Fig. 8-37). Then P_a is approximately the area under the probability-density curve from $-d$ to $+\infty$. But shifting the decision level in Fig. 8-37 from 0 (the ordinary binary case) to $-d$ is the same as decreasing A by d units, leaving the decision level at 0. With $d/\sqrt{N} = 1$, $(A - d)/\sqrt{N} = 3.2$, and we find, from Fig. 8-22 (or Fig. 6-33), $P_a \doteq 10^{-3}$. Then

$$E(n) = \frac{1}{1 - P_a} \doteq 1.001$$

a rather interesting result. On the average, only 1 binary transmission in 1,000 need be repeated!

What is the resultant effect on the probability of error? Although the error-probability curves plotted do not go below 10^{-5}, recall from Chap. 6 that for small error probability,

$$P_e \sim e^{-A^2/2N} \qquad [8\text{-}110]$$

This is due specifically to the asymptotic expression for the complementary error function given by Eq. [6-95], and repeated here:

$$\operatorname{erfc} x = \frac{2}{\sqrt{\pi}} \int_x^\infty e^{-y^2}\, dy \doteq \frac{e^{-x^2}}{x\sqrt{\pi}} \qquad x \gg 1 \qquad [8\text{-}111]$$

For bipolar transmission in gaussian noise we have

$$P_e = \frac{1}{2} \operatorname{erfc} \frac{A}{\sqrt{2N}} \sim e^{-A^2/2N} \qquad A/\sqrt{2N} \gg 1 \qquad [8\text{-}112]$$

using Eq. [8-111] and retaining just the dominant exponential term. (Compare with Eqs. [6-96] and [6-97] for coherent FSK and OOK.)

Now note from Fig. 8-37 that evaluating the error probability P in null-zone detection is equivalent to *increasing* A by d above. Then

$$P \sim e^{-(A+d)^2/2N} \qquad [8\text{-}113]$$

For the numbers used here, $(A + d)/\sqrt{N} = 5.2$, so that

$$P \sim (10^{-5})^{5.2/4.2} = 0.7 \times 10^{-6} \doteq P_e$$

The effect of introducing the null-zone region has thus been to reduce the error probability by more than an *order of magnitude*, while the average transmission rate is barely effected.

In practice one would not want to continue cycling indefinitely in the event that successive retransmissions continued to fall into the null zone. Retransmission may be halted after a predetermined number of repeats by simply closing down the null zone at that time. If the number chosen is $\gg E(n)$, there is no noticeable effect on the performance.[1]

The scheme considered here may be improved considerably by storing successive received samples and using all previous samples as well as the current one to come to a decision. It may then be shown that an optimum procedure, in the sense of minimum probability of error, corresponds to setting up a likelihood ratio, as in earlier sections of this chapter

[1] We have only stressed average retransmission time here. It is apparent that since the number of repeats is a random variable, this average number may be exceeded. Variance calculations are necessary to complete the analysis.

but now comparing this likelihood ratio to *two* numbers. If the likelihood ratio (or its logarithm) lies between the two numbers, another sample (in this case implying retransmission) is taken, a new likelihood ratio computed, and again tested. A binary decision is made depending on whether the ratio is greater than the larger number or less than the smaller one.

For the case of bipolar signals the two numbers become just the symmetric parameters $\pm d$ of Fig. 8-37. If gaussian additive noise is assumed, the logarithm of the likelihood ratio simplifies to the sum of the successive retransmitted signals, just as in the beginning of this chapter. The optimum procedure in this case then corresponds simply to storing retransmitted signals, and testing the sum to determine whether it falls within the null zone of Fig. 8-37, or outside, for a decision. This optimum technique is an example of a *sequential decision* technique, the general theory of which was first developed for quality control problems by the noted statistician Abraham Wald.[1] In the case of high signal-to-noise ratio, low probability of error, and small numbers of repeats, on the average, the improvement offered by this optimum technique with memory over the memoryless null-zone technique just discussed is negligible.[2] For since few retransmissions are required, on the average, it is apparent that most of the errors made occur on the first transmission.

For small signal-to-noise ratios, with large numbers of retransmissions required to attain the necessary low error rates, there is a marked difference in performance of the two techniques. It may be shown that the sequential test with memory provides 6-db SNR improvement over a nonsequential test with the signal simply repeated a fixed number of times. This is indicated by the curves of Fig. 8-38, which show the typical performance of the sequential test as contrasted to a fixed-sample

[1] Abraham Wald, "Sequential Analysis," Wiley, New York, 1947. See J. Bussgang and D. Middleton, Optimum Sequential Detection of Signals in Noise, *IRE Trans. Information Theory,* vol. IT-1, pp. 5-18, December, 1955, for applications to signal detection in noise. (The stress here is on large numbers of retransmissions and small SNR.) A. J. Viterbi, The Effect of Sequential Decision Feedback on Communication over the Gaussian Channel, *Information and Control,* vol. 8, pp. 80–92, February, 1965, and G. L. Turin, Signal Design for Sequential Detection Systems with Feedback, *IEEE Trans. Information Theory,* vol. IT-11, p. 401, July, 1965, analyze the binary communication problem with sequential feedback. The application of sequential techniques to radar was first proposed by M. Schwartz, A Statistical Approach to the Automatic Search Problem, Ph. D. dissertation, chap. 5, Harvard, Cambridge, Mass., 1951. Summaries of the Wald sequential procedure as applied to communication and radar, with detailed bibliography, also appear in C. W. Helstrom, in A. V. Balakrishnan (ed.), "Communication Theory," chap. 7, McGraw-Hill, New York, 1968, and J. C. Hancock and P. A. Wintz, "Signal Detection Theory," chap. 4, McGraw-Hill, New York, 1966.

[2] R. R. Boorstyn, Small Sample Sequential Detection, M. E. E. thesis, Polytechnic Institute of Brooklyn, New York, June, 1963.

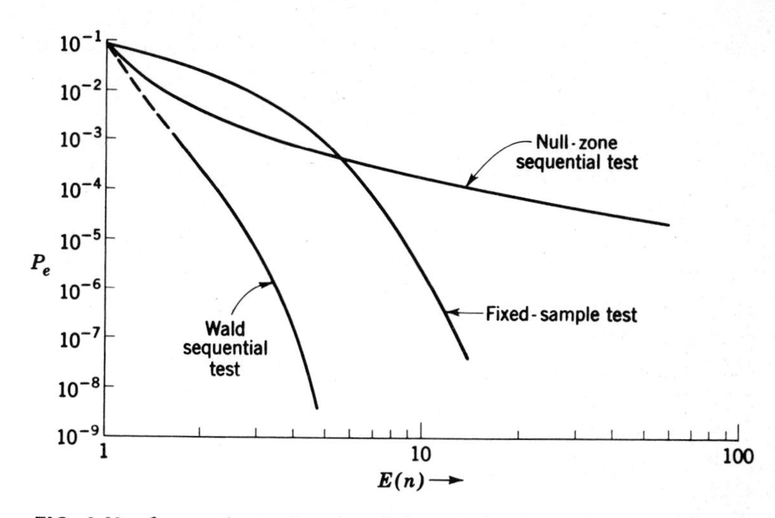

FIG. 8-38 Comparison of sequential- and fixed-sample tests (bipolar signals, additive gaussian noise, SNR = 1).

test and a null-zone test.[1] Although these curves are for 0-db SNR, similar results hold at other values.

ERROR-CORRECTION CODING[2]

The null-zone and sequential-detection procedures provide some form of error control by combining the availability of a feedback path with signal retransmission to reduce the average error rate. The schemes are essentially *error-detection* schemes with those errors that might have occurred, with the received signal plus noise falling into the null zone, possibly eliminated by retransmitting the signal.

We conclude the book by providing an introduction to coding techniques that reduce the error rate by actually correcting up to a predetermined number of errors, when they occur, by the controlled use of redundant bits inserted periodically into the message stream.

The idea of redundancy in signal transmission was first mentioned in Chap. 1. We noted there that the information content of messages such as English text, speech, or TV pictures was reduced considerably below that which might be available if the symbols used were independently generated. But redundancy serves a useful purpose—it enables

[1] M. Hecht, Non-optimum Sequential Detection Techniques, Ph.D. (E.E.) dissertation, Polytechnic Institute of Brooklyn, New York, June, 1968. See also M. Hecht and M. Schwartz, Sub-optimal Decision Feedback Communication Systems, *IEEE Trans. Commun. Technol.*, vol. COM-16, no. 6, December, 1968.

[2] The approach here is due to Professor Jack K. Wolf. The author is indebted to Professor Wolf for allowing him to use unpublished class notes in developing the material.

errors occurring to be singled out and sometimes corrected. This is the essence of introducing redundancy into a bit stream in a controlled fashion.

As an example of these ideas assume English text is used to send a telegram. The received (and garbled) message reads as follows:

CINCRATULATTONS. YUU HAV BEEM AWADDED
FELLAWSSIP TO PIB. REDLY BY JANUARE 26

It is apparent that the errors in this message are easily corrected, and the message read quite clearly. This is specifically due to the high redundancy of the English language. Certain letter (symbol) combinations just do not occur and therefore are easily corrected (providing of course that the message is not too garbled, with too many errors occurring). But what about the number 26? Is that correct? Should it have been 27, or 16, for example? (36 is obviously not possible.) No sure answer is possible here, since the numeral combinations have no redundancy and almost any numbers below 31 could have been used. (One might be wary of replying by January 1, a *holiday;* or of a date that falls on a Sunday.)

A message consisting of binary symbols generally has no built-in redundancy, however, so that errors cannot be detected. The coding techniques consist of periodically adding specified redundant bits to a binary stream so that incorrect message sequences can be so detected and corrected.

The simplest way to add redundancy to a binary message is to add a *single* binary digit to the end of each message or order of specified length, requiring the number of ones in the resultant word to be *even.* An *odd* number of errors can then be detected at the receiver. The redundant bit inserted is called a *parity check bit.* As an example, assume the parity bit is inserted every four digits. We then get the transmitted sequences shown below:

Original 4-bit sequences	Transmitted messages, with parity bit
0000	0000 0
0001	0001 1
0010	0010 1
0011	0011 0
....	.... ..
1111	1111 0

It is apparent that only $2^4 = 16$ of the possible $2^5 = 32$ message sequences 5 bits long are transmitted. The ones not transmitted (for example, 00001, 01011, etc.), are exactly those with odd numbers of 1's, and are obviously not acceptable at the receiver if so received. (An even number of errors would *not* be detected, and serves to underscore a truism of coding: it is never possible to detect and/or correct *all* errors. For if an acceptable message is transmitted, and another acceptable message received, there is no way of detecting the error.) The acceptable messages (16 of the possible 32 in this case) are called *code words.*

Parity checking is commonly used in digital computers for error checking, but provides no error correction. How does one construct code words to *correct* one, two, or more errors, as desired. The object here is to choose as code words the appropriate subset of all the possible words of a given length that provide the desired property. As an example, assume that 3 redundant bits are now added to every 4 information bits (the word) transmitted, and any single error is to be corrected. $2^4 = 16$ words of the $2^7 = 128$ possible must now be selected to provide the single-error-correcting property.

Assume that we search through all 128 possible 7-bit sequences and come up with the following sixteen 7-bit words, each one corresponding to a different one of the original message words:

Original message	Transmitted message (code words)
0000	0000000
0001	0001011
0010	0010101
0011	0011110
0100	0100110
0101	0101101
0110	0110011
0111	0111000
1000	1000111
1001	1001100
1010	1010010
1011	1011001
1100	1100001
1101	1101010
1110	1110100
1111	1111111

Note that these 16 transmitted words have the property that any two of the 7-bit messages differ in at least three positions. We now decode a received 7-bit sequence or word by comparing it to each of the 16 possible code words, as stored at the receiver, selecting the one that differs from the received word in the least number of positions. It is apparent that if one error (or none) occurs during transmission, we will select correctly.

(Why?) So this set of 16 code words provides exactly the property for which we were looking—the ability to correct one error if it occurs.

But exhaustively searching through all possible words for the desired set can be exhausting (to allow a play on words here!), particularly if the original message size becomes long. Also, extending the procedure to the correction of more than one error compounds the difficulty. Is there a more systematic way of providing the appropriate code word set? Happily a more general method, called *algebraic coding theory*, does exist for constructing code words. We shall introduce the method shortly after first evaluating the effectiveness of our 16-word single-error-correcting 7-bit code set.

For although the tabulated set corrects any *single* error, it is apparent that multiple errors may occur. The 7-bit code words must be transmitted in the same time interval as the original 4-bit word messages in order to maintain real-time transmission (Fig. 8-39). This requires

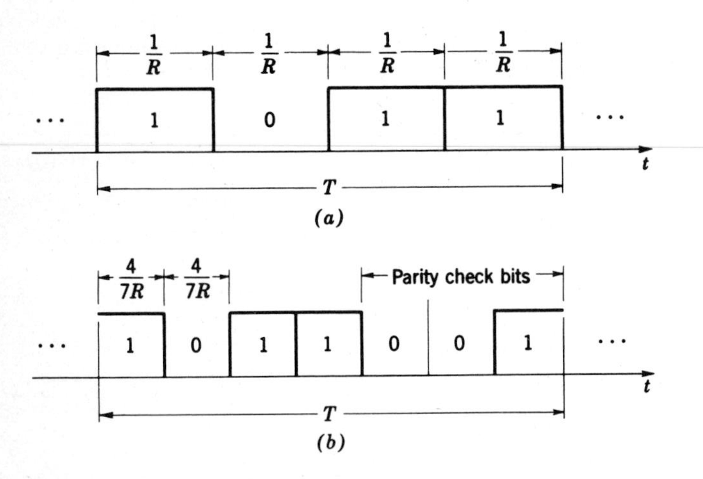

FIG. 8-39 Single-error correcting code. (a) Original binary stream; (b) encoded binary word.

reducing the binary pulses to $4/7$ of their original width, hence increasing the transmission bandwidth by $7/4$. For fixed-noise spectral density, the noise power in turn increases by $7/4$. It could very well be possible, then, that the increase in noise more than overcomes the reduction in error probability due to single-error correction, resulting in an *increase* in overall error probability! If so, encoding techniques are obviously not called for! According to Shannon's capacity theorem, however, binary codes should exist that reduce the error rate as low as possible. We shall in fact find that there is some improvement in error performance by encoding 4-bit words into 7-bit words, the increase in noise notwithstanding.[1]

To calculate the coded word error probability, assume that the uncoded error probability is 10^{-5} per bit. The probability of error in 4 bits is then

$$P_{e,\text{uncoded}} = 1 - (1 - 10^{-5})^4 \doteq 4 \times 10^{-5}$$

Furthermore, from Fig. 8-22 or Fig. 6-33, the SNR is 9.5 db. The coded SNR is then 2.4 db ($4/7$) less. Again from Fig. 8-22 or Fig. 6-33, the probability of bit error for an SNR of 7.1 db is $p = 6 \times 10^{-4}$. The coded word error probability is just the probability that any two or more bits will be in error, and is given by

$$P_{e,\text{coded}} = \sum_{j=2}^{7} \binom{7}{j} p^j (1 - p)^{7-j} \doteq 21p^2 \qquad p \ll 1$$

In particular, with $p = 6 \times 10^{-4}$,

$$P_{e,\text{coded}} \doteq 8 \times 10^{-6}$$

This compares to the uncoded word error probability of 4×10^{-5}, so that there is an improvement in error rate.

How do we now systematically determine the appropriate code words? Before generalizing the results, we go back to 4-bit message words and the particular group of 7-bit transmitted messages used. Label the individual message bits $m_1m_2m_3m_4$, and the three parity check bits used $c_1c_2c_3$. (The m's and c's can in general be 0 or 1.) Some of the group of 16 transmitted words are repeated here:

m_1	m_2	m_3	m_4	c_1	c_2	c_3
0	0	0	0	0	0	0
0	0	0	1	0	1	1
0	0	1	0	1	0	1
0	0	1	1	1	1	0
0	1	0	0	1	1	0
0	1	0	1	1	0	1
...	...	...	...	...	...	...
1	1	0	1	0	1	0
1	1	1	0	1	0	0
1	1	1	1	1	1	1

[1] An alternate way of arriving at the same $7/4$ deterioration in SNR is by assuming matched-filter detection of each bit at the receiver. As noted both in this chapter and in Chap. 6, the matched-filter output SNR is proportional to E/n_0, with E the signal energy, n_0 the noise spectral density. The energy is proportional to the bit interval, and is hence reduced by $7/4$ over noncoded transmission. The coded SNR is then again $4/7$ of the uncoded SNR.

Looking first at the column of c_1's, we note that c_1 in any row is 0 when the number of 1's in the $m_1m_2m_3$ slots is even; c_1 is 1 when the number is odd. This may be expressed algebraically by defining *modulo* 2 (or *mod* 2) additions (this is denoted by the symbol $\oplus$):

$$\begin{aligned} 0 \oplus 0 &= 0 \\ 0 \oplus 1 &= 1 \\ 1 \oplus 0 &= 1 \\ 1 \oplus 1 &= 0 \end{aligned}$$

the value of c_1 is then just the mod 2 sum of m_1, m_2, m_3. Similarly, we find c_2 determined by the mod 2 sum of m_1, m_2, m_4; and c_3 given by the sum of m_1, m_3, m_4:

$$\begin{aligned} c_1 &= m_1 \oplus m_2 \oplus m_3 \\ c_2 &= m_1 \oplus m_2 \qquad \oplus m_4 \\ c_3 &= m_1 \qquad \oplus m_3 \oplus m_4 \end{aligned} \qquad [8\text{-}114]$$

In mod 2 arithmetic, addition and subtraction are identical, so that these equations can also be rewritten as

$$\begin{aligned} 1 \cdot m_1 \oplus 1 \cdot m_2 \oplus 1 \cdot m_3 \oplus 0 \cdot m_4 \oplus 1 \cdot c_1 \oplus 0 \cdot c_2 \oplus 0 \cdot c_3 &= 0 \\ 1 \cdot m_1 \oplus 1 \cdot m_2 \oplus 0 \cdot m_3 \oplus 1 \cdot m_4 \oplus 0 \cdot c_1 \oplus 1 \cdot c_2 \oplus 0 \cdot c_3 &= 0 \\ 1 \cdot m_1 \oplus 0 \cdot m_2 \oplus 1 \cdot m_3 \oplus 1 \cdot m_4 \oplus 0 \cdot c_1 \oplus 0 \cdot c_2 \oplus 1 \cdot c_3 &= 0 \end{aligned} \qquad [8\text{-}115]$$

(The $\cdot$ represents the usual multiplication.)

Now note that the same equations may be written in matrix-vector form as

$$3\left\{\overbrace{\begin{bmatrix} 1 & 1 & 1 & 0 & 1 & 0 & 0 \\ 1 & 1 & 0 & 1 & 0 & 1 & 0 \\ 1 & 0 & 1 & 1 & 0 & 0 & 1 \end{bmatrix}}^{7}\right. \left.\overbrace{\begin{bmatrix} m_1 \\ m_2 \\ m_3 \\ m_4 \\ c_1 \\ c_2 \\ c_3 \end{bmatrix}}^{1}\right\} 7 = \begin{bmatrix} 0 \\ 0 \\ 0 \end{bmatrix} \qquad [8\text{-}116]$$

If we define a matrix H (called the parity check matrix)

$$H \equiv \begin{bmatrix} 1 & 1 & 1 & 0 & 1 & 0 & 0 \\ 1 & 1 & 0 & 1 & 0 & 1 & 0 \\ 1 & 0 & 1 & 1 & 0 & 0 & 1 \end{bmatrix} \qquad [8\text{-}117]$$

and use the symbol **C** to denote the code word written as the column vector of Eq. [8-116], it is apparent that Eq. [8-116] in shorthand form is just

$$HC = 0 \tag{8-118}$$

All code words are then solutions of this equation. In particular, if $\mathbf{C}_1$ is a code word, as is $\mathbf{C}_2$, then so is $\mathbf{C}_1 \oplus \mathbf{C}_2$, since

$$H(\mathbf{C}_1 \oplus \mathbf{C}_2) = H\mathbf{C}_1 \oplus H\mathbf{C}_2 = \mathbf{0} \oplus \mathbf{0} = \mathbf{0} \tag{8-119}$$

(By $\oplus$ here we mean mod 2 addition of the individual elements in each row.)

To see how we correct errors easily, assume that code word $\mathbf{C}$ is transmitted, but due to errors the 7-bit code word (or vector) $\mathbf{R}$ is received. If we denote vector $\mathbf{E}$ as a vector with a 1 in each position where an error occurred, and with a 0 in the position where the bit was correctly received it is apparent that we may write

$$\mathbf{R} = \mathbf{C} \oplus \mathbf{E} \tag{8-120}$$

For example, if 1101010 is the code word that was transmitted, while the received code word is 1111010 (there is an error in the third position), then $\mathbf{E}$ has the elements 0010000. To decode, multiply the received vector $\mathbf{R}$ by matrix H to get

$$H\mathbf{R} = \underbrace{H\mathbf{C}}_{=\mathbf{0}} \oplus H\mathbf{E} = H\mathbf{E} \tag{8-121}$$

If the transmitted word is correctly received, it is apparent that $H\mathbf{R}$ at the receiver is just zero. So the presence of a nonzero $H\mathbf{R} = H\mathbf{E}$ denotes an error(s). More than this, however, if $\mathbf{E}$ contains just a *single* 1 (*one* error made), the receiver output $H\mathbf{E}$ will equal the column of H corresponding to that error, so that the specific position or location of the error is pinpointed. As an example consider the received code word $\mathbf{R}$ with an error in the third position as just cited. Then

$$H\mathbf{R} = \begin{bmatrix} 1 & 1 & 1 & 0 & 1 & 0 & 0 \\ 1 & 1 & 0 & 1 & 0 & 1 & 0 \\ 1 & 0 & 1 & 1 & 0 & 0 & 1 \end{bmatrix} \begin{bmatrix} 1 \\ 1 \\ 1 \\ 1 \\ 0 \\ 1 \\ 0 \end{bmatrix} = \begin{bmatrix} 1 \\ 0 \\ 1 \end{bmatrix} \tag{8-122}$$

Since $\begin{bmatrix} 1 \\ 0 \\ 1 \end{bmatrix}$ corresponds to the third column of matrix H, the error is exactly in the third position!

How do we now generalize these results? Assume that the original message sequence contains k nonredundant 0's and 1's. Denote this message as $m_1m_2 \ldots m_k$, with the m_i's 1 or 0, depending on the particular message. We add r parity check bits $c_1c_2 \ldots c_r$, so that the total n-bit transmitted message is $m_1, m_2 \ldots m_kc_1c_2 \ldots c_r$, with $n = r + k$. The list of code words then contains 2^k of the 2^n possible sequences of length n. The r parity check bits are chosen as the solution of the following set of r algebraic equations:

$$
\begin{aligned}
h_{11}m_1 \oplus h_{12}m_2 \oplus \cdots \oplus h_{1k}m_k \oplus c_1 \qquad\qquad &= 0 \\
h_{21}m_1 \oplus h_{22}m_2 \oplus \cdots \oplus h_{2k}m_k \qquad \oplus c_2 \qquad &= 0 \\
\cdots\cdots\cdots\cdots\cdots\cdots\cdots\cdots\cdots\cdots\cdots\cdots\cdots\cdots & \\
h_{r1}m_1 \oplus h_{r2}m_2 \oplus \cdots \oplus h_{rk}m_k \qquad\qquad \oplus c_r &= 0
\end{aligned}
\tag{8-123}
$$

The h_{ij}'s are 0 or 1. What distinguishes one code from another is the choice of parameters k, r, and n, and of course the choice of the h_{ij}.

This set of r equations may again be written in matrix form as

$$HC = 0 \tag{8-124}$$

where the $(r \times n)$ matrix H is given by

$$
H \equiv \overbrace{\left.\begin{bmatrix}
h_{11}h_{12} & \cdots & h_{ik} & 1 & 0 & \cdots & 0 \\
h_{21}h_{22} & \cdots & h_{2k} & 0 & 1 & \cdots & 0 \\
\cdots & \cdots & \cdots & \cdots & \cdots & \cdots & \cdots \\
h_{r1}h_{r2} & \cdots & h_{rk} & 0 & 0 & \cdots & 1
\end{bmatrix}\right\} r}^{n}
\tag{8-125}
$$

and $\mathbf{C}$ is the word column vector given by

$$
\mathbf{C} = \overbrace{\left.\begin{bmatrix}
m_1 \\ m_2 \\ \cdot \\ \cdot \\ \cdot \\ m_k \\ c_1 \\ c_2 \\ \cdot \\ \cdot \\ \cdot \\ c_r
\end{bmatrix}\right\} n}^{1}
\tag{8-126}
$$

The object then is to choose H (which in turn determines **C**) to provide the code properties desired.

Denote the received n element vector by R again. It is apparent as a generalization of the special case ($n = 7$) considered previously that R may again be written as the mod 2 sum of the transmitted vector **C** and an error vector **E**. **E** again contains 1's in those positions where errors occurred, and 0's in those positions where bits were received correctly. Repeating the definition of E,

$$\mathbf{R} = \mathbf{C} \oplus \mathbf{E} \qquad [8\text{-}120]$$

The method of decoding algebraic codes is to again compute the vector $H\mathbf{R}$. This vector, containing r components (r the number of redundant bits), is called the *syndrome* **S** (a medical term), since it is symptomatic of the pattern of errors that actually occurred. Now if a code word is received $\mathbf{S} = 0$. (Note that this does *not* mean there were no errors. If **C** was converted during transmission into another code word of the set of 2^k possible, that is, **E** itself is a code word, there is *no* way of recovering **C** correctly.) If **S** contains any nonzero elements, however, we have not received a code word and we know that one or more errors have occurred.

If we are just interested in *detecting* errors, we merely examine the vector **S** to see if it contains any nonzero elements. If, however, we are interested in correcting errors, we must somehow surmise from **S** where the errors occurred. As previously, we have

$$\mathbf{S} \equiv H\mathbf{R} = H(\mathbf{C} + \mathbf{E}) = H\mathbf{E} \qquad [8\text{-}127]$$

Now assume that a *single error* occurred in the ith digit of the n transmitted. Then

$$\mathbf{E} = \begin{bmatrix} 0 \\ 0 \\ 0 \\ \cdot \\ \cdot \\ \cdot \\ 1 \\ 0 \\ \cdot \\ \cdot \\ \cdot \\ 0 \end{bmatrix} \begin{matrix} \\ \\ \\ \\ \\ \\ \leftarrow i\text{th position} \\ \\ \\ \\ \\ \\ \end{matrix} \qquad [8\text{-}128]$$

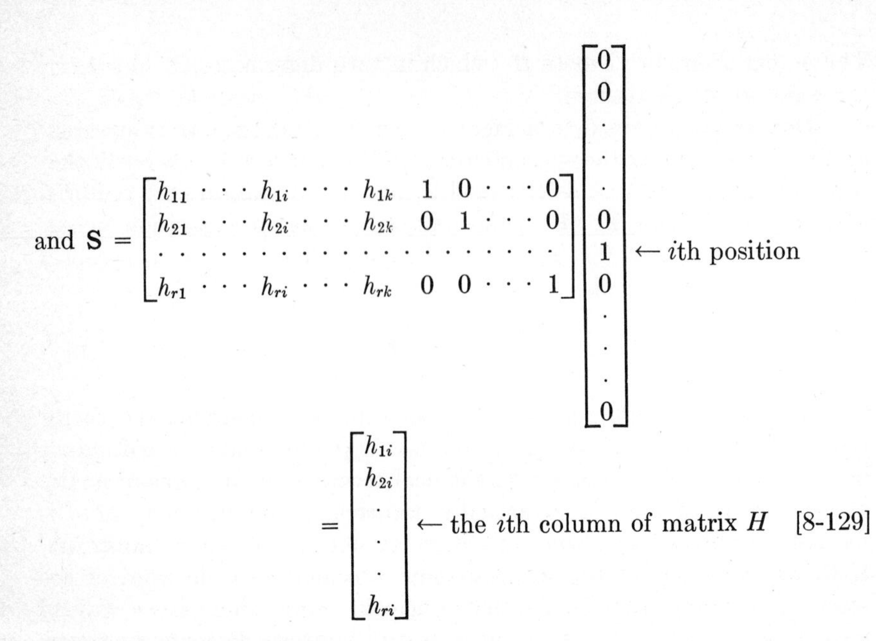

$$\text{and } \mathbf{S} = \begin{bmatrix} h_{11} & \cdots & h_{1i} & \cdots & h_{1k} & 1 & 0 & \cdots & 0 \\ h_{21} & \cdots & h_{2i} & \cdots & h_{2k} & 0 & 1 & \cdots & 0 \\ \cdot & \cdot & \cdot & \cdot & \cdot & \cdot & \cdot & \cdot & \cdot \\ h_{r1} & \cdots & h_{ri} & \cdots & h_{rk} & 0 & 0 & \cdots & 1 \end{bmatrix} \begin{bmatrix} 0 \\ 0 \\ \cdot \\ \cdot \\ \cdot \\ 0 \\ 1 \\ 0 \\ \cdot \\ \cdot \\ \cdot \\ 0 \end{bmatrix} \leftarrow i\text{th position}$$

$$= \begin{bmatrix} h_{1i} \\ h_{2i} \\ \cdot \\ \cdot \\ \cdot \\ h_{ri} \end{bmatrix} \leftarrow \text{the } i\text{th column of matrix } H \qquad [8\text{-}129]$$

This generalizes the previous result.

How do we now choose H? If we want a code that corrects all single errors, as assumed here, we must choose H so that (1) all columns are distinct (each different error then gives a distinct syndrome); and (2) no column is made up of all zeros (if an error occurs the syndrome is nonzero).

Note that the H matrix of Eq. [8-117] is an example of such a matrix that corrects single errors. An error occurring in the first bit of the seven transmitted for a particular code word would produce as the syndrome in this case

$$\mathbf{S} = \begin{bmatrix} 1 \\ 1 \\ 1 \end{bmatrix}$$

just the first column of matrix H.

If two or more errors occurred, the resultant syndrome would be the mod 2 vector sum of the corresponding columns of H where the errors occurred. As an example, if errors occurred in the first and second bit positions of the 7-bit code word, the resultant syndrome would be

$$\mathbf{S} = \begin{bmatrix} 1 \\ 1 \\ 1 \end{bmatrix} \oplus \begin{bmatrix} 1 \\ 1 \\ 0 \end{bmatrix} = \begin{bmatrix} 0 \\ 0 \\ 1 \end{bmatrix}$$

This indicates (erroneously) an error in position 7. (See Eq. [8-117].)

Since binary words are encoded and decoded as a group, it is apparent that the n bits in a given word may be reordered before transmission in any manner desired. The only stipulation, of course, is that the receiver know the manner of reordering so that it can unscramble the word and send it off in the appropriate sequence. Since this is the case, the ordering can then be chosen so that the syndrome is just the binary number corresponding to the bit position where the error occurred. For the 7-bit code of Eq. [8-117], the appropriate reordering, if so desired, is just

$$H = \begin{matrix} c_3 & c_2 & m_4 & c_1 & m_3 & m_2 & m_1 \end{matrix} \atop \begin{bmatrix} 0 & 0 & 0 & 1 & 1 & 1 & 1 \\ 0 & 1 & 1 & 0 & 0 & 1 & 1 \\ 1 & 0 & 1 & 0 & 1 & 0 & 1 \end{bmatrix} \qquad [8\text{-}130]$$

We have thus far discussed single-error correction only. It is apparent that there should be ways of designing codes to correct more than one error. The interested reader is referred to the references for detailed discussions of various types of encoding techniques.[1]

Suffice it to say that in order to correct more than a single error, not only must the column of H be distinct and different from zero but certain vector sums of column of H must be distinct. (As noted earlier, if two or more errors occur, the resultant syndrome $\mathbf{S}$ is the mod 2 sum of the corresponding columns of H. This sum should obviously not be another column of H in order for it to be distinguishable.) In particular in order to correct all patterns of t or fewer errors we must choose H such that

1. All sums of t or fewer columns are distinct from all other sums of t or fewer columns.
2. All sums of t or fewer columns do not sum to zero.

8-9 SUMMARY

In this chapter we attempted to unify discussions in previous chapters on digital communications in the presence of noise.

Starting first with binary signals we asked the question: Are there optimum binary waveshapes and optimum receiver mechanizations or

[1] W. W. Peterson, "Error-correcting Codes," Wiley, New York, 1961, and E. R. Berlekamp, "Algebraic Coding Theory," McGraw-Hill, New York, 1968, provide comprehensive, high-level treatments. Current articles on coding appear in such journals as the *IEEE Transactions on Information Theory*, and *Information and Control*. The book by J. M. Wozencraft and I. M. Jacobs, "Principles of Communication Engineering," chap. 6, Wiley, New York, 1965, devotes considerable discussion to various types of codes with particular emphasis on sequential decoders. Lucky, Salz, and Weldon, *op. cit.*, chaps. 10–12, provide a clear discussion of coding techniques with specific tables of codes and many references.

processing techniques to minimize the error probability? To answer this question we applied known techniques of statistical decision theory. Starting first with single-received-signal samples and then generalizing to multiple independent samples drawn from known probability distributions we found that the optimum processing procedure consisted of setting up a likelihood ratio and determining whether this ratio was greater than or smaller than a known constant. Alternatively, the optimum procedure consisted of subdividing the m-dimensional space of the m received signal samples into two disjoint decision regions, one corresponding to one binary signal transmitted, the other to the other signal. In most cases considered, the likelihood ratio could be simplified considerably to provide simple processing procedures for the m samples.

Specializing to the important case of additive white gaussian noise as the disturbance on the channel, we found that the optimum processor consisted of a pair of matched filters, one for each signal transmitted. (This reduces to one filter in the case of bipolar or on-off signals.) In the digital version of these filters, each received signal sample is weighted by the corresponding stored transmitted sample, all m weighted samples then being added together. The optimum signal shapes then turned out to be *any* pair of equal and opposite signals. As shown first in Chap. 6, the error probability then depends solely on the signal-energy-to-noise spectral density.

With the optimum way of transmitting and receiving then established, the next question asked was whether one could in any way improve on the signal detectability as given by the probability of error. The answer here is provided by the Shannon channel-capacity theorem. It *is* theoretically possible to drive the probability of error in the presence of additive gaussian noise to as low a value as desired by appropriate encoding and decoding operations at the transmitter and receiver, respectively. This is possible provided the binary transmission rate R in bits per second does not exceed the *channel capacity*, a number determined by the channel bandwidth, average signal power, and noise spectral density.

Shannon's theorem provides no recipe for the encoding and decoding operations. We gave two examples of systems that provide the desired error performance, however. One consists of storing RT bits in an interval T sec long and then transmitting 1 of $M = 2^{RT}$ possible orthogonal waveshapes T sec long. Such systems require bandwidths the order of M/T, increasing exponentially with T. The other scheme discussed was an information-feedback system requiring the availability of a noise-free feedback channel. This latter scheme operates with a fixed-channel bandwidth.

The significance of these different schemes and their relative performance as compared to optimum binary transmission, the meaning of channel capacity, bandwidth–SNR tradeoffs, etc., was further developed in terms of a space-communications example. Additional examples like this are included among the problems at the end of this chapter.

We concluded this chapter by examining some methods of detecting and correcting binary errors as a means of further improving the performance of binary systems. The most common schemes for error detection and correction consist of inserting check bits in the binary stream to enable a specified number of errors to be detected and/or corrected. The trick here is to choose coding schemes that provide the error performance predicted by Shannon. We also discussed a decision-feedback scheme using a null-zone region for detecting binary errors.

PROBLEMS

8-1 Consider the received-signal conditional-density functions $f(v|1)$ and $f(v|2)$ shown in Fig. P 8-1.

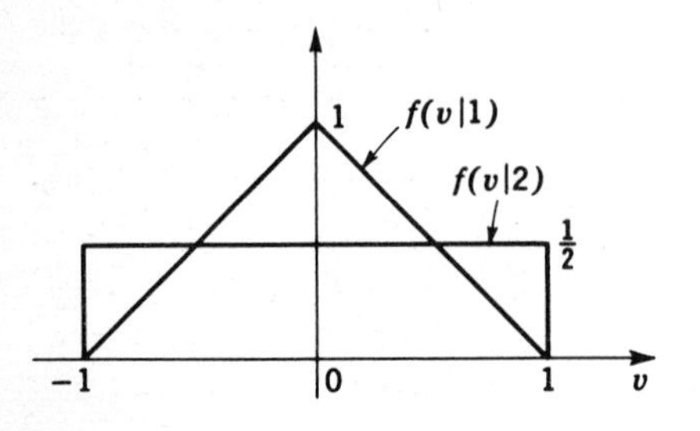

FIG. P 8-1

(a) Indicate the two decision regions V_1 and V_2 for the following values of P_1, the a priori probability of transmitting a 1: 0.3, 0.5, 0.7.
(b) Calculate the probability of error in each of the three cases of (a).

8-2 The probability-density function of a sample of received signal v corresponding to $s_1(t)$ transmitted is given by

$$f(v|1) = k_1 e^{-|v|} \qquad -\infty < v < \infty$$

while the corresponding density function corresponding to signal $s_2(t)$ is

$$f(v|2) = k_2 e^{-2|v|} \qquad -\infty < v < \infty$$

(a) Find the appropriate values of k_1 and k_2.
(b) The a priori probabilities are $P_1 = \frac{3}{4}$, $P_2 = \frac{1}{4}$. Find the values of v for which we choose s_1, and the values for which we choose s_2.

8-3 An OOK signal is transmitted over a fading medium, and gaussian noise of mean-squared value N added at the receiver. The composite signal plus noise is envelope detected before binary decisions are made. At the decision point, then, the sampled envelope r has either one of the two density functions

$$f(r|1) = \frac{re^{-r^2/2N_T}}{N_T} \qquad \text{or} \qquad f(r|2) = \frac{re^{-r^2/2N}}{N}$$

corresponding, respectively, to signal plus noise received, and to noise alone (zero signal). Here $N_T = N + S$, with S the mean signal power averaged over the fading $(0 < r < \infty)$.

(a) Show that in the case of equally likely binary signal transmission the optimum decision test consists of deciding on a 1 ("on" signal) transmitted if the envelope r exceeds a threshold $b = \sqrt{2N(1 + N/S) \log_e (1 + S/N)}$.
(b) $S/N = 10$. Calculate b and evaluate the overall probability of error. Repeat for $S/N = 1$.
(c) m independent samples r_j, $j = 1, \ldots, m$, of r are taken before a decision is

made. Show that the optimum test consists of determining whether $\sum_{j=1}^{m} r_j^2$ is greater or less than $2mN(1 + N/S) \log_e (1 + S/N)$.

8-4 A bipolar binary signal $\pm A$ is received in the presence of additive gaussian noise of variance N. Find the appropriate decision levels if one sample of signal plus noise is taken, for $P_1 = 0.3$, 0.5, and 0.7.

8-5 A bipolar binary signal of amplitude ± 1 has added to it noise $n(t)$ with density function $f(n) = \frac{3}{32}(4 - n^2)$. Find the minimum probability of error if the a priori probabilities are $P_1 = \frac{2}{5}$ and $P_{-1} = \frac{3}{5}$.

8-6 The received voltage for binary transmission has the two conditional-density functions $f(v|1)$ and $f(v|2)$ shown in the figure. Find the optimum decision rule and minimum probability of error in the three cases $P_1 = \frac{1}{2}, \frac{2}{3}, \frac{1}{3}$.

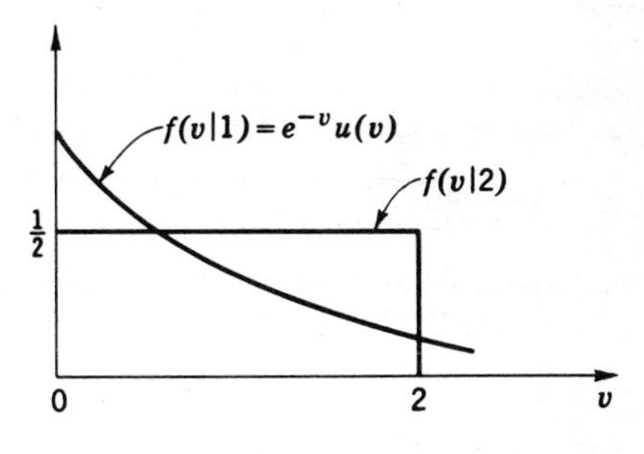

FIG. P 8-6

8-7 A received waveform $v(t)$ is of the form

$$v(t) = \begin{matrix} +2E \\ \text{or} \\ -E \end{matrix} + n(t)$$

with $n(t)$ zero-mean gaussian noise, of variance N. The a priori signal probabilities are $P(+2E) = \frac{1}{3}$, $P(-E) = \frac{2}{3}$. A single sample of $v(t)$ is taken.

(*a*) For what values of v should we choose $+2E$ in order to minimize the overall probability of error?
(*b*) Give an expression for the overall probability of error.

8-8 One of two signals is transmitted over a noisy channel. The conditional-density functions of the received random variable v are

$$f(v|1) = \frac{1}{2\pi} \qquad |v| \le \pi$$
$$= 0 \qquad \text{otherwise}$$
$$f(v|2) = \frac{1}{2\pi}(1 + \cos v) \qquad |v| \le \pi$$
$$= 0 \qquad \text{otherwise}$$

(a) With $P_1 = P_2 = \frac{1}{2}$, find the decision region of v corresponding to minimum probability of error.
(b) Find the minimum probability of error in (a).
(c) Find the values of P_1 and P_2 such that the optimum decision rule says *always* decide on signal 1. What is the probability of error in this case?

8-9 Either one of two noiselike signals is transmitted. The density functions of the received signal are

$$f(v|1) = \frac{e^{-v^2/2\sigma_1^2}}{\sqrt{2\pi\sigma_1^2}} \qquad f(v|2) = \frac{e^{-v^2/2\sigma_2^2}}{\sqrt{2\pi\sigma_2^2}}$$

$P_1 = P_2 = \frac{1}{2}$. Find the optimum receiver processing in the case of one, and then two, independent samples. Show in this latter case that signal 1 is declared present if $v_1^2 + v_2^2 > d$, d a prescribed decision level. What are the two-dimensional regions V_1 and V_2 in this latter case?

8-10 Binary signals with $P_1 = P_2 = \frac{1}{2}$ are received in additive gaussian noise of rms value of 0.5 volt. Two independent samples are used at the receiver to decide on signal s_1 or s_2. At the (known) sampling times the two-dimensional signal vectors are, respectively, $\mathbf{s}_1 = (+15 \text{ volts}, +15 \text{ volts})$, and $\mathbf{s}_2 = (-7 \text{ volts}, -7 \text{ volts})$.

(a) Find the decision rule that minimizes the probability of error.
(b) Sketch regions V_1 and V_2 in the two-dimensional plane of the received vector $\mathbf{v}$. Indicate $\mathbf{s}_1$ and $\mathbf{s}_2$ in the same sketch.
(c) Find the minimum probability of error in terms of the complementary error function erfc x defined as $1 - (2/\sqrt{\pi}) \int_0^x e^{-x^2}\, dx$.

8-11 Devise a detection scheme for *three* signals $s_1(t) = +a$, $s_2(t) = 0$, and $s_3(t) = -a$, received in additive gaussian noise of variance N. Assume that the signals are equiprobable. Find the optimum thresholds and the minimum error probability. *Hint:* Symmetry may be used in locating the optimum thresholds.

8-12 *Diversity Transmission* (use of more than one channel to improve performance). Consider the system shown in Fig. P 8-12. Bipolar signals $\pm a$ are sent out, in parallel over two channels as shown. Because of differing attenuation (or fading)

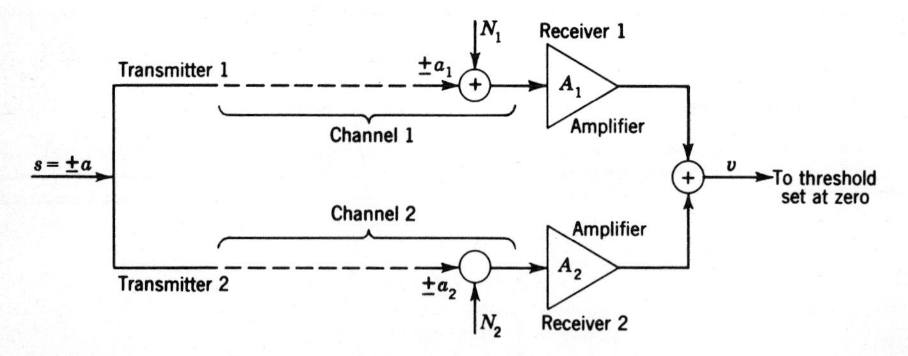

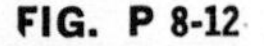
FIG. P 8-12

along the two paths, the signals arrive as $\pm a_1$ and $\pm a_2$, respectively, at each receiver. Gaussian noise of variance N_1 and N_2, respectively, is added at each receiver as shown.

(a) Show that the summed output v is a gaussian variable of expected value $\pm(A_1 a_1 + A_2 a_2)$, and variance $A_1^2 N_1 + A_2^2 N_2$.
(b) Show that the probability of error depends on the effective SNR $(a_1 + Ka_2)^2/(N_1 + K^2 N_2)$, with $K = A_2/A_1$. $K = 0$ $(A_2 = 0)$ and $K = \infty$ $(A_1 = 0)$ correspond to the single-receiver case. Show that the diversity system provides SNR, and hence error probability, improvement over the single-receiver case.
(c) Show that the optimum choice of the gain ratio K is given by

$$K_{\text{opt}} = (a_2/a_1)(N_1/N_2)$$

This is equivalent to setting $A_1 = (a_1/N_1)g$, $A_2 = (a_2/N_2)g$, g some arbitrary gain constant. The optimum diversity system hence weights each receiver input by the ratio of signal to noise (a_1/N_1) measured at that input. This type of combining is called *maximal-ratio combining*. Show that the effective SNR for this case is given by $a_1^2/N_1 + a_2^2/N_2$, the sum of the two SNR's.
(d) As a special case of diversity combining, assume that the two signal terms a_1 and a_2 represent samples in *time* of a transmitted signal. Let $N_1 = N_2 = N$ be the variance of the noise added, the two noise samples assumed independent. Show that the optimal processing of the two signal-plus-noise samples, in the sense of minimum probability of error, consists of adding them after weighting the first by a_1, the second by a_2. A little thought indicates this is the same as *matched filtering*. Compare this *matched filtering* for time diversity with *maximal-ratio combining* for diversity techniques in general.

8-13 A binary message (0 or 1) is to be transmitted in the following manner. The transmitter has two coins labeled C_0 and C_1. Coin C_0 has a probability of a head p_0 and coin C_1 has a probability of a head p_1. If message i $(i = 0$ or $1)$ is to be transmitted, coin C_i is flipped n times (independent tosses) and the sequence of heads and tails is observed by the receiver.

(a) Assuming the two messages are equally likely, find the optimum (minimum probability of error) decision rule for deciding between the two messages. Indicate one simple method for implementing this rule.
(b) Set up an expression for the resultant probability of error.

8-14 Consider the binary detection problem where we receive (after processing) the random variable v given by

$$v = s + n$$

where s, the signal, is either equal to 0 or 1 with equal prior probabilities. n is an exponential random variable with density

$$f(n) = \tfrac{1}{2} e^{-|n|} \qquad -\infty < n < \infty$$

Find:

(a) The decision rule which leads to the minimum probability of error.
(b) The resultant minimum probability of error.

8-15 One of two equally likely signals is transmitted and received in additive gaussian noise of variance 2 volts^2. The signals are $s_1(t) = 4$ volts $= -s_2(t)$, the binary interval being 1 msec long. Eight equispaced independent samples of the received signal

$$v(t) = \begin{matrix} s_1(t) \\ \text{or} \\ s_2(t) \end{matrix} + n(t)$$

are taken, and comprise an 8-vector $\mathbf{v}$.

(a) For the following sets of observed data $\mathbf{v}$ which signal would you decide was sent?
 (1) $\mathbf{v} = (4.5, 0, -1.5, 2, -6, 10, 1, -4)$.
 (2) $\mathbf{v} = (-5, -3, -4, -5, -3, 20, 15, 5)$.
(b) What is the probability of error?

8-16 $s_1(t)$ is a triangle 1 msec long with peak voltage of 4 volts. $s_2(t) = 0$. Repeat Prob. 8-15 for the following sets of observed data samples:

(a) $\mathbf{v} = (-2, -6, +1, +5, +6, +2, +1, -8)$.
(b) $\mathbf{v} = (0, -2, -4, +5, +6, +2, -4, -2)$.

8-17 One of two equally likely signals $s_1(t)$ and $s_2(t)$ is transmitted, and gaussian noise added. m independent samples of signal plus noise are taken and comprise the vector $\mathbf{v}$. Show the decision is to choose signal s_1 if the distance between $\mathbf{v}$ and $\mathbf{s}_1$ is less than the distance between $\mathbf{v}$ and $\mathbf{s}_2$. (This verifies Eq. [8-16].)

8-18 Show that the optimum processing in Prob. 8-17 is equally well given by Eq. [8-18] or [8-19].

8-19 Two independent samples for the case of Prob. 8-17 are taken. Draw the decision line dividing region V_1 from V_2. What is the effect of having unequal a priori probabilities?

8-20 Consider the integral

$$\int_{-\infty}^{\infty} \frac{\sin \pi(2Bt - k)}{\pi(2Bt - k)} \cdot \frac{\sin \pi(2Bt - m)}{\pi(2Bt - m)}\, dt$$

(See Eq. [8-41] in the text.) Show by a simple change of variables that this may be written as the convolution integral

$$\frac{1}{2B\pi} \int_{-\infty}^{\infty} \frac{\sin(\tau - x)}{(\tau - x)} \frac{\sin x}{x}\, dx$$

with $\tau \equiv (k - m)\pi$. Recalling that the Fourier transform of $(\sin ax)/\pi x$ is 1, $|\omega| \le a$; 0, $|\omega| > 0$, take Fourier transforms and show that the integral is $(1/2B)\delta_{km}$, where δ_{km} is the Kronecker delta. The $(\sin x)/x$ functions are thus examples of *orthogonal* functions.

8-21 (a) As a generalization of Prob. 8-20 above, prove that

$$\int_{-\infty}^{\infty} \frac{\sin \omega_1(t - x)}{\pi(t - x)} \frac{\sin \omega_2 x}{\pi x}\, dx = \frac{\sin \omega_1 t}{\pi t}$$

assuming $\omega_1 \le \omega_2$. *Hint:* This is already in the form of a convolution integral. Use the approach suggested in Prob. 8-20.

(b) As a special case let $\omega_1 = \omega_2 = 2\pi B$; $t = (k - m)/2B$. Show that this gives the same result as in Prob. 8-20.

8-22 *Orthogonal functions.* Consider a set of functions $\Phi_i(t)$ with the property $\int_a^b \Phi_i(t)\Phi_j(t)\,dt = \delta_{ij}$. The $\Phi_i(t)$'s then constitute a normalized orthogonal or *orthonormal* set of functions, over the integration range (a,b).

(*a*) We desire to approximate an arbitrary function $f(t)$ by a linear sum of orthonormal functions:

$$f(t) \sim \sum_{j=1}^{n} b_j\Phi_j(t) \equiv f_n(t)$$

Show that the mean-squared error between $f(t)$ and $f_n(t)$,

$$\epsilon^2 \equiv \int_a^b [f(t) - f_n(t)]^2\,dt$$

is minimized by choosing

$$b_j = \int_a^b f(t)\Phi_j(t)\,dt$$

Use the symbol a_j to denote this special case of b_j. $\sum_{j=1}^{n} a_j\Phi_j(t)$ then approximates $f(t)$ best in a least mean-squared sense. The a_j's are sometimes called the generalized Fourier coefficients.

(*b*) The orthogonal set $\Phi_j(t)$ is said to be *complete* if, using the Fourier coefficients a_j, that is,

$$f_n(t) = \sum_{j=1}^{n} a_j\Phi_j(t)$$

$\epsilon^2 \to 0$, as $n \to \infty$. Show that for this case

$$\int_a^b f^2(t)\,dt = \sum_{j=1}^{\infty} a_j{}^2$$

This is a generalized form of Parseval's theorem, first met in Chap. 2. That is, the energy in the signal equals the energy in the orthogonal functions. We then write

$$f(t) = \sum_{j=1}^{\infty} a_j\Phi_j(t)$$

where the equality is meant in this sense of equal energy.

(*c*) Let the interval (a,b) be $(-T/2, +T/2)$. Find the normalized set of sines and cosines that are orthogonal over this interval.

(*d*) According to Prob. 8-20 (and Eq. [8-41] in the text), the $(\sin x)/x$ functions are orthogonal over the interval $(-\infty, \infty)$. Normalize these functions and show how the a_j coefficients are related to the sampled values $f(j/2B)$ of a function $f(t)$ expanded in terms of the $(\sin x)/x$ functions. (See Eq. [8-40].) Show that the Parseval theorem in this case is given by

$$\int_{-\infty}^{\infty} f^2(t)\,dt = \frac{1}{2B}\sum_{j=-\infty}^{\infty} f^2\left(\frac{j}{2B}\right)$$

(*e*) Let

$$f_1(t) = \sum_{j=1}^{\infty} a_j \Phi_j(t)$$

$$a_j = \int_a^b f_1(t)\Phi_j(t)\,dt$$

$$f_2(t) = \sum_{j=1}^{\infty} b_j \Phi_j(t)$$

$$b_j = \int_a^b f_2(t)\Phi_j(t)\,dt$$

Show that

$$\int_a^b f_1(t) f_2(t)\,dt = \sum_{j=1}^{\infty} a_j b_j$$

Use this to verify Eq. [8-43] in the text, as a special case.

8-23 (*a*) Plot the capacity in bits per second versus bandwidth W of a channel with additive band-limited gaussian noise of spectral density $n_0/2$ and average power S.

(*b*) $S/n_0 = 100$. Find the maximum rate of transmission of binary information if PSK is used and a maximum probability of error of 10^{-5} is to be maintained. Repeat for $P_e = 10^{-4}$. What is the maximum rate in both cases if FSK transmission is used?

(*c*) $S/n_0 = 100$ again. The channel bandwidth is $W = 10$ Hz. If the binary digits of (*b*) above may be encoded using as complicated a digital scheme as desired, what is the maximum rate of transmission in bits per second with a probability of error as small as desired? Compare with (*b*) above. What is the SNR in this case?

(*d*) $S/n_0 = 100$. The channel bandwidth may be made as large as necessary. Repeat (*c*) and again compare with (*b*).

8-24 Digital communications for a deep-space probe (10^8 miles from earth) is to be investigated. Assume 500-MHz transmission with space vehicle and earth antenna gains of 10 db and 40 db, respectively. The transmitter power is limited to 10 watts. The receiver-noise figure is 3 db. Bit error probability is to be less than 10^{-5}.

(*a*) Find the maximum rate of binary transmission if a PSK system with synchronous detection is considered.

(*b*) The binary data are to be encoded into one of 64 orthogonal signals. Find the maximum binary rate in this case. What is the encoding or storage time required at both transmitter and receiver?

(*c*) What is the maximum possible transmission rate if an arbitrarily large bandwidth and complex encoding are allowed, and the probability of error is to be made as small as desired? Compare (*a*), (*b*), (*c*).

8-25 The antenna sizes used in Prob. 8-24 are fixed. Repeat the problem for two different frequencies: 250 MHz and 1,000 MHz.

8-26 Repeat Prob. 8-24 if a maser receiver with an effective noise temperature of 30°K is used on earth.

8-27 "Design" a radio communications system for a deep-space probe (10^8 miles) with the following constraints: (1) transmitter power is 10 watts; (2) transmission bandwidth is 1 kHz; (3) SNR at receiver input $\geq$20 db. Calculate or assume reasonable values for all other parameters.

8-28 4-MHz video information is to be transmitted from a synchronous satellite to the earth 23,000 miles away via an FM carrier at 2 GHz. The receiver, operated at room temperature, has a noise figure of 6 db. The SNR at the discriminator output is to be at least 30 db, with the CNR at the input above threshold. Select or calculate an appropriate set of realistic parameters for: (1) transmitter power level; (2) FM modulation index; (3) transmission bandwidth. The antenna gains are, respectively, 10 db and 40 db at the transmitter and receiver.

8-29 Real-time transmission of TV pictures from a lunar spacecraft to earth (400,000 km) is to be investigated. A receiver CNR of 20 db at a bandwidth of 4 MHz is required. A 500-MHz carrier is to be used, with a 2 meter $\times$ 2 meter antenna on the spacecraft and an antenna with 45-db gain on the earth. What is the power requirement of the transmitter on the space vehicle?

8-30 Investigate the possibility of maintaining voice contact with a space mission 8×10^9 km from the earth. (This represents the distance to the farthest planet of the solar system.) A receiver CNR of 20 db is desired. A low-noise receiving system with a 5°K effective noise temperature at 2 MHz is available. The spacecraft may be allowed up to 250 watts power output and can carry a large unfurlable parabolic antenna.

8-31 Show that in the information-feedback scheme of Sec. 8-7 the estimate of the number θ transmitted is given by Eq. [8-79] at the end of j transmissions.

8-32 The information-feedback scheme of Sec. 8-7 is to be considered for the deep-space probe of Prob. 8-24. Determine its transmission rate in bits per second, the transmission time T, and the number of amplitude levels M required. The system is to operate with a probability of error of 10^{-5} and the same system constraints (power antenna gains and noise figure) as in Prob. 8-24. Compare with the binary and orthogonal signal schemes considered in Prob. 8-24.

8-33 Consider the null-zone detection scheme of Sec. 8-8. With no null zone ($d = 0$) the signal amplitude to rms noise ratio $A/\sqrt{N}$ is chosen to provide an error probability of 10^{-3}. The null zone is then introduced. Plot the resultant error probability P_e versus $d/\sqrt{N}$. Plot $E(n)$, the expected number of transmissions, versus $d/\sqrt{N}$ as well. From these two curves prepare a composite curve relating P_e to $E(n)$ for fixed $A/\sqrt{N}$.

8-34 Compare probability of error for the null-zone technique of Prob. 8-33 with a bipolar scheme in which the signal is simply repeated a predetermined number of times. The samples of received signal plus noise (assumed independent) are summed and the resultant polarity used to decide on a + signal or a $-$ signal. $A/\sqrt{N}$ is the same in both cases, and $E(n)$ in the null-zone case is to be taken equal to the fixed number of repeats in the bipolar case.

8-35 Consider the binary code with parity check matrix

$$H = \begin{bmatrix} 1 & 0 & 0 & 1 & 0 & 1 \\ 0 & 1 & 0 & 0 & 1 & 1 \\ 0 & 0 & 1 & 1 & 1 & 1 \end{bmatrix}$$

(*a*) Is the word (101010) a code word?
(*b*) A code word is of the form (X01100). Is X a 0 or a 1?
(*c*) Suppose the code word (001111) is transmitted, but (000010) is received. What is the resultant syndrome? Where would this syndrome indicate an error had occurred?
(*d*) How many code words are in this code?
(*e*) What is the smallest number of errors that could change one code word into another code word? Why?

8-36 Consider a binary communication system consisting of two links as shown in Fig. P 8-36. The noise in *each* channel is such that (1) errors occur independently;

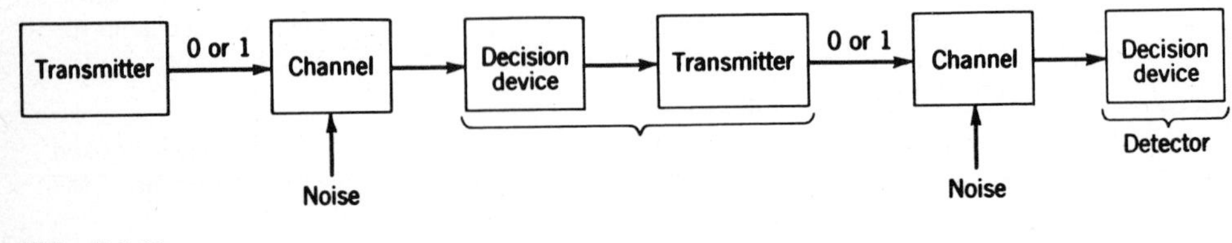

FIG. P 8-36

and (2) the probability that a 1 is received when a 0 is transmitted is p. The probability that a 0 is received when a 1 is transmitted is also p.

(*a*) Find the following four probabilities for the entire system:

A 0 is received when a 0 is transmitted
A 1 is received when a 0 is transmitted
A 0 is received when a 1 is transmitted
A 1 is received when a 1 is transmitted

(*b*) Assume that a simple coding scheme is used such that a 0 is transmitted as three successive 0's and a 1 as three successive 1's. At the detector, the following (majority) decision rule is used:

$$\begin{Bmatrix} \text{Decide 0 if 000, 001, 010, or 100 is received} \\ \text{Decide 1 if 111, 110, 101, or 011 is received} \end{Bmatrix}$$

If a 0 and 1 are equally likely, what is the probability of deciding incorrectly? Evaluate for $p = \frac{1}{3}$.

8-37 A binary message consists of words which are 5 bits long. The message words are to be encoded using a single-error-correcting code. The first 5 bits of each code word must be the message bits m_1, m_2, m_3, m_4, m_5, while the remaining bits are check bits.

(*a*) It may be shown that for a single-error-correcting (Hamming) code $2^r - 1 \geq k + r$, with k the number of message bits and r the number of check bits. What is the minimum number of check bits for $k = 5$?
(*b*) Construct an appropriate H matrix for this code.
(*c*) Find the syndrome at the receiver if there is an error in m_5.
(*d*) How does this code respond to double errors?

8-38 Consider a single-error-correcting code for 11 message bits.

(*a*) How many check bits are required? (See Prob. 8-37*a*.)
(*b*) Find a suitable H matrix.
(*c*) Find the syndrome if the single error occurs in the 7th position.

8-39 A code consists of three message digits m_1, m_2, m_3 and three check digits c_1, c_2, c_3. The transmitted sequence is $m_1c_1m_2c_2m_3c_3$. At the transmitter the check digits are formed from the following equations:

$$c_1 = m_1 \oplus m_3$$
$$c_2 = m_1 \oplus m_2$$
$$c_3 = m_1 \oplus m_2 \oplus m_3$$

(*a*) For the message $m_1 = 0$, $m_2 = 1$, $m_3 = 1$, find the transmitted sequence.
(*b*) Write down the H matrix.
(*c*) Will this code correct single errors? Why?
(*d*) Assume that the sequence 011100 is received and that no more than one error has occurred. Decode this sequence: find the location of the error and the transmitted message m_1, m_2, m_3.

8-40 Consider a binary code with three message digits and three check digits in each code word. The code word is of the form

$$m_1m_2m_3c_1c_2c_3$$

where m's are the message digits and the c_i's are the check digits. Assume that the check digits are computed from the set of equations

$$c_1 = m_1 \oplus m_2 \oplus m_3$$
$$c_2 = m_1 \oplus m_3$$
$$c_3 = m_1 \oplus m_2$$

(*a*) How many code words are there in the code?
(*b*) Find the code word which begins 110.........
(*c*) Suppose that the received word is 010111. Decode to the closest code word (i.e., the code word that differs from the received word in the fewest positions).

8-41 Calculate the probability of error for the following binary codes and compare:

(n,k)	t
(7,4)	1
(15,11)	1
(15,7)	2
(15,5)	3
(31,26)	1
(31,21)	2
(31,16)	3
(31,11)	5
(31,6)	7

where n is the total number of digits (message digits plus check digits), k is the number of check digits, and t is the error-correcting capability of the code; that is, t or less errors can be corrected.

Assume that the probability of bit error when no coding is used is 10^{-5}. Adjust the duration of the binary digits in each code so that the transmission rate (message digits per second) is constant. *Hint:*

$$P_e = \binom{n}{t+1} p^{t+1}(1-p)^{n-(t+1)} + \binom{n}{t+2} p^{t+2}(1-p)^{n-(t+2)} + \cdots + p^n$$

Why? For very small p, as true here, $P_e \doteq \binom{n}{t+1} p^{t+1}$.

INDEX

INDEX